Solving Direct and Inverse Heat Conduction Problems

Jan Taler   Piotr Duda

Solving Direct and Inverse Heat Conduction Problems

Springer

Professor Jan Taler
Dr. Piotr Duda
Cracow University Technology
Institute of Process and Power Engineering
Al. Jana Pawla II 37
31-864 Kraków, Poland

Library of Congress Control Number: 2006925848

ISBN-10 3-540-33470-X Springer Berlin Heidelberg New York
ISBN-13 978-3-540-33470-5 Springer Berlin Heidelberg New York

Springer is a part of Springer Science+Business Media
springer.com

Printed in The Netherlands

Typesetting: by the authors and techbooks using a Springer LaTeX macro package
Cover design: *design & production* GmbH, Heidelberg

Printed on acid-free paper SPIN: 11588993 89/techbooks 5 4 3 2 1 0

Preface

This book is devoted to the concept of simple and inverse heat conduction problems. The process of solving direct problems is based on the temperature determination when initial and boundary conditions are known, while the solving of inverse problems is based on the search for boundary conditions when temperature properties are known, provided that temperature is the function of time, at the selected inner points of a body.

In the first part of the book (Chaps. 1–5), we have discussed theoretical basis for thermal conduction in solids, motionless liquids and liquids that move in time. In the second part of the book, (Chapters 6–26), we have discussed at great length different engineering problems, which we have presented together with the proposed solutions in the form of theoretical and mathematical examples. It was our intention to acquaint the reader in a step-by-step fashion with all the mathematical derivations and solutions to some of the more significant transient and steady-state heat conduction problems with respect to both, the movable and immovable heat sources and the phenomena of melting and freezing. Lots of attention was paid to non-linear problems. The methods for solving heat conduction problems, i.e. the exact and approximate analytical methods and numerical methods, such as the finite difference method, the finite volume method, the finite element method and the boundary element method are discussed in great detail. Aside from algorithms, applicable computational programs, written in a FORTRAN language, were given. The accuracy of the results obtained by means of various numerical methods was evaluated by way of comparison with accurate analytical solutions.

The presented solutions not only allow to illustrate mathematical methods used in thermal conduction but also show the methods one can use to solve concrete practical problems, for example during the designing and life-time calculations of industrial machinery, combustion engines and in refrigerating and air conditioning engineering.

Many examples refer to the topic of heating and thermo-renovation of apartment buildings. The methods for solving problems involved with welding and laser technology are also discussed in great detail.

This book is addressed to undergraduate and PhD students of mechanical, power, process and environmental engineering. Due to the complexity of the heat conduction problems elaborated in this book, this edition can also serve as a reference book that can be used by nuclear, industrial and civil engineers.

Jan Taler is the author of the theoretical part of this book, mathematical exercises (excluding 12.1 & 12.3), and C, D & H attachments (found at the back of this book).

Piotr Duda wrote in the FORTRAN language all presented programs and solved with their help exercises 7.3, 11.2–11.7, 15.1, 15.2, 15.4, 15.5, 15.7, 15.8, 15.11, 15.13, 15.15, 16.5, 16.9, 16.10, 17.7, 18.5–18.8, 21.5, 21.7–21.10, 22.7, 23.3–23.7, 24.4 and 24.5. He also carried out calculations using the following programs: ANSYS (in Exercises 11.18–11.22, 12.4, 21.9 and 25.10), BETIS (in Exercise 12.4) and MathCAD (in Exercises 14.10, 16.2, 16.4, 17.6 and 25.10). Furthermore, Piotr Duda is the author of Exercises 12.1 and 12.3, and attachments A, B, E, F and G.

Kraków
June, 2005.

Jan Taler
Piotr Duda

Contents

Nomenclature

a	thermal diffusivity, m^2/s
a, b	length of rectangle sides, m
A	surface area, m^2
$\mathbf{A}$	matrix of the algebraic equation coefficients
$\mathbf{b}$	vector of the coefficients from the right side of an equation
Bi	Biot number $\alpha l/\lambda$
c	specific heat, J/(kg·K)
c_p	specific heat at constant pressure, J/(kg·K)
c_v	specific heat at constant volume, J/(kg·K)
c^*	specific heat substitute, J/(kg·K)
$\mathbf{C}$	an inverse matrix to A coefficients matrix
d, D	diameter, m
d_w	inner diameter, m
d_z	outer diameter, m
$\mathbf{d}$	directional versor
D	pulse duration, s
$\mathbf{e}$	coordinates versor
E	longitudinal elasticity modulus (Young modulus), MPa
E	distance of a temperature sensor from a solid's surface, m
f	measured temperature, in °C or K
F	dimensionless measured temperature
Fo	Fourier number (at/L^2)
g	thickness, m
[g]	column vector of temperature gradient
h	enthalpy, J/kg
h	height, m
h_{sl}	latent heat of melting (change in enthalpy due to a change from a solid to liquid phase), J/kg
H	dimensionless heat flux
i	enthalpy, J/kg
I	current, A
I_0	modified Bessel function of the first kind of order zero
I_1	modified Bessel function of the first kind of order one
J_0	Bessel function of the first kind of order zero
J_1	Bessel function of the first kind of order one
k	overall heat transfer coefficient, $W/(m^2 \cdot K)$
K	dimensionless heat transfer coefficient
K_0	modified Bessel function of the second kind of order zero
K_1	modified Bessel function of the second kind of order one

L	length, m
L	fin height, m
L_c	equivalent fin height, m
L	latent heat of melting, J/kg
L_x, L_y, L_z	length in x, y, z direction, m
m	mass, kg
m	fin parameter $(2\alpha/\lambda t)^{0.5}$, 1/m
n	outwardly directed unit normal vector to the control volume boundary
n_x, n_y, n_z	directional cosinuses
N_i^e	shape function for i- node of a finite element e
p	pressure, MPa
P	perimeter, m
P	temperature-change duratioon, s
P_t, P_1	perpendicular and linear pitch to the direction of air flow, m
q	energy per unit of surface , J/m^2
q	a variable in Laplace transform $q = \sqrt{s/a}$
q_l	energy per unit of length, J/m
$\dot{q}$	heat flux, W/m^2
$\dot{q}_m$	thermal load of heat furnace wall (absorbed heat flux), W/m^2
$\dot{q}_v$	energy generation rate per unit volume (uniform within the body), W/m^3
$\dot{Q}$	heat flow, W
r	radius, m
r_w	inner surface radius, m
r_z	outer surface radius, m
R	heat resistance, K/W
s	phase boundary position, m
s	complex variable
s	fin spacing, m
S	shape coefficient, m
Ste	Stefan number $c\,(T_m - T_0)/h_{sl}$
t	time, s
t	width, m
t	fin width, m
t	water-wall tube spacing, m
t_u	time of heat flux step-change, s
T	temperature, °C or K
T_b	fin base temperature, °C or K
T_{cz}	fluid temperature, °C or K
T_m	melting temperature, °C or K
T_n	mean temperature over the wall thickness, °C or K
T_i	temperature in i-node, °C or K
T_r	external temperature during radiation heat transfer, K
T_p	air temperature, °C or K
T_{sp}	temperature of combustion gases, °C or K

T_0	initial temperature, °C or K
$\overline{T}$	mean temperature, °C or K
$\overline{T}$	Laplace transform
u	inner energy, J/kg
u	transformed temperature, temperature difference $(T - T_0)$, K
$u(\mathbf{r}, t)$	influence function (system response to unit step function)
U	voltage, V
U	Laplace transform
v	specific volume, m^3/kg
v	velocity, m/s
v_T	temperature change rate, K/s or K/min
V	volume, m^3
w	width, m
w_x, w_y, w_z	velocity components, m/s
x, y, z	Cartesian coordinates
X	dimensionless coordinate
y	smoothed temperature, °C or K
Y_0	Bessel function of the second kind of order zero
Y_1	Bessel function of the second kind of order one

Greek symbols

α	heat transfer coefficient, $W/(m^2 \cdot K)$
β	linear or volumetric thermal expansion coefficient, 1/K
γ	unit step function (Heaviside function)
δ	depth of heat penetration, m
δ	Dirac function
Δt	time step, s
ΔT	temperature difference, K
$\Delta x, \Delta r$	spatial step in the x or r direction, m
ε	surface emissivity
ε	temperature measurement error, K
η	fin efficiency
η	similarity variable (dimensionless coordinate)
Θ	angle coordinate
Θ	dimensionless temperature
Θ	excess temperature (temperature difference between true and initial temperature or between true and ambient temperature), K
κ	thermal diffusivity, m^2/s
λ	thermal conductivity, $W/(m \cdot K)$
$\lambda_{xx}, \lambda_{yy}, \lambda_{zz}$	thermal conductivity in x, y, z direction, $W/(m \cdot K)$
Λ	thermal conductivity tensor
μ_n	n-th root of a characteristic equation
ν	Poisson ratio
ξ	a coordinate in moving coordinate system, m
ρ	density, kg/m^3
ρ	dimensionless radius
σ	Stefan-Boltzmann constant, $5.67 \cdot 10^{-8}$ $W/(m^2 \cdot K^4)$

σ_r, σ_φ, σ_z	radial, tangential and axial stresses respectively, MPa
τ	relaxation time, s
τ	time constant, s
φ, Φ	angle coordinate
Φ	dissipation function, W/m^3
ω	circular frequency, 1/s

PART I

Heat Conduction Fundamentals

Heat conduction is, aside from convection and radiation, the basic form of heat transfer. It is the only type of heat flow that occurs in non-transparent solids. In the cases of gases and fluids, heat conduction usually occurs in combination with other forms of conduction, such as convection and radiation.

1 Fourier Law

In order to describe heat conduction phenomena, one usually uses a law formulated by Fourier [2], which has the following form for one-dimensional problems:

$$\dot{q} = -\lambda \frac{\partial T}{\partial x}, \tag{1.1}$$

where, $\dot{q}$ is the heat flux expressed in W/m^2, λ – a thermal conductivity in W/(m·K), T – a temperature in °C or K, while x – a coordinate in m.

The minus sign in (1.1) testifies to the fact that heat flows in the direction of the decreasing temperature. Therefore, derivative $\partial T/\partial x$ is negative (Fig. 1.1), so in order to obtain positive value $\dot{q}$, the minus sign occurs on the right side of (1.1). In such a case, the direction of x axis and of the heat flux vector are the same.

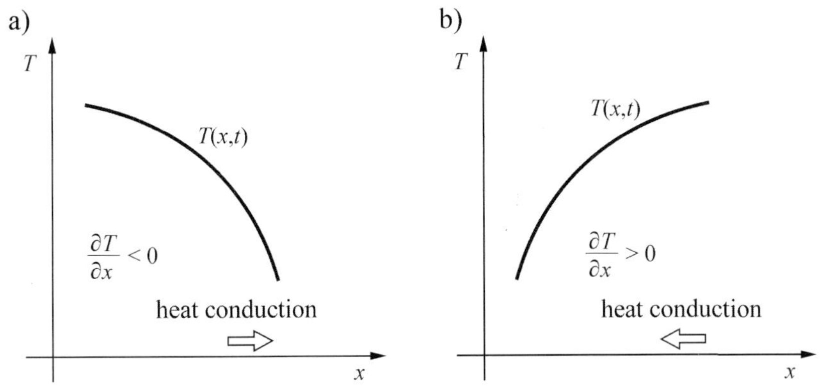

Fig. 1.1. Schematic diagram that illustrates the sign change of the first derivative of function $T(x, t)$, which describes temperature changes

If $T(x, t)$ is an increasing function (Fig. 1.1b), then one obtains a negative value of $\dot{q}$ from (1.1). If one assumes that the value of heat flux $\dot{q}$ is always positive, then the minus sign in (1.1) should be omitted in the case depicted in Fig. 1.1b.

In general, the heat flux vector is defined by *Fourier law*

$$\dot{\mathbf{q}} = -\lambda \nabla T\,, \tag{1.2}$$

where, ∇T = grad T is a temperature gradient, a right angle vector to an isothermic surface turned in the direction of a given point, where the function increases most rapidly.

Hamiltonian vectorial operator, also described as *nabla*, has the following form in the three basic coordinate systems:

- in the Cartesian coordinate system (x, y, z) (Fig. 1.2a)

$$\nabla = \mathbf{e}_x \frac{\partial}{\partial x} + \mathbf{e}_y \frac{\partial}{\partial y} + \mathbf{e}_z \frac{\partial}{\partial z}\,, \tag{1.3}$$

where, $\mathbf{e}_x = \mathbf{i}$, $\mathbf{e}_y = \mathbf{j}$, $\mathbf{e}_z = \mathbf{k}$,

- in the cylindrical coordinate system(r, θ, z) (Fig. 1.2b)

$$\nabla = \mathbf{e}_r \frac{\partial}{\partial r} + \mathbf{e}_\theta \frac{1}{r}\frac{\partial}{\partial \theta} + \mathbf{e}_z \frac{\partial}{\partial z}\,, \tag{1.4}$$

- in the spherical coordinate system (r, θ, φ) (Fig. 1.2c)

$$\nabla = \mathbf{e}_r \frac{\partial}{\partial r} + \mathbf{e}_\theta \frac{1}{r}\frac{\partial}{\partial \theta} + \mathbf{e}_\varphi \frac{1}{r\sin\theta}\frac{\partial}{\partial \varphi}\,, \tag{1.5}$$

where unit vectors $\mathbf{e}_x$, $\mathbf{e}_y$, $\mathbf{e}_z$, $\mathbf{e}_r$, $\mathbf{e}_\theta$, $\mathbf{e}_z$ and $\mathbf{e}_r$, $\mathbf{e}_\theta$, $\mathbf{e}_\varphi$, constitute an orthogonal local base in the Cartesian, cylindrical and spherical coordinate system respectively.

If Hamiltonian operator is known (1.3–1.5), it is easy to write a heat conduction equation in different coordinate systems. Thermal conductivity

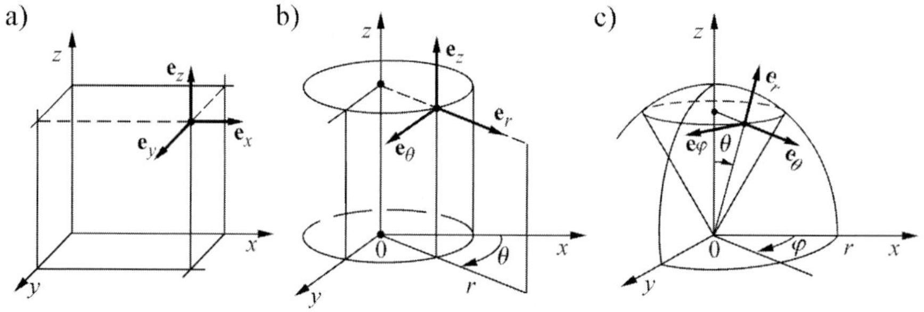

Fig. 1.2. Orthogonal coordinate systems: a) cartesian, b) cylindrical, c) spherical

λ may be temperature or location dependent. For isotropic bodies, the thermal conductivity is a scalar. In the case of anisotropic bodies, the heat conduction coefficient is a 2nd order symmetrical tensor. Fourier Law has in this case the following form:

$$\dot{\mathbf{q}} = -\Lambda \cdot \nabla T, \tag{1.6}$$

where $\mathbf{\Lambda}$ is a 2nd order symmetrical tensor

$$\Lambda = \begin{bmatrix} \lambda_{xx} & \lambda_{xy} & \lambda_{xz} \\ \lambda_{yx} & \lambda_{yy} & \lambda_{yz} \\ \lambda_{zx} & \lambda_{zy} & \lambda_{zz} \end{bmatrix}. \tag{1.7}$$

Therefore, the coordinates of the heat flux vector have in the Cartesian coordinate system the following forms:

$$\begin{aligned} \dot{q}_x &= -\left(\lambda_{xx}\frac{\partial T}{\partial x} + \lambda_{xy}\frac{\partial T}{\partial y} + \lambda_{xz}\frac{\partial T}{\partial z}\right), \\ \dot{q}_y &= -\left(\lambda_{yx}\frac{\partial T}{\partial x} + \lambda_{yy}\frac{\partial T}{\partial y} + \lambda_{yz}\frac{\partial T}{\partial z}\right), \\ \dot{q}_z &= -\left(\lambda_{zx}\frac{\partial T}{\partial x} + \lambda_{zy}\frac{\partial T}{\partial y} + \lambda_{zz}\frac{\partial T}{\partial z}\right). \end{aligned} \tag{1.8}$$

It can be proved that it is always possible to find three mutually orthogonal directions in space, such that

$$\begin{aligned} \lambda_{ij} &\neq 0 \quad \text{for} \quad i = j, \\ \lambda_{ij} &= 0 \quad \text{for} \quad i \neq j. \end{aligned} \tag{1.9}$$

These directions are called the *principal directions of an anisotropic body.* When principal directions are parallel to an axis of an assumed coordinate system, the conduction tensor is simplified to the following form:

$$\Lambda = \begin{bmatrix} \lambda_{xx} & 0 & 0 \\ 0 & \lambda_{yy} & 0 \\ 0 & 0 & \lambda_{zz} \end{bmatrix}. \tag{1.10}$$

Heat flux vector components are then defined as follow:

$$\dot{q}_x = -\lambda_{xx}\frac{\partial T}{\partial x}, \qquad \dot{q}_y = -\lambda_{yy}\frac{\partial T}{\partial y}, \qquad \dot{q}_z = -\lambda_{zz}\frac{\partial T}{\partial z}. \tag{1.11}$$

Heat flux normal component on the body's surface is defined using the following formula:

$$\dot{\mathbf{q}} \cdot \mathbf{n} = -\Lambda \cdot \nabla T \cdot \mathbf{n} = \left(\dot{q}_x n_x + \dot{q}_y n_y + \dot{q}_z n_z \right), \tag{1.12}$$

where, $n_x = \cos(n, x)$, $n_y = \cos(n, y)$, $n_z = \cos(n, z)$ are directional cosinuses of a normal to a surface.

If heat flow reaches the body surface, then the product $\dot{\mathbf{q}} \cdot \mathbf{n}$ has a negative value, since angle φ between normal $\mathbf{n}$ directed to the outside of the body and heat flow $\dot{\mathbf{q}}$ directed to the inside of the body is larger than $\pi/2$. Scalar product $\dot{\mathbf{q}} \cdot \mathbf{n}$ is then smaller than zero.

Literature

1. Bird RB, Stewart WE, Lightfoot EN (2002) Transport Phenomena. Sec. Ed., Wiley, New York
2. Fourier JB (1822) Théorie analytique de la chaleur. Paris
3. Trajdos T (1971) Tensor analysis. Mathematics. Engineer's guide (in Polish). WNT, Warszawa

2 Mass and Energy Balance Equations

In this chapter, we will discuss mass and energy conservation equations while allowing for the fact that a solid can be mobile. Such situation occurs in processes of continual steel casting, during the transport of loose materials and in number of other processes.

2.1 Mass Balance Equation for a Solid that Moves at an Assigned Velocity

Mass and energy balance will be calculated for a finite sector of a conductive area with a time-invariable volume V and surface A (Fig. 2.1).

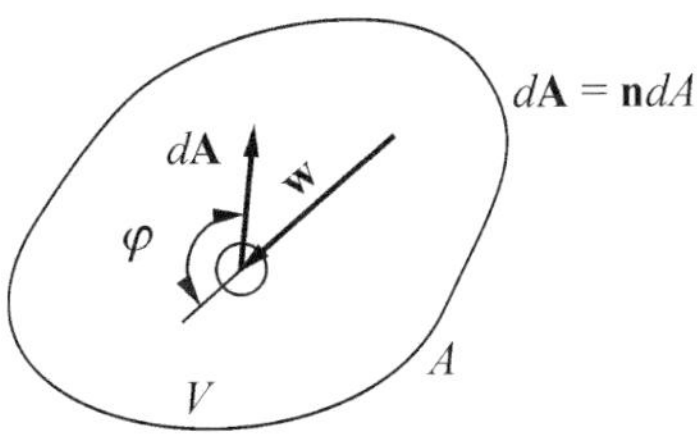

Fig. 2.1. Mass flow with velocity **w** to the inside of volume V through surface A

Only a normal component of the mass flow penetrates through the external surface to the interior

$$\dot{m}_n = \int_A \rho w_n dA\,, \tag{2.1}$$

where

$$w_n = -\mathbf{w} \cdot \mathbf{n}\,. \tag{2.2}$$

One should note the scalar product sign

$$\mathbf{w} \cdot \mathbf{n} = |\mathbf{w}|\,|\mathbf{n}| \cos\varphi = w\cos\varphi = w_n\,, \tag{2.3}$$

since $|\mathbf{n}| = 1$ (Fig. 2.2).

If mass flow is directed to the interior of the control volume V, it should have a positive value. Normal vector is a unitary vector directed to the outside. This means that for angles $\varphi > 90°$, when $\mathbf{w}$ is directed to the inside of the control volume, product $\mathbf{w} \cdot \mathbf{n}$ has a negative value. In order to obtain a positive value of $\mathbf{w} \cdot \mathbf{n}$ product, when $\mathbf{w}$ is directed to the inside, one should add a minus sign in front of this product.

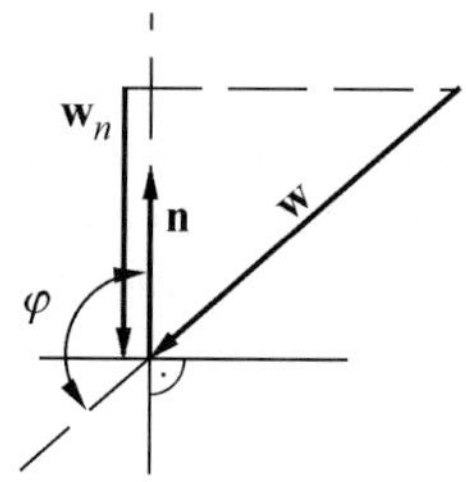

Fig. 2.2. The determination of a normal component $\mathbf{w}_n$ of $\mathbf{w}$ vector

Mass balance conservation equation has the following form:

$$\frac{\partial}{\partial t}\int_V \rho \, dV = -\int_A \rho \mathbf{w} \cdot d\mathbf{A}, \tag{2.4}$$

$$\frac{\partial}{\partial t}\int_V \rho dV = -\int_A \rho \mathbf{w} \cdot \mathbf{n} dA. \tag{2.5}$$

Gauss–Ostrogradski equation is used to transform the right-hand-side

$$\int_A \mathbf{F} \cdot \mathbf{n} dA = \int_V \nabla \cdot \mathbf{F} dV, \tag{2.6}$$

where $\mathbf{F}$ is a vector of continual partial derivatives, while ∇ is a Hamiltonian operator, formulated as

$$\nabla = \mathbf{i}\frac{\partial}{\partial x} + \mathbf{j}\frac{\partial}{\partial y} + \mathbf{k}\frac{\partial}{\partial z}. \tag{2.7}$$

After the transformation of the right-hand-side of (2.5) with (2.6), one obtains

$$\frac{\partial}{\partial t}\int_V \rho \, dV = -\int_V \nabla \cdot (\rho \mathbf{w}) \, dV.$$

Since the control volume boundaries are not time-dependent, the time derivative can be inserted in place of an integral. Additionally, after moving the right-hand-side of the equation to the left side, one obtains

$$\int_V \left[\frac{\partial \rho}{\partial t} + \nabla \cdot (\rho \mathbf{w}) \right] dV = 0 . \tag{2.8}$$

Equation (2.8) is satisfied for every volume V, including a small volume. If $V \to 0$, then one obtains from (2.8) the following:

$$\frac{\partial \rho}{\partial t} + \nabla \cdot (\rho \mathbf{w}) = 0 . \tag{2.9}$$

This is a continuity equation written in a differential form. For Cartesian coordinates, it assumes the form

$$\frac{\partial \rho}{\partial t} + \frac{\partial (\rho w_x)}{\partial x} + \frac{\partial (\rho w_y)}{\partial y} + \frac{\partial (\rho w_z)}{\partial z} = 0 . \tag{2.10}$$

Substantial derivative will be entered below in order to shorten the notation.

$$\frac{D}{Dt} = \frac{\partial}{\partial t} + w_x \frac{\partial}{\partial x} + w_y \frac{\partial}{\partial y} + w_z \frac{\partial}{\partial z} = \frac{\partial}{\partial t} + \mathbf{w} \cdot \nabla . \tag{2.11}$$

Such derivative shows how fast a parameter changes in time from the point of view of an observer, who moves along with the substance itself.

Continuity (2.9) can be expressed then in the following form:

$$\frac{D\rho}{Dt} + \rho (\nabla \cdot \mathbf{w}) = 0 . \tag{2.12}$$

2.2 Inner Energy Balance Equation

Inner energy balance equation for the control volume has the following form:

$$\begin{aligned} \frac{\partial}{\partial t} \int_V \rho u dV = & - \int_A \rho u \mathbf{w} \cdot \mathbf{n} dA - \int_V \rho p \frac{Dv}{Dt} dV - \\ & - \int_A \dot{\mathbf{q}} \cdot \mathbf{n} dA + \int_V \dot{q}_v dV + \int_V \Phi dV . \end{aligned} \tag{2.13}$$

Below, we will analyze the inner energy general balance equation for gases, liquids or solids, which move at $\mathbf{w}$ velocity, since heat conduction occurs not only in solids but also in gases and liquids. The integral can be found on the left-hand-side of the (2.13); it shows the inner energy changes that occur in time in volume V. The first term on the right-hand-side represents the inner energy flow, which moves to the control volume.

The second term represents reversible compression work of a medium contained within the control volume with respect to a unit of time. With consideration to the equation of continuity (2.12), the second expression on the right-hand-side (2.13) can be formulated in the following way:

$$-\int_V \rho p \frac{Dv}{Dt} dV = \int_V \frac{p}{\rho} \frac{D\rho}{Dt} dV = -\int_V p(\nabla \cdot \mathbf{w}) dV . \tag{2.14}$$

Reversible compression power is utilized for a density change of a medium ρ and contributes to a change in inner energy contained within the control volume V.

The third expression on the right-hand-side of the equation is the conduction-transferred energy flow. The fourth expression represents the power of volumetric heat sources of density $\dot{q}_v$, which can be the function of position $\mathbf{r}$, temperature T or time t. Dissipation function Φ is an irreversible power of internal friction forces, which influence the moving liquid particles. From Gauss–Ostrogradski theorem (2.6) applied to the first and third expression on the right-hand-side of the (2.13) and upon consideration of (2.14), one obtains

$$\int_V \left[\frac{\partial}{\partial t}(\rho u) + \nabla \cdot (\rho u \mathbf{w}) + p(\nabla \cdot \mathbf{w}) + \nabla \cdot \dot{\mathbf{q}} - \dot{q}_v - \Phi \right] dV = 0 . \tag{2.15}$$

when $V \to 0$, the (2.15) assumes the following form:

$$\frac{\partial}{\partial t}(\rho u) + \nabla \cdot (\rho u \mathbf{w}) = -\nabla \cdot \dot{\mathbf{q}} - p(\nabla \cdot \mathbf{w}) + \dot{q}_v + \Phi . \tag{2.16}$$

The left side of the (2.16) can be transformed in the following way:

$$\begin{aligned}
&\frac{\partial(\rho u)}{\partial t} + \nabla \cdot (\rho u \mathbf{w}) = \rho \frac{\partial u}{\partial t} + u \frac{\partial \rho}{\partial t} + \frac{\partial}{\partial x}(u \rho w_x) + \\
&+ \frac{\partial}{\partial y}(u \rho w_y) + \frac{\partial}{\partial z}(u \rho w_z) = \\
&= u \left[\frac{\partial \rho}{\partial t} + \frac{\partial(\rho w_x)}{\partial x} + \frac{\partial(\rho w_y)}{\partial y} + \frac{\partial(\rho w_z)}{\partial z} \right] + \\
&+ \rho \left(\frac{\partial u}{\partial t} + w_x \frac{\partial u}{\partial x} + w_y \frac{\partial u}{\partial y} + w_z \frac{\partial u}{\partial z} \right) = u \left[\frac{\partial \rho}{\partial t} + \nabla \cdot (\rho \mathbf{w}) \right] + \rho \frac{Du}{Dt} .
\end{aligned} \tag{2.17}$$

By allowing for the equation of continuity (2.9) in (2.17), one obtains

$$\frac{\partial(\rho u)}{\partial t}+\nabla\cdot(\rho u\mathbf{w})=\rho\frac{Du}{Dt}. \tag{2.18}$$

The first expression on the right-hand-side of (2.16), with respect to Fourier Law (1.6), can be written in the form

$$-\nabla\cdot\dot{\mathbf{q}}=\nabla\cdot(\Lambda\nabla T). \tag{2.19}$$

In order to transform the second expression on the right-hand-side of (2.16), the equation of continuity will be applied (2.12)

$$\frac{D\rho}{Dt}+\rho\nabla\cdot\mathbf{w}=0, \tag{2.20}$$

from which one obtains

$$\nabla\cdot\mathbf{w}=-\frac{1}{\rho}\frac{D\rho}{Dt}. \tag{2.21}$$

Term

$$-p\nabla\cdot\mathbf{w}=\frac{p}{\rho}\frac{D\rho}{Dt} \tag{2.22}$$

can be transformed in the following way:

$$\begin{aligned}-p\nabla\cdot\mathbf{w}&=\frac{p}{\rho}\frac{D\rho}{Dt}=\frac{p}{\rho}\frac{D\rho}{Dt}+\rho\frac{D}{Dt}\left(\frac{p}{\rho}\right)-\rho\frac{D}{Dt}\left(\frac{p}{\rho}\right)=\\&=\frac{p}{\rho}\frac{D\rho}{Dt}+\rho\left(\rho\frac{Dp}{Dt}-p\frac{D\rho}{Dt}\right)\frac{1}{\rho^2}-\rho\frac{D}{Dt}(pv)=\frac{Dp}{Dt}-\rho\frac{D}{Dt}(pv).\end{aligned} \tag{2.23}$$

By allowing for (2.18), (2.19) and (2.23) in the inner energy balance equation (2.16) after the performed transformations, one has

$$\rho\frac{Di}{Dt}=\nabla\cdot(\Lambda\nabla T)+\dot{q}_v+\frac{Dp}{Dt}+\Phi, \tag{2.24}$$

where, $i = u + pv$ is the specific enthalpy of a medium expressed in J/kg. Since the enthalpy differential of a substance $i = i(p, T)$ can be formulated [13, 16, 26] as

$$di=\left(\frac{\partial i}{\partial T}\right)_p dT+\left(\frac{\partial i}{\partial p}\right)_T dp=c_p dT+v(1-\beta T)\,dp, \tag{2.25}$$

the (2.24) can be written in the form

$$\rho c_p \frac{DT}{Dt} = \nabla \cdot (\Lambda \nabla T) + \dot{q}_v + \beta T \frac{Dp}{Dt} + \Phi, \tag{2.26}$$

where

$$\beta = \frac{1}{v}\left(\frac{\partial v}{\partial T}\right)_p = -\frac{1}{\rho}\left(\frac{\partial \rho}{\partial T}\right)_p \tag{2.27}$$

is the volumetric expansion coefficient.

For liquids and solids, regarded as incompressible, one can assume that

$$c_p \approx c_v = c \quad \text{and} \quad \beta = 0. \tag{2.28}$$

In the case of a perfect gas, which fulfils Clapeyron equation $pv = RT$, the volumetric expansion coefficient amounts to

$$\beta = \frac{1}{T} \tag{2.29}$$

and *energy balance equation* (2.26) has the form

$$\rho c_p \frac{DT}{Dt} = \nabla \cdot (\Lambda \nabla T) + \dot{q}_v + \frac{Dp}{Dt} + \Phi. \tag{2.30}$$

Equation (2.26) is a general energy balance equation for solids, gases or liquids, which move (flow) at **w** velocity. Thermophysical properties λ, c and ρ may be position or temperature dependent. A body can also be anisotropic, when the thermal conductivity is direction-dependent. Rate of energy generation per unit volume of heat sources $\dot{q}_v$ can be a function of position, temperature and time. On the basis of the general equation (2.26), energy balance equations will be written in a Cartesian, cylindrical and spherical coordinate system. Term Dp/Dt will be omitted, since it is very small. It will be only taken into consideration in the cases of supersonic flows.

2.2.1 Energy Balance Equations in Three Basic Coordinate Systems

Axes of a coordinate system will be selected in such a way that their directions will be the same as the general directions of an anisotropic body. Furthermore, we will present energy balance equations for different coordinate systems under the assumption that a liquid or a solid is incompressible.

Cartesian Coordinate System (Fig. 1.2a)

$$\dot{q}_x = -\lambda_{xx}\frac{\partial T}{\partial x}, \qquad \dot{q}_y = -\lambda_{yy}\frac{\partial T}{\partial y}, \qquad \dot{q}_z = -\lambda_{zz}\frac{\partial T}{\partial z}; \tag{2.31}$$

$$\begin{aligned} -\nabla\cdot\dot{\mathbf{q}} &= \nabla\cdot\left(\Lambda\nabla T\right) = -\frac{\partial \dot{q}_x}{\partial x} - \frac{\partial \dot{q}_y}{\partial y} - \frac{\partial \dot{q}_z}{\partial z} = \\ &= \frac{\partial}{\partial x}\left(\lambda_{xx}\frac{\partial T}{\partial x}\right) + \frac{\partial}{\partial y}\left(\lambda_{yy}\frac{\partial T}{\partial y}\right) + \frac{\partial}{\partial z}\left(\lambda_{zz}\frac{\partial T}{\partial z}\right), \end{aligned} \tag{2.32}$$

$$\frac{D}{Dt} = \frac{\partial}{\partial t} + w_x\frac{\partial}{\partial x} + w_y\frac{\partial}{\partial y} + w_z\frac{\partial}{\partial z}\,. \tag{2.33}$$

By allowing for (2.32) and (2.33) in (2.30), one obtains a heat balance equation

$$\begin{aligned} &\rho c_p\left(\frac{\partial T}{\partial t} + w_x\frac{\partial T}{\partial x} + w_y\frac{\partial T}{\partial y} + w_z\frac{\partial T}{\partial z}\right) = \\ &= \frac{\partial}{\partial x}\left(\lambda_{xx}\frac{\partial T}{\partial x}\right) + \frac{\partial}{\partial y}\left(\lambda_{yy}\frac{\partial T}{\partial y}\right) + \frac{\partial}{\partial z}\left(\lambda_{zz}\frac{\partial T}{\partial z}\right) + \dot{q}_v + \Phi, \end{aligned} \tag{2.34}$$

where the dissipation function is defined [20] by the following expression

$$\begin{aligned} \Phi = 2\mu\Bigg[&\left(\frac{\partial w_x}{\partial x}\right)^2 + \left(\frac{\partial w_y}{\partial y}\right)^2 + \left(\frac{\partial w_z}{\partial z}\right)^2 + \frac{1}{2}\left(\frac{\partial w_y}{\partial x} + \frac{\partial w_x}{\partial y}\right)^2 + \\ &+ \frac{1}{2}\left(\frac{\partial w_z}{\partial y} + \frac{\partial w_y}{\partial z}\right)^2 + \frac{1}{2}\left(\frac{\partial w_x}{\partial z} + \frac{\partial w_z}{\partial x}\right)^2\Bigg], \end{aligned} \tag{2.35}$$

where μ stands for a dynamic viscosity.

If a body is immovable, then $\mathbf{w} = 0$ and $\Phi = 0$. Thermophysical properties can be temperature or position-dependent. If we accept such an assumption (2.34) is simplified to a form

$$\rho c_p\frac{\partial T}{\partial t} = \frac{\partial}{\partial x}\left(\lambda_{xx}\frac{\partial T}{\partial x}\right) + \frac{\partial}{\partial y}\left(\lambda_{yy}\frac{\partial T}{\partial y}\right) + \frac{\partial}{\partial z}\left(\lambda_{zz}\frac{\partial T}{\partial z}\right) + \dot{q}_v\,. \tag{2.36}$$

Cylindrical Coordinate System (Fig. 1.2b)

$$\dot{q}_r = -\lambda_{rr}\frac{\partial T}{\partial r},\qquad \dot{q}_\theta = -\lambda_{\theta\theta}\frac{1}{r}\frac{\partial T}{\partial \theta},\qquad \dot{q}_z = -\lambda_{zz}\frac{\partial T}{\partial z};\tag{2.37}$$

$$\begin{aligned}-\nabla\cdot\dot{\mathbf{q}} = \nabla\cdot\left(\Lambda\nabla T\right) = \frac{1}{r}\frac{\partial}{\partial r}\left(r\lambda_{rr}\frac{\partial T}{\partial r}\right)+\\ +\frac{1}{r^2}\frac{\partial}{\partial\theta}\left(\lambda_{\theta\theta}\frac{\partial T}{\partial\theta}\right)+\frac{\partial}{\partial z}\left(\lambda_{zz}\frac{\partial T}{\partial z}\right),\end{aligned}\tag{2.38}$$

$$\frac{D}{Dt} = \frac{\partial}{\partial t} + w_r\frac{\partial}{\partial r} + \frac{w_\theta}{r}\frac{\partial}{\partial\theta} + w_z\frac{\partial}{\partial z}.\tag{2.39}$$

By allowing for (2.38) and (2.39) in (2.30), one obtains the equation below:

$$\begin{aligned}\rho c_p\left(\frac{\partial T}{\partial t} + w_r\frac{\partial T}{\partial r} + \frac{w_\theta}{r}\frac{\partial T}{\partial\theta} + w_z\frac{\partial T}{\partial z}\right) =\\ = \frac{1}{r}\frac{\partial}{\partial r}\left(r\lambda_{rr}\frac{\partial T}{\partial r}\right) + \frac{1}{r^2}\frac{\partial}{\partial\theta}\left(\lambda_{\theta\theta}\frac{\partial T}{\partial\theta}\right) + \frac{\partial}{\partial z}\left(\lambda_{zz}\frac{\partial T}{\partial z}\right) + \dot{q}_v + \Phi,\end{aligned}\tag{2.40}$$

where the dissipation function is defined by the following formula:

$$\begin{aligned}\Phi = 2\mu\left\{\left(\frac{\partial w_r}{\partial r}\right)^2 + \left[\frac{1}{r}\left(\frac{\partial w_\theta}{\partial\theta} + w_r\right)\right]^2 + \left(\frac{\partial w_z}{\partial z}\right)^2\right\} +\\ +\mu\left\{\left(\frac{\partial w_\theta}{\partial z} + \frac{1}{r}\frac{\partial w_z}{\partial\theta}\right)^2 + \left(\frac{\partial w_z}{\partial r} + \frac{\partial w_r}{\partial z}\right)^2 + \left[\frac{1}{r}\frac{\partial w_r}{\partial\theta} + r\frac{\partial}{\partial r}\left(\frac{w_\theta}{r}\right)\right]^2\right\}.\end{aligned}\tag{2.41}$$

For an immobile solid $\mathbf{w} = 0$ and $\Phi = 0$. The (2.40) is then simplified to a form

$$\rho c_p\frac{\partial T}{\partial t} = \frac{1}{r}\frac{\partial}{\partial r}\left(r\lambda_{rr}\frac{\partial T}{\partial r}\right) + \frac{1}{r^2}\frac{\partial}{\partial\theta}\left(\lambda_{\theta\theta}\frac{\partial T}{\partial\theta}\right) + \frac{\partial}{\partial z}\left(\lambda_{zz}\frac{\partial T}{\partial z}\right) + \dot{q}_v.\tag{2.42}$$

Spherical Coordinate System (Fig. 1.2c)

$$\dot{q}_r = -\lambda_{rr}\frac{\partial T}{\partial r},\qquad \dot{q}_\theta = -\lambda_{\theta\theta}\frac{1}{r}\frac{\partial T}{\partial\theta},\qquad \dot{q}_\varphi = -\lambda_{\varphi\varphi}\frac{1}{r\sin\theta}\frac{\partial T}{\partial\varphi};\tag{2.43}$$

$$-\nabla\cdot\mathbf{q}=\nabla\cdot\left(\Lambda\nabla T\right)=\frac{1}{r^2}\frac{\partial}{\partial r}\left(r^2\lambda_{rr}\frac{\partial T}{\partial r}\right)+$$
$$+\frac{1}{r^2\sin\theta}\frac{\partial}{\partial\theta}\left(\lambda_{\theta\theta}\sin\theta\frac{\partial T}{\partial\theta}\right)+\frac{1}{r^2\sin^2\theta}\frac{\partial}{\partial\varphi}\left(\lambda_{\varphi\varphi}\frac{\partial T}{\partial\varphi}\right),\tag{2.44}$$

$$\frac{D}{Dt}=\frac{\partial}{\partial t}+w_r\frac{\partial}{\partial r}+\frac{w_\theta}{r}\frac{\partial}{\partial\theta}+\frac{w_\varphi}{r\sin\theta}\frac{\partial}{\partial\varphi}.\tag{2.45}$$

By allowing for (2.44) and (2.45) in (2.30), one obtains

$$\rho c_p\left(\frac{\partial T}{\partial t}+w_r\frac{\partial T}{\partial r}+\frac{w_\theta}{r}\frac{\partial T}{\partial\theta}+\frac{w_\varphi}{r\sin\theta}\frac{\partial T}{\partial\varphi}\right)=$$
$$=\frac{1}{r^2}\frac{\partial}{\partial r}\left(r^2\lambda_{rr}\frac{\partial T}{\partial r}\right)+\frac{1}{r^2\sin\theta}\frac{\partial}{\partial\theta}\left(\lambda_{\theta\theta}\sin\theta\frac{\partial T}{\partial\theta}\right)+\tag{2.46}$$
$$+\frac{1}{r^2\sin^2\theta}\frac{\partial}{\partial\varphi}\left(\lambda_{\varphi\varphi}\frac{\partial T}{\partial\varphi}\right)+\dot{q}_v+\Phi,$$

$$\Phi=2\mu\left[\left(\frac{\partial w_r}{\partial r}\right)^2+\frac{1}{r^2}\left(\frac{\partial w_\theta}{\partial\theta}+w_r\right)^2+\right.$$
$$\left.+\frac{1}{r^2}\left(\frac{1}{\sin\theta}\frac{\partial w_\varphi}{\partial\varphi}+w_r+w_\theta\operatorname{ctg}\theta\right)^2\right]+$$
$$+\mu\left\{\left[r\frac{\partial}{\partial r}\left(\frac{w_\theta}{r}\right)+\frac{1}{r}\frac{\partial w_r}{\partial\theta}\right]^2+\left[\frac{1}{r\sin\theta}\frac{\partial w_r}{\partial\varphi}+r\frac{\partial}{\partial r}\left(\frac{w_\varphi}{r}\right)\right]^2+\right.\tag{2.47}$$
$$\left.+\frac{1}{r^2}\left[\sin\theta\frac{\partial}{\partial\theta}\left(\frac{w_\varphi}{\sin\theta}\right)+\frac{1}{\sin\theta}\frac{\partial w_\theta}{\partial\varphi}\right]^2\right\}.$$

For an immobile solid $\mathbf{w}=0$ and $\Phi=0$. The (2.46) is then simplified to a form

$$\rho c_p \frac{\partial T}{\partial t} = \frac{1}{r^2}\frac{\partial}{\partial r}\left(r^2 \lambda_{rr} \frac{\partial T}{\partial r}\right) + \frac{1}{r^2 \sin\theta}\frac{\partial}{\partial \theta}\left(\lambda_{\theta\theta} \sin\theta \frac{\partial T}{\partial \theta}\right) + \\ + \frac{1}{r^2 \sin^2\theta}\frac{\partial}{\partial \varphi}\left(\lambda_{\varphi\varphi} \frac{\partial T}{\partial \varphi}\right) + \dot{q}_v . \tag{2.48}$$

If a body is isotropic, then $\lambda_{rr} = \lambda_{\theta\theta} = \lambda_{\varphi\varphi} = \lambda$. Thermal conductivity λ can be a function of temperature and position. Thermophysical properties of solids are listed in Appendix B at the back of the book.

2.3 Hyperbolic Heat Conduction Equation

In the Subsect. 2.2, transient heat conduction equations were worked out for immobile solids with Cartesian (2.36), cylindrical (2.42) and spherical coordinates (2.48). These are a parabolic type of equations, based on Fourier Law, which assumes that heat diffuses at an infinitely fast rate. This means that any disturbances on the body's edges are immediately perceived in the form of temperature changes within the whole body volume. In the case of extremely rarefied gases, as well as helium and dialectric crystals at very low temperatures, the rate of heat flow propagation w_g has a finite value. In such instances, the heat flux is formulated using the following *constitutive equation* [5, 25]:

$$\dot{\mathbf{q}} + \tau \frac{\partial \dot{\mathbf{q}}}{\partial t} = -\lambda \nabla T , \tag{2.49}$$

where τ is the relaxation time. If one assumes that $\nabla T = 0$, then (2.49) becomes a homogeneous differential equation of 1st order, whose solution is proportional to expression $\exp(-t/\tau)$.

Relaxation time τ is thus an equivalent of a time constant present in an expression that defines the temporal temperature flow of a body with a concentrated heat storage capacity. Energy balance equation (2.30) has in such a case the following form:

$$\rho c_p \frac{\partial T}{\partial t} = -\nabla \cdot \dot{\mathbf{q}} + \dot{q}_v . \tag{2.50}$$

If one assumes that ρ and c_p are constant and differentiates equation with respect to time (2.50), one gets

$$\rho c_p \frac{\partial^2 T}{\partial t^2} = -\nabla \cdot \frac{\partial \dot{\mathbf{q}}}{\partial t} + \frac{\partial \dot{q}_v}{\partial t} . \tag{2.51}$$

Divergence from both sides of (2.49) is formulated as

$$\nabla \cdot \dot{\mathbf{q}} + \tau \nabla \cdot \frac{\partial \dot{\mathbf{q}}}{\partial t} = -\nabla \cdot \left(\lambda \nabla T \right). \tag{2.52}$$

By allowing for (2.50) and (2.51) in (2.52), one obtains

$$-\rho c_p \frac{\partial T}{\partial t} + \dot{q}_v - \tau \left(\rho c_p \frac{\partial^2 T}{\partial t^2} - \frac{\partial \dot{q}_v}{\partial t} \right) = -\nabla \cdot \left(\lambda \nabla T \right), \tag{2.53}$$

from where one gets, after assuming that thermal conductivity is constant

$$\frac{\partial T}{\partial t} + \tau \frac{\partial^2 T}{\partial t^2} = a \nabla^2 T + \frac{1}{\rho c_p} \left(\dot{q}_v + \tau \frac{\partial \dot{q}_v}{\partial t} \right), \tag{2.54}$$

where $a = \lambda / \rho\, c_p$, while $\nabla^2 T$ is a Laplace operator.

Equation (2.54) is a partial hyperbolic equation, which describes the thermal wave propagation with the finite velocity

$$w_q = \sqrt{\frac{a}{\tau}}. \tag{2.55}$$

Relaxation time τ is very small and for an alluminium, for example, it comes to about 10^{-11} s, while for a liquid helium to about 10^{-6} s at extremely low temperatures. Since for a liquid helium with a temperature close to an absolute zero, the diffusivity a amounts to 10 m^2/s, the velocity of thermal wave propagation is

$$w_q = \sqrt{\frac{10}{10^{-6}}} = 3162 \text{ m/s}. \tag{2.56}$$

It can be seen, therefore, that the value of w_q even in this case is large and for the majority of calculations done for transient heat conduction processes, it is assumed that $\tau = 0$ s, i.e. the velocity of heat propagation is infinitely large.

2.4 Initial and Boundary Conditions

Transient heat conduction problems are initial-boundary problems for which one is required to assign appropriate initial and boundary conditions. *Initial conditions*, also called *Cauchy conditions*, are temperature values of a body at its initial moment $t_0 = 0$ s.

$$T(\mathbf{r},t)\big|_{t=0} = T_0(\mathbf{r}). \tag{2.57}$$

If temperature distribution is written in the form of a hyperbolic equation of heat conduction (2.54), then a initial derivative value must also be given.

$$\left.\frac{\partial T(\mathbf{r},t)}{\partial t}\right|_{t=0} = \dot{T}_0(\mathbf{r}), \tag{2.58}$$

where $\mathbf{r}$ is the positional vector of the analyzed point (a field vector in a given point). Symbol $\dot{T}$ stands for the first temperature derivative with respect to time $\dot{T} = \partial T/\partial t$. In practise, one rarely makes use of the hyperbolic equation (2.54); therefore, only an assigned condition (2.57) is indispensible in order to determine transient temperature distribution.

One can distinguish four basic types of boundary conditions [15, 27], which do not describe, nevertheless, all real conditions that occur in practise, such as body heating and cooling by radiation, the melting or freezing of bodies or complex heat transfer.

2.4.1 First Kind Boundary Conditions (Dirichlet Conditions)

Temperature distribution on the edge of body A is assigned as follows

$$T(\mathbf{r},t)\big|_{A} = T_A(\mathbf{r}_A,t), \tag{2.59}$$

where $\mathbf{r}_A$ is a positional vector of a point located on the body's surface. If temperature of the body surface, $T(\mathbf{r}_A,t)$ is known from measurements taken, then the boundary conditions can be formulated as boundary conditions of the Ist order.

2.4.2 Second Kind Boundary Conditions von Neumann Conditions)

$$\Lambda\nabla T\cdot\mathbf{n}\big|_{A} = \dot{q}(\mathbf{r}_A,t). \tag{2.60}$$

If x and y axes of a coordinate system are compatible with the main anisotropic axes, then the condition (2.60) assumes the form:

$$\left.\left(\lambda_{xx}\frac{\partial T}{\partial x}n_x + \lambda_{yy}\frac{\partial T}{\partial y}n_y + \lambda_{zz}\frac{\partial T}{\partial z}n_z\right)\right|_{A} = \dot{q}(\mathbf{r}_A,t), \tag{2.61}$$

where $n_x = \cos(n, x)$, $n_y = \cos(n, y)$ and $n_z = \cos(n, z)$ are directional cosinuses of a normal to a surface.

For isotropic bodies, condition (2.60) assumes the form

$$\left(\lambda \frac{\partial T}{\partial n}\right)\bigg|_A = \dot{q}\left(\mathbf{r}_A, t\right). \tag{2.62}$$

If a surface is thermally insulated, then

$$\frac{\partial T}{\partial n}\bigg|_A = 0. \tag{2.63}$$

The boundary condition of 2nd kind is frequently set on the surface of radiated bodies, e.g. on the surface of boilers' radiant tubes. Surface temperature of the tubes, $T_A(\mathbf{r}_A,t)$ is much lower than the temperature of combustion gases T_{sp} in a furnace chamber and practically does not affect the heat flux transferred by the outer surface of the tubes. If the heat flux from a body surface is known from measurements taken, then the boundary condition of 2nd kind can be applied irrespectively of the type of heat transfer present on the body surface. Condition (2.62) is often set when solving steady-state and transient inverse heat conduction problems [9, 22, 23]. If thermophysical properties of a body c, ρ and λ are temperature invariant, then the inverse problem becomes linear, thereby easier to solve, when a boundary condition of 1st or 2nd kind is applied on the body surface.

2.4.3 Third Kind Boundary Conditions

The boundary condition of 3rd kind is also known as *Robin boundary condition* [15] or *Newton law of cooling*. Its heat penetration coefficient, also called *heat transfer coefficient* [28], expresses the intensity of convective heat exchange. Coefficient α is dependent on the type of heat exchange that occurs on a body's surface, the fluid type, and on the velocity and direction of the fluid flow with regard to body's surface. Heat transfer coefficient α is also frequently a function of surface temperature or of the difference between surface temperature T_A and factor T_{cz}, e.g. during boiling, condensation and natural convection.

$$-\left(\Lambda \nabla T \cdot \mathbf{n}\right)\big|_A = \alpha\left(\mathbf{r}_\mathbf{A}, t, T_A\right)\left[T\left(\mathbf{r}_\mathbf{A}, t\right) - T_{cz}\right]. \tag{2.64}$$

If a body is isotropic, then condition (2.64) is simplified to a form

$$-\left(\lambda\frac{\partial T}{\partial n}\right)\bigg|_A = \alpha\left(T_A - T_{cz}\right). \tag{2.65}$$

The selected values of heat transfer coefficients are listed in Table 2.1. From the analysis of this table, it is evident that in the case of droplet condensation the surface temperature of a solid is close in value to a temperature of condensing vapour.

In practise, the application of the 3rd kind boundary condition encounters difficulties with respect to the determination of spatial heat transfer coefficient changes and the medium's temperature at small flow velocity. If a liquid remains in a state of rest, then as a result of natural convection the medium moves alongside the solid's surface demonstrating, at the same time, significant temperature pulsations. Due to this reason, it is difficult to define the medium's temperature $T_{cz}(t)$. It also should be added that spatial-temporal changes in the heat transfer coefficient on the surface of a solid can be determined when a conjugated heat transfer problem in a liquid and solid is solved using a computerized fluid mechanics program, which consists of different CFD methods (abbr. *Computer Fluid Dynamics*).

Due to a shortage, however, of competent models, which would describe turbulent liquid dynamics, and experimentally-determined constants, the CFD-determined heat transfer coefficients can significantly digress from experimentally determined coefficients.

Table 2.1. Approximate values of heat transfer coefficients

Heat exchange conditions		α [W/(m^2·K)]
Forced convection	Gases	10–500
	Oils	50–1700
	Water	300–12000
	liquid (melted) metals	1000–45000
Free convection	Gases	5–30
	Oils	10–350
	Water	100–1200
	liquid metals	1000–7000
Phase change	bubble boiling	2000–50000
	film boiling	100–300
	water vapour film condensation	4000–17000
	water vapour droplet condensation	30000–120000
	condensation of vapour of organic liquids	500–2300

2.4.4 Fourth Kind Boundary Conditions

The boundary conditions of 4th kind occur at the point where two body-surfaces meet (① and ② Fig. 2.3).

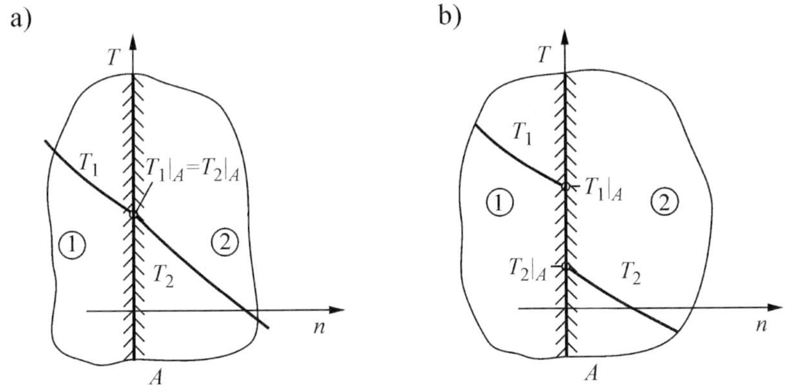

Fig. 2.3. The course of temperature on the boundary of two adjoining bodies : a) ideal contact, b) contact resistance on the boundary

If contact is ideal, then the temperature of the two bodies is identical at the point of contact. Furthermore, heat flux becomes uniform. In the case of an ideal contact, the following equalities are valid (Fig. 2.3a):

$$T_1\big|_A = T_2\big|_A, \tag{2.66}$$

$$\lambda_1 \frac{\partial T_1}{\partial n}\bigg|_A = \lambda_2 \frac{\partial T_2}{\partial n}\bigg|_A. \tag{2.67}$$

In reality, thermal resistance occurs at the point where two bodies meet and the bodies' temperatures at the point of contact are not the same (Fig. 2.3b).

Thermal resistance at the point of contact is characterized by means of the *contact heat transfer coefficient* α_{kt}, which is defined as follows

$$-\lambda_1 \frac{\partial T_1}{\partial n}\bigg|_A = \alpha_{kt}\left(T_1\big|_A - T_2\big|_A\right) = -\lambda_2 \frac{\partial T_2}{\partial n}\bigg|_A. \tag{2.68}$$

Coefficient α_{kt}, which characterizes the contact resistance, mainly depends on the roughness of a surface and on the pressure force of both bodies. The contact resistance can be significantly reduced by polishing the two touching surfaces and by moistening them with a liquid, e.g. silicon, oil or petroleum jelly.

2.4.5 Non-Linear Boundary Conditions

If a body surface is heated or cooled by radiation, then boundary condition is nonlinear

$$-\lambda \left.\frac{\partial T}{\partial n}\right|_A = \sigma F\left[\left(T|_A\right)^4 - T_r^4\right]. \tag{2.69}$$

Temperatures of surface $T|_A$ and surroundings T_r are expressed in degrees Kelvin. Symbol T_r denotes the temperature of surrounding walls, which the body can "see." It is not a temperature of a medium T_{cz} directly located by the body surface, as it is in the case of convection. Shape coefficient F is the emissivity function of the body surface and its surroundings and of mutual configuration between the body and surroundings. If a surface area of a convex or flat body A is significantly smaller than the surface area of surroundings A_r, then the heat flux on the body surface is only the emissivity function of the body surface ε and temperatures T_A and T_r:

$$-\lambda \left.\frac{\partial T}{\partial n}\right|_A = \varepsilon\sigma\left[\left(T|_A\right)^4 - T_r^4\right]. \tag{2.70}$$

Symbol $\sigma = 5.67\cdot10^{-8}$ W/(m$^2\cdot$K^4) is a Stefan-Boltzmann constant. Thermal exchange by radiation plays a large role even in the case when surface temperatures of a body $T|_A$ are relatively low, e.g. thermal exchange by radiation in central heating radiators can amount to 40% of the total heat flow transferred by radiators. In reality, thermal exchange by convection and radiation occurs simultaneously on the surfaces of bodies treated with gases and, to a smaller degree, with liquids. Heat flux on the body surface is formulated as follows:

$$-\lambda \left.\frac{\partial T}{\partial n}\right|_A = \alpha\left(T|_A - T_{cz}\right) + \sigma F\left[\left(T|_A\right)^4 - T_r^4\right].$$

If $A \ll A_r$, then on the basis of (2.70) one obtains

$$-\lambda \left.\frac{\partial T}{\partial n}\right|_A = \alpha\left(T|_A - T_{cz}\right) + \varepsilon\sigma\left[\left(T|_A\right)^4 - T_r^4\right]. \tag{2.71}$$

If $T_{cz} = T_r$, as, for example, in a room, whose temperature is identical to the temperature of the confining walls, the boundary condition (2.71) can be easily transformed into a form

$$-\lambda \left.\frac{\partial T}{\partial n}\right|_A = \left(\alpha + \alpha_r\right)\left(T|_A - T_r\right), \tag{2.72}$$

where α_r is a radiative heat transfer coefficient expressed as

Table 2.2. Emissivity of various surfaces with temperature of 300 K or with temperature from 300 to 500 K

Material and Surface Type	ε	Material and Surface Type	ε
Perfectly black body	1	Brick	
Shiny surface	0	-red	0.9–0.93
Metals		-refractory	0.9
Aluminium-smooth and polished	0.04–0.05	Wood	0.8–0.9
Aluminium-smooth and oxidized	0.11–0.12	Paints	
Aluminium-rough and oxidized	0.2–0.3	-aluminium	0.3–0.6
Aluminium-anodized	0.9–0.7	- minium	0.93
Bronze-polished	0.1	- oil, including white	0.9–0.96
Bronze-oxidized	0.6	Gypsum, marble, plaster	0.9–0.97
Chromium-polished	0.08–0.17	Burnt clay	0.91
Tin-polished	0.05	Soil	0.93–0.96
Tin-polished or galvanized	0.02–0.03	Rubber-hard	0.95
Copper-polished	0.04–0.05	Rubber-soft, gray, coarse	0.86
Copper-oxidized	0.87–0.83	Graphite	0.7
Nickel-polished	0.05–0.07	Tiles (white)	0.88
Lead-oxidized surface	0.6	Limestone	0.9–0.8
Platinum-polished	0.05	Ice (273 K)	
Mercury	0.1	- smooth	0.97
Silver-polished	0.01–0.02	-coarse	0.99
Carbon steel-polished	0.1	Mica	0.75
Carbon steel-oxidized surface	0.8	Paper	0.91–0.98
Stainless steel-polished	0.1	Sandstone	0.82
Stainless steel-oxidized surface	0.5–0.8	Cream-coloured sand	0.92
Tungsten-polished	0.03–0.05	Porcelain (glazed)	0.92
Non-metals		Rocks	0.88–0.95
Asbestos	0.96	Leather	0.95
Asphalt	0.93	Snow	0.8–0.9
Concrete	0.94	Pyrex glass	0.82–0.8
		Teflon	0.85–0.92
		Silicon carbide	0.9
		Water (a layer 1 mm thick or more)	0.96

$$\alpha_r = \varepsilon\sigma\left(T|_A + T_r\right)\left[\left(T|_A\right)^2 + T_r^2\right]. \tag{2.73}$$

It is worth to notice that $\alpha_r \neq 0$, even when $T|_A = T_r$. Coefficient α_r depends, to a large degree, on surface emissivity of the solid ε. Values of ε

for selected surfaces are listed in Table 2.2 [4, 21]. When $T_A << T_r$, one obtains from (2.70)

$$\lambda \frac{\partial T}{\partial n}\bigg|_A = \varepsilon\sigma\left(T_r\right)^4 = \dot{q}\,, \tag{2.74}$$

therefore, a non-linear boundary condition can be substituted by a 2nd kind boundary condition.

2.4.6 Boundary Conditions on the Phase Boundaries

If phase-changes occur, for instance during the process of melting or freezing, then a heat absorption or emission during the melting of a substance takes place on the phase boundary between a liquid and a solid. The boundary between the liquid and the solid phase $s(t)$ is time-variable; this is why this kind of problems are called *a free and movable boundary* (Fig. 2.4).

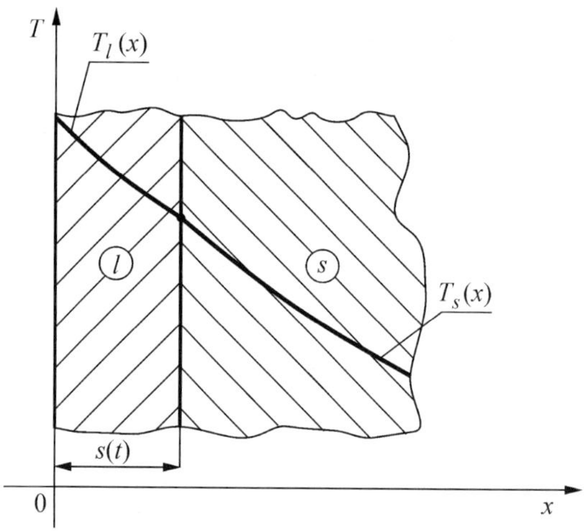

Fig. 2.4. Heat conduction problem with movable boundary

On the boundary $x = s(t)$, the following conditions are met:

$$T_l\big|_{x=s(t)} = T_s\big|_{x=s(t)} \tag{2.75}$$

and

$$\lambda_l \frac{\partial T_l}{\partial x}\bigg|_{x=s(t)} = \lambda_s \frac{\partial T_s}{\partial x}\bigg|_{x=s(t)} \pm \rho L \frac{ds}{dt}\,. \tag{2.76}$$

The plus sign refers to a process of melting, while minus sign to a process of freezing. Symbol L is the heat from substance melting (positive value), expressed in J/kg. Temperature uniformity and heat flux step change occurs on the boundary between the liquid phase (l) and solid phase (s). When solving this kind of problem, one can omit the encountered difficulties by introducing the thermal capacity substitute $c^*(T)$. Phase change takes place at T_m temperature and that elicits enthalpy $h(T)$ [3].

$$h(T) = \int_{T_{od}}^{T} c(T)\,dT + L \cdot \gamma(T - T_m), \tag{2.77}$$

where T_{od} is a randomly selected reference temperature, $c(T)$ is proper heat, while γ a unitary step function (Heaviside step function), expressed as follows:

$$\gamma(T - T_m) = \begin{cases} 1, & \text{for} \quad T \geq T_m \\ 0, & \text{for} \quad T < T_m \end{cases}. \tag{2.78}$$

Proper thermal substitute $c^*(T)$ is determined from formula

$$c^*(T) = \frac{dh}{dT} = c(T) + L\delta(T - T_m), \tag{2.79}$$

where δ is Dirac delta function. In accordance with the definition of this function, one obtains

$$\begin{aligned} &\delta(T - T_m) = 0 \qquad \text{for} \quad T \neq T_m, \\ &\delta(T - T_m) \to \infty \quad \text{for} \quad T = T_m, \\ &\int_{-\infty}^{\infty} \delta(T - T_m)\,dT = 1. \end{aligned} \tag{2.80}$$

By introducing proper thermal substitute, one eliminates step-change heat flux, which occurs in (2.76). Transient heat conduction problem in the liquid (l) and solid (s) domain is substituted by a problem in one domain, but with a specific heat $c^*(T)$ that changes abruptly on the boundary between phases. This method of analysis for problems with movable boundary is called the *enthalpy method* and is frequently employed when designating temperature fields by means of the finite element method (FEM).

Literature

1. Baehr HD, Stephan K (1993) Wärme-Stoffübertragung. Springer, Berlin
2. Bejan A (1993) Heat Transfer. Wiley, New York
3. Bonacini C, Comini G, Fasano A, Primicerio M (1973) Numerical solution of phase-change problems. Int. J. of Heat and Mass Transfer 16: 1825–1832
4. Brewster MQ (1992) Thermal Radiative Transfer and Properties. Wiley, New York
5. Cattaneo C (1958) A form of heat conduction equation which eliminates the paradox of instantaneous propagation. Compute Rendus 247: 431–433
6. Cebeci T, Bradshaw P (1984) Physical and Computational Aspects of Convective Heat Transfer. Springer, New York
7. Галин НМ, Кирилов ПЛ (1987) Тепломассообмен (в ядерной энергетике). Энергоатомиздат, Москва
8. Gersten K (1992) Single-Phase Fluid Flow. Introduction and Fundamentals: Handbook of Heat Exchanger Design. Ed. Hewitt G. F. Begell House, New York
9. Grądziel S, Taler D (2000) The method for measuring heat flux and heat transfer coefficient (in Polish). Technical Journal, 2M: 85–100
10. Hartnett JP, Rohsenow WM (1973) Basic Concepts of Heat Transfer: Handbook of Heat Transfer. McGraw-Hill, New York
11. Incropera FP, DeWitt DP (1996) Fundamentals of Heat and Mass Transfer. Ed. 4, Wiley, New York
12. Jaluria Y, Torrance KE (1986) Computational Heat Transfer. Hemisphere-Springer, Washington–Berlin.
13. Joel R (1997) Engineering Thermodynamics. Ed. 5, Longman, Harlow
14. Kakaç S, Yener Y (1995) Convective Heat Transfer. Ed. 2, CRC Press, Boca Raton
15. Kythe PK, Puri P, Schäferkotter MR (1997) Partial Differential Equations and Mathematics. CRC Press, Boca Raton
16. Look DC, Sauer HJ (1986) Engineering Thermodynamics. PWS Engineeing, Boston
17. Reddy JN, Gartling DK (1994) The Finite Element Method in Heat Transfer and Fluid Dynamics. CRC Press, Boca Raton
18. Riley KF, Hobson MP, Bence SJ (2000) Mathematical Methods for Physics and Engineering. Cambridge University Press, Cambridge
19. Saad MA (1997) Thermodynamics: Principles and Practice. Prentice Hall, Upper Saddle River
20. Slattery JC (1999) Advanced Transport Phenomena. Cambridge University Press, Cambridge
21. Sparrow EM, Cess RD (1970) Radiation Heat Transfer. Books/Cole Publishing Company, Belmont
22. Taler J (2001) The inverse methods for determining heat transfer coefficients, XI Symposium Mass and Heat Transfer. Gliwice–Szczyrk.

23. Taler J (1995) The theory and practice of heat flow processes identification (in Polish). Ossolineum, Warszawa–Wrocław–Kraków
24. Taler J, Duda P (1999) The solution of two-dimensional inverse heat conduction problems using unstructured meshes. XVII Congress of Thermodynamics. T. IV, Kraków–Zakopane: 1413–1428
25. Vernotte P (1961) La nouvelle équation de la chaleur. J. de la Transm. de la Chaleur 6
26. Wark K (1983) Thermodynamics. Ed. 4, McGraw-Hill, New York
27. Weinberger HF (1965) A First Course in Partial Differential Equations. Wiley, New York
28. Wiśniewski S, Wiśniewski T (2000) Heat transfer (in Polish). WNT, Warszawa

3 The Reduction of Transient Heat Conduction Equations and Boundary Conditions

The general transient heat conduction equation (2.34) is rather complex and often numerous difficulties are encountered while solving such equation, even when numerical methods are applied. In this chapter, we will present the methods for simplifying (2.34) and relevant boundary conditions, which help to find a solution for an initial-boundary value problem.

3.1 Linearization of a Heat Conduction Equation

If thermophysical properties of a body are temperature-dependent, then the equation for transient heat conduction in an motionless body

$$c(T)\rho(T)\frac{\partial T}{\partial t} = \nabla \cdot \left[\lambda(T)\nabla T\right] + \dot{q}_v(\mathbf{r},t) \tag{3.1}$$

is non-linear. In order to transform (3.1), one can make use of Kirchhoff transform

$$\vartheta = \frac{1}{\lambda_0}\int_0^T \lambda(T)\,dT\,. \tag{3.2}$$

Since

$$\frac{\partial \vartheta}{\partial t} = \frac{\partial \vartheta}{\partial T} \cdot \frac{\partial T}{\partial t} = \frac{\lambda(T)}{\lambda_0}\frac{\partial T}{\partial t}, \tag{3.3}$$

$$\nabla \vartheta = \frac{\partial \vartheta}{\partial T}\nabla T = \frac{\lambda(T)}{\lambda_0}\nabla T, \tag{3.4}$$

and that leads to

$$\frac{\partial T}{\partial t} = \frac{\lambda_0}{\lambda(T)}\frac{\partial \vartheta}{\partial t}, \tag{3.5}$$

$$\lambda(T)\nabla T = \lambda_0 \nabla \vartheta, \tag{3.6}$$

then (3.1) can be transformed into a form

$$\frac{1}{a(\vartheta)}\frac{\partial \vartheta}{\partial t} = \nabla^2 \vartheta + \frac{\dot{q}_v}{\lambda_0}. \tag{3.7}$$

This is still a non-linear equation.

For steady-state problems, when $\partial\vartheta/\partial t = 0$, (3.7) becomes linear

$$\nabla^2 \vartheta = -\frac{\dot{q}_v}{\lambda_0}. \tag{3.8}$$

Equation (3.8) can be solved by means of exact analytical methods under the condition that boundary conditions can be transformed into a linear form. This is possible in the case of the 1st and 2nd kind boundary conditions; one obtains then, respectively

$$T\big|_A = T_A, \tag{3.9}$$

$$\vartheta\big|_A = \frac{1}{\lambda_0}\int_0^{T_A} \lambda(T)\,dT = \vartheta_A \tag{3.10}$$

and

$$\left[\lambda(T)\frac{\partial T}{\partial n}\right]\bigg|_A = \dot{q}_A, \tag{3.11}$$

$$\lambda_0 \frac{\partial \vartheta}{\partial n}\bigg|_A = \dot{q}_A. \tag{3.12}$$

Boundary problems expressed by (3.8) and conditions (3.10) and (3.12) are linear and easier to solve than corresponding non-linear problems. Once distribution $\vartheta(\mathbf{r})$ is determined, temperature distribution $T(\mathbf{r})$ is calculated. If thermal conductivity is approximated by means of the linear function

$$\lambda(T) = \lambda_0(1+\beta T), \tag{3.13}$$

where λ_0 and β are constants, then $\vartheta(T)$ determined from (3.2) assumes the form

$$\vartheta(T) = T + \frac{1}{2}\beta T^2 . \qquad (3.14)$$

If $\vartheta(\mathbf{r})$ is known, then temperature $T(\mathbf{r})$ can be determined from (3.14)

$$T(\mathbf{r}) = \frac{-1 + \sqrt{1 + 2\beta\vartheta(\mathbf{r})}}{\beta} . \qquad (3.15)$$

If 3rd kind boundary condition is assigned on the edge of the area

$$\left[\lambda(T)\frac{\partial T}{\partial n}\right]\Big|_A = \alpha\left(T_{cz} - T|_A\right), \qquad (3.16)$$

then after the application of Kirchhoff transform (3.2), condition (3.16) continues to be non-linear

$$\lambda_0 \frac{\partial \vartheta}{\partial n}\Big|_A = \alpha\left[T_{cz} - T(\vartheta)|_A\right]. \qquad (3.17)$$

Finally, it should be emphasized here that Kirchhoff transform enables one to find accurate solutions to steady-state non-linear heat conduction problems with 1st or 2nd kind boundary conditions. In terms of transient heat conduction problems, Kirchhoff transform does not contribute much to the application and, therefore, is not recommended.

Due to the development of numerical methods for solving non-linear steady-state and transient heat conduction problems, one can refrain from using the Kirchhoff transform (3.2) and can solve non-linear problems in a direct way.

3.2 Spatial Averaging of Temperature

In many instances, temperature changes in a particular direction or in the whole body volume are minimal. Real body temperature can be approximated then by applying average temperature in the particular direction or in the whole body volume.

3.2.1 A Body Model with a Lumped Thermal Capacity

If temperature difference within a body is minimal, one can assume then that total body temperature, within its entire volume, is constant and equal to the average temperature. Such assumption can be accepted in the case of rather small-size bodies with a large thermal conductivity λ and small heat transfer coefficient $\bar{\alpha}$ on an outer surface.

One can assume for simple-shape bodies, such as the back-surface-insulated L-thick plate, a cylinder and a globe with an outer surface radius L, that as long as the following criterion is satisfied

$$Bi = \frac{\bar{\alpha}L}{\lambda} \leq 0.15, \tag{3.18}$$

the inner temperature of the body does not differ by more than 5% from the surface temperature. When criterion (3.18) is fulfilled, the body can be regarded as a body with lumped thermal capacity. When the body is heated or coolled with a liquid at temperature T_{cz} and there are no heat sources inside (Fig. 3.1), then the heat balance equation has the following form:

$$\rho V c \frac{dT}{dt} = \bar{\alpha} A \left(T_{cz} - T\right). \tag{3.19}$$

Initial condition is indispensible for solving (3.19)

$$T\big|_{t=0} = T_0. \tag{3.20}$$

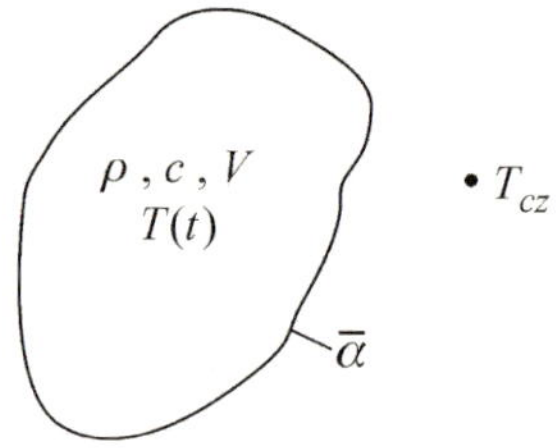

Fig. 3.1. A body model with concentrated (lumped) thermal capacity

If a temperature of a medium, which surrounds the body undergoes a step increase from temperature T_0 to temperature constant T_{cz} when $t=0$, then the initial problem (3.19)–(3.20) has a simple solution

$$\frac{T - T_{cz}}{T_0 - T_{cz}} = e^{-t/\tau}, \tag{3.21}$$

where time constant τ is formulated as

$$\tau = \frac{\rho V c}{\bar{\alpha} A}. \tag{3.22}$$

The model of a body with lumped thermal capacity is often applied in practise.

3.2.2 Heat Conduction Equation for a Simple Fin with Uniform Thickness

Fins used for the intensification of heat exchange from the gas side are usually very thin. Thus, the temperature differenced across the fin's width can be neglected [1]. Figure 3.2 shows the section of a simple fin with uniform thickness.

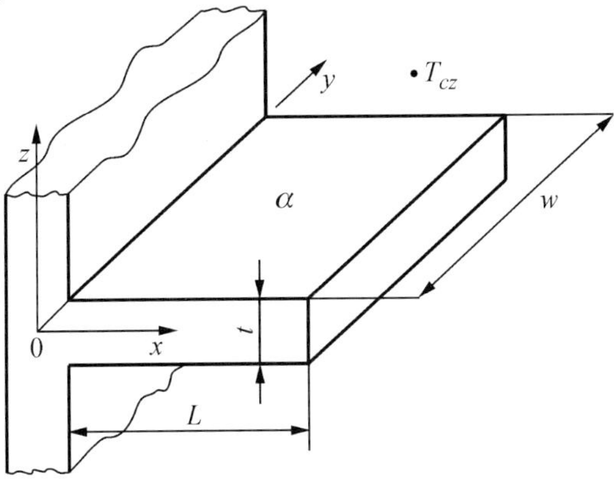

Fig. 3.2. A simple fin cross-section

A determined fin temperature distribution is expressed using the heat conduction equation

$$\frac{\partial^2 T}{\partial x^2}+\frac{\partial^2 T}{\partial y^2}+\frac{\partial^2 T}{\partial z^2}=0 \tag{3.23}$$

and the following boundary conditions:

$$T\big|_{x=0}=T_b, \tag{3.24}$$

$$-\lambda\frac{\partial T}{\partial x}\bigg|_{x=L}=\alpha\left(T\big|_{x=L}-T_{cz}\right), \tag{3.25}$$

$$\lambda\frac{\partial T}{\partial y}\bigg|_{y=0}=\alpha\left(T\big|_{y=0}-T_{cz}\right), \tag{3.26}$$

$$-\lambda\frac{\partial T}{\partial y}\bigg|_{y=W}=\alpha\left(T\big|_{y=W}-T_{cz}\right), \tag{3.27}$$

$$\lambda \frac{\partial T}{\partial z}\bigg|_{z=-t/2} = \alpha\left(T\big|_{z=-t/2} - T_{cz}\right), \tag{3.28}$$

$$-\lambda \frac{\partial T}{\partial z}\bigg|_{z=t/2} = \alpha\left(T\big|_{z=t/2} - T_{cz}\right). \tag{3.29}$$

In order to shorten (3.23), the equation will be integrated from $y = 0$ to $y = W$ and from $z = -t/2$ to $z = t/2$:

$$\int_0^W \int_{-t/2}^{t/2} \frac{\partial^2 T}{\partial x^2} dz dy + \int_0^W \int_{-t/2}^{t/2} \frac{\partial^2 T}{\partial y^2} dz dy + \int_0^W \int_{-t/2}^{t/2} \frac{\partial^2 T}{\partial z^2} dz dy = 0 . \tag{3.30}$$

Once average temperature is introduced

$$\overline{T} = \frac{1}{tW} \int_{-t/2}^{t/2} \int_0^W T dy dz \tag{3.31}$$

the following is obtained from (3.30)

$$\begin{aligned} &\frac{\partial^2 \overline{T}}{\partial x^2} + \frac{1}{tW} \int_{-t/2}^{t/2} \left(\frac{\partial T}{\partial y}\bigg|_{y=0} - \frac{\partial T}{\partial y}\bigg|_{y=W} \right) dz + \\ &+ \frac{1}{tW} \int_0^W \left(\frac{\partial T}{\partial z}\bigg|_{z=-t/2} - \frac{\partial T}{\partial z}\bigg|_{z=t/2} \right) dy = 0 . \end{aligned} \tag{3.32}$$

Once we assume that $\overline{T}(x) = T(x, y, z)$ and make allowance for boundary conditions (3.26)–(3.29), we can write (3.32) in the following form:

$$\frac{d^2 \overline{T}}{dx^2} - \frac{\alpha}{\lambda}\left(\frac{2}{W} + \frac{2}{t}\right)\left(\overline{T} - T_{cz}\right) = 0. \tag{3.33}$$

If $t \ll W$, then (3.33) is simplified to a form usually used in practical computations ((1), ex. 6.15)

$$\frac{d^2 \overline{T}}{dx^2} - m^2\left(\overline{T} - T_{cz}\right) = 0 , \tag{3.34}$$

where

$$m^2 = \frac{2\alpha}{\lambda t}. \tag{3.35}$$

Boundary conditions (3.24) and (3.25) assume the form

$$\bar{T}\Big|_{x=0} = T_b, \tag{3.36}$$

$$-\lambda \frac{\partial \bar{T}}{\partial x}\Bigg|_{x=L} = \alpha\left(\bar{T}\Big|_{x=L} - T_{cz}\right). \tag{3.37}$$

The fin temperature distribution is a solution to an (3.34) when boundary conditions are (3.36)–(3.37).

3.2.3 Heat Conduction Equation for a Circular Fin with Uniform Thickness

The section of a circular fin with uniform thickness is shown in Fig. 3.3. It is assumed that the thermal conductivity of the fin material is made of λ and heat transfer coefficient on the fin surface α can be the function of position and temperature. The fin height L is significantly larger than its width t. The determined fin temperature distribution is expressed using the heat conduction equation

$$\frac{1}{r}\frac{\partial}{\partial r}\left[\lambda(T)\, r \frac{\partial T}{\partial r}\right] + \frac{\partial}{\partial z}\left[\lambda(T)\frac{\partial T}{\partial z}\right] = 0 \tag{3.38}$$

and boundary conditions (Fig. 3.3)

$$T\Big|_{r=r_1} = T_b, \tag{3.39}$$

$$-\left[\lambda(T)\frac{\partial T}{\partial r}\right]\Bigg|_{r=r_2} = \alpha\left(T\Big|_{r=r_2} - T_{cz}\right), \tag{3.40}$$

$$\left[\lambda(T)\frac{\partial T}{\partial z}\right]\Bigg|_{z=-t/2} = \alpha\left(T\Big|_{z=-t/2} - T_{cz}\right), \tag{3.41}$$

$$-\left[\lambda(T)\frac{\partial T}{\partial z}\right]\Bigg|_{z=t/2} = \alpha\left(T\Big|_{z=t/2} - T_{cz}\right). \tag{3.42}$$

In order to reduce (3.38), the equation will be integrated from $z = -t/2$ to $z = t/2$

$$\int_{-t/2}^{t/2} \frac{1}{r}\frac{\partial}{\partial r}\left[\lambda(T) r \frac{\partial T}{\partial r}\right] dz + \int_{-t/2}^{t/2} \frac{\partial}{\partial z}\left[\lambda(T)\frac{\partial T}{\partial z}\right] dz = 0 . \tag{3.43}$$

Introducing an average temperature

$$\overline{T} = \frac{1}{t}\int_{-t/2}^{t/2} T dz \tag{3.44}$$

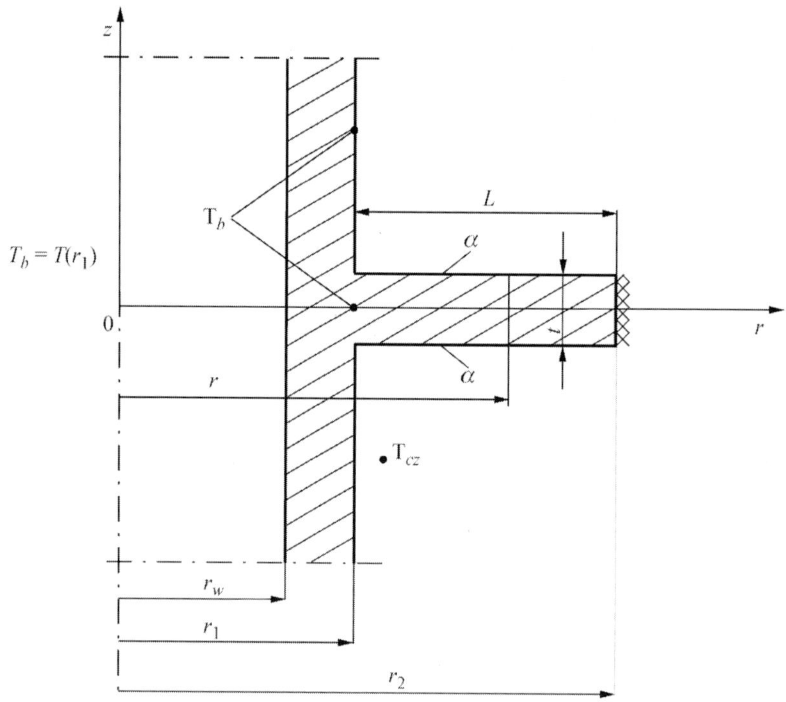

Fig. 3.3. Circular fin with uniform thickness

(Equation (3.43) can be transformed into a form

$$\frac{1}{r}\frac{\partial}{\partial r}\left[\lambda(\overline{T})\frac{\partial \overline{T}}{\partial r}\right] + \frac{1}{r}\left[\lambda(T)\frac{\partial T}{\partial z}\bigg|_{z=t/2} - \lambda(T)\frac{\partial T}{\partial z}\bigg|_{z=-t/2}\right] = 0 . \tag{3.45}$$

Taking into account the boundary conditions (3.41) and (3.42) in (3.45), one obtains

$$\frac{1}{r}\frac{d}{dr}\left[\lambda(\overline{T})\frac{d\overline{T}}{dr}\right] - \frac{2\alpha}{t}\left(\overline{T} - T_{cz}\right) = 0 . \tag{3.46}$$

Boundary conditions (3.39) and (3.40) can be integrated with respect to *dz* within the limits from $-t/2$ to $t/2$, and upon the consideration of (3.44) and the assumption that $T \approx \bar{T}$, one obtains

$$\bar{T}\Big|_{r=r_1} = T_b \,, \tag{3.47}$$

$$-\left[\lambda\left(\bar{T}\right)\frac{\partial \bar{T}}{\partial r}\right]\Bigg|_{r=r_2} = \alpha\left(\bar{T}\Big|_{r=r_2} - T_{cz}\right). \tag{3.48}$$

Due to the assumption that the temperature drop across the fin's width is insignificantly small, the two-dimensional heat conduction (3.38) was approximated by means of (3.46), which is easier to solve.

3.2.4 Heat Conduction Equation for a Circular Rod or a Pipe that Moves at Constant Velocity

In many technological processes, such as the production of wire or artificial conductors, continuous steel casting, heat rolling or during a transport of loose materials, a solid constantly moves at velocity **w**. For a conductor, which is extruded at a velocity w_z from an opening with diameter d (Fig. 3.4), the energy balance equation for cylindrical coordinates (2.40) with the exclusion of a dissipation function Φ has the following form:

$$\begin{aligned} \rho c_p\left(\frac{\partial T}{\partial t} + w_z\frac{\partial T}{\partial z}\right) &= \frac{1}{r}\frac{\partial}{\partial r}\left(r\lambda_{rr}\frac{\partial T}{\partial r}\right) + \\ &+ \frac{1}{r^2}\frac{\partial}{\partial \theta}\left(\lambda_{\theta\theta}\frac{\partial T}{\partial \theta}\right) + \frac{\partial}{\partial z}\left(\lambda_{zz}\frac{\partial T}{\partial z}\right) + \dot{q}_v \,. \end{aligned} \tag{3.49}$$

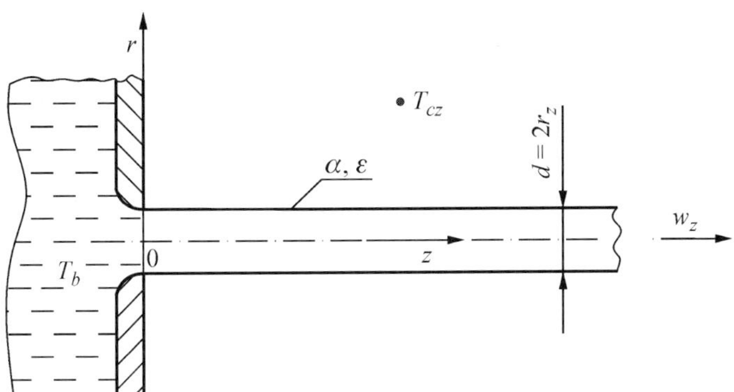

Fig. 3.4. Extrusion of a wire with diameter d from melted substance with temperature $T_b > T_{cz}$

If we additionally assume that temperature field is source-free ($\dot{q}_v = 0$) and spherically-symmetrical ($\partial T/\partial \theta = 0$), then (3.49) will be simplified to a form

$$\rho c_p \left(\frac{\partial T}{\partial t} + w_z \frac{\partial T}{\partial z} \right) = \frac{1}{r} \frac{\partial}{\partial r} \left(r \lambda_{rr} \frac{\partial T}{\partial r} \right) + \frac{\partial}{\partial z} \left(\lambda_{zz} \frac{\partial T}{\partial z} \right). \tag{3.50}$$

Temperature T_b of a melted substance is higher than the temperature of surroundings T_{cz} and the temperature of surroundings T_r, which the wire exchanges heat with by means of radiation.

If one assumes that conductor's diameter d is significantly smaller than its length, one can disregard the radially-directed temperature drop. Once (3.50) is integrated, one obtains

$$\int_0^{r_z} \rho c_p \left(\frac{\partial T}{\partial t} + w_z \frac{\partial T}{\partial z} \right) r dr = \int_0^{r_z} \left[\frac{1}{r} \frac{\partial}{\partial r} \left(r \lambda_{rr} \frac{\partial T}{\partial r} \right) + \frac{\partial}{\partial z} \left(\lambda_{zz} \frac{\partial T}{\partial z} \right) \right] r dr. \tag{3.51}$$

Once the average temperature, formulated as

$$\overline{T} = \frac{2}{r_z^2} \int_0^{r_z} r T dr \tag{3.52}$$

is introduced (3.51) can be transformed into a form

$$\frac{r_z^2}{2} \rho c_p \left(\frac{\partial \overline{T}}{\partial t} + w_z \frac{\partial \overline{T}}{\partial z} \right) = \left(r \lambda_{rr} \frac{\partial T}{\partial r} \right) \bigg|_{r_z} - \left(r \lambda_{rr} \frac{\partial T}{\partial r} \right) \bigg|_{r=0} + \frac{r_z^2}{2} \frac{\partial}{\partial z} \left(\lambda_{zz} \frac{\partial \overline{T}}{\partial z} \right). \tag{3.53}$$

In order to transform (3.53), the boundary conditions shown below are applied.

$$\lambda_{rr} \frac{\partial T}{\partial r} \bigg|_{r=0} = 0, \tag{3.54}$$

$$-\lambda_{rr} \frac{\partial T}{\partial r} \bigg|_{r=r_z} = \alpha \left(T\big|_{r=r_z} - T_{cz} \right) + \varepsilon \sigma \left(T^4\big|_{r=r_z} - T_r^4 \right). \tag{3.55}$$

By allowing for (3.54) and (3.55) in (3.53) and assuming that $T \approx \overline{T}$, (3.53) assumes the form

$$\frac{r_z^2}{2} c_p \rho \left(\frac{\partial \overline{T}}{\partial t} + w_z \frac{\partial \overline{T}}{\partial z} \right) = -\alpha r_z \left(\overline{T} - T_{cz} \right) - \\ - \varepsilon \sigma r_z \left(\overline{T}^4 - T_r^4 \right) + \frac{r_z^2}{2} \frac{\partial}{\partial z} \left(\lambda_{zz} \frac{\partial \overline{T}}{\partial z} \right), \tag{3.56}$$

from which, after simple transformations, one gets

$$\rho c_p \left(\frac{\partial \overline{T}}{\partial t} + w_z \frac{\partial \overline{T}}{\partial z} \right) = \frac{\partial}{\partial z} \left(\lambda_{zz} \frac{\partial \overline{T}}{\partial z} \right) - \frac{2\alpha}{r_z} \left(\overline{T} - T_{cz} \right) - \frac{2\varepsilon\sigma}{r_z} \left(\overline{T}^4 - T_r^4 \right), \tag{3.57}$$

where $c_p = c$ is the specific heat of a solid, while $\sigma = 5.67 \cdot 10^{-8}$ W/(m^2·K^4), a Stefan-Boltzmann constant.

The remaining boundary conditions and an initial condition can be transformed in a similar way. By doing so, one can obtain one-dimensional initial-boundary value problem, which is easier to solve than the two-dimensional problem.

Literature

1. Kraus AD (1988) Analysis of extended surface. Transactions of the ASME. J. of Heat Transfer 110: 1071–1081

$$\alpha = \bar{\alpha} f(r/r_1) = \bar{\alpha} f(\rho), \tag{4.3}$$

where

$$\varepsilon = \frac{\lambda_b - \lambda(T_{cz})}{\lambda(T_{cz})}, \quad \lambda_b = \lambda(T_b),$$

the problems (3.46)–(3.48) can be written in the form

$$\frac{d}{d\rho}\left[(1+\varepsilon\theta)\rho\frac{d\theta}{d\rho}\right] - N^2 f(\rho)\rho\theta = 0, \tag{4.4}$$

$$\theta\big|_{\rho=1} = 1, \tag{4.5}$$

$$\left[(1+\varepsilon\theta)\frac{d\theta}{d\rho}\right]\Bigg|_{\rho=k} = 0. \tag{4.6}$$

Symbols N and k denote, correspondingly

$$N = r_1\sqrt{\frac{2\bar{\alpha}}{\lambda(T_{cz})t}}, \qquad k = \frac{r_2}{r_1}. \tag{4.7}$$

Introducing the new variable

$$Q = -2\pi(1+\varepsilon\theta)\rho\frac{d\theta}{d\rho} \tag{4.8}$$

Equation (4.4) will be substituted by two-equations system

$$\frac{dQ}{d\rho} = -2\pi N^2 \rho f(\rho)\theta, \tag{4.9}$$

$$\frac{d\theta}{d\rho} = -\frac{Q}{2\pi(1+\varepsilon\theta)\rho}. \tag{4.10}$$

Boundary conditions (4.5) and (4.6) assume the form

$$\theta\big|_{\rho=1} = 1, \tag{4.11}$$

$$Q\big|_{\rho=k} = 0. \tag{4.12}$$

Equations (4.9) and (4.10) and boundary conditions (4.11) and (4.12) constitute a two-point boundary condition, which one can solve using a secant method [2]. Initial problem for system (4.9)–(4.10) is being solved at every iterative step with one of the widely available methods: the 4th order Runge-Kutt method. The solution obtained in such way has a high degree of accuracy. This method allows one to determine heat flux $\dot{q} = Q/2\pi\rho$ more accurately than the finite element method (FEM).

4.2 One-Dimensional Inverse Transient Heat Conduction Problem

For inverse problems, the substitution of a single transient heat conduction equation by two 1st order equations allows one to simplify to a significant extent the solution of an inverse problem. The transient heat conduction equation

$$c(T)\rho(T)\frac{\partial T}{\partial t} = -\frac{1}{r^m}\frac{\partial}{\partial r}\left(r^m \dot{q}\right), \tag{4.13}$$

where

$$\dot{q} = -\lambda(T)\frac{\partial T}{\partial r}, \tag{4.14}$$

can be transformed dimensionless into a equation system

$$\frac{\partial H}{\partial X} = C_1 H + C(\theta)\Omega(\theta)\frac{\partial \theta}{\partial Fo}, \tag{4.15}$$

$$\frac{\partial \theta}{\partial X} = \frac{H}{K(\theta)}, \tag{4.16}$$

where

$$C_1 = \frac{mE^*}{1 - XE^*}, \qquad E^* = \frac{E}{r_E}. \tag{4.17}$$

In (4.15) and (4.16), the following symbols were assumed:

$$C = \frac{c(T)}{c_0}, \quad \Omega = \frac{\rho(T)}{\rho_0}, \quad K = \frac{\lambda(T)}{\lambda_0}, \quad X = \frac{r_E - r}{r_E - r_w} = \frac{r_E - r}{E}, \tag{4.18a}$$

$$Fo = \frac{\kappa_0 t}{E^2}, \qquad \theta = \frac{T\lambda_0}{\dot{q}_0 E}, \qquad H = \frac{\dot{q}}{\dot{q}_0}, \tag{4.18b}$$

where $\kappa_0 = \lambda_0/(c_0\rho_0)$ is a thermal diffusivity, θ a dimensionless temperature, H a dimensionless heat flux, $\dot{q}_0 > 0$ a certain positive constant.

Figure 4.1 shows an example of an inverse transient heat conduction problem.

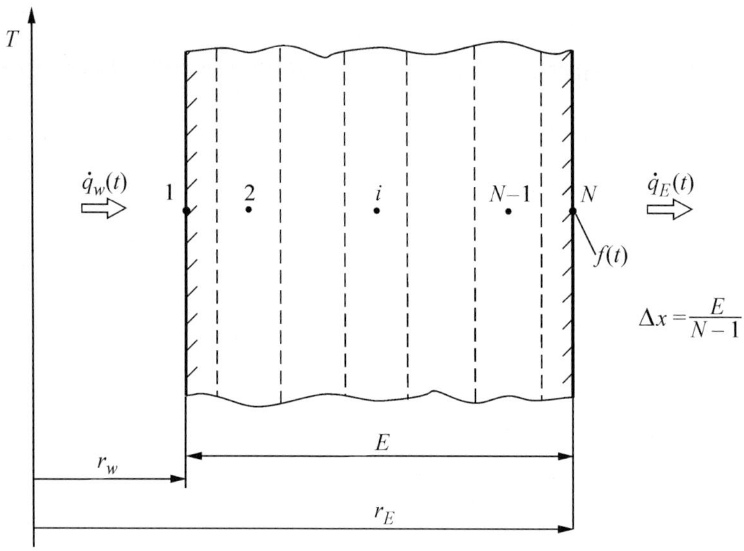

Fig. 4.1. Inverse transient heat conduction problem; temperature $f(t)$ and heat flux $\dot{q}_E(t)$ are known at point $r = r_E$

Temperature $f(t)$ is measured at point $r = r_E$. Heat flux $\dot{q}_E(t)$ is also known on this surface. If surface $r = r_E$ is thermally insulated, then $\dot{q}_E(t) = 0$. Two boundary conditions are, therefore, known on surface $r = r_E$, i.e. for $X = 0$

$$\left[K(\theta)\frac{\partial \theta}{\partial X} \right]\Big|_{X=0} = H_E(Fo), \tag{4.19}$$

$$\theta\big|_{X=0} = F(Fo), \tag{4.20}$$

where

$$H_E(Fo) = \frac{\dot{q}_E(Fo)}{\dot{q}_0}, \qquad F = \frac{f(Fo)\lambda_0}{\dot{q}_0 E}. \tag{4.21}$$

Neither temperature nor heat flux $\dot{q}_w(t)$ is known at surface $r = r_w$. Time derivative $\partial\theta/\partial Fo$ will be approximated by finite differences

$$\left.\frac{\partial\theta}{\partial Fo}\right|_{Fo_j} \approx P_j, \qquad j = 1,\ 2,\ 3,\ \ldots,\ NT\,, \tag{4.22}$$

where NT is the last temporal point.

Derivative with respect to time will be approximated by the forward and backward finite differences at the first and last temporal point, respectively

$$P_1 = \frac{T(X,\Delta Fo) - T(X,0)}{\Delta Fo}, \tag{4.23}$$

$$P_{NT} = \frac{T(X,Fo_{NT}) - T(X,Fo_{NT-1})}{\Delta Fo}, \tag{4.24}$$

where

$$Fo_j = (j-1)\Delta Fo = (j-1)\frac{\kappa_0 \Delta t}{E^2}. \tag{4.25}$$

At points $j = 2, 3, \ldots, NT - 1$ derivative with respect to time will be approximated by the central difference quotient

$$P_j = \frac{T(X,Fo_{j+1}) - T(X,Fo_{j-1})}{2\Delta Fo}. \tag{4.26}$$

Once the right-hand side of the (4.15) is approximated by quotient (4.22), one obtains

$$\frac{dH_j}{dX} = C_1 H_j + C(\theta_j)\Omega(\theta_j)P_j, \tag{4.27}$$

$$\frac{d\theta_j}{dX} = \frac{H_j}{K(\theta_j)}, \qquad j = 1,\ \ldots,\ NT\,, \tag{4.28}$$

where $H_j = H(X, Fo_j)$, $\theta_j = \theta\,(X, Fo_j)$.

Conditions (4.19) and (4.20) assume the form

$$H_j\Big|_{X=0} = H_E\left(Fo_j\right), \tag{4.29}$$

$$\theta_j\Big|_{X=0} = F\left(Fo_j\right), \qquad j = 1, \ldots, NT. \tag{4.30}$$

Thus, inverse transient heat conduction problem was reduced to an initial problem of a non-linear ordinary differential equations system. Initial problem (4.27)–(4.30) can be solved, for example, using the 4th order Runge-Kutta method.

Literature

1. Taler J (1998) Solving non-linear inverse heat conduction problems. Scientific Papers, Institute of Thermal Engineering and Fluid Mechanics PWr, 10th Heat and Mass Transfer Symposium, part. 2, Series: Conferences. No. 53, No. 9: 831–841
2. Taler J, Przybyliński P (1982) Heat transfer through circular fins with variable conductivity and non-uniform heat transfer coefficient. Chemical and Process Engineering. Vol. 3, No. 3–4: 659–673

5 Variable Change

The solution to many transient heat conduction problems, in particular problems with movable boundaries or heat sources, can be simplified by introducing new variables. Figure 5.1 shows an example of a heat source moving along x axis at velocity v, while Fig. 5.2 shows an ablation that occurs at velocity $v(t)$. One can transform transient heat conduction equation

$$\frac{\partial^2 T}{\partial x^2}+\frac{\partial^2 T}{\partial y^2}+\frac{\partial^2 T}{\partial z^2}=\frac{1}{a}\frac{\partial T}{\partial t} \tag{5.1}$$

in an motionless coordinate system with origin 0 under the condition that either the coordinate system, which moves along with the heat source, or a movable boundary will be introduced. Once the new coordinate ξ is introduced, expressed by equation (Fig. 5.1 and Fig. 5.2)

$$x=\int_0^t v(t)\,dt+\xi=s(t)+\xi \tag{5.2}$$

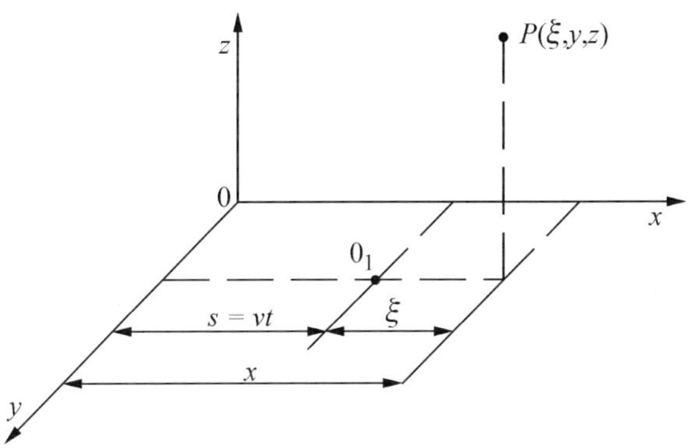

Fig. 5.1. Movable coordinate system that moves in the direction of x axis together with the heat source at velocity v

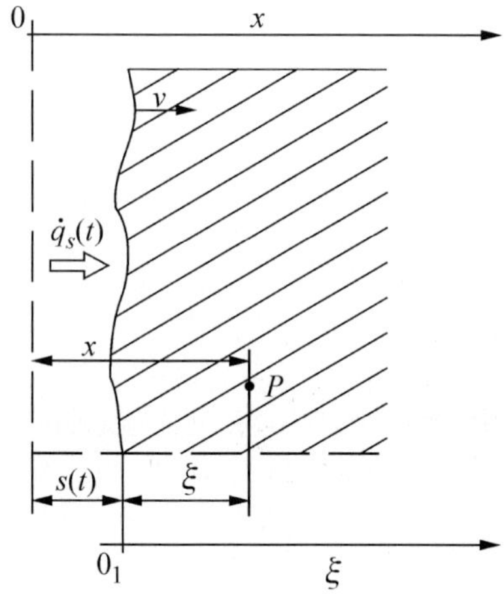

Fig. 5.2. Movable boundary (ablation) heated by a heat flow with density $\dot{q}_s(t)$

one obtains

$$\xi = x - \int_0^t v(t)\,dt = x - s(t). \tag{5.3}$$

If velocity v is independent of time, then

$$s(t) = vt, \qquad \xi = x - vt. \tag{5.4}$$

Taking into account that $T = T[\xi(t), t]$ and

$$\frac{\partial T}{\partial x} = \frac{\partial T}{\partial \xi}\frac{\partial \xi}{\partial x} = \frac{\partial T}{\partial \xi}, \tag{5.5}$$

$$\frac{\partial^2 T}{\partial x^2} = \frac{\partial^2 T}{\partial \xi^2}, \tag{5.6}$$

$$\frac{\partial T}{\partial t} = \frac{\partial T}{\partial \xi}\frac{\partial \xi}{\partial t} + \frac{\partial T}{\partial t} = -v(t)\frac{\partial T}{\partial \xi} + \frac{\partial T}{\partial t} \tag{5.7}$$

Equation (5.1) can be written in the following form:

$$\frac{\partial T}{\partial t} - v(t)\frac{\partial T}{\partial \xi} = a\left(\frac{\partial^2 T}{\partial \xi^2} + \frac{\partial^2 T}{\partial y^2} + \frac{\partial^2 T}{\partial z^2}\right). \tag{5.8}$$

In the new coordinate system (ξ, y, z) with the point of origin 0_1, a heat source or ablation boundary remain constant and do not change in time. Equation (5.8) has the same form as the heat conduction equation for a body that moves in the direction of axis ξ at velocity $-v(t)$.

Another example that testifies to the practicability of changing the variables is the determination of a transient temperature field for a semi-infinite body (Fig. 5.3).

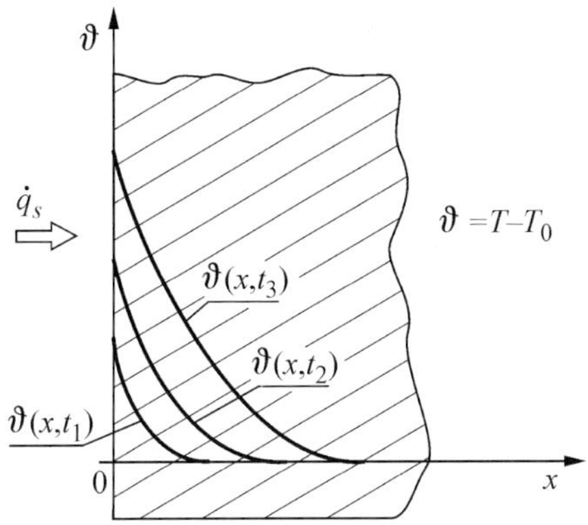

Fig. 5.3. Heating semi-infinite body with a heat flow at constant density

Temperature distribution in the semi-infinite body is formulated by the heat conduction equation

$$\frac{\partial^2 T}{\partial x^2} = \frac{1}{a}\frac{\partial T}{\partial t}, \tag{5.9}$$

using boundary conditions

$$-\lambda \left.\frac{\partial T}{\partial x}\right|_{x=0} = \dot{q}_s, \tag{5.10}$$

$$T\big|_{x\to\infty} = T_0 \tag{5.11}$$

and initial condition

$$T(x,0)=T_0\,. \tag{5.12}$$

Once dimensionless variable [1] is introduced

$$\eta=\frac{x}{\sqrt{2at}} \tag{5.13}$$

and the following solution form assumed

$$T(x,t)=T_0-\frac{\dot{q}_s x}{\lambda}\frac{f(\eta)}{\eta} \tag{5.14}$$

Equation (5.9) can be reduced to a form

$$\frac{d^2 f}{d\eta^2}+\eta\frac{df}{d\eta}-f=0\,. \tag{5.15}$$

Conditions (5.10) and (5.11) have the form

$$\left.\frac{df(\eta)}{d\eta}\right|_{\eta=0}=1\,, \tag{5.16}$$

$$f(\eta)\big|_{\eta\to\infty}=0\,. \tag{5.17}$$

Due to the fact that variable (5.13) is introduced, partial equation (5.9) was substituted by an ordinary differential equation (5.15).

Literature

1. Slattery JC (1999) Advanced Transport Phenomena. University Press, Cambridge

Part II

Exercises. Solving Heat Conduction Problems

Part two contains theoretical and numerical problems presented in the form of exercises. The topics of heat conduction and heat transfer are presented together. In this part, we discuss the solution for steady-state heat conduction problems using exact and approximate analytical and numerical methods. Transient problems are thoroughly discussed; this includes approximate and accurate analytical methods, such as the variable separation method, Laplace transform, integral heat balance method and numerical methods including, in particular, finite volume method and finite element method. Not only initial-boundary problems, called simple, are analyzed but also inverse steady-state and transient heat conduction problems. The processes of freezing and melting are discussed in depth, as well as the heat flow around stationary and movable heat sources.

6 Heat Transfer Fundamentals

In this chapter, the basics of heat conduction and transfer are discussed. The chapter contains 29 exercises, which illustrate Fourier Law, the solving of heat transfer coefficients for multi-layered flat and cylindrical partitions, the determination of a quasi-steady-state temperature field and the computation of a radiant tube temperature in boilers. Critical thickness of thermal insulation on the surface of the cylindrical tube is first determined analytically, then calculated. The methods for solving selected inverse steady-state heat conduction problems, which occur during heat flux measurement carried out by means of different types of sensors, are presented here. A great deal of attention is paid to the determination of temperature distribution and the efficiency of simple, circular, rectangular and hexagonal fins. The calculation results of efficiency in complex-shape fins, determined by means of an equivalent circular fin method and segment method, are compared with the results obtained from FEM. Examples that illustrate the computation of a heat transfer coefficient in pipes finned longitudinally and crosswise are presented here as well. Three exercises deal with the way steady-state temperature distribution is determined using control volume method. These exercises present the methods for solving problems and the computational programs used. In the last exercise of this chapter, temperature distribution and circular fin efficiency is determined under the assumption that thermal conductivity of the fin's material is temperature dependent. The problem is reduced to a two-point boundary problem for the system of two ordinary differential equations.

Exercise 6.1 Fourier Law in a Cylindrical Coordinate System

A radiant tube with an outer diameter d_z = 32 mm, wall thickness g = 5 mm and length L = 20 m is made of a steel with a thermal conductivity λ = 47 W/(m·K) (Fig. 6.1). Water-vapour mixture, heated by combustions gases that surround the tube on the outside, flows inside the tube. Inner surface temperature is T_w = 200°C, while the outer surface temperature is T_z = 250°C. The aim is to compute the heat flow transferred from the

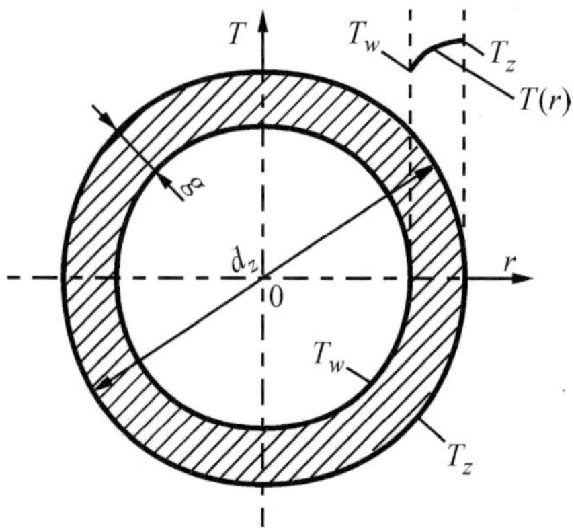

Fig. 6.1. Tube cross-section

combustion gases to the water-vapour mixture and the heat flux on the inner and outer surface.

Solution

Heat flux across the tube's wall thickness is formulated using Fourier Law

$$\dot{q}(r) = \lambda \frac{dT}{dr}. \tag{1}$$

Heat flow conducted through a flat wall can be written in the form

$$\dot{Q} = A(r)\dot{q} = A(r)\lambda \frac{dT}{dr}, \tag{2}$$

where $A(r) = 2\pi rL$.

The separation of variables in (2), gives

$$dT = \frac{\dot{Q}}{2\pi L\lambda}\frac{dr}{r}. \tag{3}$$

On the basis of known inner and outer surface temperature, one can write the boundary conditions as

$$T(r_z) = T_z, \qquad T(r_w) = T_w. \tag{4}$$

Once (3) is integrated, temperature distribution across the wall thickness of the tube is obtained

$$T = \frac{\dot{Q}}{2\pi L\lambda}\ln|r| + C\,, \tag{5}$$

where $\dot{Q}$ and C are constants computed from boundary conditions (4). Unknown value $\dot{Q}$ is equal to

$$\dot{Q} = \frac{(T_z - T_w)\cdot 2\pi L\lambda}{\ln(r_z / r_w)}\,.$$

Substituting of the data gives

$$\dot{Q} = \frac{2\pi\cdot 20\cdot 47\cdot (250-200)}{\ln\left(\frac{0.016}{0.011}\right)} = 788100\ \text{W}\,.$$

Inner surface heat flux is

$$\dot{q}_w = \frac{\dot{Q}}{2\pi r_w L} = \frac{788100}{2\pi\cdot 0.011\cdot 20} = 5.702\cdot 10^5\ \text{W/m}^2\,.$$

Outer surface heat flux is

$$\dot{q}_z = \frac{\dot{Q}}{2\pi r_z L} = \frac{788100}{2\pi\cdot 0.016\cdot 20} = 3.92\cdot 10^5\ \text{W/m}^2\,.$$

Exercise 6.2 The Equivalent Heat Transfer Coefficient Accounting for Heat Exchange by Convection and Radiation

A non-insulated tube (Fig. 6.2) with a nominal diameter d_n = 38 mm (1½″), and the following measurements: d_z = 38 mm, wall thickness g = 2.6 mm, length L = 5; the tube is kept in a room whose temperature is T_{ot} = 20°C. Water with temperature 80°C flows inside the tube. The tube's outer surface emissivity is ε = 0.8. Lets assume that outer surface temperature is identical to the temperature of a flowing medium inside and that heat transfer coefficient by means of convection is formulated as

$$\alpha_k = 5.0\sqrt[4]{\frac{T_z - T_{ot}}{T_{ot} d_z}} \quad \text{or} \quad \alpha_k = 1.21\sqrt[4]{\frac{\Delta T}{d_z}}\,,$$

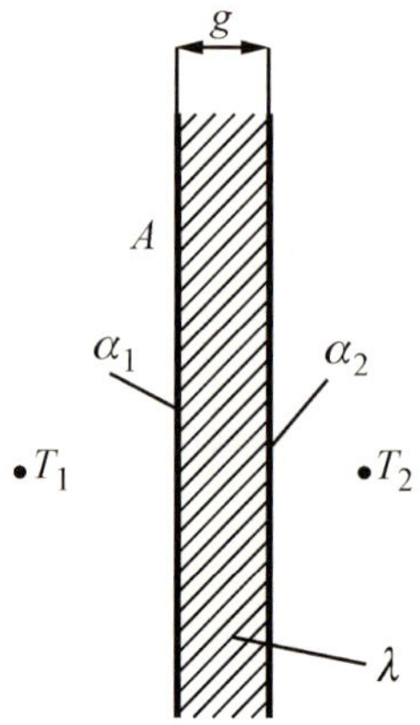

Fig. 6.3. Flat wall

Thus

$$k = \frac{1}{\frac{1}{5} + \frac{0.4}{1} + \frac{1}{15}} = 1.5 \ \mathrm{W/(m^2 \cdot K)}.$$

Heat flux transferred by the flat wall can be determined from the following formula:

$$\dot{q} = k\left(T_1 - T_2\right),$$

$$\dot{q} = 1.5 \cdot \left(20 + 20\right) = 60 \ \mathrm{W/m^2}$$

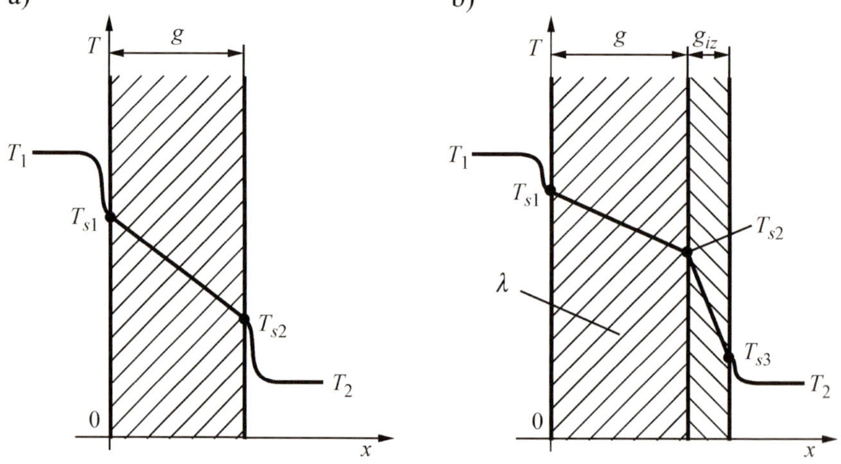

Fig. 6.4. Temperature distribution: (**a**) non-insulated wall, (**b**) insulated wall

Heat flow conducted by the flat wall can be determined as follows:

$$\dot{Q} = A \cdot k\left(T_1 - T_2\right) = 15.6 \cdot 1.5 \cdot \left(20 + 20\right) = 936 \text{ W}.$$

Temperature T_{s1} is (Fig. 6.4)

$$T_{s1} = T_1 - \frac{\dot{q}}{\alpha_1} = 20 - \frac{60}{5} = 20 - 12 = 8^\circ\text{C}.$$

Temperature T_{s2} is calculated using similar expression

$$T_{s2} = T_2 + \frac{\dot{q}}{\alpha_2} = -20 + \frac{60}{15} = -16^\circ\text{C}.$$

b) Thermally insulated wall
Heat transfer coefficient

$$k = \frac{1}{\dfrac{1}{\alpha_1} + \dfrac{g}{\lambda} + \dfrac{g_{iz}}{\lambda_{iz}} + \dfrac{1}{\alpha_2}} = \frac{1}{\dfrac{1}{5} + \dfrac{0.4}{1} + \dfrac{0.1}{0.04} + \dfrac{1}{15}} = 0.3158 \text{ W/(m}^2\text{·K)}.$$

Heat flux transferred by the wall is computed in the following way:

$$\dot{q} = k\left(T_1 - T_2\right) = 0.3158\left[20 - \left(-20\right)\right] = 12.632 \text{ W/m}^2.$$

Heat flow transferred by an insulated wall is:

$$\dot{Q} = A\dot{q} = Ak\left(T_1 - T_2\right) = 15.6 \cdot 12.632 = 197.06 \text{ W}.$$

Surface temperature T_{s1} is:

$$T_{s1} = T_1 - \frac{\dot{q}}{\alpha_1} = 20 - \frac{12.632}{5} = 17.47^\circ\text{C}.$$

Temperature T_{s2} is calculated by subtracting a temperature drop across the wall thickness from temperature T_{s1}

$$T_{s2} = T_{s1} - \frac{\dot{q}g}{\lambda} = 17.47 - \frac{12.632 \cdot 0.4}{1} = 12.42^\circ\text{C}.$$

Temperature T_{s3} is calculated as follows:

$$T_{s3} = T_2 + \frac{\dot{q}}{\alpha_2} = -20 + \frac{12.632}{15} = -19.16^\circ\text{C}.$$

One can observe that insulating a wall with a foamed polystyrene has a significant effect on the heat flow transferred by the wall and on the wall's temperature. In the case when there is a lack of insulation, heat flow $\dot{Q}$ is

$n = 936/197.06 = 4.75$ times larger than when the wall is insulated. When insulation is applied, the wall does not freeze, since its outer surface temperature increases from $T_{s2} = -16°C$ to $T_{s2} = 12.42°C$.

From the conducted analysis, one can deduce that buildings should be insulated on the outside surface, since the temperature of the walls remains then positive (is above zero).

Exercise 6.4 Overall Heat Transfer Coefficient and Heat Loss Through a Pipeline Wall

Pipeline (Fig. 6.5) with an outer diameter $d_z = 273$ mm, wall thickness $g = 16$ mm and length $L = 70$ m is made of a material whose thermal conductivity is $\lambda = 45$ W/(m·K). The pipeline is thermally insulated by a layer, which is 10 cm thick ($g_{iz} = 10$ cm) and made of a material with $\lambda_{iz} = 0.08$ W/(m·K). A medium with temperature $T_w = 400°C$ flows inside the pipeline, while on the inner surface a heat transfer coefficient is $\alpha_w = 500$ W/(m^2·K). Air temperature, which surrounds the pipeline on the outside, is $T_z = 20°C$, while a heat transfer coefficient on an outer surface is $\alpha_z = 10$ W/(m^2·K). The aim here is to compute:

1. overall heat transfer coefficient related to:
   a) outer insulation surface
   b) inner surface of the pipeline
   c) tube's length
2. heat loss

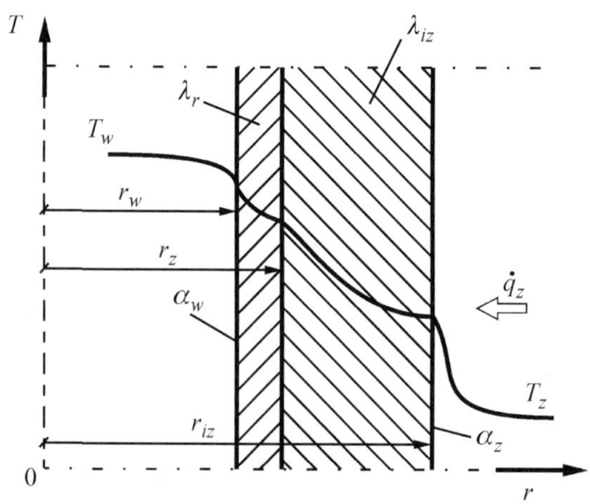

Fig. 6.5. Longitudinal cross-section of a pipeline

Solution

a) Total drop in temperature is equal to the sum of temperature decreases due to, correspondingly, convectional heat exchange on an outer surface of the insulation, the heat conduction in the insulation, the pipeline conduction and the convectional inner surface heat exchange

$$\frac{\dot{q}_z}{k_z} = \frac{\dot{q}_z}{\alpha_z} + \frac{\dot{q}_z r_{iz}}{\lambda_{iz}} \ln\frac{r_{iz}}{r_z} + \frac{\dot{q}_z r_{iz}}{\lambda_r} \ln\frac{r_z}{r_w} + \frac{\dot{q}_z}{\alpha_w}\frac{r_{iz}}{r_w}.$$

From the equation above, one can determine the overall heat transfer coefficient related to the outer surface of the tube

$$k_z = \frac{1}{\dfrac{1}{\alpha_z} + \dfrac{r_{iz}}{\lambda_{iz}} \ln\dfrac{r_{iz}}{r_z} + \dfrac{r_{iz}}{\lambda_r} \ln\dfrac{r_z}{r_w} + \dfrac{1}{\alpha_w}\dfrac{r_{iz}}{r_w}}.$$

Substitution of the numerical values gives

$$k_z = \frac{1}{\dfrac{1}{10} + \dfrac{0.2365}{0.08} \ln\dfrac{0.2365}{0.1365} + \dfrac{0.2365}{45} \ln\dfrac{0.1365}{0.1205} + \dfrac{1}{500}\dfrac{0.2365}{0.1205}} =$$

$$= 0.5782 \text{ W/(m}^2 \cdot \text{K)}.$$

Heat flow conducted by the pipeline and the insulation heat loss is given by:

$$\dot{Q} = A_{iz} k_z \left(T_w - T_z\right) = 2\pi r_{iz} L k_z \left(T_w - T_z\right).$$

Thus

$$\dot{Q} = 2 \cdot \pi \cdot 0.2365 \cdot 70 \cdot 0.5782 \cdot (400 - 20) = 22855 \text{ W}.$$

b) In order to compute the heat transfer coefficient for the pipeline's inner surface, one should begin by adding up all temperature decreases from the inner surface. Total temperature drop equals the sum of temperature decreases connected with, correspondingly, convectional inner surface heat exchange, pipeline heat conduction, heat conduction in an insulation and the insulation's convectional outer surface heat exchange:

$$\frac{\dot{q}_w}{k_w} = \frac{\dot{q}_w}{\alpha_w} + \frac{\dot{q}_w r_w}{\lambda_r} \ln\frac{r_z}{r_w} + \frac{\dot{q}_w r_w}{\lambda_{iz}} \ln\frac{r_{iz}}{r_z} + \frac{\dot{q}_w}{\alpha_z}\frac{r_w}{r_{iz}}.$$

Thus, the overall heat transfer coefficient related to inside tube surface is

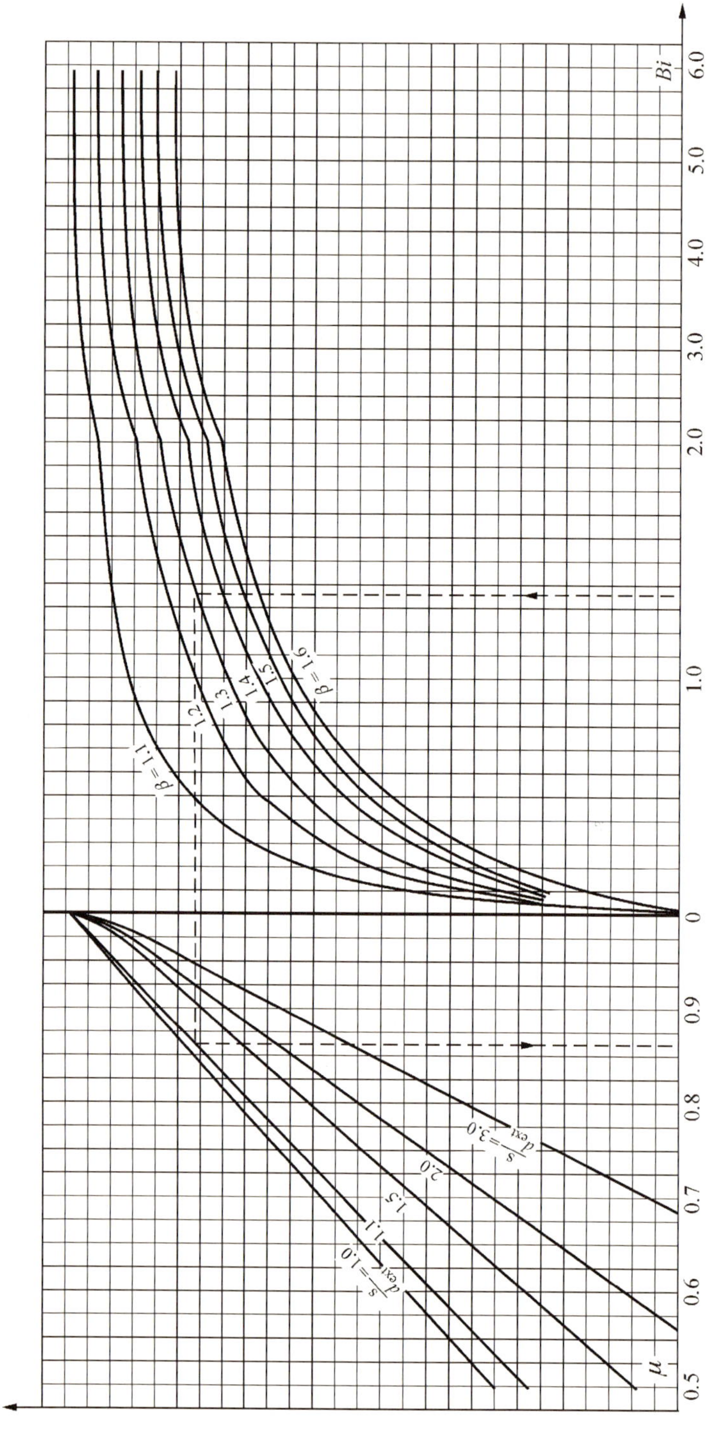

Fig. 6.9. Heat dissipation coefficient μ for radiant smooth tubes [6], where $\beta = d_z/d_w$, $s/d_{ext} = t/d_z$

The solution is obtained by integrating (1) twice in r:

$$T = C_1 \ln r + C_2 \,. \tag{4}$$

After substituting (4) for boundary condition (2) and (3) and determining constants, temperature distribution is formulated as

$$T = T_w + \frac{\dot{q} r_z}{\lambda}\left(\ln\frac{r}{r_w} + \frac{1}{Bi} \right), \tag{5}$$

where $Bi = \alpha\, r_w/\lambda$.

The tube temperature at point P is calculated by means of an approximate formula (5)

$$Bi = \frac{20000 \cdot 0.01}{40} = 5\,,$$

$$T_p = T\big|_{r=r_z} = 350\,^\circ\mathrm{C} + \frac{350000 \cdot 0.016}{40}\left(\ln\frac{0.016}{0.01} + \frac{1}{5} \right) = 443.8\,^\circ\mathrm{C}\,.$$

Real temperature at point P is lower, since heat flows from the tube's front-side to its unheated rear-side from the brickwork side. According to paper [6], the radiant tube's real temperature at point P can be calculated from the formula below

$$T_p' = T_w + \mu \dot{q} \frac{r_z}{r_w}\left(\frac{1}{\alpha_w} + \frac{2}{1 + \dfrac{r_z}{r_w}} \frac{g}{\lambda} \right),$$

where μ is a so called *heat dissipation coefficient*, which is determined from Fig. 6.9. For Bi = 5, $\beta = r_z/r_w$ = 32/20 = 1.6 and t/d_z = 1.2375, one obtains $\mu = 0.89$. Thus, temperature T_p' is

$$T_p' = 350\,^\circ\mathrm{C} + 0.89 \cdot 350000 \frac{0.016}{0.01}\left(\frac{1}{20000} + \frac{2}{1 + \dfrac{0.016}{0.01}} \cdot \frac{0.006}{40} \right) = 432\,^\circ\mathrm{C}.$$

Temperature calculated by means of the approximate formula (5) equals T_p = 443.8°C. It is, therefore, higher than the real temperature T_p' = 432°C. The difference, however $(T_p - T_p')$ is small.

Exercise 6.7 Quasi-Steady-State of Temperature Distribution and Stresses in a Pipeline Wall

The aim is to calculate the difference between inner surface temperature and average temperature across the thickness of a steel pipe wall with an outer diameter of $d_z = 324$ mm and wall thickness $g = 65$ mm, made of a ferritic steel 10CrMo910 with thermal conductivity $\lambda = 35.5$ W/(m·K) and thermal diffusivity $a = \lambda / c\rho = 7.137 \cdot 10^{-6}$ m^2/s. The outer surface of the pipe is thermally insulated. The pipe heating (the steam superheater chamber) takes place at constant temperature rate equal to $v_T = 10$ K/min. Lets assume that a quasi-stationary state forms itself in the pipe wall (Fig. 6.10) and is characterized by a stable heating rate equal to v_T. Quasi-stationary state usually occurs for $Fo = at/g^2 > 0.5$ during the heating or cooling of an element, if a temperature change rate of a medium or of an inner surface wall temperature remains constant. We will also calculate thermal stresses (axial) on an inner surface of the pipe under the assumption that the pipe ends can be easily elongated (are free). The following material constants apply for the computation: elastic modulus $E = 181600$ MPa, thermal expansion coefficient $\beta = 1.35 \cdot 10^{-5}$ 1/K, Poisson ratio $\nu = 0.301$.

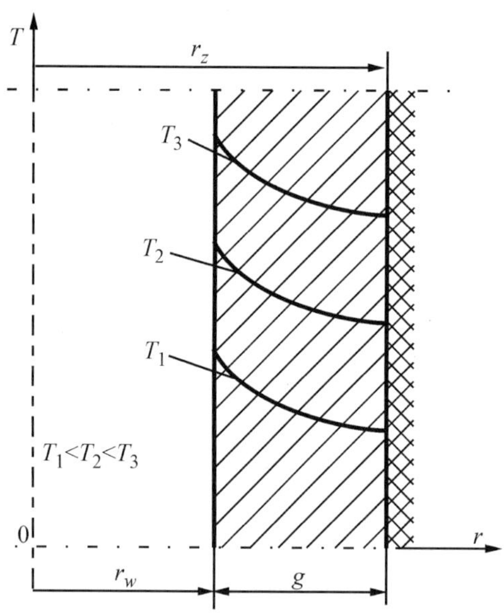

Fig. 6.10. Quasi-steady-state temperature field in the pipe wall (cylindrical chamber)

Solution

Due to a stable temperature change rate within a whole body volume equal to $\partial T/\partial t = v_T$, the heat conduction equation assumes the following form:

$$\frac{1}{r}\frac{d}{dr}\left(r\frac{dT}{dr}\right)=\frac{v_T}{a}. \tag{1}$$

Equation (1) will be solved using the following boundary conditions:

$$T\big|_{r=r_w}=T_w=v_T t, \tag{2}$$

$$\left.\frac{dT}{dr}\right|_{r=r_z}=0. \tag{3}$$

The solution is obtained by integrating (1) twice in r:

$$T=\frac{1}{4}\frac{v_T}{a}r^2+C_1\ln r+C_2. \tag{4}$$

Constants C_1 and C_2 are determined from the boundary conditions (2) and (3). Substitution of the C_1 and C_2 into (4) yields

$$T(r,t)=v_T t+\frac{v_T}{4a}\left(r^2-r_w^2-2r_z^2\ln\frac{r}{r_w}\right). \tag{5}$$

Average temperature $T_m(t)$ across the wall thickness is given by

$$T_m(t)=\frac{2}{r_z^2-r_w^2}\int_{r_w}^{r_z}rT\,dr, \tag{6}$$

from which, after substitution of (5) for (6) and subsequent integration, one gets

$$T_m(t)=v_T t+\frac{v_T}{a}\left[\frac{1}{8}\left(3r_z^2-r_w^2\right)-\frac{1}{2}r_z^4\frac{\ln\dfrac{r_z}{r_w}}{r_z^2-r_w^2}\right]. \tag{7}$$

Equation (7) can be also transformed into the following form:

$$T_m(t)=v_T t+\frac{v_T g^2}{a}\frac{\left(3u^2-1\right)\left(u^2-1\right)-4u^4\ln u}{8\left(u^2-1\right)\left(u-1\right)^2}. \tag{8}$$

The unknown temperature difference

$$\Delta T = T_m(t) - T\big|_{r=r_w} \tag{9}$$

is

$$\Delta T = \frac{v_T g^2}{a} \frac{(3u^2-1)(u^2-1)-4u^4 \ln u}{8(u^2-1)(u-1)^2},$$

where $u = r_z / r_w$.

The inner surface axial stress is given by

$$\sigma_T = \frac{E\beta}{1-\nu}\Delta T,$$

where ΔT is expressed using (9).

Temperature difference ΔT is

$$u = 324/194 = 1.6701,$$

$$\Delta T = \frac{10}{60}\frac{(0.065)^2}{7.137\cdot 10^{-6}}\frac{(3\cdot 1.67^2-1)(1.67^2-1)-4\cdot 1.67^4 \ln 1.67}{8(1.67^2-1)(1.67-1)^2} =$$

$$= -42.6444 \text{ K}.$$

The inner surface axial stress is

$$\sigma_T = \frac{E\beta}{1-\nu}\Delta T = \frac{181600\cdot 1.35\cdot 10^{-5}}{1-0.301}(-42.6444) = -149.57 \text{ MPa}.$$

Exercise 6.8 Temperature Distribution in a Flat Wall with Constant and Temperature Dependent Thermal Conductivity

The aim is to determine temperature distribution in a flat wall with thickness g heated by a heat flow with density $\dot{q}$ and cooled on the opposite side by water at temperature T_{cz} (Fig. 6.11). Heat transfer coefficient from the plate surface to water α is constant. Lets assume that the thermal conductivity of the plate material changes with temperature in a linear manner:

$$\lambda(T) = a + bT, \tag{1}$$

where a and b are constants; temperature T is expressed in °C.

We will also calculate temperature distribution with an assumption that the thermal conductivity is constant and equals

$$\lambda_m = \frac{1}{2}\left[\lambda\left(T\big|_{x=0}\right) + \lambda\left(T\big|_{x=g}\right)\right], \tag{2}$$

where $T\big|_{x=0}$ and $T\big|_{x=g}$ are front-side and rear-side surface temperatures, calculated using temperature-dependent thermal conductivity. The following values are adopted for the calculation: $g = 0.016$ m, $\dot{q}$ = 274800 W/m^2, α = 2400 W/(m^2·K), T_{cz} = 20°C, a = 14.64 W/(m·K), b = 0.0144 W/(m·K^2). Calculation results will be presented in a tabular and graphical form.

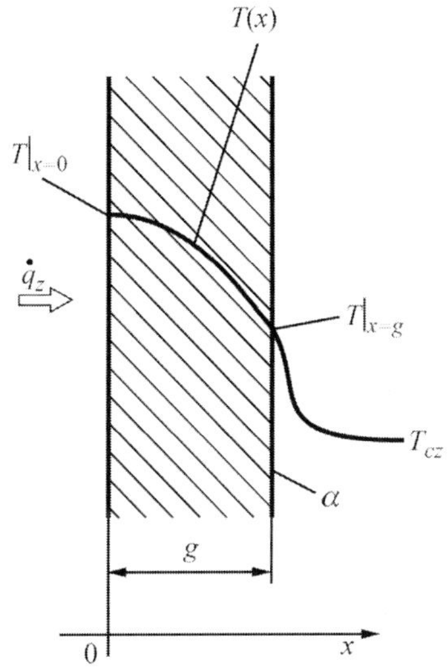

Fig. 6.11. Plate heating

Solution

First apply Fourier Law:

$$\dot{q} = -\lambda(T)\frac{dT}{dx} = -(a + bT)\frac{dT}{dx}. \tag{3}$$

Note that heat flux $\dot{q}$ is constant within the entire plate thickness, since heat flow $\dot{Q} = A\dot{q}$ is constant for steady-state heat conduction. Separation of variables in (3) gives

$$\dot{q}dx = -(a + bT)dT. \tag{4}$$

By integrating (4), one obtains a quadratic equation with respect to T

$$\dot{q}x = -aT - \frac{1}{2}bT^2 + C\,, \tag{5}$$

whose solution is the function

$$T(x) = -\frac{a}{b} + \sqrt{\left(\frac{a}{b}\right)^2 - \frac{2(\dot{q}x - C)}{b}}\,. \tag{6}$$

Constant C is determined from condition

$$T\big|_{x=g} = \frac{\dot{q}}{\alpha} + T_{cz}\,. \tag{7}$$

After substituting (7) for (5), one obtains

$$C = \dot{q}g + aT\big|_{x=g} + \frac{1}{2}b\left(T\big|_{x=g}\right)^2,$$

where $T\big|_{x=g}$ is expressed by (7).

In order to determine temperature distribution, constant C is calculated first:

$$T\big|_{x=g} = \frac{274800}{2400} + 20 = 129.5\,^{\circ}\mathrm{C},$$

$$C = 274800 \cdot 0.016 + 14.65 \cdot 129.5 + \frac{1}{2}0.0144 \cdot 129.5^2 = 6414.7208\ \mathrm{W/m}.$$

Temperature distribution is expressed by the following function:

$$T(x) = -\frac{14.65}{0.0144} + \sqrt{\left(\frac{14.65}{0.0144}\right)^2 - \frac{2(274800x - 6414.7208)}{0.0144}}\,.$$

Table 6.1 and Fig. 6.12 shows the determined $T(x)$ distribution. Mean thermal conductivity λ_m determined from (2) is:

$$T\big|_{x=0} = 370.4274\,^{\circ}\mathrm{C},\quad T\big|_{x=g} = 129.5\,^{\circ}\mathrm{C},$$

$$\lambda\left(T\big|_{x=0}\right) = \lambda\left(370.4274\,^{\circ}\mathrm{C}\right) = 19.9842\ \mathrm{W/(m \cdot K)},$$

$$\lambda\left(T\big|_{x=g}\right) = \lambda\left(129.5\,^{\circ}\mathrm{C}\right) = 16.5148\ \mathrm{W/(m \cdot K)},$$

$$\lambda_m = 0.5 \cdot (19.9842 + 16.5148) = 18.2495\ \mathrm{W/(m \cdot K)}.$$

Table 6.1. Temperature distribution $T(x)$ across the plate thickness

No.	x [m]	Non-linear Problem T [°C]	Linear Problem T [°C]
1	0.000	370.4274	370.4271
2	0.001	356.6077	355.3692
3	0.002	342.6476	340.3112
4	0.003	328.5427	325.2533
5	0.004	314.2884	310.1953
6	0.005	299.8799	295.1374
7	0.006	285.3119	280.0794
8	0.007	270.5793	265.0215
9	0.008	255.6761	246.9635
10	0.009	240.5964	234.9056
11	0.010	225.3337	219.8476
12	0.011	209.8812	204.7897
13	0.012	194.2317	189.7317
14	0.013	178.3773	174.6738
15	0.014	162.3099	159.6158
16	0.015	146.0206	144.5579
17	0.016	129.5000	129.5000

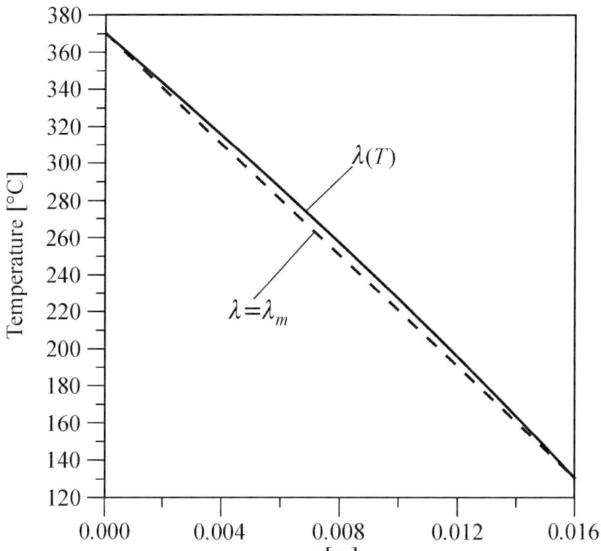

Fig. 6.12. Temperature distribution across a wall thickness for constant and temperature dependent thermal conductivity

In the case of a linear problem, when the thermal conductivity is constant and equals λ_m, temperature distribution across the plate thickness is formulated as

$$T(x)=\frac{\dot{q}(g-x)}{\lambda_m}+\frac{\dot{q}}{\alpha}+T_{cz}.$$

After substitution of the numerical values, function $T(x)$ has the form

$$T(x)=-15057.9468\cdot x+370.4271°\text{C}.$$

Temperature distribution of $T(x)$ for the linear problem is presented in Table 6.1 and Fig. 6.12. It is clear from the analysis of results presented there that the variable conductivity causes discernible temperature differences with regard to the linear problem.

Exercise 6.9 Determining Heat Flux on the Basis of Measured Temperature at Two Points Using a Flat and Cylindrical Sensor

Flat or cylindrical conductometric sensors are used to measure heat flux $\dot{q}$. The sensors operate by measuring temperature at two selected points: P_1 i P_2 (see Fig. 6.13). The known values are the thermal conductivity λ (T), coordinates of measured temperature points x_1 and x_2 (Fig. 6.13a) or (Fig. 6.13b) and measured temperatures T_1 and T_2. The aim is to determine unknown heat fluxes, assuming that the thermal conductivity of the sensor's material is a function of temperature and formulated as

$$\lambda(T)=a+bT+cT^2+dT^3, \tag{1}$$

where a, b, c and d are known coefficients.
The following values are assumed for the computation:
g = 0.016 m; x_1 = 0.002 m; x_2 = 0.012 m;
a = 14.99 W/(m·K);
$b = 1.35\cdot10^{-2}$ W/(m·K^2);
$c = -4.51\cdot10^{-6}$ W/(m·K^3);
$d = 3.59\cdot10^{-9}$ W/(m·K^4);
T_1 = 330°C; T_2 = 180°C.
In the case of a cylindrical wall:
r_2 = 0.020 m; r_1 = 0.030 m and r_z = 0.032 m.
Remaining values are the same as the values for a flat wall.

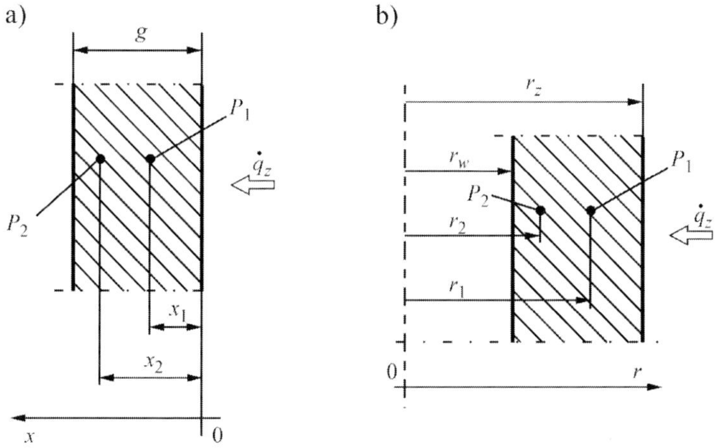

Fig. 6.13. Conductometric sensors: (**a**) flat sensor, (**b**) cylindrical sensor

Solution

a) *Flat sensor* (Fig. 6.13a)
Once the Fourier Law variables are separated

$$\dot{q} = -\lambda(T)\frac{dT}{dx}, \tag{2}$$

and subsequently the integration is carried out from x_1 to x_2, one obtains

$$\int_{x_1}^{x_2} \dot{q}dx = -\int_{T_1}^{T_2} \lambda(T)\,dT. \tag{3}$$

Introducing the mean thermal conductivity

$$\lambda_m = \frac{1}{T_1 - T_2}\int_{T_2}^{T_1} \lambda(T)\,dT, \tag{4}$$

and transforming (3), one obtains an expression which can be used to calculate heat flux $\dot{q}$

$$\dot{q} = \lambda_m \frac{T_1 - T_2}{x_2 - x_1}. \tag{5}$$

After substitution of (1) into (4) and integration, one obtains

$$\lambda_m = \frac{1}{T_1 - T_2}\left[a(T_1 - T_2) + \frac{1}{2}b\left(T_1^2 - T_2^2\right) + \frac{1}{3}c\left(T_1^3 - T_2^3\right) + \frac{1}{4}d\left(T_1^4 - T_2^4\right)\right]. \tag{6}$$

Solution

In order to derive a differential equation, which describes heat conduction in a sensor of Gardon's measuring device, the energy balance equation will be written for an elementary volume $dV = 2\ r g_c\ r$ (Fig. 6.15)

$$c(T)\rho(T)2\pi r g_c \Delta r \frac{\partial T}{\partial t} = \dot{Q}_1 - \dot{Q}_2 + \dot{Q}_3, \tag{2}$$

where

$$\dot{Q}_1 = -2\pi r g_c \left[\lambda(T)\frac{\partial T}{\partial r}\right]\Bigg|_r, \quad \dot{Q}_2 = -2\pi(r+\Delta r) g_c \left[\lambda(T)\frac{\partial T}{\partial r}\right]\Bigg|_{r+\Delta r}, \tag{3}$$

$$\dot{Q}_3 = -2\pi r \Delta r \dot{q}.$$

By substituting (3) into (2), one obtains the following for $\Delta r \to 0$

$$c(T)\rho(T)\frac{\partial T}{\partial t} = \frac{1}{r}\frac{\partial}{\partial r}\left[\lambda(T) r \frac{\partial T}{\partial r}\right] + \frac{\dot{q}}{g_c}. \tag{4}$$

In the case of steady-state problems $\partial T/\partial t = 0$, and temperature distribution is only a function of a single variable r; therefore,

$$\frac{1}{r}\frac{\partial}{\partial r}\left[\lambda(T) r \frac{\partial T}{\partial r}\right] = -\frac{\dot{q}}{g_c}. \tag{5}$$

Boundary conditions have the following form:

$$\left.\frac{dT}{dr}\right|_{r=0} = 0, \quad T\big|_{r=r_1} = T_1, \quad T\big|_{r=r_2} = T_2. \tag{6}$$

In order to linearize problem (5)–(6), Kirchhoff's transformation will be used

$$U = \int_0^T \lambda(T)\, dT. \tag{7}$$

Since

$$\frac{dU}{dr} = \frac{dU}{dT}\frac{dT}{dr} = \lambda(T)\frac{dT}{dr}, \tag{8}$$

Equation (5) and boundary conditions (6) become linear

$$\frac{1}{r}\frac{d}{dr}\left(r\frac{dU}{dr}\right) = -\frac{\dot{q}}{g_c}, \tag{9}$$

$$\left.\frac{dU}{dr}\right|_{r=0} = 0, \quad U\big|_{r=r_1} = U_1, \quad U\big|_{r=r_2} = U_2, \tag{10}$$

where

$$U_1 = \int_0^{T_1} \lambda(T)\,dT, \qquad U_2 = \int_0^{T_2} \lambda(T)\,dT. \tag{11}$$

From (9) with boundary conditions (10), one obtains

$$U = -\frac{1}{4}\frac{\dot{q}\left(r^2 - r_1^2\right)}{g_c} + U_1. \tag{12}$$

From the third boundary condition (10) it follows that

$$U_2 = -\frac{1}{4}\frac{\dot{q}\left(r_2^2 - r_1^2\right)}{g_c} + U_1. \tag{13}$$

After simple transformations of (13) we obtain a formula for heat flux $\dot{q}$

$$\dot{q} = \frac{4\left(U_1 - U_2\right)g_c}{r_2^2 - r_1^2}. \tag{14}$$

Since

$$U_1 - U_2 = \int_0^{T_1} \lambda(T)\,dT - \int_0^{T_2} \lambda(T)\,dT = \int_{T_2}^{T_1} \lambda(T)\,dT = \lambda_m\left(T_1 - T_2\right), \tag{15}$$

where

$$\lambda_m = \frac{\int_{T_2}^{T_1} \lambda(T)\,dT}{T_1 - T_2} \tag{16}$$

Equation (14) can be written in the following form:

$$\dot{q} = \frac{4\lambda_m\left(T_1 - T_2\right)g_c}{r_2^2 - r_1^2}. \tag{17}$$

If $\lambda(T) = a + bT$, then the average thermal conductivity given by (16) is

$$\lambda_m = \frac{\int_{T_2}^{T_1} (a + bT)\,dT}{T_1 - T_2} = \frac{a\left(T_1 - T_2\right) + \frac{1}{2}b\left(T_1^2 - T_2^2\right)}{T_1 - T_2} = \tag{18}$$

$$= a + \frac{1}{2}b\left(T_1 + T_2\right) = \lambda\left(T_m\right),$$

where $T_m = \frac{1}{2}\left(T_1 + T_2\right)$.

On the basis of the derived formulae, one can determine the heat flux value. Average temperature T_m measures

$$T_m = \frac{1}{2}(T_1 + T_2) = \frac{1}{2}(420 + 250) = 335\,^{\circ}\mathrm{C}.$$

After calculating the average thermal conductivity

$$\lambda_m = \lambda(T_m) = a + bT_m = 15{,}1 + 0.0136 \cdot 335 = 19.656\ \mathrm{W/(m{\cdot}K)},$$

one can calculate heat flux from (17):

$$\dot{q} = \frac{4\lambda_m (T_1 - T_2) g_c}{r_2^2 - r_1^2} = \frac{4 \cdot 19.656 \cdot (420 - 250) \cdot 0.0018}{0.010^2 - 0^2} =$$

$$= 240589.44\ \mathrm{W/m^2}.$$

Exercise 6.11 One-Dimensional Steady-State Plate Temperature Distribution Produced by Uniformly Distributed Volumetric Heat Sources

A round plate with a diameter $d_z = 200$ mm and thickness $g = 20$ mm is electrically heated. The aim is to calculate the upper and lower surface temperature of the plate under the assumption that heat is uniformly generated within the entire plate volume. The bottom and side parts of the plate surface are thermally insulated.

Heat transfer coefficient on the plate surface is $\alpha = 300$ W/(m²·K). Surrounding air temperature is $T_p = 20$°C. The plate's thermal conductivity is $\lambda = 30$ W/(m·K). The second aim is to calculate temperature distribution within the entire plate thickness under the assumption that the plate's thermal power is $\dot{Q} = 6$ kW.

Solution

The problem under consideration is shown in Fig. 6.16.

Since the power of internal heat sources remains constant

$$\dot{q}_v = \frac{\dot{Q}}{V} = \frac{4\dot{Q}}{\pi d_z^2 g}, \tag{1}$$

temperature field is determined using the heat conduction equation

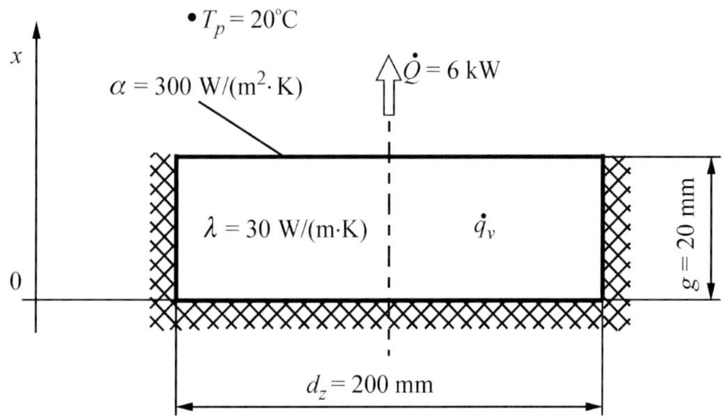

Fig. 6.16. Plate heating

$$\frac{d^2T}{dx^2} = -\frac{\dot{q}_v}{\lambda} \tag{2}$$

and boundary conditions

$$\left.\frac{dT}{dx}\right|_{x=0} = 0, \tag{3}$$

$$-\lambda \left.\frac{dT}{dx}\right|_{x=g} = \alpha\left(T|_{x=g} - T_p\right). \tag{4}$$

The solution is obtained by integrating (2) twice in x

$$T = -\frac{1}{2}\frac{\dot{q}_v}{\lambda}x^2 + C_1 x + C_2. \tag{5}$$

Once constants C_1 and C_2 are determined from boundary conditions (3) and (4) and substituted into (5), one obtains

$$T(x) = \frac{\dot{q}_v g^2}{2\lambda}\left[1 - \left(\frac{x}{g}\right)^2\right] + \frac{\dot{q}_v g}{\alpha} + T_p. \tag{6}$$

Insulated surface temperature ($x = 0$) is

$$T(0) = \frac{\dot{q}_v g^2}{2\lambda} + \frac{\dot{q}_v g}{\alpha} + T_p. \tag{7}$$

The upper part plate surface temperature ($x = g$) is

$$T(g) = \frac{\dot{q}_v g}{\alpha} + T_p. \tag{8}$$

Substitution of the numerical values into (1), (7) and (8) yields

$$\dot{q}_v = \frac{4 \cdot 6000}{\pi \cdot 0.2^2 \cdot 0,02} = 9.5493 \cdot 10^6 \text{ W/m}^3 .$$

Temperature of the lower part of the plate surface is

$$T(0) = \frac{9.5493 \cdot 10^6 \cdot (0.02)^2}{2 \cdot 30} + \frac{9.5493 \cdot 10^6 \cdot 0.02}{300} + 20 = 720.28\,^\circ\text{C} .$$

Temperature of the upper part is slightly lower:

$$T(0.02) = \frac{9.5493 \cdot 10^6 \cdot 0.02}{300} + 20 = 656.62\,^\circ\text{C} .$$

Exercise 6.12 One-Dimensional Steady-State Pipe Temperature Distribution Produced by Uniformly Distributed Volumetric Heat Sources

Electric current flows at an intensity of 300 A through a pipe made of an alloy steel pipe (Fig. 6.17) with an inner diameter d_w = 7.2 mm and outer diameter d_z = 8 mm. Thermal conductivity of this steel is λ = 18.4 W/(m·K), while its specific resistance ρ = 0.85 ($\Omega \cdot$ mm^2)/m. The aim is to calculate temperature distribution within the entire wall thickness under the assumption that the outer surface of the pipe is thermally insulated and that total heat generated by the pipe flows inside it. The inner surface temperature of the pipe is T_w = 300°C. The second aim is to calculate the heat transfer coefficient on the inner surface of the pipe, if the temperature of a medium is T_{cz} = 20°C.

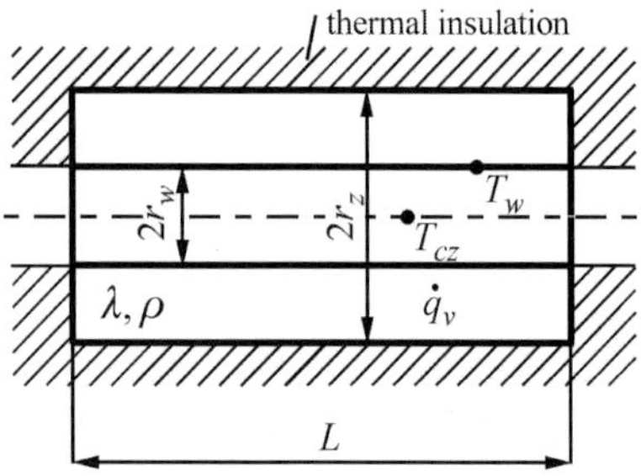

Fig. 6.17. Electrically heated steel pipe

Solution

Firstly, we will determine temperature distribution within the pipe wall under the assumption that heat is uniformly generated within the entire body volume. The heat conduction equation

$$\frac{1}{r}\frac{d}{dr}\left(r\frac{dT}{dr}\right)=-\frac{\dot{q}_v}{\lambda} \tag{1}$$

will be solved using the following boundary conditions:

$$T\big|_{r=r_w}=T_w, \tag{2}$$

$$\left.\frac{dT}{dr}\right|_{r=r_z}=0. \tag{3}$$

The solution is obtained by integrating (1) twice in r

$$\frac{dT}{dr}=-\frac{\dot{q}_v r}{2\lambda}+\frac{C_1}{r} \tag{4}$$

and

$$T(r)=-\frac{\dot{q}_v r^2}{4\lambda}+C_1\ln r+C_2. \tag{5}$$

By substituting (5) into (2) and (4) into (3), one gets two algebraic equations

$$-\frac{\dot{q}_v r_w^2}{4\lambda}+C_1\ln r_w+C_2=T_w, \tag{6}$$

$$-\frac{\dot{q}_v r_z}{2\lambda}+\frac{C_1}{r_z}=0, \tag{7}$$

whose solution are constants C_1 and C_2:

$$C_1=\frac{\dot{q}_v r_z^2}{2\lambda}, \tag{8}$$

$$C_2=\frac{\dot{q}_v}{2\lambda}\left(\frac{r_w^2}{2}-r_z^2\ln r_w\right)+T_w. \tag{9}$$

By substituting constants C_1 and C_2 into (5), one obtains the following, after simple transformations:

$$T\big|_{r=r_z} = T_z\,, \tag{2}$$

$$\left.\frac{dT}{dr}\right|_{r=r_z} = 0\,. \tag{3}$$

Integrating (1) twice, yields

$$\frac{dT}{dr} = -\frac{\dot{q}_v r}{2\lambda} + \frac{C_1}{r} \tag{4}$$

$$T(r) = -\frac{\dot{q}_v r^2}{4\lambda} + C_1 \ln r + C_2\,. \tag{5}$$

By substituting (4) into (3) and (5) into (2), one gets

$$C_1 = \frac{\dot{q}_v r_z^2}{2\lambda}\,, \tag{6}$$

$$C_2 = T_z + \frac{\dot{q}_v r_z^2}{4\lambda} - \frac{\dot{q}_v r_z^2}{2\lambda} \ln r_z\,. \tag{7}$$

Substituting (6) and (7) into (5), gives the temperature distribution $T(r)$

$$T(r) = T_z + \frac{\dot{q}_v}{2\lambda}\left(\frac{r_z^2 - r^2}{2} - r_z^2 \ln\frac{r_z}{r}\right). \tag{8}$$

Inner surface temperature is

$$T_w = T(r_w) = T_z + \frac{\dot{q}_v}{2\lambda}\left(\frac{r_z^2 - r_w^2}{2} - r_z^2 \ln\frac{r_z}{r_w}\right). \tag{9}$$

Temperature drop across within the wall thickness, determined from (9) is given by

$$T_z - T_w = \frac{\dot{q}_v}{2\lambda}\left(r_z^2 \ln\frac{r_z}{r_w} + \frac{r_w^2 - r_z^2}{2}\right). \tag{10}$$

The same result is obtained from (11) in Ex. 6.12.

Inner surface temperature determined from (9) is

$$T_w = 303.77805 + \frac{8.38715\cdot 10^8}{2\cdot 18.4}\left(-0.004^2 \ln\frac{0.004}{0.0036} + \right.$$

$$\left. + \frac{0.004^2 - 0.0036^2}{2}\right) = 300\,^\circ\mathrm{C}.$$

The remaining results are identical to the results from Ex. 6.12.

Exercise 6.14 General Equation of Heat Conduction in Fins

The aim is to derive a differential equation to describe heat transfer in fins with arbitrary shapes (Fig. 6.18) under the assumption that temperature across the fin thickness is constant. In other words, one should disregard temperature drop across the fin thickness and derive a formula for fin efficiency.

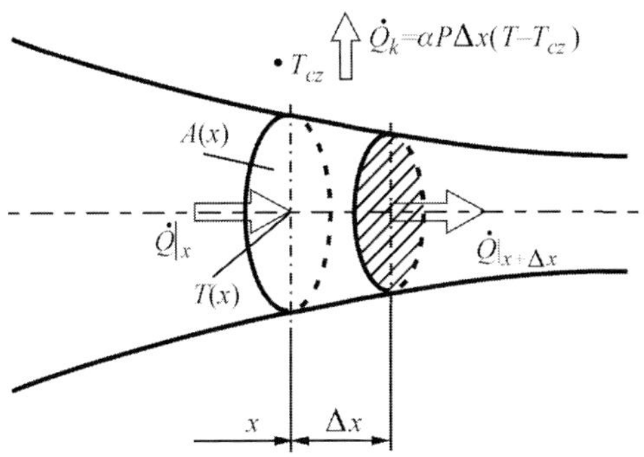

Fig. 6.18. Heat flow through fins with arbitrary shapes

Solution

By assuming that fin temperature remains constant within the fin's cross-section and changes only in the direction of x axis, the heat balance for control volume $A(x)\Delta x$ has the form

$$\dot{Q}\Big|_x = \dot{Q}\Big|_{x+\Delta x} + \dot{Q}_k \,. \tag{1}$$

Heat flows $\dot{Q}$ are expressed by the following formulas:

$$\dot{Q}\Big|_x = -\lambda A \frac{dT}{dx}\bigg|_x, \quad \dot{Q}\Big|_{x+\Delta x} = -\lambda A \frac{dT}{dx}\bigg|_{x+\Delta x}, \quad \dot{Q}_k = \alpha P \Delta x \left(T - T_{cz}\right). \tag{2}$$

By substituting (2) for (1), one obtains

$$\frac{\lambda A \dfrac{dT}{dx}\bigg|_{x+\Delta x} - \lambda A \dfrac{dT}{dx}\bigg|_x}{\Delta x} - \alpha P \left(T - T_{cz}\right) = 0\,, \tag{3}$$

where $A(x)$ is a cross-section area of a fin perpendicular to the direction of heat flow through a fin, P is the fin circumference at a point with x coordinate. If $\Delta x \to 0$, then (3) assumes the form:

$$\frac{d}{dx}\left(\lambda A\frac{dT}{dx}\right)-\alpha P\left(T-T_{cz}\right)=0\,. \tag{4}$$

For a constant thermal conductivity λ and constant cross-section A, (4) can be written in the form

$$\frac{d^2T}{dx^2}-m^2\left(T-T_{cz}\right)=0\,, \tag{5}$$

where

$$m^2=\frac{\alpha P}{\lambda A}\,. \tag{6}$$

Two boundary conditions are necessary in order to determine temperature distribution in a fin of height L. The first condition is assigned at point $x = 0$ in the base of the fin, the second at the end of the fin at point $x = L$. In practical computations, it is usually assumed that fin base temperature T_b is constant and equal to a temperature of a surface on which the fin is mounted; i.e. it is assumed that fins do not disturb temperature distribution in a construction element to which they are attached. The fin tip is usually regarded as being thermally insulated, since the surface area of the tip is considerably smaller than the area of fin's side surfaces; therefore, one can neglect the heat flow transmitted by the tip. Assuming that heat exchange takes place on the tip of the fin, the boundary conditions have the form

$$T\big|_{x=0}=T_b\,,$$

$$-\lambda\frac{dT}{dx}\bigg|_{x=L}=\alpha_w\left(T\big|_{x=L}-T_{cz}\right),$$

where α_w is the heat transfer coefficient from the tip to surroundings, while temperature T_{cz} is the temperature of a medium that surrounds the fin. It is usually assumed that $\alpha_w = \alpha$ or $\alpha_w = 0$, when a fin tip is thermally insulated.

Fin efficiency is a ratio of a heat flow $\dot{Q}$, transferred by an actual fin, to a maximal heat flow $\dot{Q}_{max}$, which the fin could transfer. Maximal heat flow $\dot{Q}_{max}$ occurs when temperature of the fin is uniform within its entire volume and is equal to the base temperature T_b. Heat flow $\dot{Q}$ can be calculated as a flow that is conducted through the base of the fin or as a dissipated flow by lateral surfaces of the fin and the tip:

$$\dot{Q} = -\left(\lambda A \frac{dT}{dx}\right)\Bigg|_{x=0} = \int_0^L \alpha\left(T - T_{cz}\right) P\, dx\,,$$

$$\dot{Q}_{max} = \int_0^L \alpha\left(T_b - T_{cz}\right) P\, dx\,.$$

Fin efficiency is formulated as

$$\eta = \frac{\dot{Q}}{\dot{Q}_{max}}\,.$$

Assuming that α, P and T_{cz} are independent of their position, formulas for efficiency in fins with standard shapes are not very complicated.

Exercise 6.15 Temperature Distribution and Efficiency of a Straight Fin with Constant Thickness

The aim is to determine temperature distribution and efficiency of a straight fin with constant thickness under the assumption that fin tip is thermally insulated. Next, to consider heat exchange through the fin tip, the fin height will be increased by half of the fin's thickness. Then, the fin temperature, heat flow dissipated by the fin and fin efficiency should also be calculated. The following values are assumed for the calculation: fin material – copper with the thermal conductivity $\lambda = 390$ W/(m·K), fin thickness $t = 0.5$ mm, height $L = 7.5$ cm, width $w = 0.7$ m, fin base temperature $T_b = 80°\text{C}$, air temperature $T_{cz} = 20°\text{C}$, heat transfer coefficient $\alpha = 10$ W/(m^2·K).

Solution

Differential (5) from Ex. 6.14, which describes thermal exchange in a fin, has the following form:

$$\frac{d^2T}{dx^2} - m^2\left(T - T_{cz}\right) = 0\,. \tag{1}$$

If fin base temperature is T_b, while fin tip is thermally insulated, the boundary conditions have the form

$$T\big|_{x=0} = T_b\,, \tag{2}$$

$$\left.\frac{dT}{dx}\right|_{x=L}=0\,. \tag{3}$$

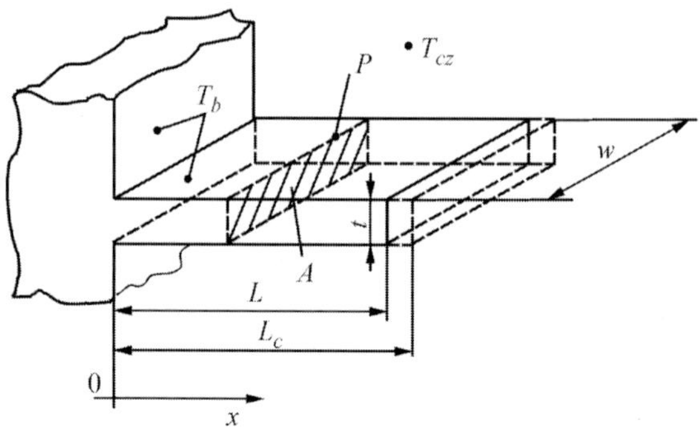

Fig. 6.19. Straight fin of constant cross-section

In the case of a straight fin, shown in Fig. 6.19, circumference P, on which thermal exchange takes place, is $P = 2(w + t)$, while the cross-section area with regard to the direction of thermal conduction is $A = wt$. Fin parameter m^2 ((6), Ex. 6.14) has, in the given case, the form

$$m^2=\frac{2\alpha\left(w+t\right)}{\lambda wt}\approx\frac{2\alpha}{\lambda t}, \tag{4}$$

since usually $t << w$. Once the new variable is introduced

$$\theta=T-T_{cz} \tag{5}$$

Equation (1) and boundary conditions (2), (3) can be rewritten in the following way

$$\frac{d^2\theta}{dx^2}-m^2\theta=0\,, \tag{6}$$

$$\left.\theta\right|_{x=0}=\theta_b,\qquad \theta_b=T_b-T_{cz}\,, \tag{7}$$

$$\left.\frac{d\theta}{dx}\right|_{x=L}=0\,. \tag{8}$$

Solution of the homogenous (6) has the following form:

$$\theta=C_1e^{mx}+C_2e^{-mx}\,. \tag{9}$$

From boundary conditions (7)and (8), two algebraic equations are obtained

$$C_1 + C_2 = \theta_b, \tag{10}$$

$$mC_1 e^{mL} - mC_2 e^{-mL} = 0. \tag{11}$$

Once constants C_1 and C_2 are determined from (10) and (11) and substituted into (9), one obtains the following after transformations:

$$\frac{\theta(x)}{\theta_b} = \frac{T(x) - T_{cz}}{T_b - T_{cz}} = \frac{\cosh m(L-x)}{\cosh mL}. \tag{12}$$

Flow $\dot{Q}$ and $\dot{Q}_{max}$ will be determined in order to define fin efficiency:

$$\dot{Q} = -\lambda A \left.\frac{dT}{dx}\right|_{x=0} = -\left[\lambda wt\left(T_b - T_{cz}\right)\frac{-m\sinh m(L-x)}{\cosh mL}\right]\Bigg|_{x=0} = \tag{13}$$
$$= \lambda wt\sqrt{\frac{2\alpha}{\lambda t}}\left(T_b - T_{cz}\right)\text{tgh } mL,$$

$$\dot{Q}_{max} = \alpha 2wL\left(T_b - T_{cz}\right). \tag{14}$$

Efficiency of a straight fin with constant cross-section is

$$\eta = \frac{\dot{Q}}{\dot{Q}_{max}} = \frac{\lambda wt\sqrt{\frac{2\alpha}{\lambda t}}\left(T_b - T_{cz}\right)\text{tgh } mL}{\alpha 2wL\left(T_b - T_{cz}\right)} = \frac{\text{tgh } mL}{mL}, \tag{15}$$

where m is formulated in (4).

Hyperbolic trigonometric functions are expressed using the following formulas:

$$\sinh x = \frac{e^x - e^{-x}}{2}, \qquad \cosh x = \frac{e^x + e^{-x}}{2}, \qquad \text{tgh } x = \frac{\sinh x}{\cosh x} = \frac{e^x - e^{-x}}{e^x + e^{-x}}.$$

In (12) and (15), fin tip heat exchange is not taken into consideration. It can be considered only when the fin height is increased by $t/2$ (Fig. 6.19). Once the fin height substitute is introduced, L_c

$$L_c = L + \frac{t}{2}$$

temperature distribution and fin efficiency are expressed as

$$\frac{\theta(x)}{\theta_b} = \frac{T(x) - T_{cz}}{T_b - T_{cz}} = \frac{\cosh m(L_c - x)}{\cosh mL_c},$$

$$\frac{\theta}{\theta_b}=\frac{T-T_{sp}}{T_b-T_{sp}}=\frac{\cosh m(L-x)}{\cosh mL}, \qquad (3)$$

where

$T_b = T_{sc}$ – wall temperature, which the casing is welded onto,

T_{sp} – temperature of combustion gases,

$T = T(x)$ – temperature of a casing within distance x from the channel wall,

m – fin parameter, defined as follows

$$m=\sqrt{\alpha_o P/\lambda_o A},$$

where α_o – heat transfer coefficient on an outer surface of the casing, $P = \pi d_z$ – outer surface of the casing,

$$A=\frac{\pi}{4}\left(d_z^2-d_w^2\right)=\pi d_{sr} g_o, \qquad d_{sr}=\frac{d_w+d_z}{2}=d_z-g.$$

Parameter m:

$$m=\sqrt{\frac{\alpha_o P}{\lambda_o A}}=\sqrt{\frac{\alpha_o \pi d_z}{\lambda_o \pi d_{sr} g_o}}=\sqrt{\frac{\alpha_o d_z}{\lambda_o d_{sr} g_o}}.$$

After substitution, one obtains

$$m=\sqrt{\frac{45\cdot 0.012}{50\cdot(0.012-0.006)\cdot 0.003}}=20\ 1/\mathrm{m}.$$

Tip temperature of the casing T_t, indicated by the thermoelement, is determined from (3) for $x = L$

$$\frac{T_t-T}{T_b-T_{sp}}=\frac{1}{\cosh mL}, \qquad T_t=T+\frac{T_b-T_{sp}}{\cosh mL}.$$

Since

$$mL = 20\cdot 0.12 = 2.4,$$

then

$$T_t=210+\frac{146.7-210}{\cosh(2.4)}=210-\frac{63.3}{5.5569}=198.61^\circ\mathrm{C}.$$

Relative temperature measurement error for combustion gases is

$$\varepsilon=\frac{T_{sp}-T_t}{T_{sp}}\cdot 100\,\%=\frac{210-198.61}{210}5.42\%.$$

Exercise 6.17 Temperature Distribution and Efficiency of a Circular Fin of Constant Thickness

The aim is to derive an equation for a circular fin of constant thickness from a general equation of heat transfer in fins (Ex. 6.14) and to determine formulas for temperature distribution in fins, for a fin-transferred heat flow and for fin efficiency. Following that calculate fin tip temperature, fin efficiency and dissipate heat flow using the following data: fin-base temperature $T_b = 90°C$, temperature of surroundings $T_{cz} = 20°C$, $r_1 = 12.5$ mm, $r_2 = 28.5$ mm, $t = 0.4$ mm, material of a fin – aluminium with thermal conductivity $\lambda = 205$ W/(m·K), heat transfer coefficient on the fin surface α = 70 W/(m^2·K). Take into account heat exchange on a fin tip by increasing fin height L to $L_c = L + t/2$.

Solution

In the case of a circular fin shown in Fig. 6.21, surface area of a fin cross-section is $A = 2\pi rt$, while circumference, on which thermal exchange occurs is $P = 4\pi r$. Parameter m is defined as

$$m = \sqrt{\frac{\alpha P}{\lambda A}} = \sqrt{\frac{\alpha 4\pi r}{\lambda 2\pi rt}} = \sqrt{\frac{2\alpha}{\lambda t}} \; . \tag{1}$$

Differential equation (4) in Ex. 6.14 assumes the following form:

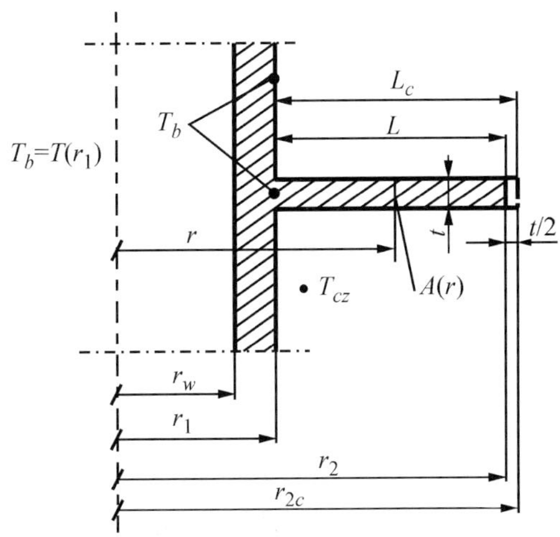

Fig. 6.21. Circular fin with constant thickness

$$\frac{1}{r}\frac{d}{dr}\left(r\frac{dT}{dr}\right)-\frac{2\alpha}{\lambda t}\left(T-T_{cz}\right)=0\,. \tag{2}$$

Once excess temperature $\theta = T - T_{cz}$ and parameter m given by (1) are introduced, (2) can be written in a form

$$\frac{d^2\theta}{dr^2}+\frac{1}{r}\frac{d\theta}{dr}-m^2\theta=0\,. \tag{3}$$

It is a modified Bessel equation, for which general solution has the form

$$\theta(r)=C_1I_0(mr)+C_2K_0(mr)\,. \tag{4}$$

Constants C_1 and C_2 will be determined from boundary conditions

$$\theta\big|_{r=r_1}=\theta_b,\ \theta_b=T_b-T_{cz}, \tag{5}$$

$$\left.\frac{d\theta}{dr}\right|_{r=r_2}=0\,. \tag{6}$$

Once constants are determined and substituted into (4), one obtains a formula for temperature distribution $\theta(r)$ in a fin:

$$\frac{\theta}{\theta_b}=\frac{T(r)-T_{cz}}{T_b-T_{cz}}=\frac{K_0(mr)I_1(mr_2)+I_0(mr)K_1(mr_2)}{I_0(mr_1)K_1(mr_2)+K_0(mr_1)I_1(mr_2)}\,. \tag{7}$$

Heat flow $\dot{Q}$ dissipated by the fin

$$\dot{Q}=-\lambda A_b\left.\frac{dT}{dr}\right|_{r=r_1}=2\pi\lambda r_1 t\theta_b m\frac{K_1(mr_1)I_1(mr_2)-I_1(mr_1)K_1(mr_2)}{K_0(mr_1)I_1(mr_2)+I_0(mr_1)K_1(mr_2)}\,. \tag{8}$$

Since maximal flow $\dot{Q}_{max}$ is

$$\dot{Q}_{max}=\alpha A_z\left(T_b-T_{cz}\right)=\alpha 2\pi\left(r_2^2-r_1^2\right)\theta_b\,, \tag{9}$$

fin efficiency, then, can be determined as:

$$\eta=\frac{\dot{Q}}{\dot{Q}_{max}}=\frac{2r_1}{m\left(r_2^2-r_1^2\right)}\frac{K_1(mr_1)I_1(mr_2)-I_1(mr_1)K_1(mr_2)}{K_0(mr_1)I_1(mr_2)+I_0(mr_1)K_1(mr_2)}\,. \tag{10}$$

Fin tip heat exchange can be taken into account by substituting radius r_2 in (7)–(10) for a slightly larger radius $r_{2c} = r_2 + t/2$. As in the case of a straight fin, fin length is larger: $L_c = L + t/2$.

After substitution of the numerical values, one obtains

$$mr_1 = 41.32 \cdot 0.0125 = 0.5165,$$

$$r_{2c} = r_2 + \frac{t}{2} = 0.0285 + 0.0002 = 0.0287 \text{ m},$$

$$mr_{2c} = 1.1859 \qquad .$$

The values of the Bessel functions are [5]

$$I_0(mr_{2c}) = I_0(1.1859) = 1.3837, \qquad K_1(mr_{2c}) = K_1(1.1859) = 0.4443,$$

$$K_0(mr_{2c}) = K_0(1.1859) = 0.3247,$$

$$I_1(mr_{2c}) = I_1(1.1859) = 0.7035, \qquad I_0(mr_1) = I_0(0.5165) = 1.0678,$$

$$K_0(mr_1) = K_0(0.5165) = 0.8977,$$

one obtains

$$\frac{T(r_{2c}) - 20}{90 - 20} = \frac{0.3247 \cdot 0.7035 + 1.3837 \cdot 0.4443}{1.0678 \cdot 0.4443 + 0.8977 \cdot 0.7035} = 0.7624,$$

$$T(r_{2c}) = 73.38^\circ \text{C}.$$

In order to determine fin efficiency, two additional Bessel functions are needed:

$$K_1(mr_1) = K_1(0.5165) = 1.5887,$$

$$I_1(mr_1) = I_1(0.5165) = 0.2670.$$

Heat flow dissipated by the fin is

$$\eta = \frac{2 \cdot 0.0125}{41.32\left(0.0287^2 - 0.0125^2\right)} \frac{1.5887 \cdot 0.7035 - 0.2670 \cdot 0.4443}{0.8977 \cdot 0.7035 + 1.0678 \cdot 0.4443} = 0.8188 .$$

Fin-diffused heat flow:

$$\dot{Q} = \eta \dot{Q}_{\max} = \eta \alpha 2\pi \left(r_{2c}^2 - r_1^2\right)\left(T_b - T_{cz}\right) =$$

$$= 0.8188 \cdot 70 \cdot 2 \cdot \pi \left(0.0287^2 - 0.0125^2\right)\left(90 - 20\right) = 16.82 \text{ W}.$$

Bessel function values, present in formulas for temperature distribution and circular fin efficiency, can be read from table [5] or calculated by means of library procedures [4, 7].

Exercise 6.18 Approximated Calculation of a Circular Fin Efficiency

Calculate circular fin efficiency from Ex. 6.17 using the following approximation formulas:

a) according to Schmidt [11, 12]

$$\eta_S = \frac{\operatorname{tgh} mL_c\varphi}{mL_c\varphi}, \tag{1}$$

where

$$\varphi = 1 + 0.35\ln\left(1 + \frac{L_c}{r_1}\right), \qquad L_c = r_{2c} - r_1, \qquad m = \sqrt{\frac{2\alpha}{\lambda t}}. \tag{2}$$

If $\eta > 0.5$, then efficiency η calculated using (1) does not differ more than ±1% from the real value.

b) according to Brandt [2]

$$\eta_B = \frac{2r_1}{2r_1 + L_c}\frac{\operatorname{tgh} mL_c}{mL_c}\left[1 + \frac{\operatorname{tgh} mL_c}{2mr_1} - C\frac{(\operatorname{tgh} mL_c)^p}{(mr_1)^n}\right], \tag{3}$$

where

$$C = 0.071882, \qquad p = 3.7482, \qquad n = 1.4810. \tag{4}$$

Maximal error from efficiency determination by means of (3) is smaller than 0.6% from an error made when determining efficiency by means of (1).

c) formula according to [3]

$$\eta_H = \frac{1}{1 + \frac{1}{3}(mL_c)^2\sqrt{r_{2c}/r_1}}. \tag{5}$$

Equation (5) gives good results, when $\eta > 0.75$. For the calculation, use the values from Ex. 6.17.

Solution

a) according to Schmidt

$$m = 41.32\ 1/\mathrm{m}, \qquad r_1 = 0.0125\ \mathrm{m}, \qquad r_{2c} = 0.0287\,\mathrm{m},$$

$$L_c = r_{2c} - r_1 = 0.0162\ \mathrm{m}, \quad \varphi = 1.2909,$$

$$\eta_S = \frac{\mathrm{tgh}(41.32 \cdot 0.0162 \cdot 1.2909)}{41.32 \cdot 0.0162 \cdot 1.2909} = 0.8082\,.$$

Since a real value of the efficiency is η = 0.8188, relative error, then, comes to

$$\varepsilon_S = \frac{(\eta_S - \eta)}{\eta} 100\% = \frac{0.8082 - 0.8188}{0.8188} \cdot 100\% = -1.295\%\,.$$

b) according to Brandt [2]

$$\eta_B = \frac{2 \cdot 0.0125}{2 \cdot 0.0125 + 0.0162} \cdot \frac{0.584575}{41.32 \cdot 0.0162} \times$$

$$\times \left(1 + \frac{0.584575}{2 \cdot 41.32 \cdot 0.0125} - 0.071882 \cdot \frac{(0.584575)^{3.7482}}{(41.32 \cdot 0.0125)^{1.4810}} \right) = 0.8162.$$

Relative error is:

$$\varepsilon_B = \frac{\eta_B - \eta}{\eta} \cdot 100\% = \frac{0.8162 - 0.8188}{0.8188} \cdot 100\% = -0.312\%\,.$$

c) formula according to [3]

$$\eta_H = \frac{1}{1 + \frac{1}{3}(41.32 \cdot 0.0162)^2 \sqrt{\frac{0.0287}{0.0125}}} = 0.8155\,.$$

Relative error is:

$$\varepsilon_H = \frac{\eta_H - \eta}{\eta} \cdot 100\% = \frac{0.8155 - 0.8188}{0.8188} \cdot 100\% = -0.409\%\,.$$

From the comparison of the results presented above, one can see that the least accurate result is given by Schmidt formula. Brandt formula allows for the most accurate calculation of circular fin efficiency; however, the amount of work required to obtain the results is not much smaller than in the case of the analytical formula. Equation (5) is both simple, yet accurate.

Exercise 6.19 Calculating Efficiency of Square and Hexagonal Fins

Calculate fin efficiency (equivalent fins) in a fin-plate exchanger made of pipes with an outer diameter of d_z = 10 mm and wall thickness g_r = 1 mm.

Distance between pipes in a hexagonal system is constant and is equal to $2s$ (Fig. 6.22b). Perpendicular and longitudinal pitch 2s, with pipes arranged in rows, is assumed constant (Fig. 6.22a). Calculations are to be made for a

a) in-line
b) hexagonal

pipe configuration for two fin-plates of different thickness: $t = 0.33$ mm and $t = 0.13$ mm, assuming that $2s = 25$ mm. Fin plates are made of aluminium alloy with thermal conductivity of $\lambda = 165$ W/(m·K). Heat transfer coefficient is $\alpha = 50$ W/(m^2·K).

Solution

Pipe configuration is shown in Fig. 6.22. Fin efficiency will be calculated using (5) from Ex. 6.18.

$$\eta = \frac{1}{1+\frac{1}{3}(mL)^2\sqrt{\frac{r_2^*}{r_1}}},$$

where $L = r_2^* - r_1$, $r_1 = d_z/2$, r_2^* – equivalent radius of a circular fin, calculated from a condition of equality of a conventional and circular fin surface area, as shown in Fig. 6.22. Fin parameter is calculated using formula

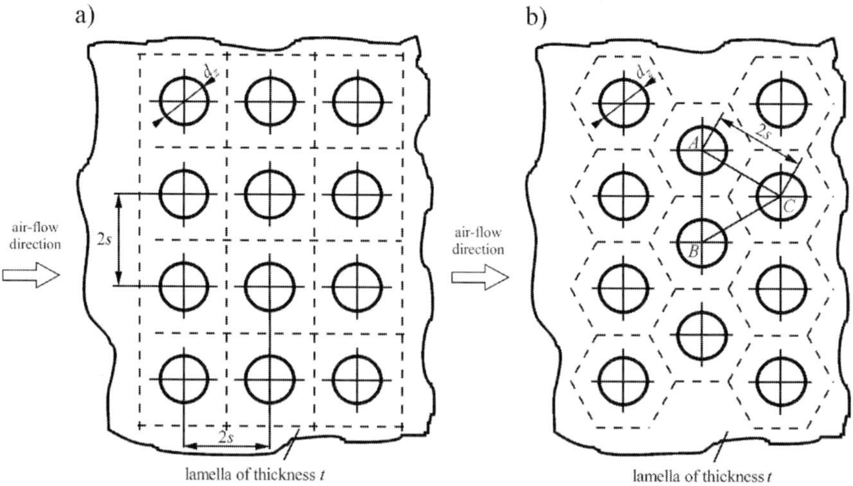

Fig. 6.22. Pipe lay-out in a fin-plate exchanger: (**a**) in-line pipe configuration, (**b**) hexagonal (staggered) pipe configuration

$$m = \sqrt{\frac{2\alpha}{\lambda t}}.$$

a) *In-line pipe configuration* is shown in Fig. 6.22a

Equivalent radius r_2^* is calculated from a condition of equality of circular and square fin surface area, whose side is $2s$

$$2s \cdot 2s - \pi r_1^2 = \pi \left(r_2^*\right)^2 - \pi r_1^2,$$

$$r_2^* = \frac{2s}{\sqrt{\pi}}.$$

After substitution, one obtains

$$r_2^* = \frac{2s}{\sqrt{\pi}} = \frac{0.025}{\sqrt{\pi}} = 0.0141 \text{ m}.$$

Parameter m:

when $t = 0.33$ mm, $m = \sqrt{\frac{2 \cdot 50}{165 \cdot 0.00033}} = 42.85 \ \ 1/\text{m}$,

when $t = 0.13$ mm, $m = \sqrt{\frac{2 \cdot 50}{165 \cdot 0.00013}} = 68.28 \ \ 1/\text{m}$.

Since,

$$L = r_2^* - r_1 = 0.0141 - 0.005 = 0.0091 \text{ m},$$

fin efficiency is

when $t = 0{,}33$ mm, $\eta = \frac{1}{1 + \frac{1}{3}(42.85 \cdot 0.0091)^2 \sqrt{\frac{0.0141}{0.005}}} = 0.9216,$

when $t = 0.13$ mm, $\eta = \frac{1}{1 + \frac{1}{3}(68.28 \cdot 0.0091)^2 \sqrt{\frac{0.0141}{0.005}}} = 0.8223.$

b) *Hexagonal pipe configuration* is shown in Fig. 6.22b

Area of triangle ABC:

$$S_{ABC} = \frac{1}{2} 2s\sqrt{(2s)^2 - s^2} = \sqrt{3}s^2.$$

Fin surface area: $S_{\dot{z}} = 2S_{ABC}$ and

$$2S_{ABC} - \pi r_1^2 = \pi \left(r_2^*\right)^2 - \pi r_1^2, \qquad 2\sqrt{3}s^2 = \pi \left(r_2^*\right)^2;$$

One obtains then

$$r_2^* = \sqrt{\frac{2\sqrt{3}}{\pi}}s = \left(\frac{2\sqrt{3}}{\pi}\right)^{1/2} s.$$

Equivalent radius:

$$r_2^* = \left(\frac{2\sqrt{3}}{\pi}\right)^{1/2} \cdot s = \left(\frac{2\sqrt{3}}{\pi}\right)^{1/2} \cdot 0.0125 = 0.0131\,\text{m}.$$

Since,

$$L = r_2^* - r_1 = 0.0131 - 0.005 = 0.0081\ \text{m},$$

one obtains the following efficiency values:

when, $t = 0.33$ mm, $$\eta = \frac{1}{1 + \frac{1}{3}(42.85 \cdot 0.0081)^2 \sqrt{\frac{0.0131}{0.005}}} = 0.9390,$$

when, $t = 0.13$ mm, $$\eta = \frac{1}{1 + \frac{1}{3}(68.28 \cdot 0.0081)^2 \sqrt{\frac{0.0131}{0.005}}} = 0.8583.$$

From the comparison of results given, it is clear that hexagonal configuration ensures greater fin efficiency; therefore, the flow of transferred heat is larger than it is in the case of a in-line pipe configuration.

Exercise 6.20 Calculating Efficiency of Hexagonal Fins by Means of an Equivalent Circular Fin Method and Sector Method

The aim is to calculate efficiency of an equivalent fin in a fin-plate exchanger with a staggered fin configuration (Fig. 6.23). Pipes with an outer diameter of d_z = 7.59 mm arranged together with the following pitches: P_t = 21 mm and P_l = 12.7 mm (Fig. 6.23). The thickness of aluminium alloy based fin plates is t = 0.115 mm. Thermal conductivity of the fin

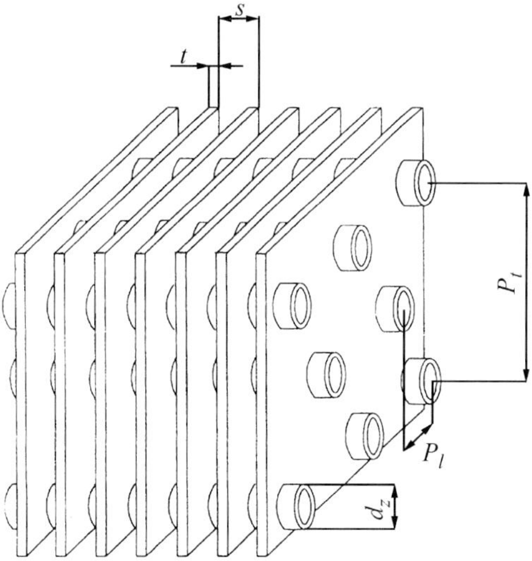

Fig. 6.23. Fin-plate heat exchanger with a staggered pipe configuration

material is $\lambda = 165$ W/(m·K). Heat transfer coefficient on a fin-plate surface is equal to α = 40 (W/m^2·K). Calculate efficiency of fins with an equivalent outer radius by means of a simplified Brandt formula and sector method.

Solution

a) Method I

First, we will calculate an equivalent outer radius of a circular fin, whose surface area equals a surface area of a equivalent fin shown in Fig. 6.24. The equivalent fin surface area is

$$A_{\dot{z}} = \left[\left(2A_{OAB} + 4A_{OBC}\right) - \pi r_1^2\right] \cdot 2,$$

where A_{OBC} is an area of triangle OBC.

The above formula took into account the fact that thermal exchange occurs on both sides of fin plates. Surface area of a circular fin A_o with an equivalent outer r_2^* and inner radius substitute and r_1, respectively, is

$$A_{\dot{z}o} = 2\pi\left[\left(r_2^*\right)^2 - r_1^2\right].$$

$$r_2^* = \sqrt{\frac{2A_{OAB} + 4A_{OBC}}{\pi}} = \sqrt{\frac{2\cdot 2.1089\cdot 10^{-5} + 4\cdot 5.6156\cdot 10^{-5}}{\pi}} =$$

$$= 9.2155\cdot 10^{-3}\,\text{m}, \quad m = \sqrt{\frac{2\alpha}{\lambda t}} = \sqrt{\frac{2\cdot 40}{165\cdot 0.000115}} = 64.93\ 1/\text{m},$$

$$L = r_2^* - r_1 = 9.2155\cdot 10^{-3} - 3.795\cdot 10^{-3} = 5.4205\cdot 10^{-5}\,\text{m},$$

$$mL = 0.35195,$$

$$\eta_e = \frac{2\cdot 3.795\cdot 10^{-3}}{2\cdot 3.795\cdot 10^{-3} + 5.4205\cdot 10^{-3}}\frac{\text{tgh}\ 0.35195}{0.35195}\times$$

$$\times\left[1 + \frac{\text{tgh}\ 0.35195}{2\cdot 64.93\cdot 3.795\cdot 10^{-3}} - 0.071882\frac{\text{tgh}\ 0.35195^{3.7482}}{\left(64.93\cdot 3.795\cdot 10^{-3}\right)^{1.481}}\right] =$$

$$= 0.5834\cdot 0.96066\cdot\left(1 + 0.68606 - 9.8256\cdot 10^{-3}\right) = 0.9394.$$

b) *Sector method*

In keeping with the sector method, a triangular fin *OAB* will be substituted by an equivalent circular fin sector (with an equivalent surface area).

$$A_{OAB} - \frac{2\gamma}{2\pi}r_1^2 = \frac{2\gamma}{2\pi}\pi r_{2,1}^2 - \frac{2\gamma}{2\pi}\pi r_1^2,$$

hence,

$$r_{2,1} = \sqrt{\frac{A_{OAB}}{\gamma}} = \sqrt{\frac{2.1089\cdot 10^{-5}}{0.189}} = 0.010563\ \text{m}.$$

Circular fin efficiency with an outer radius $r_{2,1}$ and inner radius r_1 amounts to

$$L_1 = r_{2,1} - r_1 = 0.010563 - 3.795\cdot 10^{-3} = 6.768\cdot 10^{-3}\ \text{m},$$

$$mL_1 = 0.43945,$$

$$\eta_1 = \frac{2\cdot 3.795\cdot 10^{-3}}{2\cdot 3.795\cdot 10^{-3} + 6.768\cdot 10^{-3}}\frac{\text{tgh}0.43945}{0.43945}\times$$

$$\times\left[1 + \frac{\text{tgh}0.43945}{2\cdot 64.93\cdot 3.795\cdot 10^{-3}} - 0.071882\frac{\text{tgh}0.43945^{3.7482}}{\left(64.93\cdot 3.795\cdot 10^{-3}\right)^{1.481}}\right] =$$

$$= 0.5834\cdot 0.94024\cdot\left(1 + 0.83842 - 0.020836\right) = 0.9034.$$

Radius $r_{2,2}$ of a circular fin sector with a surface area identical to a surface area of a triangular fin *OBC* is calculated in a similar way.

$$L_2 = r_{2,2} - r_1 = 5.2205 \cdot 10^{-3} \text{ m},$$

$$mL_2 = 64.93 \cdot 5.2205 \cdot 10^{-3} = 0.33897,$$

$$\eta_2 = \frac{2 \cdot 3.795 \cdot 10^{-3}}{2 \cdot 3.795 \cdot 10^{-3} + 5.2205 \cdot 10^{-3}} \frac{\text{tgh}0.33897}{0.33897} \times$$

$$\times \left[1 + \frac{\text{tgh}0.33897}{2 \cdot 64.93 \cdot 3.795 \cdot 10^{-3}} - 0.071882 \frac{(\text{tgh}0.33897)^{3.7482}}{\left(64.93 \cdot 3.795 \cdot 10^{-3}\right)^{1.481}} \right] =$$

$$= 0.59248 \cdot 0.96338 \cdot \left(1 + 0.66263 - 8.6258 \cdot 10^{-3}\right) = 0.9441.$$

Fin efficiency calculated by means of a sector method comes to

$$\eta_s = \frac{2\left(A_{OAB} - \frac{2\gamma}{2\pi}\pi r_1^2\right)\eta_1 + 4\left(A_{OBC} - \frac{2\varphi}{2\pi}\pi r_1^2\right)\eta_2}{2\left(A_{OAB} - \frac{2\gamma}{2\pi}\pi r_1^2\right) + 4\left(A_{OBC} - \frac{2\varphi}{2\pi}\pi r_1^2\right)} =$$

$$= \frac{2\left(A_{OAB} - \gamma r_1^2\right)\eta_1 + 4\left(A_{OBC} - \varphi r_1^2\right)\eta_2}{2\left(A_{OAB} - \gamma r_1^2\right) + 4\left(A_{OBC} - \varphi r_1^2\right)}.$$

After substitution of the numerical values, one obtains

$$\eta_s = \frac{2\left(2.1089 \cdot 10^{-5} - 0.189 \cdot 0.003795^2\right)0.9034 + 4\left(5.6156 \cdot 10^{-5} - 0.6909\right.}{2\left(2.1089 \cdot 10^{-5} - 0.189 \cdot 0.003795^2\right) + 4\left(5.6156 \cdot 10^{-5} - 0.6909\right.} \times$$

$$\times \frac{\left.\cdot 0.003795^2\right)0.9441}{\left.\cdot 0.003795^2\right)} = \frac{2 \cdot 1.8367 \cdot 10^{-5} \cdot 0.9034 + 4 \cdot 4.62056 \cdot 10^{-5} \cdot 0.9441}{3.6734 \cdot 10^{-5} + 1.84823 \cdot 10^{-4}} =$$

$$= 0.9373.$$

From the comparison of results, with $\eta_e = 0.9394$ and $\eta_s = 0.9373$, it is clear that method I, in which the whole fin is substituted by an equivalent circular fin, and sector method generate almost identical results. However, method I requires less computational work.

Exercise 6.21 Calculating Rectangular Fin Efficiency

The aim is to calculate efficiency of an equivalent fin in a fin-plate exchanger with a serial pipe configuration (Fig. 6.25). Pipes with outer diameters d_z = 10 mm are arranged with the following pitches: P_t = 25 mm and P_l = 22 mm. The thickness of fin-plates from aluminium alloy with thermal conductivity λ = 165 W/(m·K) is t = 0.13 mm. Heat transfer coefficient on a fin-plate surface equals α = 35 (W/m^2·K). Calculate efficiency of an equivalent circular fin with radius r_2^* by means of a simplified (5) from Ex. 6.18.

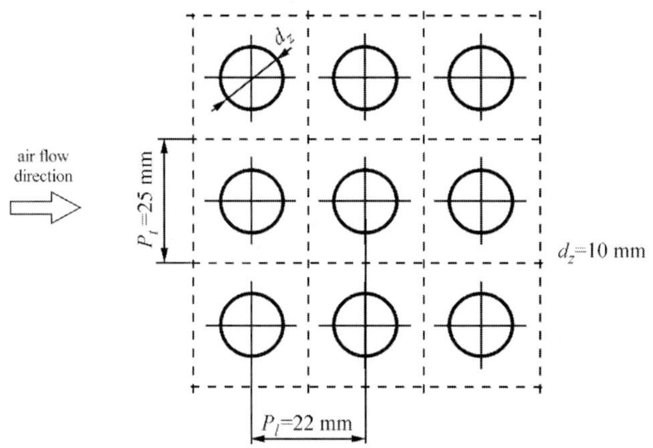

Fig. 6.25. Serial pipe configuration in a fin-plate exchanger

Solution

Equivalent radius r_2^* of a circular fin will be calculated from the condition of equality of a circular and rectangular fin surface area (Fig. 6.25):

$$2\left[\pi\left(r_2^*\right)^2 - \pi r_1^2\right] = 2\left(P_t P_l - \pi r_1^2\right).$$

From the above formula, one obtains

$$r_2^* = \sqrt{P_t P_l / \pi}\,.$$

The efficiency of a circular fin is calculated from (5), Ex. 6.18

$$\eta = \frac{1}{1 + \frac{1}{3}(mL)^2\sqrt{r_2^*/r_1}},$$

where

$$m = \sqrt{2\alpha / \lambda t}, \qquad L = r_2^* - r_1, \qquad r_1 = d_z / 2.$$

After substitution, one obtains

$$r_2^* = \sqrt{\frac{(0.025 \cdot 0.022)}{\pi}} = 0.01323 \text{ m}, \qquad r_1 = 0.5 d_1 = 0.005 \text{ m},$$

$$L = r_2^* - r_1 = 0.01323 - 0.005 = 0.008231 \text{ m},$$

$$m = \sqrt{\frac{2 \cdot 35}{165 \cdot 0.00013}} = 57.13 \, 1/\text{m}, \quad mL = 57.13 \cdot 0.008231 = 0.47023,$$

$$\eta = \frac{1}{1 + \frac{1}{3} \cdot 0.47023^2 \sqrt{\frac{0.01323}{0.005}}} = 0.8929.$$

Exercise 6.22 Heat Transfer Coefficient in Exchangers with Extended Surfaces

Derive a formula for an overall heat transfer coefficient k for thermal exchangers with extended surfaces (Fig. 6.26). Compare coefficients to an inner surface k_{rw} and outer surface k_{rz} of a smooth pipe and to an entire outer surface k_c (fins + surfaces between fins). Take into account the presence of scale on an inner fin-free surface, assuming that coefficient α_g from the finned side is determined in a way that it already accounts for the resistance from external scale.

Solution

In order to derive a formula for a heat transfer coefficient, an equivalent heat transfer coefficient α_{zr} will be introduced with regard to smooth surface area A_g, on which fins are mounted. It takes into consideration both heat transfer by fins and heat transfer by inter-fin spaces. It is assumed, at the same time, that fins do not disturb a one-dimensional temperature field in a wall of an element to which they are attached. If T_w stands for a wall surface temperature from a liquid's side, while T_z is an outer surface temperature from the gas side (Fig. 6.26), then heat flow transferred by an outer surface

$$\dot{Q}_s = \alpha_{zr} A_g \left(T_z - T_g\right) = \alpha_g A_{m\dot{z}} \left(T_z - T_g\right) + \alpha_g A_{\dot{z}} \left(T_z - T_g\right) \eta, \tag{1}$$

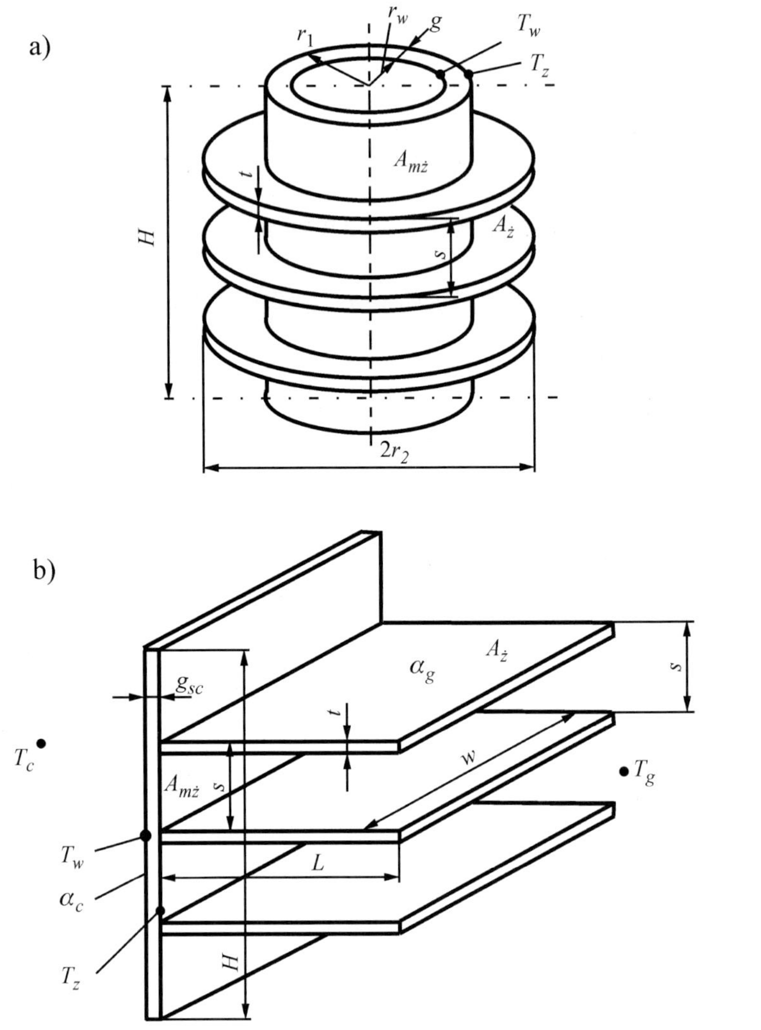

Fig. 6.26. Extended heat transfer surfaces: (**a**) finned pipe, (**b**) flat plate with fins

where $\dot{Q}_s$ stands for a heat flow that flows from a liquid of temperature T_c to a gas of temperature T_g that corresponds to a single pitch s, while $T_c > T_s$. Surface area A_g stands for a surface area of a smooth pipe or fin-free plate that corresponds to a single pitch s:

$$A_g = A_{rz} = 2\pi r_1 s \quad \text{for a smooth pipe,}$$

$$A_g = A = w \cdot s \quad \text{for a smooth plate.}$$

Surface area between fins, which corresponds to single pitch s, can be calculated in a similar way:

$$A_{m\dot{z}} = 2\pi r_1 (s - t) \quad \text{for a finned pipe,}$$

$$A_{m\dot{z}} = (s - t) w \qquad \text{for a finned plate.}$$

The surface of a single fin, which corresponds to a single pitch is given by

$$A_{\dot{z}} = 2\pi \left(r_{2c}^2 - r_1^2 \right), \quad \text{where } r_{2c} = r_2 + \frac{t}{2}, \text{ for a finned pipe,}$$

$$A_{\dot{z}} = 2wL_c, \quad \text{where } L_c = L + \frac{t}{2}, \text{ for a finned plate.}$$

From (1), one obtains a formula for an equivalent heat transfer coefficient

$$\alpha_{zr} = \alpha_g \left(\frac{A_{m\dot{z}}}{A_g} + \frac{A_{\dot{z}}}{A_g} \eta \right). \tag{2}$$

If an equivalent heat transfer coefficient is known, then the calculation of an overall heat transfer coefficient k_{rz}, with regard to a plate surface or an outer surface of a smooth pipe, is conducted in an identical way as the calculation of a smooth pipe or plate with a heat transfer coefficient equal to α_{zr}. Heat flow $\dot{Q}_s$, which corresponds to a single pitch, can be calculated for a pipe (Fig. 6.27a) using the following formula:

$$\dot{Q}_s = k_{rz} A_g \left(T_c - T_g \right) = \dot{q}_1 A_g . \tag{3}$$

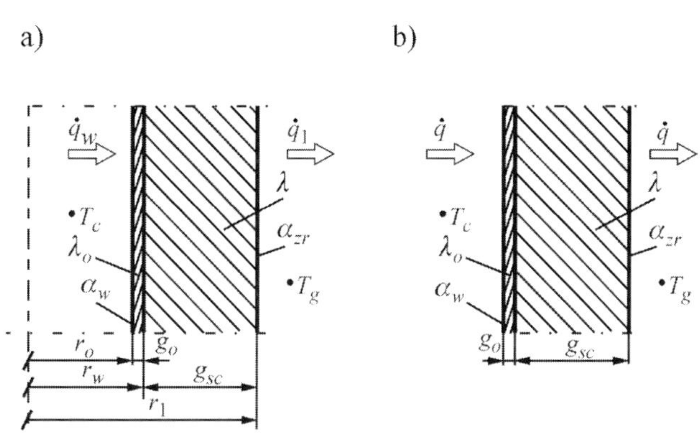

Fig. 6.27. Auxiliary diagram for determining overall heat transfer coefficient k for finned surfaces with a scale layer of thickness g_o: (**a**) pipe, (**b**) plate

Overall heat transfer coefficient k_{rz}, which pertains to a surface area of a smooth pipe with an outer radius of r_1 is calculated in the same way as the coefficient for a smooth pipe, although here one allows for the fact that four temperature decreases contribute to a temperature drop of $T_c - T_g$; these are:

• Temperature decrease between a temperature of a liquid T_c and an inner scale surface temperature T_o formulated as

$$T_c - T_o\big|_{r=r_o} = \frac{\dot{q}_o}{\alpha_c}. \tag{4}$$

Since the following equality occurs in a pipe

$$\dot{Q} = \dot{q}_o 2\pi r_o H = \dot{q}_w 2\pi r_w H = \dot{q}_1 2\pi r_1 H, \tag{5}$$

then

$$\dot{q}_o r_o = \dot{q}_w r_w = \dot{q}_1 r_1, \tag{6}$$

hence $\dot{q}_o$ can be calculated as shown below:

$$\dot{q}_o = \frac{\dot{q}_1 r_1}{r_o} \approx \frac{\dot{q}_1 r_1}{r_w}. \tag{7}$$

Equation (4) can be written then in the following way:

$$T_c - T\big|_{r=r_o} = \frac{\dot{q}_1 r_1}{r_w \alpha_c}. \tag{8}$$

• Temperature decrease inside a scale layer is formulated as

$$T\big|_{r=r_o} - T\big|_{r=r_w} \approx \frac{\dot{q}_o g_o}{\lambda_o} \approx \frac{\dot{q}_1 r_1}{r_w}\frac{g_o}{\lambda_o}. \tag{9}$$

Hence, after introducing thermal contact resistance

$$R_o = \frac{g_o}{\lambda_o} \left[(\mathrm{m}^2 \cdot \mathrm{K})/\mathrm{W} \right] \tag{10}$$

one obtains

$$T\big|_{r=r_o} - T\big|_{r=r_w} = \frac{\dot{q}_1 r_1}{r_w} R_o. \tag{11}$$

• Temperature decrease across a thickness of a pipe wall is

$$T\big|_{r=r_w} - T\big|_{r=r_1} = \frac{\dot{q}_1 r_1}{\lambda} \ln \frac{r_1}{r_w}. \tag{12}$$

• Temperature decrease between an outer surface of a pipe and gas temperature T_g is

$$T\big|_{r=r_1} - T_g = \frac{\dot{q}_1}{\alpha_{zr}}. \tag{13}$$

After summation of four temperature decreases formulated by (8)–(13), one obtains temperature difference $(T_c - T_g)$

$$T_c - T_g = \frac{\dot{q}_1 r_1}{r_w \alpha_c} + \frac{\dot{q}_1 r_1}{r_w} R_o + \frac{\dot{q}_1 r_1}{\lambda} \ln\frac{r_1}{r_w} + \frac{\dot{q}_1}{\alpha_{zr}}. \tag{14}$$

From the definition of an overall heat transfer coefficient k_{rz}, the following results from (3):

$$T_c - T_g = \frac{\dot{q}_1}{k_{rz}}. \tag{15}$$

From the dependence (14) and (15), one obtains

$$\frac{\dot{q}_1}{k_{rz}} = \frac{\dot{q}_1 r_1}{r_w \alpha_c} + \frac{\dot{q}_1 r_1}{r_w} R_o + \frac{\dot{q}_1 r_1}{\lambda} \ln\frac{r_1}{r_w} + \frac{\dot{q}_1}{\alpha_{zr}}, \tag{16}$$

hence,

$$\frac{1}{k_{rz}} = \frac{r_1}{r_w \alpha_c} + \frac{r_1}{r_w} R_o + \frac{r_1}{\lambda} \ln\frac{r_1}{r_w} + \frac{1}{\alpha_{zr}}. \tag{17}$$

One can determine overall heat transfer coefficient k_w with regard to inner surface A_w and coefficient k_c with regard to total surface $A_c = A + A_{mc}$ from the following equality:

$$\dot{Q}_s = k_{zr} A_g \left(T_c - T_g\right) = k_w A_w \left(T_c - T_g\right) = k_c A_c \left(T_c - T_g\right),$$

hence,

$$\frac{1}{k_w} = \frac{1}{k_{rz}} \frac{A_w}{A_g}, \qquad \frac{1}{k_c} = \frac{1}{k_{rz}} \frac{A_c}{A_g},$$

where: $A_w = 2\pi r_w\, s,\quad A_g = 2\pi r_1\, s,\quad A_c = A + A_m.$

In the case of a finned flat wall (Fig. 6.27b), overall heat transfer coefficient k_g is determined, with regard to a smooth surface, from a formula

$$\frac{\dot{q}}{k_g} = \frac{\dot{q}}{\alpha_w} + \frac{\dot{q} g_o}{\lambda_o} + \frac{\dot{q} g_{sc}}{\lambda} + \frac{\dot{q}}{\alpha_{zr}},$$

from which the following equality results

$$\frac{1}{k_g}=\frac{1}{\alpha_w}+R_o+\frac{g_{sc}}{\lambda}+\frac{1}{\alpha_{zr}},$$

where $R_o = g_o/\lambda_o$ [(m^2·K)/W] is a thermal contact resistance of a scale layer.

The unknown overall heat transfer coefficient with regard to total surface $A_c = A + A_m$, can be determined from the condition

$$\dot{Q}_s = k_g A_g\left(T_c - T_g\right) = k_c A_c\left(T_c - T_g\right),$$

which results in

$$\frac{1}{k_c}=\frac{1}{k_g}\frac{A_c}{A_g}.$$

Exercise 6.23 Calculating Overall Heat Transfer Coefficient in a Fin Plate Exchanger

Calculate overall heat transfer coefficient in a fin-plate exchanger analyzed in Ex. 6.20. The thickness of an aluminium made pipe wall with thermal conductivity λ_r = 165 W/(m·K) measures g_r = 1 mm. Heat transfer coefficient on an inner surface of the pipe is α_c = 5000 W/(m^2·K). The pitch of fin spacing is s = 1.6 mm. Other values remain the same as they were in Ex. 6.20. For the calculation, assume fin efficiency of $\eta = \eta_e$ = 0.9394, calculated by means of method I (Ex. 6.20). How many times does an overall heat transfer coefficient increase in comparison to a smooth pipe?

Solution

First, surface areas will be calculated:

- fin surface area (Fig. 6.24)

$$A_{\dot{z}} = \left(2A_{OAB} + 4A_{OBC}\right) = 2\left(2\cdot 2.1089\cdot 10^{-5} + 4\cdot 5.6156\cdot 10^{-5}\right) = \\ = 5.33604\cdot 10^{-4}\ \text{m}^2. \tag{1}$$

- surface area between fins

$$A_{m\dot{z}} = \pi d_z\left(s-t\right) = \pi\cdot 7.59\cdot 10^{-3}\left(1.6-0.115\right)\cdot 10^{-3} = 3.54094\cdot 10^{-5}\ \text{m}^2.$$

- smooth surface area

$$A_g = \pi d_z s = \pi\cdot 7.59\cdot 10^{-3}\cdot 1.6\cdot 10^{-3} = 3.81515\cdot 10^{-5}\,\text{m}^2.$$

- surface area of a fin and a pipe between fins

$$A_c = A_{\dot{z}} + A_{m\dot{z}} = 5.33604\cdot 10^{-4} + 3.54094\cdot 10^{-5} = 5.690134\cdot 10^{-4}\ \text{m}^2.$$

An equivalent heat transfer coefficient is

$$\alpha_{zr} = \alpha_g \left(\frac{A_{m\dot{z}}}{A_g} + \frac{A_{\dot{z}}}{A_g}\eta \right) = 40\left(\frac{3.54094\cdot 10^{-5}}{3.81515\cdot 10^{-5}} + \frac{5.33604\cdot 10^{-4}}{3.81515\cdot 10^{-5}}\cdot 0.9394 \right) = \tag{2}$$

$$= 596.58\ \mathrm{W/\left(m^2\cdot K\right)}$$

Overall heat transfer coefficient (k_{rz}), with regard to an outer surface of a smooth pipe, is calculated in the following way:

$$\frac{1}{k_{rz}} = \frac{r_1}{r_w\alpha_c} + \frac{r_1}{\lambda_r}\ln\frac{r_1}{r_w} + \frac{1}{\alpha_{zr}} = \frac{7.59\cdot 10^{-3}}{2\left(\left(7.59\cdot 10^{-3}/2\right) - 0.001\right)}\frac{1}{5000} +$$

$$+\frac{7.59\cdot 10^{-3}}{2\cdot 165}\times\ \ln\frac{7.59\cdot 10^{-3}}{2\left(\dfrac{7.59\cdot 10^{-3}}{2} - 0.001\right)} + \frac{1}{596.58} = 2.71556\cdot 10^{-4} +$$

$$+7.0346\cdot 10^{-6} + 1.676221\cdot 10^{-3} = 1.9548\cdot 10^{-3}\ \mathrm{(m^2\cdot K)/W},$$

$$k_{rz} = 511.56\ \mathrm{W/(m^2\cdot K)}.$$

Overall heat transfer coefficient (k) for smooth fin-free pipes with respect to an outer surface of a pipe is:

$$\frac{1}{k} = \frac{r_1}{r_w\alpha_c} + \frac{r_1}{\lambda_r}\ln\frac{r_1}{r_w} + \frac{1}{\alpha} = 2.71556\cdot 10^{-4} + 7.0346\cdot 10^{-6} + 0.025 =$$

$$= 0.0252786\ \mathrm{(m^2\cdot K)/W},$$

$$k = 39.559\ \ \mathrm{W/(m^2\ \cdot K)}.$$

Due to the application of flat fins (fin plates), heat transfer coefficient increased n times:

$$n = \frac{k_{rz}}{k} = \frac{511.56}{39.559} = 12.93\,.$$

Exercise 6.24 Overall Heat Transfer Coefficient for a Longitudinally Finned Pipe with a Scale Layer on an Inner Surface

A *dowtherm* was applied in a steam boiler. It flows inside a ring-shaped gap between two pipes. In order to increase heat transfer coefficient, the center pipe is longitudinally finned on its outer surface (Fig. 6.28). A water-vapour mixture flows inside the pipe. The following data is used for the calculation:

Overall heat transfer coefficient k_{rz} calculated with respect to an outer surface of a smooth pipe is formulated as

$$\frac{1}{k_{rz}} = \frac{r_1}{r_w}\frac{1}{\alpha_w} + \frac{r_1}{r_w}R_o + \frac{r_1}{\lambda}\ln\frac{r_1}{r_w} + \frac{1}{\alpha_{zr}}.$$

First coefficient k_{rz} is calculated, for the situation when an inner pipe surface is scale free. Since $R_o = 0$ and

$$r_1 = \frac{d_1}{2} = \frac{48.3\cdot 10^{-3}}{2} = 24.15\cdot 10^{-3}\,\text{m},\quad r_w = \frac{40.94\cdot 10^{-3}}{2} = 20.47\cdot 10^{-3}\,\text{m},$$

then,

$$\frac{1}{k_{rz}} = \frac{24.15\cdot 10^{-3}}{20.47\cdot 10^{-3}}\frac{1}{10000} + \frac{24.15\cdot 10^{-3}}{55}\ln\frac{24.15\cdot 10^{-3}}{20.47\cdot 10^{-3}} + \frac{1}{3198.85} =$$

$$= 1.17978\cdot 10^{-4} + 7.2592\cdot 10^{-5} + 3.1264\cdot 10^{-4} = 5.0321\cdot 10^{-4}\ (\text{m}^2\cdot\text{K})/\text{W}.$$

hence,

$$k_{rz} = 1987.2\ \text{W}/(\text{m}^2\cdot\text{K}).$$

It can be seen that here overall heat transfer coefficient is only slightly bigger than a dowtherm heat transfer coefficient, equal to α_d = 1500 W/(m^2·K). This is caused by a large heat transfer coefficient α_w, for which fin efficiency is rather small. In order to consider the effect of scale layer on a calculated coefficient k_{rz} , it is sufficient to add calculated value $1/k_{rz}$ to a scale resistance equal to

$$\frac{r_1}{r_w}R_o = \frac{r_1}{r_w}\cdot\frac{g_o}{\lambda_o} = \frac{24.15\cdot 10^{-3}}{20.47\cdot 10^{-3}}\cdot\frac{0.0001}{1} = 1.1789\cdot 10^{-4}\,(\text{m}^2\cdot\text{K})/\text{W}.$$

Overall heat transfer coefficient k'_{rz} for a pipe with a scale layer on an inner surface is

$$\frac{1}{k'_{rz}} = 5.0321\cdot 10^{-4} + 1.1798\cdot 10^{-4} = 6.2119\cdot 10^{-4}\ (\text{m}^2\cdot\text{K})/\text{W},$$

therefore,

$$k'_{rz} = 1609.82\ \text{W}/(\text{m}^2\cdot\text{K}).$$

Due to a scale accumulation on an inner surface, overall heat transfer coefficient became smaller by

$$\Delta k_{rz} = k_{rz} - k'_{rz} = 1987.23 - 1609.82 = 377.4\,\text{W}/(\text{m}^2\cdot\text{K}).$$

Exercise 6.25 Overall Heat Transfer Coefficient for a Longitudinally Finned Pipe

Calculate overall heat transfer coefficient, with respect to a smooth outer surface of a pipe, for finned pipes with an outer diameter d_1 = 16 mm and thickness g_r = 2 mm, made of aluminium with thermal conductivity λ = 205 W/(m·K). The outer diameter of a fin,with thickness t = 0.1 is d_2 = 35 mm. Pitch of the fin spacing is s = 1.1 mm. A hot water of temperature t_w = 80°C flows inside the pipe at a speed of w_1 = 0.5 m/s; it warms the air, which circulates the finned pipes crosswise. Heat transfer coefficient from the air side is equal to α = 60 W/(m^2·K). How is the overall heat transfer coefficient going to change when the velocity of flowing water increases to w_2 = 2 m/s? Calculate heat transfer coefficient from water to pipe using the following Stender and Merkel formulas

$$\alpha_w = 2040\left(1 + 0.015t_w\right) w^{0.87} d_w^{-0.13}, \tag{1}$$

in which the units are as follow :

α_w [W/(m^2·K)], t_w [°C], w [m/s], d_w [m].

Solution

A diagram of the finned pipe is presented in Fig. 6.30. First, we will calculate fin efficiency from formula

$$\eta = \frac{1}{1 + \frac{1}{3}\left(mL_c\right)^2 \sqrt{\frac{r_{2c}}{r_1}}}, \tag{2}$$

where,

$$r_1 = \frac{d_1}{2} = \frac{16 \cdot 10^{-3}}{2} = 8 \cdot 10^{-3} \text{ m},$$

$$r_2 = \frac{d_2}{2} = \frac{35 \cdot 10^{-3}}{2} = 17.5 \cdot 10^{-3} \text{ m},$$

$$r_{2c} = r_2 + \frac{t}{2} = 17.5 \cdot 10^{-3} + 0.05 \cdot 10^{-3} = 17.55 \cdot 10^{-3} \text{ m},$$

$$L_c = L + \frac{t}{2} = r_2 - r_1 + \frac{t}{2} = 17.5 \cdot 10^{-3} - 8 \cdot 10^{-3} + \frac{0.1 \cdot 10^{-3}}{2} = 9.55 \cdot 10^{-3} \text{ m}.$$

Exercise 6.26 Determining One-Dimensional Temperature Distribution in a Flat Wall by Means of Finite Volume Method

Determine temperature distribution and heat flux on a slab surface, in which temperature field is described using the following differential equation and boundary conditions:

$$\frac{d^2T}{dx^2} = -\frac{\dot{q}_v}{\lambda}, \qquad 0 \le x \le L, \tag{1}$$

$$\left.\frac{dT}{dx}\right|_{x=0} = 0, \tag{2}$$

$$-\lambda \left.\frac{dT}{dx}\right|_{x=L} = \alpha\left[\left.T\right|_{x=L} - T_{cz}\right]. \tag{3}$$

Determine temperature distribution by means of finite volume method for the following data: λ = 15 W/(m·K), α = 300 W/(m^2·K), L = 0.06 m, $\dot{q}_v$ = 1·10^6 W/m^3, T_{cz} = 20°C. Compare determined temperature distribution with the accurate analytical solution (Ex. 6.11)

$$T(x) = T_{cz} + \frac{\dot{q}_v L}{\alpha} + \frac{\dot{q}_v L^2}{2\lambda}\left[1 - \left(\frac{x}{L}\right)^2\right], \tag{4}$$

when the wall thickness is divided into five finite volumes (Fig. 6.31).

Solution

Heat balance equation for the first finite volume has the form

$$\frac{\lambda}{\Delta x}\left(T_2 - T_1\right) + \dot{q}_v \frac{\Delta x}{2} = 0, \tag{5}$$

from which one obtains

$$2T_2 - 2T_1 = -\frac{\dot{q}_v \left(\Delta x\right)^2}{\lambda}, \tag{6}$$

where $\Delta x = L/4$.

Heat balance equation for finite volumes with numbers i = 2, 3, 4 has the form

$$\frac{\lambda}{\Delta x}\left(T_{i-1}-T_i\right)+\frac{\lambda}{\Delta x}\left(T_{i+1}-T_i\right)+\dot{q}_v\Delta x=0\,, \tag{7}$$

hence,

$$T_{i-1}-2T_i+T_{i+1}=-\frac{\dot{q}_v\left(\Delta x\right)^2}{\lambda}, \qquad i=2,\,3,\,4. \tag{8}$$

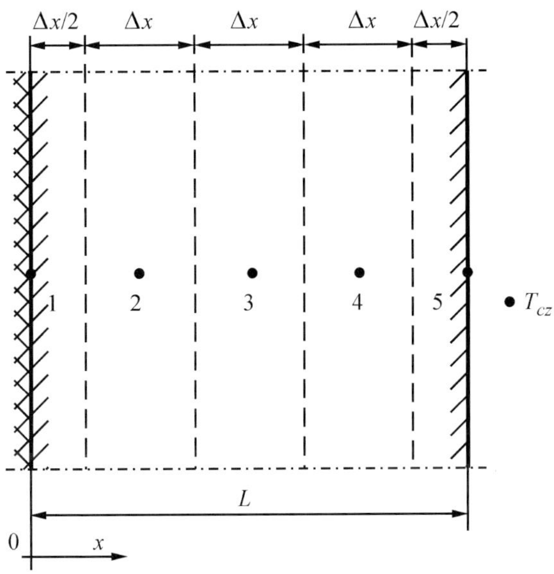

Fig. 6.31. Wall division into control volumes

Heat balance equation for control volume no. 5 has the form

$$\frac{\lambda}{\Delta x}\left(T_4-T_5\right)+\alpha\left(T_{cz}-T_5\right)+\dot{q}_v\frac{\Delta x}{2}=0\,, \tag{9}$$

hence, one obtains

$$2T_4-\left(2+\frac{2\alpha\Delta x}{\lambda}\right)T_5=-\frac{\dot{q}_v\left(\Delta x\right)^2}{\lambda}-\frac{2\alpha\Delta x}{\lambda}T_{cz}\,. \tag{10}$$

By substituting the numerical values in (6), (8) and (10), one obtains the following system:

$$\begin{aligned} -2T_1 + 2T_2 &= -15 \\ T_1 - 2T_2 + T_3 &= -15 \\ T_2 - 2T_3 + T_4 &= -15 \\ T_3 - 2T_4 + T_5 &= -15 \\ 2T_4 - 2.6T_5 &= -27. \end{aligned} \tag{11}$$

This system can also be written in the matrix form

$$\mathbf{AT} = \mathbf{b}, \tag{12}$$

where

$$\mathbf{A} = \begin{bmatrix} -2 & 2 & 0 & 0 & 0 \\ 1 & -2 & 1 & 0 & 0 \\ 0 & 1 & -2 & 1 & 0 \\ 0 & 0 & 1 & -2 & 1 \\ 0 & 0 & 0 & 2 & -2.6 \end{bmatrix}, \quad \mathbf{T} = \begin{Bmatrix} T_1 \\ T_2 \\ T_3 \\ T_4 \\ T_5 \end{Bmatrix}, \quad \mathbf{b} = \begin{Bmatrix} -15 \\ -15 \\ -15 \\ -15 \\ -27 \end{Bmatrix}.$$

Equation system (12) has been solved using Gauss elimination method. Solution can be written in the form

$$\mathbf{T} = \mathbf{A}^{-1}\mathbf{b},$$

where

$$\mathbf{T} = \begin{Bmatrix} 340.0 \\ 332.5 \\ 310.0 \\ 272.5 \\ 220.0 \end{Bmatrix}, \quad \mathbf{A}^{-1} = \begin{bmatrix} -3.67 & -6.33 & -5.33 & -4.33 & -1.67 \\ -3.17 & -6.33 & -5.33 & -4.33 & -1.67 \\ -2.67 & -5.33 & -5.33 & -4.33 & -1.67 \\ -2.17 & -4.33 & -4.33 & -4.33 & -1.67 \\ -1.67 & -3.33 & -3.33 & -3.33 & -1.67 \end{bmatrix}.$$

It is important to know the inverse matrix $\mathbf{A}^{-1}$, since it allows to define the relationship between temperature in a given node and the heat source unit power $\dot{q}_v$ and the parameters α and T_{cz}, which prevail in the boundary condition (3). In order to compare obtained temperature values with analytical solution, temperature values in nodes were also calculated using (4) (Table 6.2).

Heat flux on a slab surface is

$$\dot{q}\big|_{x=L} = \alpha\left(T\big|_{x=L} - T_{cz}\right).$$

Table 6.2. Comparison of temperature values calculated numerically nodes with exact values

Node no.	Coordinate x [m]	Finite volume method T [°C]	Exact solution T [°C]
1	0	340.0	340.0
2	0.015	332.5	332.5
3	0.030	310.0	310.0
4	0.045	272.5	272.5
5	0.060	220.0	220.0

Accurate value:

$$\dot{q}\Big|_{x=L} = 300\left(220-20\right) = 60000 \text{ W/m}^2 .$$

Approximated value $\dot{q}_p(L)$ calculated from formula

$$\dot{q}_p\Big|_{x=L} = \alpha\left(T_5 - T_{cz}\right)$$

is

$$\dot{q}_p\Big|_{x=L} = 300\left(220-20\right) = 60000 \text{ W/m}^2 .$$

Approximated value $\dot{q}_p\Big|_{x=L}$ can also be determined by means of Fourier Law, while first derivative $dT/dx\Big|_{x=L}$ will be calculated by means of a backward finite difference differential quotient with an accuracy of 2nd order

$$\dot{q}_p\Big|_{x=L} = -\lambda \frac{dT}{dx}\Big|_{x=L} \approx -\lambda \left(\frac{3T_N - 4T_{N-1} + T_{N-2}}{2\Delta x} \right)\Bigg|_{N=5} .$$

After substituting node temperature values calculated by means of the finite volume method, one obtains

$$\dot{q}_p\Big|_{x=L} = -15 \frac{3 \cdot 220 - 4 \cdot 272.5 + 310}{2 \cdot 0.015} = 60000 \text{ W/m}^2 .$$

From the analysis of the obtained results, it is clear that the accuracy of the finite volume method is very good.

Program for calculating temperature distribution in slabs

```
 program mat
dimension a(50,50),b(50),c(50,50),t(50)
open(unit=1,file='mat.in')
```

```
      open(unit=2,file='mat.out')
      read(1,*)n
       read(1,*) ((a(i,j),j=1,n),i=1,n), (b(i),i=1,n)
      write(2,*)'A'
      write(2,66) ((a(i,j),j=1,n),i=1,n)
      write(2,*)'B'
      write(2,66) (b(i),i=1,n)
      call matinv(a,n,c)
      write(2,*)'A^-1'
      write(2,66) ((c(i,j),j=1,n),i=1,n)
      do 20 i=1,n
      sum=0.0
      do 10 j=1,n
   10 sum=sum+c(i,j)*b(j)
   20 t(i)=sum
      write(2,*)''
      write(2,50) (i,t(i),i=1,n)
      stop    7
   50 format(5('T(',i2,')=',f8.2,2x))
   66 format(5f8.2)
      end

data(mat.in)
5
-2. 2. 0. 0. 0.
1. -2. 1. 0. 0.
0. 1. -2. 1. 0.
0. 0. 1. -2. 1.
0. 0. 0. 2. -2.6
-15. -15. -15. -15. -27.
results(mat.out)
A
   -2.00    2.00     .00     .00     .00
    1.00   -2.00    1.00     .00     .00
     .00    1.00   -2.00    1.00     .00
     .00     .00    1.00   -2.00    1.00
     .00     .00     .00    2.00   -2.60
B
  -15.00  -15.00  -15.00  -15.00  -27.00
A^-1
   -3.67   -6.33   -5.33   -4.33   -1.67
   -3.17   -6.33   -5.33   -4.33   -1.67
   -2.67   -5.33   -5.33   -4.33   -1.67
   -2.17   -4.33   -4.33   -4.33   -1.67
   -1.67   -3.33   -3.33   -3.33   -1.67

T( 1)=  340.00  T( 2)=  332.50  T( 3)=  310.00
T( 4)=  272.50  T( 5)=  220.00
```

Exercise 6.27 Determining One-Dimensional Temperature Distribution in a Cylindrical Wall By Means of Finite Volume Method

Solve the problem formulated in Ex. 6.12 using the finite volume method. Calculate tube wall temperature in five uniformly spaced points (Fig. 6.32). Compare the obtained numerical solution with the accurate analytical solution.

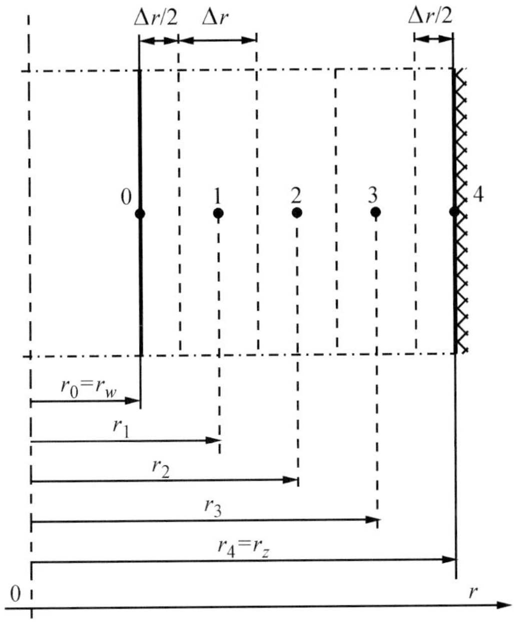

Fig. 6.32. Division of a cylindrical wall into finite volumes

Solution

Heat balance equation for finite volumes, adjacent to the inner surface, is not required, since the inner surface temperature is known: $T_0 = T_w$. Heat balance equations for nodes i = 1, 2, 3 can be written in the condensed form

$$2\pi\left(r_i - \frac{\Delta r}{2}\right)\lambda\frac{T_{i-1}-T_i}{\Delta r} + 2\pi\left(r_i + \frac{\Delta r}{2}\right)\lambda\frac{T_{i+1}-T_i}{\Delta r} + 2\pi r_i \Delta r \dot{q}_v = 0, \tag{1}$$

$i = 1, 2, 3,$

where $\Delta r = (r_z - r_w)/4$.

On the basis of (1), one obtains the following equations for nodes numbered 1, 2 and 3:

$$\frac{r_1 - \frac{\Delta r}{2}}{\Delta r}\left(T_0 - T_1\right) + \frac{r_1 + \frac{\Delta r}{2}}{\Delta r}\left(T_2 - T_1\right) + \frac{\dot{q}_v r_1 \Delta r}{\lambda} = 0\,, \tag{2}$$

$$\frac{r_2 - \frac{\Delta r}{2}}{\Delta r}\left(T_1 - T_2\right) + \frac{r_2 + \frac{\Delta r}{2}}{\Delta r}\left(T_3 - T_2\right) + \frac{\dot{q}_v r_2 \Delta r}{\lambda} = 0\,, \tag{3}$$

$$\frac{r_3 - \frac{\Delta r}{2}}{\Delta r}\left(T_2 - T_3\right) + \frac{r_3 + \frac{\Delta r}{2}}{\Delta r}\left(T_4 - T_3\right) + \frac{\dot{q}_v r_3 \Delta r}{\lambda} = 0\,. \tag{4}$$

An equation for node 4 has the form

$$2\pi\left(r_4 - \frac{\Delta r}{2}\right)\lambda\frac{T_3 - T_4}{\Delta r} + \pi\left[r_4^2 - \left(r_4 - \frac{\Delta r}{2}\right)^2\right]\dot{q}_v = 0\,. \tag{5}$$

After simple transformations (2)–(5) can be written in the following form:

$$-2T_1 + \frac{r_1 + \frac{\Delta r}{2}}{r_1}T_2 = -\frac{r_1 - \frac{\Delta r}{2}}{r_1}T_0 - \frac{\dot{q}_v\left(\Delta r\right)^2}{\lambda}\,, \tag{6}$$

$$\frac{r_2 - \frac{\Delta r}{2}}{r_2}T_1 - 2T_2 + \frac{r_2 + \frac{\Delta r}{2}}{r_2}T_3 = -\frac{\dot{q}_v\left(\Delta r\right)^2}{\lambda}\,, \tag{7}$$

$$\frac{r_3 - \frac{\Delta r}{2}}{r_3}T_2 - 2T_3 + \frac{r_3 + \frac{\Delta r}{2}}{r_3}T_4 = -\frac{\dot{q}_v\left(\Delta r\right)^2}{\lambda}\,, \tag{8}$$

$$2T_3 - 2T_4 = -\frac{r_4 - \frac{\Delta r}{4}}{r_4 - \frac{\Delta r}{2}}\frac{\dot{q}_v\left(\Delta r\right)^2}{\lambda}\,. \tag{9}$$

After substitution of the numerical values

$$\Delta r = \frac{r_z - r_w}{4} = \frac{(8-7.2)\cdot 10^{-3}}{2\cdot 4} = 1\cdot 10^{-4}\ \text{m},$$

$$r_1 = r_w + \Delta r = 3.6\cdot 10^{-3} + 1\cdot 10^{-4} = 3.7\cdot 10^{-3}\ \text{m},$$

$$r_2 = r_w + 2(\Delta r) = 3.6\cdot 10^{-3} + 2\cdot 1\cdot 10^{-4} = 3.8\cdot 10^{-3}\ \text{m},$$

$$r_3 = r_w + 3(\Delta r) = 3.6\cdot 10^{-3} + 3\cdot 1\cdot 10^{-4} = 3.9\cdot 10^{-3}\ \text{m}, \tag{10}$$

$$\dot{q}_v = 8.38715\cdot 10^{8}\ \text{W/m}^3,$$

$$\lambda = 18.4\ \text{W/(m}\cdot\text{K)},$$

$$T_0 = T_w = 300\,^\circ\text{C},$$

the equation system (7)–(10) assumes the form

$$\begin{aligned}
&-2T_1 + 1.0135135T_2 = -296.4017693,\\
&0.986842T_1 - 2T_2 + 1.013157 89T_3 = -\ 0.4558234,\\
&0.9871795T_2 - 2T_3 + 1.0128205T_4 = -\ 0.4558234,\\
&2T_3 - 2T_4 = -0.458708.
\end{aligned} \tag{11}$$

Equation system (11) can be written in the matrix form

$$\mathbf{AT} = \mathbf{b}, \tag{12}$$

where

$$\mathbf{A} = \begin{bmatrix} -2 & 1.0135135 & 0 & 0 \\ 0.986842 & -2 & 1.01315789 & 0 \\ 0 & 0.9871795 & -2 & 1.0128205 \\ 0 & 0 & 2 & -2 \end{bmatrix},$$

$$\mathbf{T} = \begin{Bmatrix} T_1 \\ T_2 \\ T_3 \\ T_4 \end{Bmatrix}, \quad \mathbf{b} = \begin{Bmatrix} -296.4017693 \\ -0.4558234 \\ -0.4558234 \\ -0.458708 \end{Bmatrix}. \tag{13}$$

The solution of the equation system (12), obtained by means of the Gauss elimination method, has the form

$$\mathbf{T} = \mathbf{A}^{-1}\mathbf{b}, \tag{14}$$

where

$$\mathbf{T} = \begin{Bmatrix} 301.6718 \\ 302.8493 \\ 303.5464 \\ 303.7757 \end{Bmatrix}, \mathbf{A}^{-1} = \begin{bmatrix} -1.0137 & -1.0411 & -1.0685 & -0.5411 \\ -1.0137 & -2.0544 & -2.1085 & -1.0678 \\ -1.0137 & -2.0544 & -3.1215 & -1.5807 \\ -1.0137 & -2.0544 & -3.1215 & -2.0807 \end{bmatrix}. \quad (15)$$

The comparison of the obtained numerical solution with analytical solution from Ex. 6.12

$$T(r) = T_0 + \frac{\dot{q}_v}{2\lambda}\left(r_z^2 \ln\frac{r}{r_w} + \frac{r_w^2 - r^2}{2} \right) \quad (16)$$

is presented in Table 6.3. Heat flux on the inner surface of the pipe will be determined from formula

$$\dot{q}_w = \lambda \frac{dT}{dr}\bigg|_{r=r_w} \approx \lambda \frac{-3T_0 + 4T_1 - T_2}{2\Delta r}. \quad (17)$$

Table 6.3. The comparison of numerically and analytically calculated node temperature values

Node no.	Coordinate x [m]	Finite Volume Method T [°C]	Exact Method T [°C]
1	0.0036	300	300.0
2	0.0037	301.6718	301.6725
3	0.0038	302.8493	302.8506
4	0.0039	303.5464	303.5482
5	0.0040	303.7757	303.7781

Heat transfer coefficient on an inner surface of the pipe will be determined from formula

$$\alpha = \frac{\dot{q}_w}{T\big|_{r=r_w} - T_{cz}} = \frac{\dot{q}_w}{T_0 - T_{cz}},$$

where $\dot{q}_w$ is expressed using (17).

After substitution, one obtains

$$\dot{q}_w = 18.4 \frac{-3 \cdot 300 + 4 \cdot 301.6718 - 302.8493}{2 \cdot 1 \cdot 10^{-4}} = 353086.8 \text{ W/m}^2$$

and

$$\alpha = \frac{353086.8}{300 - 20} = 1261.024 \text{ W/(m}^2\text{·K)}.$$

It is clear that the obtained results are almost identical to the results obtained by means of the analytical formulas from Ex. 6.12.

Exercise 6.28 Inverse Steady-State Heat Conduction Problem for a Pipe Solved by Space-Marching Method

Electrical current with an intensity of 300 A flows through a stainless steel pipe with an inner diameter d_w = 7.2 mm and outer diameter d_z = 8 mm. Thermal conductivity of this steel is λ = 18.4 W/(m·K), while its specific resistance ρ = 0.85 ($\Omega \cdot$ mm^2)/m. Assuming that the outer surface of the pipe is thermally insulated and the temperature of this surface is known and equals T_z = 303.77805°C, calculate temperature distribution across the wall thickness. Assume that total heat rate generated inside the wall flows to the interior of the pipe. Calculate heat transfer coefficient on the inner surface of the pipe, if the temperature of a medium measures T_{cz} = 20°C. Perform calculations using finite volume method. Analytical solution of the problem formulated above is presented in Ex. 6.13.

Solution

A division of the wall into finite volumes is shown in Fig. 6.32. An equation system for temperature in finite volume nodes has the same form as (6)–(9) in Ex. 6.27, therefore

$$\begin{aligned}
&\frac{r_1 - \Delta r/2}{r_1} T_0 - 2T_1 + \frac{r_1 + \Delta r/2}{r_1} T_2 = -\frac{\dot{q}_v (\Delta r)^2}{\lambda}, \\
&\frac{r_2 - \Delta r/2}{r_2} T_1 - 2T_2 + \frac{r_2 + \Delta r/2}{r_2} T_3 = -\frac{\dot{q}_v (\Delta r)^2}{\lambda}, \\
&\frac{r_3 - \Delta r/2}{r_3} T_2 - 2T_3 + \frac{r_3 + \Delta r/2}{r_3} T_4 = -\frac{\dot{q}_v (\Delta r)^2}{\lambda}, \\
&2T_3 - 2T_4 = -\frac{r_4 - \Delta r/4}{r_4 - \Delta r/2} \frac{\dot{q}_v (\Delta r)^2}{\lambda}.
\end{aligned} \tag{1}$$

unknown. One can also see that calculated node temperatures differ slightly from the "exact" values (calculated using analytical formulas) given in Table 6.3. This is mainly due to the rounding calculation errors made in this exercise. In inverse problems, rounding errors or temperature measurement errors, especially in transient problems, significantly influence the accuracy of the obtained results.

Exercise 6.29 Temperature Distribution and Efficiency of a Circular Fin with Temperature-Dependent Thermal Conductivity

Determine efficiency of a circular fin with constant thickness t by assuming that thermal conductivity of the fin material λ is a linear temperature function of temperature:

$$\lambda(T)=\lambda(T_{cz})(1+\varepsilon\theta),$$

where $\varepsilon=\left[\lambda(T_1)-\lambda(T_{cz})\right]/\lambda(T_{cz}),\quad \theta=(T-T_{cz})/(T_1-T_{cz})$.

Symbols T_1 and T_{cz} stand for fin base temperature and medium's temperature, respectively. Draw graphs show the fin efficiency as the function of the parameter M:

$$k=r_2/r_1,\qquad M=L^{3/2}\sqrt{\alpha/\left[\lambda(T_{cz})Lt\right]},$$

where r_1 and r_2 are the base and fin tip radius, $L=r_2-r_1$ fin height, α – heat transfer coefficient on the fin surface. Show results in a tabular form for k = 2 as a function of parameter M.

Solution

Equations and boundary conditions, which describe fin temperature field, have the following form (see reference [12] and equation (4.8)–(4.12) in Part 1 of this book):

$$\frac{dQ}{d\rho}=-2\pi N^2\rho\theta, \tag{1}$$

$$\frac{d\theta}{d\rho}=-\frac{Q}{2\pi(1+\varepsilon\theta)\rho}, \tag{2}$$

$$\theta\big|_{\rho=1}=1, \tag{3}$$

$$Q\big|_{\rho=k} = 0, \tag{4}$$

$$Q = -2\pi(1+\varepsilon\theta)\rho\frac{d\theta}{d\rho}. \tag{5}$$

Symbol $\rho = r/r_1$ is a dimensionless radius. The following relation occurs between parameters N and M:

$$N^2 = \frac{2}{(k-1)^2}M^2. \tag{6}$$

Fin efficiency η, defined as a ratio of real-fin-dissipated heat flow to isothermal-fin-dissipated heat flow, is formulated as

$$\eta_{\dot{z}} = \frac{-\left[2\pi t r \lambda(T_{cz})(1+\varepsilon\theta)\frac{dT}{dr}\right]\Bigg|_{r=r_1}}{\int\limits_{r_1}^{r_2} 4\pi\alpha(T_1 - T_\infty) r\,dr}. \tag{7}$$

Once (5) is substituted in (7) and subsequently transformed, one obtains

$$\eta_{\dot{z}} = \frac{Q\big|_{\rho=1}}{\pi N^2(k^2-1)} = \frac{(k-1)Q\big|_{\rho=1}}{2\pi(k+1)M^2}. \tag{8}$$

By solving two-point boundary value problem (1)–(4), one is able to determine $Q|_{\rho=1}$. Two-point boundary value problem will be solved iteratively using secant method, also called *Newton-Raphson method* [1, 8]. Boundary problem (1)–(4) will be substituted by the initial problem under the assumption that value Q is given at the base of the fin

$$Q\big|_{\rho=1} = \beta. \tag{9}$$

If we assume a certain numerical value β, we will be able to solve the initial problem, formulated by (1) and (2) and by initial conditions (3) and (9), at a given iterative step. *Runge-Kutta Method of 4th order* was applied to solve the initial problem. Value β must be chosen in such way that condition (4) is satisfied. Variable $Q|_{\rho=k}$ is, therefore, a function of parameter β

$$Q\big|_{\rho=k} = \bar{Q}(\beta). \tag{10}$$

One should find such value of parameter β^* for which $\bar{Q}(\beta^*) = 0$. Therefore, the solution of two-point boundary problems (1)–(4) is reduced to the determination of the root of the following algebraic equation:

$$\bar{Q}(\beta) = 0\,. \tag{11}$$

Such equation will be solved by means of the secant method, according to which (Fig. 6.33)

$$\frac{\beta_n - \beta_{n-1}}{\bar{Q}(\beta_n) - \bar{Q}(\beta_{n-1})} = \frac{\beta_{n+1} - \beta_n}{\bar{Q}(\beta_{n+1}) - \bar{Q}(\beta_n)}\,. \tag{12}$$

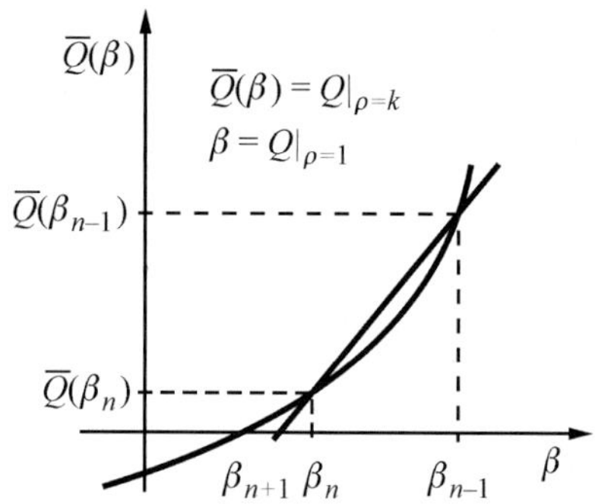

Fig. 6.33. Determining the root of a non-linear algebraic equation $\bar{Q}(\beta)$ by means of the secant method

Next, by taking into account the following condition in (12)

$$\bar{Q}(\beta_{n+1}) = 0 \tag{13}$$

one obtains,

$$\frac{\beta_n - \beta_{n-1}}{\bar{Q}(\beta_n) - \bar{Q}(\beta_{n-1})} = \frac{\beta_{n+1} - \beta_n}{0 - \bar{Q}(\beta_n)}\,, \tag{14}$$

from which it follows that

$$\beta_{n+1} = \beta_n - \frac{(\beta_n - \beta_{n-1})\bar{Q}(\beta_n)}{\bar{Q}(\beta_n) - \bar{Q}(\beta_{n-1})}\,, \qquad n = 0, 1, \ldots \tag{15}$$

At the beginning of the calculation, two values $\beta_0 > 0$ and $\beta_1 > 0$ are assumed, and following that roots β_2, β_3, ... are calculated using (15). Iterative process is continued until the following condition is met

$$|\beta_{n+1} - \beta_n| < e,$$

where e is the assigned tolerance of the calculation and equals $e = 0.001$. The results of fin efficiency calculation are presented in Fig. 6.34–6.37 for different values of $k = r_2/r_1$, ε and M.

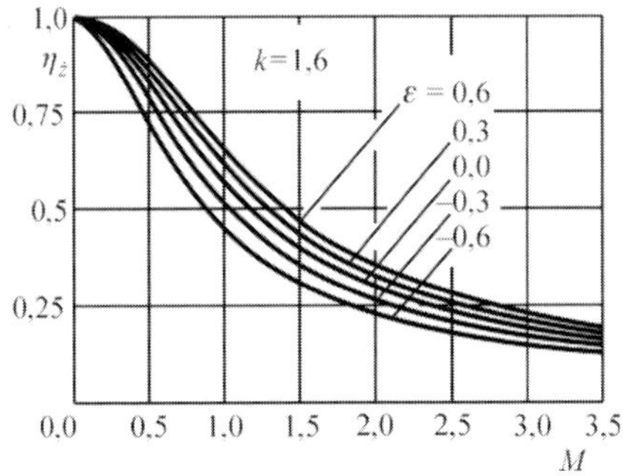

Fig. 6.34. Efficiency of a fin with constant thickness and variable thermal conductivity for $k = 1.6$

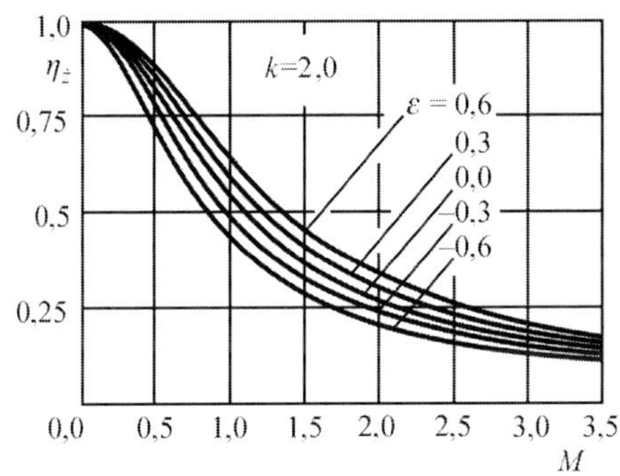

Fig. 6.35. Efficiency of a circular fin with constant thickness and variable thermal conductivity for $k = 2.0$

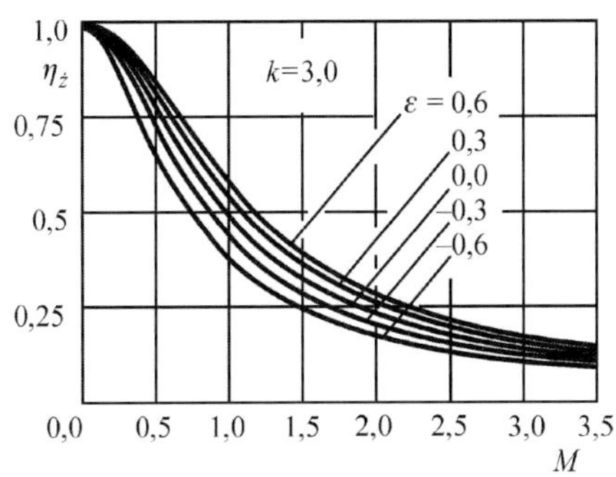

Fig. 6.36. Efficiency of a circular fin of constant thickness and variable thermal conductivity for $k = 3.0$

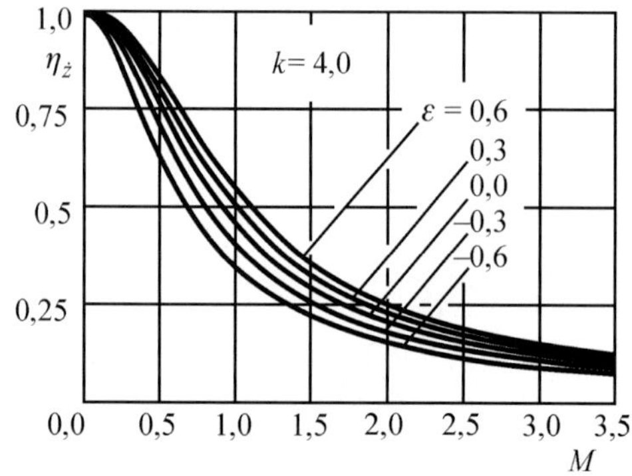

Fig. 6.37. Efficiency of a circular fin of constant thickness and variable thermal conductivity for $k = 4.0$

From the comparisons presented in Table 6.4, it is evident that the given method for calculating fin efficiency is highly accurate. In paper [12], the method described above was also used to determine circular fin efficiency with position-dependent heat transfer coefficient α.

Table 6.4. Efficiency $\eta_{\dot{z}}$ of a circular fin with constant thickness for $k = 2$; value $\eta_{\dot{z}}$ for $\varepsilon = 0$ (constant thermal conductivity) calculated by means of the analytical formula ((10), Ex. 6.17) is given in brackets

ε	Parameter M											
	0	0.25	0.5	0.75	1.0	1.25	1.5	1.75	2.0	2.25	2.5	2.75
–0.6	1.0	0.8831	0.6915	0.5351	0.4217	0.3410	0.2829	0.2401	0.2078	0.1827	0.1628	0.1468
–0.3	1.0	0.9241	0.7648	0.6095	0.4868	0.3959	0.3292	0.2795	0.2419	0.2126	0.1894	0.1707
0.0	1.0	0.9445	0.8133	0.6674	0.5418	0.4440	0.3704	0.3149	0.2725	0.2395	0.2133	0.1921
	(1.0)	(.9445)	(0.8133)	(0.6674)	(0.5418)	(0.4440)	(0.3704)	(0.3149)	(0.2725)	(0.2395)	(.2133)	(0.1921)
0.3	1.0	0.9565	0.8465	0.7128	0.5885	0.4867	0.4077	0.3473	0.3007	0.2643	0.2353	0.2119
0.6	1.0	0.9642	0.8702	0.7486	0.6282	0.5247	0.4419	0.3772	0.3269	0.2874	0.2558	0.2303

Literature

1. Bjőrck A, Dahlquist G (1987) Numerical methods. PWN, Warsaw
2. Brandt F (1985) Wärmeűbertragung in Dampferzeugern und Wärmeaustauschern, FDBR Fachverband Dampfkessel, Behälter- und Rohrleitungsbau E.V. Vulkan Verlag, Essen
3. Gnielinski V, Žukauskas A, Skrinska A (1992) Banks of plain and finned tubes. Chapter 2.5.3.Hewitt G. F.: Handbook of Heat Exchanger Design. Begell House, New York: 2.5.3–1 – 2.5.3–16
4. IMSL Math/Library (1994) Special Functions, Visual Numerics. Houston, Texas, USA.

5. Janke E, Emde F, Lösch F (1960) Tafeln höherer Funktionen. B. G. Teubner, Stuttgart
6. Lokshin VA, Peterson F, Schwarz AL (1988) Standard Methods of Hydraulic Design for Power Boilers. Hemisphere-Springer, Berlin-Washington
7. MathCAD 7 Professional (1997). MathSoft Inc, Cambridge Massachusetts, USA
8. Press WH, Teukolsky SA, Vetterling WT, Flannery BP (1996) Numerical Recipies in Fortran 77. Sec. Ed. Cambridge University Press
9. Recknagel H, Sprenger E, Hönmann W, Schramek ER (1994) Handbook, Heating and air conditioning, including refrigeration engineering and hot water supply. EWFE
10. Schmidt ThE (1950) Die Wärmeleistung von berippten Oberflächen. Abh. Deutsch. Kältetechn. Verein. Nr. 4, C.F. Műller, Karlsruhe
11. Schmidt ThE (1949) Heat transfer calculations for extended surfaces. Refig. Eng., Apr.: 351–357
12. Taler J, Przybyliński P (1982) Heat transfer by circular fins of variable conduction and non-uniform heat transfer coefficient. Chemical and Process Engineering 3, No 3–4: 659–676
13. Weber RL (1994) Principles of Enhanced Heat Transfer. Wiley & Sons, New York

7 Two-Dimensional Steady-State Heat Conduction. Analytical Solutions

In order to solve steady-state heat conduction problems, we have employed in this chapter a well-known separation of variables method, which is an analytical method. We have derived formulas for two-dimensional temperature distribution in fins of an infinite and finite length and in the radiant tubes of boilers. A computational program was developed for determining temperature and heat flux in finite-length-fins.

Exercise 7.1 Temperature Distribution in an Infinitely Long Fin with Constant Thickness

Determine temperature distribution in an infinitely long fin, shown in Fig. 7.1, by means of separation of variables method.

Fin base temperature is T_b, while the temperature of a fin-surrounding medium is T_{cz}. Heat transfer coefficient α on the fin surface is constant.

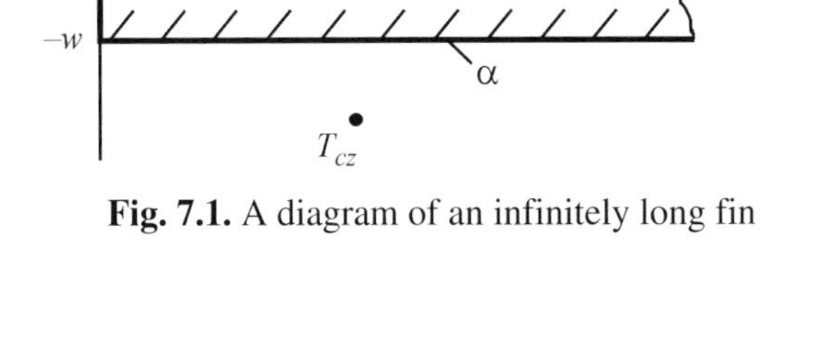

Fig. 7.1. A diagram of an infinitely long fin

Solution

Due to a symmetry of the temperature field, only an upper half of the fin will be examined here. Temperature distribution is described by the following differential equation

$$\frac{\partial^2 T}{\partial x^2}+\frac{\partial^2 T}{\partial y^2}=0 \tag{1}$$

and boundary conditions

$$T(0,y)=T_b\,, \tag{2}$$

$$T(\infty,y)=T_{cz}\,, \tag{3}$$

$$\frac{\partial T}{\partial y}(x,0)=0\,, \tag{4}$$

$$-\lambda\frac{\partial T}{\partial y}(x,w)=\alpha\left[T(x,w)-T_{cz}\right]. \tag{5}$$

Once dimensionless variables are introduced, such as

- temperature

$$\Theta=\frac{T-T_{cz}}{T_b-T_{cz}}\,, \tag{6}$$

- coordinates

$$X=\frac{x}{w},\qquad Y=\frac{y}{w}\,, \tag{7}$$

and Biot number

$$Bi=\frac{\alpha w}{\lambda}\,, \tag{8}$$

problem (1)–(5) can be written in the dimensionless form:

$$\frac{\partial^2 \Theta}{\partial X^2}+\frac{\partial^2 \Theta}{\partial Y^2}=0, \tag{9}$$

$$\Theta(0,Y)=1, \tag{10}$$

$$\Theta(\infty,Y)=0, \tag{11}$$

$$\Theta(0,Y)=1, \tag{10}$$

$$\frac{\partial \Theta}{\partial X}(L,Y)=0, \tag{11}$$

$$\frac{\partial \Theta}{\partial Y}(X,0)=0, \tag{12}$$

$$\frac{\partial \Theta}{\partial Y}(X,1)+Bi\cdot\Theta(X,1)=0. \tag{13}$$

In accordance with the separation of variables method, the solution of problems (9)–(13) is searched for in the form

$$\Theta(X,Y)=U(X)\cdot V(Y). \tag{14}$$

Function $V(Y)$ has the same form as it does in Ex. 7.1. Function $U(X)$ is, much like in Ex. 7.1, determined from equation

$$\frac{\partial^2 U}{\partial X^2}-\mu^2 U=0, \tag{15}$$

when boundary conditions are

$$UV\big|_{X=0}=1, \tag{16}$$

$$\frac{\partial U}{\partial X}(L)=0\text{, where } L=l/w. \tag{17}$$

The solution for (15), which satisfies condition (17), is the function

$$U=C\cosh\mu(L-X). \tag{18}$$

Since the characteristic equation (25) from Ex. 7.1 has an infinite number of positive elements μ_n, an infinite number of functions exist

$$U_n=C_n\cosh\mu_n(L-X),\qquad n=1,2,\ldots \tag{19}$$

which satisfy (15) and boundary condition (17). The solution of problems (9)–(13) will be searched for in the form

$$\Theta(X,Y)=\sum_{n=1}^{\infty}U_nV_n=\sum_{n=1}^{\infty}C_n\cosh\mu_n(L-X)\cdot\cos\mu_nY. \tag{20}$$

Constant C_n in expression (20) is determined from the heterogeneous boundary condition (10), which yields the following result:

$$\sum_{n=1}^{\infty}C_n\cosh\mu_nL\cdot\cos\mu_nY=1. \tag{21}$$

After multiplying both sides of the equation by $\cos \mu_m Y$ and after integrating dY from 0 to 1, one gets

$$\sum_{n=1}^{\infty} \int_0^1 C_n \cosh \mu_n L \cdot \cos \mu_n Y \cdot \cos \mu_m Y = \int_0^1 \cos \mu_m Y dY \,. \tag{22}$$

From orthogonal condition of function $\cos \mu_n Y \cdot \cos \mu_m Y$ ((33), Ex. 7.1), one obtains

$$C_n \cosh \mu_n L = \frac{\int_0^1 \cos \mu_n Y dY}{\int_0^1 \cos^2 \mu_n Y dY} \,, \tag{23}$$

hence

$$C_n = \frac{2 \sin \mu_n}{\cosh \mu_n L \left(\mu_n + \sin \mu_n \cos \mu_n \right)} \,. \tag{24}$$

By substituting constant C_n formulated in (24) into expression (20), temperature distribution is formulated as

$$\Theta(X,Y) = 2 \sum_{n=1}^{\infty} \frac{\sin \mu_n}{\cosh \mu_n L \left(\mu_n + \sin \mu_n \cos \mu_n \right)} \times \\ \times \cosh \mu_n \left(L - X \right) \cdot \cos \mu_n Y, \tag{25}$$

where μ_n are positive elements of the characteristic transcendental equation

$$\mathrm{ctg} \mu = \frac{1}{Bi} \mu \,. \tag{26}$$

Exercise 7.3 Calculating Temperature Distribution and Heat Flux in a Straight Fin with Constant Thickness and Insulated Tip

Calculate temperature distribution in a fin on the basis of a formula derived in Ex. 7.2. Calculate fin temperature in points shown in Fig. 7.3. Also calculate heat flux at the fin base in points (0,0) and (0,w). Determine a formula for averaged temperature and heat flux across the fin thickness. Compare mean temperature values across the entire fin length and heat flux at the fin base with one-dimensional solution. Assume the following values for the calculation: w = 0.003m, l = 0.024 m, α = 100 W/(m^2·K), T_b = 95°C, T_{cz} = 20°C, Δx = 0.003 m, λ = 50 W/(m·K).

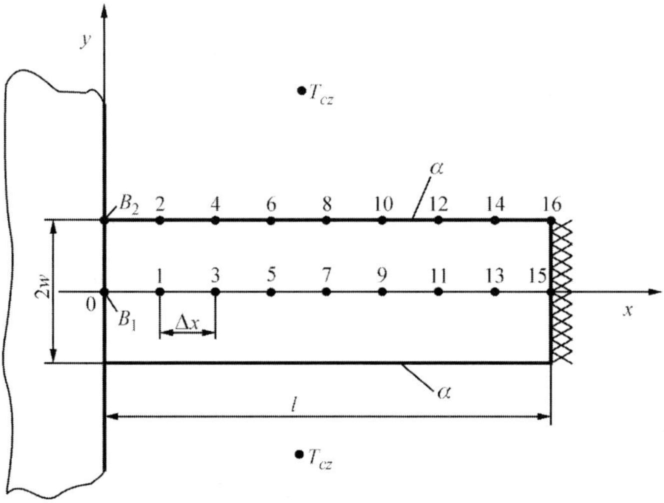

Fig. 7.3. A fin diagram with marked nodes, in which temperature is calculated

Solution

On the basis of (25) from Ex. 7.2, temperature distribution $T(X,Y)$ will be calculated from formula

$$T = T_{cz} + \left(T_b - T_{cz}\right)\Theta\,, \tag{1}$$

where

$$\Theta\left(X,Y\right) = 2\sum_{n=1}^{\infty}\frac{\sin\mu_n}{\left(\mu_n + \sin\mu_n\cdot\cos\mu_n\right)}\frac{\cosh\mu_n\left(L-X\right)}{\cosh\mu_n L}\cos\mu_n Y\,, \tag{2}$$

where: $X = x/w$, $Y = y/w$, $Bi = \alpha w/\lambda$.

Elements of the characteristic equation

$$\text{ctg}\mu = \frac{1}{Bi}\mu \tag{3}$$

will be determined by means of the interval halving method; one should note, however, that the infinite element μ_n lies in an interval between $\mu_{n,\min} = (n-1)\pi$ and $\mu_{n,\max} = (n-1/2)\pi$. Values $\mu_{n,\max}$ are characteristic values of the Strum-Liouville problem in an instance when $Bi \to \infty$, i.e. when constant temperature is assigned on the fin surface.

Mean temperature across the fin thickness is determined from formula

$$\overline{T}\left(x\right) = \frac{1}{w}\int_0^w\left[T_{cz} + \left(T_b - T_{cz}\right)\theta\right]dy\,, \tag{4}$$

from where, one obtains

$$\overline{T} = T_{cz} + 2\left(T_b - T_{cz}\right)\sum_{n=1}^{\infty}\frac{\sin\mu_n}{\mu_n\left(\mu_n + \sin\mu_n \cdot \cos\mu_n\right)}\frac{\cosh\mu_n\left(L - X\right)}{\cosh\mu_n L}\sin\mu_n . \quad (5)$$

Heat flux in the direction of x axis comes to

$$\dot{q}_x = -\lambda\frac{\partial T}{\partial x} = \frac{2\lambda\left(T_b - T_{cz}\right)}{w}\sum_{n=1}^{\infty}\frac{\mu_n \sin\mu_n}{\left(\mu_n + \sin\mu_n \cdot \cos\mu_n\right)}\frac{\sinh\mu_n\left(L - X\right)}{\cosh\mu_n L}\times \times\cos\mu_n Y . \quad (6)$$

Mean heat flux value across the fin thickness is the function of x coordinate

$$\overline{\dot{q}_x} = \frac{1}{w}\int_0^w \dot{q}_x\left(x, y\right)dy = \frac{2\lambda\left(T_b - T_{cz}\right)}{w}\sum_{n=1}^{\infty}\frac{\sin\mu_n}{\left(\mu_n + \sin\mu_n \cdot \cos\mu_n\right)}\times \times\frac{\sinh\mu_n\left(L - X\right)}{\cosh\mu_n L}\sin\mu_n . \quad (7)$$

Fin temperature distribution T_{1d}, determined under the assumption that temperature decrease within the fin thickness is negligibly small, is expressed by function (Ex. 6.15)

$$T_{1d}\left(x\right) = T_{cz} + \left(T_b - T_{cz}\right)\frac{\cosh m\left(l - x\right)}{\cosh ml} , \quad (8)$$

where $m = \sqrt{\alpha/\lambda w}$.

In the given case

$$m = \sqrt{\frac{50}{100 \cdot 0.003}} = 12.90994 \text{ 1/m} .$$

In the case of the one-dimensional solution, heat flux is formulated as

$$\dot{q}_{1d}\left(x\right) = -\lambda\frac{dT_{1d}}{dx} = \left(T_b - T_{cz}\right)\lambda m\frac{\sinh m\left(l - x\right)}{\cosh ml} . \quad (9)$$

Allowing that $Bi = \alpha w/\lambda = 100 \cdot 0.003/50 = 0.006$, the first ten elements of the characteristic equation (3) were determined:

μ_1	μ_2	μ_3	μ_4	μ_5	μ_6	μ_7	μ_8	μ_9	μ_{10}
0.0774	3.1435	6.2841	9.4254	12.5668	15.7083	18.8499	21.9914	25.1330	28.2745

Next, the elements were applied to (2). Temperature $T(x, y)$ in nodes shown in Fig. 7.3 was calculated by means of the FORTRAN program, which comes with this exercise. Proper values μ_n were calculated by means of the sub-program presented in paper [1]. Mean temperature distribution $\overline{T}(x)$ and temperature $T_{1d}(x)$ were also calculated.

Heat flux was calculated at two points: B_1 and B_2 (Fig. 7.3). Mean heat flux $\overline{\dot{q}_x}$ at the fin base ($X = 0$) was calculated on the basis of (7). For comparison purposes $\dot{q}_{1d}$ was also calculated for $x = 0$ by means of (9). Temperature calculation results are shown in Table 7.1.

Table 7.1. Calculation results

x [m]	Node no.	Temperature [°C]	Node no.	Temperature [°C]	Node no.	$\overline{T}(x)$ [°C]	$T_{1d}(x)$ [°C]
0	B_1	95.00	B_2	94.99	B_1, B_2	95.00	95.00
0.003	1	92.09	2	91.88	1, 2	92.02	92.02
0.006	3	89.55	4	89.34	3, 4	89.48	89.47
0.009	5	87.42	6	87.22	5, 6	87.35	87.34
0.012	7	85.70	8	85.50	7, 8	85.63	85.62
0.015	9	84.37	10	84.18	9, 10	84.30	84.29
0.018	11	83.42	12	83.23	11, 12	83.36	83.34
0.021	13	82.86	14	82.67	13, 14	82.80	82.78
0.024	15	82.67	16	82.48	15, 16	82.61	82.59

Calculated heat flux measures:

- at point B_1

  $$\dot{q}_x(0,0) = 49764 \text{ W/m}^2,$$

- at point B_2

  $$\dot{q}_x(0,w) = 66646 \text{ W/m}^2,$$

- mean heat flux at the fin base $\overline{\dot{q}_x}$

  $$\overline{\dot{q}_x}(0) = 53253 \text{ W/m}^2,$$

- heat flux $\dot{q}_{1d}$ at the fin base

  $$\dot{q}_{1d}(0) = 53341 \text{ W/m}^2.$$

From the analysis of the obtained results, it is evident that there is a small temperature decrease across the fin thickness. Also, heat flux $\dot{q}_x$ varies in points B_1 and B_2. A good accuracy of results is evident in $\overline{T}(x)$ and $T_{1d}(x)$,

```
      dimension eigen(*)
      teta=0.
      x_b=x/w
      dlug_b=dlug/w
      do i=1,ne
      s=eigen(i)
      teta=teta+sin(s)*cosh(s*(dlug_b-x_b))*sin(s)/
     &s/(s+sin(s)*cos(s))/cosh(s*dlug_b)
      enddo
      temperature_sr=t_cz+(t_b-t_cz)*2.*teta
      end function

      function temperature(x,y,t_cz,t_b,dlug,w,ne,eigen)
      dimension eigen(*)
      teta=0.
      x_b=x/w
      y_b=y/w
      dlug_b=dlug/w
      do i=1,ne
         s=eigen(i)
      teta=teta+sin(s)*cosh(s*(dlug_b-x_b))*cos(s*y_b)/
     &(s+sin(s)*cos(s))/cosh(s*dlug_b)
      enddo
      temperature=t_cz+(t_b-t_cz)*2.*teta
      end function

c     procedure calculates elements of characteristic eq.
c     x*tan(x)=bi where bi is Biot number, ne calculated
c     element quantity, eigen vector with recorded calculated
c     elements
      subroutine equation_elements(bi,ne,eigen)
      dimension eigen(*)
      pi=3.141592654
      do i=1,ne
         xi=(float(i)-1.)*pi
         xf=pi*(float(i)-.5)
         do while (abs(xf-xi).ge.5.E-06)
            xm=(xi+xf)/2.
            y=xm*sin(xm)/cos(xm)-bi
            if (y.lt.0.) then
               xi=xm
            else
               xf=xm
            endif
         enddo
        eigen(i)=xm
      enddo
      return
```

```
      end
data(fin.in)
10 0.006
20. 95. 0.024 0.003  50. 100.
results(fin.out)
CALCULATING TWO-DIMENSIONAL FIN TEMPERATURE FIELD

DATA ENTERED
ne      =        10
Biot number=.60000E-02
t_cz    =.20000E+02 [C]
t_b     =.95000E+02 [C]
dlug    =.24000E-01 [m]
w       =.30000E-02 [m]
lambda  =.50000E+02 [W/mK]
alfa    =.10000E+03 [W/m2K]

CALCULATION OF FIRST 10 EQUATION ELEMENTS X*TAN(X)=BI

CALCULATED EQUATION ELEMENTS
Lp          mi
 1     .773851E-01
 2     .314350E+01
 3     .628414E+01
 4     .942542E+01
 5     .125668E+02
 6     .157083E+02
 7     .188499E+02
 8     .219914E+02
 9     .251330E+02
10     .282745E+02

CALCULATED TEMPERATURE [C]
  x[m]    T(x,B1)      T(x,B2)     T_sr(x)     T_1d(x)
 .000   .95000E+02  .94990E+02 .95000E+02 .95000E+02
 .003   .92095E+02  .91887E+02 .92026E+02 .92021E+02
 .006   .89554E+02  .89346E+02 .89485E+02 .89475E+02
 .009   .87426E+02  .87224E+02 .87359E+02 .87346E+02
 .012   .85702E+02  .85506E+02 .85637E+02 .85621E+02
 .015   .84372E+02  .84180E+02 .84308E+02 .84290E+02
 .018   .83428E+02  .83238E+02 .83365E+02 .83345E+02
 .021   .82863E+02  .82675E+02 .82801E+02 .82781E+02
 .024   .82676E+02  .82488E+02 .82613E+02 .82593E+02
 .027   .82863E+02  .82675E+02 .82801E+02 .82781E+02

CALCULATED HEAT FLUX [W/m2]
q_x(0,0)=.49764E+05
q_x(0,w)=.66646E+05
q_x_sr(0)=.53253E+05
q_x_1d(0)=.53341E+05
```

Exercise 7.4 Temperature Distribution in a Radiant Tube of a Boiler

Determine formula for temperature distribution in the boiler's radiant tube (Fig. 7.4) by means of the separation of variables method. Assuming that heat flux $\dot{q}_m$ (thermal load of the water-wall) transferred by the water-wall is known (calculated with reference to a wall regarded as a plane), as well as the temperature of a medium that flows inside the tube T_{cz} and heat transfer coefficient α on an inner surface of the tube, determine temperature field in the function of coordinates r and φ. Also calculate the inner and outer surface tube temperature for angle $\varphi = 0$ and $\varphi = \pi$ rad; use the following values for the calculation:

- outer surface tube radius $r_z = 0.019$ m,
- inner surface tube radius $r_w = 0.015$ m,
- scale of radiant tube spacing $s = 0.042$ m,
- thermal load of the water-wall $\dot{q}_m = 300000$ W/m^2;
- heat transfer coefficient on the inner surface of the tube $\alpha = 15\ 000$ W/(m^2·K),
- temperature of a medium $T_{cz} = 330$°C,
- heat conduction coefficient of the steel which the tube is made of $\lambda = 45$ W/(m·K).

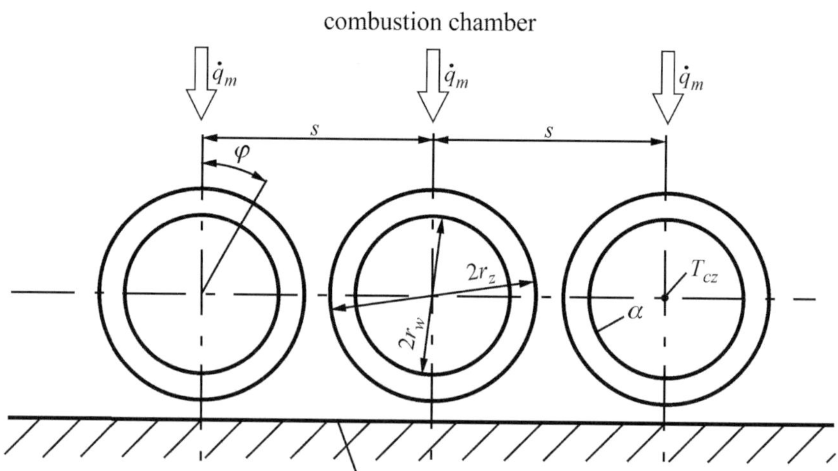

Fig. 7.4. A diagram of a radiant tube spacing in a combustion chamber

Heat flux on the outer surface of the tube is expressed by function [12]

$$\dot{q}(\varphi) = \dot{q}_m \left(0.3649 + 0.4777\cos\varphi + 0.1574\cos 2\varphi\right) . \tag{1}$$

Solution

Tube temperature distribution is expressed by heat conduction equation

$$\frac{1}{r}\frac{\partial}{\partial r}\left(r\frac{\partial T}{\partial r}\right) + \frac{1}{r^2}\frac{\partial^2 T}{\partial \varphi^2} = 0 \tag{2}$$

and boundary conditions

$$\lambda \left.\frac{\partial T}{\partial r}\right|_{r=r_z} = \dot{q}_0 + \sum_{n=1}^{\infty} \dot{q}_n \cos(n\varphi) , \tag{3}$$

$$\lambda \left.\frac{\partial T}{\partial r}\right|_{r=r_w} = \alpha T\Big|_{r=r_w} , \tag{4}$$

where T is the temperature excess of the tube ζ above the temperature of the medium ζ_{cz}, i.e. $T = \zeta - \zeta_{cz}$.

In conformity with the separation of variables method, the solution is searched for in the form

$$T(r,\varphi) = U(r) \cdot V(\varphi) . \tag{5}$$

By substituting (5), one obtains equation

$$r^2 U''V + rU'V + UV'' = 0 . \tag{6}$$

After a division of (6) by UV and the separation of variables, one obtains

$$\frac{r^2 U'' + rU'}{U} = -\frac{V''}{V} . \tag{7}$$

Since r and φ are independent variables, equality (7) occurs only when its both sides are equal to the same constant. If the constant were negative, the solution $V(\varphi)$ would then contain exponential functions, which would unable one to satisfy periodic boundary condition (3) written in the Fourier series form. Separated constant, therefore, must be either a positive integral number or zero. If one assumes that both sides of (7) are equal to n^2, one obtains

$$r^2U'' + rU' - n^2U = 0, \tag{8}$$

$$V'' + n^2V = 0, \qquad n = 0, 1, \ldots \tag{9}$$

In the case of a circular-symmetrical load only $\dot{q}_0 \neq 0$, whereas $\dot{q}_1 = \dot{q}_2 = \cdots = 0$. For $n = 0$, the solution of (8) and (9) has the form

$$U(r) = A_0' + B_0' \ln r \tag{10}$$

and

$$V(\varphi) = C_0' + D_0'\varphi \,. \tag{11}$$

Due to the circular-symmetrical load $D_0' = 0$, the product $U(r)V(\varphi)$ can be written in the form

$$U(r)V(\varphi) = A_0 + B_0 \ln r, \qquad n = 0\,, \tag{12}$$

where, $A_0 = A_0'C'$, $B_0 = B_0'C_0'$.

For $n \geq 1$, the solution of (8)–(9) has the forms

$$U(r) = A_n' r^n + B_n' r^{-n}\,, \tag{13}$$

$$V(\varphi) = C_n' \cos n\varphi + D_n' \sin n\varphi\,. \tag{14}$$

Due to the symmetry of tube heating (condition (3)) with respect to the plane, which is perpendicular to the water-wall and which crosses the tube axis, constant $D_n' = 0$. Product $U(r)V(\varphi)$ can be written in the form

$$U(r)V(\varphi) = \left(C_n r^n + D_n r^{-n}\right)\cos n\varphi, \qquad n \geq 1, \tag{15}$$

where, $C_n = A_n' C_n'$ and $D_n = B_n' C_n'$.

Expression (5), which describes the distribution of excess temperature in the tube, has the form

$$T(r,\varphi) = A_0 + B_0 \ln r + \sum_{n=1}^{\infty}\left(C_n r^n + D_n r^{-n}\right)\cos n\varphi\,. \tag{16}$$

After substituting (16) into boundary conditions (3) and (4), one can determine constants, which can be written after transformation in the following form:

$$A_0 = \frac{\dot{q}_0 r_z}{\lambda}\left(\frac{1}{Bi} - \ln r_w\right),$$

$$B_0 = \frac{\dot{q}_0 r_z}{\lambda},$$

$$C_n = \frac{\dot{q}_n r_z}{\lambda} \frac{\frac{1}{n} u^n (Bi + n) \frac{1}{r_w^n}}{Bi\left(u^{2n} + 1\right) + n\left(u^{2n} - 1\right)},$$

$$D_n = -\frac{\dot{q}_n r_z}{\lambda} \frac{\frac{1}{n} u^n (Bi - n) r_w^n}{Bi\left(u^{2n} + 1\right) + n\left(u^{2n} - 1\right)},$$

where, $u = r_z / r_w$, $Bi = \alpha\, r_w / \lambda$.

Since in this exercise, the heat flux on the outer surface of the pipe is defined by (1), then

$$\dot{q}_0 = 0.3649 \dot{q}_m,$$
$$\dot{q}_1 = 0.4777 \dot{q}_m,$$
$$\dot{q}_2 = 0.1574 \dot{q}_m,$$

and

$$\dot{q}_3 = \dot{q}_4 = \ldots = 0,$$

It is easy to calculate tube temperature, when only 2 terms are accounted for in the series (16). Once the following is calculated

$$u = \frac{r_z}{r_w} = \frac{0.019}{0.015} = 1.2667,$$

$$Bi = \frac{\alpha r_w}{\lambda} = \frac{15\,000 \cdot 0.015}{45} = 5.0$$

and substituted into solution (16), one obtains

$$T(r_z, 0) = 52.28°\text{C},$$
$$T(r_w, 0) = 23.53°\text{C},$$
$$T(r_z, \pi) = 2.31°\text{C},$$
$$T(r_w, \pi) = 1.04°\text{C}.$$

These are temperatures above the medium's temperature. Corresponding pipe temperatures are:

$$\zeta(r_z, 0) = T_{cz} + T(r_z, 0) = 330 + 52.28 = 382.28°\text{C},$$
$$\zeta(r_w, 0) = T_{cz} + T(r_w, 0) = 330 + 23.53 = 353.53°\text{C},$$

$$\zeta(r_z,\pi)=T_{cz}+T(r_z,\pi)=330+2.31=332.31°\text{C},$$
$$\zeta(r_w,\pi)=T_{cz}+T(r_w,\pi)=330+1.04=331.04°\text{C}.$$

It is evident that temperature $\zeta(r_z,0)$ is the maximum temperature across the whole cross-section of the tube. Provided that this temperature is known, one can correctly chose the right type of steel for the radiant tube of a boiler.

Literature

1. Becker M (1986) Heat Transfer. A Modern Approach. Plenum Press, New York
2. Gdula SJ et. al. (1984) Heat conduction. Group work. PWN, Warsaw: 134–138
3. Irey RK (1968) Errors in the one-dimensional fin solution. Transactions of the ASME, J. Heat Transfer 90: 175–178
4. Lau W, Tan CW (1973) Errors in one-dimensional heat transfer analyses in straight and annular fins. Transactions of the ASME, J. Heat Transfer 95: 549–551
5. Look DC (1995) Fin on pipe (insulated tip): minimum conditions for fin to be beneficial. Heat Transfer Engineering 16, No. 3: 65–75
6. Look DCJr, Kang HS (1991) Effects of variation in root temperature on heat lost from a thermally non-symmetric fin. Int. J. Heat Mass Transfer 34, No. 4/5: 1059–1065
7. Look DCJr (1999) Fin (on pipe) effectiveness: one dimensional and two dimensional. Transactions of the ASME, J. Heat Transfer 121: 227–230
8. Look DCJr (1988) Two-dimensional fin performance: Bi (top surface) > Bi (bottom surface). Transactions of the ASME, J. Heat Transfer 110: 780–782
9. Ma SW, Behbahani AI, Tsuei YG (1991) Two-dimensional rectangular fin with variable heat transfer coefficient. Int. J. Heat and Mass Transfer 34, No 1: 79–85
10. Młynarski F, Taler J (1979) Analytische Untersuchung der Temperatur – und Spannungsverteilung in strahlungsbeheizten Kesselrohren unter der Berücksichtigung der wasserseitigen Ablagerungen. VGB Kraftwerkstechnik 5: 440–447
11. Taler J, Kulesza L (1982) Calculating temperature field in finned, bimetallic and homogenous screen pipes with a layer of residue on an inner surface. Archives of Power Engineering 1: 3–23
12. Taler J (1980) Weighed residues method and its application for temperature field calculation in boilers' elements. Monograph 14, Pub. Krakow University of Technology, Kraków
13. (1999) The Heat Transfer Problem Solver, Problem 3–5. Research and Education Association. Piscataway, New Jersey USA: 142–145
14. Unal HC (1988) The effect of the boundary condition at a fin tip on the performance of the fin with and without internal heat generation. Int. J. Heat Mass Transfer 31, No. 7: 1483–1496
15. Kraus AD, Aziz A, Welty J (2001) Extended Surface Heat Transfer. Wiley & Sons, New York

8 Analytical Approximation Methods. Integral Heat Balance Method

In this chapter an integral heat balance method, which is an approximate method [1–8], was applied to solve various engineering problems. It ensures fast obtaining of an approximated solution, marked by simplicity of form. Accuracy of approximated solution is higher when heat flow is significantly larger in one direction than it is in others. For this reason, integral heat balance method is recommended for determining steady-state temperature fields, which are not much different from one-dimensional steady-state temperature field.

Exercise 8.1 Temperature Distribution within a Rectangular Cross-Section of a Bar

Determine temperature distribution in an infinitely long bar of a rectangular cross-section, heated by thermal sources with constant power density $\dot{q}_v$. Bar temperature T_b (Fig. 8.1) at perimeter is constant. Determine temperature distribution by means of integral heat balance method.

Solution

Temperature distribution $T(x,y)$ is expressed by heat conduction equation

$$\frac{\partial^2 T}{\partial x^2} + \frac{\partial^2 T}{\partial y^2} = -\frac{\dot{q}_v}{\lambda} \tag{1}$$

and by boundary conditions

$$T\big|_{x=\pm\frac{b}{2}} = T_b \, , \tag{2}$$

$$T\big|_{y=\pm\frac{h}{2}} = T_b \, . \tag{3}$$

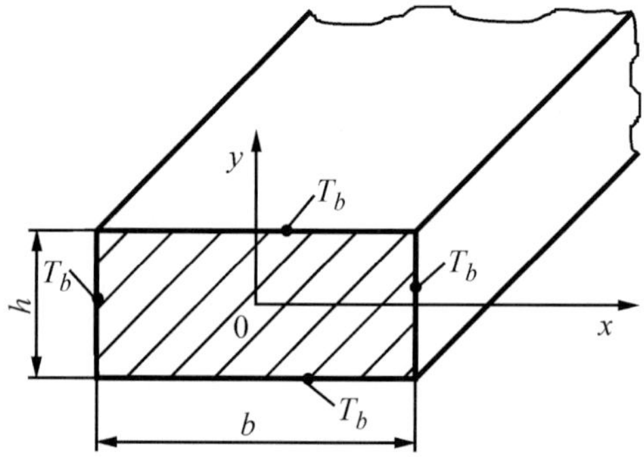

Fig. 8.1. A bar of a rectangular cross-section

In accordance with integral heat balance method, temperature distribution will be approximated by the following function, which satisfies boundary conditions and temperature field symmetry conditions with respect to axis x and y:

$$T(x,y)=T_b+\left(T_{\max}-T_b\right)\left[1-\left(\frac{x}{b/2}\right)^2\right]\left[1-\left(\frac{y}{h/2}\right)^2\right]. \tag{4}$$

If we assume in (4) that $x = 0$ and $y = 0$, then $T(0,0) = T_{max}$. Constant $T_{max,}$ which is present in function (4) and approximates real temperature distribution, is determined through an integration of (1) within an entire rectangle area: $-b/2 \le x \le b/2$, $-h/2 \le y \le h/2$. Approximate temperature distribution (4) is replaced by an integral

$$\int_{-h/2}^{h/2}\int_{-b/2}^{b/2}\left(\frac{\partial^2 T}{\partial x^2}+\frac{\partial^2 T}{\partial y^2}+\frac{\dot{q}_v}{\lambda}\right)dxdy=0 \tag{5}$$

and after transformations, one obtains

$$T_{\max}=T_b+\frac{3\dot{q}_v}{16\lambda}\frac{b^2h^2}{b^2+h^2}\ . \tag{6}$$

Solution (6) is a good approximation of an accurate solution, when $b >> h$ or $b << h$. In both cases, temperature distribution is almost one-dimensional. The lowest accuracy is obtained in the case of a square cross-section, when $b = h$. Then difference $(T_{max} - T_b)$ determined from (6) is by 27% larger than a difference obtained from an exact analytical solution.

Exercise 8.2 Temperature Distribution in an Infinitely Long Fin of Constant Thickness

Determine temperature distribution in a fin depicted in Fig. 8.2. Fin base temperature T_b is constant. Determine temperature distribution by means of approximate method: integral heat balance method.

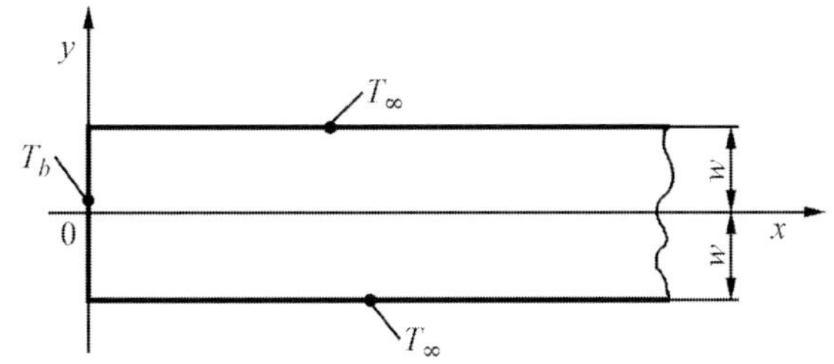

Fig. 8.2. Two-dimensional steady-state temperature field in an infinitely long fin

Solution

Temperature distribution is expressed by a differential equation of steady-state heat conduction

$$\frac{\partial^2 T}{\partial x^2}+\frac{\partial^2 T}{\partial y^2}=0 , \tag{1}$$

and by boundary conditions

$$T(0,y)=T_{max}\left[1-\left(\frac{y}{w}\right)^2\right], \qquad T(x,\pm w)=T_\infty, \qquad T(\infty,y)=T_\infty . \tag{2}$$

Once new variable is introduced $\Theta = T - T_\infty$, (1) and conditions (2) can be written in the following form:

$$\frac{\partial^2 \Theta}{\partial x^2}+\frac{\partial^2 \Theta}{\partial y^2}=0 , \tag{3}$$

$$\Theta(0,y)=\Theta_{max}\left[1-\left(\frac{y}{w}\right)^2\right], \qquad \Theta(x,\pm w)=0, \qquad \Theta(\infty,y)=0 , \tag{4}$$

where $\Theta_{max} = T_{max} - T_\infty$.

Approximate temperature distribution will be searched for in a form

$$\Theta(x,y)=\left(w^2-y^2\right)U(x) . \tag{5}$$

Function (5) is selected in such a way that it satisfies boundary condition $\Theta(y, \pm w) = 0$ and the temperature field symmetry condition with respect to axis x, i.e.

$$\left.\frac{\partial \Theta}{\partial y}\right|_{y=0} = 0 . \tag{6}$$

In accordance with heat balance method, heat conduction equation (3) is integrated within the area under analysis (due to a symmetry of temperature field after division in two parts)

$$\int_0^w \int_0^\infty \left(\frac{\partial^2 \Theta}{\partial x^2} + \frac{\partial^2 \Theta}{\partial y^2} \right) dxdy = 0 . \tag{7}$$

By substituting (5) into (7), one obtains

$$\int_0^w \int_0^\infty \left[\left(w^2 - y^2 \right) U'' - 2U \right] dxdy = 0 ,$$

from which, after integration in the direction of y-axis, we have

$$\int_0^\infty \left[\left(w^2 y - \frac{y^3}{3} \right) U'' - 2Uy \right]_0^w dx = 0 . \tag{8}$$

Applying the boundary conditions (4) in (8), one obtains

$$\int_0^\infty \left[\frac{2}{3} w^3 U'' - 2wU \right] dx = 0 , \tag{9}$$

which yields differential equation

$$U'' - \frac{3}{w^2} U = 0 . \tag{10}$$

Boundary conditions for U result from the first and third boundary condition in (4). From the first condition, one obtains

$$\Theta_{\max} \left[1 - \left(\frac{y}{w} \right)^2 \right] = \left(w^2 - y^2 \right) U(0) , \tag{11}$$

from which, after simple transformations, one has

$$U(0) = \frac{\Theta_{\max}}{w^2} . \tag{12}$$

From third boundary condition in (4), one obtains a second boundary condition for $U(x)$

$$U(\infty) = 0 \, . \tag{13}$$

Solution of (10) has the form

$$U(x) = C_1 \exp\left(-\sqrt{3}\frac{x}{w}\right) + C_2 \exp\left(\sqrt{3}\frac{x}{w}\right) . \tag{14}$$

By substituting (14) into condition (12) and (13), one is able to determine constant C_1 and C_2

$$C_1 = \frac{\Theta_{\max}}{w^2}, \qquad C_2 = 0 \, . \tag{15}$$

Function $U(x)$ is given by

$$U(x) = \frac{\Theta_{\max}}{w^2} \exp\left(-\sqrt{3}\frac{x}{w}\right) . \tag{16}$$

By allowing for (16) in (5), the expression for approximate temperature distribution has the form

$$\Theta(x, y) = \Theta_{\max} \frac{w^2 - y^2}{w^2} \exp\left(-\sqrt{3}\frac{x}{w}\right) . \tag{17}$$

Exercise 8.3 Determining Temperature Distribution in a Boiler's Water-Wall Tube by Means of Functional Correction Method

Solved the problem formulated in Ex. 7.4 using functional correction method.

Solution

First we will briefly characterize *functional correction method* [6], which is also an analytical approximate method. If heat flux is much larger in one direction than it is in another, then temperature field resembles one-dimensional temperature field. Functional correction method is very effective for determining such temperature fields. A typical example of such problem is the conduction of heat in a fin, in which the flow is much larger in the direction of x axis than it is in the direction of y axis (Ex. 7.2, Fig. 7.2). Heat conduction equation

$$\frac{\partial^2 T}{\partial x^2}+\frac{\partial^2 T}{\partial y^2}=0 \tag{1}$$

will be approximated by differential equation, in which derivative $\partial^2 T/\partial y^2$ will be approximated by mean value within the fin thickness

$$\frac{\partial^2 T}{\partial y^2}=f(x), \tag{2}$$

where

$$f(x)=-\frac{1}{w}\int_0^w \frac{\partial^2 T}{\partial x^2}dy. \tag{3}$$

As follows from (2), it has been assumed that second derivative $\partial^2 T/\partial y^2$ is a function of coordinate x and doesn't depend, on coordinate y.

Temperature distribution is determined in the following way.

Once (2) is integrated twice, one obtains

$$T(x,y)=\frac{1}{2}y^2 f(x)+C_1 y+C_2, \tag{4}$$

where C_1 and C_2 are constants, which are determined from appropriate boundary conditions in the direction of y axis. Once (4) is substituted into (3) and all mathematical operations carried out, one obtains ordinary differential equation of 2nd order for $f(x)$. After integrating this equation in the presence of appropriate boundary conditions at the base and tip of a fin, one obtains temperature distribution in the fin.

In order to determine temperature distribution in a water-wall tube, whose temperature field is expressed by equation

$$\frac{1}{r}\frac{\partial}{\partial r}\left(r\frac{\partial T}{\partial r}\right)+\frac{1}{r^2}\frac{\partial^2 T}{\partial \varphi^2}=0 \tag{5}$$

and boundary conditions

$$\lambda \left.\frac{\partial T}{\partial r}\right|_{r=r_z}=\dot{q}_0+\sum_{n=1}^{\infty}\dot{q}_n \cos n\varphi, \tag{6}$$

$$\lambda \left.\frac{\partial T}{\partial r}\right|_{r=r_w}=\alpha \left.T\right|_{r=r_w}, \tag{7}$$

where $T = \zeta - \zeta_{cz}$ is excess temperature of tube $\zeta(r,\varphi)$ above a temperature of medium ζ_{cz}, functional correction method will be applied. In a water-wall tube heat flux in circumferential direction is much smaller than in radial direction. One could determine, therefore, tube temperature distribution under the assumption that temperature is only a function of a single variable r, while ignore tube heat flow in the circumferential direction. Maximum tube temperature $T_{max} = T(r_z, 0)$ determined in such way is to large, since an outflow of heat from frontal part of the pipe to its cooler rear side is not taken into consideration. One can make more accurate determination of temperature distribution in a tube by means of functional correction method, according to which (5) can be substituted by approximate equation

$$\frac{1}{r}\frac{\partial}{\partial r}\left(r\frac{\partial T}{\partial r}\right) = f(\varphi), \tag{8}$$

where

$$f(\varphi) = -\frac{2}{r_z^2 - r_w^2}\int_{r_w}^{r_z}\frac{1}{r^2}\frac{\partial^2 T}{\partial \varphi^2} r\,dr. \tag{9}$$

Once (9) is integrated twice, one obtains

$$T(r,\varphi) = \frac{1}{4}f(\varphi)r^2 + C_1 \ln r + C_2. \tag{10}$$

After constants C_1 and C_2 are determined from boundary condition (6) and (7) and substituted into (10), one obtains

$$T = \left[\frac{1}{4}\left(r^2 - r_w^2\right) - \frac{1}{2}r_z^2 \ln\frac{r}{r_w} + \frac{1}{2Bi}\left(r_w^2 - r_z^2\right)\right]f(\varphi) + \\ + \frac{\dot{q}(\varphi) r_z}{\lambda}\left(\ln\frac{r}{r_w} + \frac{1}{Bi}\right), \tag{11}$$

where $Bi = \dfrac{\alpha r_w}{\lambda}$, $\dot{q}(\varphi) = \dot{q}_0 + \sum_{n=1}^{\infty}\dot{q}_n \cos n\varphi$.

Once (11) is substituted into (9) and all computational operations conducted, the following differential equation is obtained

$$\frac{d^2 f}{d\varphi^2} - m^2 f = -\frac{2m^2}{r_z^2 - r_w^2}\left(\frac{1}{2}\frac{r_z}{\lambda}\ln^2 u + \frac{r_z}{\lambda}\frac{1}{Bi}\ln u\right)\sum_{n=1}^{\infty}\dot{q}_n n^2 \cos n\varphi, \tag{12}$$

where $u = r_z / r_w$,

$$m^2 = \frac{4Bi\left(u^2 - 1\right)}{-Bi\left(u^2 - 1\right) + 2Biu^2 \ln^2 u + 4\left(u^2 - 1\right)\ln u + 2Bi \ln u}. \tag{13}$$

Boundary conditions for function $f(\varphi)$ result from temperature field symmetry conditions

$$\left.\frac{\partial T}{\partial \varphi}\right|_{\varphi=0} = \left.\frac{\partial T}{\partial \varphi}\right|_{\varphi=\pi}, \tag{14}$$

from which follows that

$$\left.\frac{\partial f}{\partial \varphi}\right|_{\varphi=0} = \left.\frac{\partial f}{\partial \varphi}\right|_{\varphi=\pi} = 0 . \tag{15}$$

The solution of (12) with boundary conditions (15) is function

$$f(\varphi) = \frac{2m^2}{r_z^2 - r_w^2}\left(\frac{1}{2}\frac{r_z}{\lambda}\ln^2 u + \frac{r_z}{\lambda}\frac{1}{Bi}\ln u\right)\sum_{n=1}^{\infty}\frac{\dot{q}_n n^2}{m^2+n^2}\cos n\varphi . \tag{16}$$

From (11), (13) and (16), tube temperature was calculated on an outer and inner tube surface for angles $\varphi = 0$ and $\varphi = \pi$ as a function of Biot number Bi. The obtained approximate values were compared with values, which were calculated by means of analytical formula (Ex. 7.4, (16)). The comparison is presented in Table 8.1.

Calculation results were given in a dimensionless form $\Theta = \dfrac{T\lambda}{\dot{q}_m r_z} = \dfrac{(\zeta - \zeta_{cz})\lambda}{\dot{q}_m r_z}$ as a function of Biot number $Bi = \alpha r_w/\lambda$. From the comparison of results presented in Table 8.1, one can conclude that the accuracy of approximate method is very good. Also, in the functional correction method computational formulas have a simpler form than they do in the analytical method.

Table 8.1. Comparison of temperature $\Theta = T\lambda/\dot{q}_m r_z$ calculated by means of approximate formula (11) and analytical formula ((16), Ex. 7. 4); in top rows values Θ were given, calculated by means of exact analytical formulas, while in bottom rows approximate values were given

φ [rad]	r \ Bi	1.0	1.5	2.0	3.0	5.0	8.0	15.0	20.0
0	r_z	1.0310	0.7893	0.6608	0.5260	0.4127	0.3466	0.2938	0.2784
		1.0304	0.7890	0.6607	0.5261	0.4131	0.3471	0.2945	0.2792
	r_w	0.8172	0.5709	0.4397	0.3018	0.1858	0.1180	0.06373	0.04798
		0.8173	0.5709	0.4397	0.3018	0.1858	0.1179	0.06372	0.04797
π	r_z	0.07604	0.04564	0.03403	0.02452	0.01821	0.01504	0.01267	0.01203
		0.07576	0.04576	0.03410	0.02456	0.01824	0.01508	0.01274	0.01208
	r_w	0.06287	0.03433	0.02331	0.01427	0.008192	0.00507	0.00272	0.00204
		0.06289	0.03434	0.02331	0.01426	0.008178	0.00506	0.00271	0.00204

Literature

1. Burmeister LC (1979) Triangular fin performance by the heat balance integral method. Transactions of the ASME, J. of Heat Transfer 101: 562–564
2. Chung BTF (1978) The heat balance integral in steady-state conduction. Transactions of the ASME, J. of Heat Transfer 100: 651 (Discussion with Sfeirem AA)
3. Sfeir AA (1976) The heat balance integral in steady-state conduction. Transactions of the ASME, J. of Heat Transfer 98: 466–469
4. Sfeir AA, Clumpner JA (1977) Continuous casting of cylindrical ingots. Transactions of the ASME, J. of Heat Transfer 99: 29–34
5. Taler J (1979) Analytical, approximate solution of heat conduction problems. Engineering Treaties 27, Nr 2: 351–357
6. Taler J (1980) Weighed residues method and its application for temperature field calculation in boilers' elements (in Polish). Monograph 14, Pub. Krakow University of Technology, Kraków
7. Taler J (1978) The application of least square method to transient heat conduction problems (in Polish). Heavy Machinery Construction Archive 25, No 3: 483–502
8. Thorpe JF (1965) Axial heat conduction in reactor fuel elements. Nuclear Science and Engineering 23: 329–334
9. Yuen WW, Wessel RA (1981) Application of the integral method to two-dimensional transient heat conduction problems. Transactions of the ASME, J. of Heat Transfer 103: 397–399

9 Two-Dimensional Steady-State Heat Conduction. Graphical Method

In this chapter authors demonstrate that temperature gradient determines the direction of its largest increment and that constant temperature lines (isotherms) are perpendicular to a constant heat flux line. They also discuss graphical methods used for determining isotherms and constant heat flux lines. Furthermore, they graphically determine temperature distribution and heat flow in a chimney cross-section with a square and circular opening.

Exercise 9.1 Temperature Gradient and Surface-Transmitted Heat Flow

Assuming that temperature gradient is known, determine body surface heat flux in the normal direction n to the aforementioned surface and in the arbitrary direction marked by versor **l**. Determine heat flow, which is transfered through a body surface.

Solution

In order to determine heat flow $\dot{Q}$ exchanged between a body surface with an area A and its surroundings, one needs to know what the normal heat flux component on the body surface is

$$\dot{q}_n = \lambda \frac{\partial T}{\partial n}, \tag{1}$$

normal derivative $\partial T/\partial n$, however, can also be formulated as shown below

$$\frac{\partial T}{\partial n} = \mathbf{n} \cdot \nabla T, \tag{2}$$

where **n** is the unit normal-to-surface vector directed to the outside.

Heat flow $\dot{Q}$ is determined from formula

$$\dot{Q} = \int_A \lambda \frac{\partial T}{\partial n} dA = -\int_A \dot{\mathbf{q}} \cdot d\mathbf{A} = -\int_A \dot{\mathbf{q}} \cdot \mathbf{n} dA = \int_A \lambda \mathbf{n} \cdot \nabla T dA , \tag{3}$$

where ∇T is the temperature gradient formulated as

$$\nabla T = \mathbf{i}\frac{\partial T}{\partial x} + \mathbf{j}\frac{\partial T}{\partial y} + \mathbf{k}\frac{\partial T}{\partial z} .$$

Heat flux vector $\dot{\mathbf{q}}$ is expressed by Fourier Law

$$\dot{\mathbf{q}} = -\lambda \nabla T .$$

Heat flux component, which is tangent to a body surface and occurs in the case of non-isothermal body surfaces does not influence surface-to-surroundings-exchanged heat flow $\dot{Q}$. Heat flux in the direction l is formulated as

$$\dot{q}_l = \lambda \frac{\partial T}{\partial l} = \lambda \mathbf{l} \cdot \nabla T ,$$

where $\partial T/\partial l$ is the temperature directional derivative in direction l, while $\mathbf{l}$ a directional unit vector, directed to the outside of the body. If the outer body surface is isothermal (Fig. 9.1), then the heat flux vector is perpendicular to the body surface and $\dot{q} = \dot{q}_n$, i.e. the tangent component of the heat flux vector is not present. In such a case, heat flux in the direction l is formulated as

$$\dot{q}_l = \lambda \nabla T \cdot \mathbf{l} = \dot{q} \cdot l \cdot \cos\varphi = \dot{q}_n \cos\varphi .$$

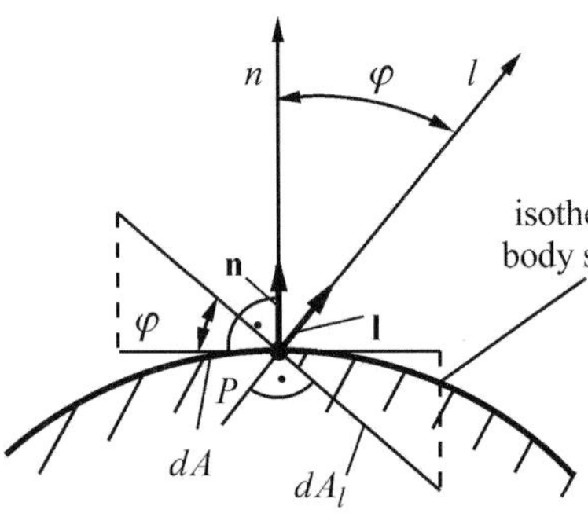

Fig. 9.1. A diagram that shows how heat flux is determined on an isothermal body surface

The element of surface dA_l is

$$dA_l = \frac{dA}{\cos\varphi} .$$

A heat flow that moves through surface A is formulated as

$$\dot{Q} = \int_{A_l} \dot{q}_l dA_l = \int_{A_l} \dot{q}_n \cos\varphi \frac{dA}{\cos\varphi} = \int_{A_l} \dot{q}_n dA ,$$

it is, therefore, identical to (3).

Exercise 9.2 Orthogonality of Constant Temperature Line and Constant Heat Flux

Demonstrate that a direction of a temperature gradient is the direction of the fastest temperature increment and that constant temperature lines (isotherms) are orthogonal to a constant heat flux line.

Solution

The concept of direction will be applied in this exercise and will be briefly characterized. Take a straight line; every parallel vector to this line is called the *direction vector* for this line. Every line has many direction vectors, since a given direction vector multiplied by a scalar other than zero can serve as a direction vector. Let $P_0 = P_0(x_0, y_0, z_0)$ be an arbitrary point and $\mathbf{d} = (a, b, c)$, a given direction vector. Therefore, there is only one line, which passes through point P_0 at direction $\mathbf{d}$. Let $\mathbf{v}_0 = (x_0, y_0, z_0)$ and $\mathbf{v} = (x, y, z)$ be position vectors of point P_0 and P , respectively (Fig. 9.2).

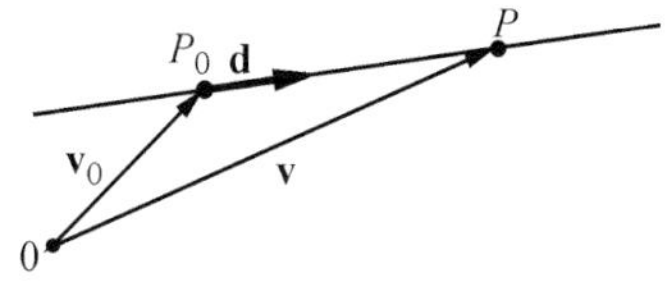

Fig. 9.2. A diagram showing the concept of direction

Therefore $\mathbf{v} = \mathbf{v}_0 + \overrightarrow{P_0P}$ and point P lies on the straight line, if vector $\overrightarrow{P_0P}$ is parallel to $\mathbf{d}$, i.e. when $\overrightarrow{P_0P} = t\mathbf{d}$, where t is a scalar. Therefore, $\mathbf{v}$ is a position vector of the point that lies on the straight line, if

$$\mathbf{v} = \mathbf{v}_0 + t\mathbf{d} . \tag{1}$$

Vector **d** can be a unit vector. In order to demonstrate that a direction of the temperature gradient direction is a direction of the largest temperature increment, while the opposite direction (gradient with minus sign), a direction of the fastest temperature decrease, one assumes that a dislocation from point **x** = (x_1, x_2, x_3) to point **x** = h**d**, is carried out where **d** is the direction, while h a step whose length is assigned. Such dislocation, therefore, is executed from point (x_1, x_2, x_3) to point $(x_1 + \delta x_1, x_2 + \delta x_2, x_3 + \delta x_3)$, where

$$\delta x_i = h d_i \ , \tag{2}$$

while d_i stands for the directional cosines of vector **d**, such that

$$\sum_{i=1}^{3} d_i^2 = 1 \ . \tag{3}$$

One can determine temperature change dT when function $T(x_1 + \delta x_1, x_2 + \delta x_2, x_3 + \delta x_3)$ is written in the form of Taylor series and when the computation is limited to the linear terms

$$\begin{aligned} dT &= T\left(x_1 + \delta x_1, x_2 + \delta x_2, x_3 + \delta x_3\right) - T\left(x_1, x_2, x_3\right) = \\ &= \frac{\partial T}{\partial x_1}\delta x_1 + \frac{\partial T}{\partial x_2}\delta x_2 + \frac{\partial T}{\partial x_3}\delta x_3 , \end{aligned} \tag{4}$$

where derivatives $\partial T/\partial x_i$, *and* =1, 2, 3 are calculated at point **x**. In order to determine the direction cosines d_i, which satisfy (3) and contribute to the largest temperature increment dT, Lagrange multiplier method will be applied.

Lagrange function has the form

$$F\left(d_1, d_2, d_3\right) = dT + \mu\left(\sum_{i=1}^{3} d_i^2 - 1\right) , \tag{5}$$

where μ is the Lagrange multiplier, while dT is expressed by (4). From the necessary condition for existing maximum of the function F, i.e. from the condition of zero setting of the first derivatives from function F:

$$\frac{\partial F}{\partial d_j} = 0, \quad j = 1, 2, 3 , \tag{6}$$

where

$$\frac{\partial F}{\partial d_j} = h\frac{\partial T}{\partial x_j} + 2\mu d_j, \quad j = 1, 2, 3 , \tag{7}$$

one obtains

$$d_j = \frac{-h}{2\mu}\frac{\partial T}{\partial x_j} \ . \tag{8}$$

From (8), follow the following equalities

$$\frac{d_1}{\partial T/\partial x_1} = \frac{d_2}{\partial T/\partial x_2} = \frac{d_3}{\partial T/\partial x_3} \ . \tag{9}$$

One can see, therefore, that $d_i = \partial T/\partial x_i$, $i = 1, 2, 3$ and direction $\mathbf{d}$ is parallel to gradient $\nabla T(\mathbf{x})$ at point $\mathbf{x}$. The fastest local temperature increment for the small step h occurs when direction $\mathbf{d}$ is equal to gradient ∇T. The fastest local temperature decrease occurs, therefore, in the direction $[-\nabla T(\mathbf{x})]$. Hence, we can conclude that the heat flux vector, expressed by Fourier Law

$$\dot{\mathbf{q}} = -\lambda \nabla T \tag{10}$$

is parallel to the direction of the largest temperature decrement at point $\mathbf{x}$.

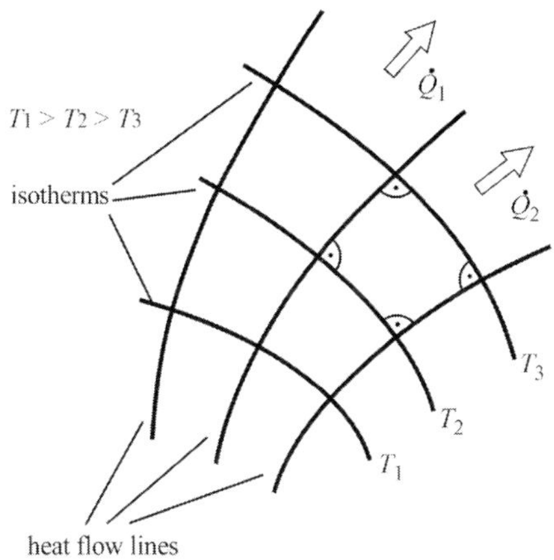

Fig. 9.3. Orthogonality of heat flow lines and isotherms

In order to demonstrate that heat flow lines are orthogonal to isotherms, a temperature difference between points $(\mathbf{x} + d\mathbf{x})$ and $\mathbf{x}$ can be expressed in the following way:

$$dT = \left|\nabla T(\mathbf{x})\right|\left|d\mathbf{x}\right|\cos\varphi \, . \tag{11}$$

If we assume in (11) that $d\mathbf{x} = h\mathbf{d}$, then

$$dT = h\sum_{i=1}^{3}\frac{\partial T}{\partial x_i}d_i$$

Is identical with expression (4). If we assume that $\varphi = 180°$ in (11), then direction $d\mathbf{x}$ is the same as the direction of the largest temperature decrease $\nabla T(\mathbf{x})$. However, for $\varphi = 90°$, the function increment expressed by (11) is equal to zero. Therefore, heat flux vector $\dot{\mathbf{q}}$ at an arbitrary point $\mathbf{x}$ is perpendicular (orthogonal) to an isotherm, which passes through that point (Fig. 9.3).

If body surface is adiabatic (extremely well insulated thermally), then isotherms are perpendicular to such surface.

Exercise 9.3 Determining Heat Flow between Isothermal Surfaces

Illustrate a graphical method for determining heat flow between isothermal surfaces in two-dimensional heat conduction problems.

Solution

Graphical method can be applied to two-dimensional problems, which involve only isothermal and adiabatic surfaces. In order to graphically illustrate two-dimensional heat conduction, two curve categories are drawn: the first one consists of constant temperature lines (isotherms), the second, lines parallel to a heat flux vector (heat flow lines). If constant temperature difference ΔT exists between isotherms, then a distance between two adjacent isotherms indicates the heat flux value. If a distance between isotherms at a set value ΔT is small, then temperature gradient is large. In steady-state heat conduction problems, the transferred heat flow $\dot{Q}$ is constant. What this means is that if the cross-section the heat flow $\dot{Q}$ passes through becomes smaller, then heat flux and, consequently, temperature gradient becomes larger. If the cross-section is larger, then heat flux and temperature gradient are smaller, while the distance between the isotherms becomes bigger. Since heat transfer occurs in the presence of a temperature gradient, heat does not flow along the isotherms. Therefore, the isotherm-orthogonal heat flow line can be regarded as an insulated boundary through which the heat conduction does not take place. Two adjacent heat flow lines form, therefore, a path (trajectory), which the specific heat flow passes through.

Three isotherms are presented in Fig. 9.4; they differ from each other with respect to ΔT and two heat flow lines separated in between by heat flow $\dot{Q}_i$.

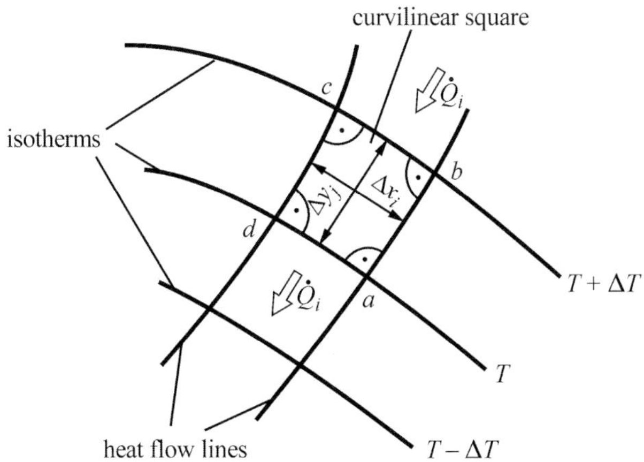

Fig. 9.4. An illustration of a graphical method used when determining heat flow in two-dimensional problems

If one assumes that the thickness of the analyzed element is L in the direction perpendicular to the surface and heat flux is approximated by formula

$$\dot{q}_i = -\lambda \frac{T-(T+\Delta T)}{\Delta y_j} = \frac{\lambda \Delta T}{\Delta y_j}, \tag{1}$$

where

$$\Delta y_j = \frac{ab+cd}{2}, \tag{2}$$

then heat flow $\dot{Q}_i$ is formulated as

$$\dot{Q}_i = A_i \dot{q}_i = L \Delta x_i \dot{q}_i = \lambda L \frac{\Delta x_i}{\Delta y_j} \Delta T, \tag{3}$$

where

$$\Delta x_i = \frac{ad+bc}{2}. \tag{4}$$

If isotherms and heat flow lines, which intercept each other at right angles, are drawn in a way that curvilinear tetragons *abcd* become curvilinear squares, i.e.

$$\Delta x_i \approx \Delta y_j, \tag{5}$$

then (3) assumes the following form:

$$\dot{Q}_i = \lambda L \Delta T. \tag{6}$$

If heat passes between two isothermal surfaces with temperature T_g and T_z, when $T_g > T_z$ and area is divided into M paths traversed by the heat flow itself and temperature difference between isotherms is calculated from

$$\Delta T = \frac{T_g - T_z}{N}, \quad \text{where } N \text{ is the integer number,} \tag{7}$$

then the total heat flow measures

$$\dot{Q} = \sum_{i=1}^{M} \dot{Q}_i = M\dot{Q}_i , \tag{8}$$

where M stands for the number of paths (it does not have to be an integer number). By substituting (6) into (8) and accounting for (7), one obtains

$$\dot{Q} = \frac{ML}{N} \lambda \left(T_g - T_z\right) . \tag{9}$$

Equality (9) is applied when it is necessary to define shape coefficient S

$$\dot{Q} = S\lambda \left(T_g - T_z\right) , \tag{10}$$

where

$$S = \frac{ML}{N} . \tag{11}$$

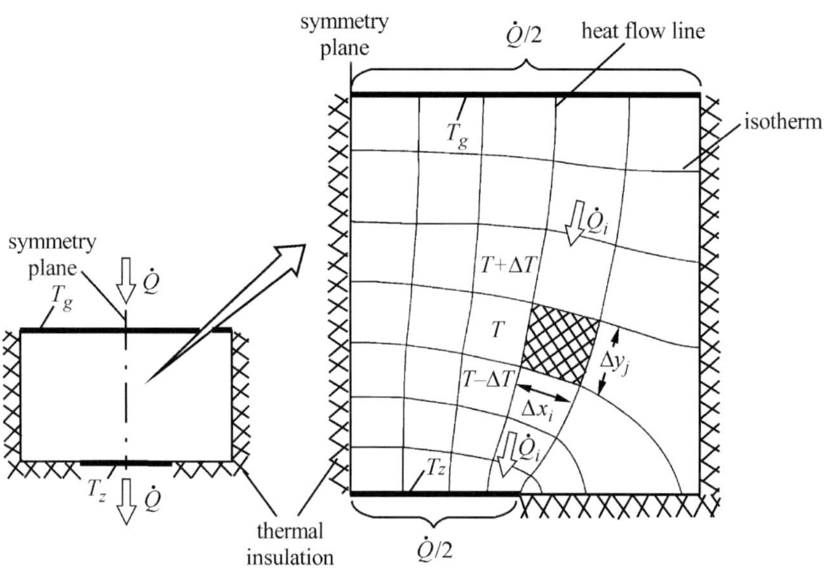

Fig. 9.5. An illustration of a graphical method for determining heat flow $\dot{Q}$ between isothermal surfaces with temperature T_g and $T_z < T_g$

Shape coefficients S are tabularized for various isothermal surface configurations that one can come across in engineering. In the graphical method, heat flow is defined by means of (10).

When drawing isotherms and heat flow lines, one should note that symmetry lines, which come from shape symmetry of an analyzed area and from symmetry of boundary conditions, are adiabatic lines. As a result, the analysis can be limited to a repeatable part of the total area, for e.g. in the case of an area shown in Fig. 9.5, one analyzes only one of its halves [1].

If boundary surfaces are thermally insulated, isotherms then are perpendicular to such surfaces. Examples of the practical application of the graphical method can be found in papers [1–5].

Exercise 9.4 Determining Heat Loss Through a Chimney Wall; Combustion Channel (Chimney) with Square Cross-Section

A chimney made of stainless steel with a square cross-section and a side equal to b = 14 cm is insulated on an outer surface by a glass wool whose heat conduction coefficient is λ = 0.05 W/(m·K). An outer dimension of the square chimney is a = 38 cm. Combustion gases from a boiler fired by a natural gas flow inside the chimney. Due to the condensation of water vapour on the inner surface of the chimney, one can assume that the surface is isothermal, with the temperature measures T_g = 40°C. Outer surface temperature is T_z = 0°C. Calculate chimney-to-surroundings heat loss per 1m of chimney length. Compare calculation results with the computed heat flow value by means of the shape coefficient S.

Solution

Chimney cross-section is shown in Fig. 9.6a. A division of 1/8 of the chimney cross-section into curvilinear squares is shown in Fig. 9.6b. Since the number of paths, traversed by the heat flow, amounts to M = 5, while the number of sections ΔT to N = 6, heat flow is ((9), Ex. 9.3)

$$\frac{\dot{Q}}{L} = 8\frac{M}{N}\lambda\left(T_g - T_z\right) = 8\cdot\frac{5}{6}\cdot 0.05\cdot\left(40-0\right) = 13.33 \text{ W/m}\,. \tag{1}$$

Number 8 in (1) allows for the number of repeatable elements in the chimney cross-section.

Multiplier 8 appears because there are 8 identic elements, which the chimney cross-section is divided to. If $M = 3$ and $N = 6$, then from (1), one obtains

$$\frac{\dot{Q}}{L} = 8 \cdot \frac{3}{6} \cdot 1 \cdot (100 - 30) = 280 \text{ W/m}.$$

The same heat flow $\dot{Q}/L$ can be calculated with the use of shape coefficient S

$$\frac{\dot{Q}}{L} = \frac{S}{L} \lambda \left(T_g - T_z \right), \tag{2}$$

where S is formulated as (Appendix D)

$$S = \frac{2\pi L}{\ln\left(1.08 \frac{a}{d}\right)}. \tag{3}$$

By substituting (3), one obtains

$$S = \frac{2\pi L}{\ln\left(1.08 \frac{0.52}{0.13}\right)} = 4.294L.$$

If S is known, one can calculate $\dot{Q}/L$ from (2)

$$\frac{\dot{Q}}{L} = 4.294 \cdot 1 \cdot (100 - 30) = 300.58 \text{ W/m}.$$

Heat loss estimation error with the graphical method is

$$\Delta \dot{Q} = \frac{300.58 - 280}{300.58} \cdot 100\% = 6.85\%.$$

Literature

1. Bejan A (1993) Heat Transfer. Wiley, New York
2. Incropera FP, DeWitt DP (1996) Fundamentals of Heat and Mass Transfer. Wiley, New York
3. Jakob M (1949) Heat Transfer 1. Wiley, New York
4. Welty JR (1974) Engineering Heat Transfer. New York
5. White FM (1991) Heat and Mass Transfer. Reading. Addison-Wesley

10 Two-Dimensional Steady-State Problems. The Shape Coefficient

The subject of discussion in this chapter is the determination of heat flow between isothermal surfaces by means of a shape coefficient. The concept of shape coefficient initially appeared in 1913 [5]. Shape coefficients for different geometric systems of isothermal surfaces, which one can come up against in practice, are compiled in Appendix D, at the back of this book. The appendix contains 38 different configurations of isothermal surfaces.

Exercise 10.1 Buried Pipe-to-Ground Surface Heat Flow

Discuss the method for determining heat flow $\dot{Q}$ between two isothermal surfaces using shape coefficient S or thermal resistance R. Calculate heat flow per 1 m of pipe length transmitted from a non-insulated pipe, with an outer surface diameter $d_z = 325$ mm and temperature $T_1 = 110°\text{C}$, to ground surface at temperature $T_2 = 3°\text{C}$. The pipe is buried in the ground (Fig. 10.1) at a depth of $h = 130$ cm. Assume that thermal conductivity of the ground is $\lambda = 1.2$ W/(m · K).

Solution

Heat flow $\dot{Q}$, between two isothermal surfaces at temperatures T_1 and T_2 that are separated by a medium with a thermal conductivity λ, is expressed by (10) in Ex. 9.3:

$$\dot{Q} = S\lambda\left(T_1 - T_2\right) . \qquad (1)$$

Shape coefficients are usually tabularized [1–4, 6, 7]. Shape coefficient formulas for bodies with various shapes are compiled in Appendix D. In some scientific papers, e.g. in [4], heat flow $\dot{Q}$ is formulated as

$$\dot{Q} = \frac{T_1 - T_2}{R} \,, \tag{2}$$

where R is the heat conduction resistance. From the comparison of (1) and (2), one can see that

$$R = \frac{1}{\lambda S} \,. \tag{3}$$

For the problem formulated in this exercise, shape coefficient S has the form (Fig. 10.1)

$$S = \frac{2\pi L}{\ln\left[\frac{h}{r} + \sqrt{\left(\frac{h}{r}\right)^2 - 1}\right]} \,, \qquad h > r \,, \tag{4}$$

where $r = d_z/2$ stands for an outer surface radius of the pipe, while L for the pipe length.

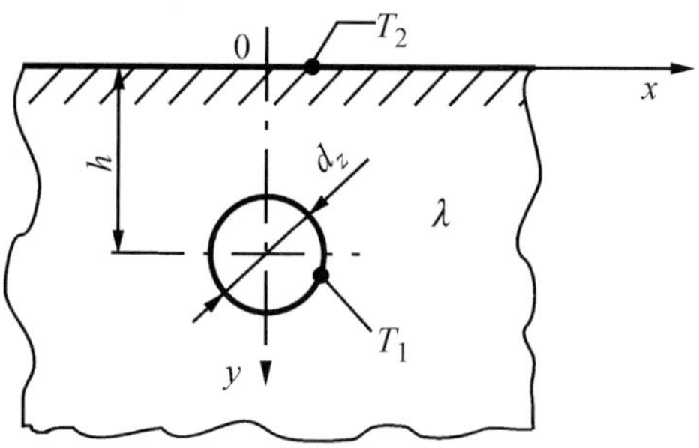

Fig. 10.1. A diagram of a pipe buried underground at depth h

Ground temperature can be determined from formula

$$\frac{T - T_2}{T_1 - T_2} = \frac{1}{2\ln\left[\frac{h}{r} + \sqrt{\left(\frac{h}{r}\right)^2 - 1}\right]} \ln \frac{x^2 + \left(y + \sqrt{h^2 - r^2}\right)^2}{x^2 - \left(y - \sqrt{h^2 - r^2}\right)^2} \,. \tag{5}$$

Equations (4) and (5) were derived under the assumption that ground temperature equals temperature T_2 on the entire surface ($y = 0$) from $x = 0$ to an infinite distance from the pipe axis, i.e. $T\,(0,0) = T\,(\infty, 0) = T_2$. By substituting (4) into (1), one can determine heat flow within 1m long pipe length:

$$\frac{\dot{Q}}{L}=\frac{S\lambda}{L}\left(T_1-T_2\right)=\frac{2\pi\lambda\left(T_1-T_2\right)}{\ln\left[\frac{h}{r}+\sqrt{\left(\frac{h}{r}\right)^2-1}\right]} . \tag{6}$$

After substitution of the numerical values, one obtains

$$\frac{\dot{Q}}{L}=\frac{2\pi\cdot 1.2\cdot\left(110-3\right)}{\ln\left[\frac{1.3}{0.1625}+\sqrt{\left(\frac{1.3}{0.1625}\right)^2-1}\right]}=291.4\ \text{W/m} . \tag{7}$$

In order to decrease heat loss, i.e. reduce quotient $\dot{Q}/L$, the pipe should be insulated to lower temperature T_1.

Exercise 10.2 Floor Heating

Pipes with diameter d_z = 18 mm were used for floor heating. They were spaced at s = 120 mm from each other at a depth of h = 30 mm. Outer surface temperature of a pipe is T_1 = 40°C, while floor temperature T_2 = 25°C. Assume that thermal conductivity of the floor is λ= 1.0 W/(m · K). Assuming that there is no additional thermal insulation from the ground, i.e. the floor is regarded as a semi-infinite body, calculate the heat flow transferred from an individual pipe to the floor surface within the length of 1 m.

Solution

Please, refer to Fig. 10.2.

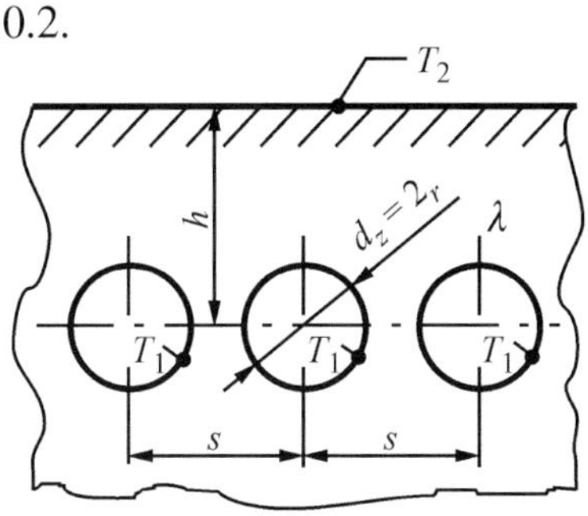

Fig. 10.2. Floor heating

Heat flow transferred by an individual pipe, with respect to its length, is given by

$$\frac{\dot{Q}}{L}=\frac{S\lambda}{L}\left(T_1-T_2\right), \tag{1}$$

where the shape coefficient S is formulated as (Appendix D, item 7)

$$S=\frac{2\pi L}{\ln\left(\dfrac{s}{\pi r}\sinh\dfrac{2\pi h}{s}\right)}. \tag{2}$$

By substituting (2) into (1), one obtains

$$\frac{\dot{Q}}{L}=\frac{2\pi\lambda\left(T_1-T_2\right)}{\ln\left(\dfrac{s}{\pi r}\sinh\dfrac{2\pi h}{s}\right)}. \tag{3}$$

Substituting the numerical values into (3) gives

$$\frac{\dot{Q}}{L}=\frac{2\pi\cdot 1.0\cdot\left(40-25\right)}{\ln\left(\dfrac{0.12}{\pi\cdot 0.009}\sinh\dfrac{2\pi\cdot 0.03}{0.12}\right)}=41.35\ \text{W/m}.$$

In order to increase heat flow $\dot{Q}/L$, one should insulate pipes from below in order to decrease the heat flow transfer to the ground.

Exercise 10.3 Temperature of a Radioactive Waste Container Buried Underground

Radioactive waste is stored in a spherical container whose outer diameter is 1.6 m; it is located 11 m below the ground level. Surface temperature of the ground is T_2 = 10°C. Assuming that waste generated heat flow is 5 kW, and the ground thermal conductivity equals λ = 1.8 W/(m · K), calculate outer surface temperature of the container.

Solution

Outer surface temperature of the container (Fig. 10.3) is calculated using formula

$$\dot{Q}=S\lambda\left(T_1-T_2\right), \tag{1}$$

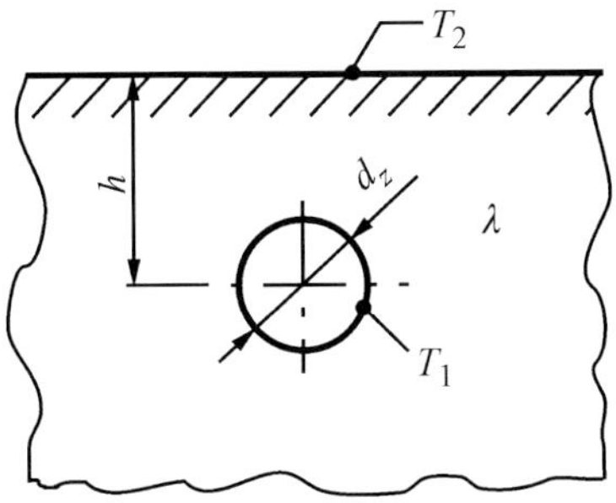

Fig. 10.3. A diagram of a spherical container stored underground

hence,

$$T_1 = T_2 + \frac{\dot{Q}}{S\lambda} . \qquad (2)$$

Shape coefficient (Appendix D, item 18) equals

$$S = \frac{4\pi r}{1 - \dfrac{r}{2h}} = \frac{4\pi \cdot 0.8}{1 - \dfrac{0.8}{2 \cdot 11}} = 10.43 .$$

By substituting S and other numerical values into (2), one obtains

$$T_1 = 10 + \frac{5000}{10.43 \cdot 1.8} = 276.3\,°\mathrm{C} .$$

Literature

1. Andrews RV (1955) Solving conductive heat transfer problems with electrical-analogue shape factors. Chem. Eng. Progres. 51. No. 2: 67–71
2. Hahne E, Grigull U (1975) Formfaktor und Formwiderstand der stationären mehrdimensionalen Wärmeleitung. Int. J. Heat Mass Transfer 18: 751–767
3. Hassani AV, Hollands KG (1990) Conduction shape factor for a region of uniform thickness surrounding a three-dimensional body of arbitrary shape. J. Heat Transfer 112: 492–495
4. Кутателадзе СС (1990) Теплопередача и гидродинамическое сопротивление. Справочное пособие. Энергоатомиздат, Москва
5. Langmuir I, Adams EQ, Meikle GS (1913) Flow of heat through furnace walls: the shape factor. Transact. Am. Electrochem. Soc. 24: 53–81
6. Sunderland JE, Johnson KR (1964) Shape factors for heat conduction through bodies with isothermal or convective boundary conditions. Trans. ASHRAE. 70: 237–241
7. (1984) VDI – Wärmeatlas 4. VDI-Verlag, Düsseldorf

11 Solving Steady-State Heat Conduction Problems by Means of Numerical Methods

This chapter is devoted to numerical methods, which are used to determine steady-state temperature fields. It contains detailed description of the following numerical methods: finite-difference method, finite-volume method (control volume), finite element method (FEM) and pseudo-transient method for solving stationary problems, based on the method of lines. Linear and non-linear problems, both simple and inverse, are solved here. Specific computational programs are developed for determining steady-state temperature fields, while Gauss elimination method, Gauss-Seidel iterative method or over-relaxation method are applied to integrate an algebraic equation system. Ordinary differential equation system in the pseudo-transient method is solved using Rung-Kutta method of 4th order. Finite element method, based on Galerkin method, is discussed in great detail, as well as the two methods for creating global equation system in FEM. Basic matrixes and vectors, which occur in FEM for one-dimensional and two-dimensional triangular and rectangular elements, are also developed. Furthermore, authors present their own solutions to FEM problems. The obtained results are compared with analytical solutions or the solutions acquired by means of finite volume method. The application of the ANSYS program is presented in Ex. 11.20, 11.21 and 11.22. Hexagonal fin efficiency is determined in Ex. 11.21, while the effect the shape of pins on the heating surface of the cast iron heating boiler has on the temperature distribution and pin-transferred heat flow is analyzed in Ex. 11.22.

Exercise 11.1 Description of the Control Volume Method

Describe how transient heat conduction problems are solved by means of the control volume method; assume that thermal conductivity can be temperature dependent. Write heat balance equation for control volume in the Cartesian and cylindrical coordinate system for two-dimensional problems.

Solution

Control volume method, also called elementary balance method or finite volume method, is a universal and effective method for solving heat conduction problems. If the thickness of an analyzed area is d and thermal properties c, ρ, λ and power densityof internal heat sources $\dot{q}_v$ are temperature dependent, then heat conduction equation can be written in the form

$$c(T)\rho(T)\frac{\partial T}{\partial t} = -\mathrm{div}\,\dot{\mathbf{q}} + \dot{q}_v . \tag{1}$$

The area is divided into control volumes, which have the following dimensions: Δx, Δy and d in the Cartesian coordinate system (Fig. 11.1) or Δr, $\Delta\varphi$ and d (Fig. 11.2) in the cylindrical coordinate system. Once (1) is integrated over the control volume, the following equation for a single cell (control volume) is obtained:

$$\int_{CV} c(T)\rho(T)\frac{\partial T}{\partial t}dV = -\int_{CV} \mathrm{div}\,\dot{\mathbf{q}}\,dV + \int_{CV} \dot{q}_v\,dV , \tag{2}$$

where *CV* stands for the *control volume.*

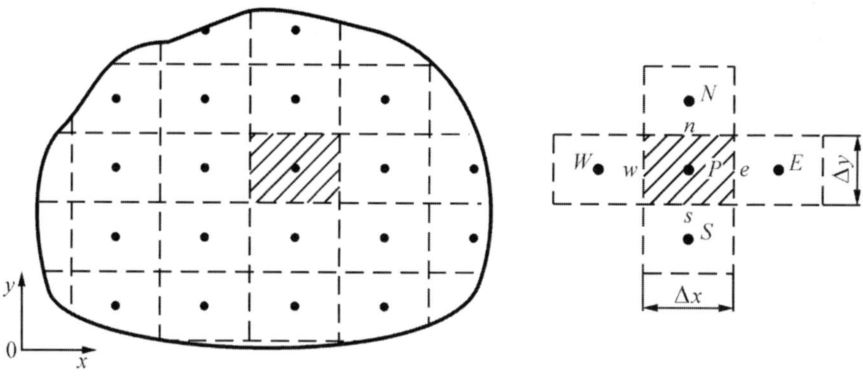

Fig. 11.1. A diagram of an area divided into finite volumes in the Cartesian coordinate system

If we apply Green-Gauss-Ostrogradski theorem to the first term on the right-hand-side of (2), the equation will assume the form

$$\int_{CV} c(T)\rho(T)\frac{\partial T}{\partial t}dV = -\int_{S} \mathbf{n}\cdot\dot{\mathbf{q}}dS + \int_{CV} \dot{q}_v\,dV , \tag{3}$$

where S is the control volume surface, while $\mathbf{n}$ a normal unit surface vector directed to the outside of the control volume. From expression

$$\mathbf{n} \cdot \dot{\mathbf{q}} = 1 \cdot |\dot{\mathbf{q}}| \cos(\mathbf{n}, \dot{\mathbf{q}}) = \dot{q}_n \tag{4}$$

it is evident that when the heat flows up to the control volume, the heat flux vector $\dot{\mathbf{q}}$ is directed to the inside of the control volume and the angle between vector $\mathbf{n}$ and $\dot{\mathbf{q}}_\mathbf{n}$ is 180°. The scalar product (4) is then negative, while the surface integral in (3) is positive. If ΔV denotes the volume of a control cell, then individual terms in (3) can be approximated in the following way:

$$\int_{CV} c(T)\rho(T)\frac{\partial T}{\partial t}dV \approx \Delta V c(T_P)\rho(T_P)\frac{dT_P}{dt}, \tag{5}$$

$$-\int_S \mathbf{n} \cdot \dot{\mathbf{q}}\, dS = \sum_{i=1}^{4} \dot{Q}_i, \tag{6}$$

$$\int_{CV} \dot{q}_v dV = \Delta V\, \dot{q}_v(T_P), \tag{7}$$

where $\dot{Q}_i$ is the heat flow that flows in from the neighbouring cell. Substituting equalities (5)–(7) in (3), one obtains the following heat balance equation

$$\Delta V c(T_P)\rho(T_P)\frac{dT_P}{dt} = \sum_{i=1}^{4} \dot{Q}_i + \Delta V \dot{q}_v(T_P), \tag{8}$$

which will be written in a greater detail in the Cartesian and cylindrical coordinate system.

a) *Heat balance equation- Cartesian coordinates*

A division of an area into control volumes and a control volume are shown in Fig.11.1. The volume of a single cell is $\Delta V = (\Delta x)(\Delta y)d$. Heat flows, which inflow from nodes W, N, E and S to node P are expressed by the following formulas:

$$\dot{Q}_{W-P} = (\Delta y)d\,\dot{q}_{W-P} = (\Delta y)d\frac{\lambda(T_W) + \lambda(T_P)}{2} \cdot \frac{T_W - T_P}{\Delta x}, \tag{9}$$

$$\dot{Q}_{N-P} = (\Delta x)d\,\dot{q}_{N-P} = (\Delta x)d\frac{\lambda(T_N) + \lambda(T_P)}{2} \cdot \frac{T_N - T_P}{\Delta y}, \tag{10}$$

$$\dot{Q}_{E-P} = (\Delta y) d\, \dot{q}_{E-P} = (\Delta y) d \frac{\lambda(T_E) + \lambda(T_P)}{2} \cdot \frac{T_E - T_P}{\Delta x}, \tag{11}$$

$$\dot{Q}_{S-P} = (\Delta x) d\, \dot{q}_{S-P} = (\Delta x) d \frac{\lambda(T_S) + \lambda(T_P)}{2} \cdot \frac{T_S - T_P}{\Delta y}. \tag{12}$$

Substituting the expressions (9)–(12) in (8), one obtains

$$\begin{aligned}(\Delta x)(\Delta y) d\, c(T_P)\rho(T_P)\frac{dT_P}{dt} &= (\Delta y) d \frac{\lambda(T_W) + \lambda(T_P)}{2} \cdot \frac{T_W - T_P}{\Delta x} + \\ &+ (\Delta x) d \frac{\lambda(T_N) + \lambda(T_P)}{2} \cdot \frac{T_N - T_P}{\Delta y} + (\Delta y) d \frac{\lambda(T_E) + \lambda(T_P)}{2} \cdot \\ &\cdot \frac{T_E - T_P}{\Delta x} + (\Delta x) d \frac{\lambda(T_S) + \lambda(T_P)}{2} \cdot \frac{T_S - T_P}{\Delta y} + (\Delta x)(\Delta y) d \cdot \dot{q}_V(T_P).\end{aligned} \tag{13}$$

Assuming constant properties

$$\lambda_P = \lambda(T_P), \quad c_P = c(T_P), \quad \ldots, \quad a_P = \frac{\lambda_P}{c_P \cdot \rho_P}$$

Equation (13) is simplified to the following equation

$$\begin{aligned}\frac{dT_P}{dt} = a_P \Bigg[&\frac{\lambda_W + \lambda_P}{2\lambda_P} \cdot \frac{T_W - T_P}{(\Delta x)^2} + \frac{\lambda_N + \lambda_P}{2\lambda_P} \cdot \frac{T_N - T_P}{(\Delta y)^2} + \\ &+ \frac{\lambda_E + \lambda_P}{2\lambda_P} \cdot \frac{T_E - T_P}{(\Delta x)^2} + \frac{\lambda_S + \lambda_P}{2\lambda_P} \cdot \frac{T_S - T_P}{(\Delta y)^2} \Bigg] + \frac{\dot{q}_{v,P}}{c_P \cdot \rho_P}.\end{aligned} \tag{14}$$

When steady-state problem is analyzed, $dT_p/dt = 0$. For a uniform grid $\Delta x = \Delta y$ and for constant and temperature independent thermal properties and heat source power, (14) is simplified to a form

$$T_W + T_N + T_E + T_S - 4T_P + \frac{\dot{q}_v (\Delta x)^2}{\lambda} = 0. \tag{15}$$

b) *Heat balance equation-cylindrical coordinates*

Heat balance (8) can be transformed into a form similar to (14) after calculating of the following quantities (Fig. 11.2):

$$\Delta V = \pi \left(r_n^2 - r_s^2 \right) \frac{\Delta\varphi}{2\pi} d = \frac{\left(r_n^2 - r_s^2 \right)\Delta\varphi}{2} d, \tag{16}$$

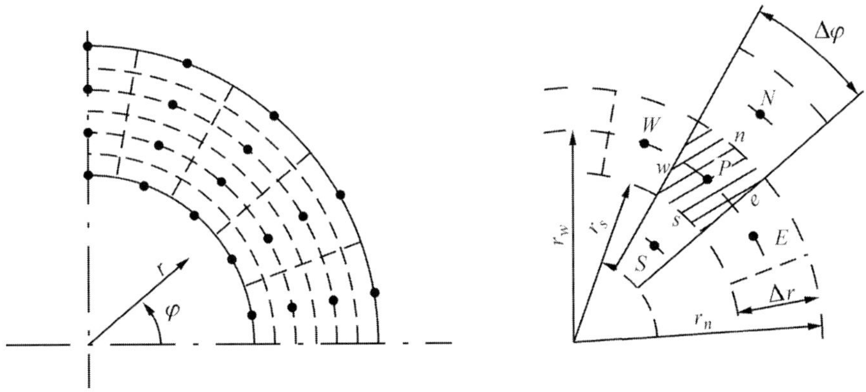

Fig. 11.2. A division of an area into finite volumes in the cylindrical coordinate system

$$\dot{Q}_{W-P} = (\Delta r)\, d\, \frac{\lambda_W + \lambda_P}{2} \cdot \frac{T_W - T_P}{r_w (\Delta\varphi)}, \tag{17}$$

$$\dot{Q}_{N-P} = r_n (\Delta\varphi)\, d\, \frac{\lambda_N + \lambda_P}{2} \cdot \frac{T_N - T_P}{\Delta r}, \tag{18}$$

$$\dot{Q}_{E-P} = (\Delta r)\, d\, \frac{\lambda_E + \lambda_P}{2} \cdot \frac{T_E - T_P}{r_w (\Delta\varphi)}, \tag{19}$$

$$\dot{Q}_{S-P} = r_s (\Delta\varphi)\, d\, \frac{\lambda_S + \lambda_P}{2} \cdot \frac{T_S - T_P}{\Delta r}, \tag{20}$$

From (8), one obtains

$$\begin{aligned} \frac{dT_P}{dt} = \frac{2a_P}{\left(r_n^2 - r_s^2\right) \cdot \Delta\varphi} \Bigg[& \frac{\lambda_W + \lambda_P}{2\lambda_P} \cdot \frac{\Delta r}{r_w (\Delta\varphi)} \left(T_W - T_P\right) + \\ & + \frac{\lambda_N + \lambda_P}{2\lambda_P} \cdot \frac{r_n (\Delta\varphi)}{\Delta r} \left(T_N - T_P\right) + \frac{\lambda_E + \lambda_P}{2\lambda_P} \cdot \frac{\Delta r}{r_w (\Delta\varphi)} \left(T_E - T_P\right) + \\ & + \frac{\lambda_S + \lambda_P}{2\lambda_P} \cdot \frac{r_s \cdot \Delta\varphi}{\Delta r} \left(T_S - T_P\right) \Bigg] + \frac{\dot{q}_{v,P}}{c_P \cdot \rho_P}. \end{aligned} \tag{21}$$

In the case of steady-state problems, one should assume that $dT_p/dt = 0$. Heat balance (14) in the Cartesian coordinates or (21) in cylindrical coordinates is written for all nodes, including the nodes in the control volumes that abut to a boundary. Appropriate boundary conditions should be allowed for in the equations for boundary-adjacent control volumes. In order to determine node temperature in the cases of transient problems, one should solve the ordinary differential equation system by means of the Runge-Kutta method, for instance. In steady-state problems, one can obtain an algebraic equation system, which can be solved by direct methods, e.g. Gauss elimination method, or by iterative methods like Gauss-Seidel method.

Exercise 11.2 Determining Temperature Distribution in a Square Cross-Section of a Long Rod by Means of the Finite Volume Method

Determine temperature distribution in a square cross-section of an infinitely long rod with prescribed temperature on lateral surfaces (Fig. 11.3). In order to solve the problem, apply the control volume method, while the obtained algebraic equation system solve by means of the iterative Gauss-Seidel method. Write a computational program for the determination of temperature in nodes 1–4.

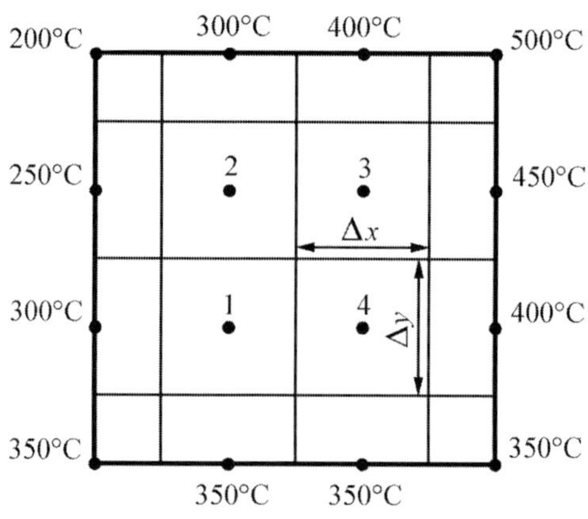

Fig. 11.3. Square cross-section of an infinitely long rod with prescribed surface temperature

Solution

Equation (15) from Ex. 11.1 will be used to solve the above stated problem; in this case, it has the form

$$T_W + T_N + T_E + T_S - 4T_P = 0 \;. \tag{1}$$

From the equation above, we have, respectively, for nodes 1 to 4:

- node 1

$$\begin{aligned} &300 + T_2 + T_4 + 350 - 4T_1 = 0\,, \\ &4T_1 - T_2 - T_4 = 650; \end{aligned} \tag{2}$$

- node 2

$$\begin{aligned} &250 + 300 + T_3 + T_1 - 4T_2 = 0\;, \\ &-T_1 + 4T_2 - T_3 = 550; \end{aligned} \tag{3}$$

- node 3

$$\begin{aligned} &T_2 + 400 + 450 + T_4 - 4T_3 = 0\,, \\ &-T_2 + 4T_3 - T_4 = 850; \end{aligned} \tag{4}$$

- node 4

$$\begin{aligned} &T_1 + T_3 + 400 + 350 - 4T_4 = 0\;, \\ &-T_1 - T_3 + 4T_4 = 750. \end{aligned} \tag{5}$$

According to the Gauss-Seidel method, (2)–(5) are transformed in a way that the temperature in the first node is determined from the first equation, in the second node from the second equation, in the third node from the third equation and in the fourth node from the fourth equation:

$$T_1 = \frac{1}{4}\left(650 + T_2 + T_4\right)\,, \tag{6}$$

$$T_2 = \frac{1}{4}\left(550 + T_1 + T_3\right)\,, \tag{7}$$

$$T_3 = \frac{1}{4}\left(850 + T_2 + T_4\right)\,, \tag{8}$$

$$T_4 = \frac{1}{4}\left(750 + T_1 + T_3\right)\,. \tag{9}$$

Next, initial approximation is assumed, e.g.

$T_1^{(0)} = 250°\text{C}$, $T_2^{(0)} = 250°\text{C}$, $T_3^{(0)} = 250°\text{C}$, $T_4^{(0)} = 250°\text{C}$

and, in turn, individual temperatures are determined from (6)–(9). Temperature determined in this way is substituted into the subsequent equation, i.e.

$$T_1^{(1)} = \frac{1}{4}\left(650 + T_2^{(0)} + T_4^{(0)}\right) = \frac{1}{4}(650 + 250 + 250) = 287.5\,^\circ\mathrm{C},$$

$$T_2^{(1)} = \frac{1}{4}\left(550 + T_1^{(1)} + T_3^{(0)}\right) = \frac{1}{4}(550 + 287.5 + 250) = 271.875\,^\circ\mathrm{C},$$

$$T_3^{(1)} = \frac{1}{4}\left(850 + T_2^{(1)} + T_4^{(0)}\right) = \frac{1}{4}(850 + 271.875 + 250) = 342.969\,^\circ\mathrm{C},$$

$$T_4^{(1)} = \frac{1}{4}\left(750 + T_1^{(1)} + T_3^{(1)}\right) = \frac{1}{4}(750 + 287.5 + 342.969) = 345.117\,^\circ\mathrm{C}.$$

The second approximation is done in a similar way

$$T_1^{(2)} = \frac{1}{4}\left(650 + T_2^{(1)} + T_4^{(1)}\right) = \frac{1}{4}(650 + 271.875 + 345.117) = 316.748\,^\circ\mathrm{C},$$

$$T_2^{(2)} = \frac{1}{4}\left(550 + T_1^{(2)} + T_3^{(1)}\right) = \frac{1}{4}(550 + 316.748 + 342.969) = 302.429\,^\circ\mathrm{C},$$

$$T_3^{(2)} = \frac{1}{4}\left(850 + T_2^{(2)} + T_4^{(1)}\right) = \frac{1}{4}(850 + 302.429 + 345.117) = 374.387\,^\circ\mathrm{C},$$

$$T_4^{(2)} = \frac{1}{4}\left(750 + T_1^{(2)} + T_3^{(2)}\right) = \frac{1}{4}(750 + 316.748 + 374.387) = 360.284\,^\circ\mathrm{C}.$$

From the third approximation, one obtains

$$T_1^{(3)} = \frac{1}{4}\left(650 + T_2^{(2)} + T_4^{(2)}\right) = \frac{1}{4}(650 + 302.429 + 360.284) = 328.178\,^\circ\mathrm{C},$$

$$T_2^{(3)} = \frac{1}{4}\left(550 + T_1^{(3)} + T_3^{(2)}\right) = \frac{1}{4}(550 + 328.178 + 374.387) = 313.141\,^\circ\mathrm{C},$$

$$T_3^{(3)} = \frac{1}{4}\left(850 + T_2^{(3)} + T_4^{(2)}\right) = \frac{1}{4}(850 + 313.141 + 360.284) = 380.856\,^\circ\mathrm{C},$$

$$T_4^{(3)} = \frac{1}{4}\left(750 + T_1^{(3)} + T_3^{(3)}\right) = \frac{1}{4}(750 + 328.178 + 380.856) = 364.759\,^\circ\mathrm{C}.$$

After fourth iteration, one has

$$T_1^{(4)} = \frac{1}{4}\left(650 + T_2^{(3)} + T_4^{(3)}\right) = \frac{1}{4}(650 + 313.141 + 364.759) = 331.975\,^\circ\mathrm{C},$$

$$T_2^{(4)} = \frac{1}{4}\left(550 + T_1^{(4)} + T_3^{(3)}\right) = \frac{1}{4}(550 + 331.975 + 380.856) = 315.708^\circ\text{C},$$

$$T_3^{(4)} = \frac{1}{4}\left(850 + T_2^{(4)} + T_4^{(3)}\right) = \frac{1}{4}(850 + 315.708 + 364.759) = 382.617^\circ\text{C},$$

$$T_4^{(4)} = \frac{1}{4}\left(750 + T_1^{(4)} + T_3^{(4)}\right) = \frac{1}{4}(750 + 331.975 + 382.617) = 366.148^\circ\text{C}.$$

Following that, iterative calculations are conducted in a way that satisfies the inequality below:

$$\left|T_i^{(k+1)} - T_i^{(k)}\right| < \varepsilon; \qquad i = 1, 2, 3, 4; \qquad k = 0, 1\ldots \tag{10}$$

For ε = 0.00001°C after k = 14 iterations, the following temperature values are obtained:

T_1 = 333.333°C, T_2 = 316.667°C,

T_3 = 383.333°C,

T_4 = 366.667°C.

Calculations were carried out by means of the FORTRAN program. In spite of the fact that a large number of iterations was done, calculation time is very short, since the formulas are very simple in form.

Program for temperature determination in nodes 1–4

```
c      Calculating two-D temperature field in a flat rod
c      by means of Gauss-Seidel method
       program seidel
       dimension t(50),tt(50)
       logical inaccurate
       open(unit=1,file='seidel.in')
       open(unit=2,file='seidel.out')
       read(1,*)n,toler,niter,t_pocz
       write(2,'(a)') "CALCULATING TWO-DIMENSIONAL TEMPERATURE
      &FIELD IN A FLAT & ROD "
       write(2,'(/a)') "DATA ENTERED"
       write(2,'(a,i10)')     "equation number  n=",n
       write(2,'(a,e10.5,a)')"calcul. toler=",toler,"[C]"
       write(2,'(a,i10)') "max. iteration number niter=",niter
       write(2,'(a,e10.5,a)')"init.temp.t_pocz=",t_pocz,"  [C]"
       do i=1,n
          t(i)=t_pocz
```

```
         tt(i)=t_pocz
      enddo
      i=0
      inaccurate=.true.
      do while ((i.le.niter).and.inaccurate)
         t(1)=(650.+t(2)+t(4))/4.
         t(2)=(550.+t(1)+t(3))/4.
         t(3)=(850.+t(2)+t(4))/4.
         t(4)=(750.+t(1)+t(3))/4.
         inaccurate=.false.
         do j=1,n
            if (abs(tt(j)-t(j)).gt.toler) inaccurate=.true.
         enddo
         if  (inaccurate) then
            do j=1,n
               tt(j)=t(j)
            enddo
         endif
         i=i+1
      enddo
......write(2,'(/a)')"CALCULATED TEMPERATURE"
      write(2,'(a)')"   Lp     T[C] "
      do j=1,n
         write(2,'(i5,3x,e11.6)')j,t(j)
      enddo
      write(2,'(a,i10)') "final iteration number=",i
      end  program seidel

data(seidel.in)
4 0.00001 10000000 250.

results(seidel.out)
CALCULATING TWO-DIMENSIONAL TEMPERATURE FIELD IN A FLAT ROD

DATA ENTERED
equation number n= 4
calcul. toler=.10000E-04 [C]
max. iteration number niter=  10000000
init.temp.t_pocz=.25000E+03 [C]

CALCULATED TEMPERATURE
   Lp     T[C]
    1   .333333E+03
    2   .316667E+03
    3   .383333E+03
    4   .366667E+03
final iteration number=  14
```

Exercise 11.3 A Two-Dimensional Inverse Steady-State Heat Conduction Problem

Solve an inverse heat conduction problem. Temperature is measured at a point inside a body. The unknown is the temperature of a node, which lies on the edge of that body. Consider two cases (Fig. 11.4):

a) Temperature is measured in node 1, while the unknown is the temperature in node B, which lies on the body edge.

b) Temperature is measured in node 3, while the unknown is the temperature in node B, which lies on the body edge.

As measurement values f_1 and f_3 adopt temperatures determined in the previous exercise (Ex. 11.2), for nodes 1 and 3, respectively, i.e.

$$f_1 = T_1 = 333.333°\text{C},$$

$$f_3 = T_3 = 383.333°\text{C}.$$

How the calculation results are going to change, if measurement values contain a measurement error $\Delta T = +1.0°\text{C}$ i.e.

$$f_1 = T_1 + \Delta T = 333.333 + 1.0 = 334.333°\text{C},$$

$$f_3 = T_3 + \Delta T = 383.333 + 1.0 = 384.333°\text{C}.$$

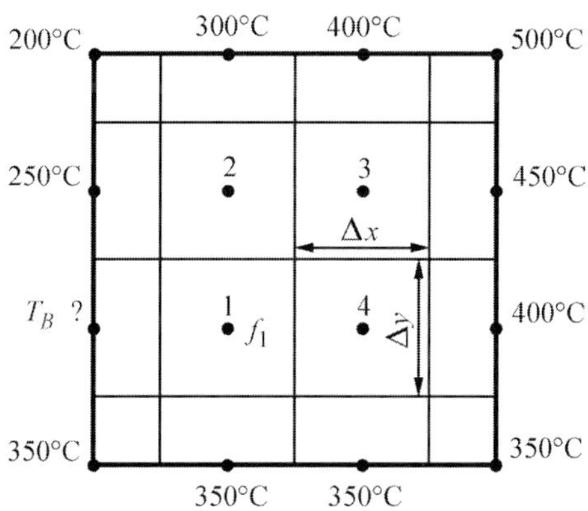

Fig. 11.4. Inverse heat conduction problem; temperature f_1 is measured in node 1, while the unknown temperature at point T_B lies on the body edge

Solution

In general, temperatures in volume nodes or finite elements are formulated by the equation system

$$\begin{aligned} a_{11}T_1 + a_{12}T_2 + a_{13}T_3 + \ldots + a_{1n}T_n &= b_1 \\ a_{21}T_1 + a_{22}T_2 + a_{23}T_3 + \ldots + a_{2n}T_n &= b_2 \\ \ldots\ldots\ldots\ldots\ldots\ldots\ldots\ldots\ldots\ldots\ldots\ldots\ldots\ldots \\ a_{n1}T_1 + a_{n2}T_2 + a_{n3}T_3 + \ldots + a_{nn}T_n &= b_n \end{aligned} \tag{1}$$

Parameters that appear in the boundary conditions are expressed by the terms on the right side of the system, i.e. in vector $\mathbf{b} = (b_1, b_2, \ldots, b_n)^T$, while coefficients a_{ij}, $i = 1, \ldots, n$, $j = 1, \ldots, n$, i.e. the coefficient matrix **A** is known. If the equation system (1) is written in the matrix form

$$\mathbf{AT} = \mathbf{b}, \tag{2}$$

where

$$\mathbf{A} = \begin{bmatrix} a_{11} & a_{12} & a_{13} & \ldots & a_{1n} \\ a_{21} & a_{22} & a_{23} & \ldots & a_{2n} \\ \ldots & \ldots & \ldots & \ldots & \ldots \\ a_{n1} & a_{n2} & a_{n3} & \ldots & a_{nn} \end{bmatrix}, \quad \mathbf{T} = \begin{bmatrix} T_1 \\ T_2 \\ \ldots \\ T_n \end{bmatrix}, \quad \mathbf{b} = \begin{bmatrix} b_1 \\ b_2 \\ \ldots \\ b_n \end{bmatrix}, \tag{3}$$

then the solution of the system (2) has the form

$$\mathbf{T} = \mathbf{A}^{-1}\mathbf{b}, \tag{4}$$

where $\mathbf{A}^{-1}$ is the inverse matrix to **A.**

Once we determine the inverse matrix

$$\mathbf{C} = \mathbf{A}^{-1} = \begin{bmatrix} c_{11} & c_{12} & c_{13} & \ldots & c_{1n} \\ c_{21} & c_{22} & c_{23} & \ldots & c_{2n} \\ \ldots & \ldots & \ldots & \ldots & \ldots \\ c_{n1} & c_{n2} & c_{n3} & \ldots & c_{nn} \end{bmatrix} \tag{5}$$

we can determine node temperatures from (4)

$$\begin{aligned} T_1 &= c_{11}b_1 + c_{12}b_2 + c_{13}b_3 + \ldots + c_{1n}b_n \\ T_2 &= c_{21}b_1 + c_{22}b_2 + c_{23}b_3 + \ldots + c_{2n}b_n \\ &\ldots\ldots\ldots\ldots\ldots\ldots\ldots\ldots\ldots\ldots\ldots\ldots \\ T_i &= c_{i1}b_1 + c_{i2}b_2 + c_{i3}b_3 + \ldots + c_{i,j-1}b_{j-1} + c_{i,j}b_j + c_{i,j+1}b_{j+1} + \ldots + c_{in}b_n \\ &\ldots\ldots\ldots\ldots\ldots\ldots\ldots\ldots\ldots\ldots\ldots\ldots \\ T_n &= c_{n1}b_1 + c_{n2}b_2 + c_{n3}b_3 + \ldots + c_{nn}b_n \end{aligned} \tag{6}$$

If temperature f_i in node i is known from measurements taken, while coefficient b_j is unknown, then from the equality condition of measured temperature f_i and calculated T_i, one obtains

$$f_i = T_i, \tag{7}$$

from where, after accounting for (6), one is able to determine coefficient b_j

$$b_j = \frac{f_j - c_{i1}b_1 - c_{i2}b_2 - c_{i3}b_3 - \ldots - c_{i,j-1}b_{j-1} - c_{i,j+1}b_{j+1} - \ldots - c_{in}b_n}{c_{i,j}}. \tag{8}$$

If the measurement data contains an error, then (8) assumes the form

$$b_j = \frac{\left(f_j + \Delta T\right) - c_{i1}b_1 - c_{i2}b_2 - c_{i3}b_3 - \ldots - c_{i,j-1}b_{j-1} - c_{i,j+1}b_{j+1} - \ldots - c_{in}b_n}{c_{i,j}}.$$

Should the problem formulated in this exercise appear (Fig. 11.4), then the balance equation system has the following form

$$\begin{aligned}
&4T_1 - T_2 - T_4 = T_B + 350\\
&-T_1 + 4T_2 - T_3 = 550\\
&-T_2 + 4T_3 - T_4 = 850\\
&-T_1 - T_3 + 4T_4 = 750
\end{aligned}$$

Hence, the coefficient matrix has the form

$$\mathbf{A} = \begin{bmatrix} 4 & -1 & 0 & -1 \\ -1 & 4 & -1 & 0 \\ 0 & -1 & 4 & -1 \\ -1 & 0 & -1 & 4 \end{bmatrix}.$$

Inverse matrix, determined by means of MATINV program (see Appendix E), is

$$\mathbf{C} = \mathbf{A}^{-1} = \begin{bmatrix} c_{11} & c_{12} & c_{13} & \ldots & c_{1n} \\ c_{21} & c_{22} & c_{23} & \ldots & c_{2n} \\ \ldots & \ldots & \ldots & \ldots & \ldots \\ c_{n1} & c_{n2} & c_{n3} & \ldots & c_{nn} \end{bmatrix} = \begin{bmatrix} 0.292 & 0.083 & 0.042 & 0.083 \\ 0.083 & 0.292 & 0.083 & 0.042 \\ 0.042 & 0.083 & 0.292 & 0.083 \\ 0.083 & 0.042 & 0.083 & 0.292 \end{bmatrix}.$$

(7) assumes the form

$$f_1 = c_{11}\left(T_B + 350\right) + c_{12} \cdot 550 + c_{13} \cdot 850 + c_{14} \cdot 750 \qquad (i = 1)$$

Temperature T_B is

$$T_B = \frac{f_1 - 350c_{11} - 550c_{12} - 850c_{13} - 750c_{14}}{c_{11}} =$$

$$= \frac{333.333 - 350 \cdot 0.292 - 550 \cdot 0.083 - 850 \cdot 0.042 - 750 \cdot 0.083}{0.292} =$$

$$= 299.770\,^\circ\text{C}$$

In an instance when data is burdened with errors, one gets

$$T'_B = \frac{(f_1 + \Delta T) - 350c_{11} - 550c_{12} - 850c_{13} - 750c_{14}}{c_{11}} =$$

$$= \frac{334.333 - 350 \cdot 0.292 - 550 \cdot 0.083 - 850 \cdot 0.042 - 750 \cdot 0.083}{0.292} =$$

$$= 303.195\,^\circ\text{C}.$$

If temperature is measured in node 3, then (7) assumes the form

$$f_3 = c_{31}(T_B + 350) + c_{32} \cdot 550 + c_{33} \cdot 850 + c_{34} \cdot 750,$$

hence,

$$T_B = \frac{f_3 - 350c_{31} - 550c_{32} - 850c_{33} - 750c_{34}}{c_{31}} =$$

$$= \frac{383.333 - 350 \cdot 0.042 - 550 \cdot 0.083 - 850 \cdot 0.292 - 750 \cdot 0.083}{0.042} =$$

$$= 298.405\,^\circ\text{C}.$$

For measured temperature f_3 with error ΔT, temperature T'_B is

$$T'_B = \frac{f_3 - 350c_{31} - 550c_{32} - 850c_{33} - 750c_{34}}{c_{31}} =$$

$$= \frac{384.333 - 350 \cdot 0.042 - 550 \cdot 0.083 - 850 \cdot 0.292 - 750 \cdot 0.083}{0.042} =$$

$$= 322.214\,^\circ\text{C}.$$

From the analysis of the given results, one can see that in the case of temperature measured at point 1, the accuracy of the obtained results is greater for both, the accurate temperature f_1 and the error-burdened data $(f_1 + \Delta T)$. Temperature changes in T_B have a larger effect on temperature T_1. There seems to be a distinct cause and effect relationship between T_B and T_1 than there is between T_B and T_3. If temperature changes in T_B affect, to a small degree, temperature changes at the point of measurement, as it happens in the case of node 3, it is difficult to accurately determine temperature T_B on the basis of measured temperature f_3. Small temperature measurement error in f_3 triggers a very large change in temperature T_B. The explanation for this is that the inverse problem has been ill conditioned. In practice the problem can be avoided, if the sensor or temperature measuring sensors are placed within a close proximity to a surface on which the boundary condition is being determined. This is, however, not always possible. If the temperature sensor is located far from the surface, on which the boundary conditions are identified, one should expect the obtained results to be far less accurate. This is precisely what this exercise has demonstrated; adding error ΔT = 1°C to the "accurate" measurement value at point 3 causes the determined temperature T'_B = 322.214°C to be significantly different from the real temperature T_B = 300°C.

Program inv

```
C      Inverse matrix calculation
       program inv
       dimension a(50,50),c(50,50)
       open(unit=1,file='inv.in')
       open(unit=2,file='inv.out')
       read(1,*)n
       read(1,*) ((a(i,j),j=1,n),i=1,n)
       write(2,'(a)')"INVERSE MATRIX CALCULATION"
       write(2,'(a)') "DATA ENTERED"
       write(2,'(a,i10)')"matrix A dimension  n=",n
       write(2,'(a)') "matrix A"
       write(2,'(4f8.2)') ((a(i,j),j=1,n),i=1,n)
       call matinv(a,n,c)
       write(2,'(a)') "CALCULATED MATRIX A^-1"
       write(2,'(4f9.3)') ((c(i,j),j=1,n),i=1,n)
       end  program inv
```

```
data(inv.in)
4
 4. -1.  0. -1.
-1.  4. -1.  0.
 0. -1.  4. -1.
-1.  0. -1.  4.
```

```
results(inv.out)
INVERSE MATRIX CALCULATION
DATA ENTERED
matrix A dimension  n=         4
matrix A
    4.00   -1.00     .00   -1.00
   -1.00    4.00   -1.00     .00
     .00   -1.00    4.00   -1.00
   -1.00     .00   -1.00    4.00
```

```
CALCULATED MATRIX A^-1
      .292      .083      .042      .083
      .083      .292      .083      .042
      .042      .083      .292      .083
      .083      .042      .083      .292
```

Exercise 11.4 Gauss-Seidel Method and Over-Relaxation Method

Describe Gauss-Seidel method and over-relaxation method, which are frequently employed when solving a system of algebraic equations obtained from the control volume method. Write a computational program in the FORTRAN language for the calculation using the over-relaxation method. Show how the equation system obtained in Ex. 11.2 can be solved by means of this program.

Solution

In the Gauss-Seidel method, the system of algebraic equations, which are the heat balance equations for the control volume,

$$\begin{aligned} &a_{11}T_1 + a_{12}T_2 + a_{13}T_3 + \ldots + a_{1n}T_n = b_1 \\ &a_{21}T_1 + a_{22}T_2 + a_{23}T_3 + \ldots + a_{2n}T_n = b_2 \\ &\ldots\ldots\ldots\ldots\ldots\ldots\ldots\ldots\ldots\ldots \\ &a_{n1}T_1 + a_{n2}T_2 + a_{n3}T_3 + \ldots + a_{nn}T_n = b_n, \end{aligned} \tag{1}$$

is transformed to a form

$$\begin{aligned} T_1 &= \frac{1}{a_{11}} \cdot \left(b_1 - a_{12}T_2 - a_{13}T_3 - \ldots - a_{1n}T_n\right) \\ T_2 &= \frac{1}{a_{22}} \cdot \left(b_2 - a_{21}T_1 - a_{23}T_3 - \ldots - a_{2n}T_n\right) \\ &\ldots\ldots\ldots\ldots\ldots\ldots\ldots\ldots\ldots\ldots \\ T_n &= \frac{1}{a_{nn}} \cdot \left(b_n - a_{n1}T_1 - a_{n2}T_2 - \ldots - a_{n,n-1}T_{n-1}\right). \end{aligned} \tag{2}$$

Gauss-Seidel method is an iterative method. One begins calculations by selecting the initial approximation first: $T_1^{(0)}$, $T_2^{(0)}$, ..., $T_n^{(0)}$; more often than not it is assumed that all temperature values equal zero. Quite often, moreover, temperatures $T_i^{(0)} = b_i/a_{ii}$, $i = 1, \ldots, n$ are selected as initial values. By substituting $T_2 = T_3 = \ldots = T_n = 0$ into the first equation in the system (2), one is able to calculate the first approximation of $T_1^{(1)}$. Temperature $T_1^{(1)}$ is automatically taken into account in the second equation of the system (2).

The remaining temperature values, which are present on the right-hand-side of the second equation, are assumed to be as follow: $T_3 = T_4 = \ldots = T_n = 0$. This is how $T_2^{(1)}$ is calculated from the second equation. By using the same method to determine temperature in the remaining nodes, the following approximation is obtained: $T_1^{(1)}, T_2^{(1)}, \ldots, T_n^{(1)}$. The determination of node temperature in the iterative k-step progresses as follows:

$$T_1^{(k+1)} = \frac{1}{a_{11}} \cdot \left(b_1 - a_{12} T_2^{(k)} - a_{13} T_3^{(k)} - \ldots - a_{1n} T_n^{(k)} \right)$$

$$T_2^{(k+1)} = \frac{1}{a_{22}} \cdot \left(b_2 - a_{21} T_1^{(k+1)} - a_{23} T_3^{(k)} - \ldots - a_{2n} T_n^{(k)} \right)$$

$$\ldots\ldots\ldots\ldots\ldots\ldots\ldots\ldots\ldots\ldots\ldots\ldots$$

$$T_i^{(k+1)} = \frac{1}{a_{ii}} \cdot \left(b_i - a_{i1} T_1^{(k+1)} - a_{i2} T_2^{(k+1)} - \ldots - a_{i,i-1} T_{i-1}^{(k+1)} - a_{i,i+1} T_{i+1}^{(k)} - \ldots - a_{in} T_n^{(k)} \right) \qquad (3)$$

$$\ldots\ldots\ldots\ldots\ldots\ldots\ldots\ldots\ldots\ldots\ldots\ldots$$

$$T_n^{(k+1)} = \frac{1}{a_{nn}} \cdot \left(b_n - a_{n1} T_1^{(k+1)} - a_{n2} T_2^{(k+1)} - \ldots - a_{n,n-1} T_{n-1}^{(k+1)} \right).$$

This calculation method, expressed by the equations in (3), was applied in a program presented in Ex.11.2. It does not require of one to use coefficients a_{ij}, $i = 1, \ldots, n$, $j = 1, \ldots, n$ in the calculation. The calculation process is the same when coefficients a_{ij} are temperature dependent, if the thermal conductivity, for instance, is temperature dependent. A drawback to this method is the fact that one is forced to rewrite all balance equations anew, when the new problem must be analyzed.

In order to make the program more universally applicable, a formula for $T_i^{(k+1)}$ in the system (3) will be used for the calculation; it can be written in the slightly different form

$$T_i^{(k+1)} = \frac{1}{a_{ii}} \left(b_i - \sum_{j=1}^{i-1} a_{ij} T_j^{(k+1)} - \sum_{j=i+1}^{n} a_{ij} T_j^{(k)} \right), \qquad i = 1, \ldots, n. \qquad (4)$$

Over-relaxation method is a form of modification of the Gauss-Seidel method; it aims to accelerate the iterative process

$$T_i^{(k+1)} = T_i^{(k)} + \frac{\omega}{a_{ii}} \left(b_i - \sum_{j=1}^{i-1} a_{ij} T_j^{(k+1)} - \sum_{j=i}^{n} a_{ij} T_j^{(k)} \right), \qquad (5)$$

where $1 \le \omega \le 2$ is an over-relaxation coefficient. If $\omega = 1$, then over-relaxation method is identical to Gauss-Seidel method. Both, Gauss-Seidel method and over-relaxation method are convergent when

$$\sum_{\substack{j=1 \\ j \ne i}}^{n} \left| a_{ij} \right| < \left| a_{ii} \right|, \qquad i = 1, 2, \ldots, n. \qquad (6)$$

Iterative process is continued until the criterion below is satisfied

$$\left|T_i^{(k+1)} - T_i^{(k)}\right| < \varepsilon, \qquad i = 1, ..., n, \tag{7}$$

where ε is the assigned calculation tolerance or

$$\left\|\mathbf{T}^{(k+1)} - \mathbf{T}^{(k)}\right\| < \varepsilon_1, \tag{8}$$

where ε_1 is the assigned calculation tolerance, e.g. ε_1 = 0.001 K. Square norm is calculated from formula

$$\left\|\mathbf{T}^{(k+1)} - \mathbf{T}^{(k)}\right\| = \left[\sum_{i=1}^{n}\left(T_i^{(k+1)} - T_i^{(k)}\right)^2\right]^{\frac{1}{2}}. \tag{9}$$

Sub-program SOR for solving equation system (1) by means of the over-relaxation method is shown in Appendix F. Equation system (2)–(5) in Ex. 11.2 has the following form:

$$\begin{bmatrix} 4 & -1 & 0 & -1 \\ -1 & 4 & -1 & 0 \\ 0 & -1 & 4 & -1 \\ -1 & 0 & -1 & 4 \end{bmatrix} \begin{Bmatrix} T_1 \\ T_2 \\ T_3 \\ T_4 \end{Bmatrix} = \begin{Bmatrix} 650 \\ 550 \\ 850 \\ 750 \end{Bmatrix}, \tag{10}$$

The broken brackets are column vectors.

A program for solving system (10) with the help of sub-program SOR is shown below. The value of over-relaxation coefficient is assumed to equal 1,2. Obtained results are the same as they are in Ex. 11.2.

Program content, data and the solution of equation system (10)

```
C      Solution of an equation system by means of over-
C      relaxation method and sub-program SOR
       program nadrel
       dimension a(50,51),xi(50)
       nmax=50
       mmax=nmax+1
       open(unit=1,file='nadrel.in')
       open(unit=2,file='nadrel.out')
       read(1,*)n, w, niter, toler
       m=n+1
       read(1,*)((a(i,j),j=1,m),i=1,n)
       read(1,*)(xi(i),i=1,n)
       write(2,'(a)') "SOLUTION OF EQUATION SET
      &BY OVER-RELAXATION METHOD"
       write(2,'(/a)') "DATA ENTERED"
       write(2,'(a,i10)')     "equation number  n=",n
       write(2,'(a,e10.5)')   "relaxation coefficient w=",w
```

```
      write(2,'(a,i10)')    "max. iter. number niter=",niter
      write(2,'(a,e10.5,a)') "calc.toler.toler=",toler,"[C]"
      write(2,'(a)') "matrix A"
      write(2,'(5f8.2)') ((a(i,j),j=1,m),i=1,n)
      write(2,'(a)') "initial vector XI"
      write(2,'(4f8.2)') (xi(i),i=1,n)
      call sor(a,nmax,mmax,n,xi,w,niter,toler,k)
      write(2,'(/a)')"CALCULATION RESULTS"
      write(2,'(a)')"   Lp     X "
      do j=1,n
         write(2,'(i5,3x,4e11.6)')j,xi(j)
      enddo
      write(2,'(a,i10)') "final iteration number=",k
      end  program nadrel
```

```
data(nadrel.in)
 4  1.2  30  1.0E-3
 4. -1.  0. -1.  650.
-1.  4. -1.  0.  550.
 0. -1.  4. -1.  850.
-1.  0. -1.  4.  750.
 0.  0.  0.  0.
```

```
results(nadrel.out)
SOLUTION OF EQUATION SET BY OVER-RELAXATION METHOD
DATA ENTERED
equation number   n=        4
relaxation coefficient w=.12000E+01
max. iteration number niter=        30
calc.toler.toler=.10000E-02 [C]
matrix A
    4.00   -1.00     .00   -1.00  650.00
   -1.00    4.00   -1.00     .00  550.00
     .00   -1.00    4.00   -1.00  850.00
   -1.00     .00   -1.00    4.00  750.00
initial vector XI
     .00     .00     .00     .00
CALCULATION RESULTS
   Lp     X
    1   .333333E+03
    2   .316667E+03
    3   .383333E+03
    4   .366667E+03
final iteration number=        11
```

Exercise 11.5 Determining Two-Dimensional Temperature Distribution in a Straight Fin with Uniform Thickness by Means of the Finite Volume Method

Determine temperature distribution in a fin presented in Fig. 11.5 by means of the control volume method. For the calculation adopt the values given in Ex. 7.3. Also calculate heat flow at the fin base and its efficiency.

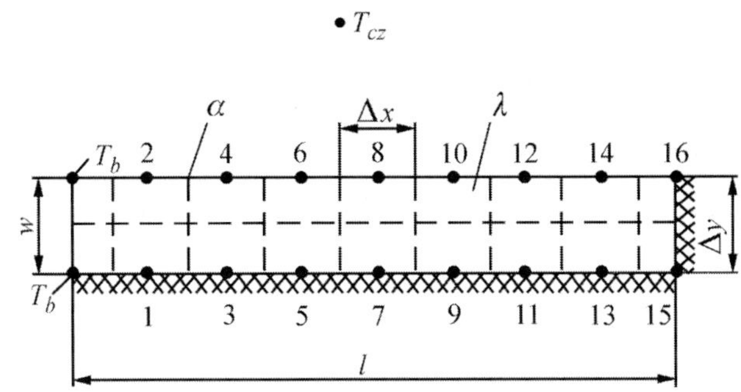

Fig. 11.5. A division of half of the fin into control volumes

Solution

Heat balance equations for control volumes have the form:

• node 1

$$\lambda \frac{\Delta y}{2} \frac{T_b - T_1}{\Delta x} + \lambda \frac{\Delta y}{2} \frac{T_3 - T_1}{\Delta x} + \lambda \Delta x \frac{T_2 - T_1}{\Delta y} = 0, \tag{1}$$

hence,

$$\left(\frac{\Delta y}{\Delta x} + \frac{\Delta x}{\Delta y} \right) T_1 - \frac{\Delta x}{\Delta y} T_2 - \frac{\Delta y}{2\Delta x} T_3 = \frac{\Delta y}{2\Delta x} T_b. \tag{2}$$

• node 2

$$\lambda \frac{\Delta y}{2} \frac{T_b - T_2}{\Delta x} + \lambda \frac{\Delta y}{2} \frac{T_4 - T_2}{\Delta x} + \lambda \Delta x \frac{T_1 - T_2}{\Delta y} + \alpha \Delta x \left(T_{cz} - T_2 \right) = 0, \tag{3}$$

from where, after transformations, one obtains

$$-\frac{\Delta x}{\Delta y} T_1 + \left(\frac{\Delta y}{\Delta x} + \frac{\Delta x}{\Delta y} + \frac{\alpha \Delta x}{\lambda} \right) T_2 - \frac{\Delta y}{2\Delta x} T_4 = \frac{\Delta y}{2\Delta x} T_b + \frac{\alpha \Delta x}{\lambda} T_{cz}. \tag{4}$$

- nodes 3, 5, 7, 9, 11, 13

Heat balance equation for i-node has the form

$$\lambda \frac{\Delta y}{2} \frac{T_{i-2} - T_i}{\Delta x} + \lambda \frac{\Delta y}{2} \frac{T_{i+2} - T_i}{\Delta x} + \lambda \Delta x \frac{T_{i+1} - T_i}{\Delta y} = 0, \tag{5}$$

hence,

$$-\frac{\Delta y}{2\Delta x} T_{i-2} + \left(\frac{\Delta y}{\Delta x} + \frac{\Delta x}{\Delta y} \right) T_i - \frac{\Delta x}{\Delta y} T_{i+1} - \frac{\Delta y}{2\Delta x} T_{i+2} = 0, \tag{6}$$

$i = 3, 5, 7, 9, 11, 13$

- nodes 4, 6, 8, 10, 12, 14

Heat balance equation for i-node has the form

$$\lambda \frac{\Delta y}{2} \frac{T_{i-2} - T_i}{\Delta x} + \lambda \frac{\Delta y}{2} \frac{T_{i+2} - T_i}{\Delta x} + \lambda \Delta x \frac{T_{i-1} - T_i}{\Delta y} + \alpha \Delta x \left(T_{cz} - T_i \right) = 0, \tag{7}$$

hence,

$$-\frac{\Delta y}{2\Delta x} T_{i-2} - \frac{\Delta x}{\Delta y} T_{i-1} + \left(\frac{\Delta y}{\Delta x} + \frac{\Delta x}{\Delta y} + \frac{\alpha \Delta x}{\lambda} \right) T_i - \frac{\Delta y}{2\Delta x} T_{i+2} = \frac{\alpha \Delta x}{\lambda}, \tag{8}$$

$i = 2, 4, 6, 8, 10, 12.$

- node 15

$$\lambda \frac{\Delta y}{2} \frac{T_{13} - T_{15}}{\Delta x} + \lambda \frac{\Delta x}{2} \frac{T_{16} - T_{15}}{\Delta y} = 0, \tag{9}$$

from where, after simple transformations, one obtains

$$-\frac{\Delta y}{\Delta x} T_{13} + \left(\frac{\Delta y}{\Delta x} + \frac{\Delta x}{\Delta y} \right) T_{15} - \frac{\Delta x}{\Delta y} T_{16} = 0. \tag{10}$$

- node 16

$$\lambda \frac{\Delta y}{2} \frac{T_{14} - T_{16}}{\Delta x} + \lambda \frac{\Delta x}{2} \frac{T_{15} - T_{16}}{\Delta y} + \alpha \frac{\Delta x}{2} \left(T_{cz} - T_{16} \right) = 0; \tag{11}$$

from where, one obtains

$$-\frac{\Delta y}{\Delta x} T_{14} - \frac{\Delta x}{\Delta y} T_{15} + \left(\frac{\Delta y}{\Delta x} + \frac{\Delta x}{\Delta y} + \frac{\alpha \Delta x}{\lambda} \right) T_{16} = \frac{\alpha \Delta x}{\lambda} T_{cz}. \tag{12}$$

After substitution of $T_b = 95°C$, $T_{cz} = 20°C$, $\Delta x = \Delta y = 0.003$ m, $\lambda = 50$ W/(m·K), $\alpha = 100$ W/(m²·K), (2), (4), (6), (8), (10) and (12) assume the form

$$
\begin{aligned}
&2T_1 - T_2 - 0.5T_3 = 47.5 \qquad\qquad -T_1 + 2.006T_2 - 0.5T_4 = 47.62\\
&-0.5T_{i-2} + 2T_i - T_{i+1} - 0.5T_{i+2} = 0; \qquad i = 3, 5, 7, 9, 11, 13\\
&-0.5T_{i-2} - T_{i-1} + 2.006T_i - 0.5T_{i+2} = 0.12; \qquad i = 2, 4, 6, 8, 10, 12, 14\\
&-T_{13} + 2T_{15} - T_{16} = 0 \qquad\qquad -T_{14} - T_{15} + 2.006T_{16} = 0.12.
\end{aligned}
\tag{13}
$$

Gauss-Seidel method will be used to solve equation system (13). A program similar to the one in Ex. 11.2 will be used for the calculation. For that reason, one should rewrite (13), so that one could determine temperature T_i from i-equation. Equation system (13) assumes the form

Node no.	Equation
1	$T_1 = 0.5(T_2 + 0.5T_3 + 47.5)$
2	$T_2 = \frac{1}{2.006}(T_1 + 0.5T_4 + 47.62)$
3	$T_3 = 0.5(0.5T_1 + T_4 + 0.5T_5)$
4	$T_4 = \frac{1}{2.006}(0.5T_2 + T_3 + 0.5T_6 + 0.12)$
5	$T_5 = 0.5(0.5T_3 + T_6 + 0.5T_7)$
6	$T_6 = \frac{1}{2.006}(0.5T_4 + T_5 + 0.5T_8 + 0.12)$
7	$T_7 = 0.5(0.5T_5 + T_8 + 0.5T_9)$
8	$T_8 = \frac{1}{2.006}(0.5T_6 + T_7 + 0.5T_{10} + 0.12)$
9	$T_9 = 0.5(0.5T_7 + T_{10} + 0.5T_{11})$
10	$T_{10} = \frac{1}{2.006}(0.5T_8 + T_9 + 0.5T_{12} + 0.12)$
11	$T_{11} = 0.5(0.5T_9 + T_{12} + 0.5T_{13})$
12	$T_{12} = \frac{1}{2.006}(0.5T_{10} + T_{11} + 0.5T_{14} + 0.12)$
13	$T_{13} = 0.5(0.5T_{11} + T_{14} + 0.5T_{15})$
14	$T_{14} = \frac{1}{2.006}(0.5T_{12} + T_{13} + 0.5T_{16} + 0.12)$
15	$T_{15} = 0.5(T_{13} + T_{16})$
16	$T_{16} = \frac{1}{2.006}(T_{14} + T_{15} + 0.12)$.

(14)

The system was solved with the assumed tolerance that equals ε = 0.00001°C. As an initial solution, the following was assumed:

$$T_1^{(0)} = T_2^{(0)} = \ \dots \ = T_{16}^{(0)} = 20\,^{\circ}\mathrm{C}\,. \tag{15}$$

Computational Program Content

```
C     Calculating two-dimensional fin temp. field (Fig.11.5)
C     by means of control volume method equation system
C     solved by Gauss-Seidel method
      program seidel2
      dimension t(50),tt(50)
      logical inaccurate
      open(unit=1,file='seidel2.in')
      open(unit=2,file='seidel2.out')
      read(1,*)n,toler,niter,t_pocz
      write(2,'(a)')
     &"CALCULATING TWO-DIMENSIONAL FIN TEMPERATURE FIELD "
      write(2,'(/a)') "DATA ENTERED"
      write(2,'(a,i10)')     "equation number  n=",n
      write(2,'(a,e10.5,a)') "cal.toler toler=",toler," [C]"
      write(2,'(a,i10)') "max. iteration number niter=",niter
      write(2,'(a,e10.5,a)')"init temp. t_pocz=",t_pocz,"[C]"
      do i=1,n
         t(i)=t_pocz
         tt(i)=t_pocz
      enddo

      i=0
      inaccurate=.true.
      do while ((i.le.niter).and.inaccurate)
      t(1)=(t(2)+0.5*t(3)+47.5)*0.5
      t(2)=(t(1)+0.5*t(4)+47.62)/2.006
      t(3)=(0.5*t(1)+t(4)+0.5*t(5))*0.5
      t(4)=(0.5*t(2)+t(3)+0.5*t(6)+0.12)/2.006
      t(5)=(0.5*t(3)+t(6)+0.5*t(7))*0.5
      t(6)=(0.5*t(4)+t(5)+0.5*t(8)+0.12)/2.006
      t(7)=(0.5*t(5)+t(8)+0.5*t(9))*0.5
      t(8)=(0.5*t(6)+t(7)+0.5*t(10)+0.12)/2.006
      t(9)=(0.5*t(7)+t(10)+0.5*t(11))*0.5
      t(10)=(0.5*t(8)+t(9)+0.5*t(12)+0.12)/2.006
      t(11)=(0.5*t(9)+t(12)+0.5*t(13))*0.5
      t(12)=(0.5*t(10)+t(11)+0.5*t(14)+0.12)/2.006
      t(13)=(0.5*t(11)+t(14)+0.5*t(15))*0.5
      t(14)=(0.5*t(12)+t(13)+0.5*t(16)+0.12)/2.006
      t(15)=(t(13)+t(16))*0.5
      t(16)=(t(14)+t(15)+0.12)/2.006
      inaccurate=.false.
```

```
do j=1,n
    if (abs(tt(j)-t(j)).gt.toler) inaccurate=.true.
enddo
  if (inaccurate) then
    do j=1,n
      tt(j)=t(j)
    enddo
  endif
 i=i+1
enddo
write(2,'(/a)')"CALCULATED TEMPERATURE"
write(2,'(a)')"   Lp      T[C]  "
do j=1,n
 write(2,'(i5,3x,e11.6)')j,t(j)
enddo
write(2,'(a,i10)') "final iteration number=",i
end  program seidel2

data(seidel.in)
16 0.00001 100000 20.

results(seidel.out)
DATA ENTERED
equation number  n=          16
cal.toler toler =.10000E-04 [C]
max. iteration number niter=     100000
init temp. t_pocz=.20000E+02 [C]
CALCULATED TEMPERATURE
Lp      T[C]
 1    .921152E+02
 2    .919377E+02
 3    .895855E+02
 4    .893836E+02
 5    .874594E+02
 6    .872585E+02
 7    .857353E+02
 8    .855386E+02
 9    .844046E+02
10    .842117E+02
11    .834597E+02
12    .832696E+02
13    .828950E+02
14    .827066E+02
15    .827072E+02
16    .825193E+02
final iteration number=        536
```

Table 11.1. Temperatures in control volume nodes shown in Fig. 11.5

Node no.	Temperature		Node no.	Temperature	
	Control Volume Method	Analytical Method		Control Volume Method	Analytical Method
1	92.11	92.09	9	84.40	84.37
2	91.94	91.88	10	84.21	84.18
3	89.58	89.55	11	83.46	83.42
4	89.38	89.34	12	83.27	83.23
5	87.46	87.42	13	82.89	82.86
6	87.26	87.22	14	82.71	82.67
7	85.73	85.70	15	82.71	82.67
8	85.54	85.50	16	82.52	82.48

Calculation results are presented in Table 11.1.

One can see from the table above that temperatures calculated by means of control volume method are almost the same as the values determined by means of the analytical formula ((2), Ex.7.3). On the basis of temperature distribution, one can calculate fin-base heat flow, which equals the fin-to-surroundings transferred heat flow. If a fin perpendicular to the diagram surface (fin length) measures 1 m in length, then the heat flow at the fin-base is expressed by formula

$$\dot{Q}_b = 2\left(\frac{w}{2}\cdot 1\cdot\lambda\frac{T_b - T_2}{\Delta x} + \frac{w}{2}\cdot 1\cdot\lambda\frac{T_b - T_1}{\Delta x}\right) =$$

$$= \frac{2w\lambda}{2\Delta x}\left(2T_b - T_1 - T_2\right) = 2\lambda\left(T_b - \frac{T_1 + T_2}{2}\right),$$

hence,

$$\dot{Q}_b = 2\cdot 50\left(95 - \frac{92.11 + 91.94}{2}\right) = 297.5 \text{ W}.$$

Multiplier 2 was placed in front of the square brackets, since the heat flow given off by the fin is twice as large; only half of the fin was taken into consideration in Fig.11.5. In order to calculate efficiency, one needs to know what the value of heat flow $\dot{Q}_{\max}$ is, given off by an isothermal fin with temperature T_b within its entire volume

$$\dot{Q}_{\max} = 2\cdot l\cdot 1\cdot\alpha\left(T_b - T_{cz}\right) = 2\cdot 0.024\cdot 1\cdot 100\cdot\left(95 - 20\right) = 360 \text{ W}.$$

Therefore, fin efficiency determined by means of the control volume method is

$$\eta = \frac{\dot{Q}}{\dot{Q}_{\max}} = \frac{297.5}{360} = 0.826 .$$

Fin efficiency η_e calculated by means of the analytical method is formulated as

$$\eta_e = \frac{\dot{Q}_e}{\dot{Q}_{\max}} = \frac{2w\overline{\dot{q}_x}(0)}{\dot{Q}_{\max}} = \frac{2w\overline{\dot{q}_x}(0)}{360} .$$

Hence,

$$\eta_e = \frac{2 \cdot 0.003 \cdot 53253}{360} = 0.887 .$$

Relative error from the efficiency calculation above is

$$\Delta\eta = \frac{\eta - \eta_e}{\eta_e} \cdot 100\% = \frac{0.826 - 0.887}{0.887} \cdot 100\% = -6.8\%.$$

In spite of the coarse control volume grid, a good agreement was established between temperature distribution (Table 11.1) and two-dimensional analytical solution. Calculation of heat flux at the fin-base, as indicated by the calculated efficiency value, is less accurate.

In order to improve accuracy, fin efficiency will be calculated in a different way. Fin-dissipated heat flow can also be calculated by determining heat flow received by the lateral surfaces of the fin first:

$$\dot{Q} = 2\alpha\Delta x\left[\frac{1}{2}(T_b - T_{cz}) + (T_2 - T_{cz}) + (T_4 - T_{cz}) + (T_6 - T_{cz}) + (T_8 - T_{cz}) + \right.$$

$$\left. + (T_{10} - T_{cz}) + (T_{12} - T_{cz}) + (T_{14} - T_{cz}) + \frac{1}{2}(T_{16} - T_{cz})\right] =$$

$$= 2\alpha\Delta x\left[\frac{1}{2}T_b + T_2 + T_4 + T_6 + T_8 + T_{10} + T_{12} + T_{14} + \frac{1}{2}T_{16} - 8T_{cz}\right] =$$

$$= 2 \cdot 100 \cdot 0.003\left(\frac{95}{2} + 91.94 + 89.38 + 87.26 + 85.54 + 84.21 + 83.27 + \right.$$

$$\left. + 82.71 + \frac{82.52}{2} - 8 \cdot 20\right) = 319.842\ \text{W/m}.$$

Fin efficiency, then, determined by means of the control volume method is

$$\eta = \frac{\dot{Q}}{\dot{Q}_{\max}} = \frac{319.842}{360} = 0.888 .$$

Relative error is at

$$\Delta\eta = \frac{\eta - \eta_e}{\eta_e} \cdot 100\% = \frac{0.888 - 0.887}{0.887} \cdot 100\% = 0.112\% .$$

Exercise 11.6 Determining Two-Dimensional Temperature Distribution in a Square Cross-Section of a Chimney

Determine temperature distribution in a chimney cross-section presented in Fig. 11.6. External dimensions of the chimney are $2b \times 2b$. Internal canal has a square cross-section and the length of its side is $2a$. For the calculation assume that $b = 0.375$ m and $a = 0.125$ m. Thermal conductivity of the chimney's material is $\lambda = 1.25$ W/(m·K). Heat transfer coefficient from emissions to inner surface is $\alpha_w = 60$ W/(m²·K), while from outer surface to surroundings is $\alpha_z = 20$ W/(m²·K). Emissions temperature measures $T_w = 250$°C, while air temperature of surroundings $T_z = 10$°C. Solve the problem using control volume method.

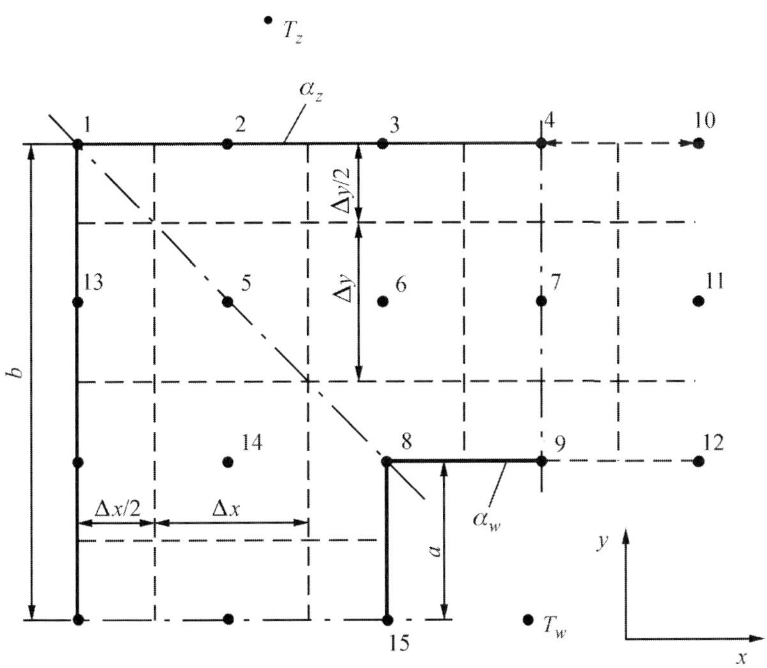

Fig. 11.6. A division of ¼ of a chimney into control volumes

Solution

Due to the symmetry of temperature field, only 1/8 of the chimney cross-section will be analyzed below. Temperature will be determined for nodes from node 1 to 9 (Fig. 11.6). Control volumes are squares with a side that measures $\Delta x = \Delta y = (b - a)/2 = (0.375 - 0.125)/2 = 0.125$ m.

Heat balance equations for control volumes have the following form:

- Node 1

$$\alpha_z\left(\frac{\Delta x}{2}+\frac{\Delta y}{2}\right)\left(T_z - T_1\right) + \lambda\frac{\Delta x}{2}\frac{T_{13}-T_1}{\Delta y} + \lambda\frac{\Delta y}{2}\frac{T_2 - T_1}{\Delta x} = 0, \tag{1}$$

from where, one obtains (when $\Delta y = \Delta x$)

$$T_1 = \frac{1}{1+\Delta Bi_z}\left[\frac{T_2 + T_{13}}{2} + \left(\Delta Bi_z\right)T_z\right]. \tag{2}$$

When $T_{13} = T_2$, (2) assumes the form

$$T_1 = \frac{1}{1+\Delta Bi_z}\left[T_2 + \left(\Delta Bi_z\right)T_z\right], \tag{3}$$

where $\Delta Bi_z = \alpha(\Delta x)/\lambda$.

- Node 2

$$\alpha_z\Delta x\left(T_z - T_2\right) + \lambda\frac{\Delta y}{2}\frac{T_1 - T_2}{\Delta x} + \lambda\frac{\Delta y}{2}\frac{T_3 - T_2}{\Delta x} + \lambda\Delta x\frac{T_5 - T_2}{\Delta y} = 0, \tag{4}$$

hence,

$$T_2 = \frac{1}{2+\Delta Bi_z}\left[T_5 + \frac{T_1 + T_3}{2} + \left(\Delta Bi_z\right)T_z\right]. \tag{5}$$

- Node 3

$$\alpha_z\Delta x\left(T_z - T_3\right) + \lambda\frac{\Delta y}{2}\frac{T_2 - T_3}{\Delta x} + \lambda\frac{\Delta y}{2}\frac{T_4 - T_3}{\Delta x} + \lambda\Delta x\frac{T_6 - T_3}{\Delta y} = 0, \tag{6}$$

where from, after transformations, one obtains

$$T_3 = \frac{1}{2+\Delta Bi_z}\left[T_6 + \frac{T_2 + T_4}{2} + \left(\Delta Bi_z\right)T_z\right]. \tag{7}$$

- Node 4

$$\alpha_z\Delta x\left(T_z - T_4\right) + \lambda\frac{\Delta y}{2}\frac{T_3 - T_4}{\Delta x} + \lambda\frac{\Delta y}{2}\frac{T_{10} - T_4}{\Delta x} + \lambda\Delta x\frac{T_7 - T_4}{\Delta y} = 0, \tag{8}$$

hence, when $T_{10} = T_3$, one obtains

$$T_4 = \frac{1}{2 + \Delta Bi_z}\left[T_7 + T_3 + \left(\Delta Bi_z\right)T_z\right]. \tag{9}$$

• Node 5

$$\lambda\Delta x \frac{T_2 - T_5}{\Delta y} + \lambda\Delta x \frac{T_{14} - T_5}{\Delta y} + \lambda\Delta y \frac{T_{13} - T_5}{\Delta x} + \lambda\Delta y \frac{T_6 - T_5}{\Delta x} = 0\,, \tag{10}$$

hence, when $T_{13} = T_2$ and $T_{14} = T_6$, one gets

$$T_5 = \frac{1}{2}\left(T_2 + T_6\right). \tag{11}$$

• Node 6

$$\lambda\Delta y \frac{T_5 - T_6}{\Delta x} + \lambda\Delta y \frac{T_7 - T_6}{\Delta x} + \lambda\Delta x \frac{T_3 - T_6}{\Delta y} + \lambda\Delta x \frac{T_8 - T_6}{\Delta y} = 0\,, \tag{12}$$

hence, one obtains

$$T_6 = \frac{1}{4}\left(T_3 + T_5 + T_7 + T_8\right). \tag{13}$$

• Node 7

$$\lambda\Delta y \frac{T_6 - T_7}{\Delta x} + \lambda\Delta y \frac{T_{11} - T_7}{\Delta x} + \lambda\Delta x \frac{T_4 - T_7}{\Delta y} + \lambda\Delta x \frac{T_9 - T_7}{\Delta y} = 0\,, \tag{14}$$

hence, when $T_{11} = T_6$, one obtains

$$T_7 = \frac{1}{4}\left(T_4 + 2T_6 + T_9\right). \tag{15}$$

• Node 8

$$\begin{aligned}&\alpha_w\left(\frac{\Delta x}{2} + \frac{\Delta y}{2}\right)\left(T_w - T_8\right) + \lambda\Delta x \frac{T_6 - T_8}{\Delta y} + \lambda\Delta y \frac{T_{14} - T_8}{\Delta x} + \lambda \frac{\Delta y}{2}\frac{T_9 - T_8}{\Delta x} + \\ &+\lambda \frac{\Delta x}{2}\frac{T_{15} - T_8}{\Delta y} = 0,\end{aligned} \tag{16}$$

hence, after transformations and when $T_{15} = T_6$ and $T_{14} = T_9$, one has

$$T_8 = \frac{1}{3 + \Delta Bi_z}\left[2T_6 + T_9 + \left(\Delta Bi_w\right)T_w\right], \tag{17}$$

where $\Delta Bi_z = \alpha(\Delta x)/\lambda$.

• Node 9

$$\alpha_w \Delta x\left(T_w - T_9\right) + \lambda \frac{\Delta y}{2}\frac{T_8 - T_9}{\Delta x} + \lambda \frac{\Delta y}{2}\frac{T_{12} - T_9}{\Delta x} + \lambda \Delta x \frac{T_7 - T_9}{\Delta y} = 0,$$

hence, after transformations and when $T_{12} = T_8$, one has

$$T_9 = \frac{1}{2 + \Delta Bi_z}\left[T_7 + T_8 + \left(\Delta Bi_w\right)T_w\right].$$

One should emphasize that due to temperature field symmetry, planes 1-5-8 and 4-7-9 are thermally insulated (are adiabatic). There is no need to take additional nodes located outside area 1-4-9-8-1 into consideration, if one takes into consideration that symmetry planes are thermally insulated. The heat balance for node 8, with the plane 1-5-8 thermally insulated, has the form

$$\alpha_w \frac{\Delta x}{2}\left(T_w - T_8\right) + \lambda \Delta x \frac{T_6 - T_8}{\Delta y} + \lambda \frac{\Delta y}{2}\frac{T_9 - T_8}{\Delta x} = 0,$$

hence, after transformations, (17) is obtained. Equations for nodes located in symmetry planes 1, 5, 4, 7 and 9 can be derived in a similar way.

After substitution, one has

$$\Delta Bi_z = \frac{\alpha_z \Delta x}{\lambda} = \frac{20 \cdot 0.125}{1.25} = 2.0,$$

$$\Delta Bi_w = \frac{\alpha_w \Delta x}{\lambda} = \frac{60 \cdot 0.125}{1.25} = 6.0.$$

Heat balance equations for nodes 1 to 9 have the form

$$T_1 = \frac{1}{3}\left(T_2 + 20\right),\ T_2 = \frac{1}{4}\left(T_5 + \frac{T_1 + T_3}{2} + 20\right),\ T_3 = \frac{1}{4}\left(T_6 + \frac{T_2 + T_4}{2} + 20\right),$$

$$T_4 = \frac{1}{4}\left(T_7 + T_3 + 20\right),\qquad T_5 = \frac{1}{2}\left(T_2 + T_6\right),\qquad T_6 = \frac{1}{4}\left(T_3 + T_5 + T_7 + T_8\right),$$

$$T_7 = \frac{1}{4}\left(T_4 + 2T_6 + T_9\right),\ T_8 = \frac{1}{9}\left(2T_6 + T_9 + 1500\right),\ T_9 = \frac{1}{9}\left(T_7 + T_8 + 1500\right).$$

The equation system above will be solved by Gauss-Seidel method, when $\varepsilon = 0.00001$°C. The printout of the program in FORTRAN language is presented below.

Computational program in FORTRAN language used for determining temperature distribution in a chimney cross-section

```
C     Calculating two-dim. temperature field in a chimney
C     cross-section (Fig. 11.6) by means of control volume
C     method, equation system solved by Gauss-Seidel method
      program seidel3
      dimension t(50),tt(50)
      logical inaccurate
      open(unit=1,file='seidel3.in')
      open(unit=2,file='seidel3.out')
      read(1,*)n,toler,niter,t_pocz
      write(2,'(a)') "CALCULATING TWO-DIMENSIONAL TEMPERATURE
     &FIELD IN CHIMNEY CROSS-SECTION"
      write(2,'(/a)') "DATA ENTERED"
      write(2,'(a,i10)')    "equation number  n=",n
      write(2,'(a,e10.5,a)') "calculation tolerance toler=",
     &toler, " [C]"
      write(2,'(a,i10)')    "max.iteration number niter=",
     &niter
      write(2,'(a,e10.5,a)')"initial temp. t_pocz=",t_pocz,
     &" [C]"
      do i=1,n
         t(i)=t_pocz
         tt(i)=t_pocz
      enddo
      i=0
      inaccurate=.true.
      do while ((i.le.niter).and.inaccurate)
         t(1)=(t(2)+20.)/3.
         t(2)=(t(5)+0.5*t(1)+0.5*t(3)+20.)/4.
         t(3)=(t(6)+0.5*t(2)+0.5*t(4)+20.)/4.
         t(4)=(t(7)+t(3)+20.)/4.
         t(5)=(t(2)+t(6))/2.
         t(6)=(t(3)+t(5)+t(7)+t(8))/4.
         t(7)=(t(4)+2.*t(6)+t(9))/4.
         t(8)=(2.*t(6)+t(9)+1500.)/9.
         t(9)=(t(7)+t(8)+1500.)/8.
         inaccurate=.false.
         do j=1,n
             if (abs(tt(j)-t(j)).gt.toler) inaccurate=.true.
         enddo
             if  (inaccurate) then
            do j=1,n
               tt(j)=t(j)
            enddo
         endif
         i=i+1
      enddo
```

```
write(2,'(/a)')"CALCULATED TEMPERATURE"
      write(2,'(a)')"    Lp      T[C]  "
      do j=1,n
         write(2,'(i5,3x,e11.6)')j,t(j)
      enddo
write(2,'(a,i10)') "final iteration number=",i
      end  program seidel3
```

```
data (seidel3.in)
9 0.00001 100000 10.
```

```
results(seidel3.out)
CALCULATING TWO-DIMENSIONAL TEMPERATURE FIELD IN CHIMNEY
CROSS-SECTION

DATA ENTERED
equation number  n=9
calculation tolerance  toler=.10000E-04 [C]
max. iteration number niter=    100000
initial temp.    t_pocz=.10000E+02 [C]

CALCULATED TEMPERATURE
   Lp      T[C]
    1    .169665E+02
    2    .308994E+02
    3    .437345E+02
    4    .477868E+02
    5    .732471E+02
    6    .115595E+03
    7    .127413E+03
    8    .217985E+03
    9    .230675E+03
final iteration number= 26
```

The following initial values were assumed: $T_1^{(0)} = T_2^{(0)} = \ldots = T_9^{(0)} =$ 10°C. After $n = 26$ iterations, the following temperature values were obtained:

$$T_1 = 16.96\,^\circ\text{C}, \qquad T_2 = 30.90\,^\circ\text{C}, \qquad T_3 = 43.73\,^\circ\text{C},$$
$$T_4 = 47.78\,^\circ\text{C}, \qquad T_5 = 73.25\,^\circ\text{C}, \qquad T_6 = 115.59\,^\circ\text{C},$$
$$T_7 = 127.41\,^\circ\text{C}, \qquad T_8 = 217.98\,^\circ\text{C}, \qquad T_9 = 230.67\,^\circ\text{C}.$$

The accuracy of this solution can be evaluated if one calculates the heat flow, which is dissipated through the outer and inner chimney surface on the length of 1 m. Outer surface heat flow is

$$\dot{Q}_z = 8\alpha_z\left[\frac{\Delta x}{2}(T_1 - T_z) + \Delta x(T_2 - T_z) + \Delta x(T_3 - T_z) + \frac{\Delta x}{2}(T_4 - T_z)\right] =$$

$$= 8\cdot 20\cdot 0.125\left(\frac{1}{2}16.96 + 30.90 + 43.73 + \frac{1}{2}47.78 - 3\cdot 10\right) = 1540 \text{ W}.$$

Outer surface heat flow can be calculated from the formula below

$$\dot{Q}_w = 8\left[\alpha_w\frac{\Delta x}{2}(T_w - T_8) + \alpha_w\frac{\Delta x}{2}(T_w - T_9)\right],$$

where from, after transformations, one has

$$\dot{Q}_w = 8\alpha_w(\Delta x)\left(T_w - \frac{T_8 + T_9}{2}\right) = 8\cdot 60\cdot 0.125\left(250 - \frac{217.98 + 230.67}{2}\right) =$$

$$= 1540.5 \text{ W}.$$

Heat flows $\dot{Q}_z$ and $\dot{Q}_w$ should be equal, since the heat conduction is steady-state. Relative difference $\Delta\dot{Q}$ is at

$$\Delta\dot{Q} = \frac{\dot{Q}_w - \dot{Q}_z}{\dot{Q}_w}\cdot 100\% = \frac{1540.5 - 1540}{1540.5}\cdot 100\% = 0.03\%.$$

The difference between $\dot{Q}_z$ *i* $\dot{Q}_w$ is attributed to a rather small number of control volumes.

Exercise 11.7 Pseudo-Transient Determination of Steady-State Temperature Distribution in a Square Cross-Section of a Chimney; Heat Transfer by Convection and Radiation on an Outer Surface of a Chimney

Determine steady-state temperature distribution in a cross-section of a chimney presented in Fig. 11.6; allow for both, heat transfer by convection and heat transfer by radiation. Assume that the equivalent emissivity of the chimney's interior is $\varepsilon_w = 0.9$, while the outer surface $\varepsilon_z = 0.8$. Other values remain the same as they are in Ex. 11.6. Use control volume method to determine temperature distribution.

Solution

The presence of radiation renders this problem to be non-linear. If the problem is solved as a steady-state problem, one obtains a non-linear algebraic equation system for node temperature, which can be solved by Newton-Raphson method or by other iteration methods. The problem in question can be also solved as a transient problem, since there are a number of well developed methods, which can be used for solving non-linear ordinary differential equation systems, for e.g., Rung- Kutta method. Temperature distribution is determined after a sufficiently long duration; that is, the unknown steady-state temperature distribution is determined. Heat balance equations for individual control volumes (Fig. 11.6) have the form (when $\Delta x = \Delta y$):

- Node 1

$$\frac{1}{2}(\Delta x)\frac{1}{2}(\Delta x)c\rho\frac{dT_1}{dt} = \alpha_z \cdot (\Delta x)(T_z - T_1) + \varepsilon_z\sigma \cdot (\Delta x)\left(T_z^4 - T_1^4\right) + \\ +\lambda\frac{\Delta x}{2}\frac{T_{13} - T_1}{\Delta x} + \lambda\frac{\Delta x}{2}\frac{T_2 - T_1}{\Delta x}, \tag{1}$$

hence, one obtains

$$\frac{dT_1}{dt} = \frac{4a}{(\Delta x)^2}\left[(\Delta Bi_z)T_z - \frac{\varepsilon_z\sigma \cdot (\Delta x)}{\lambda}\left(T_z^4 - T_1^4\right) - (1 + \Delta Bi_z)T_1 + \right. \\ \left. + \frac{T_2}{2} + \frac{T_{13}}{2}\right], \tag{2}$$

following that, when $T_{13} = T_2$

$$\frac{dT_1}{dt} = \frac{4a}{(\Delta x)^2}\left[(\Delta Bi_z)T_z - \frac{\varepsilon_z\sigma \cdot (\Delta x)}{\lambda}\left(T_z^4 - T_1^4\right) - (1 + \Delta Bi_z)T_1 + T_2\right], \tag{3}$$

where temperatures are expressed in Kelvin, while $\sigma = 5.67 \cdot 10^{-8}$ W/(m^2·K^4) is the Stefan-Boltzmann constant.

- Node 2

$$\frac{1}{2}(\Delta x)(\Delta x)c\rho\frac{dT_2}{dt} = \alpha_z \cdot (\Delta x)(T_z - T_2) + \varepsilon_z\sigma \cdot (\Delta x)\left(T_z^4 - T_2^4\right) + \\ +\lambda\frac{\Delta x}{2}\frac{T_1 - T_2}{\Delta x} + \lambda\frac{\Delta x}{2}\frac{T_3 - T_2}{\Delta x} + \lambda(\Delta x)\frac{T_5 - T_2}{\Delta x}, \tag{4}$$

which results in

$$\frac{dT_2}{dt} = \frac{2a}{(\Delta x)^2}\left[(\Delta Bi_z)T_z + \frac{\varepsilon_z\sigma \cdot (\Delta x)}{\lambda}\left(T_z^4 - T_2^4\right) - (2 + \Delta Bi_z)T_2 + \right. \\ \left. + \frac{T_1}{2} + \frac{T_3}{2} + T_5\right]. \tag{5}$$

• Node 3

$$\frac{1}{2}(\Delta x)(\Delta x)c\rho\frac{dT_3}{dt}=\alpha_z\cdot(\Delta x)(T_z-T_3)+\varepsilon_z\sigma\cdot(\Delta x)\left(T_z^4-T_3^4\right)+ \\ +\lambda\frac{\Delta x}{2}\frac{T_2-T_3}{\Delta x}+\lambda\frac{\Delta x}{2}\frac{T_4-T_3}{\Delta x}+\lambda\cdot(\Delta x)\frac{T_6-T_3}{\Delta x}, \tag{6}$$

$$\frac{dT_3}{dt}=\frac{2a}{(\Delta x)^2}\left[(\Delta Bi_z)T_z+\frac{\varepsilon_z\sigma\cdot(\Delta x)}{\lambda}\left(T_z^4-T_3^4\right)-(2+\Delta Bi_z)T_3+ \right. \\ \left. +\frac{T_2}{2}+\frac{T_4}{2}+T_6\right]. \tag{7}$$

• Node 4

$$(\Delta x)\frac{1}{2}(\Delta x)c\rho\frac{dT_4}{dt}=\alpha_z(\Delta x)(T_z-T_4)+\varepsilon_z\sigma\cdot(\Delta x)\left(T_z^4-T_4^4\right)+ \\ +\lambda\frac{\Delta x}{2}\frac{T_{10}-T_4}{\Delta x}+\lambda\frac{\Delta x}{2}\frac{T_3-T_4}{\Delta x}+\lambda\cdot(\Delta x)\frac{T_7-T_4}{\Delta x}, \tag{8}$$

hence, when $T_{10}=T_3$,

$$\frac{dT_4}{dt}=\frac{2a}{(\Delta x)^2}\left[(\Delta Bi_z)T_z+\frac{\varepsilon_z\sigma\cdot(\Delta x)}{\lambda}\left(T_z^4-T_4^4\right)-(2+\Delta Bi_z)T_4+T_3+T_7\right]. \tag{9}$$

• Node 5

$$(\Delta x)(\Delta x)c\rho\frac{dT_5}{dt}=\lambda\cdot(\Delta x)\frac{T_2-T_5}{\Delta x}+\lambda\cdot(\Delta x)\frac{T_6-T_5}{\Delta x}+ \\ +\lambda\cdot(\Delta x)\frac{T_{13}-T_5}{\Delta x}+\lambda\cdot(\Delta x)\frac{T_{14}-T_5}{\Delta x}, \tag{10}$$

where from, after transformations and when $T_{13}=T_2$ and $T_{14}=T_6$, one has

$$\frac{dT_5}{dt}=\frac{2a}{(\Delta x)^2}(T_2+T_6-2T_5), \tag{11}$$

• Node 6

$$(\Delta x)^2c\rho\frac{dT_6}{dt}=\lambda\cdot(\Delta x)\frac{T_3-T_6}{\Delta x}+\lambda\cdot(\Delta x)\frac{T_5-T_6}{\Delta x}+ \\ +\lambda\cdot(\Delta x)\frac{T_7-T_6}{\Delta x}+\lambda\cdot(\Delta x)\frac{T_8-T_6}{\Delta x}, \tag{12}$$

where from, after transformations, one obtains

$$\frac{dT_6}{dt} = \frac{2a}{(\Delta x)^2}\left(T_3 + T_5 + T_7 + T_8 - 4T_6\right). \tag{13}$$

- Node 7

$$(\Delta x)^2 c\rho \frac{dT_7}{dt} = \lambda \cdot (\Delta x)\frac{T_4 - T_7}{\Delta x} + \lambda \cdot (\Delta x)\frac{T_6 - T_7}{\Delta x} + \lambda \cdot (\Delta x)\frac{T_9 - T_7}{\Delta x} + \\ + \lambda \cdot (\Delta x)\frac{T_{11} - T_7}{\Delta x}, \tag{14}$$

hence, when $T_{11} = T_6$, one has

$$\frac{dT_7}{dt} = \frac{a}{(\Delta x)^2}\left(T_4 + 2T_6 + T_9 - 4T_7\right), \tag{15}$$

- Node 8

$$\frac{3}{4}(\Delta x)^2 c\rho \frac{dT_8}{dt} = \varepsilon_w \sigma \cdot (\Delta x)\left(T_w^4 - T_8^4\right) + \alpha_w \cdot (\Delta x)\left(T_w - T_8\right) + \\ + \lambda \cdot (\Delta x)\frac{T_6 - T_8}{\Delta x} + \lambda \frac{\Delta x}{2}\frac{T_9 - T_8}{\Delta x} + \lambda \frac{\Delta x}{2}\frac{T_{18} - T_8}{\Delta x} + \lambda \cdot (\Delta x)\frac{T_{14} - T_8}{\Delta x}, \tag{16}$$

hence, after transformations and when $T_{14} = T_6$ and $T_{15} = T_9$, one has

$$\frac{dT_8}{dt} = \frac{4}{3}\frac{a}{(\Delta x)^2}\left[\left(\Delta Bi_w\right)T_w + 2T_6 + T_9 - \left(3 + \Delta Bi_w\right)T_8 \right. \\ \left. + \frac{\varepsilon_w \sigma \cdot (\Delta x)}{\lambda}\left(T_w^4 - T_8^4\right)\right]. \tag{17}$$

- Node 9

$$\frac{1}{2}(\Delta x)^2 c\rho \frac{dT_9}{dt} = \varepsilon_w \sigma \cdot (\Delta x)\left(T_w^4 - T_9^4\right) + \alpha_w \cdot (\Delta x)\left(T_w - T_9\right) + \\ + \lambda \cdot (\Delta x)\frac{T_7 - T_9}{\Delta x} + \lambda \frac{\Delta x}{2}\frac{T_8 - T_9}{\Delta x} + \lambda \frac{\Delta x}{2}\frac{T_{12} - T_9}{\Delta x}, \tag{18}$$

hence, when $T_{12} = T_8$, one gets

$$\frac{dT_9}{dt} = \frac{2a}{(\Delta x)^2}\left[\left(\Delta Bi_w\right)T_w + \frac{\varepsilon_w \sigma \cdot (\Delta x)}{\lambda}\left(T_w^4 - T_9^4\right) + T_7 + T_8 - \left(2 + \Delta Bi_w\right)T_9\right] \tag{19}$$

After assuming for $\Delta x = \Delta y = (b - a)/2 = (0.375 - 0.125)/2 = 0.125$ m, $\lambda = 1.25$ W/(m·K), $\alpha_w = 60$ W/(m²·K), $\alpha_z = 20$ W/(m²·K), $T_w = 250 + 273.15 = 523.15$ K, $T_z = 10 + 273.15 = 283.15$ K , one can calculate ΔBi_w and ΔBi_z:

$$\Delta Bi_w = \frac{\alpha_w \cdot (\Delta x)}{\lambda} = \frac{60 \cdot 0.125}{1.25} = 6.0,$$

$$\Delta Bi_z = \frac{\alpha_z \cdot (\Delta x)}{\lambda} = \frac{20 \cdot 0.125}{1.25} = 2.0$$

and

$$\frac{\varepsilon_w \sigma \cdot (\Delta x)}{\lambda} = \frac{0.9 \cdot 5.67 \cdot 10^{-8} \cdot 0.125}{1.25} = 5.103 \cdot 10^{-9}\ 1/\mathrm{K}^3,$$

$$\frac{\varepsilon_z \sigma \cdot (\Delta x)}{\lambda} = \frac{0.8 \cdot 5.67 \cdot 10^{-8} \cdot 0.125}{1.25} = 4.536 \cdot 10^{-9}\ 1/\mathrm{K}^3.$$

In order to calculate temperature distribution, one needs to know what the value of heat diffusivity $a = \lambda / c\rho$ is. To quickly reach a steady-state, a should have a large value, e.g. $a = 1.5625 \cdot 10^{-5}$ m/s²; then $a/(\Delta x)^2 = 1.5625 \cdot 10^{-5}/(0.125)^2 = 0.001$ 1/s . If we were to assume that $a = 5.2 \cdot 10^{-7}$ m²/s, we would significantly lengthen the whole calculation, since the transient state (chimney heating) would last longer. At an initial moment when $t = 0$ s, chimney temperature is uniform and is of 10°C, thus $T_1(0) = T_2(0) = \ldots = T_9(0) = 283.15$ K. Once all the data is taken into consideration, (3), (5), (7), (9), (11), (13), (15), (17) and (19) assume, respectively, the following forms:

$$\frac{dT_1}{dt} = 0.004\left[566.3 + 4.536 \cdot 10^{-9}\left(283.15^4 - T_1^4\right) - 3T_1 + T_2\right], \tag{20}$$

$$\frac{dT_2}{dt} = 0.002\left[566.3 + 4.536 \cdot 10^{-9}\left(283.15^4 - T_2^4\right) - 4T_2 + \frac{T_1}{2} + \frac{T_3}{2} + T_5\right], \tag{21}$$

$$\frac{dT_3}{dt} = 0.002\left[566.3 + 4.536 \cdot 10^{-9}\left(283.15^4 - T_3^4\right) - 4T_3 + \frac{T_2}{2} + \frac{T_4}{2} + T_6\right], \tag{22}$$

$$\frac{dT_4}{dt} = 0.002\left[566.3 + 4.536 \cdot 10^{-9}\left(283.15^4 - T_4^4\right) - 4T_4 + T_7 + T_3\right], \tag{23}$$

$$\frac{dT_5}{dt} = 0.002\left(T_2 - 2T_5 + T_6\right), \tag{24}$$

$$\frac{dT_6}{dt} = 0.001\left(T_3 + T_5 + T_7 + T_8 - 4T_6\right), \tag{25}$$

$$\frac{dT_7}{dt} = 0.001\left(T_4 + 2T_6 + T_9 - 4T_7\right), \tag{26}$$

$$\frac{dT_8}{dt} = \frac{0.004}{3}\left[3138.9 + 5.103 \cdot 10^{-9}\left(523.15^4 - T_8^4\right) + 2T_6 + T_9 - 9T_8\right], \tag{27}$$

$$\frac{dT_9}{dt} = 0.002\left[3138.9 + 5.103 \cdot 10^{-9}\left(523.15^4 - T_9^4\right) + T_7 + T_8 - 8T_9\right]. \tag{28}$$

Initial conditions have the form

$$T_1(0) = T_2(0) = \ \ldots \ = T_9(0) = 283.15 \text{ K} \tag{29}$$

Problem (20)–(29) will be solved by the Rung-Kutta method. A subprogram for the integration of the equation system by means of the Rung-Kutta method is presented in Appendix G. If we assume that temperature is already in a steady-state after t = 9120 s, then the following temperature values will be obtained:

$T_1 = 288.36 \text{ K} = 15.21^\circ\text{C}$, $T_2 = 300.99 \text{ K} = 27.84^\circ\text{C}$,

$T_3 = 312.55 \text{ K} = 39.40^\circ\text{C}$, $T_4 = 315.99\ K = 42.84^\circ\text{C}$,

$T_5 = 345.27 \text{ K} = 72.12^\circ\text{C}$, $T_6 = 389.56 \text{ K} = 116.41^\circ\text{C}$,

$T_7 = 401.19 \text{ K} = 128.04^\circ\text{C}$, $T_8 = 499.22 \text{ K} = 226.07^\circ\text{C}$,

$T_9 = 509.66 \text{ K} = 236.51^\circ\text{C}$.

The integration step in Rung-Kutta method was assumed to equal Δt = 60 s. Heat flow transferred by an inner surface of the chimney within the length of 1 m is at

$$\dot{Q}_w = 4\Big[\alpha_w \cdot (\Delta x)(T_w - T_8) + \varepsilon_w \sigma \cdot (\Delta x)\left(T_w^4 - T_8^4\right) + \alpha_w \cdot (\Delta x)(T_w - T_9) +$$
$$+\varepsilon_w \sigma \cdot (\Delta x)\left(T_w^4 - T_9^4\right)\Big] = 4\big[60 \cdot 0.125(523.15 - 499.22) + 0.9 \cdot 5.67 \times$$
$$\times 10^{-8} \cdot 0.125 \cdot \left(523.15^4 - 499.22^4\right) + 60 \cdot 0.125(523.15 - 509.66) +$$
$$+0.9 \cdot 5.67 \cdot 10^{-8} \cdot 0.125 \cdot \left(523.15^4 - 509.66^4\right)\big] = 1638.65 \text{ W}.$$

Heat flow given off by an outer surface is

$$\dot{Q}_z = 8\Bigg[\alpha_z \frac{\Delta x}{2}(T_1 - T_z) + \frac{1}{2}\varepsilon_z \sigma \cdot (\Delta x)\left(T_1^4 - T_z^4\right) + \alpha_z \cdot (\Delta x)(T_2 - T_z) +$$
$$+\varepsilon_z \sigma \cdot (\Delta x)\left(T_2^4 - T_z^4\right) + \alpha_z \cdot (\Delta x)(T_3 - T_z) + \varepsilon_z \sigma \cdot (\Delta x)\left(T_3^4 - T_z^4\right) +$$
$$+\alpha_z \frac{\Delta x}{2}(T_4 - T_z) + \frac{1}{2}\varepsilon_z \sigma \cdot (\Delta x)\left(T_4^4 - T_z^4\right)\Bigg] = 8\Bigg[\alpha_z \cdot (\Delta x) \cdot \left[\frac{T_1}{2} + T_2 +\right.$$
$$\left. + T_3 + \frac{T_4}{2} - 3T_z\right] + \varepsilon_z \sigma \cdot (\Delta x)\left(\frac{T_1^4}{2} + T_2^4 + T_3^4 + \frac{T_2^4}{2} - 3T_z^4\right)\Bigg],$$

hence, after substitution of the numerical values, one has

$$\dot{Q}_z = 8\Bigg[20 \cdot 0.125 \cdot \left(\frac{288.36}{2} + 300.99 + 312.55 + \frac{315.99}{2} - 3 \cdot 283.15\right) +$$
$$+0.8 \cdot 5.67 \cdot 10^{-8} \cdot 0.125\left(\frac{288.36^4}{2} + 300.99^4 + 312.55^4 +\right.$$
$$\left.\frac{300.99^4}{2} - 3 \cdot 283.15^4\right)\Bigg] = 1598.71 \text{ W}$$

Heat flows $\dot{Q}_w$ and $\dot{Q}_z$ should be equal. Relative difference

$$\Delta\dot{Q} = \frac{\dot{Q}_w - \dot{Q}_z}{\dot{Q}_w} \cdot 100\% = \frac{1638.65 - 1598.71}{1638.65} \cdot 100\% = 2.44\%$$

results from the small number of control volumes. By increasing the number of control volumes, one can improve the accuracy of the obtained results and at the same time decrease the difference between $\dot{Q}_w$ and $\dot{Q}_z$. If heat exchange by radiation is neglected in (20)–(28), i.e. when $\varepsilon_w = \varepsilon_z = 0$, then the following node temperature values are obtained (after $t = 9780$ s):

$T_1 = 290.12 \text{ K} = 16.97^\circ\text{C}$, $T_2 = 304.05 \text{ K} = 30.90^\circ\text{C}$,
$T_3 = 316.88 \text{ K} = 43.73^\circ\text{C}$, $T_4 = 320.94 \text{ K} = 47.79^\circ\text{C}$,
$T_5 = 346.40 \text{ K} = 73.25^\circ\text{C}$, $T_6 = 388.74 \text{ K} = 115.59^\circ\text{C}$,
$T_7 = 400.56 \text{ K} = 127.41^\circ\text{C}$, $T_8 = 491.13 \text{ K} = 217.98^\circ\text{C}$,
$T_9 = 503.82 \text{ K} = 230.67^\circ\text{C}$,

They are almost identical to temperatures calculated in Ex.11.6. Due to the fact that steady-state was treated as a particular case of transient state, when $t \to \infty$, temperature distribution was calculated in a relatively simple way, as it was not necessary to apply iteration and to select approximate, initial temperature values at the beginning of the iteration process.

Computational program in the FORTRAN language used for determining temperature distribution in a cross-section of a chimney

```
c     Calculating two-dim. temperature field in a chimney
c     cross-section (Fig. 11.6) by means of control volume
c     method, equation system solved by Runge- Kutta method
      program rkutta
      integer co_ile_druk
      dimension y(6000),f(6000)
      open(unit=1,file='rkutta.in')
      open(unit=2,file='rkutta.out')
      read(1,*) t,dt,m,n_row,n_time
      read(1,*) t_init
      read(1,*) co_ile_druk
      write(2,'(a)')"CALCULATING 2-D TEMP FIELD IN A CHIMNEY"
      write(2,'(/a)') "DATA ENTERED"
      write(2,'(a,e10.5,a)')  "initial time=",t," [s]"
      write(2,'(a,e10.5,a)')  "time step=",dt," [s]"
      write(2,'(a,i10)')    "parameter m=",m
      write(2,'(a,i10)')    "equation number  n_row=",n_row
      write(2,'(a,i10)')"time step number  n_time=",n_time
      write(2,'(a,e10.5,a)')"initial temp. t_init=",t_init,
     &" [C]"
      write(2,'(a,i10)')  "printing frequency=", co_ile_druk
      z_w=5.103E-9
      z_z=4.536E-9
      write(2,'(/a)')"CALCULATED TEMPERATURE [K]"
      write(2,'(a,a)')"t[s]   T(1)    T(2)    T(3)    T(4)",
     &"    T(5)    T(6)    T(7)    T(8)    T(9)"
      numerator=0
      kolejny=1
      to i=1,n_row
      y(i)= t_init
      enddo
```

```
      write(2,'(f9.0,9f8.3)') t,((y(i)),i=1,n_row)

    8 if( (kolejny-n_time).le.0.0 ) then
              kolejny= kolejny+1
    6    licznik = licznik+1
              call runge (n_row,y,f,t,dt,m,k)
              goto (10,20),k
   10 f(1)=0.004*(566.3+z_z*(283.15**4-y(1)**4)-3.*y(1)+y(2))
      f(2)=0.002*(566.3+z_z*(283.15**4-y(2)**4)-4.*y(2)+
     &0.5*y(1)+0.5*y(3)+y(5))
      f(3)=0.002*(566.3+z_z*(283.15**4-y(3)**4)-4.*y(3)+
     &0.5*y(2)+0.5*y(4)+y(6))
      f(4)=0.002*(566.3+z_z*(283.15**4-y(4)**4)-
     &4.*y(4)+y(7)+y(3))
      f(5)=0.002*(y(2)-2.*y(5)+y(6))
      f(6)=0.001*(y(3)+y(5)+y(7)+y(8)-4.*y(6))
      f(7)=0.001*(y(4)+2.*y(6)+y(9)-4.*y(7))
      f(8)=0.004*(3138.9+z_w*(523.15**4-y(8)**4)+2.*y(6)+
     &y(9)-9.*y(8))/3.
      f(9)=0.002*(3138.9+z_w*(523.15**4-y(9)**4)+y(7)+
     &y(8)-8.*y(9))
         goto 6
   20    continue
      if((float(licznik / co_ile_druk)* co_ile_druk)
     & .ne. licznik) goto 6
         write(2,'(f9.0,9f8.3)') t,((y(i)),i=1,n_row)
              goto 8
       endif
      stop
      end
```

```
data(rkutta.in)
0.0 60. 0 9 200
283.15
1
```

```
results(rkutta.out - set part)
CALCULATING 2-D TEMP FIELD IN A CHIMNEY

DATA ENTERED
initial time=.00000E+00 [s]
time step=.60000E+02 [s]
parameter m=          0
equation number  n_row=         9
time step number  n_time=       200
initial temp.    t_init=.28315E+03 [C]
```

```
printing frequency=         1

CALCULATED TEMPERATURE [K]
t[s]    T(1)     T(2)     T(3)     T(4)     T(5)     T(6)     T(7)
T(8)     T(9)
0. 283.150 283.150 283.150 283.150 283.150 283.150 283.150
283.150 283.150
60. 283.150 283.158 283.268 283.306 283.277 286.477 287.810
387.704 418.660
120. 283.155 283.217 283.892 284.131 283.975 293.481 296.517
439.175 469.283
180. 283.178 283.394 285.038 285.567 285.357 301.487 305.863
463.512 488.055
240. 283.231 283.722 286.546 287.393 287.323 309.332 314.766
475.076 495.437
     .................
9000. 288.361 300.990 312.555 315.994 345.275 389.560 401.193
499.218 509.656
9060. 288.361 300.990 312.555 315.994 345.275 389.560 401.193
499.218 509.656
9120. 288.361 300.990 312.555 315.995 345.275 389.560 401.193
499.218 509.656
9180. 288.361 300.990 312.555 315.995 345.275 389.560 401.193
499.218 509.656
9240. 288.361 300.990 312.555 315.995 345.275 389.560 401.193
499.218 509.656
     .................
11820. 288.361 300.990 312.555 315.995 345.275 389.560
401.193 499.218 509.656
11880. 288.361 300.990 312.555 315.995 345.275 389.560
401.193 499.218 509.656
11940. 288.361 300.990 312.555 315.995 345.275 389.560
401.193 499.218 509.656
12000. 288.361 300.990 312.555 315.995 345.275 389.560
401.193 499.218 509.656
```

Exercise 11.8 Finite Element Method

Describe the procedure for calculating temperature fields by means of the finite element method (FEM). List main advantages and disadvantages of FEM.

Historical Development of FEM

The precursor of the FEM method was a mathematician by the name of Courant, who in 1943 employed the segmental approximation by polynomial method in combination with the variational method in order to solve the torsion problem [3]. The method was developed and its present name, namely the *finite element method* appeared in 1950s [2, 7]. Traditional analytical approximation methods, such as variational methods or Galerkin methods [5] have many limitations, which arise from the approximation of solution within the entire analyzed area by means of a single function. It is almost impossible to apply analytical approximation method in an instance when the shape of an analyzed region is complex and its boundary conditions change in a time and position. Similar limitations, with respect to shape, characterize the classical finite difference method, in which partial derivatives in differential equations are approximated by means of differentce quotients. Finite difference method allows one to analyze different boundary conditions; the shape of a body, however, should be isometric, e.g. a rectangle, prism, cylinder, sphere, or a flat, cylindrical or spherical wall. The universal applicability of the finite difference method, such as the *control volume method*, also known as *finite volume method*, allows to find a solution in construction elements or in the complex shape regions [1]. Capabilities of this method are very similar to the capabilities of FEM. A region, whose temperature distribution we are trying to establish, can be divided into control volumes (cells) of arbitrary shapes; due to this reason, one can analyze curvilinear boundaries or other complex-shape boundaries.

The application of finite element method in heat transfer and fluid mechanics also has certain limitations. At the boundary of a given element, heat or mass that flows from one element can differ from the heat that flows towards adjacent element, in spite of the fact that the same section of the boundary is in both instances analyzed. This is due to the fact that a discontinuity of heat flux occurs in FEM at the boundary of adjacent elements. To eliminate this problem, a so called *finite element balance method* was developed. Another difference between FEM and the control (finite) volume method is the approximation method for a temperature derivative after time in transient problems. In FEM, thermal capacity of an element is distributed among the element nodes at appropriate weights; in finite volume method, however, thermal capacity of a cell (element) is concentrated in a single node that lies inside the cell. From the comparison of calculation results obtained by means of FEM and finite volume method, it is evident that the concentration of thermal capacity in a single point not only does not tamper with calculation accuracy but increases it. In inverse problems, concentration of thermal capacity in a single point, which lies inside the control volume, improves the solution stability.

Finally, one can conclude that although finite difference method and FEM were treated initially as two separate methods, the discrepancy is almost invisible between the finite volume method, which derives from finite difference method, and the FEM balance method, with a concentrated thermal capacity in finite elements.

Also the functions, which interpolate temperature distribution (or other unknown quantities) inside the finite volume or a finite element can be employed in both methods.

Solution

FEM consists of the following calculation steps:

1. Division of an area into finite elements (a grid generation), Fig. 11.7.
2. Mathematical formulation of Galerkin or variational method (Ritz method) for the analyzed boundary or initial-boundary problem within the area of a single element.

3. Selection of functions, which interpolate temperature distribution inside the element (shape function).
4. Determination of an algebraic equation system for a steady-state problem or of an ordinary differential equation system for transient problems in a single element by means of Galerkin or variational method, formulated in step 2. The equation number equals the node number in a given element, since node temperatures are the unknown quantities in the element.
5. Summing up of equation systems for individual elements, with an aim to create a single universal node-temperature equation system for the whole analyzed region.
6. Allowing for the parameters present in the boundary conditions of the global equation system.
7. Solving the algebraic equation system in the case of a steady-state problem or the ordinary differential equation system in the case of a transient problem.
8. Calculation of heat flux, heat flow and other secondary quantities and graphical representation of the calculation results (*post-processing*).

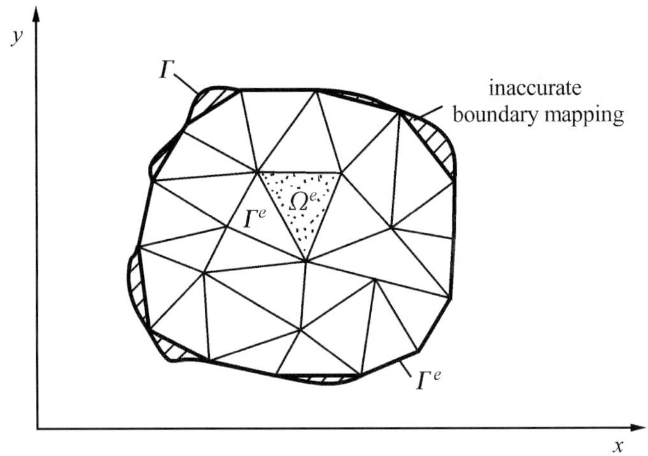

Fig. 11.7. Division of an area into finite elements

The procedure outlined above is typical of large commercial programs, designed to provide solutions to problems from different disciplines. In terms of individual solutions to specific problems and a development of ones own computational program, the procedure steps can have a different sequence; for instance, boundary conditions can already be accounted for when creating algebraic equations for a given element, i.e. at point 4. Also the global equation system can be created in a different way by summing up, for instance, algebraic equations (or ordinary differential equations) for node i, for example, obtained for elements with the same shape coefficient N_i around the node in Galerkin method.

Assuming that few elements share a common node i (Fig. 11.8), the shape coefficient N_i assumed in Galerkin method occurs only in elements with common node i. In other nodes, the value of such coefficient equals zero. This is the reason why we can add up the equations obtained for all elements with the same shape coefficient N_i around node i when creating a global equation system.

If the same is done with respect to all other nodes, a global algebraic equation system or ordinary differential equation system is obtained for node temperatures that can be solved by means of different methods.

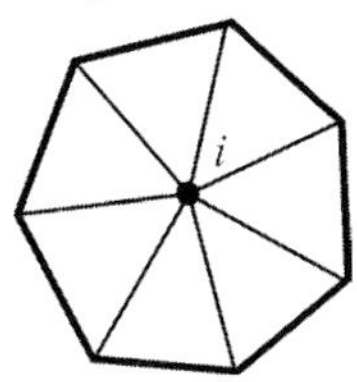

Fig. 11.8. Elements taken into consideration when determining an algebraic equation for steady-state problems or an ordinary differential equation for transient problems in a single node

The second method for creating a global equation system is the same as in the finite volume method, in which heat balance is written for node i.

Finite element method has the following advantages:

- It is suited for problem analysis in complex shape bodies.
- Boundary conditions can be non-linear and time and location-dependent .
- There are number of very good software programs that enable one to quickly solve numerous problems, including problems related to deformation mechanics, heat transfer and fluid mechanics.
- One can solve non-linear problems, when thermo-physical properties of a material are temperature-dependent, and problems in heterogeneous bodies with location-dependent properties in, for instance, composite or anisotropic materials.
- A division of an area into finite elements is automatically carried out (using commonly available software programs), which makes it easier to evaluate accuracy of obtained results by increasing density of an element grid.
- Calculation results are obtained in a graphical and numerical form, which make it easier for a user to quickly analyze obtained results (majority of software programs can be installed in personal computers). The software costs are gradually decreasing.

Finite element method is not, however, free of drawbacks; its main drawbacks are as follow:

- The initial installation costs of the FEM software program are very high.
- Source programs are usually not included in the software set; thus, there is no possibility for program modification or improvement.
- If the problem under analysis is part of a larger problem, it is difficult then to combine one software set with one's own programs or with other sets, especially if the problem is to be solved in an on-line mode.
- Particular attention should be paid to the accuracy of results obtained by means of FEM. The apparent ease, which the results are obtained with in a graphical form is rather deceptive. Even when boundary conditions are prescribed incorrect, e.g. when the end conditions for a construction element are incorrectly set during the determination of thermal stresses, the obtained results seem to be correct at a first glance.

Exercise 11.9 Linear Functions That Interpolate Temperature Distribution (Shape Functions) Inside Triangular and Rectangular Elements

Describe the simplest forms of temperature-distribution-interpolating functions inside triangular and rectangular elements (*shape functions*).

Solution

First, we will discuss triangular elements (Fig. 11.9).

Temperature distribution in a triangular element will be approximated by a linear function

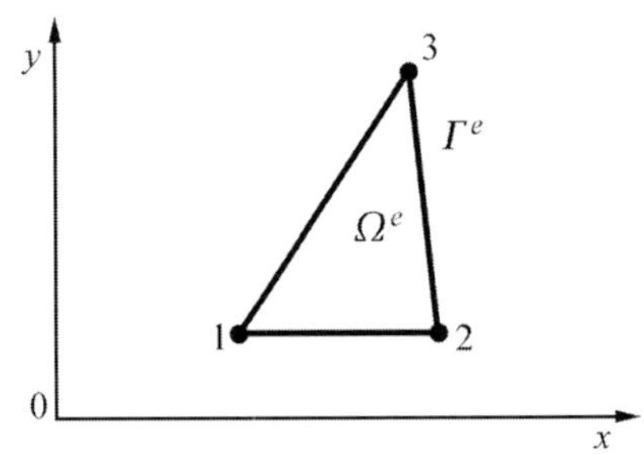

Fig. 11.9. Triangular finite element

$$T^e(x,y)=\alpha_1^e+\alpha_2^e x+\alpha_3^e y. \tag{1}$$

Constants α_1^e, α_2^e, α_3^e will be determined from conditions

$$T^e(x_1,y_1)=T_1^e, \qquad T^e(x_2,y_2)=T_2^e, \qquad T^e(x_3,y_3)=T_3^e. \tag{2}$$

By substituting (1) into (2), one obtains the following equation system

$$\begin{aligned}
\alpha_1^e+\alpha_2^e x_1+\alpha_3^e y_1&=T_1^e\\
\alpha_1^e+\alpha_2^e x_2+\alpha_3^e y_2&=T_2^e\\
\alpha_1^e+\alpha_2^e x_3+\alpha_3^e y_3&=T_3^e,
\end{aligned} \tag{3}$$

where from, one has

$$\begin{aligned}
\alpha_1^e&=\frac{1}{2A^e}\left[(x_2y_3-x_3y_2)T_1^e+(x_3y_1-x_1y_3)T_2^e+(x_1y_2-x_2y_1)T_3^e\right],\\
\alpha_2^e&=\frac{1}{2A^e}\left[(y_2-y_3)T_1^e+(y_3-y_1)T_2^e+(y_1-y_2)T_3^e\right],\\
\alpha_3^e&=\frac{1}{2A^e}\left[(x_3-x_2)T_1^e+(x_1-x_3)T_2^e+(x_2-x_1)T_3^e\right],
\end{aligned} \tag{4}$$

where

$$2A^e=\begin{vmatrix}1 & x_1 & y_1\\ 1 & x_2 & y_2\\ 1 & x_3 & y_3\end{vmatrix} \quad (A^e\text{–a surface area of triangle 1-2-3 from Fig. 11.9}). \tag{5}$$

After substituting (4) into (1) and ordering, one has

$$T^e(x,y)=N_1^e\cdot T_1^e+N_2^e\cdot T_2^e+N_3^e\cdot T_3^e, \tag{6}$$

where N_1^e, N_2^e, N_3^e are, so called, *shape functions*, formulated as

$$\begin{aligned}
N_1^e&=\frac{1}{2A^e}\left(a_1^e+b_1^e x+c_1^e y\right),\\
N_2^e&=\frac{1}{2A^e}\left(a_2^e+b_2^e x+c_2^e y\right),\\
N_3^e&=\frac{1}{2A^e}\left(a_3^e+b_3^e x+c_3^e y\right),
\end{aligned} \tag{7}$$

where

$$\begin{aligned}
&a_1^e=x_2y_3-x_3y_2,\quad b_1^e=y_2-y_3,\quad c_1^e=x_3-x_2,\\
&a_2^e=x_3y_1-x_1y_3,\quad b_2^e=y_3-y_1,\quad c_2^e=x_1-x_3,\\
&a_3^e=x_1y_2-x_2y_1,\quad b_3^e=y_1-y_2,\quad c_3^e=x_2-x_1.
\end{aligned} \tag{8}$$

On the basis of temperature distribution inside the element, one can determine heat flux vector

$$\dot{\mathbf{q}} = -\left(\lambda_x \frac{\partial T}{\partial x} \mathbf{i} + \lambda_y \frac{\partial T}{\partial y} \mathbf{j} \right). \tag{9}$$

Derivative $\partial T/\partial x$ is formulated as

$$\frac{\partial T}{\partial x} = \frac{\partial N_1^e}{\partial x} T_1^e + \frac{\partial N_2^e}{\partial x} T_2^e + \frac{\partial N_3^e}{\partial x} T_3^e, \tag{10}$$

hence, after substituting into (7), one gets

$$\begin{aligned} \frac{\partial T}{\partial x} &= \frac{1}{2A^e}\left(b_1^e \cdot T_1^e + b_2^e \cdot T_2^e + b_3^e \cdot T_3^e \right) = \frac{1}{2A^e}\left[\left(y_2 - y_3 \right) T_1^e + \right. \\ &\left. + \left(y_3 - y_1 \right) T_2^e + \left(y_1 - y_2 \right) T_3^e \right]. \end{aligned} \tag{11}$$

Derivative $\partial T/\partial y$ can be calculated in a similar way

$$\frac{\partial T}{\partial y} = \frac{1}{2A^e}\left[\left(x_3 - x_2 \right) T_1^e + \left(x_1 - x_3 \right) T_2^e + \left(x_2 - x_1 \right) T_3^e \right]. \tag{12}$$

It is evident, thus, that components of the heat flux vector

$$\dot{q}_x = -\lambda_x \frac{\partial T}{\partial x} \quad \text{and} \quad \dot{q}_y = -\lambda_y \frac{\partial T}{\partial y} \tag{13}$$

are constant and position-independent inside the element. It seems, therefore, that heat flux equality does not occur on the element boundary, since heat flux is constant yet different in every element. The condition of temperature continuity, however, is preserved. It is easy to demonstrate that the calculated temperature transient is the same for two adjacent elements with common side, regardless of the element in which the transient is calculated. The lack of continuity on the element boundary impairs the accuracy of the solution obtained by means of FEM. In order to calculate heat flux at a specific point inside the analyzed region or to determine heat flow using a boundary segment with an assigned temperature, one should employ a denser element mesh so that a satisfactory result in terms of accuracy could be obtained.

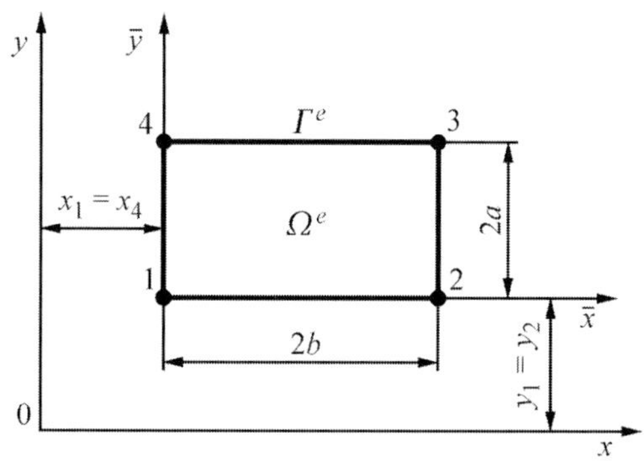

Fig. 11.10. Linear rectangular finite element

A linear tetragonal element, presented in Fig. 11.10, will be discussed below.

Temperature distribution inside the element will be approximated by function

$$T^e = \beta_1^e + \beta_2^e \cdot \bar{x} + \beta_3^e \cdot \bar{y} + \beta_4^e \cdot \bar{x} \cdot \bar{y}. \tag{14}$$

Constants $\beta_1^e, \dots, \beta_4^e$ will be determined from conditions

$$\begin{aligned} &T^e(0,0) = T_1^e, \qquad T^e(2b,0) = T_2^e, \\ &T^e(2b,2a) = T_3^e, \qquad T^e(0,2a) = T_4^e, \end{aligned} \tag{15}$$

from which, once (14) is substituted, the following equation system is obtained:

$$\begin{aligned} &\beta_1^e = T_1^e, \qquad \beta_1^e + 2b\beta_2^e = T_2^e, \\ &\beta_1^e + 2b\beta_2^e + 2a\beta_3^e + 4ab\beta_4^e = T_3^e, \qquad \beta_1^e + 2a\beta_3^e = T_4^e. \end{aligned} \tag{16}$$

Once the above equation system is solved, obtains

$$\begin{aligned} &\beta_1^e = T_1^e, \qquad \beta_2^e = \frac{1}{2b}\left(T_2^e - T_1^e\right), \\ &\beta_3^e = \frac{1}{2a}\left(T_4^e - T_1^e\right), \qquad \beta_4^e = \frac{1}{4ab}\left(T_1^e - T_2^e + T_3^e - T_4^e\right). \end{aligned} \tag{17}$$

After substituting (17) into (14) and after transformations, one obtains

$$T^e = N_1^e \cdot T_1^e + N_2^e \cdot T_2^e + N_3^e \cdot T_3^e + N_4^e \cdot T_4^e, \tag{18}$$

where shape functions are formulated as follow:

$$N_1^e = \left(1 - \frac{\bar{x}}{2b}\right)\left(1 - \frac{\bar{y}}{2a}\right), \qquad N_2^e = \frac{\bar{x}}{2b}\left(1 - \frac{\bar{y}}{2a}\right),$$
$$N_3^e = \frac{\bar{x} \cdot \bar{y}}{4ab}, \qquad N_4^e = \frac{\bar{y}}{2a}\left(1 - \frac{\bar{x}}{2b}\right). \tag{19}$$

Shape functions N_i^e have the following properties:

- $N_i^e\left(x_j, y_j\right) = \delta_{i,j}$ ($i, j = 1, 2, 3$) for a triangular element (20)

and

- $N_i^e\left(\bar{x}_j, \bar{y}_j\right) = \delta_{i,j}$ ($i, j = 1, 2, 3, 4$) for a tetragonal element (21)

where $\delta_{i,j}$ is the Kronecker delta, which satisfies

$$\delta_{i,j} = \begin{cases} 1, & \text{for} \quad i = j \\ 0, & \text{for} \quad i \neq j\,. \end{cases} \tag{22}$$

From properties (20) and (21), it follows that

$$\sum_{i=1}^{n} N_i^e = 1, \tag{23}$$

where n stands for the number of nodes in an element.

Therefore, in i-node the shape function $N_i^e = 1$, in other nodes, however, it equals zero. Aside from linear functions discussed above, one can apply other interpolation functions, for instance, the square functions.

Exercise 11.10 Description of FEM Based on Galerkin Method

Derive basic equations in FEM for a single element using Galerkin method. Assume that two-dimensional temperature field is source-based, while three different boundary conditions of 1st, 2nd and 3rd order are assigned on the body's edge. Allow for the fact that the medium is anisotropic, i.e. $\lambda_x \neq \lambda_y$.

Solution

We need to find a solution for the heat conduction equation

$$\frac{\partial}{\partial x}\left(\lambda_x \frac{\partial T}{\partial x}\right) + \frac{\partial}{\partial y}\left(\lambda_y \frac{\partial T}{\partial y}\right) + \dot{q}_v = 0 \tag{1}$$

when boundary conditions are (Fig. 11.11):

$$T\big|_{\Gamma_T} = T_b, \tag{2}$$

$$\left(\lambda_x \frac{\partial T}{\partial x} n_x + \lambda_y \frac{\partial T}{\partial y} n_y\right)\Bigg|_{\Gamma_q} = \dot{q}_B \tag{3}$$

and

$$\left(\lambda_x \frac{\partial T}{\partial x} n_x + \lambda_y \frac{\partial T}{\partial y} n_y\right)\Bigg|_{\Gamma_\alpha} = \alpha\left(T_{cz} - T\big|_{\Gamma_\alpha}\right). \tag{4}$$

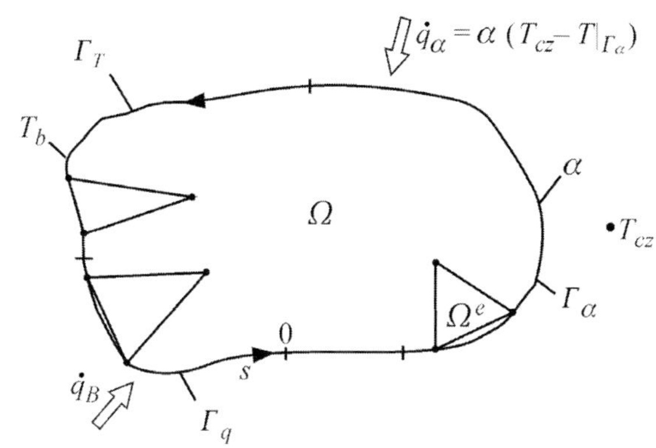

Fig. 11.11. A diagram with different boundary conditions

In (1)–(4), the following symbols are used:

λ_x – thermal conductivity of a material in *x-axis* direction
λ_y – thermal conductivity of a material in *y-axis* direction
T_b – temperature set on the body boundary Γ_T,
$\dot{q}_B$ – heat flux on the body boundary Γ_q,
α – heat transfer coefficient on the body boundary Γ_α,
T_{cz} – temperature of a medium.

In order to make these calculations more widely applicable, three different types of boundary conditions are assumed:

- 1st order kind condition, section Γ_T,
- 2nd order kind condition, section Γ_q,
- 3rd order kind condition, section Γ_α.

It is also assumed that heat flux $\dot{q}$ in (3) is positive, i.e. the body is being heated. Normal to boundary **n** is a unit vector directed to the outside of the region, while its components are equal to directional cosinuses

$$n_x = \cos\varphi, \qquad n_y = \cos\left(\frac{\pi}{2} - \varphi\right) = \sin\varphi, \tag{5}$$

where φ is the slope angle of normal to a horizontal plane.

Boundary conditions (3) and (4) can be written in a slightly different form, if conductivity matrix is entered into the equation

$$[\lambda] = \begin{bmatrix} \lambda_x & 0 \\ 0 & \lambda_y \end{bmatrix} \tag{6}$$

and column vector of temperature gradient

$$\{g\} = \begin{Bmatrix} \dfrac{\partial T}{\partial x} \\ \dfrac{\partial T}{\partial y} \end{Bmatrix}. \tag{7}$$

Heat flux components $\dot{q}_x$ and $\dot{q}_y$ can be written then in the form

$$\begin{Bmatrix} \dot{q}_x \\ \dot{q}_y \end{Bmatrix} = -[\lambda]\{g\}. \tag{8}$$

If we take into account, moreover, that

$$\dot{\mathbf{q}} = -\lambda_x \frac{\partial T}{\partial x}\mathbf{i} - \lambda_y \frac{\partial T}{\partial y}\mathbf{j} = \dot{q}_x\mathbf{i} + \dot{q}_y\mathbf{j}, \qquad \mathbf{n} = n_x\mathbf{i} + n_y\mathbf{j} \tag{9}$$

and

$$-\left(-\lambda_x \frac{\partial T}{\partial x} n_x - \lambda_y \frac{\partial T}{\partial y} n_y\right) = -\,\dot{\mathbf{q}} \cdot \mathbf{n} = \dot{q}_n, \tag{10}$$

then boundary conditions (3) and (4) can be correspondingly written in the form

$$\dot{q}_n\big|_{\Gamma_q^e} = \dot{q}_B, \tag{11}$$

$$\dot{q}_n\big|_{\Gamma_\alpha^e} = \alpha\left(T_{cz} - T\big|_{\Gamma_\alpha^e}\right), \tag{12}$$

where $\dot{q}_n$ is the component value of the normal heat flux. One also assumes that the body thickness in the direction perpendicular to the plane of the diagram is of 1m.

Boundary problem (1)–(4) was formulated for the whole region Ω. In FEM, Galerkin method is first formulated for a single element Ω^e. It is assumed that three types of boundary conditions are assigned, as they are for the whole region, on the boundary of a single element. One needs to apply such a formulation to elements adjacent to body boundary (Fig. 11.11). It is not necessary, however, to consider boundary conditions for elements, which lie inside the body. Temperature distribution inside the element Ω^e is approximated by function

$$T^e(x,y)=\sum_{j=1}^{n}T_j^e\cdot N_j^e(x,y)=\left[N^e\right]\left\{T^e\right\}, \tag{13}$$

where n is the number of nodes in the element, T_j^e – temperature in j-node and $N_j^e(x, y)$ the shape function (interpolation function).

Galerkin method will be used to determine an approximate temperature T_j^e in nodes, j = 1, ..., n.

$$\int_{\Omega^e}\left[\frac{\partial}{\partial x}\left(\lambda_x\frac{\partial T^e}{\partial x}\right)+\frac{\partial}{\partial y}\left(\lambda_y\frac{\partial T^e}{\partial y}\right)+\dot{q}_v\right]N_i^e(x,y)\,dxdy=0. \tag{14}$$

Green theorem will be applied in order to transform (14):

$$\int_{\Omega^e}\left(\frac{\partial G}{\partial x}-\frac{\partial F}{\partial y}\right)dxdy=\oint_{\Gamma^e}\left(Fdx+Gdx\right). \tag{15}$$

Integration on the boundary Γ^e is anti-clockwise.

If one assumes that

$$F=-\lambda_y\frac{\partial T^e}{\partial y}N_i^e \quad\text{and}\quad G=\lambda_x\frac{\partial T^e}{\partial x}N_i^e, \tag{16}$$

then on the basis of (15), one obtains

$$\begin{aligned}&\int_{\Omega^e}\left[\frac{\partial}{\partial x}\left(\lambda_x\frac{\partial T^e}{\partial x}N_i^e\right)+\frac{\partial}{\partial y}\left(\lambda_y\frac{\partial T^e}{\partial y}N_i^e\right)\right]dxdy=\\&=\int_{\Gamma^e}N_i^e\left(-\lambda_y\frac{\partial T^e}{\partial y}dx+\lambda_x\frac{\partial T^e}{\partial x}dy\right).\end{aligned} \tag{17}$$

Once the left-hand-side of (17) is transformed, the equation can be written in the form

$$\int_{\Omega^e}\left[\frac{\partial}{\partial x}\left(\lambda_x\frac{\partial T^e}{\partial x}\right)+\frac{\partial}{\partial y}\left(\lambda_y\frac{\partial T^e}{\partial y}\right)\right]N_i^e dxdy=-\int_{\Omega^e}\left(\lambda_x\frac{\partial T^e}{\partial x}\frac{\partial N_i^e}{\partial x}+\right.$$
$$\left.+\lambda_y\frac{\partial T^e}{\partial y}\frac{\partial N_i^e}{\partial y}\right)dxdy+\int_{\Gamma^e}N_i^e\left(-\lambda_y\frac{\partial T^e}{\partial y}dx+\lambda_x\frac{\partial T^e}{\partial x}dy\right). \quad (18)$$

By substituting (18) into (14), one has

$$\int_{\Omega^e}\left(\lambda_x\frac{\partial T^e}{\partial x}\frac{\partial N_i^e}{\partial x}+\lambda_y\frac{\partial T^e}{\partial y}\frac{\partial N_i^e}{\partial y}\right)dxdy=\int_{\Omega^e}N_i^e\dot{q}_v dxdy+$$
$$+\int_{\Gamma^e}N_i^e\left(-\lambda_y\frac{\partial T^e}{\partial y}dx+\lambda_x\frac{\partial T^e}{\partial x}dy\right). \quad (19)$$

Because of (Fig. 11.12)

$$-dx = ds\cdot\sin\varphi = n_y ds\,, \quad (20)$$

$$dy = ds\cdot\cos\varphi = n_x ds \quad (21)$$

and on the basis of (10), the expression in the brackets in the curvilinear integral in (19) can be transformed in the following way:

$$-\lambda_y\frac{\partial T^e}{\partial y}dx+\lambda_x\frac{\partial T^e}{\partial x}dy=\lambda_y\frac{\partial T}{\partial y}n_y ds+\lambda_x\frac{\partial T}{\partial x}n_x ds=-\dot{\mathbf{q}}\cdot\mathbf{n}=\dot{q}_n ds\,. \quad (22)$$

Hence, from the above and the boundary conditions (11) and (12) in (19), one gets

$$\int_{\Omega^e}\left(\lambda_x\frac{\partial T^e}{\partial x}\frac{\partial N_i^e}{\partial x}+\lambda_y\frac{\partial T^e}{\partial y}\frac{\partial N_i^e}{\partial y}\right)dxdy=\int_{\Omega^e}N_i^e\dot{q}_v dxdy+\int_{\Gamma_q^e}N_i^e\dot{q}_B ds+$$
$$+\int_{\Gamma_\alpha^e}N_i^e\alpha\left(T_{cz}-T^e\right)ds. \quad (23)$$

After substituting (13) into (23), one gets

$$\int_{\Omega^e}\left(\lambda_x\frac{\partial N_i^e}{\partial x}\sum_{j=1}^{n}T_j^e\frac{\partial N_j^e}{\partial x}+\lambda_y\frac{\partial N_i^e}{\partial y}\sum_{j=1}^{n}T_j^e\frac{\partial N_j^e}{\partial y}\right)dxdy=\int_{\Omega^e}N_i^e\dot{q}_v dxdy+$$
$$+\int_{\Gamma_q^e}N_i^e\dot{q}_B ds-\int_{\Gamma_\alpha^e}N_i^e\alpha\sum_{j=1}^{n}T_j^e N_j^e ds+\int_{\Gamma_\alpha^e}N_i^e\alpha T_{cz} ds\,. \quad (24)$$

Equation (24) for i-node can be written in the form

$$\sum_{j=1}^{n}\left(K_{c,ij}^{e}+K_{\alpha,ij}^{e}\right)\cdot T_{j}^{e}=f_{Q,i}^{e}+f_{q,i}^{e}+f_{\alpha,i}^{e}\,, \tag{25}$$

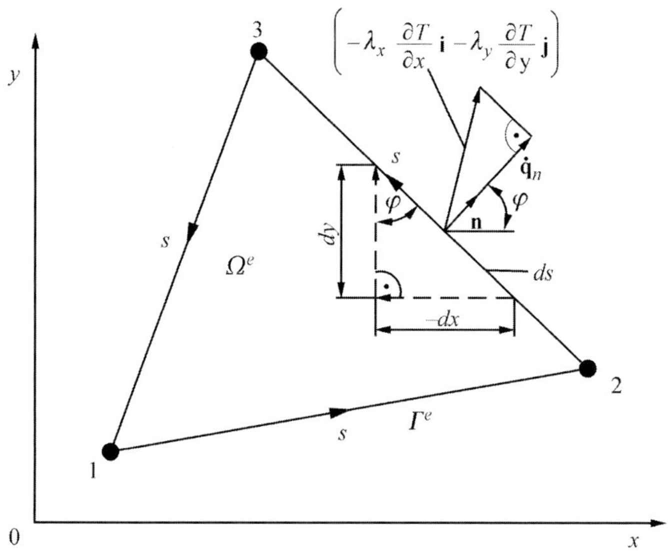

Fig. 11.12. A diagram with a calculation of a curvilinear integral on the element's perimeter (in an anti-clockwise direction)

where

$$K_{c,ij}^{e}=\int_{\Omega^{e}}\left(\lambda_{x}\frac{\partial N_{i}^{e}}{\partial x}\frac{\partial N_{j}^{e}}{\partial x}+\lambda_{y}\frac{\partial N_{i}^{e}}{\partial y}\frac{\partial N_{j}^{e}}{\partial y}\right)dxdy\,, \tag{26}$$

$$K_{\alpha,ij}^{e}=\int_{\Gamma_{\alpha}^{e}}\alpha N_{i}^{e}N_{j}^{e}ds\,, \tag{27}$$

$$f_{Q,i}^{e}=\int_{\Omega^{e}}\dot{q}_{v}N_{i}^{e}dxdy\,, \tag{28}$$

$$f_{q,i}^{e}=\int_{\Gamma_{q}^{e}}\dot{q}_{B}N_{i}^{e}ds \tag{29}$$

and

$$f_{\alpha,i}^{e} = \int_{\Gamma_{\alpha}^{e}} \alpha T_{cz} N_{i}^{e} ds \,. \tag{30}$$

If in (14), and by that in (25) N_j^e is assumed to be the shape function for the consecutive nodes of a finite element, then one obtains for a given element a system of n equations, which can be written in the form

$$\left(\left[K_c^e\right]+\left[K_\alpha^e\right]\right)\left\{T^e\right\}=\left\{f_Q^e\right\}+\left\{f_q^e\right\}+\left\{f_\alpha^e\right\}, \tag{31}$$

where matrix coefficient and elements of column vectors are expressed by (26)–(30), while vector $\{T^e\}$ has the form $\{T^e\} = [\, T_1^e, \ldots , T_n^e]$, where n is the number of nodes in an element. In contrast to other exercises where bold type designates matrixes and column vectors, the traditional notation used in FEM is preserved in (31). [ ] stands for a matrix or row vector, while { } a column vector. Matrix $[K_c^e]$ is called *stiffnes* or *conductivity matrix*. Matrix $[K_c^e]$ is symmetrical, since $K_{c,ij}^{e} = K_{c,ji}^{e}$. Equation (31) forms the basis of FEM for (1).

Equation system (31) is frequently written in a slightly different form. Once (13) is substituted into (7), temperature gradient vector can be written in the form

$$\{g\}=\begin{bmatrix} \dfrac{\partial N_1^e}{\partial x} & \dfrac{\partial N_2^e}{\partial x} & \cdots & \dfrac{\partial N_n^e}{\partial x} \\ \dfrac{\partial N_1^e}{\partial y} & \dfrac{\partial N_2^e}{\partial y} & \cdots & \dfrac{\partial N_n^e}{\partial y} \end{bmatrix}\left\{T^e\right\} . \tag{32}$$

If we denote by [B] the matrix in the square bracket:

$$[B]=\begin{bmatrix} \dfrac{\partial N_1^e}{\partial x} & \dfrac{\partial N_2^e}{\partial x} & \cdots & \dfrac{\partial N_n^e}{\partial x} \\ \dfrac{\partial N_1^e}{\partial y} & \dfrac{\partial N_2^e}{\partial y} & \cdots & \dfrac{\partial N_n^e}{\partial y} \end{bmatrix}, \tag{33}$$

Then $\{g\}$ can be expressed in a shortened form

$$\{g\}=\left[B^e\right]\left\{T^e\right\}. \tag{34}$$

Conductivity matrix $[K_c^e]$ can be written then in the form

$$\left[K_c^e\right]= \int_{\Omega^e}\left[B^e\right]^T [\lambda]\left[B^e\right]dxdy\,. \tag{35}$$

The remaining matrixes and column vectors present in (31) can be expressed in the following way:

$$\left[K_\alpha^e\right] = \int_{\Gamma_\alpha^e} \alpha \left[N^e\right]^T \left[N^e\right] ds,$$
$$\left\{f_Q^e\right\} = \int_{\Omega^e} \dot{q}_v \left[N^e\right]^T dxdy\,, \tag{36}$$

$$\left\{f_q^e\right\} = \int_{\Gamma_q^e} \dot{q}_B \left[N^e\right]^T d\Gamma$$
$$\left\{f_\alpha^e\right\} = \int_{\Gamma_\alpha^e} \alpha T_{cz} \left[N^e\right]^T ds\,, \tag{37}$$

where $\left[N^e\right] = \left[N_1^e,\, N_2^e,\, \ldots,\, N_n^e\right]$

$$\left[N^e\right]^T = \begin{Bmatrix} N_1^e \\ N_2^e \\ \vdots \\ N_n^e \end{Bmatrix},$$ while n is the number of nodes in an element.

Exercise 11.11 Determining Conductivity Matrix for a Rectangular and Triangular Element

Determine conductivity matrix $[K_c^e]$ for a rectangular and triangular element.

Solution

Equation (26) from Ex. 11.10 and formulas for shape functions shown in Ex. 11.9. will be used to calculate the elements of a conductivity matrix.

a) *Conductivity matrix* $[K_c^e]$ *for a finite rectangular element*

Matrix elements $[K_c^e]$ are expressed by (26) in Ex. 11.10

$$K_{c,ij}^e = \int_{\Omega^e} \left(\lambda_x \frac{\partial N_i^e}{\partial x} \frac{\partial N_j^e}{\partial x} + \lambda_y \frac{\partial N_i^e}{\partial y} \frac{\partial N_j^e}{\partial y} \right) dxdy\,. \tag{1}$$

Only two matrix elements will be calculated from (19) in Ex. 11.9 for the shape functions $K^e_{c,11}$ and $K^e_{c,12}$. Let x, y be local coordinates.

After determining derivatives

$$\frac{\partial N_1^e}{\partial x} = -\frac{1}{2b}\left(1-\frac{y}{2a}\right) \text{ and } \frac{\partial N_1^e}{\partial y} = -\frac{1}{2a}\left(1-\frac{x}{2b}\right) \tag{2}$$

and

$$\frac{\partial N_2^e}{\partial x} = \frac{1}{2b}\left(1-\frac{y}{2a}\right) \text{ and } \frac{\partial N_2^e}{\partial y} = -\frac{x}{4ab}\ , \tag{3}$$

element $K^e_{c,11}$ will be calculated first

$$\begin{aligned} K^e_{c,11} &= \int_{\Omega^e}\left(\lambda_x \frac{\partial N_1^e}{\partial x}\frac{\partial N_1^e}{\partial x} + \lambda_y \frac{\partial N_1^e}{\partial y}\frac{\partial N_1^e}{\partial y}\right) dxdy = \\ &= \lambda_x \int_0^{2a}\left[\int_0^{2b}\frac{1}{4b^2}\left(1-\frac{y}{2a}\right)^2 dx\right]dy + \lambda_y \int_0^{2a}\left[\int_0^{2b}\frac{1}{4a^2}\left(1-\frac{x}{2b}\right)^2 dx\right]dy = \\ &= \frac{\lambda_x}{3}\frac{a}{b} + \frac{\lambda_y}{3}\frac{b}{a}\ . \end{aligned} \tag{4}$$

Element $K^e_{c,12}$ is calculated in a similar way:

$$\begin{aligned} K^e_{c,12} &= \int_{\Omega^e}\left(\lambda_x \frac{\partial N_1^e}{\partial x}\frac{\partial N_2^e}{\partial x} + \lambda_y \frac{\partial N_1^e}{\partial y}\frac{\partial N_2^e}{\partial y}\right) dxdy = \\ &= -\lambda_x \int_0^{2a}\left[\int_0^{2b}\frac{1}{4b^2}\left(1-\frac{y}{2a}\right)^2 dx\right]dy + \lambda_y \int_0^{2a}\left[\int_0^{2b}\frac{1}{8a^2b}\left(x-\frac{x^2}{2b}\right) dx\right]dy = \\ &= -\frac{\lambda_x}{3}\frac{a}{b} + \frac{\lambda_y}{6}\frac{b}{a}\ . \end{aligned} \tag{5}$$

Also the remaining elements of the conductivity matrix $[K_c^e]$ can be determined in a similar way, namely

$$\left[K_c^e\right]=\frac{\lambda_x}{6}\frac{a}{b}\begin{bmatrix}2 & -2 & -1 & 1\\ -2 & 2 & 1 & -1\\ -1 & 1 & 2 & -2\\ 1 & -1 & -2 & 2\end{bmatrix}+\frac{\lambda_y}{6}\frac{b}{a}\begin{bmatrix}2 & 1 & -1 & -2\\ 1 & 2 & -2 & -1\\ -1 & -2 & 2 & 1\\ -2 & -1 & 1 & 2\end{bmatrix}. \tag{6}$$

b) *Conductivity matrix [K_c^e] for a finite triangular element*

Matrix [B] for a triangular element is formulated as

$$\left[B^e\right]=\begin{bmatrix}\frac{\partial N_1^e}{\partial x} & \frac{\partial N_2^e}{\partial x} & \frac{\partial N_3^e}{\partial x}\\ \frac{\partial N_1^e}{\partial y} & \frac{\partial N_2^e}{\partial y} & \frac{\partial N_3^e}{\partial y}\end{bmatrix}=\frac{1}{2A^e}\begin{bmatrix}b_1^e & b_2^e & b_3^e\\ c_1^e & c_2^e & c_3^e\end{bmatrix}, \tag{7}$$

where A^e is the surface area of a triangle, while coefficients b_i^e, c_i^e, $i = 1, 2, 3$, are expressed by formulas in (8), Ex. 11.9. Since the coefficients in matrix [B^e] are constants and λ_x and λ_y are material constants independent of position and temperature inside the element, the conductivity matrix can be easily determined, since

$$\left[K_c^e\right]=\int_{\Omega^e}\left[B^e\right]^T\left[\lambda^e\right]\left[B^e\right]dxdy=\left[B^e\right]^T\left[\lambda^e\right]\left[B^e\right]\int_{\Omega^e}dxdy$$

or

$$\left[K_c^e\right]=\left[B^e\right]^T\left[\lambda^e\right]\left[B^e\right]A^e. \tag{8}$$

Once (7) is substituted into (8) and the appropriate operations carried out, one obtains

$$\left[K_c^e\right]=\frac{\lambda_x^e}{4A^e}\begin{bmatrix}\left(b_1^e\right)^2 & b_1^e b_2^e & b_1^e b_3^e\\ b_1^e b_2^e & \left(b_2^e\right)^2 & b_2^e b_3^e\\ b_1^e b_3^e & b_2^e b_3^e & \left(b_3^e\right)^2\end{bmatrix}+\frac{\lambda_x^e}{4A^e}\begin{bmatrix}\left(c_1^e\right)^2 & c_1^e c_2^e & c_1^e c_3^e\\ c_1^e c_2^e & \left(c_2^e\right)^2 & c_2^e c_3^e\\ c_1^e c_3^e & c_2^e c_3^e & \left(c_3^e\right)^2\end{bmatrix}. \tag{9}$$

It can easily verify that the same results are obtained when calculating matrix coefficient with (1). When a body is isotropic, i.e. $\lambda_x^e = \lambda_x^e = \lambda^e$, the conductivity matrix [K_c^e] for a triangle expressed by (9) can be written in a simpler form, by introducing the notation shown in Fig. 11.13.

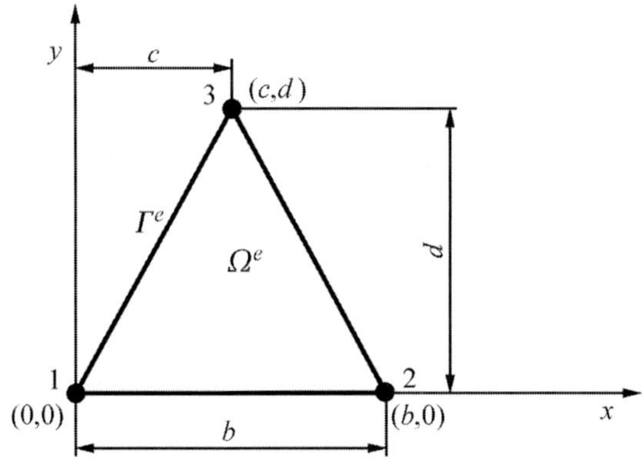

Fig. 11.13. Triangular finite element

On the basis of formulas (8) from Ex. 11.9 and the notations from Fig. 11.13, (9) for $\lambda_x^e = \lambda_x^e = \lambda^e$ can be written in the form

$$\left[K_c^e\right] = \frac{\lambda^e}{4A^e}\begin{bmatrix} d^2+(c-b)^2 & -d^2-c(c-b) & b(c-b) \\ -d^2-c(c-b) & d^2+c^2 & -cb \\ b(c-b) & -cb & b^2 \end{bmatrix}. \quad (10)$$

In a case when a triangle is rectangular in shape (Fig. 11.14), conductivity matrix assumes the form

$$\left[K_c^e\right] = \frac{\lambda^e}{2}\begin{bmatrix} \dfrac{d}{b} & -\dfrac{d}{b} & 0 \\ -\dfrac{d}{b} & \dfrac{d}{b}+\dfrac{b}{d} & -\dfrac{b}{d} \\ 0 & -\dfrac{b}{d} & \dfrac{b}{d} \end{bmatrix}. \quad (11)$$

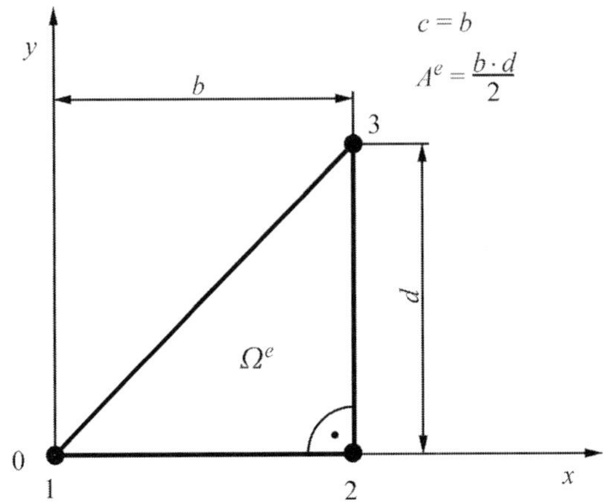

Fig. 11.14. Finite element in a rectangular triangle form

Exercise 11.12 Determining Matrix $[K_\alpha^e]$ in Terms of Convective Boundary Conditions for a Rectangular and Triangular Element

Determine matrix $[K_\alpha^e]$ for a rectangular and triangular element.

Solution

Matrix $[K_\alpha^e]$ present in (25) [Ex. 11.10], whose coefficients are expressed by (27) [Ex. 11.10], arises from 3rd order boundary conditions assigned on the boundary of an element. Matrix $[K_\alpha^e]$ can be also determined by means of (36) from Ex. 11.10. Coefficients $K^e_{\alpha,ij}$ will be calculated by means of (27) from Ex. 11.10

$$K^e_{\alpha,ij} = \int_{\Gamma^e_\alpha} \alpha N_i^e N_j^e ds \,. \qquad (1)$$

The determination of integrals in FEM is discussed, among others, in articles [4, 6].

a) *Rectangular finite element*

If convective heat transfer is assigned on all sides of an element with a heat transfer coefficient α, then matrix $[K_\alpha^e]$ is formulated as

$$\left[K_\alpha^e\right]=\int_{\Gamma_\alpha^e}\alpha\begin{bmatrix}\left(N_1^e\right)^2 & N_1^eN_2^e & N_1^eN_3^e & N_1^eN_4^e\\ N_1^eN_2^e & \left(N_2^e\right)^2 & N_2^eN_3^e & N_2^eN_4^e\\ N_1^eN_3^e & N_2^eN_3^e & \left(N_3^e\right)^2 & N_3^eN_4^e\\ N_1^eN_4^e & N_2^eN_4^e & N_3^eN_4^e & \left(N_4^e\right)^2\end{bmatrix}ds\,. \tag{2}$$

In practice, convective heat transfer is usually set on one or two element sides, which constitute a fragment of a body boundary. If convective heat transfer takes place on the side 1-2 of a rectangular element (Fig. 11.15), then in (2) one should assume that $N_3 = N_4 = 0$. Equation (2) assumes the form

$$\left[K_\alpha^e\right]=\int_0^{2b}\alpha\begin{bmatrix}\left(N_1^e\right)^2 & N_1^eN_2^e & 0 & 0\\ N_1^eN_2^e & \left(N_2^e\right)^2 & 0 & 0\\ 0 & 0 & 0 & 0\\ 0 & 0 & 0 & 0\end{bmatrix}ds\,. \tag{3}$$

Since $ds = dx$ and on the basis of (19) from Ex. 11.9, individual integrals in (3) are

$$\int_0^{2b}\left(N_1^e\right)^2dx=\frac{2b}{3}=\frac{L_{12}}{3}\,, \tag{4}$$

where L_{12} is the length of the side 1-2 of the element in question.

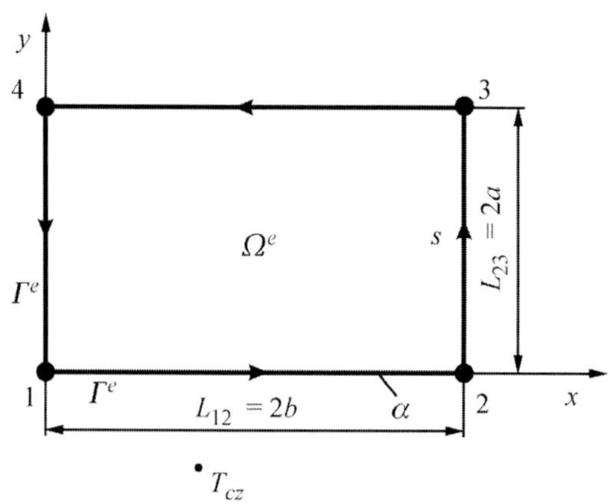

Fig. 11.15. Convective heat transfer is prescribed on the boundary 1-2 of a rectangular element

Furthermore, once the following is determined

$$\int_0^{2b} N_1^e N_2^e dx = \frac{2b}{6} = \frac{L_{12}}{6} \tag{5}$$

and

$$\int_0^{2b} \left(N_2^e\right)^2 dx = \frac{2b}{3} = \frac{L_{12}}{3} \tag{6}$$

matrix (3) assumes the form

$$\left[K_\alpha^e\right] = \frac{\alpha L_{12}}{6}\begin{bmatrix} 2 & 1 & 0 & 0 \\ 1 & 2 & 0 & 0 \\ 0 & 0 & 0 & 0 \\ 0 & 0 & 0 & 0 \end{bmatrix}. \tag{7}$$

Similar results are obtained for the remaining sides of the element

$$\left[K_\alpha^e\right] = \frac{\alpha L_{23}}{6}\begin{bmatrix} 0 & 0 & 0 & 0 \\ 0 & 2 & 1 & 0 \\ 0 & 1 & 2 & 0 \\ 0 & 0 & 0 & 0 \end{bmatrix}, \tag{8}$$

$$\left[K_\alpha^e\right] = \frac{\alpha L_{34}}{6}\begin{bmatrix} 0 & 0 & 0 & 0 \\ 0 & 0 & 0 & 0 \\ 0 & 0 & 2 & 1 \\ 0 & 0 & 1 & 2 \end{bmatrix}, \tag{9}$$

$$\left[K_\alpha^e\right] = \frac{\alpha L_{41}}{6}\begin{bmatrix} 2 & 0 & 0 & 1 \\ 0 & 0 & 0 & 0 \\ 0 & 0 & 0 & 0 \\ 1 & 0 & 0 & 2 \end{bmatrix}, \tag{10}$$

where L_{23}, L_{34}, L_{41} are the lengths of the sides on which the convective heat transfer takes place.

b) *Triangular finite element*

For a triangular element, matrix $[K_\alpha^e]$ has the form

$$\left[K_\alpha^e\right] = \int_{\Gamma_\alpha^e} \alpha \begin{bmatrix} \left(N_1^e\right)^2 & N_1^e N_2^e & N_1^e N_3^e \\ N_2^e N_1^e & \left(N_2^e\right)^2 & N_2^e N_3^e \\ N_3^e N_1^e & N_3^e N_2^e & \left(N_3^e\right)^2 \end{bmatrix} ds . \tag{11}$$

The above matrix refers to a case when convective heat transfer takes place on all three sides of a triangular element. When heat transfer occurs only on the side 1-2, one assumes in (11) that $N_3^e = 0$, while after integration, one has

$$\left[K_\alpha^e\right] = \frac{\alpha L_{12}}{6} \begin{bmatrix} 2 & 1 & 0 \\ 1 & 2 & 0 \\ 0 & 0 & 0 \end{bmatrix} . \tag{12}$$

Formulas for the remaining sides are obtained in a similar way

$$\left[K_\alpha^e\right] = \frac{\alpha L_{23}}{6} \begin{bmatrix} 0 & 0 & 0 \\ 0 & 2 & 1 \\ 0 & 1 & 2 \end{bmatrix} , \tag{13}$$

$$\left[K_\alpha^e\right] = \frac{\alpha L_{31}}{6} \begin{bmatrix} 2 & 0 & 1 \\ 0 & 0 & 0 \\ 1 & 0 & 2 \end{bmatrix} , \tag{14}$$

where L_{12}, L_{23}, L_{31} are the respective side lengths of the triangular element.

When calculating curvilinear integrals, present in (11), for a triangular element, needed in order to determine (12)–(14), (1) was used:

$$\int_0^L N_i^m(s) \cdot N_j^n(s)\, ds = L \frac{m!n!}{(m+n+1)!} . \tag{15}$$

It is easy to calculate the integrals in (11) by means of (15); e.g. to calculate integral

$$\int_0^{L_{12}} \left(N_i^e\right)^2 ds ,$$

in (15), one assumes that $m = 2$, $n = 0$, hence

$$\int_0^{L_{12}} \left(N_i^e\right)^2 ds = L_{12} \frac{2!0!}{(2+0+1)!} = L_{12} \frac{1 \cdot 2 \cdot 1}{3 \cdot 2 \cdot 1} = \frac{L_{12}}{3} .$$

Exercise 11.13 Determining Vector $\{f_Q^e\}$ with Respect to Volumetric and Point Heat Sources in a Rectangular and Triangular Element

Determine vector $\{f_Q^e\}$ for a rectangular and triangular element, when unit heat source power is constant within the area of the element and constant for a point heat source.

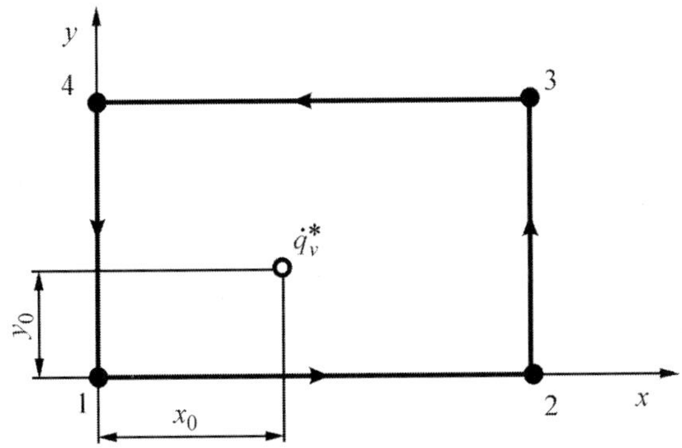

Fig. 11.16. Point heat source inside a rectangular element

Solution

Vector components $\{f_Q^e\}$ will be calculated according to (28) from Ex. 11.10

$$f_{Q,i}^e = \int_{\Omega^e} \dot{q}_v N_i^e dxdy . \tag{1}$$

a) *Rectangular element*

If power density of a heat source is constant, it is easy to calculate $\left\{f_{Q,i}^e\right\}$

$$f_{Q,i}^e = \int_0^{2a}\left(\int_0^{2b} \dot{q}_v N_i^e dx\right)dy ; \tag{2}$$

hence, after substituting into (19) from Ex. 11.9, one obtains

$$\left\{f_Q^e\right\} = \frac{\dot{q}_v A_e}{4}\begin{Bmatrix}1\\1\\1\\1\end{Bmatrix} = \dot{q}_v ab\begin{Bmatrix}1\\1\\1\\1\end{Bmatrix}, \tag{3}$$

where A_e is the surface area of an element, equal to $4ab$.

It follows from (3) that 1/4 of total body heat flow is allotted to every node in a tetragonal element.

In the case of the point heat source (Fig. 11.16), (1) assumes the form

$$\dot{q}_v = \dot{q}_v^* \delta\left(x - x_0\right)\delta\left(y - y_0\right), \tag{4}$$

where $\dot{q}_v^*$ [W/m] is the heat flow emitted at point (x_0, y_0), with respect to a unit of length as the heat source is infinitely long in the direction perpendicular to the diagram plane. Function δ is a Dirac delta, which approaches infinity at point (x_0, y_0); at the remaining points, however, it equals zero.

By substituting (4) into (1), one has

$$\left\{f_Q^e\right\} = \dot{q}_v^* \begin{Bmatrix} N_1^e\left(x_0, y_0\right) \\ N_2^e\left(x_0, y_0\right) \\ N_3^e\left(x_0, y_0\right) \\ N_4^e\left(x_0, y_0\right) \end{Bmatrix}. \tag{5}$$

b) *Triangular element*

If density $\dot{q}_v$ is constant, then from (1), one obtains

$$f_{Q,i}^e = \dot{q}_v \int_{\Omega^e} N_i^e dxdy\,. \tag{6}$$

In order to calculate the integral on the surface of a triangular element, a formula from reference [4] will be used here:

$$\int_{\Omega^e}\left(N_1^e\right)^l\left(N_2^e\right)^m\left(N_3^e\right)^n dA = \frac{l!m!n!}{\left(l+m+n+2\right)!}2A_e\,. \tag{7}$$

Since in the given case $m = 0$, $n = 0$, $l = 1$, then from (6), one has

$$f_{Q,i}^e = \frac{\dot{q}_v A_e}{3}\,. \tag{8}$$

Therefore, vector $\{f_Q^e\}$ has the form

$$\left\{f_Q^e\right\} = \frac{\dot{q}_v A_e}{3}\begin{Bmatrix}1\\1\\1\end{Bmatrix}. \tag{9}$$

It follows from (9) that 1/3 of the total heat flow in an element is allotted to every node in that element. In the case of point heat source, vector $\{f_Q^e\}$ has the form

$$\left\{f_Q^e\right\} = \dot{q}_v^* \begin{Bmatrix}N_1^e\left(x_0, y_0\right)\\N_2^e\left(x_0, y_0\right)\\N_3^e\left(x_0, y_0\right)\end{Bmatrix}. \tag{10}$$

Equations (5) and (10) refer to a case when the point heat source is located inside an element.

When a heat source is located in a node common to several elements (Fig. 11.17), then source power $\dot{q}_v^*$ per unit of length can be divided among individual elements proportionally to angle φ at the tip of a given element. For a triangular element, vector $\{f_Q^e\}$ has the form

$$\left\{f_Q^e\right\} = \frac{\varphi\, \dot{q}_v^*}{2\pi}\begin{Bmatrix}0\\1\\0\end{Bmatrix}, \tag{11}$$

where angle φ is expressed in radians.

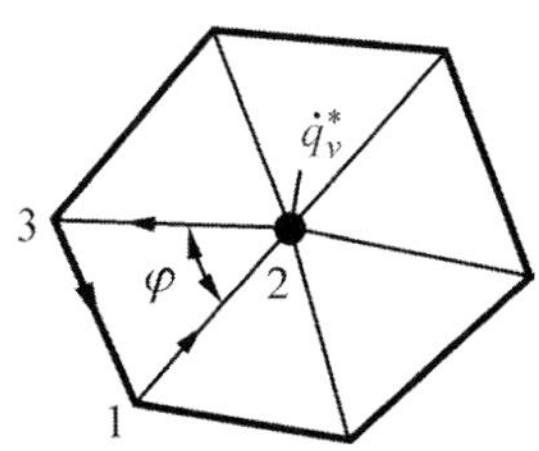

Fig. 11.17. Point heat source in a node common to several elements

In practice, the location of a heat source is of no great significance, since in the global equation system, with the heat balance equations for individual nodes, total power $\dot{q}_v^*$ is present in this equation of a node, which has a point heat source inside.

Exercise 11.14 Determining Vectors $\{f_q^e\}$ and $\{f_\alpha^e\}$ with Respect to Boundary Conditions of 2nd and 3rd Kind on the Boundary of a Rectangular or Triangular Element

Determine vectors $\{f_q^e\}$ and $\{f_\alpha^e\}$ for a finite rectangular and triangular element.

Solution

Elements of column vectors $\{f_q^e\}$ and $\{f_\alpha^e\}$ are determined from (29) and (30), Ex. 11.10

$$f_{q,i}^e = \int_{\Gamma_q^e} \dot{q}_B N_i^e ds , \tag{1}$$

$$f_{\alpha,i}^e = \int_{\Gamma_\alpha^e} \alpha T_{cz} N_i^e ds , \tag{2}$$

therefore, from almost identical integrals. If we assume that $\dot{q}_B = \alpha T_{cz}$ in the first integral, then we obtain (2). This is why only vector $\{f_q^e\}$ will be determined below.

a) *Finite rectangular element*

If heat flux is given on the boundary 1–2 of a finite element (Fig. 11.18) with thickness 1, then vector $\{f_q^e\}$ is formulated as

$$\left\{f_q^e\right\} = \int_{\Gamma_q^e} \dot{q}_B \left[N^e\right]^T ds = \int_0^{2b} \dot{q}_B \begin{Bmatrix} N_1^e \\ N_2^e \\ N_3^e \\ N_4^e \end{Bmatrix} dx . \tag{3}$$

Since also $N_3^e = N_4^e = 0$ on the side 1-2, (3) assumes the form

$$\left\{f_q^e\right\} = \dot{q}_B \int_0^{2b} \begin{Bmatrix} N_1^e(x,0) \\ N_2^e(x,0) \\ 0 \\ 0 \end{Bmatrix} dx , \tag{4}$$

where shape functions N_1^e and N_2^e are expressed by (19), in Ex. 11.9.

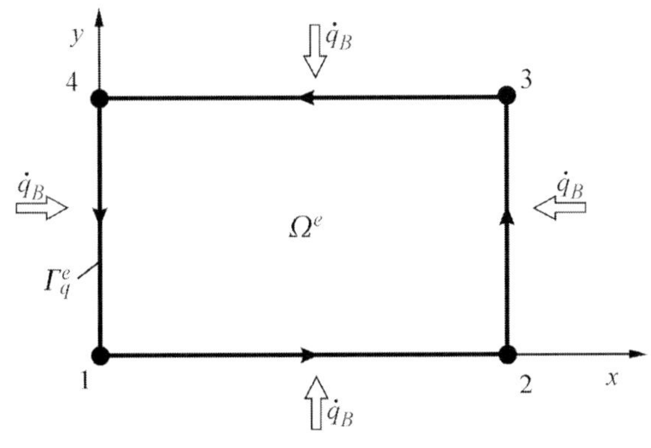

Fig. 11.18. Rectangular element heated by heat flux $\dot{q}_B$

Once the integrals are calculated

$$\int_0^{2b} N_1^e(x,0)dx = \int_0^{2b}\left(1-\frac{x}{2b}\right)dx = x - \frac{x^2}{4b}\bigg|_0^{2b} = b = \frac{L_{12}}{2}\,, \tag{5}$$

$$\int_0^{2b} N_2^e(x,0)dx = \int_0^{2b}\frac{x}{2b}dx = \frac{x^2}{4b}\bigg|_0^{2b} = b = \frac{L_{12}}{2}\,, \tag{6}$$

vector $\{f_q^e\}$ can be written in the form

$$\left\{f_q^e\right\} = \frac{\dot{q}_B L_{12}}{2}\begin{Bmatrix}1\\1\\0\\0\end{Bmatrix}. \tag{7}$$

It is evident from the analysis of (7) that the term $\dot{q}_B\, L_{12}/2$ for node no. 1 will appear on the right-hand-side of the (25), Ex. 11.10, as it will for node no. 2. This means that half of the heat, which flows through the lateral surface of an element with length L_{12} and thickness 1, flows to node no. 1. The second half flows to node no. 2. Vector $\{f_q^e\}$ can be calculated in a similar way when the heat inflows into the element through surfaces 2-3, 3-4 and 4-1; one then obtains, respectively

$$\left\{f_q^e\right\} = \frac{\dot{q}_B L_{23}}{2}\begin{Bmatrix}0\\1\\1\\0\end{Bmatrix}, \tag{8}$$

a) equation systems obtained for individual finite elements [Method I],
b) algebraic equations obtained for different elements that share, nevertheless, a common node (as an analogy to finite volume method) [Method II].

Are the temperature continuity conditions and heat flux conditions satisfied on the boundary between elements?

Solution

a) In order to create a global equation system, conductivity matrix $[K_c^e]$ and the matrix that comes from the assigned 3rd kind boundary conditions $[K_\alpha^e]$, derived for individual elements, must be summated, i.e.

$$[K]=\sum_{e=1}^{N}\left(\left[K_c^e\right]+\left[K_\alpha^e\right]\right) \tag{1}$$

That includes the summation of vectors $\{f_Q^e\}$, $\{f_q^e\}$, $\{f_\alpha^e\}$, present on the right-hand-side of the (31) in Ex. 11.10

$$\{f_Q\}=\sum_{e=1}^{N}\{f_Q^e\}\,, \tag{2}$$

$$\{f_q\}=\sum_{e=1}^{N}\{f_q^e\} \tag{3}$$

and

$$\{f_\alpha\}=\sum_{e=1}^{N}\{f_\alpha^e\}\,, \tag{4}$$

where N is the finite elements integral number, which the entire analyzed region was divided to. The global equation system for temperature in element nodes has the form

$$[K]\{T\}=\{f_Q\}+\{f_q\}+\{f_\alpha\}\,, \tag{5}$$

where $\{T\}$ is the column vector of size N, which comprises of unknown temperatures in element nodes. Next, one has to account for parameters present in the boundary conditions in the above created global equation system (5).

The method for creating matrix $[K]$, which is sparse, should be discussed in greater detail, since only some of the coefficients present in it are other than zero. It is assumed that the flat region is divided into triangular elements (Fig. 11.19).

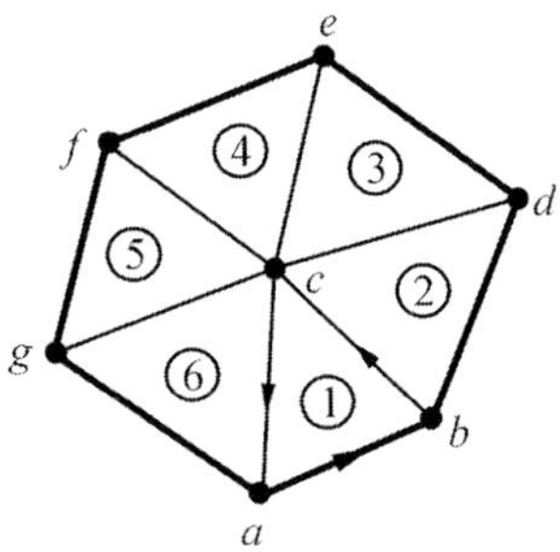

Fig. 11.19. A division of a flat region into finite triangular elements; element numbers and global node numbers are marked: ①–⑥ – finite element numbers, *a*–*g* – global node numbers

If element ① lies inside the analyzed region, thereby $[K_\alpha^{1}]$ can be disregarded, then matrix $[K^1] = [K_c^{1}]$ for the first element (Fig. 11.20) can be written in the following way:

$$\begin{array}{c} \\ \langle 1\rangle \\ \langle 2\rangle \\ \langle 3\rangle \\ \\ \end{array} \begin{array}{c} \begin{array}{ccc} a & b & c \end{array} \\ \begin{bmatrix} K_{11}^{1} & K_{12}^{1} & K_{13}^{1} \\ K_{21}^{1} & K_{22}^{1} & K_{23}^{1} \\ K_{31}^{1} & K_{32}^{1} & K_{33}^{1} \end{bmatrix} \\ \begin{array}{ccc} \langle 1\rangle & \langle 2\rangle & \langle 3\rangle \end{array} \end{array} . \tag{6}$$

If global node numbers of a triangular element are marked as *a*, *b*, and *c*, while local node numbers as ⟨1⟩, ⟨2⟩, ⟨3⟩, then one can see that coefficient K^1_{aa} corresponds to coefficient K^1_{11} in matrix $[K^1]$ (6), coefficient K^1_{bc} corresponds to coefficient K^1_{23}, etc. When creating a matrix of coefficients $[K]$ according to (1), one should be guided by global indexes, i.e. one should add coefficient K^e_{aa} that occurs in the matrix of element ⓔ to coefficient K^1_{aa} that occurs in the conductivity matrix of element ①. Coefficients that have the same global indexes in conductivity matrixes of other elements are added together. Such common coefficients appear in conductivity matrixes of elements, which share a common node, e.g. in the case of elements in Fig. 11.19, the node common to six elements is node *c*.

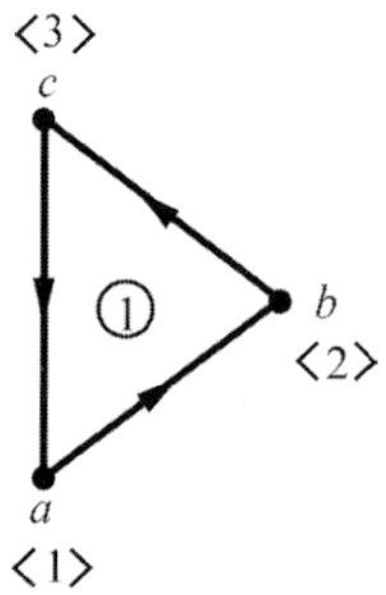

Numeration of nodes in element no. ①	
Global node number	Local node number
a	⟨1⟩
b	⟨2⟩
c	⟨3⟩

Fig. 11.20. A diagram of global (*a*, *b* and *c*) and local (⟨1⟩, ⟨2⟩, ⟨3⟩) node numeration in a triangular element

A global equation system for node temperatures can be created in another way, which resembles the way heat balance equations are developed using the control volume method. One can also create control volume in FEM around node *c* (Fig. 11.19), common to surrounding elements, by linking centers of gravity of triangular elements with the midpoints of triangle sides (Fig . 11.21). The equation number equal to the number of nodes in an element is assigned to every element. There are three equations in the case of a triangular element. When creating a global equation for node *c* (Fig. 11.19), only those equations are considered in which the shape function was selected as a weight function in the Galerkin method at point *c*. For element ① when local nodes are positioned as shown in Fig. 11.20, the third equation is the equation in question; it results from the application of Galerkin method when weight coefficient equals $N_3^{1}(x, y)$.

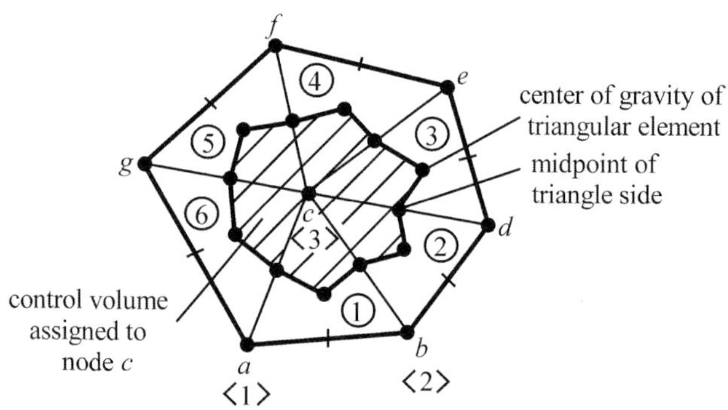

Fig. 11.21. Control volume in FEM with a region divided into triangular elements; linear functions interpolate temperature distribution inside the element

If similar local node numeration is assumed for the remaining elements shown in Fig. 11.19, then we must account for the third equation in every equation system for a given element, since in every element local node ⟨3⟩

corresponds to node c. In Galerkin method, function $N_3^e(x,y)$ plays a role of a weight function for element ⓔ.

b) The second method for creating global equation system, based on the formulation of an appropriate equation for a given node, indicates that there is a close relationship between FEM and the control (finite) volume method.

The following conditions should be retained when aggregating (summating) elements (Fig. 11.22):

- continuity of temperature field, including boundaries between elements;
- continuity of heat flow, also on the boundary between elements.

The first condition is satisfied in FEM; the second condition, however, is not completely satisfied. On the boundary between elements, the following temperature continuity takes place

$$T_3^1 = T_3^2 = T_4$$

and

$$T_2^1 = T_1^2 = T_2 \ . \tag{7}$$

The above indexes (7) are the element numbers. Temperature equality on the boundary between elements follows from the equation of temperature in nodes and linear character of temperature distribution inside the elements.

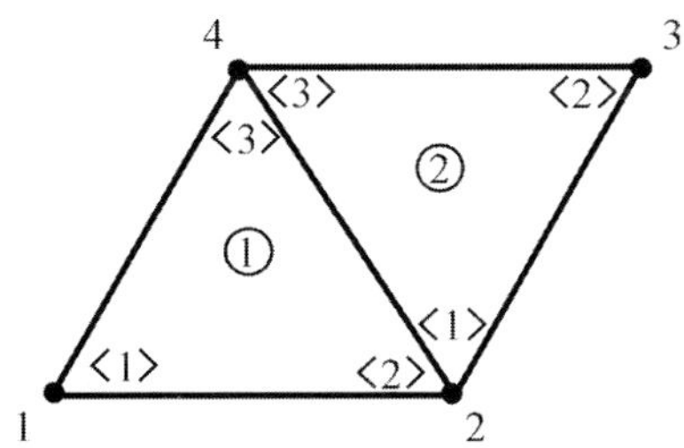

Fig. 11.22. Global and local numeration of nodes in finite elements

In agreement with (29), Ex. 11.10, the equality of integrals takes place on the boundary between elements:

$$\int_{L_{23}^1} \dot{q}_n^1 N_2^1 ds = - \int_{L_{13}^2} \dot{q}_n^2 N_1^2 ds \ , \tag{8}$$

$$\int_{L_{23}^1} \dot{q}_n^1 N_3^1 ds = - \int_{L_{13}^2} \dot{q}_n^2 N_3^2 ds \ . \tag{9}$$

Therefore, only in the case of very small elements, when the side common to both elements is very short in length, the heat flux equality is ensured on the boundary where two elements meet. Furthermore, heat flux inside an element is constant when temperature distribution is interpolated inside the element by linear functions. Therefore, heat flux step-change occurs at the point of contact of two elements. At such point, there is also no continuity among the first derivatives of function $T^e(x, y)$. This lack of continuity at the point of contact between elements negatively affects the accuracy of solution. In order to determine heat flux at a given point in an analyzed region or to calculate heat flow transmitted by a body boundary, the region should be divided into very small elements, so that the accuracy of the determined heat flux values is satisfactory.

Exercise 11.16 Determining Temperature Distribution in a Square Cross-Section of an Infinitely Long Rod by Means of FEM, in which the Global Equation System is Constructed using Method I (from Ex. 11.15)

Determine temperature distribution in a square cross-section of an infinitely long rod (Fig. 11.23). Upper and lower surfaces are thermally insulated. Left vertical surface is heated by a heat flow whose density is $\dot{q}_B = 200\ 000$ W/m^2, while the surface on the opposite side is cooled by water at temperature $T_{cz} = 20°C$ with a heat transfer coefficient equal to $\alpha = 1000$ W/(m^2·K). Thermal conductivity of the rod's material $\lambda_x = \lambda_y = 50$ W/(m·K). The length of the side within the square cross-section of the rod is $a = 2$ cm.

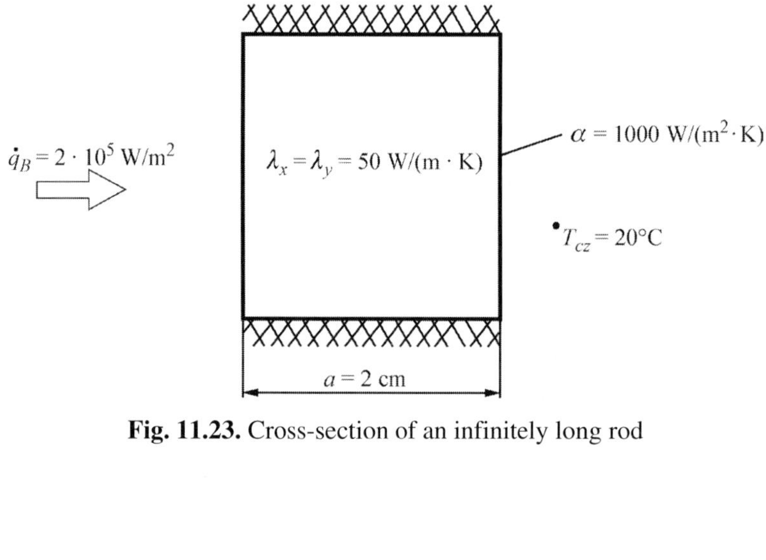

Fig. 11.23. Cross-section of an infinitely long rod

Solution

Temperature field will be treated as two-dimensional. Cross-section of the rod will be divided into four elements. Local and global node numeration is given in Fig. 11.24. and Table 11.2.

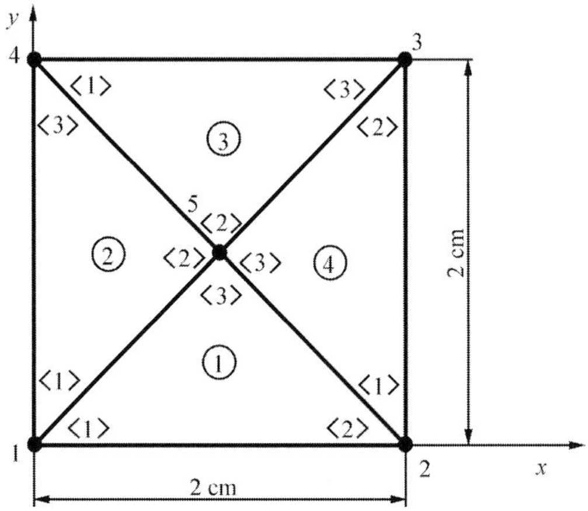

Fig. 11.24. A division of a rod's cross-section into four elements with a marked global and local node numeration; local numeration is given in brackets ⟨ ⟩

Table 11.2. Local and global node numeration

Element No.	Local Node Number	Global Node Number
	⟨1⟩	1
①	⟨2⟩	2
	⟨3⟩	5
	⟨1⟩	1
②	⟨2⟩	5
	⟨3⟩	4
	⟨1⟩	4
③	⟨2⟩	5
	⟨3⟩	3
	⟨1⟩	2
④	⟨2⟩	3
	⟨3⟩	5

First we will determine the quantities in individual elements of the conductivity matrix $[K_c^e]$ and in the element of matrix element $[K_\alpha^4]$, since a convection heat transfer is assigned on the side 2-3 (⟨1⟩-⟨2⟩). Conductivity matrix for a triangular element is formulated by (9) in Ex. 11.11, which in the given case for $\lambda_x = \lambda_y = \lambda$ has the form

$$\left[K_c^e\right]=\frac{\lambda}{4A^e}\begin{bmatrix}\left(b_1^e\right)^2+\left(c_1^e\right)^2 & b_1^e b_2^e+c_1^e c_2^e & b_1^e b_3^e+c_1^e c_3^e \\ b_1^e b_2^e+c_1^e c_2^e & \left(b_2^e\right)^2+\left(c_2^e\right)^2 & b_2^e b_3^e+c_2^e c_3^e \\ b_1^e b_3^e+c_1^e c_3^e & b_2^e b_3^e+c_2^e c_3^e & \left(b_3^e\right)^2+\left(c_3^e\right)^2\end{bmatrix}, \tag{1}$$

where matrix coefficients are formulated in (8), Ex. 11.9.

Matrix $[K_\alpha^4]$ for element ④ is formulated in (12), Ex. 11.12

$$\left[K_\alpha^4\right]=\frac{\alpha L_{12}}{6}\begin{bmatrix}2 & 1 & 0\\ 1 & 2 & 0\\ 0 & 0 & 0\end{bmatrix}, \tag{2}$$

where $L_{12} = 0$ is the length of the side in the square cross-section. From matrices $[K_c^e]$ and $[K_\alpha^4]$, one obtains:

- Element ①

$x_1^1 = 0\text{ m}, \qquad y_1^1 = 0\text{ m},$

$x_2^1 = 2\cdot 10^{-2}\text{ m}, \qquad y_2^1 = 0\text{ m},$

$x_3^1 = 1\cdot 10^{-2}\text{ m}, \qquad y_3^1 = 1\cdot 10^{-2}\text{ m},$

$b_1^1 = 0 - 1\cdot 10^{-2} = -1\cdot 10^{-2}\text{ m}, \qquad b_2^1 = 1\cdot 10^{-2} - 0 = 1\cdot 10^{-2}\text{ m},$

$b_3^1 = 0 - 0 = 0\text{ m},$

$c_1^1 = 1\cdot 10^{-2} - 2\cdot 10^{-2} = -1\cdot 10^{-2}\text{ m}, \qquad c_2^1 = 0 - 1\cdot 10^{-2} = -1\cdot 10^{-2}\text{ m},$

$c_3^1 = 2\cdot 10^{-2} - 0 = 2\cdot 10^{-2}\text{ m},$

$A^1 = \frac{1}{2}\cdot 0.02\cdot 0.01 = 1\cdot 10^{-4}\text{ m}^2.$

Conductivity matrix $[K_c^1]$ calculated with (1) is as follows:

$$\left[K_c^1\right]=50\begin{bmatrix}0.5 & 0 & -0.5\\ 0 & 0.5 & -0.5\\ -0.5 & -0.5 & 1.0\end{bmatrix}\begin{matrix}1\\2,\\5\end{matrix} \quad \text{(columns: 1, 2, 5)} \tag{3}$$

The numbers above and next to the matrix are the global node numbers.

• Element ②

$x_1^2 = 0 \text{ m}, \quad y_1^2 = 0 \text{ m}, \quad x_2^2 = 0.01 \text{ m}, \quad y_2^2 = 0.01 \text{ m},$
$x_3^2 = 0.0 \text{ m}, \quad y_3^2 = 0.02 \text{ m},$
$b_1^2 = 0.01 - 0.02 = -1 \cdot 10^{-2} \text{ m}, \quad b_2^2 = 0.02 - 0 = 2 \cdot 10^{-2} \text{ m},$
$b_3^2 = 0 - 0.01 = -1 \cdot 10^{-2} \text{ m},$
$c_1^2 = 0 - 0.01 = -1 \cdot 10^{-2} \text{ m}, \quad c_2^2 = 0 - 0 = 0 \text{ m},$
$c_3^2 = 0.01 - 0 = 1 \cdot 10^{-2} \text{ m},$
$A^2 = \frac{1}{2} \cdot 0.02 \cdot 0.01 = 1 \cdot 10^{-4} \text{ m}^2.$

Conductivity matrix is

$$\begin{matrix} & \begin{matrix} 1 & \quad 5 & \quad 4 \end{matrix} & \\ \left[K_c^2\right] = 50 & \begin{bmatrix} 0.5 & -0.5 & 0 \\ -0.5 & 1.0 & -0.5 \\ 0.0 & -0.5 & 0.5 \end{bmatrix} & \begin{matrix} 1 \\ 5. \\ 4 \end{matrix} \end{matrix} \tag{4}$$

• Element ③

$x_1^3 = 0 \text{ m}, \quad y_1^3 = 2 \cdot 10^{-2} \text{ m},$
$x_2^3 = 1 \cdot 10^{-2} \text{ m}, \quad y_2^3 = 1 \cdot 10^{-2} \text{ m},$
$x_3^3 = 2 \cdot 10^{-2} \text{ m}, \quad y_3^3 = 2 \cdot 10^{-2} \text{ m},$
$b_1^3 = 1 \cdot 10^{-2} - 2 \cdot 10^{-2} = -1 \cdot 10^{-2} \text{ m}, \quad b_2^3 = 2 \cdot 10^{-2} - 2 \cdot 10^{-2} = 0 \text{ m},$
$b_3^3 = 2 \cdot 10^{-2} - 1 \cdot 10^{-2} = 1 \cdot 10^{-2} \text{ m},$
$c_1^3 = 2 \cdot 10^{-2} - 1 \cdot 10^{-2} = 1 \cdot 10^{-2} \text{ m}, \quad c_2^3 = 0 - 2 \cdot 10^{-2} = -2 \cdot 10^{-2} \text{ m},$
$c_3^3 = 1 \cdot 10^{-2} - 0 = 1 \cdot 10^{-2} \text{ m}.$

Matrix $[K_c^3]$ is as follows:

$$\begin{matrix} & \begin{matrix} 4 & \quad 5 & \quad 3 \end{matrix} & \\ \left[K_c^3\right] = 50 & \begin{bmatrix} 0.5 & -0.5 & 0 \\ -0.5 & 1.0 & -0.5 \\ 0 & -0.5 & 0.5 \end{bmatrix} & \begin{matrix} 4 \\ 5. \\ 3 \end{matrix} \end{matrix} \tag{5}$$

• Element ④

$x_1^4 = 2\cdot 10^{-2}$ m, $y_1^4 = 0$ m, $x_2^4 = 2\cdot 10^{-2}$ m, $y_2^4 = 2\cdot 10^{-2}$ m,

$x_3^4 = 1\cdot 10^{-2}$ m, $y_3^4 = 1\cdot 10^{-2}$ m,

$b_1^4 = 2\cdot 10^{-2} - 1\cdot 10^{-2} = 1\cdot 10^{-2}$ m, $b_2^4 = 1\cdot 10^{-2} - 0 = 1\cdot 10^{-2}$ m,

$b_3^4 = 0 - 2\cdot 10^{-2} = -2\cdot 10^{-2}$ m,

$c_1^4 = 1\cdot 10^{-2} - 2\cdot 10^{-2} = -1\cdot 10^{-2}$ m, $c_2^4 = 2\cdot 10^{-2} - 1\cdot 10^{-2} = 1\cdot 10^{-2}$ m,

$c_3^4 = 2\cdot 10^{-2} - 2\cdot 10^{-2} = 0$ m.

Matrix $[K_c^4]$ is as follows:

$$\begin{array}{c} \begin{matrix} 2 & \quad 3 & \quad 5 \end{matrix} \\ \left[K_c^4\right] = 50\begin{bmatrix} 0.5 & 0 & -0.5 \\ 0 & 0.5 & -0.5 \\ -0.5 & -0.5 & 1.0 \end{bmatrix} \begin{matrix} 2 \\ 3. \\ 5 \end{matrix} \end{array} \tag{6}$$

Matrix $[K_\alpha^4]$, which results from the heat transfer on the side 2-3, has the form

$$\begin{array}{c} \begin{matrix} 2 & 3 \end{matrix} \\ \left[K_\alpha^4\right] = 3.33(3)\begin{bmatrix} 2 & 1 & 0 \\ 1 & 2 & 0 \\ 0 & 0 & 0 \end{bmatrix} \begin{matrix} 2 \\ 3, \\ \end{matrix} \end{array} \tag{7}$$

or

$$\begin{array}{c} \begin{matrix} 2 & \quad 3 \end{matrix} \\ \left[K_\alpha^4\right] = \begin{bmatrix} 6.667 & 3.333 & 0 \\ 3.333 & 6.667 & 0 \\ 0 & 0 & 0 \end{bmatrix} \begin{matrix} 2 \\ 3. \\ \end{matrix} \end{array} \tag{8}$$

The sum of matrix $[K_c^4] + [K_\alpha^4]$ is

$$\begin{array}{c} \begin{matrix} 2 & \quad 3 & \quad 5 \end{matrix} \\ \left[K_c^4\right] + \left[K_\alpha^4\right] = \begin{bmatrix} 31.667 & 3.333 & -25.0 \\ 3.333 & 31.667 & -25.0 \\ -25.0 & -25.0 & 50 \end{bmatrix} \begin{matrix} 2 \\ 3. \\ 5 \end{matrix} \end{array} \tag{9}$$

Coefficient matrix in the global equation system can be obtained by adding coefficients in matrices (3)–(6) and (9):

$$K_{11} = K_{11}^1 + K_{11}^2 = 25 + 25 = 50 \text{ W/(m} \cdot \text{K)}, \quad K_{12} = 0, \quad K_{13} = 0, \quad K_{14} = 0,$$
$$K_{15} = K_{15}^1 + K_{15}^2 = -25 + (-25) = -50 \text{ W/(m} \cdot \text{K)},$$
$$K_{21} = 0, K_{22} = K_{22}^1 + \left(K_{22}^4 + K_{\alpha,22}^4\right) = 25 + 31.6667 = 56.6667 \text{ W/(m} \cdot \text{K)},$$
$$K_{23} = \left(\left[K_{23}^4\right] + \left[K_{\alpha,23}^4\right]\right) = 3.333 \text{ W/(m} \cdot \text{K)}, \quad K_{24} = 0,$$
$$K_{25} = K_{25}^1 + K_{25}^4 = -25 + (-25) = -50 \text{ W/(m} \cdot \text{K)},$$
$$K_{31} = 0, \ K_{32} = \left(K_{32}^4 + K_{\alpha,32}^4\right) = 3.333 \text{ W/(m} \cdot \text{K)},$$
$$K_{33} = K_{33}^3 + \left(K_{33}^4 + K_{\alpha,33}^4\right) = 25 + 31.6667 = 56.6667 \text{ W/(m} \cdot \text{K)},$$
$$K_{34} = K_{34}^3 = 0 \text{ W/(m} \cdot \text{K)}, K_{35} = K_{35}^3 + K_{35}^4 = -25 + (-25) = -50 \text{ W/(m} \cdot \text{K)},$$
$$K_{41} = K_{41}^2 = 0, K_{42} = 0, K_{43} = 0, K_{44} = K_{44}^2 + K_{44}^3 = 25 + 25 = 50 \text{ W/(m} \cdot \text{K)},$$
$$K_{45} = K_{45}^2 + K_{45}^3 = -25 + (-25) = -50 \text{ W/(m} \cdot \text{K)},$$
$$K_{51} = K_{51}^1 + K_{51}^2 = -25 + (-25) = -50 \text{ W/(m} \cdot \text{K)},$$
$$K_{52} = K_{52}^1 + K_{52}^4 = -25 + (-25) = -50 \text{ W/(m} \cdot \text{K)},$$
$$K_{53} = K_{53}^3 + K_{53}^4 = -25 + (-25) = -50 \text{ W/(m} \cdot \text{K)},$$
$$K_{54} = K_{54}^2 + K_{54}^3 = -25 + (-25) = -50 \text{ W/(m} \cdot \text{K)},$$
$$K_{55} = K_{55}^1 + K_{55}^2 + K_{55}^3 + K_{55}^4 = 50 + 50 + 50 + 50 = 200 \text{ W/(m} \cdot \text{K)}.$$

If all coefficients are known, one can write then the global conductivity matrix

$$[K] = \begin{bmatrix} 50 & 0 & 0 & 0 & -50 \\ 0 & 56.667 & 3.333 & 0 & -50 \\ 0 & 3.333 & 56.667 & 0 & -50 \\ 0 & 0 & 0 & 50 & -50 \\ -50 & -50 & -50 & -50 & 200 \end{bmatrix} \text{W/(m} \cdot \text{K)}. \tag{10}$$

Following that, vector $\{f_q^2\}$ is defined according to (17), Ex. 11.14

$$\left\{f_q^2\right\} = \frac{\dot{q}_B L_{31}}{2} \begin{Bmatrix} 1 \\ 0 \\ 1 \end{Bmatrix} = \begin{Bmatrix} f_1 \\ f_5 \\ f_4 \end{Bmatrix}, \tag{11}$$

$$\left\{f_q^2\right\} = \begin{Bmatrix} f_1 \\ f_5 \\ f_4 \end{Bmatrix} = \begin{Bmatrix} 2000 \\ 0 \\ 2000 \end{Bmatrix}. \tag{12}$$

Vector $\{f_\alpha^4\}$ will be calculated according to (16), Ex. 11.14

$$\left\{f_\alpha^4\right\} = \frac{\alpha T_{cz}}{2} \begin{Bmatrix} 1 \\ 1 \\ 0 \end{Bmatrix} = \begin{Bmatrix} f_2 \\ f_3 \\ f_5 \end{Bmatrix}, \tag{13}$$

where from, after substitution, one obtains

$$\left\{f_\alpha^4\right\} = \begin{Bmatrix} f_2 \\ f_3 \\ f_5 \end{Bmatrix} = \begin{Bmatrix} 10000 \\ 10000 \\ 0 \end{Bmatrix} \text{ W/m}^2. \tag{14}$$

The right-hand-side vector has the form

$$\{f\} = \begin{Bmatrix} f_1 \\ f_2 \\ f_3 \\ f_4 \\ f_5 \end{Bmatrix} = \begin{Bmatrix} 2000 \\ 10000 \\ 10000 \\ 2000 \\ 0 \end{Bmatrix} \text{ W/m}^2. \tag{15}$$

The equation system (31) from Ex. 11.10, from which node temperatures will be determined, assumes in this case the following form

$$\begin{bmatrix} 50 & 0 & 0 & 0 & -50 \\ 0 & 56.667 & 3.333 & 0 & -50 \\ 0 & 3.333 & 56.667 & & -50 \\ 0 & 0 & 0 & 50 & -50 \\ -50 & -50 & -50 & -50 & 200 \end{bmatrix} \begin{Bmatrix} T_1 \\ T_2 \\ T_3 \\ T_4 \\ T_5 \end{Bmatrix} = \begin{Bmatrix} 2000 \\ 10000 \\ 10000 \\ 2000 \\ 0 \end{Bmatrix}. \tag{16}$$

Since the equation system (16) is solved by means of the Gauss elimination method and the program shown in Ex. 6.26, the following is obtained: $T_1 = 280°\text{C}$, $T_2 = 200°\text{C}$, $T_3 = 200°\text{C}$, $T_4 = 280°\text{C}$, $T_5 = 240°\text{C}$. Due to thermal insulation of lateral surfaces 1-2 and 3-4, temperature field is one-dimensional in the cross-section of the rod.

Temperature can be calculated from analytical formulas

$$T_1 = T_4 = \frac{\dot{q}_B a}{\lambda} + \frac{\dot{q}_B}{\alpha} = \frac{200000 \cdot 0.02}{50} + \frac{200000}{1000} = 280°\text{C},$$

$$T_5 = \frac{\dot{q}_B a}{2\lambda} + \frac{\dot{q}_B}{\alpha} = \frac{200000 \cdot 0.02}{2 \cdot 50} + \frac{200000}{1000} = 240°\text{C},$$

$$T_2 = T_3 = \frac{\dot{q}_B}{\alpha} = \frac{200000}{1000} = 200°\text{C}.$$

It is clear, therefore, that the results obtained by means of FEM and the analytical formulas are identical to each other.

Exercise 11.17 Determining Temperature Distribution in an Infinitely Long Rod with Square Cross-Section by Means of FEM, in which the Global Equation System is Constructed using Method II (from Ex. 11.15)

Solve the problem formulated in Ex. 11.16 by means of FEM; namely, the equation (of heat balance) for individual nodes. Use Method II discussed in Ex. 11.15.

Solution

• Node 1

Elements ① and ② have node 1 in common (Fig. 11.24). Equation system for element ① has the form

$$\left[K^1\right]\left\{T^1\right\} = \left\{f^1\right\}, \tag{1}$$

where $[K^1]$ is formulated in (3), Ex. 11.16. Because $\{f^1\} = [0, 0, 0]^T$, equation system (1) has the form

$$\begin{bmatrix} 25 & 0 & -25 \\ 0 & 25 & -25 \\ -25 & -25 & 50 \end{bmatrix} \begin{Bmatrix} T_1 \\ T_2 \\ T_5 \end{Bmatrix} = \begin{Bmatrix} 0 \\ 0 \\ 0 \end{Bmatrix}. \tag{2}$$

Node 1 in the global numeration is also ⟨1⟩ in the local numeration of element ①; therefore, only the first equation is taken into consideration in (2)

$$25T_1 - 25T_5 = 0. \tag{3}$$

The algebraic equation system for element ② has the form

$$\left[K^2\right]\left\{T^2\right\}=\left\{f^2\right\},$$

where $[K^2]$ is formulated in (4) while vector $\{f^2\}$ in (12), Ex. 11.16. The equation system for element ② assumes the form then

$$\begin{bmatrix} 25 & 25 & 0 \\ -25 & 50 & -25 \\ 0 & -25 & 25 \end{bmatrix}\begin{Bmatrix} T_1 \\ T_5 \\ T_4 \end{Bmatrix}=\begin{Bmatrix} 2000 \\ 0 \\ 2000 \end{Bmatrix}. \tag{4}$$

Node 1 in the global numeration is node ⟨1⟩ in the local numeration in element ②; therefore, only the first equation in (4) is taken into consideration

$$25T_1-25T_5=2000. \tag{5}$$

Once (3) is added to (5), an algebraic equation for node 1 is obtained in global numeration

$$50T_1-50T_5=2000. \tag{6}$$

- Node 2

Node 2 is shared by element ① and ④. Node 2 in the global numeration is also node ⟨2⟩ in the local numeration. Therefore, in the equation system (2), second equation is taken into consideration

$$25T_2-25T_5=0. \tag{7}$$

Equation system for element ④ has the form

$$\left[K^4\right]\left\{T^4\right\}=\left\{f^4\right\}, \tag{8}$$

where $[K^4]$ is formulated in (6), Ex. 11.16, while $\{f^4\}$ in (14), Ex. 11.16. Thus, the equation system (8) has the form

$$\begin{bmatrix} 31,667 & 3,333 & -25 \\ 3,333 & 31,667 & -25 \\ -25 & -25 & 50 \end{bmatrix}\begin{Bmatrix} T_2 \\ T_3 \\ T_5 \end{Bmatrix}=\begin{Bmatrix} 10000 \\ 10000 \\ 0 \end{Bmatrix}. \tag{9}$$

Node 2 in the global numeration is node ⟨1⟩ in the local numeration in element ④. In the equation system (9) only the first equation is taken into consideration

$$31.667\,T_2+3.333\,T_3-25\,T_5=10000. \tag{10}$$

Once (7) and (10) are added together, the equation for node 2 is obtained (in global numeration)

$$56.667\,T_2+3.333\,T_3-50\,T_5=10000. \tag{11}$$

• Node 3

Node 3 has a local number ⟨2⟩ in element ④. The second equation in the system (9) has the form

$$3.333T_2 + 31.667T_3 - 25T_5 = 10000\,. \tag{12}$$

An equation system for element ③, to which node 3 belongs, has the form

$$\left[K^3\right]\left\{T^3\right\} = \left\{f^3\right\}, \tag{13}$$

where $[K^3]$ is expressed in (4), Ex. 11.16, while $\{f^3\} = [0,0,0]^T$. The equation system (13) assumes the form

$$\begin{bmatrix} 25 & -25 & 0 \\ -25 & 50 & -25 \\ 0 & -25 & 25 \end{bmatrix} \begin{Bmatrix} T_4 \\ T_5 \\ T_3 \end{Bmatrix} = \begin{Bmatrix} 0 \\ 0 \\ 0 \end{Bmatrix}. \tag{14}$$

Only the third equation from above (14) is allowed for, since node 3 has the local number ⟨3⟩

$$25T_3 - 25T_5 = 0\,. \tag{15}$$

Once corresponding sides of (12) and (15) are added, a global equation for node 3 is obtained

$$3.333T_2 + 56.667T_3 - 50T_5 = 10\,000\,. \tag{16}$$

• Node 4

Node 4 is shared by element ② and ③. In the equation system for element ②, the third equation is taken into consideration, since the analyzed node has a local number ⟨3⟩ ②. From (4), one obtains

$$25T_4 - 25T_5 = 2000\,. \tag{17}$$

In the equation system for element ③, the first equation is taken into consideration, since node 4 has the local number equal to ⟨1⟩ in element ③. From (14) one has

$$25T_4 - 25T_5 = 0\,. \tag{18}$$

Once corresponding sides of (17) and (18) are added, one obtains a global equation for node 4

$$50T_4 - 50T_5 = 2000\,. \tag{19}$$

• Node 5

This is a node common to all elements. In the equation system (2) for element ①, the third equation is accounted for, since the analyzed node has a local number ⟨3⟩ in element ①. From (2), one gets

$$-25T_1 - 25T_2 + 50T_5 = 0. \tag{20}$$

In (4) for element ②, the second equation is considered

$$-25T_1 - 25T_4 + 50T_5 = 0, \tag{21}$$

since node 5 has the local number ⟨2⟩ in this element. In the equation system for element ③, the second equation is accounted for

$$-25T_3 - 25T_4 + 50T_5 = 0, \tag{22}$$

since node 5 has the local number ⟨2⟩ in this element. In the equation system (9) for element ④, the third equation is accounted for

$$-25T_2 - 25T_3 + 50T_5 = 0, \tag{23}$$

since node 5 has the local number ⟨3⟩ in element ④. Once corresponding sides of (20), (21), (22) and (23) are added, one has

$$-50T_1 - 50T_2 - 50T_3 - 50T_4 + 200T_5 = 0. \tag{24}$$

Equations (6), (11), (16), (19) and (24) form a global equation system.

$$\begin{aligned}
&50T_1 - 50T_5 = 2000\\
&56.6667T_2 + 3.333T_3 - 50T_5 = 10000\\
&3.333T_2 + 56.667T_3 - 50T_5 = 10000\\
&50T_4 - 50T_5 = 2000\\
&-50T_1 - 50T_2 - 50T_3 - 50T_4 + 200T_5 = 0.
\end{aligned} \tag{25}$$

Equation system (25) and system (16) from Ex. 11.16 are the same. It is clear, therefore, that regardless of how the global equation system is created, it always remains the same.

Exercise 11.18 Determining Temperature Distribution by Means of FEM in an Infinitely Long Rod with Square Cross-Section, in which Volumetric Heat Sources Operate

Determine steady-state temperature distribution in a square region whose side is $2a = 2$ cm in length. Assume that the thermal conductivity of a medium is $\lambda = 42$ W/(m·K). Heat source power per unit of volume is $\dot{q}_v = 1 \cdot 10^7$ W/m^3. Boundary conditions are illustrated in Fig.11.25. Assume the following values for the calculation: $\dot{q}_B = 200000$ W/m^2, $\alpha = 60$ W/(m^2·K), $T_{cz} = 20$°C, $T_s = 100$°C.

Solution

Boundary conditions can be written in the following way:

$$-\lambda \left.\frac{\partial T}{\partial x}\right|_{x=0} = \dot{q}_B, \tag{1}$$

$$\left.T\right|_{x=2a} = T_s, \tag{2}$$

$$-\lambda \left.\frac{\partial T}{\partial y}\right|_{y=0} = \alpha\left(T_{cz} - \left.T\right|_{y=0}\right), \tag{3}$$

$$-\lambda \left.\frac{\partial T}{\partial y}\right|_{y=2a} = 0. \tag{4}$$

Temperature distribution will be determined by means of FEM with a division depicted in Fig.11.25. Local and global node numeration is presented in Table 11.3.

Conductivity matrix (rigidity) in the case of the square element and constant thermal conductivity, when $\lambda_x = \lambda_y = \lambda$, has the following form ((6), Ex. 11.11).

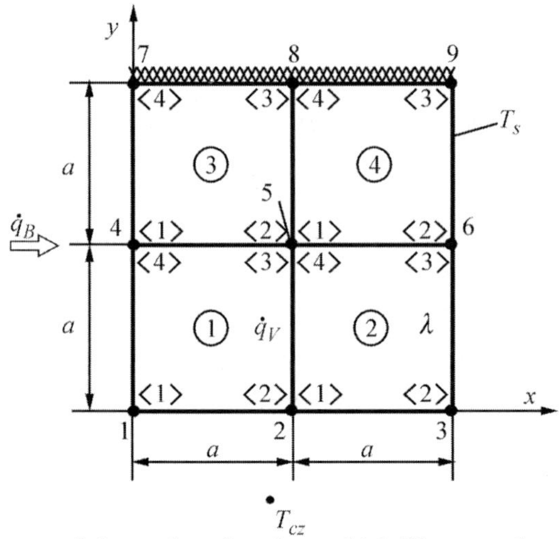

Fig. 11.25. Diagram of the analyzed region, which illustrates boundary conditions and division of an area into finite elements

Table 11.3. Local and global node numeration

Element No.	Local Node No.	Global Node No.
①	⟨1⟩	1
	⟨2⟩	2
	⟨3⟩	5
	⟨4⟩	4
②	⟨1⟩	2
	⟨2⟩	3
	⟨3⟩	6
	⟨4⟩	5
③	⟨1⟩	4
	⟨2⟩	5
	⟨3⟩	8
	⟨4⟩	7
④	⟨1⟩	5
	⟨2⟩	6
	⟨3⟩	9
	⟨4⟩	8

$$\left[K_c^e\right]=\frac{\lambda}{6}\begin{bmatrix}4 & -1 & -2 & -1\\ -1 & 4 & -1 & -2\\ -2 & -1 & 4 & -1\\ -1 & -2 & -1 & 4\end{bmatrix}. \tag{5}$$

Next, matrixes of conductivity (stiffness) $[K_c^e]$ will be written for individual elements. Global node numeration will be used.

- Element ①

$$\left[K_c^1\right]=7\begin{array}{c} \begin{matrix}\;\;1 & \;\;2 & \;\;5 & \;\;4\end{matrix}\\ \begin{bmatrix}4 & -1 & -2 & -1\\ -1 & 4 & -1 & -2\\ -2 & -1 & 4 & -1\\ -1 & -2 & -1 & 4\end{bmatrix}\end{array}\begin{matrix}\\1\\2\\5\\4\end{matrix}, \tag{6}$$

- Element ②

$$\left[K_c^2\right]=7\begin{array}{c} \begin{matrix}\;\;2 & \;\;3 & \;\;6 & \;\;5\end{matrix}\\ \begin{bmatrix}4 & -1 & -2 & -1\\ -1 & 4 & -1 & -2\\ -2 & -1 & 4 & -1\\ -1 & -2 & -1 & 4\end{bmatrix}\end{array}\begin{matrix}\\2\\3\\6\\5\end{matrix}, \tag{7}$$

- Element ③

$$\left[K_c^3\right]=7\begin{array}{c} \begin{matrix}\;\;4 & \;\;5 & \;\;8 & \;\;7\end{matrix}\\ \begin{bmatrix}4 & -1 & -2 & -1\\ -1 & 4 & -1 & -2\\ -2 & -1 & 4 & -1\\ -1 & -2 & -1 & 4\end{bmatrix}\end{array}\begin{matrix}\\4\\5\\8\\7\end{matrix}, \tag{8}$$

- Element ④

$$\left[K_c^4\right]=7\begin{array}{c} \begin{matrix}\;\;5 & \;\;6 & \;\;9 & \;\;8\end{matrix}\\ \begin{bmatrix}4 & -1 & -2 & -1\\ -1 & 4 & -1 & -2\\ -2 & -1 & 4 & -1\\ -1 & -2 & -1 & 4\end{bmatrix}\end{array}\begin{matrix}\\5\\6\\9\\8\end{matrix}, \tag{9}$$

Global conductivity matrix (stiffness) $[K_c]$ results from the summation of matrixes for individual elements. By doing so, one should also pay attention to coefficients with the same global indexes, as they should be summated as well. Coefficients of the global conductivity matrix (index c was not included in the designations of coefficients $K_{c,ij}$ in order to shorten the notation) are as follow:

$$K_{11} = K_{11}^1 = 28, \quad K_{12} = K_{12}^1 = -7, \quad K_{14} = K_{14}^1 = -7, \quad K_{15} = K_{15}^1 = -14,$$

$$K_{21} = K_{21}^1 = -7, \quad K_{22} = K_{22}^1 + K_{22}^2 = 7(4+4) = 56, \quad K_{23} = K_{23}^2 = -7,$$

$$K_{24} = K_{24}^1 = -14, \quad K_{25} = K_{25}^1 + K_{25}^2 = 7(-1+(-1)) = -14,$$

$$K_{26} = K_{26}^2 = -14, \quad K_{32} = K_{32}^2 = -7, \quad K_{33} = K_{33}^2 = 28, \quad K_{35} = K_{35}^2 = -14,$$

$$K_{36} = K_{36}^2 = -7, \quad K_{41} = K_{41}^1 = -7, \quad K_{42} = K_{42}^1 = -14,$$

$$K_{44} = K_{44}^1 + K_{44}^3 = 56, \quad K_{45} = K_{45}^1 + K_{45}^3 = -7 + (-7) = -14,$$

$$K_{47} = K_{47}^3 = -7, \quad K_{48} = K_{48}^3 = -14,$$

$$K_{48} = K_{48}^3 = -14, \quad K_{51} = K_{51}^1 = -14, \quad K_{52} = K_{52}^1 + K_{52}^2 = -7 + (-7) = -14,$$

$$K_{53} = K_{53}^2 = -14, \quad K_{54} = K_{54}^1 + K_{54}^3 = -7 + (-7) = -14,$$

$$K_{55} = K_{55}^1 + K_{55}^2 + K_{55}^3 + K_{55}^4 = 28 + 28 + 28 + 28 = 112,$$

$$K_{56} = K_{56}^2 + K_{56}^4 = -7 + (-7) = -14, \quad K_{57} = K_{57}^3 = -14,$$

$$K_{58} = K_{58}^3 + K_{58}^4 = -7 + (-7) = -14, \quad K_{59} = K_{59}^4 = -14, \quad K_{62} = K_{62}^2 = -14,$$

$$K_{63} = K_{63}^2 = -7, \quad K_{65} = K_{65}^2 + K_{65}^4 = -7 + (-7) = -14,$$

$$K_{66} = K_{66}^2 + K_{66}^4 = 28 + 28 = 56,$$

$$K_{68} = K_{68}^4 = -14, \quad K_{69} = K_{69}^4 = -7, \quad K_{74} = K_{74}^3 = -7, \quad K_{75} = K_{75}^3 = -14,$$

$$K_{77} = K_{77}^3 = 28,$$

$$K_{78} = K_{78}^3 = -7, \quad K_{84} = K_{84}^3 = -14, \quad K_{85} = K_{85}^3 + K_{85}^4 = -7 + (-7) = -14,$$

$$K_{86} = K_{86}^4 = -14,$$

$$K_{87} = K_{87}^3 = -7, \quad K_{88} = K_{88}^3 + K_{88}^4 = 28 + 28 = 56, \quad K_{89} = K_{89}^4 = -7,$$

$$K_{95} = K_{95}^4 = -14, \quad K_{96} = K_{96}^4 = -7, \quad K_{98} = K_{98}^4 = -7, \quad K_{99} = K_{99}^4 = 28.$$

The remaining coefficients equal zero. Global conductivity matrix $[K_c]$ has the form

$$\left[K_c\right]=\begin{bmatrix} 28 & -7 & 0 & -7 & -14 & 0 & 0 & 0 & 0 \\ -7 & 56 & -7 & -14 & -14 & -14 & 0 & 0 & 0 \\ 0 & -7 & 28 & 0 & -14 & -7 & 0 & 0 & 0 \\ -7 & -14 & 0 & 56 & -14 & 0 & -7 & -14 & 0 \\ -14 & -14 & -14 & -14 & 112 & -14 & -14 & -14 & -14 \\ 0 & -14 & -7 & 0 & -14 & 56 & 0 & -14 & -7 \\ 0 & 0 & 0 & -7 & -14 & 0 & 28 & -7 & 0 \\ 0 & 0 & 0 & -14 & -14 & -14 & -7 & 56 & -7 \\ 0 & 0 & 0 & 0 & -14 & -7 & 0 & -7 & 28 \end{bmatrix}. \qquad (10)$$

Next, matrix $[K_\alpha]$ is determined

$$\left[K_\alpha\right]=\left[K_\alpha^1\right]+\left[K_\alpha^2\right], \qquad (11)$$

since heat transfer takes place on the lateral surface 1-2 of element ① and on 2-3 of element ②. Matrix $[K_\alpha^1]$ will be calculated using (7), from Ex. 11.12

$$\left[K_\alpha^1\right]=\frac{\alpha L_{12}}{6}\begin{array}{c}\begin{matrix}1 & 2 & 5 & 4\end{matrix}\\ \begin{bmatrix} 2 & 1 & 0 & 0 \\ 1 & 2 & 0 & 0 \\ 0 & 0 & 0 & 0 \\ 0 & 0 & 0 & 0 \end{bmatrix}\end{array}\begin{matrix}\\ 1\\2\\5\\4\end{matrix}=\frac{60\cdot 0.01}{6}\begin{array}{c}\begin{matrix}1 & 2 & 5 & 4\end{matrix}\\ \begin{bmatrix} 2 & 1 & 0 & 0 \\ 1 & 2 & 0 & 0 \\ 0 & 0 & 0 & 0 \\ 0 & 0 & 0 & 0 \end{bmatrix}\end{array}\begin{matrix}\\ 1\\2\\5\\4\end{matrix}= \qquad (12)$$

$$=\begin{array}{c}\begin{matrix}1 & 2 & 5 & 4\end{matrix}\\ \begin{bmatrix} 0.2 & 0.1 & 0 & 0 \\ 0.1 & 0.2 & 0 & 0 \\ 0 & 0 & 0 & 0 \\ 0 & 0 & 0 & 0 \end{bmatrix}\end{array}\begin{matrix}\\ 1\\2\\5\\4\end{matrix}.$$

Matrix $[K_\alpha^2]$ will be calculated in a similar way.

$$\left[K_\alpha^2\right]=\frac{\alpha L_{12}}{6}\begin{array}{c}\begin{matrix}2 & 3 & 6 & 5\end{matrix}\\ \begin{bmatrix} 2 & 1 & 0 & 0 \\ 1 & 2 & 0 & 0 \\ 0 & 0 & 0 & 0 \\ 0 & 0 & 0 & 0 \end{bmatrix}\end{array}\begin{matrix}\\ 2\\3\\6\\5\end{matrix}=\begin{array}{c}\begin{matrix}2 & 3 & 6 & 5\end{matrix}\\ \begin{bmatrix} 0.2 & 0.1 & 0 & 0 \\ 0.1 & 0.2 & 0 & 0 \\ 0 & 0 & 0 & 0 \\ 0 & 0 & 0 & 0 \end{bmatrix}\end{array}\begin{matrix}\\ 2\\3\\6\\5\end{matrix}. \qquad (13)$$

Once matrix coefficients $[K_\alpha^1]$ and $[K_\alpha^2]$, with the same global indexes, are added together, the following is obtained (index α in $K_{\alpha,ij}$ was omitted in order to simplify the notation)

$$K_{11} = K_{11}^1 = 0.2;\ K_{12} = K_{12}^1 = 0.1;\ K_{21} = K_{21}^1 = 0.1;$$
$$K_{22} = K_{22}^1 + K_{22}^2 = 0.2 + 0.2 = 0.4;\ K_{23} = K_{23}^2 = 0.1;$$
$$K_{32} = K_{32}^2 = 0.1;\ K_{33} = K_{33}^2 = 0.2.$$

Matrix $[K_\alpha]$, which results from boundary conditions on the boundary 1--2-3, has the form

$$\left[K_\alpha\right] = \begin{bmatrix} 0.2 & 0.1 & 0 & 0 & 0 & 0 & 0 & 0 & 0 \\ 0.1 & 0.4 & 0.1 & 0 & 0 & 0 & 0 & 0 & 0 \\ 0 & 0.1 & 0.2 & 0 & 0 & 0 & 0 & 0 & 0 \\ 0 & 0 & 0 & 0 & 0 & 0 & 0 & 0 & 0 \\ 0 & 0 & 0 & 0 & 0 & 0 & 0 & 0 & 0 \\ 0 & 0 & 0 & 0 & 0 & 0 & 0 & 0 & 0 \\ 0 & 0 & 0 & 0 & 0 & 0 & 0 & 0 & 0 \\ 0 & 0 & 0 & 0 & 0 & 0 & 0 & 0 & 0 \\ 0 & 0 & 0 & 0 & 0 & 0 & 0 & 0 & 0 \end{bmatrix} \mathrm{W/(m\cdot K)}. \tag{14}$$

Matrix $[K]$ is obtained as a result of adding matrix $[K_c]$ formulated in (10) and matrix $[K_\alpha]$ formulated in (14)

$$[K] = \begin{bmatrix} 28.2 & -6{,}9 & 0 & -7 & -14 & 0 & 0 & 0 & 0 \\ -6.9 & 56.4 & -6.9 & -14 & -14 & -14 & 0 & 0 & 0 \\ 0 & -6.9 & 28.2 & 0 & -14 & -7 & 0 & 0 & 0 \\ -7 & -14 & 0 & 56 & -14 & 0 & -7 & -14 & 0 \\ -14 & -14 & -14 & -14 & 112 & -14 & -14 & -14 & -14 \\ 0 & -14 & -7 & 0 & -14 & 56 & 0 & -14 & -7 \\ 0 & 0 & 0 & -7 & -14 & 0 & 28 & -7 & 0 \\ 0 & 0 & 0 & -14 & -14 & -14 & -7 & 56 & -7 \\ 0 & 0 & 0 & 0 & -14 & -7 & 0 & -7 & 28 \end{bmatrix} \frac{\mathrm{W}}{\mathrm{mK}}. \tag{15}$$

Next, one calculates the vectors on the right-hand-side of (31) from Ex. 11.10. Vector $\{f_Q^e\}$ is formulated in (3) from Ex. 11.13; it assumes the following form for the individual elements:

$$\left\{f_Q^1\right\} = \frac{\dot{q}_v a^2}{4}\begin{Bmatrix}1\\1\\1\\1\end{Bmatrix}\begin{matrix}1\\2\\5\\4\end{matrix}\ \text{W/m},\qquad \left\{f_Q^2\right\} = \frac{\dot{q}_v a^2}{4}\begin{Bmatrix}1\\1\\1\\1\end{Bmatrix}\begin{matrix}2\\3\\6\\5\end{matrix}\ \text{W/m},$$

$$\left\{f_Q^3\right\} = \frac{\dot{q}_v a^2}{4}\begin{Bmatrix}1\\1\\1\\1\end{Bmatrix}\begin{matrix}4\\5\\8\\7\end{matrix}\ \text{W/m},\qquad \left\{f_Q^4\right\} = \frac{\dot{q}_v a^2}{4}\begin{Bmatrix}1\\1\\1\\1\end{Bmatrix}\begin{matrix}5\\6\\9\\8\end{matrix}\ \text{W/m}. \tag{16}$$

Global vector $\{f_Q\}$ is obtained as a result of summing up the elements of vectors with the same global numbers.

$$\left\{f_Q\right\} = \frac{\dot{q}_v a^2}{4}\begin{Bmatrix}1\\2\\1\\2\\4\\2\\1\\2\\1\end{Bmatrix} = \begin{Bmatrix}250\\500\\250\\500\\1000\\500\\250\\500\\250\end{Bmatrix}\begin{matrix}1\\2\\3\\4\\5\\6\\7\\8\\9\end{matrix}\ \text{W/m}. \tag{17}$$

Vector $\{f_\alpha\}$ will be calculated in compliance with (2) from Ex. 11.14. Allowing for the fact that convection heat transfer is assigned on the surface 1-2 of element ① and on the surface 2-3 of element ②, vectors $\{f_\alpha^1\}$ and $\{f_\alpha^2\}$ have the form

$$\left\{f_\alpha^1\right\} = \frac{\alpha T_{cz} a}{2}\begin{Bmatrix}1\\1\\0\\0\end{Bmatrix}\begin{matrix}1\\2\\ \\ \end{matrix} = \begin{Bmatrix}6\\6\\0\\0\end{Bmatrix}\begin{matrix}1\\2\\ \\ \end{matrix}\ \text{W/m}, \tag{18}$$

$$\left\{f_\alpha^2\right\} = \frac{\alpha T_{cz} a}{2}\begin{Bmatrix}1\\1\\0\\0\end{Bmatrix}\begin{matrix}2\\3\\ \\ \end{matrix} = \begin{Bmatrix}6\\6\\0\\0\end{Bmatrix}\begin{matrix}2\\3\\ \\ \end{matrix}\ \text{W/m}. \tag{19}$$

As result of summing up the elements of vectors (18) and (19), with the same global numbers, one obtains

$$\{f_\alpha\} = \begin{Bmatrix} 6 \\ 12 \\ 6 \\ 0 \\ 0 \\ 0 \\ 0 \\ 0 \\ 0 \end{Bmatrix} \begin{matrix} 1 \\ 2 \\ 3 \\ 4 \\ 5 \\ 6 \\ 7 \\ 8 \\ 9 \end{matrix} \ \text{W/m} \ . \tag{20}$$

Heat flux $\dot{q}_B$ is assigned on surfaces 7-4 (⟨4⟩-⟨1⟩) and 4-1 (⟨4⟩-⟨1⟩), of the element ③ and ①, respectively. Vectors $\{f_q^1\}$ and $\{f_q^2\}$ calculated according to (10) from Ex. 11.14 are

$$\{f_q^1\} = \frac{\dot{q}_B a}{2} \begin{Bmatrix} 1 \\ 0 \\ 0 \\ 1 \end{Bmatrix} = \begin{Bmatrix} 1000 \\ 0 \\ 0 \\ 1000 \end{Bmatrix} \begin{matrix} 1 \\ \\ \\ 4 \end{matrix} \ \text{W/m} , \tag{21}$$

$$\{f_q^3\} = \frac{\dot{q}_B a}{2} \begin{Bmatrix} 1 \\ 0 \\ 0 \\ 1 \end{Bmatrix} = \begin{Bmatrix} 1000 \\ 0 \\ 0 \\ 1000 \end{Bmatrix} \begin{matrix} 4 \\ \\ \\ 7 \end{matrix} \ \text{W/m} . \tag{22}$$

Once these vectors are summed up (one should pay attention to the fact that the elements with the same global numbers should be summed up), one has

$$\{f_q\} = \begin{Bmatrix} 1000 \\ 0 \\ 0 \\ 2000 \\ 0 \\ 0 \\ 1000 \\ 0 \\ 0 \end{Bmatrix} \begin{matrix} 1 \\ 2 \\ 3 \\ 4 \\ 5 \\ 6 \\ 7 \\ 8 \\ 9 \end{matrix} \ \text{W/m} \ . \tag{23}$$

The sum of vectors

$$\{f\}=\{f_Q\}+\{f_\alpha\}+\{f_q\} \tag{24}$$

iso

$$\{f\}=\begin{Bmatrix}250\\500\\250\\500\\1000\\500\\250\\500\\250\end{Bmatrix}+\begin{Bmatrix}6\\12\\6\\0\\0\\0\\0\\0\\0\end{Bmatrix}+\begin{Bmatrix}1000\\0\\0\\2000\\0\\0\\1000\\0\\0\end{Bmatrix}=\begin{Bmatrix}1256\\512\\256\\2500\\1000\\500\\1250\\500\\250\end{Bmatrix}\begin{matrix}1\\2\\3\\4\\5\\6\\7\\8\\9\end{matrix}\ \text{W/m}\ . \tag{25}$$

Global equation system $[K]\{T\} = \{f\}$ assumes the form

$$\begin{bmatrix}28.2 & -6.9 & 0 & -7 & -14 & 0 & 0 & 0 & 0\\ -6.9 & 56.4 & -6.9 & -14 & -14 & -14 & 0 & 0 & 0\\ 0 & -6.9 & 28.2 & 0 & -14 & -7 & 0 & 0 & 0\\ -7 & -14 & 0 & 56 & -14 & 0 & -7 & -14 & 0\\ -14 & -14 & -14 & -14 & 112 & -14 & -14 & -14 & -14\\ 0 & -14 & -7 & 0 & -14 & 56 & 0 & -14 & -7\\ 0 & 0 & 0 & -7 & -14 & 0 & 28 & -7 & 0\\ 0 & 0 & 0 & -14 & -14 & -14 & -7 & 56 & -7\\ 0 & 0 & 0 & 0 & -14 & -7 & 0 & -7 & 28\end{bmatrix}\begin{Bmatrix}T_1\\T_2\\T_3\\T_4\\T_5\\T_6\\T_7\\T_8\\T_9\end{Bmatrix}=\begin{Bmatrix}1256\\512\\256\\2500\\1000\\500\\1250\\500\\250\end{Bmatrix}. \tag{26}$$

The equation system (26) will be transformed with the boundary condition (2), from which it follows that $T_3 = T_6 = T_9 = 100°\text{C}$

$$\begin{bmatrix}28.2 & -6.9 & 0 & -7 & -14 & 0 & 0 & 0 & 0\\ -6.9 & 56.4 & 0 & -14 & -14 & 0 & 0 & 0 & 0\\ 0 & 0 & 1 & 0 & 0 & 0 & 0 & 0 & 0\\ -7 & -14 & 0 & 56 & -14 & 0 & -7 & -14 & 0\\ -14 & -14 & 0 & -14 & 112 & 0 & -14 & -14 & 0\\ 0 & 0 & 0 & 0 & 0 & 1 & 0 & 0 & 0\\ 0 & 0 & 0 & -7 & -14 & 0 & 28 & -7 & 0\\ 0 & 0 & 0 & -14 & -14 & 0 & -7 & 56 & 0\\ 0 & 0 & 0 & 0 & 0 & 0 & 0 & 0 & 1\end{bmatrix}\begin{Bmatrix}T_1\\T_2\\T_3\\T_4\\T_5\\T_6\\T_7\\T_8\\T_9\end{Bmatrix}=\begin{Bmatrix}1256\\2602\\100\\2500\\5200\\100\\1250\\2600\\100\end{Bmatrix}. \tag{27}$$

Global equation system (27) was solved using the Gauss elimination method and the following was obtained:

$$\begin{aligned} &T_1 = 238.63^\circ\text{C}, && T_2 = 180.20^\circ\text{C}, \\ &T_3 = 100^\circ\text{C}, && T_4 = 240.68^\circ\text{C}, \\ &T_5 = 181.80^\circ\text{C}, && T_6 = 100^\circ\text{C}, \\ &T_7 = 241.27^\circ\text{C}, && T_8 = 182.21^\circ\text{C}, \\ &T_9 = 100^\circ\text{C}. \end{aligned} \tag{28}$$

Temperature distribution was also calculated by means of ANSYS program, while the region was divided into 2500 elements. The following temperatures were obtained for nodes, which correspond to nodes 1-9:

$$\begin{aligned} &T_1 = 238.51^\circ\text{C}, && T_2 = 180.15^\circ\text{C}, \\ &T_3 = 100^\circ\text{C}, && T_4 = 240.53^\circ\text{C}, \\ &T_5 = 181.67^\circ\text{C}, && T_6 = 100^\circ\text{C}, \\ &T_7 = 241.09^\circ\text{C}, && T_8 = 182.08^\circ\text{C}, \\ &T_9 = 100^\circ\text{C}. \end{aligned} \tag{29}$$

The isotherm map is shown in Fig. 11.26.

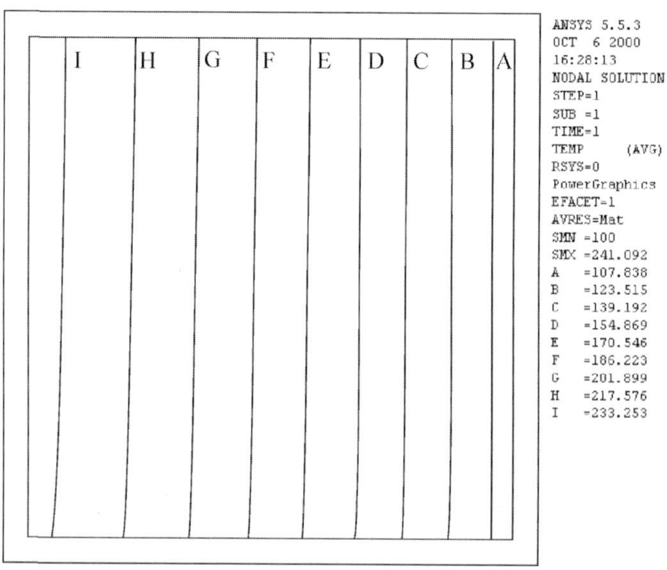

Fig. 11.26. Calculation results; calculations conducted by means of the ANSYS program

Exercise 11.19 Determining Two-Dimensional Temperature Distribution in a Straight Fin with Constant Thickness by Means of FEM

Determine temperature distribution in a fin by means of FEM. Assume the following values from Ex. 7.3 for the calculation: a = 0.003 m, λ = 50 W/(m·K), α = 100 W/(m^2·K), T_b = 95°C, T_{cz} = 20°C. Determine heat flow at the fin base and fin efficiency.

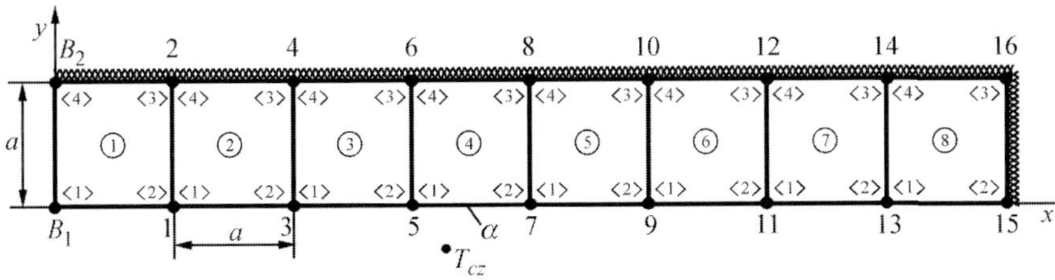

Fig. 11.27. A fin division into finite elements

Solution

Temperature distribution will be determined at the mid-point of the fin due to the symmetry of temperature field with respect to axis x. Table 11.4 contains local and global node numeration. Method II, discussed in Ex. 11.15, will be employed in the construction of the global equation system; the method is based on the summation of weighted residuals for elements with a common node. Due to the fact that the equations for individual nodes have similar structure, the equation for node three will be created and, subsequently, applied to nodes 1, 5, 7, 9, 11 and 13, while the equation for node 4 will be applied to nodes 2, 6, 8, 10, 12 and 14. Separate equations will be created for nodes 15 and 16.

• Node 3

Node 3 is shared by elements ② and ③ (Fig. 11.27). Thermal conduction matrixes $[K_c^2]$ and $[K_c^3]$ will be calculated according to (6) from Ex. 11.11, while matrixes $[K_\alpha^2]$ and $[K_\alpha^3]$ according to (7) from Ex. 11.12. Matrixes $[K_c^2]$ and $[K_\alpha^2]$ are expressed as follow:

$$\begin{matrix} & \langle 1\rangle & \langle 2\rangle & \langle 3\rangle & \langle 4\rangle & \\ \left[K_c^2\right]=\dfrac{\lambda}{6} & \begin{bmatrix} 4 & -1 & -2 & -1 \\ -1 & 4 & -1 & -2 \\ -2 & -1 & 4 & -1 \\ -1 & -2 & -1 & 4 \end{bmatrix} & & & & \begin{matrix}\langle 1\rangle\\ \langle 2\rangle\\ \langle 3\rangle\\ \langle 4\rangle\end{matrix} \end{matrix}, \tag{1}$$

Table 11.4. Local and global node numeration

Element No.	Local Node No.	Global Node No.
①	⟨1⟩	B_1
	⟨2⟩	1
	⟨3⟩	2
	⟨4⟩	B_2
②	⟨1⟩	1
	⟨2⟩	3
	⟨3⟩	4
	⟨4⟩	2
③	⟨1⟩	3
	⟨2⟩	5
	⟨3⟩	6
	⟨4⟩	4
④	⟨1⟩	5
	⟨2⟩	7
	⟨3⟩	8
	⟨4⟩	6
⑤	⟨1⟩	7
	⟨2⟩	9
	⟨3⟩	10
	⟨4⟩	8
⑥	⟨1⟩	9
	⟨2⟩	11
	⟨3⟩	12
	⟨4⟩	10
⑦	⟨1⟩	11
	⟨2⟩	13
	⟨3⟩	14
	⟨4⟩	12
⑧	⟨1⟩	13
	⟨2⟩	15
	⟨3⟩	16
	⟨4⟩	14

$$\left[K_\alpha^2\right]=\frac{\alpha a}{6}\begin{array}{c}\begin{matrix}\langle 1\rangle & \langle 2\rangle & \langle 3\rangle & \langle 4\rangle\end{matrix}\\ \begin{bmatrix}2 & 1 & 0 & 0\\ 1 & 2 & 0 & 0\\ 0 & 0 & 0 & 0\\ 0 & 0 & 0 & 0\end{bmatrix}\end{array}\begin{matrix}\langle 1\rangle\\ \langle 2\rangle\\ \langle 3\rangle\\ \langle 4\rangle\end{matrix}. \tag{2}$$

Stiffness matrix $[K]$ is obtained by way of summing up matrices (1) and (2) (coefficients with the same indexes are added up)

$$\left[K^2\right]=\left[K_c^2\right]+\left[K_\alpha^2\right]=\begin{array}{c}\begin{matrix}\quad\langle 1\rangle & \qquad\quad\langle 2\rangle & \qquad\langle 3\rangle & \langle 4\rangle\end{matrix}\\ \begin{bmatrix} \frac{2\lambda}{3}+\frac{\alpha a}{3} & -\frac{\lambda}{6}+\frac{\alpha a}{6} & -\frac{\lambda}{3} & -\frac{\lambda}{6} \\ -\frac{\lambda}{6}+\frac{\alpha a}{6} & \frac{2\lambda}{3}+\frac{\alpha a}{3} & -\frac{\lambda}{6} & -\frac{\lambda}{3} \\ -\frac{\lambda}{3} & -\frac{\lambda}{6} & \frac{2\lambda}{3} & -\frac{\lambda}{6} \\ -\frac{\lambda}{6} & -\frac{\lambda}{3} & -\frac{\lambda}{6} & \frac{2\lambda}{3} \end{bmatrix}\end{array}\begin{matrix}\\ \langle 1\rangle\\ \langle 2\rangle\\ \langle 3\rangle\\ \langle 4\rangle\end{matrix}. \qquad (3)$$

Matrixes $[K_c^3]$ and $[K_\alpha^3]$ are calculated in a similar way.

$$\left[K_c^3\right]=\frac{\lambda}{6}\begin{array}{c}\begin{matrix}\langle 1\rangle & \langle 2\rangle & \langle 3\rangle & \langle 4\rangle\end{matrix}\\ \begin{bmatrix} 4 & -1 & -2 & -1 \\ -1 & 4 & -1 & -2 \\ -2 & -1 & 4 & -1 \\ -1 & -2 & -1 & 4 \end{bmatrix}\end{array}\begin{matrix}\\ \langle 1\rangle\\ \langle 2\rangle\\ \langle 3\rangle\\ \langle 4\rangle\end{matrix}, \qquad (4)$$

$$\left[K_\alpha^3\right]=\frac{\alpha a}{6}\begin{array}{c}\begin{matrix}\langle 1\rangle & \langle 2\rangle & \langle 3\rangle & \langle 4\rangle\end{matrix}\\ \begin{bmatrix} 2 & 1 & 0 & 0 \\ 1 & 2 & 0 & 0 \\ 0 & 0 & 0 & 0 \\ 0 & 0 & 0 & 0 \end{bmatrix}\end{array}\begin{matrix}\\ \langle 1\rangle\\ \langle 2\rangle\\ \langle 3\rangle\\ \langle 4\rangle\end{matrix}. \qquad (5)$$

Rigidity matrix for element ③ is as follows:

$$\left[K^3\right]=\left[K_c^3\right]+\left[K_\alpha^3\right]=\begin{array}{c}\begin{matrix}\quad\langle 1\rangle & \qquad\quad\langle 2\rangle & \qquad\langle 3\rangle & \langle 4\rangle\end{matrix}\\ \begin{bmatrix} \frac{2\lambda}{3}+\frac{\alpha a}{3} & -\frac{\lambda}{6}+\frac{\alpha a}{6} & -\frac{\lambda}{3} & -\frac{\lambda}{6} \\ -\frac{\lambda}{6}+\frac{\alpha a}{6} & \frac{2\lambda}{3}+\frac{\alpha a}{3} & -\frac{\lambda}{6} & -\frac{\lambda}{3} \\ -\frac{\lambda}{3} & -\frac{\lambda}{6} & \frac{2\lambda}{3} & -\frac{\lambda}{6} \\ -\frac{\lambda}{6} & -\frac{\lambda}{3} & -\frac{\lambda}{6} & \frac{2\lambda}{3} \end{bmatrix}\end{array}\begin{matrix}\\ \langle 1\rangle\\ \langle 2\rangle\\ \langle 3\rangle\\ \langle 4\rangle\end{matrix}. \qquad (6)$$

Next, vectors $\{f_\alpha^2\}$ and $\{f_\alpha^3\}$ will be determined. According to (7) from Ex. 11.14, one has

$$\left\{f_\alpha^2\right\} = \frac{\alpha T_{cz} a}{2}\begin{Bmatrix}1\\1\\0\\0\end{Bmatrix}\begin{matrix}\langle 1\rangle\\\langle 2\rangle\\\langle 3\rangle\\\langle 4\rangle\end{matrix} = \begin{Bmatrix}\alpha T_{cz} a/2\\\alpha T_{cz} a/2\\0\\0\end{Bmatrix}\begin{matrix}\langle 1\rangle\\\langle 2\rangle\\\langle 3\rangle\\\langle 4\rangle\end{matrix}, \tag{7}$$

$$\left\{f_\alpha^3\right\} = \begin{Bmatrix}\alpha T_{cz} a/2\\\alpha T_{cz} a/2\\0\\0\end{Bmatrix}\begin{matrix}\langle 1\rangle\\\langle 2\rangle\\\langle 3\rangle\\\langle 4\rangle\end{matrix}. \tag{8}$$

Node 3, i.e. the node with global number 3, corresponds to local node ⟨2⟩ in element ② and to node ⟨1⟩ in element ③. Therefore, in the equation system for element ②

$$\left[K^2\right]\begin{Bmatrix}T_1\\T_3\\T_4\\T_2\end{Bmatrix} = \begin{Bmatrix}\alpha T_{cz} a/2\\\alpha T_{cz} a/2\\0\\0\end{Bmatrix} \tag{9}$$

only the second equation is taken into consideration; it has the form

$$\left[-\frac{\lambda}{6}+\frac{\alpha a}{6},\ \frac{2\lambda}{3}+\frac{\alpha a}{3},\ -\frac{\lambda}{6},\ -\frac{\lambda}{3}\right]\begin{Bmatrix}T_1\\T_3\\T_4\\T_2\end{Bmatrix} = \frac{\alpha T_{cz} a}{2}, \tag{10}$$

$$\left(-\frac{\lambda}{6}+\frac{\alpha a}{6}\right)T_1 + \left(\frac{2\lambda}{3}+\frac{\alpha a}{3}\right)T_3 - \frac{\lambda}{6}T_4 - \frac{\lambda}{3}T_2 = \frac{\alpha T_{cz} a}{2}. \tag{11}$$

In the equation system for element ③

$$\left[K^3\right]\begin{Bmatrix}T_3\\T_5\\T_6\\T_4\end{Bmatrix} = \begin{Bmatrix}\alpha T_{cz} a/2\\\alpha T_{cz} a/2\\0\\0\end{Bmatrix} \tag{12}$$

only the first equation is taken into consideration; it has the form

$$\left[\frac{2\lambda}{3}+\frac{\alpha a}{3},\ -\frac{\lambda}{6}+\frac{\alpha a}{6},\ -\frac{\lambda}{3},\ -\frac{\lambda}{6}\right]\begin{Bmatrix}T_3\\T_5\\T_6\\T_4\end{Bmatrix} = \frac{\alpha T_{cz} a}{2}, \tag{13}$$

$$\left(\frac{2\lambda}{3}+\frac{\alpha a}{3}\right)T_3+\left(-\frac{\lambda}{6}+\frac{\alpha a}{6}\right)T_5-\frac{\lambda}{3}T_6-\frac{\lambda}{6}T_4=\frac{\alpha T_{cz}a}{2}. \quad (14)$$

Once (11) and (14) are added, the equation for node 3 is obtained

$$\left(-\frac{\lambda}{6}+\frac{\alpha a}{6}\right)T_1-\frac{\lambda}{3}T_2+2\left(\frac{2\lambda}{3}+\frac{\alpha a}{3}\right)T_3-\frac{\lambda}{3}T_4+\left(-\frac{\lambda}{6}+\frac{\alpha a}{6}\right)T_5-\frac{\lambda}{3}T_6= \quad (15)$$
$$=\alpha T_{cz}a.$$

Analogically, one can write an equation for node 1

$$\left(-\frac{\lambda}{6}+\frac{\alpha a}{6}\right)T_b-\frac{\lambda}{3}T_b+2\left(\frac{2\lambda}{3}+\frac{\alpha a}{3}\right)T_1-\frac{\lambda}{3}T_2+\left(-\frac{\lambda}{6}+\frac{\alpha a}{6}\right)T_3-\frac{\lambda}{3}T_4= \quad (16)$$
$$=\alpha T_{cz}a,$$

which results in

$$2\left(\frac{2\lambda}{3}+\frac{\alpha a}{3}\right)T_1-\frac{\lambda}{3}T_2+\left(-\frac{\lambda}{6}+\frac{\alpha a}{6}\right)T_3-\frac{\lambda}{3}T_4=\left(\frac{\lambda}{2}-\frac{\alpha a}{6}\right)T_b+\alpha T_{cz}a. \quad (17)$$

• Nodes 5, 7, 9, 11, 13

Equation (15) can be applied to nodes that lie on the surface, which remains in contact with the medium; that excludes, however, node 15

$$\left(-\frac{\lambda}{6}+\frac{\alpha a}{6}\right)T_{i-2}-\frac{\lambda}{3}T_{i-1}+2\left(\frac{2\lambda}{3}+\frac{\alpha a}{3}\right)T_i-\frac{\lambda}{3}T_{i+1}+\left(-\frac{\lambda}{6}+\frac{\alpha a}{6}\right)T_{i+2}- \quad (18)$$
$$-\frac{\lambda}{3}T_{i+3}=\alpha aT_{cz}; \qquad i=3,\ 5,\ 7,\ 9,\ 11,\ 13.$$

• Node 15

The equation for node 15 is obtained in a similar way as the equation for node 3 in element ②. Analogically to (10), one can obtain the equation for node 15

$$\left[-\frac{\lambda}{6}+\frac{\alpha a}{6},\ \frac{2\lambda}{3}+\frac{\alpha a}{3},\ -\frac{\lambda}{6},\ -\frac{\lambda}{3}\right]\begin{Bmatrix}T_{13}\\T_{15}\\T_{16}\\T_{14}\end{Bmatrix}=\frac{\alpha T_{cz}a}{2}, \quad (19)$$

from which, one obtains

$$\left(-\frac{\lambda}{6}+\frac{\alpha a}{6}\right)T_{13}+\left(\frac{2\lambda}{3}+\frac{\alpha a}{3}\right)T_{15}-\frac{\lambda}{6}T_{16}-\frac{\lambda}{3}T_{14}=\frac{\alpha T_{cz}a}{2}. \quad (20)$$

• Node 4

Convection heat transfer does not occure on the surface of elements ② and ③. The algebraic equation system for element ② has the form

$$\left[K_c^2\right]\begin{Bmatrix} T_1 \\ T_3 \\ T_4 \\ T_2 \end{Bmatrix} = 0\,, \tag{21}$$

where $[K_c^2]$ is formulated in (1).

The equation for node 4 (the third equation in the system (21)) has the form

$$\frac{\lambda}{6}\left[-2,\ \ -1,\ \ 4,\ \ -1\right]\begin{Bmatrix} T_1 \\ T_3 \\ T_4 \\ T_2 \end{Bmatrix} = 0\,, \tag{22}$$

$$-\frac{\lambda}{3}T_1 - \frac{\lambda}{6}T_3 + \frac{2\lambda}{3}T_4 - \frac{\lambda}{6}T_2 = 0\,. \tag{23}$$

The equation system for element ③ has the form

$$\left[K_c^3\right]\begin{Bmatrix} T_3 \\ T_5 \\ T_6 \\ T_4 \end{Bmatrix} = 0\,, \tag{24}$$

where $[K_c^2]$ is formulated in (4).

The equation for node 4 (the fourth equation in the system (24)) has the form

$$\frac{\lambda}{6}\left[-1,\ \ -2,\ \ -1,\ \ 4\right]\begin{Bmatrix} T_3 \\ T_5 \\ T_6 \\ T_4 \end{Bmatrix} = 0\,, \tag{25}$$

from which, one obtains

$$-\frac{\lambda}{6}T_3 - \frac{\lambda}{3}T_5 - \frac{\lambda}{6}T_6 + \frac{2\lambda}{3}T_4 = 0\,. \tag{26}$$

Once (23) and (26) are added together, an algebraic equation (heat balance) for node 4 is obtained

$$-\frac{\lambda}{3}T_1-\frac{\lambda}{6}T_2-\frac{\lambda}{3}T_3+\frac{4\lambda}{3}T_4-\frac{\lambda}{3}T_5-\frac{\lambda}{6}T_6=0. \tag{27}$$

Equation for node 2 can be written analogically (indexes are reduced by 2 and the boundary conditions are accounted for)

$$-\frac{\lambda}{3}T_b-\frac{\lambda}{6}T_b-\frac{\lambda}{3}T_1+\frac{4\lambda}{3}T_2-\frac{\lambda}{3}T_3-\frac{\lambda}{6}T_4=0; \tag{28}$$

hence, one gets

$$-\frac{\lambda}{3}T_1+\frac{4\lambda}{3}T_2-\frac{\lambda}{3}T_3-\frac{\lambda}{6}T_4=\frac{\lambda}{2}T_b. \tag{29}$$

- Nodes 6, 8, 10, 12, 14

One can write a general equation for nodes 4, 6, 8, 10, 12 and 14 on the basis of (27)

$$-\frac{\lambda}{3}T_{i-3}-\frac{\lambda}{6}T_{i-2}-\frac{\lambda}{3}T_{i-1}+\frac{4\lambda}{3}T_i-\frac{\lambda}{3}T_{i+1}-\frac{\lambda}{6}T_{i+2}=0, \quad i=4,\ 6,\ 8,\ 10,\ 12,\ 14. \tag{30}$$

The equation system for element ⑧ has the form

$$\left[K_c^2\right]\begin{Bmatrix}T_{13}\\T_{15}\\T_{16}\\T_{14}\end{Bmatrix}=0, \tag{31}$$

since $[K_c^8] = [K_c^2]$. $[K_c^2]$ is formulated in (1).

- Node 16

The equation for node 16 (the third equation in the system (31)) has the form

$$\frac{\lambda}{6}\left[-2,\ \ -1,\ \ 4,\ \ -1\right]\begin{Bmatrix}T_{13}\\T_{15}\\T_{16}\\T_{14}\end{Bmatrix}=0, \tag{32}$$

hence, one has

$$-\frac{\lambda}{3}T_{13}-\frac{\lambda}{6}T_{15}+\frac{2\lambda}{3}T_{16}-\frac{\lambda}{6}T_{14}=0. \tag{33}$$

The equation system made of (17), (18), (20), (29), (30) and (33) defines node temperature distribution. Such system has the form

$$\begin{aligned}
&2\left(\frac{2\lambda}{3}+\frac{\alpha a}{3}\right)T_1-\frac{\lambda}{3}T_2+\left(-\frac{\lambda}{6}+\frac{\alpha a}{6}\right)T_3-\frac{\lambda}{3}T_4=\left(\frac{\lambda}{2}-\frac{\alpha a}{6}\right)T_b+\alpha T_{cz}a,\\
&\left(-\frac{\lambda}{6}+\frac{\alpha a}{6}\right)T_{i-2}-\frac{\lambda}{3}T_{i-1}+2\left(\frac{2\lambda}{3}+\frac{\alpha a}{3}\right)T_i-\frac{\lambda}{3}T_{i+1}+\left(-\frac{\lambda}{6}+\frac{\alpha a}{6}\right)T_{i+2}-\\
&-\frac{\lambda}{3}T_{i+3}=\alpha a T_{cz},\quad i=3,5,7,9,11,13,\\
&\left(-\frac{\lambda}{6}+\frac{\alpha a}{6}\right)T_{13}+\left(\frac{2\lambda}{3}+\frac{\alpha a}{3}\right)T_{15}-\frac{\lambda}{6}T_{16}-\frac{\lambda}{3}T_{14}=\frac{\alpha T_{cz}a}{2}\\
&-\frac{\lambda}{3}T_1+\frac{4\lambda}{3}T_2-\frac{\lambda}{3}T_3-\frac{\lambda}{6}T_4=\frac{\lambda}{2}T_b,\\
&-\frac{\lambda}{3}T_{i-3}-\frac{\lambda}{6}T_{i-2}-\frac{\lambda}{3}T_{i-1}+\frac{4\lambda}{3}T_i-\frac{\lambda}{3}T_{i+1}-\frac{\lambda}{6}T_{i+2}=0,\\
&i=4,6,8,10,12,14\\
&-\frac{\lambda}{3}T_{13}-\frac{\lambda}{6}T_{15}+\frac{2\lambda}{3}T_{16}-\frac{\lambda}{6}T_{14}=0.
\end{aligned} \tag{34}$$

The equation system (34) will be solved using the Gauss-Seidel method; due to this reason, it will be written in the form

$$T_1=\frac{1}{8+4Bi}\left[2T_2+(1-Bi)T_3+2T_4+(3-Bi)T_b+6BiT_{cz}\right]$$

$$T_i=\frac{1}{8+4Bi}\left[(1-Bi)T_{i-2}+2T_{i-1}+2T_{i+1}+(3-Bi)T_{i+2}+2T_{i+3}+6BiT_{cz}\right], \tag{35}$$

$$i=3,5,7,9,11,13$$

$$\begin{aligned}
&T_{15}=\frac{1}{4+2Bi}\left[(1-Bi)T_{13}+2T_{14}+T_{16}+3BiT_{cz}\right]\\
&T_2=\frac{1}{8}\left(2T_1+2T_3+T_4+3T_b\right)\\
&T_i=\frac{1}{8}\left(2T_{i-3}+T_{i-2}+2T_{i-1}+2T_{i+1}+T_{i+2}\right),\qquad i=4,6,8,10,12,14\\
&T_{16}=\frac{1}{4}\left(2T_{13}+T_{14}+T_{15}\right),
\end{aligned} \tag{35}$$

where $Bi = \alpha a/\lambda$ is the Biot number.

This system will be solved using the same data that was given in Ex.7.3: a = 0.003m, λ = 50 W/(m·K), α = 100 W/(m^2·K), T_b = 95°C, T_{cz} = 20°C, $Bi = \alpha a/\lambda$ = (100·0.003)/50 = 0.006. If we assume that calculation tolerance δ = 0.00001 K in Gauss-Seidel method is the solution to system (35), we will obtain the results, which are shown in Table 11.5.

Table 11.5. Temperature in control volume nodes shown in Fig. 11.27

Node No.	Temperature		Node No.	Temperature	
	FEM	Analytical Method		FEM	Analytical Method
1	91.90	91.88	9	84.20	84.18
2	92.14	92.09	10	84.39	84.37
3	89.38	89.34	11	83.26	83.23
4	89.58	89.55	12	83.45	83.42
5	87.25	87.22	13	82.70	82.67
6	87.45	87.42	14	82.88	82.86
7	85.53	85.50	15	82.51	82.48
8	85.73	85.70	16	82.70	82.67

Next, we will calculate heat flow

$$\dot{Q} = 2\int_0^a \left(-\lambda \frac{\partial T^1}{\partial x} \right) \bigg|_{x=0} dy = -2\lambda \int_0^a \frac{\partial T^1}{\partial x} \bigg|_{x=0} dy\,, \tag{36}$$

where $T^1(x, y)$ stands for the temperature distribution in element ①, formulated as

$$T^1(x, y) = N_1^1 T_b + N_2^1 T_1 + N_3^1 T_2 + N_4^1 T_b\,, \tag{37}$$

where,

$$N_1^1 = \left(1 - \frac{x}{a}\right)\left(1 - \frac{y}{a}\right),$$
$$N_2^1 = \frac{x}{a}\left(1 - \frac{y}{a}\right), \tag{38}$$
$$N_3^1 = \frac{xy}{a^2}, \quad N_4^1 = \frac{y}{a}\left(1 - \frac{x}{a}\right).$$

Derivative $\partial T^1/\partial x$ is

$$\frac{\partial T^1}{\partial x} = \left(\frac{\partial N_1^1}{\partial x} + \frac{\partial N_4^1}{\partial x} \right) T_b + \frac{\partial N_2^1}{\partial x} T_1 + \frac{\partial N_3^1}{\partial x} T_2, \tag{39}$$

where

$$\frac{\partial N_1^1}{\partial x} = -\frac{1}{a}\left(1 - \frac{y}{a}\right), \qquad \frac{\partial N_4^1}{\partial x} = -\frac{y}{a^2},$$
$$\frac{\partial N_2^1}{\partial x} = \frac{1}{a}\left(1 - \frac{y}{a}\right), \qquad \frac{\partial N_3^1}{\partial x} = \frac{y}{a^2}. \tag{40}$$

By substituting (40) into (39), one obtains

$$\frac{\partial T^1}{\partial x} = \frac{T_1 - T_b}{a} + \frac{T_2 - T_1}{a^2} y, \tag{41}$$

while after substitution of (41) into (36) and integration, one has

$$\dot{Q} = -2\lambda \left[T_1 - T_b + \frac{1}{2}(T_2 - T_1) \right] = 2\lambda \left(T_b - \frac{T_1 + T_2}{2} \right) \text{W/m}. \tag{42}$$

Maximum heat flow given off by an isothermal fin with the base temperature T_b is

$$\dot{Q}_{\max} = 2 \cdot 8\alpha a (T_b - T_{cz}) = 16\alpha a (T_b - T_{cz}). \tag{43}$$

Fin efficiency is defined as

$$\eta = \frac{\dot{Q}}{\dot{Q}_{\max}} = \frac{2\lambda \left(T_b - \frac{T_1 + T_2}{2} \right)}{16\alpha a (T_b - T_{cz})} = \frac{1}{8Bi} \frac{T_b - \frac{T_1 + T_2}{2}}{T_b - T_{cz}}, \tag{44}$$

where $Bi = \alpha a / \lambda$.

After substitution of the numerical values, one obtains

$$\eta = \frac{1}{8 \cdot 0.006} \cdot \frac{95 - \frac{91.90 + 92.14}{2}}{95 - 20} = 0.828. \tag{45}$$

As one can see, the determined fin efficiency differs from the value $\eta_e =$ 0.887 obtained by means of the analytical formula (Ex. 7.3). Relative error is at

$$\Delta\eta = \frac{\eta - \eta_e}{\eta_e} \cdot 100\% = \frac{0.828 - 0.887}{0.887} \cdot 100\% = -6.6\%. \tag{46}$$

Rather large error $\Delta\eta$ arises from the approximation of medium temperature gradient within the width of fin thickness by means of difference quotient

$$\left.\frac{\partial \overline{T}}{\partial x}\right|_{x=0} = \frac{1}{a} \int_0^a \left.\frac{\partial T}{\partial x}\right|_{x=0} dy = \frac{1}{a} \left(T_b - \frac{T_1 + T_2}{2} \right)$$

with an accuracy of 1st order.

One should add that heat flow $\dot{Q}$ at the fin base can be also calculated from formula

$$\dot{Q} = 2a\dot{q}_b, \tag{47}$$

where $\dot{q}_b$ is the heat flux at the fin base. Aside from the given fin base temperature, assigned at points B_1 and B_2 (Fig. 11.27), temperatures in nodes 1 and 2 (global numeration) are known from the FEM calculations. The equation system for element ① has the form

$$\left[K_c^1\right]\begin{Bmatrix} T_b \\ T_1 \\ T_2 \\ T_b \end{Bmatrix} = \left\{f_q\right\}, \tag{48}$$

where $\{f_q\}$ follows from the heat flux $\dot{q}_b$ assigned at the fin-base. Because $\dot{q}_b$ is assigned on the side ⟨4⟩-⟨1⟩ of element ①, vector $\{f_q\}$ has the form then

$$\left\{f_q\right\} = \begin{Bmatrix} \dot{q}_b a/2 \\ 0 \\ 0 \\ \dot{q}_b a/2 \end{Bmatrix}. \tag{49}$$

The first equation in the system (48) has the form

$$\frac{\lambda}{6}\begin{bmatrix} 4 & -1 & -2 & -1 \end{bmatrix}\begin{Bmatrix} T_b \\ T_1 \\ T_2 \\ T_b \end{Bmatrix} = \frac{\dot{q}_b a}{2}, \tag{50}$$

hence, we have

$$\frac{4\lambda}{6}T_b - \frac{\lambda}{6}T_1 - \frac{2\lambda}{6}T_2 - \frac{\lambda}{6}T_b = \frac{\dot{q}_b a}{2}. \tag{51}$$

Heat flux $\dot{q}_b$ at point B_1 determined from (51) is

$$\dot{q}_b = \frac{\lambda}{a}\left(T_b - \frac{T_1 + 2T_2}{3}\right). \tag{52}$$

The fourth equation in the system (49) has the form

$$\frac{\lambda}{6}\begin{bmatrix} -1 & -2 & -1 & 4 \end{bmatrix}\begin{Bmatrix} T_b \\ T_1 \\ T_2 \\ T_b \end{Bmatrix} = \frac{\dot{q}_b a}{2}, \tag{53}$$

hence, we obtain

$$-\frac{\lambda}{6}T_b - \frac{2\lambda}{6}T_1 - \frac{\lambda}{6}T_2 + \frac{4\lambda}{6}T_b = \frac{\dot{q}_b a}{2}. \tag{54}$$

Heat flux $\dot{q}_b$ at point B_2 determined from (54) is

$$\dot{q}_b = \frac{\lambda}{a}\left(T_b - \frac{2T_1 + T_2}{3}\right). \tag{55}$$

Arithmetic average of the heat flux in nodes B_1 and B_2 given by (52) and (55) is

$$\dot{q}_b = \frac{\lambda}{a}\left(T_b - \frac{T_1 + T_2}{2}\right). \tag{56}$$

By substituting (56) into (47), (42) is obtained. Both methods for calculating heat flux give identical results.

In order to improve accuracy, fin efficiency will be calculated using a different method.

Heat flow given off by the fin can be expressed in the following way:

$$\dot{Q} = 2\alpha a\left[\frac{1}{2}(T_b - T_{cz}) + (T_1 - T_{cz}) + (T_3 - T_{cz}) + (T_5 - T_{cz}) + (T_7 - T_{cz}) + \right.$$
$$\left. + (T_9 - T_{cz}) + (T_{11} - T_{cz}) + (T_{13} - T_{cz}) + \frac{1}{2}(T_{15} - T_{cz})\right], \tag{57}$$

$$\dot{Q} = 2\alpha\Delta x\left[\frac{1}{2}T_b + T_1 + T_3 + T_5 + T_7 + T_9 + T_{11} + T_{13} + \frac{1}{2}T_{15} - 8T_{cz}\right], \tag{58}$$

$$\dot{Q} = 2\cdot 100\cdot 0.003\left(\frac{95}{2} + 91.90 + 89.38 + 87.25 + 85.53 + \right.$$
$$\left. + 84.20 + 83.26 + 82.70 + \frac{82.51}{2} - 8\cdot 20\right) = 319.785 \text{ W/m}.$$

Fin efficiency determined by means of FEM is

$$\eta = \frac{\dot{Q}}{\dot{Q}_{\max}} = \frac{319.785}{360} = 0.888. \tag{59}$$

Relative error from the determination of efficiency is

$$\Delta\eta = \frac{\eta - \eta_e}{\eta_e}\cdot 100\% = \frac{0.888 - 0.887}{0.887}\cdot 100\% = 0.112\%,$$

This method, therefore, is much more accurate than the earlier presented method in which $\dot{Q}$ is determined from (42).

Exercise 11.20 Determining Two-Dimensional Temperature Distribution by Means of FEM in a Straight Fin with Constant Thickness (ANSYS Program)

Determine temperature distribution and efficiency of a fin presented in Fig. 11.28. For the calculation, adopt the values from Ex. 7.3: $w = 0.003$ m, $l = 0.024$m, $\alpha = 100$ W/(m ·K)^2, $T_b = 95$°C, $T_{cz} = 20$°C, $\lambda = 50$ W/(m·K). Calculate fin efficiency for cases a) and b) presented in Fig. 11.28, i.e. when the fin tip is thermally insulated and heat transfer occurs at the tip. Furthermore, for the case a) determine fin efficiency by means of the analytical formula; make use of the results obtained in Ex. 7.3.

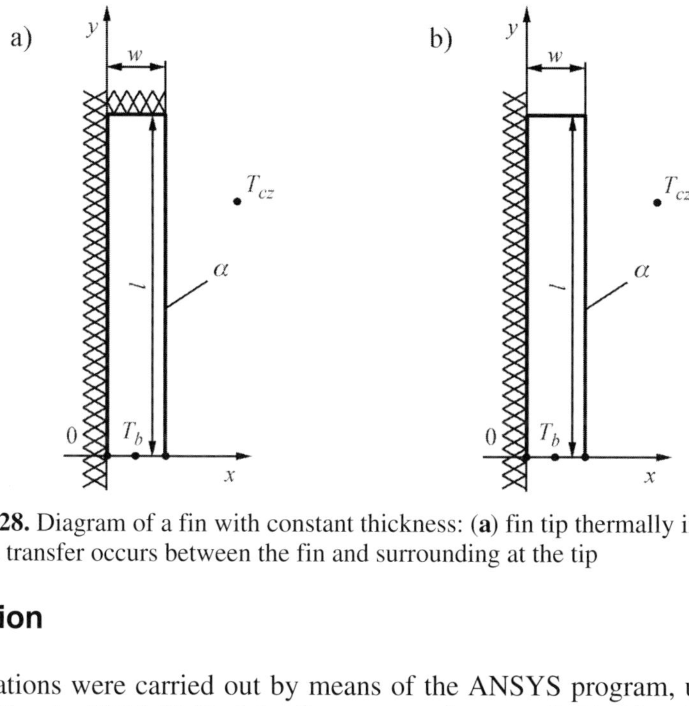

Fig. 11.28. Diagram of a fin with constant thickness: (**a**) fin tip thermally insulated, (**b**) heat transfer occurs between the fin and surrounding at the tip

Solution

Calculations were carried out by means of the ANSYS program, used for calculating by FEM. Half of the fin cross-section was divided into 288 elements. Temperature was calculated in 343 nodes (Fig. 11.29). In a case when the fin tip is thermally insulated (Fig. 11.28a), temperatures of the tip calculated by means of the analytical formulas are (Ex. 7.3)

$$T(0, l) = 82.67°\text{C} \qquad T(w, l) = 82.48°\text{C}.$$

Corresponding approximate temperatures obtained by means of FEM are

$$T(0, l) = 82.67°\text{C} \qquad T(w, l) = 82.49°\text{C}.$$

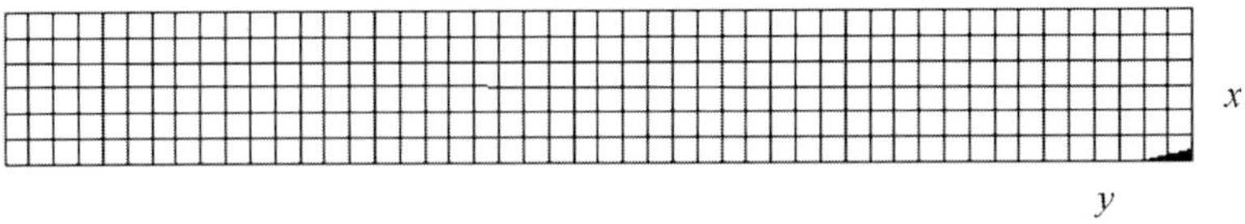

Fig. 11.29. Division of the half of fin cross-section into finite elements

The results obtained by means of the analytical solution and FEM are in a very good agreement.

Fin efficiency will be calculated from formula

$$\eta = \frac{\dot{Q}}{\dot{Q}_{\max}}, \tag{1}$$

where $\dot{Q}$ is the fin-to-surroundings transferred heat flow, which is formulated as

$$\dot{Q} = 2 \cdot \bar{\dot{q}}_x \Big|_{x=0} w \,. \tag{2}$$

In the case of the analytical solution, the mean heat flux at the fin base $\bar{\dot{q}}_x \Big|_{x=0}$ calculated by means of (7) from Ex. 7.3 is at $\bar{\dot{q}}_x \Big|_{x=0}$ = 53253 W/m^2. Therefore,

$$\dot{Q} = 2 \cdot 53253 \cdot 0.003 = 319.52 \text{ W/m} \,. \tag{3}$$

Maximum heat flow $\dot{Q}_{\max}$, i.e. heat flow transferred by an isothermal fin with fin base temperature T_b is formulated as

$$\dot{Q}_{\max} = 2\alpha l \left(T_b - T_{cz}\right) = 2 \cdot 100 \cdot 0.024 \left(95 - 20\right) = 360 \text{ W/m} \,. \tag{4}$$

Fin efficiency calculated by means of the analytical solution is

$$\eta = \frac{\dot{Q}}{\dot{Q}_{\max}} = \frac{319.52}{360} = 0.8876 \,. \tag{5}$$

Heat flow at the fin base determined by means of FEM for the fin shown in Fig. 11.28a is

$$\dot{Q} = \dot{Q}_b = 2 \cdot 157.88 = 315.76 \text{ W/m} \,. \tag{6}$$

Fin efficiency calculated by means of FEM is

$$\eta = \frac{\dot{Q}}{\dot{Q}_{\max}} = \frac{315.76}{360} = 0.8771 \,. \tag{7}$$

and is very close to the value of $\eta = 0.8876$ obtained by means of the analytical solution.

One can specify the heat flow given off by the fin, if the heat flow transferred by the lateral fin surfaces is determined first.

$$\dot{Q} = 2\int_0^l \alpha\left[T(w,y) - T_{cz}\right]dy = 319.5 \text{ W/m};$$

Next, efficiency $\eta = 0.8875$ is obtained from (7) and is almost identical to the one obtained from the analytical formula. It is evident, therefore, that node temperatures are more accurately calculated in FEM than the boundary heat flux.

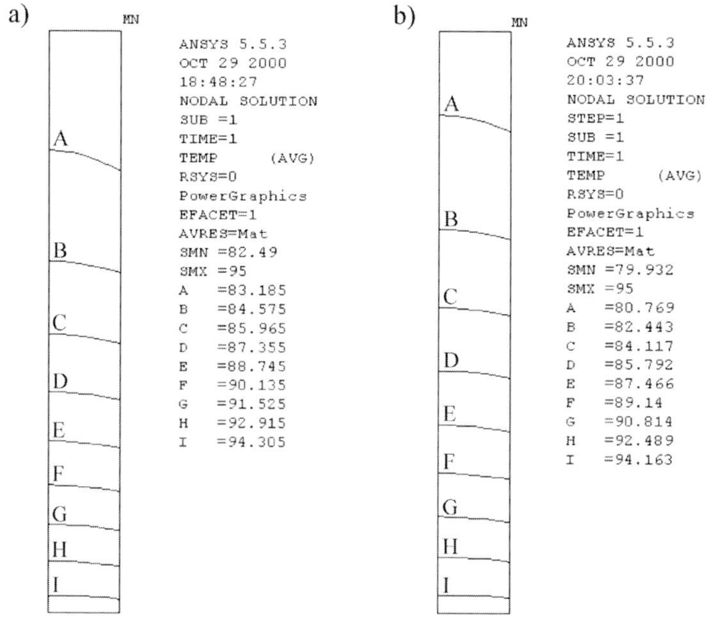

Fig. 11.30. Layout of isotherms in the fin cross-section: (**a**) thermally insulated fin tip; $T(0, l) = 82.67$°C, $T(w, l) = 82.49$°C; (**b**) heat transfer at the fin tip; $T(0, l) = 80.11$°C, $T(w, l) = 79.93$°C

Next, temperature distribution and fin efficiency were calculated, while a consideration was given to a tip heat transfer. The layout of isotherms in the fin cross-section is presented in Fig.11.30b. As one can see, temperatures $T(0, l) = 80.11$°C, $T(w, l) = 79.93$°C are slightly lower than they are in the case when the fin tip is thermally insulated. Also the maximum flow $\dot{Q}_{max}$ calculated from formula

$$\dot{Q}_{max} = 2\alpha(l+w)(T_b - T_{cz}) = 2 \cdot 100(0.024 + 0.003)(95 - 20) = 405 \text{ W/m}$$

is larger due to the fin tip heat transfer. Heat flow at the fin base calculated by means of FEM is at $\dot{Q} = \dot{Q}_b = 345.84$ W/m. Fin efficiency, therefore, is

$$\eta = \frac{\dot{Q}}{\dot{Q}_{max}} = \frac{345.84}{405} = 0.8539 .$$

The above value approximates the obtained value, while the heat transfer at the fin tip is neglected (7). In spite of the fact that fin efficiency is low, the fin-diffused heat flow is large, since the heat transfer takes place at the fin tip.

Exercise 11.21 Determining Two-Dimensional Temperature Distribution by Means of FEM in a Hexagonal Fin with Constant Thickness (ANSYS Program)

Determine temperature distribution and fin efficiency by means of FEM. Fin diagram, which results from plate fin division, is presented in Fig. 11.31. For the calculation, assume the values from Ex. 6.20: thickness of the plate-fin t = 0.000115 m, d = 0.00759 m, λ = 165 W/(m·K), α = 40 W/(m^2·K); T_b = 100°C; T_{cz} = 0°C. Compare calculated fin efficiency value with the values obtained in Ex. 6.20.

Solution

Calculations were carried out by means of FEM and ANSYS programs. The analyzed region was divided into 1377 elements (Fig. 11.32). Temperature was determined in 2934 nodes. The layout of isotherms on the fin surface is shown in Fig. 11.33. Note that temperature on the outer fin boundary is non-uniform due to the irregular shape of the fin. Maximum temperature on the outer boundary is T_{max} = 93.162°C, while minimum T_{min}= 90.379°C. Temperatures at characteristic points marked in Fig. 11.31 are, correspondingly: T_1 = 91.56°C, T_2 = 90.47°C, T_3 = 90.38°C, T_4= 100°C, T_5= 100°C .

Fin efficiency will be calculated from formula

$$\eta = \frac{\dot{Q}}{\dot{Q}_{max}}, \tag{1}$$

where $\dot{Q}_{max}$ is the heat flow transferred by the isothermal fin whose base temperature is $T_b = 100°C$; the heat flow is formulated as

$$\dot{Q}_{\max} = \alpha A_{\dot{z}} \left(T_b - T_{cz} \right), \tag{2}$$

while (Fig. 6.24, Ex. 6.20)

$$A_{\dot{z}} = 2\left(2A_{OAB} + 4A_{OBC} - \frac{\pi d^2}{4} \right) = 2\left(2 \cdot 2.1089 \cdot 10^{-5} + \right. \\ \left. + 4 \cdot 5.6156 \cdot 10^{-5} - 4.52452 \cdot 10^{-5} \right) = 4.43 \cdot 10^{-4} \text{ m}^2. \tag{3}$$

Maximum heat flow $\dot{Q}_{max}$ is at

$$\dot{Q}_{\max} = 40 \cdot 4.43 \cdot 10^{-4} \cdot (100 - 0) = 1.772 \text{ W}.$$

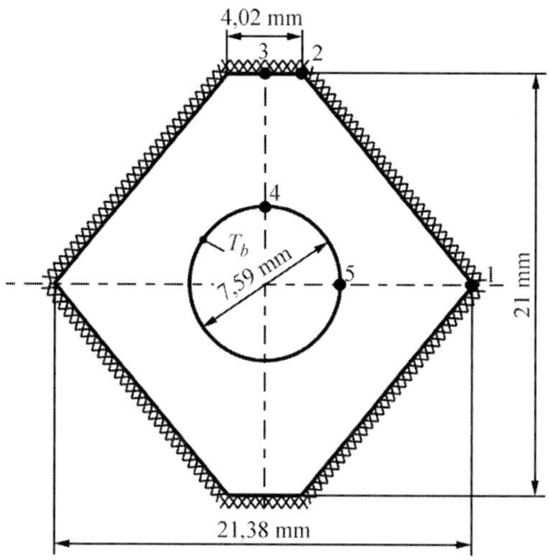

Fig. 11.31. Diagram of a conventional fin after plate-fin division

Fin-transferred heat flow $\dot{Q}$ can be determined from formula

$$\dot{Q} = \int_{A_{\dot{z}}} \alpha \left(T_s - T_{cz} \right) dA = \alpha \left(T_{sr} - T_{cz} \right) A_{\dot{z}}, \tag{4}$$

where $A_{\dot{z}}$ is the lateral area of the fin surface that exchanges heat with surroundings, while T_{sr} is the average temperature of the fin surface formulated as

$$T_{sr} = \frac{\int_{A_{\dot{z}}} T_s dA}{A_{\dot{z}}} \approx \frac{\sum_{e=1}^{N_e} T_{sr,e} \cdot A_e}{A_{\dot{z}}}. \tag{5}$$

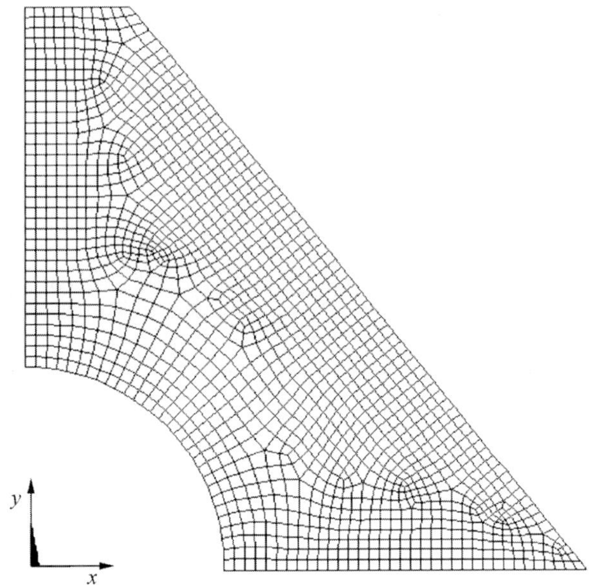

Fig. 11.32. Division of 1/8 of a fin cross-section into finite elements

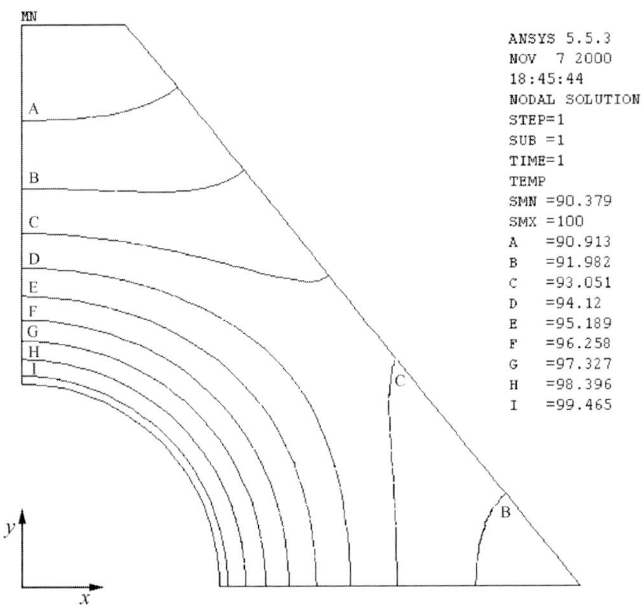

Fig. 11.33. The layout of isotherms on the fin surface

Temperature T_s is the temperature of the fin outer surface, which is in contact with surroundings. Temperature $T_{sr,e}$ is the average temperature of the element's surface exposed to surroundings, while A_e is the element's surface area from the side exposed to surroundings. Symbol N_e stands for the number of elements, which the analyzed region was divided to.

Average temperature determined from (5) by means of the ANSYS program measures $T_{sr} = 93.80°\text{C}$. Such method of determining fin-transferred heat flow $\dot{Q}$ is more accurate than the method that uses formula

$$\dot{Q} = \dot{Q}_b = \pi d t \left(-\lambda \frac{dT}{dr} \right) \Bigg|_{r=d/2}, \tag{6}$$

since an accurate determination of $\lambda(dT/dr)$ in FEM enforces the need to divide the region into a very large number of elements.

Fin-transferred heat flow $\dot{Q}$ determined by means of (4) with the help of the ANSYS program comes to

$$\dot{Q} = 1.6622 \text{ W} .$$

It is a heat flow transferred by lateral fin surfaces. Fin efficiency is at

$$\eta = \frac{\dot{Q}}{\dot{Q}_{\max}} = \frac{1.6622}{1.772} = 0.9380 .$$

Calculated efficiency differs insignificantly from the efficiency of the equivalent circular fin $\eta_e = 0.9394$ and from the fin efficiency determined by means of the segment method, equal to $\eta_e = 0.9373$.

Exercise 11.22 Determining Axisymmetrical Temperature Distribution in a Cylindrical and Conical Pin by Means of FEM (ANSYS Program)

Determine temperature distribution in cylindrical and conical pins, shown in Fig. 11.34, by means of FEM and with the use of the ANSYS program. Pins of this kind are used in gas-fired cast-iron heating boilers with an aim to increase the heat flow transferred from combustion gases to water. Pin dimensions are given in Fig.11.34. Both pins are almost identical in volume. Assume the following values for the calculation: water temperature $T_w = 75°\text{C}$, temperature of combustion gases $T_{sp} = 400°\text{C}$. Thermal conductivity of the material from which the wall and pins are made of is $\lambda =$

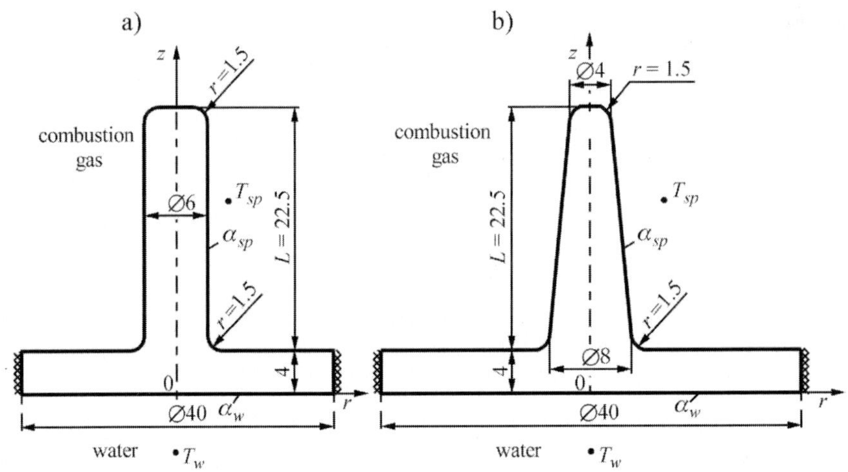

Fig. 11.34. Pinned heating surfaces: (**a**) cylindrical pin, (**b**) conical pin

48 W/(m·K). Heat transfer coefficients on the water and combustion gases side are, respectively $\alpha_w = 1000$ W/(m^2·K) and $\alpha_{sp} = 80$ W/(m^2·K).

Draw the layout of isotherms in the longitudinal cross-section of the pins and determine maximum temperatures. Also calculate pin-transferred heat flows from combustion gases to a boiler wall by determining heat flow at the base of the pins $\dot{Q}_b$ for the coordinate $z = 0.004$ m. Which of the pins ensures a larger flow of transferred heat when maximum temperature is decreased? Calculate temperature distribution and heat flux at the base of the cylindrical pin by means of the formulas obtained when a radial temperature drop is neglected.

Solution

Temperature in the cylindrical pin was determined in 3401 nodes when longitudinal cross-section was divided into 3195 elements (Fig. 11.35a), while in the conical pin in 3647 nodes when longitudinal cross-section was divided into 3439 elements (Fig. 11.35b).

Maximum temperature of the cylindrical pin at $T_{w,max} = 186.65$°C is larger than the maximum temperature of the conical pin at $T_{s,max} = 150.747$°C (Fig. 11.36). A lower maximum temperature of the conical pin is due to the fact that the pin has a more advantageous shape, since the surface area of the cross-section becomes larger as the heat flow, which is conducted through the pin's cross-section increases too.

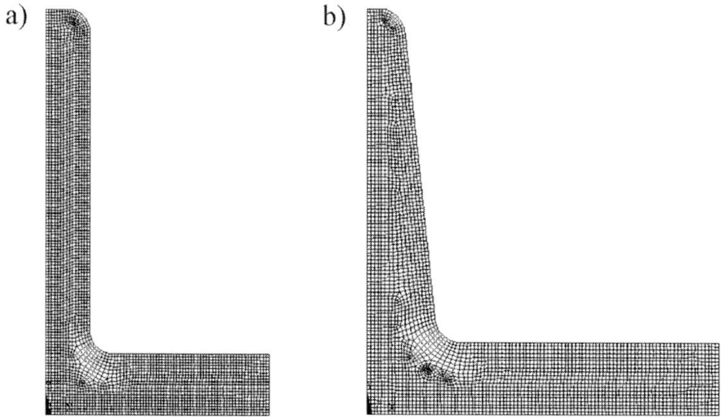

Fig. 11.35. Half of the pin's longitudinal cross-section divided into finite elements: (**a**) cylindrical pin, (**b**) conical pin

In the case of the cylindrical peg, constituent heat flux $\dot{q}_z = -\lambda \partial T/\partial z$ is much larger near the base than anywhere else and that contributes to a large increase in pin temperature within this region. Following that, heat flow at the pin base $z = 0.004$ m were calculated from formula

$$\dot{Q}_b = -2\pi \int_0^{r_b + r} \lambda \frac{\partial T}{\partial z} r dr,$$

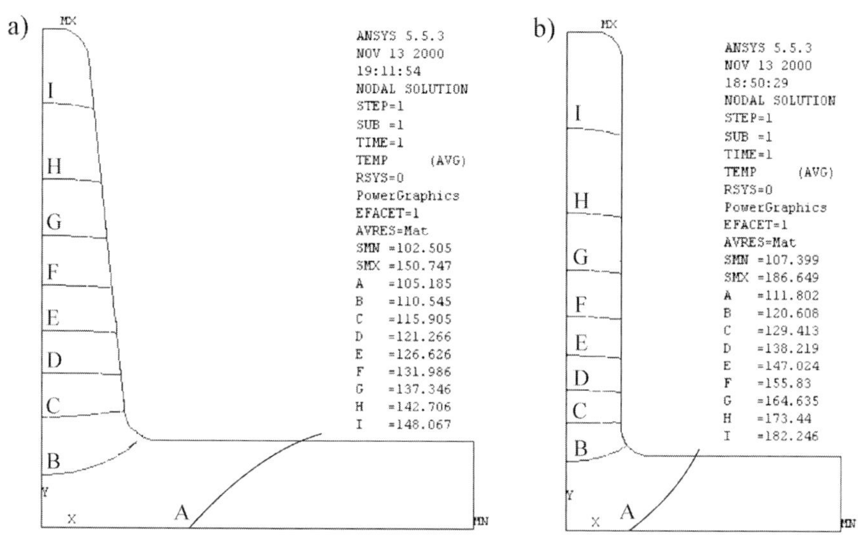

Fig. 11.36. Layout of isotherms on the pin surface: (**a**) cylindrical pin, (**b**) conical pin

where $r_b = 0.003$ m for cylindrical pin, $r_b = 0.004$ m for conical pin and $r = 0.0015$ m.

For cylindrical pin $\dot{Q}_{b,w} = 8.892$ W, while for conical pin $\dot{Q}_{b,s} = 8.456$ W. As one can see, the shape of the conical pin is a very advantageous, since in spite of the fact that the flow of transferred heat is almost the same, the temperature at the tip is much lower than it is at the tip of the cylindrical pin. Temperature of the cylindrical pin can be approximately calculated from the formula below (while disregarding radius at the curves and radial temperature drop):

$$T(z) = T_{sp} + \left(T_b - T_{sp}\right) \frac{\cosh\left[m\left(z - L - 0.004\right)\right]}{\cosh mL}, \tag{1}$$

where T_b is an average temperature at the peg base determined by means of FEM. This temperature measures approximately $T_b \approx 118.469$°C. Parameter m formulated as

$$m = \sqrt{\frac{4\alpha_{sp}}{\lambda d}}, \tag{2}$$

where d is the peg's diameter, measures

$$m = \sqrt{\frac{4 \cdot 80}{48 \cdot 0.006}} = 33.33(3)\,1/\text{m}.$$

Therefore, the tip temperature of the cylindrical pin is

$$T(L + 0.004) = T_{w,\max} = 400 + (118.469 - 400)\frac{1}{\cosh(33.3333 \cdot 0.0225)} =$$

$$= 182.548°\text{C}.$$

As one can see, this temperature is close to the temperature $T_{w,\max} = 186.65$°C obtained by means of FEM. Heat flow at the pin base can be calculated from formula

$$\dot{Q} = -\frac{\pi d^2}{4}\lambda \left.\frac{\partial T}{\partial z}\right|_{z=0.004} = -\lambda \frac{\pi d^2}{4} m\left(T_b - T_{sp}\right) \operatorname{tgh} mL = -48\frac{\pi 0.006^2}{4} \times \tag{3}$$

$$\times 33.3333(118.469 - 400)\operatorname{tgh}(33.3333 \cdot 0.0225) = 8.0893 \text{ W}.$$

The obtained value approximates the value determined by means of FEM, which is equal to $\dot{Q}_{b,w} = 8.892$ W. However, from the calculations carried out with the use of FEM, it is clear that pin-base-temperature is higher than the pin-free wall temperature from the combustion-gases-side, i.e. wall temperature for $z = 0.004$ m at a significant distance from the pin axis for, e.g. $r > 2d$. Due to the application of FEM, one can use the actual dimensions of the pins shape in the calculation, e.g. the curved edges or the two-dimensional character of the temperature field in the pin and the wall, to which the pin is attached.

Literature

1. Anderson JD (1995) Computational Fluid Dynamics. The Basics with Applications. McGraw-Hill, New York
2. Clough RW (1980) The finite element method after twenty-five years: a personal view. Computers & Structures 12, No. 4: 361–370
3. Courant R (1943) Variational methods for the solution of problems of equilibrium and vibrations. Bull. of the American Mathematical Society 49: 1–23
4. Eisenberg MA, Lawrence EM (1973) On finite element integration in natural coordinates. International J. for Numerical Methods in Engineering 7: 574–575
5. Fletcher CAJ (1984) Computational Galerkin Methods. Springer, New York
6. Huebner KH (1975) The Finite Element Method for Engineers. Wiley, New York
7. Turner MJ, Clough RW, Martin HC, Topp LJ (1956) Stiffness and deflection analysis of complex structures. J. of the Aeronautical Sciences 2, No. 9: 805–823

12 Finite Element Balance Method and Boundary Element Method

In this chapter, the finite element balance method (FEBM) and boundary element method (BEM) are discussed in depth. Also examples are given to demonstrate how two-dimensional steady-state temperature distributions can be determined using FEBM and BEM. The obtained results are compared with the results from FEM calculations, which were carried out by means of the ANSYS program with a highly dense finite element mesh.

Exercise 12.1 Finite Element Balance Method

Describe finite element balance method.

Solution

Transient heat conduction phenomenon can be described by the following equation

$$c(T)\rho(T)\frac{\partial T}{\partial t} = -\nabla \cdot \dot{\mathbf{q}}\,, \tag{1}$$

where $\dot{\mathbf{q}}$ is a heat flux vector formulated by Fourier Law

$$\dot{\mathbf{q}} = -\lambda(T)\nabla T\,. \tag{2}$$

Thermo-physical properties such as specific heat c, thermal conductivity λ and density ρ are known temperature functions. Numerical solution will be carried out using finite element balance method. Balance methods, based on finite elements, allow solving heat conduction problems in complex shape bodies. They increasingly become ever more popular.

Lets consider finite volume V with a boundary surface S. Heat balance equation, which allows for the variability of thermo-physical properties and which is integrated within volume V, has the form

$$\int_V c(T)\rho(T)\frac{\partial T}{\partial t}dV = -\int_V \nabla \cdot \dot{\mathbf{q}} dV . \tag{3}$$

By applying mean value theorem to the left-hand-side of (3), while theorem of divergence to the right-hand-side, one obtains

$$Vc(\bar{T})\rho(\bar{T})\frac{d\bar{T}}{dt} = -\int_S \dot{\mathbf{q}} \cdot \mathbf{n} dS , \tag{4}$$

where the upper dash stands for the appropriate mean value in the finite volume V.

For the purpose of simplification, finite element equations will be presented in two dimensions and in Cartesian coordinates. Equation (1) assumes the following form:

$$c(T)\rho(T)\frac{\partial T}{\partial t} = \frac{\partial}{\partial x}\left[\lambda(T)\frac{\partial T}{\partial x}\right] + \frac{\partial}{\partial y}\left[\lambda(T)\frac{\partial T}{\partial y}\right]. \tag{5}$$

An arbitrarily selected region, in which the heat flow is analyzed, was divided into triangular tri-nodal elements, marked by continuous lines (Fig. 12.1). Next, the center point of a triangle was connected with the center points of sides. This is the way in which continuous lines define the boundary of the entire region and of the individual elements, while broken lines define the control volumes. Striped regions are the examples of three control volumes. One of them is assigned to an internal node, while two to boundary nodes. In this discretisation, broken lines approximate boundary curves.

Two-dimensional mesh in finite volumes can be also created by means of the Voronoi polynomials. This is when the broken lines that define convex control volumes intersect continuous lines, which define the elements, at right angles. One triangular element was evaluated; it is shown in Fig.12.2. Integral equation similar to (5) can be written when the energy preservation principle for finite volume V_{1aoc} is applied. Equation (4) can be written in the following form

$$\left[\int_a^o \dot{\mathbf{q}} \cdot \mathbf{n} ds + \int_o^c \dot{\mathbf{q}} \cdot \mathbf{n} ds\right] = -V_{1aoc}c(T_1)\rho(T_1)\frac{dT_1}{dt}, \tag{6}$$

where $\mathbf{n}$ is a normal unit vector directed to the outside of the element surface ds, while $\dot{\mathbf{q}}$ is heat flux

$$\dot{\mathbf{q}} = -\lambda(T)\nabla T .$$

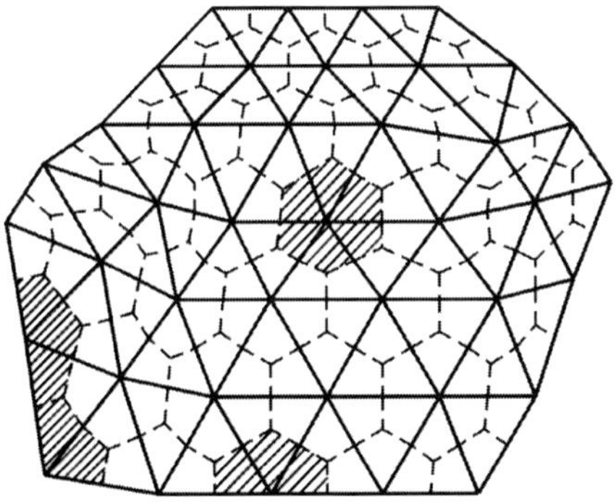

Fig. 12.1. Irregular region divided into triangular elements and finite volumes

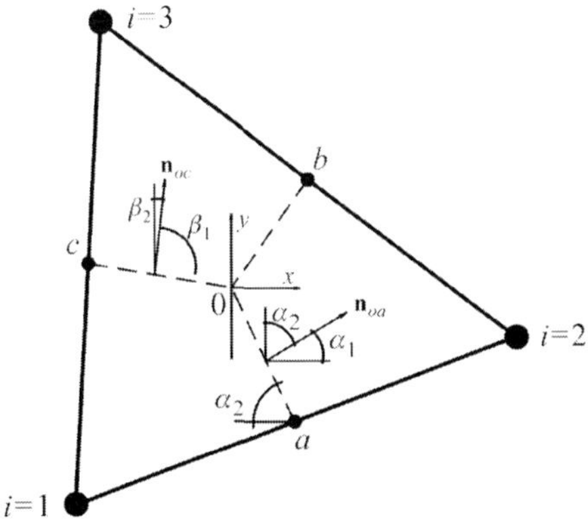

Fig. 12.2. Typical tri-nodal element with an assumed local coordinate system

To solve (6), one is required to define the functions that interpolate temperature T, thermal conductivity λ, density ρ and specific heat c.

Linear function was selected for temperature interpolation within the triangular element. Each element node has one degree of freedom, i.e. unknown temperature T_i^e. Temperature at an arbitrary point with x, y coordinates inside the element is formulated as follows:

$$T^e = a_1 + a_2 \cdot x + a_3 \cdot y. \tag{7}$$

Assuming that

$$T^e(x_1, y_1) = T_1^e, \qquad T^e(x_2, y_2) = T_2^e, \qquad T^e(x_3, y_3) = T_3^e \tag{8}$$

one obtains

$$\begin{aligned} &T_1^e = a_1 + a_2 \cdot x_1 + a_3 \cdot y_1, \qquad T_2^e = a_1 + a_2 \cdot x_2 + a_3 \cdot y_2, \\ &T_3^e = a_1 + a_2 \cdot x_3 + a_3 \cdot y_3. \end{aligned} \tag{9}$$

Once (9) are solved with respect to a_1, a_2 and a_3, one obtains

$$\begin{aligned} a_1 &= \frac{1}{2A}\left[(x_2 y_3 - x_3 y_2)T_1^e + (x_3 y_1 - x_1 y_3)T_2^e + (x_1 y_2 - x_2 y_1)T_3^e\right], \\ a_2 &= \frac{1}{2A}\left[(y_2 - y_3)T_1^e + (y_3 - y_1)T_2^e + (y_1 - y_2)T_3^e\right], \end{aligned} \tag{10}$$

where R_ε is the region around the heat source. The outer surface of this region is S_ε.

Constant C will be determined for the two-dimensional problem (Fig. 12.5), in which the integration region is a cylinder with radius ε and height $L = 1$ m. Once the theorem of divergence is applied to the first integral in (15), one obtains

$$\int_{R_\varepsilon} \nabla^2 v dV = \int_{S_\varepsilon} (\nabla v) \cdot \mathbf{n} dS = \int_{S_\varepsilon} \frac{\partial v}{\partial n} dS = \left.\frac{\partial v}{\partial r}\right|_{r=\varepsilon} S_\varepsilon . \tag{16}$$

By substituting (16) into (15), one obtains

$$S_\varepsilon \left.\frac{\partial v}{\partial r}\right|_{r=\varepsilon} = -1, \tag{17}$$

where

$$S_\varepsilon = 2\pi\varepsilon . \tag{18}$$

Once derivative of function (13) is calculated

$$\left.\frac{\partial v}{\partial r}\right|_{r=\varepsilon} = -\frac{C}{\varepsilon}$$

from (17), one obtains

$$2\pi\varepsilon \cdot \left(-\frac{C}{\varepsilon}\right) = -1 ,$$

hence

$$C = \frac{1}{2\pi}. \tag{19}$$

Therefore, the fundamental solution for the two-dimensional problem has the form

$$v = \frac{1}{2\pi} \ln \frac{1}{r} , \tag{20}$$

where $r = \sqrt{(x - x_i)^2 + (y - y_i)^2}$.

From the standpoint of physics, one can use the concept of heat exchange by conduction to explain in a straight-forward way how constant C is determined from Gauss-Ostrogradski theorem (16). Linear heat source with a power $\dot{Q}/L = 1 \mathrm{W/m}$ is located at the point (x_i, y_i); therefore, power $\dot{Q} = 1$ W is generated within the length of $L = 1$ m. Heat source power per unit of volume is

$$\dot{q}_v = \frac{\dot{Q}}{\pi\, r_z^2 L}, \tag{21}$$

where $\dot{q}_v$ is expressed in W/m^3. If the outer radius r_z of a cylindrical heat source becomes smaller, then $\dot{q}_v$ becomes larger, i.e.

$$\dot{q}_v \to \infty, \quad \text{gdy} \quad r_z \to 0. \tag{22}$$

Power $\dot{Q}$ generated by the heat source outflows to the surrounding space. Function $v(r)$ that shows in such case temperature distribution within the area whose heat conduction coefficient equals λ = 1 W/(m·K), should satisfy the heat balance equation. Source-generated heat flow must be passing through a lateral surface of the cylinder (Fig. 12.5)

$$\dot{Q} = -2\pi\varepsilon L\lambda \left.\frac{\partial v}{\partial r}\right|_{r=\varepsilon}, \tag{23}$$

hence, one gets

$$\frac{\dot{Q}}{L} = -2\pi\varepsilon\lambda \left.\frac{\partial v}{\partial r}\right|_{r=\varepsilon}. \tag{24}$$

Once we substitute $\dot{Q}/L$ = 1 W/m, λ = 1 W/(m·K) and account for $\partial v/\partial r|_{r=\varepsilon} = -C/\varepsilon$, z (23), we have (19). On the basis of divergence theorem (15), one can also determine constant C in (14) for the spatial problem. Constants C and C_1 in (14) are $C = 1/4\pi$, $C_1 = 0$.

If the fundamental solution (20) is known, one can calculate v and $\partial v/\partial n$, which appears on the right-hand-side of (9). By substituting (1) and (10) into (9), one has for $(x_i, y_i) \in R$ and $(x_i, y_i) \notin S$

$$\int_R \left(u\nabla^2 v - v\nabla^2 u\right) dxdy = -\int_R u\delta^i dxdy = -u_i\,. \tag{25}$$

By substituting (25) into (9), one obtains

$$-u_i = \int_s \left(u\frac{\partial v}{\partial n} - v\frac{\partial u}{\partial n}\right) ds\,. \tag{26}$$

On the basis of (26), one can calculate temperature u_i at any point (x_i, y_i) within region R, if temperature u and the derivative in normal direction $\partial v/\partial n$ are known for the whole boundary s. According to boundary conditions (2) and (3), temperature and normal derivative are not known for the whole boundary, but only for the part of that boundary. Equation (26), therefore, can be used to calculate temperature u_i once temperature u on the boundary section s_q and normal derivative $\partial u/\partial n$ on the boundary

$$\begin{aligned}
&-\frac{1}{8\cdot A_1}\Big[(T_1-T_5)\big\{2\lambda\big[(y_2-y_5)(y_1+y_5)-(x_5-x_2)(x_1+x_5)\big]-\\
&-2\lambda\big[(y_2-y_5)(y_2+y_5)-(x_5-x_2)(x_2+x_5)\big]\big\}\times\\
&\times(T_2-T_5)\big\{2\lambda\big[(y_5-y_1)(y_1+y_5)-(x_1-x_5)(x_1+x_5)\big]-\\
&-2\lambda\big[(y_5-y_1)(y_2+y_5)-(x_1-x_5)(x_2+x_5)\big]\big\}\Big]-\\
&-\frac{1}{8\cdot A_2}\Big[(T_2-T_5)\big\{2\lambda\big[(y_3-y_5)(y_2+y_5)-(x_5-x_3)(x_2+x_5)\big]-\\
&-2\lambda\big[(y_3-y_5)(y_3+y_5)-(x_5-x_3)(x_3+x_5)\big]\big\}\times\\
&\times(T_3-T_5)\big\{2\lambda\big[(y_5-y_2)(y_2+y_5)-(x_2-x_5)(x_2+x_5)\big]-\\
&-2\lambda\big[(y_5-y_2)(y_3+y_5)-(x_2-x_5)(x_3+x_5)\big]\big\}\Big]-\\
&-\frac{1}{8\cdot A_3}\Big[(T_3-T_5)\big\{2\lambda\big[(y_4-y_5)(y_3+y_5)-(x_5-x_4)(x_3+x_5)\big]-\\
&-2\lambda\big[(y_4-y_5)(y_4+y_5)-(x_5-x_4)(x_4+x_5)\big]\big\}\times\\
&\times(T_4-T_5)\big\{2\lambda\big[(y_5-y_3)(y_3+y_5)-(x_3-x_5)(x_3+x_5)\big]-\\
&-2\lambda\big[(y_5-y_3)(y_4+y_5)-(x_3-x_5)(x_4+x_5)\big]\big\}\Big]-\\
&-\frac{1}{8\cdot A_4}\Big[(T_4-T_5)\big\{2\lambda\big[(y_1-y_5)(y_4+y_5)-(x_5-x_1)(x_4+x_5)\big]-\\
&-2\lambda\big[(y_1-y_5)(y_1+y_5)-(x_5-x_1)(x_1+x_5)\big]\big\}\times\\
&\times(T_1-T_5)\big\{2\lambda\big[(y_5-y_4)(y_4+y_5)-(x_4-x_5)(x_4+x_5)\big]-\\
&-2\lambda\big[(y_5-y_4)(y_1+y_5)-(x_4-x_5)(x_1+x_5)\big]\big\}\Big]=0,
\end{aligned}\tag{1}$$

where

$$A_1=\frac{1}{2}\det\begin{bmatrix}1 & x_5 & y_5\\ 1 & x_1 & y_1\\ 1 & x_2 & y_2\end{bmatrix}=0.0001\,\mathrm{m}^2 .$$

By substituting the numerical values into (1), the following equation for node 5 is obtained:

$$-\frac{42\cdot 2}{8\cdot 0{,}0001}\Big[\big\{(T_1-T_5)(0-0{,}0002)+(T_2-T_5)(0{,}0002-0{,}0004)\big\}+$$
$$+\big\{(T_2-T_5)(0{,}0004-0{,}0006)+(T_3-T_5)(-0{,}0002-0)\big\}+$$
$$+\big\{(T_3-T_5)(0-0{,}0002)+(T_4-T_5)(-0{,}0006-(-0{,}0004))\big\}+$$
$$+\big\{(T_4-T_5)(-0{,}0004-(-0{,}0002))+(T_1-T_5)(-0{,}0002-0)\big\}\Big]=0,$$
$$-\Big[\big\{(T_1-T_5)(-2)+(T_2-T_5)(-2)\big\}+\big\{(T_2-T_5)(-2)+(T_3-T_5)(-2)\big\}+$$
$$+\big\{(T_3-T_5)(-2)+(T_4-T_5)(-2)\big\}+\big\{(T_4-T_5)(-2)+(T_1-T_5)(-2)\big\}\Big]=0.$$

By substituting T_2= 100°C and T_3 = 100°C, one obtains

$$T_1+T_4-4\cdot T_5=-200\,. \tag{2}$$

Heat balance equation for the finite volume, which surrounds node 4 can be written in the following form (n = 4; i = 1, 5, 3):

$$-\frac{1}{8\cdot A_4}\Big[(T_1-T_4)\big\{2\lambda\big[(y_5-y_4)(y_1+y_4)-(x_4-x_5)(x_1+x_4)\big]-$$
$$-2\lambda\big[(y_5-y_4)(y_5+y_4)-(x_4-x_5)(x_5+x_4)\big]\big\}\times$$
$$\times(T_5-T_4)\big\{2\lambda\big[(y_4-y_1)(y_1+y_4)-(x_1-x_4)(x_1+x_4)\big]-$$
$$-2\lambda\big[(y_4-y_1)(y_5+y_4)-(x_1-x_4)(x_5+x_4)\big]\big\}\Big]-$$
$$-\frac{1}{8\cdot A_3}\Big[(T_5-T_4)\big\{2\lambda\big[(y_3-y_4)(y_5+y_4)-(x_4-x_3)(x_5+x_4)\big]- \tag{3}$$
$$-2\lambda\big[(y_3-y_4)(y_3+y_4)-(x_4-x_3)(x_3+x_4)\big]\big\}\times$$
$$\times(T_3-T_4)\big\{2\lambda\big[(y_4-y_5)(y_5+y_4)-(x_5-x_4)(x_5+x_4)\big]-$$
$$-2\lambda\big[(y_4-y_5)(y_3+y_4)-(x_5-x_4)(x_3+x_4)\big]\big\}\Big]+\dot{q}_B\frac{a}{2}=0,$$

where from, after substitution of the numerical values, one has

$$-\frac{42\cdot 2}{8\cdot 0.0001}\Big[\{(T_1-T_4)(-0.0002-(-0.0002))+(T_5-T_4)\times$$
$$\times(0.0004-0.0006)\}+\{(T_5-T_4)(0.0002-0.0004)+(T_3-T_4)\times$$
$$\times(0.0002-0.0002)\}\Big]+200000\cdot\frac{0.02}{2}=0,$$

$$-\frac{42}{4}\Big[\{(T_1-T_4)(0)+(T_5-T_4)(-2)\}+$$
$$+\{(T_5-T_4)(-2)+(T_3-T_4)\cdot 0\}\Big]+2000=0,$$

$$T_5-T_4=-\ 47.6\,. \tag{4}$$

Heat balance equation for the finite volume, which surrounds node 1 can be written in the following form ($n = 1$; $i = 2, 5, 4$):

$$-\frac{1}{8\cdot A_1}\Big[(T_2-T_1)\{2\lambda[(y_5-y_1)(y_2+y_1)-(x_1-x_5)(x_2+x_1)]-$$
$$-2\lambda[(y_5-y_1)(y_5+y_1)-(x_1-x_5)(x_5+x_1)]\}\times$$
$$\times(T_5-T_1)\{2\lambda[(y_1-y_2)(y_2+y_1)-(x_2-x_1)(x_2+x_1)]- \tag{5}$$
$$-2\lambda[(y_1-y_2)(y_5+y_1)-(x_2-x_1)(x_5+x_1)]\}\Big]-$$
$$-\frac{1}{8\cdot A_4}\Big[(T_5-T_1)\{2\lambda[(y_4-y_1)(y_5+y_1)-(x_1-x_4)(x_5+x_1)]-$$
$$-2\lambda[(y_4-y_1)(y_4+y_1)-(x_1-x_4)(x_4+x_1)]\}\times$$
$$\times(T_4-T_1)\{2\lambda[(y_1-y_5)(y_5+y_1)-(x_5-x_1)(x_5+x_1)]-$$
$$-2\lambda[(y_1-y_5)(y_4+y_1)-(x_5-x_1)(x_4+x_1)]\}\Big]+\dot{q}_B\frac{a}{2}+\dot{Q}_\alpha=0,$$

$$\dot{Q}_\alpha=\int_0^{a/2}\alpha\left(T_{cz}-T(x,0)\right)dx=$$
$$=\alpha\left\{\left(T_{cz}-\frac{1}{2A}\Big[(x_2y_3-x_3y_2)T_1+(x_3y_1-x_1y_3)T_2+(x_1y_2-x_2y_1)T_5\Big]\right)\frac{a}{2}-\right.$$
$$\left.-\frac{1}{2A}\Big[(y_2-y_3)T_1+(y_3-y_1)T_2+(y_1-y_2)T_5\Big]\frac{a^2}{8}\right\}=$$
$$=\alpha\left\{\left(T_{cz}-\frac{1}{2a^2/4}\Big[a\frac{a}{2}T_1\Big]\right)\frac{a}{2}-\frac{1}{2a^2/4}\Big[\left(-\frac{a}{2}\right)T_1+\left(\frac{a}{2}\right)T_2\Big]\frac{a^2}{8}\right\}.$$

After substitution of the numerical values, one has

$$\dot{Q}_{\alpha}=\alpha\left[\left(T_{cz}-T_{1}\right)\frac{a}{2}-\left[T_{2}-T_{1}\right]\frac{a}{8}\right]=60\cdot 20\cdot\frac{0.02}{2}-60\cdot\frac{3\cdot 0.02}{8}\cdot T_{1}-$$

$$-60\cdot\frac{0.02}{8}\cdot T_{2}=12-0.45\cdot T_{1}-0.15\cdot T_{2},$$

$$-\frac{42\cdot 2}{8\cdot 0.0001}\left[\left\{\left(T_{2}-T_{1}\right)\left(0.0002-0.0002\right)+\left(T_{5}-T_{1}\right)\times\right.\right.$$

$$\left.\times\left(-0.0004-\left(-0.0002\right)\right)\right\}+\left\{\left(T_{5}-T_{1}\right)\left(0.0002-0.0004\right)+\right.$$

$$\left.\left.+\left(T_{4}-T_{1}\right)\left(-0.0002-\left(-0.0002\right)\right)\right\}\right]+200000\cdot\frac{0.02}{2}+$$

$$+\left(12-0.45\cdot T_{1}-0.15\cdot T_{2}\right)=0,$$

$$-\frac{42}{4}\left[2\left(T_{5}-T_{1}\right)\left(-2\right)\right]+2000+\left(12-0.45\cdot T_{1}-0.15\cdot T_{2}\right)=0,$$

$$-42.45\cdot T_{1}+42\cdot T_{5}=-1997. \qquad (6)$$

The solution for equation systems (2), (4) and (6) is

$$T_{1}=192.035^{\circ}\mathrm{C}; \qquad T_{4}=194.145^{\circ}\mathrm{C}; \qquad T_{5}=146.545^{\circ}\mathrm{C}.$$

Calculations by means of the ANSYS program were also carried out and the following results were obtained:

$$T_{1}=191.676^{\circ}\mathrm{C}; \qquad T_{4}=193.92^{\circ}\mathrm{C}; \qquad T_{5}=146.545^{\circ}\mathrm{C}.$$

Exercise 12.4 Determining Temperature Distribution in a Square Region using Boundary Element Method

Determine temperature distribution in a two-dimensional region, which measures 6 × 6 m (Fig. 12.9). The thermal conductivity of the material is λ = 50 W/(m·K). Boundary conditions are presented in Fig. 12.9. Temperatures are assigned on lateral surfaces, while heat flux on an upper and lower surface. Determine temperature distribution by means of BETIS program [1]. Compare the obtained results with the values determined by means of FEM.

Solution

Temperature distribution within a two-dimensional region shown in Fig. 12.9 is governed by the heat conduction equation

$$\frac{\partial^2 T}{\partial x^2}+\frac{\partial^2 T}{\partial y^2}=0 \tag{1}$$

and by boundary conditions

$$T\big|_{x=0}=300^\circ\mathrm{C}, \qquad 0\le y\le 6\ \mathrm{m}, \tag{2}$$

$$T\big|_{x=6m}=0^\circ\mathrm{C}, \qquad 0\le y\le 6\ \mathrm{m}, \tag{3}$$

$$-\lambda\left.\frac{\partial T}{\partial y}\right|_{y=0}=10000\ \mathrm{W/m^2}, \qquad 0\le x\le 6\ \mathrm{m}, \tag{4}$$

$$-\lambda\left.\frac{\partial T}{\partial y}\right|_{y=6m}=0\ \mathrm{W/m^2}, \qquad 0\le x\le 6\ \mathrm{m}. \tag{5}$$

Boundary conditions (4) and (5) will be transformed into a form

$$\left.\frac{\partial T}{\partial y}\right|_{y=0}=-\frac{10000}{\lambda}=-200\ \mathrm{K/m}, \tag{6}$$

$$\left.\frac{\partial T}{\partial y}\right|_{y=6m}=0. \tag{7}$$

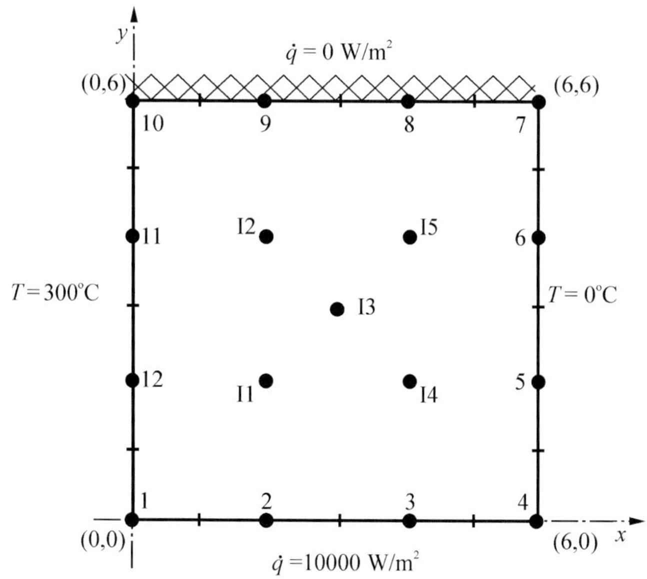

Fig. 12.9. A diagram that illustrates boundary conditions and boundary division into finite elements in BEM

Table 12.1. Node temperature values (Fig. 12.10) calculated by means of BEM and FEM

Node no.	Temperature [°C]		Node no.	Temperature [°C]	
	BEM	FEM		BEM	FEM
1	300.00	300.00	10	300.00	300.00
2	607.38	612.91	11	300.00	300.00
3	507.38	512.91	12	300.00	300.00
4	0.00	0.00	I1	336.71	350.09
5	0.00	0.00	I2	252.98	258.24
6	0.00	0.00	I3	246.05	255.04
7	0.00	0.00	I4	236.71	250.09
8	134.17	136.37	I5	152.98	158.24
9	234.17	236.37			

$$Bi^* = \frac{\alpha\left(\frac{V}{A_s}\right)}{\lambda} \le 0.05\,, \qquad (1)$$

If the above condition is met, one can neglected the temperature drop inside the body. In (1) Bi^* is the Biot number, V a body volume, while A_s an outer surface area of the body. Quotient $L^* = V/A_s$ is a characteristic body dimension. A characteristic dimension of a sphere with radius R is

$$L^* = \frac{V}{A_s} = \frac{4}{3}\frac{\pi R^3}{4\pi R^2} = \frac{R}{3}\,. \qquad (2)$$

Once we substitute (2) into (1), the condition under which we assume that the heat conduction model with lumped mass has the following form for a sphere:

$$\frac{\alpha R}{\lambda} \le 0.15\,. \qquad (3)$$

Heat balance equation will be written for an arbitrary shape body. Body volume is V, while A_s is the outer surface of the body (Fig. 13.1). Thermal conductivity λ, density ρ and specific heat c are constant and temperature invariant. Assuming that initial body temperature is

$$T\big|_{t=0} = T_0 \qquad (4)$$

and that temperature of the medium $T_{cz}(t) > T_0$, i.e. the body is heated, the energy balance has the form

$$m\frac{du}{dt} = \alpha A_s\left[T_{cz}(t) - T(t)\right], \qquad (5)$$

where $u = c_v T$ is a unitary internal temperature, while $m = \rho V$ a body mass.

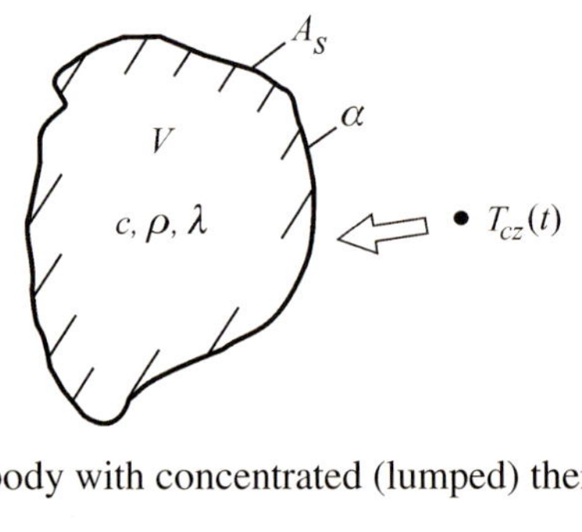

Fig. 13.1. A body with concentrated (lumped) thermal capacity

Because specific heat is practically the same for a solid when pressure and volume is constant, i.e. $c_p = c_v = c$, the energy balance (5) assumes the form

$$c\rho V\frac{dT(t)}{dt}+\alpha A_sT(t)=\alpha A_sT_{cz}(t). \tag{6}$$

Once time constant is introduced

$$\tau=\frac{c\rho V}{\alpha A_s}, \tag{7}$$

Equation (6) can be transformed into a form

$$\tau\frac{dT}{dt}+T=T_{cz}(t). \tag{8}$$

It is a heterogeneous differential equation of the first order. If temperature of the medium undergoes a step-change (Fig. 13.2), one can easily determine the solution for (6), if it is assumed that

$$\theta=T-T_{cz}. \tag{9}$$

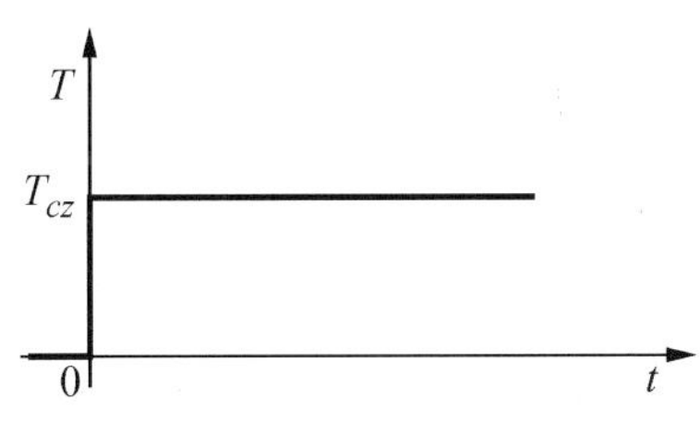

Fig. 13.2. Temperature step-change in a medium

Equation (8) and initial condition (4) have the form then

$$\tau\frac{d\theta}{dt}+\theta=0, \tag{10}$$

$$\theta\big|_{t=0}=\theta_0=T_0-T_{cz}. \tag{11}$$

After separation of variables in (10)

$$\frac{d\theta}{\theta}=-\frac{1}{\tau}dt \tag{12}$$

one obtains $\ln\theta = -t/\tau + C$ or

$$\theta = e^{-t/\tau + C} = C_1 e^{-t/\tau}. \tag{13}$$

Once we account for boundary condition (11), we can determine constant C_1: $C_1 = \theta_0$. Solution (13) assumes the form then

$$\theta(t) = \theta_0 e^{-t/\tau}, \tag{14}$$

where time constant τ is given by (7).

Heat flow delivered to the body at time t iso

$$\dot{Q}(t) = \alpha A_s \left[T_{cz} - T(t)\right]. \tag{15}$$

The quantity of body-transferred heat in time from 0 to t is

$$Q(t) = \int_0^t \dot{Q}(t)\,dt = \alpha A_s \left(T_{cz} - T_0\right)\tau\left(1 - e^{-t/\tau}\right). \tag{16}$$

Exercise 13.2 Heat Exchange between a Body with Lumped Thermal Capacity and Surroundings with Time-Dependent Temperature

Write general formula for temperature of a body with concentrated mass (small thermal resistance), if temperature of a medium T_{cz} changes in time. Derive a formula for body temperature, if the medium's temperature $T_{cz}(t) = a + bt$ changes at constant rate, equal to b, where a is a constant.

Solution

Differential equation, which describes body temperature changes, has the form (Ex. 13.1, (6))

$$c\rho V \frac{dT}{dt} + \alpha(t) A_s T(t) = \alpha(t) A_s T_{cz}(t) \tag{1}$$

when initial condition is

$$T(t)\big|_{t=0} = T_0 \,. \tag{2}$$

Once (1) is divided by $c\rho V$, one obtains

$$\frac{dT}{dt} + p(t)T = q(t), \tag{3}$$

where

$$p(t) = \frac{\alpha(t) A_s}{c\rho V},$$
$$q(t) = \frac{\alpha(t) A_s}{c\rho V} T_{cz}(t). \tag{4}$$

It is a heterogeneous differential equation of the first order. First we will determine the solution for homogeneous equation. In (3) we assume that $q(t) = 0$, following that we obtain, using variable separation method

$$T = C_1 e^{-\int p(t)dt}, \quad \text{where } C_1 \text{ is a constant.} \tag{5}$$

According to the variation of constant method, solution (5) is written in the form

$$T(t) = C_1(t) e^{-\int p(t)dt}. \tag{6}$$

Once (6) is substituted into (3) and mathematical operation carried out, one obtains

$$C_1' e^{-\int p(t)dt} - C_1(t) p(t) e^{-\int p(t)dt} + p(t) C_1(t) e^{-\int p(t)dt} = q(t), \tag{7}$$

where

$$C_1'(t) = q(t) e^{\int p(t)dt}, \tag{8}$$

$$C_1' = \frac{dC_1}{dt}.$$

Once (8) is integrated, one has

$$C_1(t) = \int q(t) e^{\int p(t)dt} + C, \tag{9}$$

where C is a constant. By substituting (9) into (6), we has

$$T(t) = Ce^{-\int p(t)dt} + e^{-\int p(t)dt} \int q(t) e^{\int p(t)dt} dt. \tag{10}$$

Constant C is determined from the initial condition (2).

If the medium's temperature changes according to formula

$$T_{cz}(t) = a + bt, \tag{11}$$

where a and b are constants and coefficient α is constant too, then from (4) we have

$$p(t)=\frac{\alpha A_s}{c\rho V}=\frac{1}{\tau}, \qquad q(t)=\frac{\alpha A_s(a+bt)}{c\rho V}=\frac{a+bt}{\tau}. \tag{12}$$

By substituting (12) into (10), one obtains

$$T(t)=Ce^{-t/\tau}+e^{-t/\tau}\int\frac{a+bt}{\tau}e^{t/\tau}dt. \tag{13}$$

Then we determine the integral below by integrating by parts

$$\int bte^{t/\tau}dt=\tau bte^{t/\tau}-\int b\left(\tau e^{t/\tau}\right)d\tau=\left(b\tau t-b\tau^2\right)e^{t/\tau}=b\tau(t-\tau)e^{t/\tau} \tag{14}$$

and calculate the remaining integrals. Equation (13) then assumes the form

$$T(t)=Ce^{-t/\tau}+e^{-t/\tau}\left[ae^{t/\tau}+\frac{b\tau(t-\tau)e^{t/\tau}}{\tau}\right],$$

$$T(t)=Ce^{-t/\tau}+(a+bt)-b\tau. \tag{15}$$

From the initial condition (2), we get

$$T_0=C+a-b\tau,$$

hence,

$$C=T_0-a+b\tau. \tag{16}$$

Once (16) is substituted into (15), (15) assumes the form

$$T(t)=T_{cz}(t)+(T_0-a)e^{-t/\tau}-b\tau\left(1-e^{-t/\tau}\right), \tag{17}$$

where $T_{cz}(t)$ is expressed by (11).

From the analysis of (17) it follows that for $t\to\infty$ one obtains

$$T(t)\Big|_{t\to\infty}=T_{cz}(t)-b\tau, \tag{18}$$

i.e. body temperature is lower than the medium's temperature by $b\tau$. The greater the time constant τ, the greater the difference in $b\tau$.

Exercise 13.3 Determining Temperature Distribution of a Body with Lumped Thermal Capacity, when the Temperature of a Medium Changes Periodically

Write a formula for temperature of a body with concentrated mass (small thermal resistance), if the medium's temperature T_{cz} changes in time. Derive a formula for the body temperature, if the medium's temperature undergoes cyclic changes according to formula

$$T_{cz}(t) = T_m + \Delta T \sin \omega t \,, \tag{1}$$

where $\omega = 2\pi/t_0$ is the circular frequency of temperature changes, while t_0 *a* change period. Heat transfer coefficient remains constant.

Solution

Body temperature changes in time are formulated in (1), Ex. 13.2. By accounting for

$$p(t) = \frac{\alpha A_s}{c\rho V} = \frac{1}{\tau} \quad \text{and} \quad q(t) = \frac{\alpha A_s}{c\rho V} T_{cz}(t) = \frac{\alpha A_s}{c\rho V} \frac{T_m + \Delta T \cos \omega t}{\tau} \tag{2}$$

the solution of (1) (Ex. 13.2) is expressed by (10) (Ex.13.2). Once expressions $p(t)$ and $q(t)$ given by (2) are substituted into (10), Ex. 13.2, and mathematical operations are carried out, one obtains

$$T(t) = Ce^{-t/\tau} + T_m + \frac{\Delta T}{\sqrt{1 + \omega^2 \tau^2}} \sin\left(\omega t - \operatorname{arctg} \omega \tau\right). \tag{3}$$

From initial condition

$$T\big|_{t=0} = T_m \tag{4}$$

one has

$$C + T_m + \frac{\Delta T}{\sqrt{1 + \omega^2 \tau^2}} \sin\left(-\operatorname{arctg} \omega \tau\right) = T_m \,, \tag{5}$$

hence,

$$C = -\frac{\Delta T}{\sqrt{1 + \omega^2 \tau^2}} \sin\left(-\operatorname{arctg} \omega \tau\right) = \frac{\omega \tau \Delta T}{1 + \omega^2 \tau^2} \,. \tag{6}$$

By substituting (6) into (3), one obtains an expression, which describes body temperature $T(t)$:

$$T(t) = T_m + \frac{\omega\tau\Delta T}{1+\omega^2\tau^2} e^{-t/\tau} + \frac{\Delta T}{\sqrt{1+\omega^2\tau^2}} \sin(\omega t - \text{arctg}\,\omega\tau). \tag{7}$$

The second term in (7) for time $t >> \tau$ can be neglected; one obtains then the "steady-state" temperature oscillations

$$T(t) = T_m + \frac{\Delta T}{\sqrt{1+\omega^2\tau^2}} \sin(\omega t - \text{arctg}\,\omega\tau). \tag{8}$$

Body temperature oscillates around mean temperature T_m, but at a lower amplitude, equal to

$$\Delta T' = \frac{\Delta T}{\sqrt{1+\omega^2\tau^2}}. \tag{9}$$

Furthermore, a phase shift occurs, which equals

$$\varphi = \text{arctg}\,\omega\tau. \tag{10}$$

Exercise 13.4 Inverse Problem: Determining Temperature of a Medium on the Basis of Temporal Thermometer-Indicated Temperature History

Determine temperature of a medium $T_{cz}(t)$ on the basis of temperature history indicated by a thermometer with known time-constant τ. Assume that the thermometer is a body with a concentrated mass and "measurement data" is expressed by function $T(t)$.

Solution

This *inverse problem* is the kind of problem whose output signal $T(t)$ is known in the analyzed medium-thermometer set-up; the input signal $T_{cz}(t)$ is the unknown. Differential equation, which describes temperature changes in the thermometer, will be used to determine $T_{cz}(t)$.

$$T_{cz}(t) = T + \tau \frac{dT}{dt} \tag{1}$$

The main difficulty in determining temperature $T_{cz}(t)$ by means of (1) lies in the accurate calculation of derivative dT/dt, especially when the readings are affected by random errors.

If the measured signal $T(t)$ is approximated locally by a polynomial of the 3rd degree with respect to time using the least-squares method, one can eliminate then, to a large extent, the effect of random errors on the value of derivative dT/dt, since the derivative is calculated when the polynomial of the 3rd degree is differentiated with respect to time and not by means of difference quotients, which are based on the readings affected by random errors. Using the polynomial of the 3rd degree and seven subsequent measurement points, the following formula is obtained for derivative dT_i/dt (Fig. 13.3):

$$\frac{dT_i}{dt} = \frac{22T_{i-3} - 67T_{i-2} - 58T_{i-1} + 58T_{i+1} + 67T_{i+2} - 22T_{i+3}}{252(\Delta t)}. \tag{2}$$

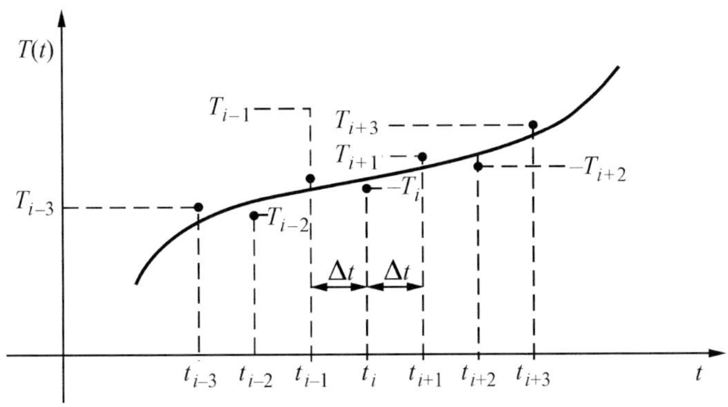

Fig. 13.3. A diagram for the calculation of the first derivative dT/dt by means of the seven-point digital filter $t_i = i(\Delta t)$

For time t_0 in (2) the following temperatures appear: T_{-3}, T_{-2} and T_{-1}. They are not known from the readings taken. One assumes then that (Fig. 13.4)

$$T_{-3} = T_0, \qquad T_{-2} = T_0, \qquad T_{-1} = T_0. \tag{3}$$

Derivatives dT_i/dt for the first three measurement points are formulated as

$$\begin{aligned} \frac{dT_0}{dt} &= \frac{22T_0 - 67T_0 - 58T_0 + 58T_1 + 67T_2 - 22T_3}{252(\Delta t)}, \\ \frac{dT_0}{dt} &= \frac{-103T_0 + 58T_1 + 67T_2 - 22T_3}{252(\Delta t)}; \\ \frac{dT_1}{dt} &= \frac{22T_0 - 67T_0 - 58T_0 + 58T_2 + 67T_3 - 22T_4}{252(\Delta t)}, \end{aligned} \tag{4}$$

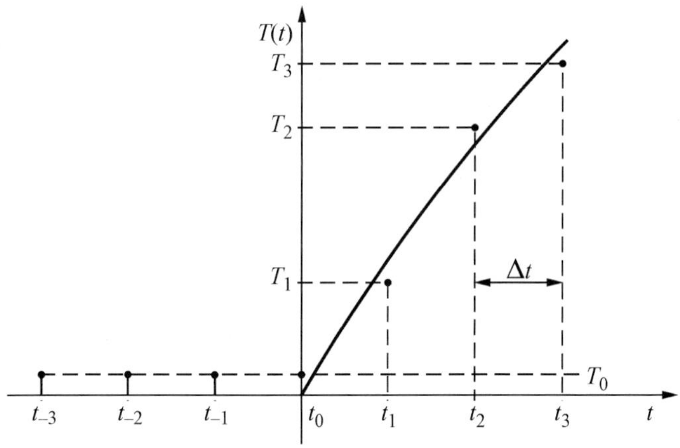

Fig. 13.4. A diagram that illustrates the calculation of derivative dT/dt for $i = 0, 1, 2$

$$\frac{dT_1}{dt} = \frac{-103T_0 + 58T_2 + 67T_3 - 22T_4}{252(\Delta t)}; \tag{5}$$

$$\frac{dT_2}{dt} = \frac{22T_0 - 67T_0 - 58T_1 + 58T_3 + 67T_4 - 22T_5}{252(\Delta t)},$$

$$\frac{dT_2}{dt} = \frac{-45T_0 - 58T_1 + 58T_3 + 67T_4 - 22T_5}{252(\Delta t)}. \tag{6}$$

At other points $i = 3, 4,...,$ derivative dT_i/dt is calculated by means of (2). Temperature of the medium $T_{cz}(t_i)$ is determined from (1)

$$T_{cz}(t_i) = T_i + \tau \frac{dT_i}{dt}, \qquad i = 0, 1,..., \tag{7}$$

where derivatives dT_i/dt are formulated in (4), (5), (6) and (2).

Exercise 13.5 Calculating Dynamic Temperature Measurement Error by Means of a Thermocouple

The weld in a thermocouple copper-constantan (Cu-Konst) has a shape of a sphere whose diameter is $d = 2$ mm. Air temperature is measured by inserting the thermocouple into the passing airflow. How long should the measuring of temperature last, so that dynamic measurement error will be smaller than 3%? Air-to-surface heat transfer coefficient of the thermocouple is at α = 60 W/(m^2·K). Assume the following thermo-physical properties of the thermocouple for the calculation: density ρ = 8900 kg/m^3, specific heat c = 415 J/(kg·K).

Solution

Temperature of the thermocouple $T(t)$ is expressed by formula ((14) from Ex. 13.1)

$$\frac{T_{cz} - T(t)}{T_{cz} - T_0} = e^{-t/\tau}, \tag{1}$$

where $\tau = c\rho V/\alpha A_s$ is the time constant.

Time t_x, after which measurement error is at ε, i.e.

$$\varepsilon = \frac{T_{cz} - T(t_x)}{T_{cz} - T_0}, \tag{2}$$

is calculated from (1): $\varepsilon = e^{-t_x/\tau}$, hence, one has

$$\ln \varepsilon = -\frac{t_x}{\tau} \qquad t_x = -\tau \ln \varepsilon. \tag{3}$$

Time constant of the thermocouple is

$$\tau = \frac{c\rho \cdot \frac{4}{3}\pi R^3}{\alpha 4\pi R^2} = \frac{c\rho R}{3\alpha} = \frac{c\rho d}{6\alpha}, \tag{4}$$

where from, after substitution, one has

$$\tau = \frac{415 \cdot 8900 \cdot 0.002}{6 \cdot 60} = 20.52 \text{ s}.$$

Measurement duration, after which the dynamic measurement error equals 3 %, is

$$t_x = -20.52 \ln 0.03 = 71.95 \text{ s}.$$

If duration exceeds 71.95 s, then the dynamic measurement error becomes smaller.

Exercise 13.6 Determining the Time It Takes to Cool Body Down to a Given Temperature

A single grape with a diameter of d = 12 mm and initial temperature of T_0 = 21°C was placed in a freezer whose temperature was T_{cz} = 8°C. Assuming that heat transfer coefficient between the surrounding air and grape surface equals α = 15 W/(m^2·K). Determine the time it takes the grape to reach the temperature of $T(t_x)$ = 10°C. Assume that $\rho = 1100$ kg/m^3 and c = 4200 J/(kg·K). Calculate how much heat is taken away from the grape during the process of cooling.

Solution

The temperature of the grape can be calculated by means of (14) from Ex. 13.1

$$\frac{T - T_{cz}}{T_0 - T_{cz}} = e^{-t/\tau}, \tag{1}$$

where

$$\tau = \frac{c\rho V}{\alpha A_s}. \tag{2}$$

Once a logarithm is found for both sides of (1), one has

$$\ln\frac{T(t_x) - T_{cz}}{T_0 - T_{cz}} = -\frac{t_x}{\tau},$$

hence,

$$t_x = -\tau \ln\frac{T(t_x) - T_{cz}}{T_0 - T_{cz}}. \tag{3}$$

The grape's time constant is

$$\tau = \frac{c\rho \cdot \frac{4}{3}\pi R^3}{\alpha 4\pi R^2} = \frac{c\rho R}{3\alpha} = \frac{c\rho d}{6\alpha}, \tag{4}$$

where from, after substitution, one has

$$\tau = \frac{4200 \cdot 1100 \cdot 0.012}{6 \cdot 15} = 616 \text{ s}.$$

Time t_x is calculated from (3):

$$t_x = -616 \ln \frac{10-8}{21-8} = 1153 \text{ s} = 19 \text{ min } 13 \text{ s}.$$

The amount of heat carried away from the grape during the cooling period is

$$Q(t_x) = \alpha A_s (T_{cz} - T_0)\tau\left(1 - e^{-t_x/\tau}\right) = 15 \cdot 4 \cdot \pi (0.006)^2 (8-21) \times$$
$$\times\, 616\left(1 - e^{-1153/616}\right) = -45.98 \text{ J}$$

Minus sign means that the heat was carried away from the grape.

In order to check the validity of results, grape heat removal can also be calculated in the following way:

$$Q = mc\left[T(t_x) - T_0\right] = \frac{4}{3}\pi R^3 \rho c\left[T(t_x) - T_0\right],$$

where from, after substitution, one obtains

$$Q = \frac{4}{3}\pi \cdot 0.006^3 \cdot 1100 \cdot 4200(10-21) = -45.98 \text{ J},$$

therefore, the same results are obtained.

Exercise 13.7 Temperature Measurement Error of a Medium whose Temperature Changes at Constant Rate

Water temperature increases at a rate of b = 15 K/min. Assuming that water heating lasted 6 minutes, calculate temperature measurement error at the end of the heating phase, if water temperature was measured by means of a three different sensors:

a) a sheathed thermocouple Ni-NiCr with the jacket's diameter of 1.5 mm and time constant of $\tau = 0.73$ s,

b) a sheathed thermocouple Ni-NiCr with the jacket's diameter of 3 mm and time constant of $\tau = 1.2$ s,

c) thermo-electric industrial sensor PTTK-SW12 (produced by KFAP Kraków) with a casing whose diameter is d_z = 18 mm and sensor length L = 140 mm. The time constant of the thermometer is $\tau = 10.8$ s.

Solution

For time $t \gg \tau$ temperature measurement error ε is ((18), Ex. 13.2)

Solution

The assumed coordinate system is presented in Fig. 14.1.

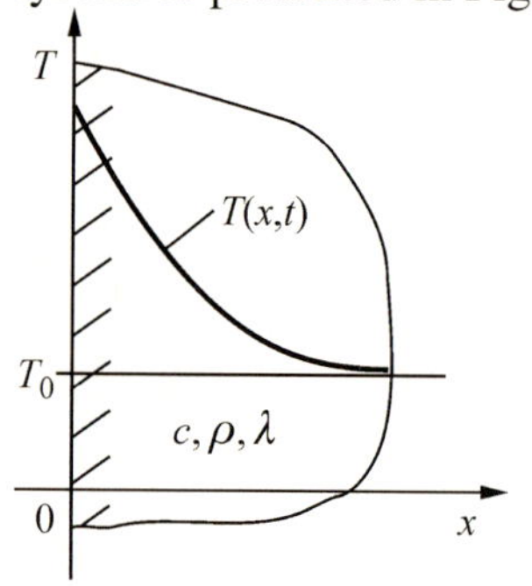

Fig. 14.1. Heated half-space with a step increase in surface temperature

Temperature distribution is described by equation

$$\frac{\partial T}{\partial t} = a\frac{\partial^2 T}{\partial x^2}, \tag{1}$$

by boundary conditions

$$T(0,t) = T_s, \tag{2}$$

$$T(\infty,t) = T_0 \tag{3}$$

and by initial condition

$$T(x,0) = T_0. \tag{4}$$

Once excess temperature surplus is introduced

$$u = T - T_0 \tag{5}$$

the initial-boundary problem (1)–(4) can be expressed in the following way:

$$\frac{\partial u}{\partial t} = a\frac{\partial^2 u}{\partial x^2}, \tag{6}$$

$$u(0,t) = u_s, \tag{7}$$

$$u(\infty,t) = 0, \tag{8}$$

$$u(x,0)=0\,. \tag{9}$$

The initial-boundary problem (6)–(9) will be solved using Laplace transform. Transformation of (6) is carried out when the initial condition is given by (9); hence, we obtain

$$\frac{d^2U(x,s)}{dx^2}-\frac{s}{a}U(x,s)=0\,. \tag{10}$$

The solution of (10) has the form

$$U(x,s)=C_1e^{\sqrt{s/a}\cdot x}+C_2e^{-\sqrt{s/a}\cdot x}\,. \tag{11}$$

By accounting for condition (8), from which it follows that $C_1 = 0$, solution (11) assumes the form

$$U(x,s)=C_2e^{-\sqrt{s/a}\cdot x}\,. \tag{12}$$

Once constant C_2 is determined from boundary condition (7), which after Laplace transformation assumes the form

$$U(0,s)=\frac{u_s}{s}\,, \tag{13}$$

one obtains

$$U(x,s)=\frac{u_s}{s}e^{-\sqrt{s/a}\cdot x}\,. \tag{14}$$

Inverse Laplace transformation is carried out using the fourth formula from Table 14.2; hence, we obtain

$$u(x,t)=u_s\operatorname{erfc}\left(\frac{x}{2\sqrt{at}}\right)=u_s\left[1-\operatorname{erf}\left(\frac{x}{2\sqrt{at}}\right)\right], \tag{15}$$

$$\frac{T(x,t)-T_s}{T_0-T_s}=\operatorname{erf}\left(\frac{x}{2\sqrt{at}}\right). \tag{16}$$

Appendix A contains a table with function erf x.

Heat flux is formulated as

$$\dot{q}(x,t)=-\lambda\left(T_0-T_s\right)\frac{\partial}{\partial x}\left[\operatorname{erf}\left(\frac{x}{2\sqrt{at}}\right)\right] \tag{17}$$

or

$$\dot{q}(x,t) = \lambda\left(T_s - T_0\right)\frac{2}{\sqrt{\pi}} \cdot e^{-x^2/4at} \cdot \frac{1}{2\sqrt{at}} = \sqrt{\frac{\lambda c\rho}{\pi t}}\left(T_s - T_0\right)e^{-x^2/4at} . \quad (18)$$

Heat flux on the body surface is

$$\dot{q}(0,t) = \sqrt{\frac{\lambda c\rho}{\pi t}}\left(T_s - T_0\right). \quad (19)$$

Exercise 14.3 Formula Derivation for Temperature Distribution in a Half-Space with a Step Increase in Heat Flux

Derive a formula for temperature distribution in a semi-infinite body with a step increase in heat flux $\dot{q}_s$ on the surface. Initial body temperature T_0 is constant and measures.

Solution

Temperature field in a half-space is defined by a differential equation

$$\frac{\partial u}{\partial t} = a\frac{\partial^2 u}{\partial x^2}, \quad (1)$$

by initial condition

$$u(x,0) = 0, \quad (2)$$

and by boundary conditions

$$-\lambda \left.\frac{\partial u}{\partial x}\right|_{x=0} = \dot{q}_s, \quad (3)$$

$$u(\infty,t) = 0, \quad (4)$$

where

$$u(x,t) = T(x,t) - T_0 . \quad (5)$$

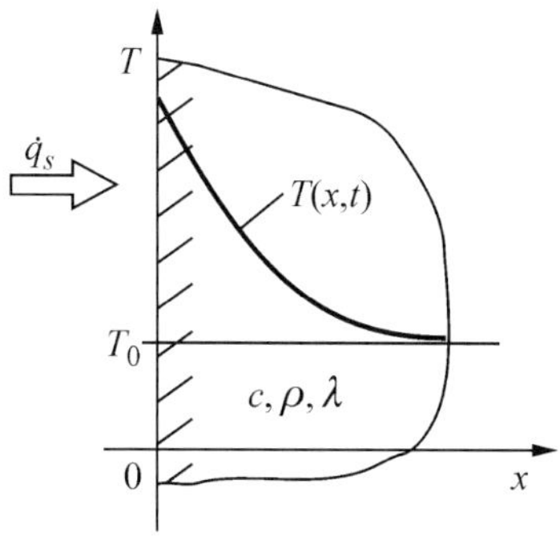

Fig. 14.2. Diagram of a half-space heated by heat flux $\dot{q}_s$

Once the Laplace transform is applied to (1) and initial condition (2) is accounted for, one gets

$$\frac{d^2U(x,s)}{dx^2} - \frac{s}{a}U(x,s) = 0. \tag{6}$$

The solution of (6) has the form

$$U(x,s) = C_1 e^{\sqrt{s/a}\cdot x} + C_2 e^{-\sqrt{s/a}\cdot x}. \tag{7}$$

From condition (4), it follows that constant C_1 in solution (7) equals zero. Constant C_2 is determined from boundary condition (3), to which Laplace transform was applied

$$-\lambda \frac{\partial U}{\partial x}\bigg|_{x=0} = \frac{\dot{q}_s}{s}. \tag{8}$$

By substituting (7) into (8) and accounting for the fact that $C_1 = 0$, one obtains

$$C_2 = \frac{\dot{q}_s}{\lambda s q}, \text{ where } q = \sqrt{s/a}. \tag{9}$$

By substituting (9) into (7), one has

$$U(x,s) = \frac{\dot{q}_s}{\lambda}\frac{e^{-qx}}{sq}. \tag{10}$$

Once inverse Laplace transformation is carried out (Table 14.2, Ex. 14.1), an expression for temperature distribution is obtained:

$$u = T - T_0 = \frac{\dot{q}_s}{\lambda}\left(2\sqrt{\frac{at}{\pi}}e^{-x^2/(4at)} - x\,\mathrm{erfc}\frac{x}{2\sqrt{at}}\right). \tag{11}$$

Surface temperature of the half-space is expressed as

$$u\big|_{x=0} = T\big|_{x=0} - T_0 = \frac{2\dot{q}_s}{\lambda}\sqrt{\frac{at}{\pi}} = \frac{2\dot{q}_s\sqrt{t}}{\sqrt{\pi}\sqrt{\lambda c \rho}}. \tag{12}$$

Exercise 14.4 Formula Derivation for Temperature Distribution in a Half-Space with a Step Increase in Temperature of a Medium

Derive a formula for temperature distribution in a semi-infinite body with a convective boundary condition.

Solution

Half-space is heated or cooled by a medium with temperature T_{cz} and an assigned heat transfer coefficient α. Half-space temperature field is defined by the heat conduction equation

$$\frac{\partial u}{\partial t} = a\frac{\partial^2 u}{\partial x^2}, \tag{1}$$

by initial condition

$$u(x,0) = 0, \tag{2}$$

and by boundary conditions

$$-\lambda\frac{\partial u}{\partial x}\bigg|_{x=0} = \alpha\left(u_{cz} - u\big|_{x=0}\right), \tag{3}$$

$$u(\infty,t) = 0, \tag{4}$$

where $u = T - T_0$.

Once Laplace transformation is carried out, solution $U(x,s)$ has the form identical to the (12) in Ex. 14.2:

$$U(x,s) = C_2 e^{-qx}, \quad \text{where} \quad q = \sqrt{s/a}. \tag{5}$$

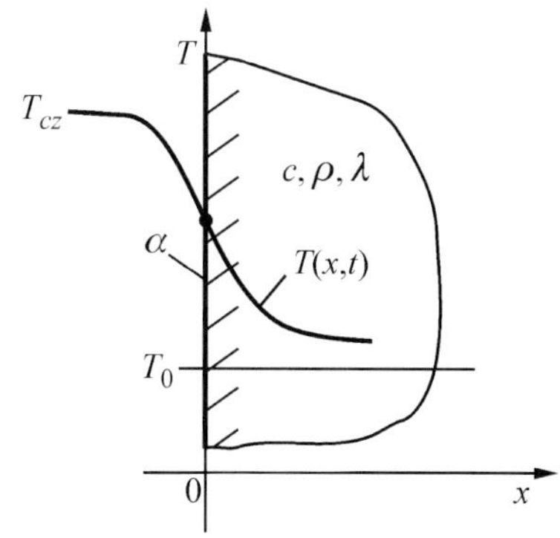

Fig. 14.3. Temperature distribution in a half-space with a convective boundary condition

In order to determine constant C_2, Laplace transformation will be done for boundary condition (3):

$$-\lambda \frac{\partial U}{\partial x}\bigg|_{x=0} = \alpha\left(\frac{u_{cz}}{s} - U\big|_{x=0}\right). \tag{6}$$

Once (5) is substituted into (6) and subsequently transformed, constant C_2 is determined

$$C_2 = \left(T_{cz} - T_0\right)\frac{\alpha/\lambda}{s\left(q + \alpha/\lambda\right)}. \tag{7}$$

Once (7) is substituted into (5), one has

$$U\left(x,s\right) = \left(T_{cz} - T_0\right)\frac{\left(\alpha/\lambda\right)\cdot e^{-qx}}{s\left(q+\alpha/\lambda\right)}. \tag{8}$$

Inverse Laplace transformation is carried out using the last formula from Table 14.2 (Ex. 14.1), hence, we obtain a formula for temperature distribution

$$u = T - T_0 = \left(T_{cz} - T_0\right)\Bigg[\operatorname{erfc}\frac{x}{2\sqrt{at}} - \exp\left(\frac{\alpha\, x}{\lambda} + \frac{\alpha^2 at}{\lambda^2}\right)\times$$
$$\times\operatorname{erfc}\left(\frac{x}{2\sqrt{at}} + \frac{\alpha}{\lambda}\sqrt{at}\right)\Bigg], \qquad 0 \le x \le \infty. \tag{9}$$

Once the symbols are introduced

$$Fo = \frac{at}{x^2}, \qquad Bi = \frac{\alpha\, x}{\lambda}, \tag{10}$$

and relations accounted for

$$\begin{aligned} &\frac{x}{2\sqrt{at}} = \frac{1}{2\sqrt{Fo}}, \\ &\frac{\alpha^2 at}{\lambda^2} = Bi^2 Fo, \\ &\frac{\alpha\sqrt{at}}{\lambda} = Bi\sqrt{Fo} \end{aligned} \tag{11}$$

solution (9) can be written in the form

$$\begin{aligned} \theta = \frac{T(x,t) - T_0}{T_{cz} - T_0} = 1 - \operatorname{erf}\left(\frac{1}{2\sqrt{Fo}}\right) - \exp\left(Bi + Bi^2 Fo\right) \times \\ \times \left[1 - \operatorname{erf}\left(\frac{1}{2\sqrt{Fo}} + Bi\sqrt{Fo}\right)\right]. \end{aligned} \tag{12}$$

Function (12) describes temperature distribution in the half-space with an initial temperature T_0, whose surface is subjected to (for time $t > 0$) a liquid with temperature $T_{cz} \neq T_0$. The half-space can be heated or cooled by a fluid with a temperature of T_{cz} = const. Half-space surface temperature $x = 0$ is defined by the following expression obtained from (12)

$$T(0,t) = T_0 + (T_{cz} - T_0)\left[1 - \operatorname{erfc}\left(\frac{\alpha}{\lambda}\sqrt{at}\right)\exp\left(\frac{\alpha^2 at}{\lambda^2}\right)\right] \tag{13}$$

or

$$T(0,t) = T_0 + (T_{cz} - T_0)\left\{1 - \exp\left(\frac{\alpha^2 at}{\lambda^2}\right)\left[1 - \operatorname{erf}\left(\frac{\alpha}{\lambda}\sqrt{at}\right)\right]\right\}. \tag{14}$$

Expression (14) is also used for the experimental determination of the heat transfer coefficient α on the basis of measured surface temperature at a given time point t_p.

Half-space temperature distribution in a semi-logarithmic and Cartesian coordinate system is shown in Figs. 14.4 and 14.5.

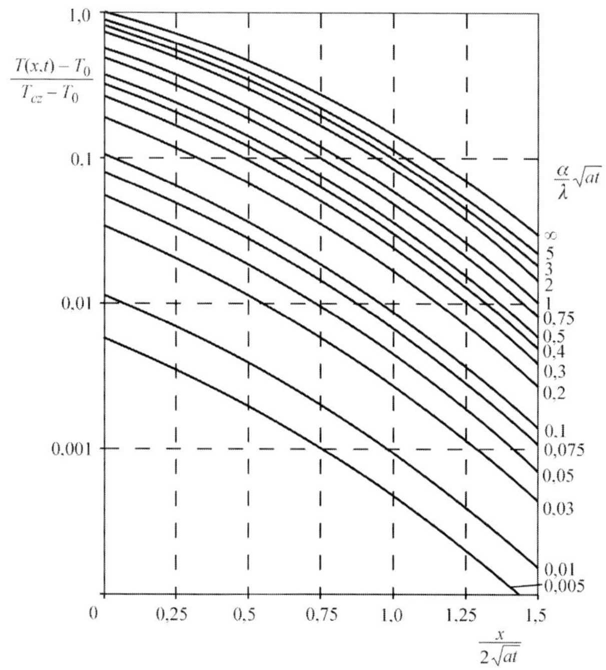

Fig. 14.4. Temperature distribution in a half-space with a convective heat transfer on the surface

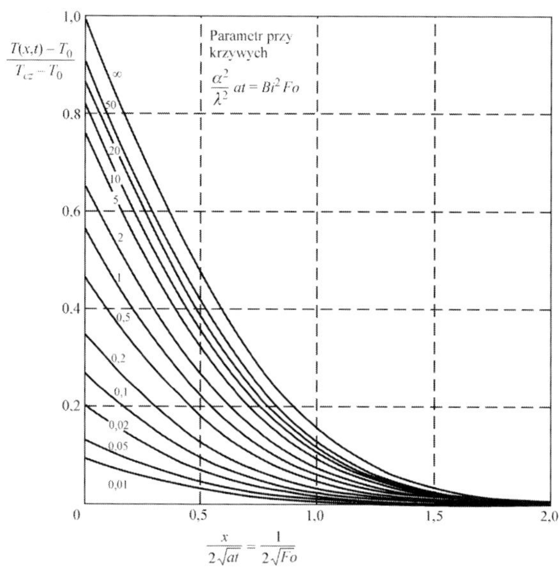

Fig. 14.5. Temperature distribution in a half-space with a convective heat transfer on the surface

By using a formula for derivative

$$\frac{d}{dx}(\operatorname{erf} x)=\frac{2}{\sqrt{\pi}}e^{-x^2}, \tag{17}$$

heat flux (16) can be expressed in the following way:

$$\dot{q}(x,t)=-\lambda\int_0^t\frac{d\varphi(t)}{dt}\left[-\frac{2}{\sqrt{\pi}}\cdot\exp\left(-\frac{x^2}{4a(t-\tau)}\right)\cdot\frac{1}{2\sqrt{a(t-\tau)}}\right]d\tau, \tag{18}$$

$$\dot{q}(x,t)=\sqrt{\frac{\lambda c\rho}{\pi}}\int_0^t\frac{1}{\sqrt{t-\tau}}\exp\left(-\frac{x^2}{4a(t-\tau)}\right)\cdot\frac{d\varphi(\tau)}{d\tau}d\tau.$$

Heat flux on the half-space surface is

$$\dot{q}(x,t)\Big|_{x=0}=\sqrt{\frac{\lambda c\rho}{\pi}}\int_0^t\frac{d\varphi(\tau)}{d\tau}\frac{1}{\sqrt{t-\tau}}d\tau. \tag{19}$$

Equation (19) is used for measuring heat flux on the basis of the half-space measured temperature $\varphi(t)$.

Exercise 14.6 Formula Derivation for a Quasi-Steady State Temperature Field in a Half-Space when Surface Temperature Changes Periodically

Derive a formula for the quasi-steady state temperature field in a half-space, when surface temperature changes periodically (Fig. 14.7). Discuss your findings.

Solution

Temperature field is expressed by equation

$$\frac{\partial T}{\partial t}=a\frac{\partial^2 T}{\partial x^2}, \tag{1}$$

by initial condition

$$T\Big|_{t=0}=\overline{T}, \qquad 0\le x<\infty \tag{2}$$

and by boundary conditions

$$T\big|_{x=0} = \overline{T} + \Delta T\cos\omega t, \qquad t > 0, \tag{3}$$

$$T\big|_{x\to\infty} = \overline{T}, \qquad t > 0, \tag{4}$$

where $\omega = 2\pi/t_0$ is the circular frequency, while t_0 a circular period.

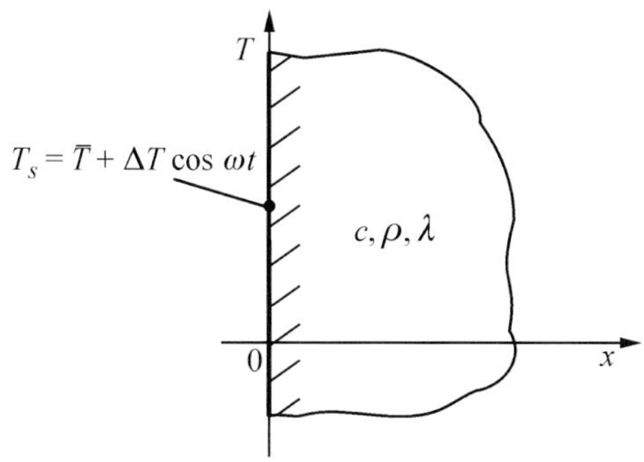

Fig. 14.7. Diagram of the half-space with periodically variable temperature

Because function $u = T - \overline{T}$, which describes excessive temperature surplus distribution has a periodic character, the solution of problem (1)–(4) will be searched for in the form

$$u = T(x,t) - \overline{T} = Ae^{kx}e^{i\omega t}, \tag{5}$$

where function $e^{i\omega t}$ is expressed by the Euler formula

$$e^{i\omega t} = \cos\omega t + i\sin\omega t. \tag{6}$$

By substituting $T(x,t)$ from expression (5) into (1), one has

$$Ae^{kx}i\omega e^{i\omega t} = aAk^2e^{kx}e^{i\omega t}, \tag{7}$$

hence, the equation

$$k^2 - \frac{i\omega}{a} = 0, \tag{8}$$

whose solution has the form

$$k_{1,2} = \pm\sqrt{i\omega/a}. \tag{9}$$

Taking into account for the identity

$$i = (i+1)^2/2 \tag{10}$$

and the decreasing temperature, as the distance from the surface increases, we assume for the further calculation that

$$k = -\sqrt{i\omega / a} = -(i+1)\sqrt{\omega / 2a}\,. \tag{11}$$

By substituting (11) into (5), one has

$$u = Ae^{-(i+1)\sqrt{\omega/2a}\cdot x}e^{i\omega t} = Ae^{-\sqrt{\omega/2a}\cdot x}\cdot e^{i\left(\omega t-\sqrt{\omega/2a}\cdot x\right)}. \tag{12}$$

Using Euler's Equation (6), solution (12) can be written in the form

$$u = T(x,t) - \bar{T} = Ae^{-x\sqrt{\omega/2a}}\left[\cos\left(\omega t - x\sqrt{\frac{\omega}{2a}}\right) + i\sin\left(\omega t - x\sqrt{\frac{\omega}{2a}}\right)\right]. \tag{13}$$

Constant A in (13) determined from condition (3) is

$$A = \Delta T. \tag{14}$$

By substituting (14) into (13) and taking only the real part of the solution, one has

$$T(x,t) = \bar{T} + \Delta T e^{-x\sqrt{\omega/2a}}\cdot\cos\left(\omega t - x\sqrt{\frac{\omega}{2a}}\right). \tag{15}$$

This is the equation that describes steady-state periodic changes in the half-space, caused by periodic temperature changes on the half-space surface.

The analysis of solution (15) leads to the following conclusions:

1. The amplitude of temperature fluctuations becomes smaller with depth.
2. Low frequency temperature fluctuations penetrate materials deeper than the high frequency fluctuations.
3. Steady-state temperature fluctuations transpire inside a body at a frequency equal to the frequency of temperature changes on the half-space surface.
4. Time points t, in which temperatures at a given point $x = \delta$ reach its maximum or minimum, are determined from equation

$$\frac{2\pi}{t_0}t - \delta\sqrt{\frac{2\pi}{t_0}\frac{1}{2a}} = n\pi, \quad n = 1, 2, 3 \ldots. \tag{16}$$

The even values of n, however, correspond to the maximum values $T(\delta, t)$, while odd values to the minimum values.

Next, we will determine the length of the temperature wave Λ. Assuming that temperature distribution is analyzed at a given time point $t = t_p$, from (15), one has

$$T\left(x,t_p\right)=\overline{T}+\Delta T e^{-x\sqrt{\omega/2a}}\cos\left(\omega t_p - x\sqrt{\frac{\omega}{2a}}\right) \tag{17}$$

where the coordinates of maximum temperature values will be determined from.

Function $\cos\varphi$ assumes maximum values equal to 1 for $\varphi = \pm n2\pi$, n = 0, 1, ... For the subsequent calculations, negative angle values φ are assumed, so that the maximums x_n of function $T(x, t_p)$ will lie on the positive semi-axis x. The position of x_n maximums in function $T(x, t_p)$ is determined from equation

$$\omega t_p - x\sqrt{\frac{\omega}{2a}} = -n2\pi, \qquad n=0, 1, 2, \ldots . \tag{18}$$

From the solution of (18), one has

$$x_n = \sqrt{\frac{2a}{\omega}}\left(\omega t_p + 2n\pi\right), \qquad n=0, 1, 2, \ldots . \tag{19}$$

Hence, for n = 0 from (19), we obtain

$$x_0 = t_p\sqrt{2\omega a} = 2t_p\sqrt{\frac{\pi a}{t_0}}, \tag{20}$$

while for n = 1

$$\omega t_p - x_1\sqrt{\frac{\omega}{2a}} = -2\pi, \tag{21}$$

$$x_1 = \sqrt{\frac{2a}{\omega}}\left(\omega t_p + 2\pi\right). \tag{22}$$

Wave length Λ is obtained from the calculation of a distance between two subsequent maximums

$$\Lambda = x_{n+1} - x_n = \sqrt{\frac{2a}{\omega}}\left[\omega t_p + 2\left(n+1\right)\pi\right] - \sqrt{\frac{2a}{\omega}}\left(\omega t_p + 2n\pi\right) = 2\pi\sqrt{\frac{2a}{\omega}}, \tag{23}$$

hence, by taking into account that $\omega = 2\pi/t_0$, one gets

$$\Lambda = x_{n+1} - x_n = 2\sqrt{\pi a t_0}. \tag{24}$$

The amplitude of temperature changes $(T_m(x, t) - \overline{T})$ at a distance x from the body surface is determined from (15), when

Solution

Here, we will make use of (16) from Ex. 14.2 for temperature distribution in the half-space

$$\frac{T(x,t)-T_s}{T_0-T_s}=\mathrm{erf}\left(\frac{x}{2\sqrt{at}}\right). \tag{1}$$

From the definition of the depth of heat penetration, we have

$$T_0-T(\delta,t)=0.01(T_0-T_s), \tag{2}$$

from where, we get

$$\frac{T(\delta,t)-T_s}{(T_0-T_s)}=0.99\,.$$

From table in Appendix A for

$$\mathrm{erf}\left(\frac{\delta}{2\sqrt{at}}\right)=0.99 \tag{3}$$

we obtain

$$\frac{\delta}{2\sqrt{at}}=1.82\,. \tag{4}$$

The depth of heat penetration is determined from (4)

$$\delta=3.64\sqrt{at}. \tag{5}$$

Time t_L is determined from (5) after substituting $\delta = L$

$$L=3.64\sqrt{at_L}\,,$$

$$t_L=\frac{1}{a}\left(\frac{L}{3.64}\right)^2=\frac{1}{1.1\cdot10^{-5}}\left(\frac{0.2}{3.64}\right)^2=274.5\text{ s}.$$

Temperature at a distance $x = \delta_1$ after time t_L is

$$T(\delta_1,t_L)=T_s+(T_0-T_s)\mathrm{erf}\left(\frac{\delta_1}{2\sqrt{at_L}}\right)=$$

$$=20+(350-20)\,\mathrm{erf}\left(\frac{0.05}{2\sqrt{1.1\cdot10^{-5}\cdot274.5}}\right)=$$

$$=20+330\,\mathrm{erf}(0.455)=20+330\cdot0.48007=178.4^\circ\text{C}.$$

Exercise 14.9 Calculating Plate Surface Temperature Under the Assumption that the Plate is a Semi-Infinite Body

Thick steel plate was quickly placed into an industrial furnace at high temperature. Furnace temperature T_p is much higher than the plate surface temperature T_s, i.e. $T_p >> T_s$. One can assume, therefore, that the plate surface was suddenly warmed by a heat flow with constant density $\dot{q}_s$, since $T_p^4 >> T_s^4$. Treat the plate as a semi-infinite body. Calculate plate surface temperature and temperature at point x = 0,03 m under the plate surface. Assume that heat flux is $\dot{q}_s$ = 500000 W/m^2, initial plate temperature T_0 = 30°C, λ = 40 W/(m·K), c = 460 J/(kg·K), ρ = 7770 kg/m^3. Calculate temperature after time t = 30 s, from the moment the charge is placed in the furnace.

Solution

Surface temperature is expressed by (12) in Ex. 14.3:

$$T_s = T\big|_{x=0} = T_0 + \frac{2\dot{q}_s\sqrt{t}}{\sqrt{\pi\lambda c\rho}} = 30 + \frac{2\cdot 500000\sqrt{30}}{\sqrt{\pi\cdot 40\cdot 460\cdot 7770}} = 288.4\,^\circ\text{C}.$$

Temperature at the point with a coordinate x is given by (11), Ex. 14.3:

$$T(x,t) = T_0 + \frac{\dot{q}_s}{\lambda}\left[2\sqrt{\frac{at}{\pi}}e^{-x^2/(4at)} - x\left(1-\operatorname{erf}\left(\frac{x}{2\sqrt{at}}\right)\right)\right], \text{ where } a = \frac{\lambda}{c\rho}.$$

After substitution of the numerical values, one obtains

$$a = \frac{\lambda}{c\rho} = \frac{40}{460\cdot 7770} = 1.11913\cdot 10^{-5}\ \text{m}^2/\text{s},$$

$$\frac{x}{2\sqrt{at}} = \frac{0.03}{2\sqrt{1.11913\cdot 10^{-5}\cdot 30}} = 0.81863\,.$$

From Appendix A, one has

$$\operatorname{erf}(0.81863) = 0.753.$$

After substituting data into temperature formula, one gets

$$T(0.03\ \text{m};\ 30\ \text{s}) = 30 + \frac{500000}{40}\left[2\sqrt{\frac{1.11913\cdot 10^{-5}\cdot 30}{\pi}}\cdot e^{-0.81863^2} - \right.$$

$$\left. -0.03(1-0.753)\right] = 30 + 39.6 = 69.6\,^\circ\text{C}.$$

Exercise 14.12 Calculating Quasi-Steady-State Ground Temperature at a Specific Depth when Surface Temperature Changes Periodically

In the summer, ground surface temperature changes from 35°C to 10°C within 24 hours. Assuming that similar temperature changes occur over a longer period of time, so that steady-state temperature fluctuations are formed underground, calculate temperature change intervals at a depth of a) $x_1 = 0.9$ m and b) $x_2 = 1.2$ m. Also determine phase shift and time-lag of temperature changes at a depth x_1 and x_2 in relation to temperature changes on the ground surface. Assume the following thermo-physical properties of the ground (clay) for the calculation: $\lambda = 1.28$ W/(m·K), $c = 880$ J/(kg·K), $\rho = 1500$ kg/m^3.

Solution

Formulas derived in Ex. 14.6. will be used to solve this exercise. Amplitude of temperature changes is given by (26), Ex. 14.6

$$\frac{T_m(x,t)-\overline{T}}{\Delta T}=e^{-x\sqrt{\omega/2a}}, \tag{1}$$

where:

$\Delta T=\dfrac{T_{max}-T_{min}}{2}$ – amplitude of ground surface temperature changes,

T_{max} – maximum temperature of ground surface in °C,

T_{min} – minimum temperature of ground surface in °C,

$\overline{T}=\dfrac{T_{max}+T_{min}}{2}$ – average ground temperature.

Ground temperature fluctuations at a depth x occur in the interval

$$\overline{T}-\Delta T e^{-x\sqrt{\omega/2a}}\le T_m(x,t)\le \overline{T}+\Delta T e^{-x\sqrt{\omega/2a}}. \tag{12}$$

A phase shift in temperature fluctuations at a depth x is ((13) from Ex. 14.6))

$$\varphi = x\sqrt{\frac{\omega}{2a}}, \tag{3}$$

while time-lag Δt, which corresponds to angle φ is

$$\Delta t=\frac{\varphi}{\omega}=\frac{x}{\omega}\sqrt{\frac{\omega}{2a}}=x\sqrt{\frac{1}{2\,\omega a}}=\frac{x}{2}\sqrt{\frac{t_0}{\pi a}}, \tag{4}$$

where $\omega = 2\pi/t_0$ stands for temperature circular frequency, t_0 – temperature change period.

Calculations will be done separately for both x_1 and x_2.

a) $x = x_1 = 0.9$ m.

$$a = \frac{\lambda}{c\rho} = \frac{1.28}{880 \cdot 1500} = 9.697 \cdot 10^{-7} \text{ m}^2\text{/s},$$

$$\Delta T = \frac{35-10}{2} = \frac{25}{2} = 12.5^\circ\text{C},$$

$$\overline{T} = \frac{35+10}{2} = 22.5^\circ\text{C},$$

$$\omega = \frac{2\pi}{t_0} = \frac{2\pi}{24 \cdot 3600} = 7.2722 \cdot 10^{-5} \text{ rad/s},$$

$$\Delta T e^{-x_1\sqrt{\omega/2a}} = 12.5 \exp\left(-0.9\sqrt{\frac{7.2722 \cdot 10^{-5}}{2 \cdot 9.697 \cdot 10^{-7}}}\right) = 0.05^\circ\text{C}.$$

Temperature fluctuations at a depth $x_1 = 0.9$ m are in the interval

$$22.45^\circ\text{C} \le T_m(x_1,t) \le 22.55^\circ\text{C}.$$

Time-lag in temperature fluctuations with respect to temperature changes on the ground surface amounts to

$$\Delta t = \frac{0.9}{2}\sqrt{\frac{24 \cdot 3600}{\pi \cdot 9.697 \cdot 10^{-7}}} = 75783.7 \text{ s} = 21 \text{ h } 3\text{min}.$$

b) $x = x_2 = 1.2$ m,

$$\Delta T e^{-x_2\sqrt{\omega/2a}} = 12.5 \exp\left(-1.2\sqrt{\frac{7.2722 \cdot 10^{-5}}{2 \cdot 9.697 \cdot 10^{-7}}}\right) = 0.008^\circ\text{C}.$$

Temperature fluctuations at a depth $x_2 = 1.2$ m are in the interval

$$22.492^\circ\text{C} \le T_m(x_2,t) \le 22.508^\circ\text{C}.$$

Time-lag Δt amounts to

$$\Delta t = \frac{1.2}{2}\sqrt{\frac{24 \cdot 3600}{\pi \cdot 9.697 \cdot 10^{-7}}} = 101044.9 \text{ s} = 28 \text{ h } 4\text{min}.$$

From the analysis of results obtained for both cases, one can discern that temperature fluctuations, which occur on the ground surface are very quickly suppressed. Temperature changes that occur underground at a depth of $x_1=0.9$ m and $x_2=1.2$ m are considerably delayed in comparison with the changes that occur on the ground surface.

Exercise 14.13 Calculating Surface Temperature at the Contact Point of Two Objects

Calculate surface temperature at the point of contact of a chamotte brick heated to a temperature of 200°C with **a)** a cast iron object, **b)** a wooden object (oak), with a temperature of 20°C. Lets assume that the objects are suddenly placed against the heated chamotte brick. Assume that the contact between the two objects is ideal, i.e. the temperature and heat flux on the contact surface are identical. Use the following thermo-physical properties of the materials for the calculation:

- chamotte brick:
  $\lambda = 0.9$ W/(m·K), $c = 835$ J/(kg·K), $\rho = 1800$ kg/m^3,
- cast iron:
  $\lambda = 53$ W/(m·K), $c = 545$ J/(kg·K), $\rho = 7200$ kg/m^3,
- oak wood:
  $\lambda = 0.19$ W/(m·K), $c = 2400$ J/(kg·K), $\rho = 700$ kg/m^3.

Solution

Temperature of the surfaces in contact with each other will be calculated from (3), Ex. 14.7

$$T_s = \frac{\sqrt{\lambda_1 c_1 \rho_1} \cdot T_{1,0} + \sqrt{\lambda_2 c_2 \rho_2} \cdot T_{2,0}}{\sqrt{\lambda_1 c_1 \rho_1} + \sqrt{\lambda_2 c_2 \rho_2}}. \tag{1}$$

After substitution, of the numerical values we have

a) chamotte brick–cast iron

$$T_s = \frac{\sqrt{0.9 \cdot 835 \cdot 1800} \cdot 200 + \sqrt{53 \cdot 545 \cdot 7200} \cdot 20}{\sqrt{0.9 \cdot 835 \cdot 1800} + \sqrt{53 \cdot 545 \cdot 7200}} = 33.43^\circ\text{C},$$

b) chamotte brick–oak wood

$$T_s = \frac{\sqrt{0.9\cdot 835\cdot 1800}\cdot 200+\sqrt{0.19\cdot 2400\cdot 700}\cdot 20}{\sqrt{0.9\cdot 835\cdot 1800}+\sqrt{0.19\cdot 2400\cdot 700}} = 141.15^{\circ}\text{C}.$$

From the comparison of the obtained results, it is clear that the contact temperature is closer to the initial temperature of the body with a higher product value $\lambda c\rho$. In the case of **a)** contact temperature T_s is closer to the initial temperature of the cast iron, while in the case of **b)** to the initial temperature of the chamotte brick.

Literature

1. Carslaw HS, Jaeger JC (1986) Conduction of Heat in Solids. Clarendon Press, Oxford
2. Janke E, Emde F, Lösch F (1960) Tafeln höherer Funktionen. Teubner, Stuttgart
3. Kulish VV, Dallas JL (2000) Fractional – diffusion solutions for transient local temperature and heat flux. Transactions of the ASME, Journal of Heat Transfer 122: 372–376
4. Tautz H (1971) Wärmeleitung und Temperaturausgleich. Verlag Chemie, Weinheim
5. (1999) The Heat Transfer Problem Solver. Research and Education Association. New Jersey, Piscataway
6. Thomson WJ (1997) Atlas for Computing Mathematical Functions. Wiley-Interscience Publication, John-Wiley and Sons, New York
7. Диткин ВА, Прудников АП (1975) Операцонное Исчисление. Высшая Школа, Москва
8. Лыков АВ (1967) Теория теплопроводности. Высшая Школа, Минск

15 Transient Heat Conduction in Simple-Shape Elements

This chapter analyzes the phenomenon of transient heat conduction in simple-shape bodies. It presents twenty exercises, which contain both, theory and computational problems. Using the separation of variables method and Laplace transform, the authors derive formulas for temperature distribution in a plate, cylinder and sphere with boundary conditions of 1st, 2nd and 3rd kind. Also develop computational programs as well as graphs and diagrams, which enable one to calculate roots of characteristic equations, temperature distribution, temperature change rate and average temperature. Dimensionless teperature values in the function of time are listed for boundary conditions of 2nd kind in tables provided. Derived formulas are applied to the calculation of temperature transients and thermal stresses.

Exercise 15.1 Formula Derivation for Temperature Distribution in a Plate with Boundary Conditions of 3rd Kind

Derive a formula for temperature distribution in a plate with a thickness $2L$, when convective heat transfer occurs between both plate surfaces and the surroundings, with temperature T_{cz}, when heat transfer coefficient α is constant. Initial plate temperature is constant and is T_0 (Fig. 15.1). Assume constant material properties for the calculation: λ, c and ρ.

Solution

Plate temperature distribution is described by the heat conduction equation

$$\frac{\partial T}{\partial t} = a\frac{\partial^2 T}{\partial x^2}, \qquad 0 \le t, \qquad 0 \le x \le L, \tag{1}$$

boundary conditions

$$\left.\frac{\partial T}{\partial x}\right|_{x=0} = 0, \qquad 0 \le t\,, \tag{2}$$

$$-\lambda \left.\frac{\partial T}{\partial x}\right|_{x=L} = \alpha\left(\left.T\right|_{x=L} - T_{cz}\right), \qquad 0 \le t \tag{3}$$

and by initial condition

$$\left.T(x,t)\right|_{t=0} = T_0, \qquad 0 \le x \le L\,. \tag{4}$$

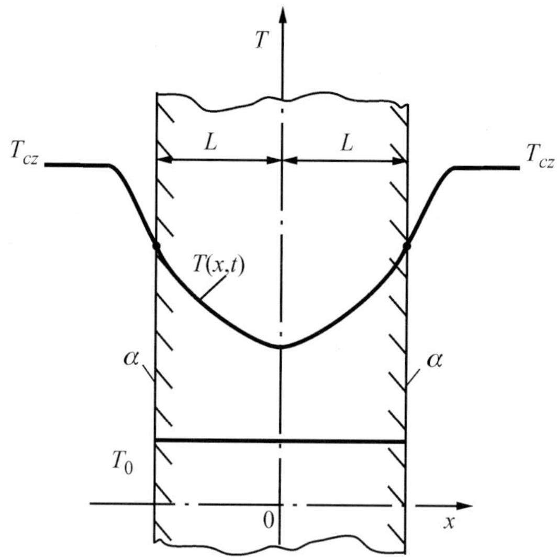

Fig. 15.1. Heating an infinitely long uniformly thick plate

Due to the symmetry of the problem, only region $0 \le x \le L$ (a half of the plate's thickness) is analyzed here. In order to reduce the boundary condition (3) to a homogeneous condition, a new variable is introduced

$$\theta\left(x,t\right) = T\left(x,t\right) - T_{cz}\,. \tag{5}$$

The initial-boundary problem (1)–(4) can be written then in the following way:

$$\frac{\partial \theta}{\partial t} = a\frac{\partial^2 \theta}{\partial x^2}\,, \tag{6}$$

$$\left.\frac{\partial \theta}{\partial x}\right|_{x=0} = 0\,, \tag{7}$$

$$-\lambda \frac{\partial \theta}{\partial x}\bigg|_{x=L} = \alpha \theta\big|_{x=L}, \tag{8}$$

$$\theta\big|_{t=0} = \theta_0 . \tag{9}$$

The separation of variables method is used to solve problems (6)–(9); according to this method, the solution has a form

$$\theta(x,t) = \varphi(t)\psi(x). \tag{10}$$

Once (10) are substituted into (6), one obtains

$$\frac{1}{a}\psi \frac{d\varphi}{dt} = \varphi \frac{d^2\psi}{dx^2}. \tag{11}$$

Once both sides of (11) are divided by (ψ,φ) can be written in the form

$$\frac{1}{a}\frac{1}{\varphi}\frac{d\varphi}{dt} = \frac{1}{\psi}\frac{d^2\psi}{dx^2}. \tag{12}$$

Due to the fact that the equality (12) should occur for any value of x and t, both sides of the equation should be equal to the constant, which should, in turn, have a negative value due to a finite temperature value in time. Once the constant is marked as $-k^2$, one has

$$\frac{1}{a}\frac{1}{\varphi}\frac{d\varphi}{dt} = \frac{1}{\psi}\frac{d^2\psi}{dx^2} = -k^2, \tag{13}$$

from which two equations follow:

$$\frac{d\varphi}{dt} + ak^2\varphi = 0, \tag{14}$$

$$\frac{d^2\psi}{d^2t} + k^2\psi = 0. \tag{15}$$

General solutions to (14) and (15) are functions

$$\varphi = C_1 e^{-ak^2 t}, \tag{16}$$

$$\psi = C_2 \cos(kx) + C_3 \sin(kx). \tag{17}$$

Once (16) and (17) are substituted into (10), one obtains

$$\theta = \varphi(t)\psi(x) = e^{-ak^2t}\left[A\cos(kx) + B\sin(kx)\right], \tag{18}$$

where $A = C_1C_2$ and $B = C_1C_3$.

From boundary condition (7), one obtains

$$\left.\frac{\partial \theta}{\partial x}\right|_{x=0} = e^{-ak^2t}k\left(-A\sin 0 + B\cos 0\right) = 0\,,$$

hence, $B = 0$. Therefore, the solution of (18) has the following form

$$\theta\left(x,t\right) = Ae^{-ak^2t}\cos\left(kx\right). \tag{19}$$

Once (19) is substituted into boundary condition (8), one has

$$\lambda kAe^{-ak^2t}\sin\left(kL\right) = \alpha Ae^{-ak^2t}\cos\left(kL\right),$$

hence the equation

$$\text{ctg}\left(kL\right) = \frac{kL}{\dfrac{\alpha L}{\lambda}}. \tag{20}$$

Once we denote $\alpha L/\lambda = Bi$ and $kL = \mu$, transcendental (20) can be written in the form

$$\text{ctg}\mu = \frac{\mu}{Bi}. \tag{21}$$

Equation (21) has an infinite number of roots, which can be determined using one of the methods for solving non-linear algebraic equations, either, for instance, the interval halving method or Muller method [1, 2]. These are iterative methods; when using them, one is required to give an approximate starting value for each element or interval in which a particular element is found. The interval with element μ_i can be determined and presented in a graphical form.

Once we denote $y_1 = \text{ctg}\,\mu$ and $y_2 = \mu/Bi$, we can easily determine roots of (21), if we define intersection points of functions $y_1(\mu)$ and $y_2(\mu)$ first (Fig. 15.2).

One can see that a different set of roots corresponds to every value of Biot number Bi

$$\mu_1 < \mu_2 < \mu_3 < \ldots < \mu_n < \ldots$$

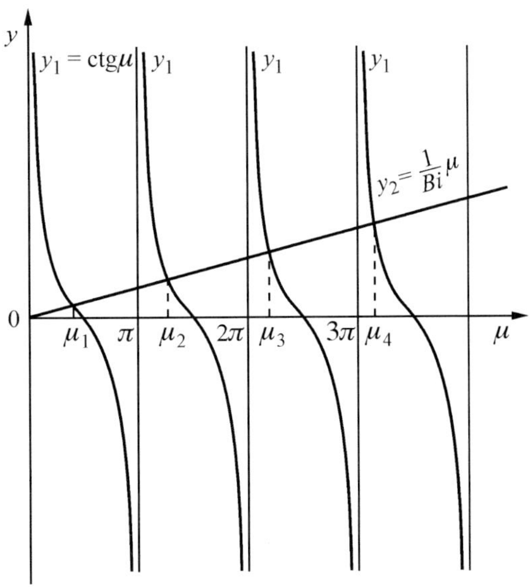

Fig. 15.2. Graphical determination of roots of characteristic equation (21): ctg $\mu = \mu/Bi$

For number $Bi \to \infty$ line $y_2 = \mu/Bi$ overlaps the x-axis and the roots of (21) are

$$\mu_1 = \frac{1}{2}\pi\,,\ \mu_2 = \frac{3}{2}\pi\,,\ \mu_3 = \frac{5}{2}\pi\,, \ldots,\ \mu_n = (2n-1)\pi/2, \quad n = 1, 2, 3\ldots .$$

For $Bi \to 0$ line $y_2 = \mu/Bi$ overlaps the y-axis and the roots of (21) are $\mu_1 = 0$, $\mu_2 = \pi$, $\mu_3 = 2\pi$ ... $\mu_n = (n-1)\pi$, $n = 1, 2, 3$... It is clear, therefore, that i-element lies in the interval

$$(i-1)\pi < \mu_i < \left(i - \frac{1}{2}\right)\pi, \qquad i = 1, 2, \ldots \tag{22}$$

Program in FORTRAN language for calculating roots of the characteristic equation (21) by means of interval halving method

```
C      Calculating roots of characteristic equation
       program p15_1
       dimension eigen(50)
       open(unit=1,file='p15_1.in')
       open(unit=2,file='p15_1.out')
       read(1,*)ne
```

```
       write(2,'(a)') "CALCULATING ROOTS OF CHARACTERISTIC
      &EQUATION"
       write(2,'(/a)') "DATA ENTERED"
       write(2,'(a,i10)')    "ne      =",ne
       write(2,'(/a,i3,a)') "CALCULATIONS OF PRIMARIES",ne,
      &" ROOTS OF EQUATION X*TAN(X)=BI"
       write(2,'(/a)')"CALCULATED EQUATION ROOTS"
       write(2,'(a,a)')"   Bi        mi1    mi2     mi3    ",
      &" mi4      mi5    mi6 "
       Bi=0.
       call equation_roots(Bi,ne,eigen)
       write(2,'(f7.3,5x,6f9.4)') Bi,(eigen(i),i=1,ne)
       Bi=0.001
       variable=0.001
       to k=1,5
          to j=1,9
             call equation_roots(Bi,ne,eigen)
             write(2,'(f7.3,5x,6f9.4)') Bi,(eigen(i),i=1,ne)
             Bi=Bi+variable
          enddo
       variable=variable*10.
       enddo
       call equation_roots(Bi,ne,eigen)
       write(2,'(f7.3,5x,6f9.4)') Bi,(eigen(i),i=1,ne)
       end program p15_1

c      procedure calculates roots of characteristic equation
c      x*tan(x)=Bi where Bi is Biot
c      number, ne the number of calcul. roots, eigen vector
c      with recorded calculated roots
       subroutine equation_roots(Bi,ne,eigen)
       dimension eigen(*)
       pi=3.141592654
       to i=1,ne
          xi=(float(i)-1.)*pi
          xf=pi*(float(i)-.5)
          to while (abs(xf-xi).ge.5.E-06)
             xm=(xi+xf)/2.
             y=xm*sin(xm)/cos(xm)-Bi
             if (y.lt.0.) then xi=xm
          else
                xf=xm
         endif
          enddo
          eigen(i)=xm
       enddo
       return
       end
```

The first six roots of (21) are presented in Table 15.1.
Every root μ_i corresponds to a single (19), which has the following form, if we take into account that $\mu_i = k_i L$:

$$\theta_i(x,t) = A_i \cos\left(\mu_i \frac{x}{L}\right) e^{-\mu_i^2 \frac{at}{L^2}}, \qquad i = 1,\ 2... \tag{23}$$

The obtained solution (23) satisfies differential equation (1) for any i, but does not satisfy the initial condition, since for $t = 0$ (23) has the form

$$\theta_i(x,0) = A_i \cos\left(\mu_i \frac{x}{L}\right), \qquad i = 1,\ 2\ ... \tag{24}$$

It is easy to satisfy the initial condition $T(x,0) = T_0(x)$, if we assume that the solution for temperature distribution is the sum of partial solutions (24)

$$\theta(x,t) = \sum_{n=1}^{\infty} A_n \cos\left(\mu_n \frac{x}{L}\right) e^{-\mu_n^2 \frac{at}{L^2}}. \tag{25}$$

Constants A_n will be determined from the initial condition. Once both sides of the (25) are multiplied, for $t = 0$, by $\cos\left(\lambda_m \frac{x}{L}\right)$ and then integrated in the interval $0 \le x \le L$, one obtains

$$\theta_0 \int_0^L \cos\left(\mu_m \frac{x}{L}\right) dx = \int_0^L \sum_{n=1}^{\infty} A_n \cos\left(\mu_n \frac{x}{L}\right) \cos\left(\mu_m \frac{x}{L}\right) dx. \tag{26}$$

Noting that

$$\int_0^L \cos\left(\mu_n \frac{x}{L}\right) \cos\left(\mu_m \frac{x}{L}\right) dx = 0, \quad \text{gdy} \quad n \ne m, \tag{27}$$

then from (26), one has

$$\theta_0 \int_0^L \cos\left(\mu_n \frac{x}{L}\right) dx = A_n \int_0^L \cos^2\left(\mu_n \frac{x}{L}\right) dx, \tag{28}$$

$$\theta_0 \frac{L}{\mu_n} \sin \mu_n = A_n \frac{L}{2}\left(1 + \frac{1}{2\mu_n} \sin 2\mu_n\right), \tag{29}$$

Table 15.1. Roots of characteristic equation ctg$\mu = \mu/Bi$

Bi	μ_1	μ_2	μ_3	μ_4	μ_5	μ_6
0.000	0.0000	3.1416	6.2832	9.4248	12.5664	15.7080
0.001	0.0316	3.1419	6.2833	9.4249	12.5665	15.7080
0.002	0.0447	3.1422	6.2835	9.4250	12.5665	15.7081
0.003	0.0547	3.1425	6.2837	9.4251	12.5666	15.7082
0.004	0.0632	3.1429	6.2838	9.4252	12.5667	15.7082
0.005	0.0706	3.1432	6.2840	9.4253	12.5668	15.7083
0.006	0.0774	3.1435	6.2841	9.4254	12.5668	15.7083
0.007	0.0836	3.1438	6.2843	9.4255	12.5669	15.7084
0.008	0.0893	3.1441	6.2845	9.4256	12.5670	15.7085
0.009	0.0947	3.1445	6.2846	9.4257	12.5671	15.7085
0.010	0.0998	3.1448	6.2848	9.4258	12.5672	15.7086
0.020	0.1409	3.1479	6.2864	9.4269	12.5680	15.7092
0.030	0.1723	3.1511	6.2880	9.4280	12.5688	15.7099
0.040	0.1987	3.1543	6.2895	9.4290	12.5696	15.7105
0.050	0.2218	3.1574	6.2911	9.4301	12.5703	15.7111
0.060	0.2425	3.1606	6.2927	9.4311	12.5711	15.7118
0.070	0.2615	3.1637	6.2943	9.4322	12.5719	15.7124
0.080	0.2791	3.1668	6.2959	9.4333	12.5727	15.7131
0.090	0.2956	3.1700	6.2975	9.4343	12.5735	15.7137
0.100	0.3111	3.1731	6.2991	9.4354	12.5743	15.7143
0.200	0.4328	3.2039	6.3148	9.4460	12.5823	15.7207
0.300	0.5218	3.2341	6.3305	9.4565	12.5902	15.7270
0.400	0.5932	3.2635	6.3461	9.4670	12.5981	15.7334
0.500	0.6533	3.2923	6.3616	9.4775	12.6060	15.7397
0.600	0.7051	3.3204	6.3770	9.4879	12.6139	15.7461
0.700	0.7506	3.3477	6.3923	9.4983	12.6218	15.7524
0.800	0.7910	3.3744	6.4074	9.5087	12.6296	15.7587
0.900	0.8274	3.4003	6.4224	9.5190	12.6375	15.7650
1.000	0.8603	3.4256	6.4373	9.5293	12.6453	15.7713
2.000	1.0769	3.6436	6.5783	9.6296	12.7223	15.8336
3.000	1.1925	3.8088	6.7040	9.7240	12.7967	15.8945
4.000	1.2646	3.9352	6.8140	9.8119	12.8678	15.9536
5.000	1.3138	4.0336	6.9096	9.8928	12.9352	16.0107
6.000	1.3496	4.1116	6.9924	9.9667	12.9988	16.0654
7.000	1.3766	4.1746	7.0640	10.0339	13.0584	16.1177
8.000	1.3978	4.2264	7.1263	10.0949	13.1141	16.1675
9.000	1.4149	4.2694	7.1806	10.1502	13.1660	16.2147
10.000	1.4289	4.3058	7.2281	10.2003	13.2142	16.2594
20.000	1.4961	4.4915	7.4954	10.5117	13.5420	16.5864
30.000	1.5202	4.5615	7.6057	10.6543	13.7085	16.7691
40.000	1.5325	4.5979	7.6647	10.7334	13.8048	16.8794
50.000	1.5400	4.6202	7.7012	10.7832	13.8666	16.9519
60.000	1.5451	4.6353	7.7259	10.8172	13.9094	17.0026
70.000	1.5487	4.6461	7.7438	10.8419	13.9406	17.0400
80.000	1.5514	4.6543	7.7573	10.8606	13.9644	17.0686
90.000	1.5535	4.6606	7.7679	10.8753	13.9830	17.0911
100.000	1.5552	4.6658	7.7764	10.8871	13.9981	17.1093

hence, one can easily determine constant A_n

$$A_n = \theta_0 \frac{2\sin\mu_n}{\mu_n + \sin\mu_n \cos\mu_n}. \tag{30}$$

By substituting (30) into (25), the expression that defines temperature distribution is obtained

$$\theta(x,t) = \theta_0 \sum_{n=1}^{\infty} \frac{2\sin\mu_n}{\mu_n + \sin\mu_n \cos\mu_n} \cos\left(\mu_n \frac{x}{L}\right) e^{-\mu_n^2 \frac{at}{L^2}}. \tag{31}$$

If we introduce a dimensionless coordinate $X = x/L$ and Fourier number $Fo = at/L^2$, then we can write the (31) in the following form

$$\frac{\theta}{\theta_0} = \frac{T(x,t) - T_{cz}}{T_0 - T_{cz}} = 2\sum_{n=1}^{\infty} \frac{\sin\mu_n \cos(\mu_n X)}{\mu_n + \sin\mu_n \cos\mu_n} e^{-\mu_n^2 Fo}. \tag{32}$$

When calculating thermal stresses and accumulated or given off energy during the processes of heating and cooling, respectively, one needs to know what the average temperature across the plate thickness is

$$\frac{\bar{\theta}}{\theta_0} = \frac{\bar{T} - T_{cz}}{T_0 - T_{cz}} = \frac{1}{L\theta_0} \int_0^L \theta dx, \tag{33}$$

where $\theta(x,t)$ is given by (32).

Once (31) is substituted into (33) and subsequently integrated, one has

$$\frac{\bar{\theta}}{\theta_0} = \sum_{n=1}^{\infty} \frac{2\sin^2\mu_n}{\mu_n(\mu_n + \sin\mu_n \cos\mu_n)} e^{-\mu_n^2 Fo}. \tag{34}$$

The case of an infinitely large heat transfer coefficient α

When heat transfer coefficient α is very large, the plate surface temperature is very close to the temperature of a medium. In the case when $\alpha \to \infty$, then $T|_{x=\pm L} = T_{cz}$. If the Biot number $Bi = \alpha L/\lambda > 100$, one can assume that the heat transfer coefficient is infinitely large. For $\alpha \to \infty$ the roots of characteristic equation (21) are (Fig. 15.1)

$$\mu_n = (2n-1)\frac{\pi}{2}, \qquad n = 1,\ 2,\ \ldots \tag{35}$$

Every root μ_i corresponds to a one (14), which, if we take into account that $\mu_i = k_i r_z$, has the following form:

$$\theta_i(r,t) = A_i J_0\left(\mu_i R\right) e^{-\mu_i^2 \frac{at}{r_z^2}}, \tag{24}$$

where $R = r/r_z$.

The program in FORTRAN language helps to calculate roots of the characteristic equation (18). The first six roots of (18) are given in Table 15.2.

Program for calculating roots of the characteristic equation $J_0(\mu)/J_1(\mu) = \mu / \mathrm{Bi}$, $Bi = \alpha r_z / \lambda$

```
C      Program for calculating roots of the charact. (18)
       program p15_4
       dimension eigen(50),zero(100)
c      file p15_4.in is attached to the program
       open(unit=1,file='p15_4.in')
       open(unit=2,file='p15_4.out')
       read(1,*)ne
       read(1,*)nc
       read(1,*)(zero(i),i=1,2*nc,2)
       read(1,*)(zero(i),i=2,2*nc,2)
       write(2,'(a)')
      & "CALCULATION OF ROOTS OF THE CHARACTERISTIC EQUATION"
       write(2,'(/a)') "INPUT DATA"
       write(2,'(a,i10)')    "ne       =",ne
       write(2,'(/a,i3,a)') "CALCULATION OF FIRST",ne,
      & " ROOTS OF EQUATION X*J1(X)/J0(X)=BI"
       write(2,'(/a)')"CALCULATED ROOTS"
       write(2,'(a,a)')"     Bi          mi1      mi2      mi3      ",
      & " mi4        mi5     mi6 "
       Bi=0.
       call equation_roots_cyl(Bi,zero,ne,eigen)
       write(2,'(f7.3,5x,6f9.4)') Bi,(eigen(i),i=1,ne)
       Bi=0.001
       zmienna=0.001
       do k=1,5
          do j=1,9
             call equation_roots_cyl(Bi,zero,ne,eigen)
             write(2,'(f7.3,5x,6f9.4)') Bi,(eigen(i),i=1,ne)
             Bi=Bi+zmienna
          enddo
       zmienna=zmienna*10.
       enddo
       call equation_roots_cyl(Bi,zero,ne,eigen)
```

```
      write(2,'(f7.3,5x,6f9.4)') Bi,(eigen(i),i=1,ne)
      end  program p15_4

c   procedure calculates roots of the characteristic equation
c   x*J1(x)/J0(x)=Bi where Bi is Biot number, ne is number of
c   calculated roots, eigen is output vector with calculated
c   roots zero() is output vector with funct. J1(x) and J0(x)
c   zeroing points see Fig.15.11
      subroutine equation_roots_cyl(Bi,zero,ne,eigen)
      dimension zero(*),eigen(*)
      pi=3.141592654
      j=0
      do i=1,ne
         xi=zero(i+j)
         xf=zero(i+1+j)
         do while (abs(xf-xi).ge.1.E-05)
            xm=(xi+xf)/2.
            y=xm*bessj1(xm)/bessj0(xm)-Bi
            if (y.lt.0.) then
               xi=xm
            else
               xf=xm
            endif
         enddo
         eigen(i)=xm
      j=j+1
      enddo
      return
      end

      FUNCTION bessj0(x)
      REAL bessj0,x
      REAL ax,xx,z
      DOUBLE PRECISION p1,p2,p3,p4,p5,q1,q2,q3,q4,q5,
     & r1,r2,r3,r4,r5,r6,s1,s2,s3,s4,s5,s6,y
      SAVE
p1,p2,p3,p4,p5,q1,q2,q3,q4,q5,r1,r2,r3,r4,r5,r6,s1,s2,
     & s3,s4,s5,s6
      DATA p1,p2,p3,p4,p5/1.d0,-.1098628627d-2,.2734510407d-4
     &,-.2073370639d-5,.2093887211d-6/, q1,q2,q3,q4,q5/-
     &.1562499995d-1,.1430488765d-3,-.6911147651d-5,
     &.7621095161d-6,-.934945152d-7/
      DATA r1,r2,r3,r4,r5,r6/57568490574.d0,-13362590354.d0,
     &651619640.7d0,-11214424.18d0,77392.33017d0,
     &-184.9052456d0/,s1,s2,s3,s4,s5,s6
     &/57568490411.d0,1029532985.d0,9494680.718d0,
     &59272.64853d0,267.8532712d0,1.d0/
```

```
 if(abs(x).lt.8.)then
   y=x**2
 bessj0=(r1+y*(r2+y*(r3+y*(r4+y*(r5+y*r6)))))/
&(s1+y*(s2+y*(s3+y*(s4+y*(s5+y*s6)))))
 else
   ax=abs(x)
   z=8./ax
   y=z**2
   xx=ax-.785398164
   bessj0=sqrt(.636619772/ax)*(cos(xx)*(p1+y*(p2+y*
&(p3+y*(p4+y*p5))))-
& z*sin(xx)*(q1+y*(q2+y*(q3+y*(q4+y*q5)))))
 endif
 return
 END

 FUNCTION bessj1(x)
 REAL bessj1,x
 REAL ax,xx,z
 DOUBLE PRECISION p5,q1,q2,q3,q4,q5,r1,r2,r3,r4,
& p1,p2,p3,p4,r5,r6,s1,s2,s3,s4,s5,s6,y
 SAVE p1,p2,p3,p4,p5,q1,q2,q3,q4,q5,r1,r2,r3,
& r4,r5,r6,s1,s2,s3,s4,s5,s6
 DATA r1,r2,r3,r4,r5,r6/72362614232.d0,-7895059235.d0,
&242396853.1d0,-2972611.439d0,15704.48260d0,
&-30.16036606d0/,s1,s2,s3,s4,s5,s6/144725228442.d0,
&2300535178.d0,18583304.74d0,99447.43394d0,
&376.9991397d0,1.d0/
 DATA p1,p2,p3,p4,p5/1.d0,.183105d-2,-.3516396496d-4,
&.2457520174d-5,-.240337019d-6/, q1,q2,q3,q4,q5
&/.04687499995d0,-.2002690873d-3,.8449199096d-5,
&-.88228987d-6,.105787412d-6/
 if(abs(x).lt.8.)then
   y=x**2
   bessj1=x*(r1+y*(r2+y*(r3+y*(r4+y*(r5+y*r6)))))/
&(s1+y*(s2+y*(s3+y*(s4+y*(s5+y*s6)))))
 else
   ax=abs(x)
   z=8./ax
   y=z**2
   xx=ax-2.356194491
   bessj1=sqrt(.636619772/ax)*(cos(xx)*
&(p1+y*(p2+y*(p3+y*
&(p4+y*p5))))-z*sin(xx)*(q1+y*(q2+y*(q3+y*(q4+y*q5)))))
&*sign(1.,x)
 endif
 return
 END
```

Table 15.2. Roots of the characteristic equation $J_0(\mu)/J_1(\mu) = \mu/Bi$, $Bi = \alpha\, r_z/\lambda$

Bi	μ_1	μ_2	μ_3	μ_4	μ_5	μ_6
0.000	0.0000	3.8317	7.0156	10.1735	13.3237	16.4706
0.001	0.0447	3.8320	7.0157	10.1736	13.3238	16.4707
0.002	0.0632	3.8322	7.0159	10.1737	13.3238	16.4708
0.003	0.0774	3.8325	7.0160	10.1738	13.3239	16.4708
0.004	0.0894	3.8327	7.0162	10.1739	13.3240	16.4709
0.005	0.0999	3.8330	7.0163	10.1740	13.3241	16.4709
0.006	0.1095	3.8333	7.0164	10.1741	13.3241	16.4710
0.007	0.1182	3.8335	7.0166	10.1742	13.3242	16.4711
0.008	0.1264	3.8338	7.0167	10.1743	13.3243	16.4711
0.009	0.1340	3.8341	7.0169	10.1744	13.3244	16.4712
0.010	0.1412	3.8343	7.0170	10.1745	13.3244	16.4712
0.020	0.1995	3.8369	7.0184	10.1754	13.3252	16.4718
0.030	0.2440	3.8395	7.0199	10.1764	13.3259	16.4725
0.040	0.2814	3.8421	7.0213	10.1774	13.3267	16.4731
0.050	0.3143	3.8447	7.0227	10.1784	13.3274	16.4737
0.060	0.3438	3.8473	7.0241	10.1794	13.3282	16.4743
0.070	0.3709	3.8499	7.0256	10.1803	13.3289	16.4749
0.080	0.3960	3.8525	7.0270	10.1813	13.3297	16.4755
0.090	0.4195	3.8551	7.0284	10.1823	13.3305	16.4761
0.100	0.4417	3.8577	7.0298	10.1833	13.3312	16.4767
0.200	0.6170	3.8835	7.0440	10.1931	13.3387	16.4828
0.300	0.7465	3.9091	7.0582	10.2029	13.3462	16.4888
0.400	0.8516	3.9344	7.0723	10.2127	13.3537	16.4949
0.500	0.9408	3.9594	7.0864	10.2225	13.3612	16.5009
0.600	1.0184	3.9841	7.1004	10.2322	13.3686	16.5070
0.700	1.0873	4.0085	7.1144	10.2419	13.3761	16.5130
0.800	1.1490	4.0325	7.1282	10.2516	13.3835	16.5191
0.900	1.2048	4.0562	7.1421	10.2613	13.3910	16.5251
1.000	1.2558	4.0795	7.1558	10.2710	13.3984	16.5312
2.000	1.5994	4.2910	7.2884	10.3658	13.4719	16.5910
3.000	1.7887	4.4634	7.4103	10.4566	13.5434	16.6498
4.000	1.9081	4.6019	7.5201	10.5423	13.6125	16.7073
5.000	1.9898	4.7131	7.6177	10.6223	13.6786	16.7630
6.000	2.0490	4.8033	7.7039	10.6964	13.7414	16.8168
7.000	2.0937	4.8772	7.7797	10.7646	13.8008	16.8684
8.000	2.1286	4.9384	7.8464	10.8271	13.8566	16.9179
9.000	2.1566	4.9897	7.9051	10.8842	13.9090	16.9650
10.000	2.1795	5.0332	7.9569	10.9363	13.9580	17.0099
20.000	2.2880	5.2568	8.2534	11.2677	14.2983	17.3442
30.000	2.3261	5.3410	8.3771	11.4222	14.4748	17.5348
40.000	2.3455	5.3846	8.4432	11.5081	14.5774	17.6508
50.000	2.3572	5.4112	8.4840	11.5621	14.6433	17.7272
60.000	2.3651	5.4291	8.5116	11.5990	14.6889	17.7807
70.000	2.3707	5.4419	8.5316	11.6258	14.7222	17.8201
80.000	2.3750	5.4516	8.5466	11.6461	14.7475	17.8502
90.000	2.3783	5.4592	8.5584	11.6620	14.7674	17.8739
100.000	2.3809	5.4652	8.5678	11.6747	14.7834	17.8931

If we account for the characteristic equation (18), we can write the (38) in the form

$$\frac{\bar{\theta}}{\theta_0}=\frac{\bar{T}(t)-T_{cz}}{T_0-T_{cz}}=\sum_{n=1}^{\infty}\frac{4Bi^2}{\mu_n^{\,2}\left(Bi^2+\mu_n^{\,2}\right)}e^{-\mu_n^{\,2}Fo}. \tag{39}$$

A case of an infinitely large heat transfer coefficient α

When heat transfer coefficient α is large, the surface temperature of a cylinder is close to the temperature of a medium. In the case when $\alpha \to \infty$, then $T|_{r=r_z}=T_{cz}$. The temperature of the cylinder is formulated in this case as

$$\frac{\theta}{\theta_0}=\sum_{n=1}^{\infty}\frac{2}{\mu_n J_1\left(\mu_n\right)}J_0\left(\mu_n\frac{r}{r_z}\right)e^{-\mu_n^{\,2}Fo}, \tag{40}$$

where μ_n, $n = 1, 2 \ldots$ are the roots of characteristic equation (22). Average temperature is formulated as

$$\frac{\bar{\theta}}{\theta_0}=\frac{2}{r_z^{\,2}\theta_0}\int_0^{r_z}\theta r dr=\sum_{n=1}^{\infty}\frac{4J_1\left(\mu_n\right)}{\mu_n^{\,2}J_1\left(\mu_n\right)}e^{-\mu_n^{\,2}Fo}=\sum_{n=1}^{\infty}\frac{4}{\mu_n^{\,2}}e^{-\mu_n^{\,2}Fo}, \tag{41}$$

where μ_n are the roots of $J_0(\mu) = 0$.

Equations (40) and (41) express plate temperature with boundary conditions of 1st kind, when surface temperature of the cylinder undergoes a step increase from initial temperature T_0 at the moment $t = 0$ to temperature T_{cz} for time $t > 0$.

Exercise 15.5 A Program for Calculating Temperature Distribution and Its Change Rate in an Infinitely Long Cylinder with Boundary Conditions of 3rd Kind

Write a program in FORTRAN language for calculating temperature distribution in an infinitely long cylinder, using the formulas derived in Ex. 15.4. The program should also allow to calculate average temperature and temperature change rate at any point in the cylinder. By means of the developed program, calculate temperature transient on the surface and in the center of the cylinder. Also calculate temperature change rate on the surface and in the center of the cylinder and average temperature transient in time. Assume the following values for the calculation: $r_z = 0.025$ m, $\lambda = 50$ W/(m·K), $a = 1\cdot10^{-5}$ m²/s, $\alpha = 2000$ W/(m²·K), $T_0 = 20$°C, $T_{cz} = 100$°C. Apply formulas derived in Ex. 15.4.

Solution

Temperature change rate is obtained once function $T(r,t)$, formulated in (35), Ex.15.4, is differentiated with respect to time. By accounting that $T(r,t) = \theta(r,t)(T_0 - T_{cz}) + T_{cz}$, one gets

$$\frac{dT(r,t)}{dr} = \frac{2a(T_{cz} - T_0)}{r_z^2} \sum_{n=1}^{\infty} \frac{\mu_n J_1(\mu_n) J_0(\mu_n R)}{J_0^2(\mu_n) + J_1^2(\mu_n)} e^{-\mu^2{}_n Fo} . \tag{1}$$

A program for calculating temperature distribution *T*(*r*,*t*), average temperature $\bar{T}(t)$ and temperature change rate *dT*(*r*,*t*)/*dt* in an infinitely long convectively heated or cooled cylinder

```
C   program for calculating temperature distribution, average
C   temperature and temp. change rate in an infinitely long
C   convectively heated or cooled cylinder
      program p15_5
      dimension eigen(50),zero(100)
      open(unit=1,file='p15_5.in')
      open(unit=2,file='p15_5.out')
      read(1,*)ne
      read(1,*)t_o,t_cz,s_rz,s_a
      read(1,*)s_lam,s_alfa
      read(1,*)nc
      read(1,*)(zero(i),i=1,2*nc,2)
      read(1,*)(zero(i),i=2,2*nc,2)
      write(2,'(a)')
     & " CALCULATING CYLINDER TEMPERATURE DISTRIBUTION "
      write(2,'(/a)') "INPUT DATA"
      write(2,'(a,i10)')      "ne     =",ne
      write(2,'(a,e10.5,a)') "t_o    =",t_o, " [C]"
      write(2,'(a,e10.5,a)') "t_cz   =",t_cz," [C]"
      write(2,'(a,e10.5,a)') "r_z    =",s_rz, " [m]"
      write(2,'(a,e10.5,a)') "a      =",s_a,  " [m^2/s]"
      write(2,'(a,e10.5,a)') "lambda =",s_lam," [W/mK]"
      write(2,'(a,e10.5,a)') "alfa   =",s_alfa," [W/m2K]"
      bi=s_alfa*s_rz/s_lam
      call equation_roots_cyl(bi,zero,ne,eigen)
      write(2,'(/a)')"CALCULATED TEMPERATURE [C]"
      write(2,'(a,a)')"  t[s]    T(0,t)     T(r_z,t)      T_sr(t)
     &      dT/dt(0,t)",        "       dT/dt(r_z,t)"
      t=0.
      do while (t.le.400.)
        write(2,'(f5.0,5(3x,e10.3))')t,
```

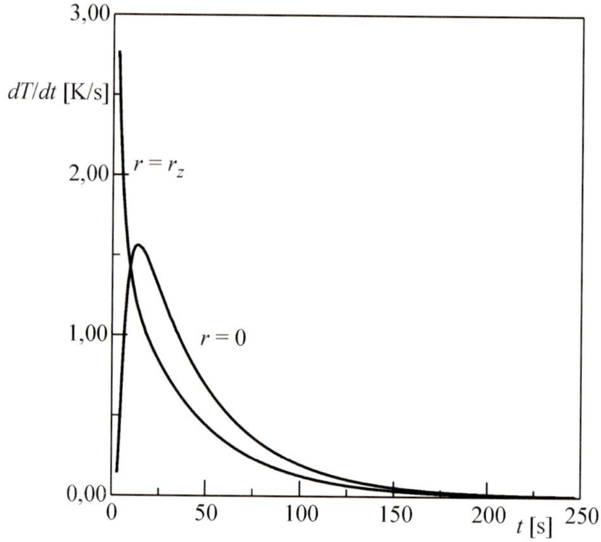

Fig. 15.13. Heating rate history $dT(r,t)/dt$ of an outer surface ($r = r_z$) and the center of a cylinder ($r = 0$) in the function of time

Exercise 15.6 Calculating Temperature in an Infinitely Long Cylinder using the Annexed Diagrams

By means of the annexed diagrams [1, 2] determine surface temperature in the center line of a steel-made infinitely long cylinder with an outer surface radius $r_z = 0.05$ m (Fig. 15.14). Also calculate the cylinder's average temperature. The surface of the cylinder, which has been heated to the initial temperature of $T_0 = 100°C$, is suddenly cooled by a compressed airflow at the temperature of $T_{cz} = 20°C$. Assume the following values for the calculation: $\alpha = 200$ W/(m^2·K), $\lambda = 50$ W/(m·K), $a = \lambda/c\rho = 1 \cdot 10^{-5}$ m^2/s, $t = 250$ s. Furthermore, determine temperature in the center line of the cylinder, if the heat transfer coefficient is infinitely large, i.e $\alpha \to \infty$. Also calculate the unknown temperatures by means of the program developed in Ex. 15.5, by accounting for 20 terms in the infinite series.

Solution

In order to find the appropriate temperatures in the diagrams given, the Biot and Fourier numbers have to be calculated first

$$Bi = \frac{\alpha r_z}{\lambda} = \frac{200 \cdot 0.05}{50} = 0.2 \ ,$$

$$Fo = \frac{at}{r_z^2} = \frac{1 \cdot 10^{-5} \cdot 250}{0.05} = 1.$$

From the diagram in Fig. 15.15, one has $\theta / \theta_0 = 0{,}66$, i.e.

$$\frac{T(r_z,t) - T_{cz}}{T_0 - T_{cz}} = \frac{T(r_z,t) - 20}{100 - 20} = 0.66,$$

hence, follows that $T(r_z,t)$ = 72.8°C. From diagram in Fig. 15.16, the dimensionless temperature can be determined in the center line of the cylinder $\theta(0, Fo)/\theta_0$

$$\frac{\theta(0, Fo)}{\theta_0} = \frac{T(0,t) - 20}{100 - 20} = 0.72 \, ,$$

hence, we obtain $T(0,t)$ = 77.6°C.

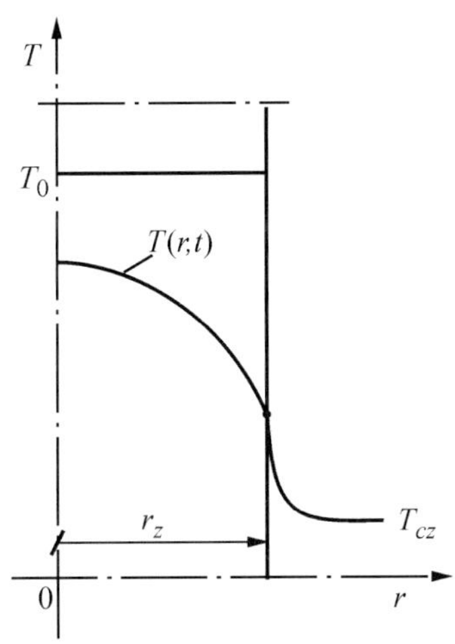

Fig. 15.14. The cooling process of a steel cylinder

Average temperature $\overline{T}(t)$ is read from the diagram in Fig. 15.17

$$\frac{\overline{\theta}}{\theta_0} = \frac{\overline{T}(t) - T_{cz}}{T_0 - T_{cz}} = 0.68 \ ,$$

hence, $\overline{T}(t) = (100 - 20) \cdot 0.68 + 20 = 74.4$°C.

When $\alpha \to \infty$, the dimensionless temperature in the center line of the cylinder cannot be read from the Fig. 15.18. As a result of the calculations conducted by means of the program from Ex.15.5, one has

$$\left.\begin{aligned} T(r_z,t) &= 72.0^{\circ}\mathrm{C} \\ T(0,t) &= 77.3^{\circ}\mathrm{C} \\ \overline{T}(t) &= 74.6^{\circ}\mathrm{C} \end{aligned}\right\} \quad \text{for} \quad \alpha = 200\,\mathrm{W}/\left(\mathrm{m}^2\cdot\mathrm{K}\right)$$

and

$$T(0,t) = 20.4^{\circ}\mathrm{C} \qquad \text{for} \quad \alpha \to \infty \quad \left(\alpha = 1\cdot 10^{6}\,\mathrm{W}/\left(\mathrm{m}^2\cdot\mathrm{K}\right)\right)$$

Similar results were obtained, with an exception of the case $\alpha \to \infty$. It is easy to determine temperature when diagrams are used, but such method is less accurate, since not every value of the Biot number has a corresponding curve in the diagrams from Figs. 15.15–15.18.

Appendix 15.2. Diagrams for the calculation of temperature in an infinitely long cylinder with a convective boundary condition

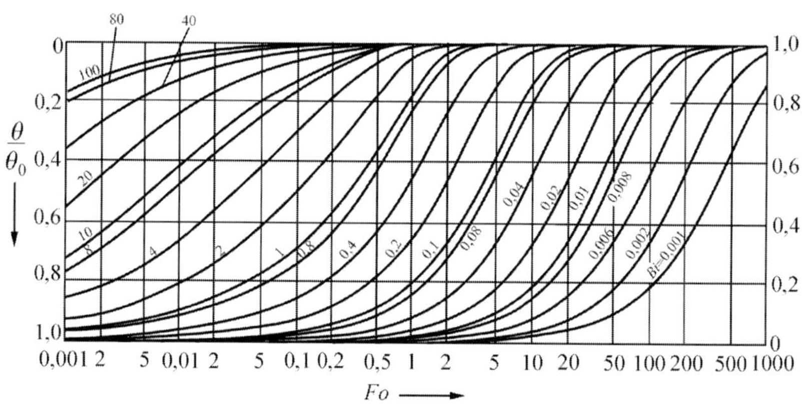

Fig. 15.15. A change in the dimensionless surface temperature of a cylinder $\theta(1, Fo)/\theta_0 = [T(r_z,t) - T_{cz}]/(T_0 - T_{cz})$ in the Fourier number function $Fo = at/r_z^2$, $Bi = \alpha r_z/\lambda$

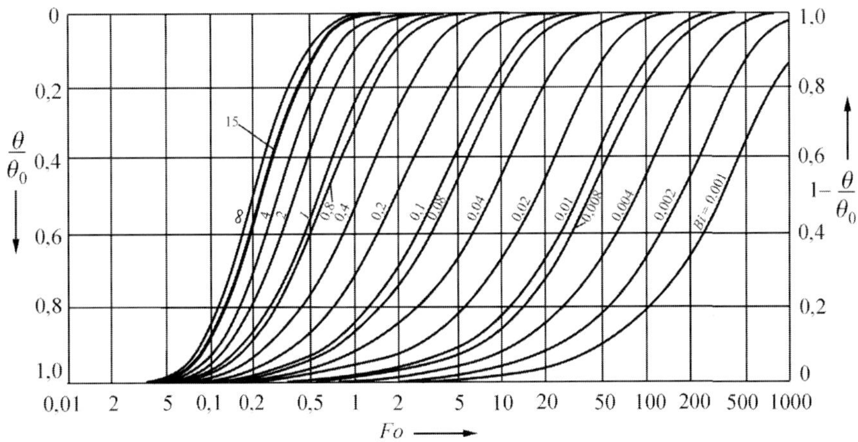

Fig. 15.16. Temperature changes in the center line of a cylinder $\theta(0,Fo)/\theta_0$ = $[T(0,t) - T_{cz}]/(T_0 - T_{cz})$ in the Fourier number function $Fo = at/r_z^2$, $Bi = \alpha r_z/\lambda$

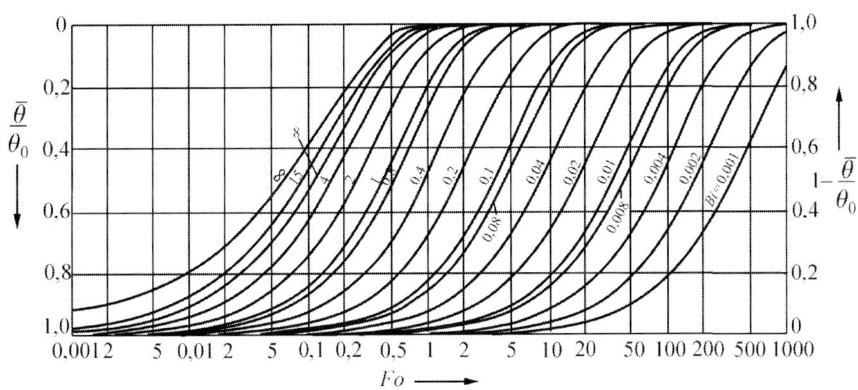

Fig. 15.17. Changes in average dimensionless temperature $\overline{\theta}/\theta_0 = [\overline{T}(Fo) - T_{cz}]/(T_0 - T_{cz})$ in the Fourier number function $Fo = at/r_z^2$, $Bi = \alpha r_z/\lambda$

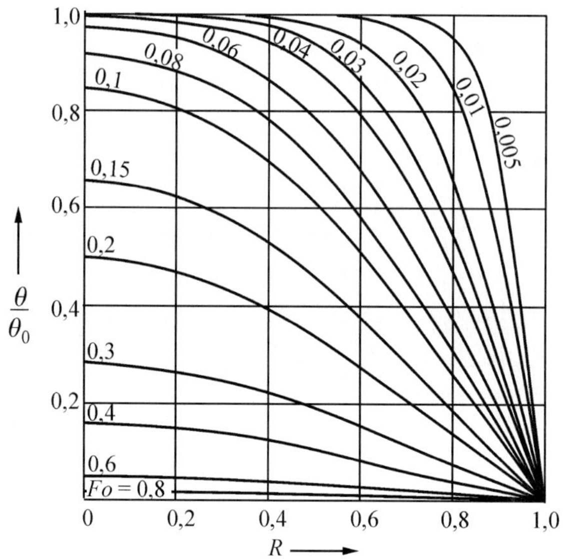

Fig. 15.18. Temperature changes in a cylinder $\theta/\theta_0 = [T(r,t) - T_{cz}]/(T_0 - T_{cz})$ in the radial coordinate function $R = r/r_z$ with a step change in the cylinder surface temperature from temperature T_0 to temperature T_{cz}

Exercise 15.7 Formula Derivation for a Temperature Distribution in a Sphere with Boundary Conditions of 3rd Kind

Derive a formula for temperature distribution in a sphere with the outer surface radius r_z and the convective boundary condition. Initial temperature of the sphere T_0 is uniform. Temperature of the medium increases step-like from the initial temperature T_0 to temperature T_{cz} (Fig. 15.19). Also derive a formula for the average temperature $\overline{T}(t)$. Consider a specific case, when $\alpha \to \infty$.

Solution

Temperature distribution is described by the heat conduction equation

$$\frac{\partial \theta}{\partial t} = a\left(\frac{\partial^2 \theta}{\partial r^2} + \frac{2}{r}\frac{\partial \theta}{\partial r}\right), \tag{1}$$

by boundary conditions

$$-\lambda \frac{\partial \theta}{\partial r}\bigg|_{r=r_z} = \alpha\, \theta\big|_{r=r_z} , \tag{2}$$

$$\frac{\partial \theta}{\partial r}\bigg|_{r=0} = 0 \tag{3}$$

and by initial condition

$$\theta\big|_{t=0} = \theta_0 , \tag{4}$$

where

$\theta = T - T_{cz}$, $\theta_0 = T_0 - T_{cz}$

If a new variable is introduced

$$\vartheta = \theta r , \tag{5}$$

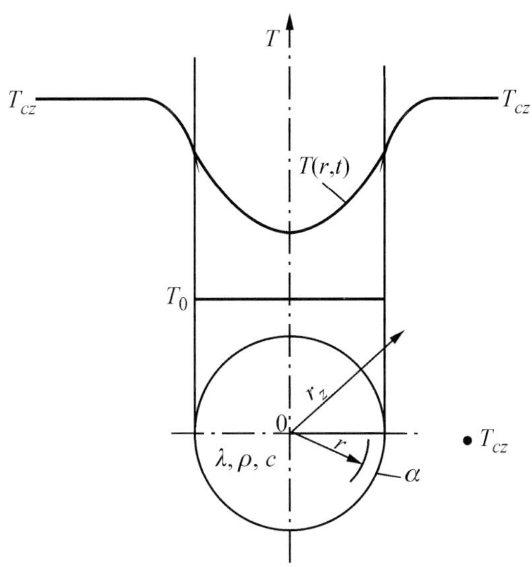

Fig. 15.19. Heating a sphere with a medium, whose temperature undergoes a step-change from initial temperature T_0 to temperature T_{cz}

the initial-boundary problem (1)–(4) can be written as follows:

$$\frac{\partial \vartheta}{\partial t} = a \frac{\partial^2 \vartheta}{\partial r^2}, \tag{6}$$

$$-\lambda \left. \frac{\partial \vartheta}{\partial r} \right|_{r=r_z} = \left(\alpha - \frac{\lambda}{r_z} \right) \vartheta \bigg|_{r=r_z}, \tag{7}$$

$$\vartheta|_{r=0} = 0, \tag{8}$$

$$\vartheta|_{t=0} = r\theta_0. \tag{9}$$

Once the separation of variables method is applied, as in the case of the flat wall (Ex. 15.1) the following solution of (6) is obtained that accounts for the boundary condition (8)

$$\vartheta = Ae^{-ak^2 t} \sin(kr). \tag{10}$$

By substituting (10) into (7), one obtains

$$\begin{aligned} -\lambda Ak \cos(k r_z) &= \left(\alpha - \frac{\lambda}{r_z} \right) A \sin(k r_z), \\ \operatorname{tg}(k r_z) &= -\frac{k r_z}{Bi - 1}. \end{aligned} \tag{11}$$

If notation $\mu = kr_z$ is introduced, characteristic equation (11) can be written in the form

$$\operatorname{tg}\mu = \frac{\mu}{1 - Bi}, \tag{12}$$

where $Bi = \alpha r_z / \lambda$.

The first six roots of (12) are presented in Table 15.3.

Graphical method for determining roots of the characteristic equation (12) is presented in Fig. 15.20. On the basis of this diagram, it is easy to determine the intervals, which contain roots of (12). If $(1 - Bi) > 0$, then n-th root of (12) lies in the interval

$$(n-1)\pi \le \mu_n \le \frac{\pi}{2} + (n-1)\pi, \quad n = 1, 2, \ldots, \quad \text{gdy } (1 - Bi) > 0. \tag{13}$$

Table 15.3. First six roots of the characteristic (12) tgμ = $\mu/(1 - Bi)$ determined by means of the program for Ex. 15.7

Bi	μ_1	μ_2	μ_3	μ_4	μ_5	μ_6
0.000	0.0000	4.4934	7.7253	10.9041	14.0662	17.2208
0.001	0.0548	4.4936	7.7254	10.9042	14.0663	17.2208
0.002	0.0774	4.4939	7.7255	10.9043	14.0663	17.2209
0.003	0.0948	4.4941	7.7256	10.9044	14.0664	17.2209
0.004	0.1095	4.4943	7.7258	10.9045	14.0665	17.2210
0.005	0.1224	4.4945	7.7259	10.9046	14.0666	17.2211
0.006	0.1341	4.4947	7.7260	10.9047	14.0666	17.2211
0.007	0.1448	4.4950	7.7262	10.9048	14.0667	17.2212
0.008	0.1548	4.4952	7.7263	10.9049	14.0668	17.2212
0.009	0.1642	4.4954	7.7264	10.9049	14.0668	17.2213
0.010	0.1730	4.4956	7.7265	10.9050	14.0669	17.2213
0.020	0.2445	4.4979	7.7278	10.9060	14.0676	17.2219
0.030	0.2991	4.5001	7.7291	10.9069	14.0683	17.2225
0.040	0.3450	4.5023	7.7304	10.9078	14.0690	17.2231
0.050	0.3854	4.5045	7.7317	10.9087	14.0697	17.2237
0.060	0.4217	4.5068	7.7330	10.9096	14.0705	17.2242
0.070	0.4551	4.5090	7.7343	10.9105	14.0712	17.2248
0.080	0.4860	4.5112	7.7356	10.9115	14.0719	17.2254
0.090	0.5150	4.5134	7.7369	10.9124	14.0726	17.2260
0.100	0.5423	4.5157	7.7382	10.9133	14.0733	17.2266
0.200	0.7593	4.5379	7.7511	10.9225	14.0804	17.2324
0.300	0.9208	4.5601	7.7641	10.9316	14.0875	17.2382
0.400	1.0528	4.5822	7.7770	10.9408	14.0946	17.2440
0.500	1.1656	4.6042	7.7899	10.9499	14.1017	17.2498
0.600	1.2644	4.6261	7.8028	10.9591	14.1088	17.2556
0.700	1.3525	4.6479	7.8156	10.9682	14.1159	17.2614
0.800	1.4320	4.6696	7.8284	10.9774	14.1230	17.2672
0.900	1.5044	4.6911	7.8412	10.9865	14.1301	17.2730
1.000	1.5708	4.7124	7.8540	10.9956	14.1372	17.2788
2.000	2.0288	4.9132	7.9787	11.0855	14.2074	17.3364
3.000	2.2889	5.0870	8.0962	11.1727	14.2764	17.3932
4.000	2.4556	5.2329	8.2045	11.2560	14.3433	17.4490
5.000	2.5704	5.3540	8.3029	11.3348	14.4080	17.5034
6.000	2.6537	5.4544	8.3913	11.4086	14.4699	17.5562
7.000	2.7165	5.5378	8.4703	11.4773	14.5288	17.6072
8.000	2.7654	5.6078	8.5406	11.5408	14.5846	17.6562
9.000	2.8044	5.6669	8.6031	11.5993	14.6374	17.7032
10.000	2.8363	5.7173	8.6587	11.6532	14.6869	17.7481
20.000	2.9857	5.9783	8.9831	12.0029	15.0384	18.0887
30.000	3.0372	6.0766	9.1201	12.1691	15.2245	18.2869
40.000	3.0632	6.1273	9.1933	12.2618	15.3334	18.4085
50.000	3.0788	6.1582	9.2384	12.3200	15.4034	18.4888
60.000	3.0893	6.1788	9.2690	12.3599	15.4518	18.5450
70.000	3.0967	6.1937	9.2909	12.3887	15.4872	18.5864
80.000	3.1023	6.2048	9.3075	12.4106	15.5140	18.6181
90.000	3.1067	6.2135	9.3204	12.4276	15.5352	18.6431
100.000	3.1102	6.2204	9.3308	12.4414	15.5521	18.6632

When $(1 - Bi) < 0$, then

$$\frac{\pi}{2}+(n-1)\pi \le \mu_n \le \pi+(n-1)\pi, \quad n=1,\ 2,\ ..., \quad \text{gdy } (1-Bi)<0. \tag{14}$$

If $Bi = 1$, then

$$\mu_n = (2n-1)\frac{\pi}{2}, \quad n=1,\ 2,\ ... \tag{15}$$

When $Bi \to \infty$, then

$$\mu_n = n\pi, \qquad n=1, 2, ... \tag{16}$$

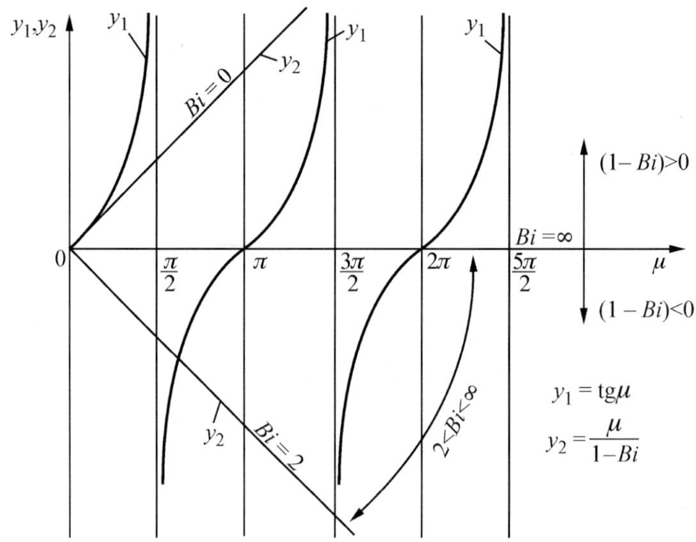

Fig. 15.20. Graphical method for determining roots of the characteristic equation $\text{tg}\mu = \mu/(1-Bi)$

Computational program in the FORTRAN language was developed to determine roots of the characteristic equation (12). The interval-halfing method was applied, using (13) and (14).

A program for calculating the first six roots of the characteristic equation (12): $\text{tg}\mu = \mu/(1 - Bi)$

```
C     program for calculating the first six roots
C     of the characteristic equation (12) tg(mi)=mi/(1-Bi)
      program p15_7
      dimension eigen(50)
      open(unit=1,file='p15_7.in')
```

```
      open(unit=2,file='p15_7.out')
      read(1,*)ne
      write(2,'(a)')
     &"CALCULATION OF ROOTS OF THE CHARACTERISTIC EQUATION"
      write(2,'(/a)') "INPUT DATA"
      write(2,'(a,i10)')   "ne      =",ne
      write(2,'(/a,i3,a)') "CALCULATION OF FIRST",ne,
     &      "ROOTS OF EQUATION X*COT(X)=1-BI"
      write(2,'(/a)')"CALCULATED ROOTS"
      write(2,'(a,a)')"    Bi         mi1    mi2    mi3     ",
     & " mi4       mi5    mi6 "
      Bi=0.

      call equation_roots_sph(Bi,ne,eigen)
      write(2,'(f7.3,5x,6f9.4)') Bi,(eigen(i),i=1,ne)
      Bi=0.001
      zmienna=0.001
      do k=1,5
         do j=1,9
            call equation_roots_sph(Bi,ne,eigen)
            write(2,'(f7.3,5x,6f9.4)') Bi,(eigen(i),i=1,ne)
            Bi=Bi+zmienna
         enddo
      zmienna=zmienna*10.
      enddo

      call equation_roots_sph(Bi,ne,eigen)
      write(2,'(f7.3,5x,6f9.4)') Bi,(eigen(i),i=1,ne)
      end  program p15_7

c     procedure calculates roots of the characteristic
c     equation x*cot(x)=1-Bi where Bi is Biot number,
c     ne is number of calculated roots, eigen is output
c     vector with calculated roots
      subroutine equation_roots_sph(Bi,ne,eigen)
      dimension eigen(*)
      pi=3.141592654
      if ((Bi.eq.1.).or.(Bi.gt.10000.)) then
       do i=1,ne
          if(Bi.eq.1.) eigen(i)=(2.*float(i)-1.)*pi/2.
          if(Bi.gt.10000.) eigen(i)=float(i)*pi
       enddo
      else
       h=1.-Bi
       if (h.lt.0.) then
          h1=pi/2.
          h2=pi
       else
```

```
         h1=0.
         h2=pi/2.
      endif
      do i=1,ne
         xi=h1+(float(i)-1.)*pi
         xf=h2+(float(i)-1.)*pi
         do while (abs(xf-xi).ge.5.E-06)
            xm=(xi+xf)/2.
            y=xm*cos(xm)/sin(xm)-h
            if (y.lt.0.) then
               xf=xm
            else
               xi=xm
            endif
         enddo
         eigen(i)=xm
     enddo
     endif
     return
   end
```

In order to satisfy the initial condition (5), the solution of (10) must be modified to a form

$$\vartheta = \sum_{n=1}^{\infty} A_n \sin\left(\mu_n \frac{r}{r_z}\right) e^{-\mu_n^2 \frac{at}{r_z^2}} . \tag{17}$$

By substituting expression (17) into (9), one obtains

$$r\left(T_0 - T_{cz}\right) = \sum_{n=1}^{\infty} A_n \sin\left(\mu_n \frac{r}{r_z}\right). \tag{18}$$

After multiplying (18) by $\sin\left(\mu_m \frac{r}{r_z}\right)$ and integrating it within the interval from $r = 0$ to $r = r_z$, one has

$$\left(T_0 - T_{cz}\right) \int_0^{r_z} r \sin\left(\mu_m \frac{r}{r_z}\right) dr = \sum_{n=1}^{\infty} A_n \int_0^{r_z} \sin\left(\mu_n \frac{r}{r_z}\right) \sin\left(\mu_m \frac{r}{r_z}\right) dr . \tag{19}$$

The integral on the right-hand-side of (19) is

$$\int_0^{r_z} \sin\left(\mu_n \frac{r}{r_z}\right) \sin\left(\mu_m \frac{r}{r_z}\right) dr = \frac{r_z\left(\mu_n \sin\mu_m \cos\mu_n - \mu_m \sin\mu_n \cos\mu_m\right)}{\mu_m^2 - \mu_n^2}, \tag{20}$$

while, after accounting for (12), it has the form

$$\int_0^{r_z} \sin\left(\mu_n \frac{r}{r_z}\right) \sin\left(\mu_m \frac{r}{r_z}\right) dr = \begin{cases} 0 \quad \text{dla} \quad m \neq n \\ \frac{r_z}{2\mu_n}\left(\mu_n - \sin\mu_n \cos\mu_n\right) \quad \text{dla} \quad m = n \end{cases} \tag{21}$$

Once we account for (21) in (19) and calculate the integral on the left-hand-side, we have the following from (19)

$$A_n = \left(T_0 - T_{cz}\right) \frac{2\left(\sin\mu_n - \mu_n \cos\mu_n\right)}{\mu_n - \sin\mu_n \cos\mu_n}. \tag{22}$$

If we substitute (22) into (17) and account for (5), then, after transformations, we obtain

$$\frac{\theta}{\theta_0} = \frac{T(r,t) - T_{cz}}{T_0 - T_{cz}} = \sum_{n=1}^{\infty} \frac{2\left(\sin\mu_n - \mu_n \cos\mu_n\right)}{\mu_n - \sin\mu_n \cos\mu_n} \cdot \frac{\sin\left(\mu_n R\right)}{\mu_n R} e^{-\mu_n^2 Fo}, \tag{23}$$

where $R = r/r_z$, $Fo = at/r_z^2$.

Dimensionless average temperature is defined as

$$\frac{\bar{\theta}}{\theta_0} = \frac{\bar{T}(t) - T_{cz}}{T_0 - T_{cz}} = \frac{3}{r_z^3} \int_0^{r_z} \frac{\theta}{\theta_0} r^2 dr = 3 \int_0^1 \frac{\theta}{\theta_0} R^2 dR. \tag{24}$$

Once we substitute (23) into (24) and account for

$$\int_0^1 R \sin\left(\mu_n R\right) dR = \frac{1}{\mu_n^2}\left(\sin\mu_n - \mu_n \cos\mu_n\right) \tag{25}$$

we have

$$\frac{\bar{\theta}}{\theta_0} = \sum_{n=1}^{\infty} \frac{6}{\mu_n^3} \frac{\left(\sin\mu_n - \mu_n \cos\mu_n\right)^2}{\mu_n - \sin\mu_n \cos\mu_n} e^{-\mu_n^2 Fo}. \tag{26}$$

The case of an infinitely large heat transfer coefficient α

When heat transfer coefficient α is large, then surface temperature of a sphere is close to a temperature of the medium, i.e. $T|_{r=r_z} = T_{cz}$. If the Biot number $Bi = \alpha r_z / \lambda > 100$, one can assume then that the heat transfer coefficient is infinitely large. For $\alpha \to \infty$, the roots of characteristic equation (12) are

$$\mu_n = n\pi, \qquad n = 1,\ 2,\ \ldots \tag{27}$$

By substituting the roots of (27) into (23), one has

$$\frac{\theta}{\theta_0}=\frac{T(r,t)-T_{cz}}{T_0-T_{cz}}=\sum_{n=1}^{\infty}2(-1)^{n+1}\frac{\sin(n\pi R)}{n\pi R}e^{-(n\pi)^2 Fo}, \tag{28}$$

where T_{cz} is, in the given case, a surface temperature of the sphere.

Average temperature

$$\frac{\overline{\theta}}{\theta_0}=\frac{\overline{T}(t)-T_{cz}}{T_0-T_{cz}}=3\int_0^1 R^2\frac{\theta}{\theta_0}dR \tag{29}$$

is given by

$$\frac{\overline{\theta}}{\theta_0}=\sum_{n=1}^{\infty}\frac{6}{(n\pi)^2}e^{-(n\pi)^2 Fo}. \tag{30}$$

Equations (29) and (30) express temperature of the sphere with the boundary condition of 1st kind, when surface temperature of the sphere increases step-like from initial temperature T_0 to temperature T_{cz}.

Exercise 15.8 A Program for Calculating Temperature Distribution and Its Change Rate in a Sphere with Boundary Conditions of 3rd Kind

Write a program in FORTRAN language for calculating temperature distribution in a sphere, using the formulas derived in Ex.15.7. The program should also allow to calculate average temperature and the temperature change rate at a random point in the sphere. Using the developed program, calculate temperature transient of the sphere surface and center. Also determine temperature change rate on the surface and in the center of the sphere and average temperature transient in time. Assume the following values for the calculation (steel quenching): r_z = 18.75 mm, $a = \lambda/c\rho$ = $1.0\cdot10^{-5}$ m^2/s, λ = 36 W/(m·K), α = 500 W/(m^2·K), T_0 = 850°C, T_{cz} = 40°C.

Solution

Temperature change rate is obtained by differentiating function $T(r,t)$ with respect to time, formulated by (23) in Ex. 15.7

$$\frac{dT(r,t)}{dt}=\frac{a(T_{cz}-T_0)}{r_z^2}\sum_{n=1}^{\infty}\frac{2\mu_n(\sin\mu_n-\mu_n\cos\mu_n)}{\mu_n-\sin\mu_n\cos\mu_n}\cdot\frac{\sin(\mu_n R)}{R}e^{-\mu_n^2 Fo} \tag{1}$$

where $R = r/r_z$, $Fo = at/r_z^2$.

A program for calculating temperature distribution *T*(*r*,*t*), average temperature $\overline{T}(t)$ and temperature change rate *dT*(*r*,*t*)/*dt* in the heated or convectively cooled sphere

```
C      program for calculating temperature distribution,
C      average temperature and temperature change rate
C      in the heated and convectively cooled sphere
       program p15_08
       dimension eigen(50)
       open(unit=1,file='p15_08.in')
       open(unit=2,file='p15_08.out')
       read(1,*)ne
       read(1,*)t_o,t_cz,s_rz,s_a
       read(1,*)s_lam,s_alfa
       write(2,'(a)')
      & "CALCULATING SPHERE TEMPERATURE DISTRIBUTION"
       write(2,'(/a)') "INPUT DATA"
       write(2,'(a,i10)')       "ne     =",ne
       write(2,'(a,e10.5,a)')   "t_o    =",t_o,  " [C]"
       write(2,'(a,e10.5,a)')   "t_cz   =",t_cz," [C]"
       write(2,'(a,e10.5,a)')   "r_z    =",s_rz," [m]"
       write(2,'(a,e10.5,a)')   "a      =",s_a,  " [m^2/s]"
       write(2,'(a,e10.5,a)')   "lambda =",s_lam," [W/mK]"
       write(2,'(a,e10.5,a)')   "alfa   =",s_alfa," [W/m2K]"
       bi=s_alfa*s_rz/s_lam
       write(*,*)bi
       call equation_roots_sph(bi,ne,eigen)
       write(*,*)eigen
       write(2,'(/a)')" CALCULATED TEMPERATURE [C]"
       write(2,'(a,a)')"  t[s]    T(0,t)     T(r_z,t)    T_sr(t)
      &     dT/dt(0,t)","     dT/dt(r_z,t)"
       t=0.
       do while (t.le.2000.)
          write(2,'(f5.0,5(3x,e10.3))')t,
      &        temperature(1.E-7,t,t_cz,t_o,s_rz,s_a,ne,eigen),
      &        temperature(s_rz,t,t_cz,t_o,s_rz,s_a,ne,eigen),
      &        temperature_sr(t,t_cz,t_o,s_rz,s_a,ne,eigen),
      &  temperature_szyb(1.E-7,t,t_cz,t_o,s_rz,s_a,ne,eigen),
      &  temperature_szyb(s_rz,t,t_cz,t_o,s_rz,s_a,ne,eigen)
          t=t+5.
       enddo
       end program p15_08

c      according to equation (1)
       function temperature_szyb(r,t,t_cz,t_o,s_rz,
      &s_a,ne,eigen)
       dimension eigen(*)
```

```
      teta=0.
      do i=1,ne
         s=eigen(i)
         teta=teta+s*(sin(s)-s*cos(s))*sin(s*r/s_rz)*
     &    exp(-s**2*s_a*t/s_rz**2)/(s-sin(s)*cos(s))/(r/s_rz)
      enddo
      temperature_szyb=2.*s_a*(t_cz-t_o)*teta/s_rz**2
      end function
c     according to equation (26) in 15.7
      function temperature_sr(t,t_cz,t_o,s_rz,s_a,ne,eigen)
      dimension eigen(*)
      teta=0.
      do i=1,ne
         s=eigen(i)
         teta=teta+(sin(s)-s*cos(s))**2*
     &    exp(-s**2*s_a*t/s_rz**2)/(s-sin(s)*cos(s))/(s**3)
      enddo
      temperature_sr=t_cz+(t_o-t_cz)*6.*teta
      end function
c     according to equation (23) in 15.7
      function temperature(r,t,t_cz,t_o,s_rz,s_a,ne,eigen)
      dimension eigen(*)
      teta=0.
      do i=1,ne
         s=eigen(i)
         teta=teta+(sin(s)-s*cos(s))*sin(s*r/s_rz)*exp(-s**2*
     &    s_a*t/s_rz**2)/(s-sin(s)*cos(s))/(s*r/s_rz)
      enddo
      temperature=t_cz+(t_o-t_cz)*2.*teta
      end function
c     procedure calculates roots of the characteristic
c     equation x*cot(x)=1-Bi where Bi is Biot number,
c     ne is number of calculated roots, eigen is
c     output vector with calculated roots
      subroutine equation_roots_sph(bi,ne,eigen)
      dimension eigen(*)
      pi=3.141592654
      if ((bi.eq.1.).or.(bi.gt.10000.)) then
      do i=1,ne
          if(bi.eq.1.) eigen(i)=(2.*float(i)-1.)*pi/2.
          if(bi.gt.10000.) eigen(i)=float(i)*pi
      enddo
      else
      h=1.-bi
       if (h.lt.0.) then
          h1=pi/2.
          h2=pi
```

```
else
    h1=0.
    h2=pi/2.
 endif
 do i=1,ne
    xi=h1+(float(i)-1.)*pi
    xf=h2+(float(i)-1.)*pi
    do while (abs(xf-xi).ge.5.E-06)
       xm=(xi+xf)/2.
       y=xm*cos(xm)/sin(xm)-h
       if (y.lt.0.) then
          xf=xm
       else
          xi=xm
    endif
    enddo
    eigen(i)=xm
 enddo
endif
return
end
```

Results from the calculation of temperature and temperature change rate are presented in Figs. 15.21 and 15.22.

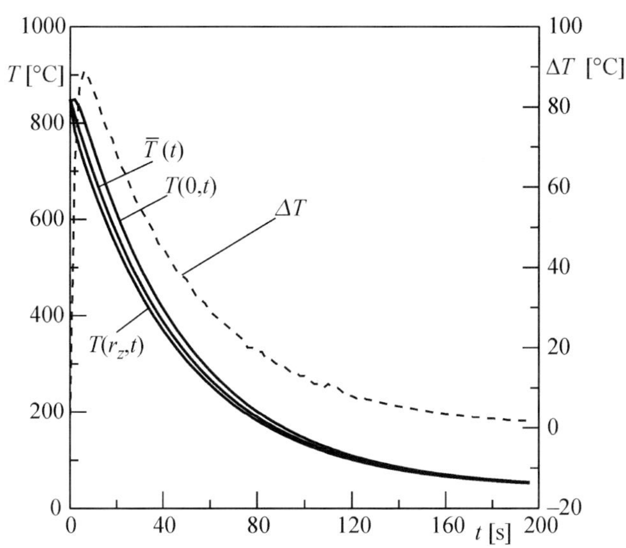

Fig. 15.21. Temperature transient of the sphere's outer surface $T(r_z,t)$, center $T(0,t)$, average temperature $\overline{T}(t)$ and temperature difference $\Delta T = T(0,t) - T(r_z,t)$ in the function of time (steel sphere quenching)

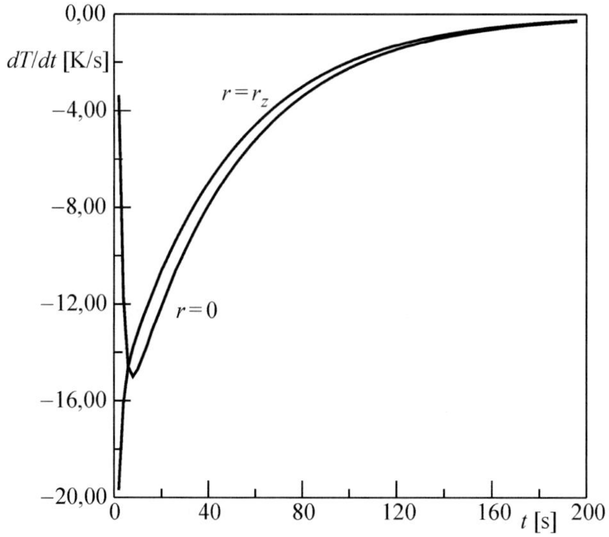

Fig. 15.22. Temperature change rate transient dT/dt on the outer surface of a sphere ($r = r_z$) and in the sphere center ($r = 0$) in the function of time (steel sphere quenching)

Exercise 15.9 Calculating Temperature of a Sphere using the Diagrams Provided

On the basis of the given diagrams [1, 2] determine temperature on an outer surface of a sphere and the temperature at its center (Fig. 15.23). Also calculate average temperature of the sphere. Outer surface of the sphere, heated to a uniform initial temperature of $T_0 = 850°C$ was suddenly cooled in a quenching bath with temperature of $T_{cz} = 40°C$. Assume the following values for the calculation: $r_z = 0.01875$ m, $\alpha = 384$ W/(m^2·K), $\lambda = 36$ W/(m·K), $a = \lambda/cp = 1\cdot10^{-5}$ m^2/s, $t = 35$ s. Also calculate the temperature of the sphere center for $t = 7$ s, if the heat transfer coefficient is infinitely large, i.e. $\alpha \rightarrow \infty$. In addition, determine the unknown temperatures by means of the program developed in Ex. 15.8.

Solution

In order to find the appropriate temperatures in diagrams provided, the Biot and Fourier numbers should be calculated first

$$Bi = \frac{\alpha r_z}{\lambda} = \frac{200\cdot 0.05}{50} = 0.2\,, \qquad Fo = \frac{at}{r_z^2} = \frac{1\cdot 10^{-5}\cdot 35}{0.01875^2} = 0.9956\,.$$

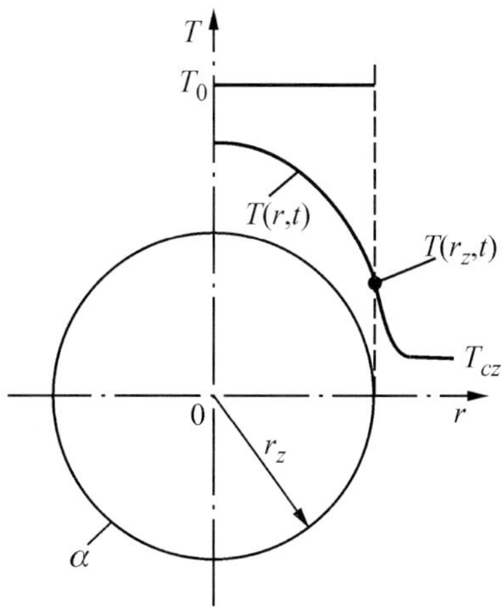

Fig. 15.23. The cooling of a steel sphere in a quenching bath

From the diagram in Fig. 15.24, one has

$$\theta = \frac{T(r_z,t) - T_{cz}}{T_0 - T_{cz}} = \frac{T(r_z,35\,\mathrm{s}) - 40}{850 - 40} = 0.54,$$

hence, $T(r_z,t)$ = 477.4°C. From Fig. 15.25, one can determine dimensionless temperature of the sphere center

$$\frac{\theta(0,Fo)}{\theta_0} = \frac{T(0,t) - 40}{850 - 40} = 0.6,$$

hence, one obtains $T(0,t) = (850 - 40) - 0.6 + 40 = 526$°C.

Dimensionless average temperature obtained from Fig. 15.26 is

$$\frac{\bar{\theta}(Fo)}{\theta_0} = \frac{\bar{T}(t) - T_{cz}}{T_0 - T_{cz}} = 0.58,$$

hence,

$$\bar{T}(t) = (850 - 40) \cdot 0.58 + 40 = 509.8\,^\circ\mathrm{C}.$$

In the case when $\alpha \to \infty$, one obtains from Fig. 15.27 for t = 7 s ($Fo = 0.199$)

$$\frac{\theta(0,Fo)}{\theta_0} = \frac{T(0,t) - T_{cz}}{T_0 - T_{cz}} = 0.28,$$

hence,

$$T(0,t) = (T_0 - T_{cz}) \cdot 0.28 + T_{cz} = (850 - 40) \cdot 0.28 + 40 = 266.8^{\circ}\mathrm{C}\,.$$

When $\alpha \to \infty$, the cooling of the sphere is very rapid.

By conducting calculations by means of the program from Ex. 15.8, one has

$$\left.\begin{aligned} T(r_z,t) &= 478.1^{\circ}\mathrm{C} \\ T(0,t) &= 523.2^{\circ}\mathrm{C} \\ \bar{T}(t) &= 496.0^{\circ}\mathrm{C} \end{aligned}\right\} \quad \text{for} \quad t = 35\ \mathrm{s} \quad \text{and} \quad \alpha = 384\ \mathrm{W/(m^2 \cdot K)}$$

and

$$T(0,t) = 266.6^{\circ}\mathrm{C} \qquad \text{for} \quad t = 7s \quad \text{and} \quad \alpha \to \infty\left(\alpha = 1 \cdot 10^7\ \mathrm{W}/\left(\mathrm{m}^2 \cdot \mathrm{K}\right)\right).$$

In both cases similar results are obtained. The diagrams help to determine temperature; the method, however, is less accurate, since not every value of the Biot number Bi has a corresponding curve in the diagram (Figs. 15.24–15.27)

Appendix 15.3. Diagrams for the calculation of temperature in a sphere with a convective boundary condition

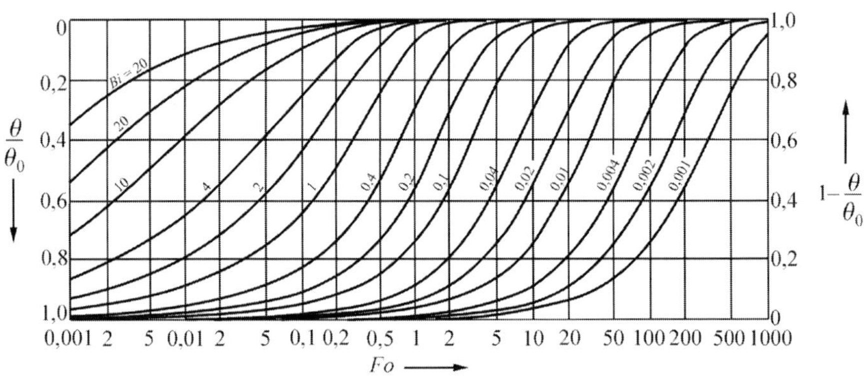

Fig. 15.24. Change in the dimensionless surface temperature of a sphere $\theta(1,F_0)/\theta_0 = [T(r_z,t) - T_{cz}]/(T_0 - T_{cz})$ in the Fourier number function $Fo = at/r_z^2$, $Bi = \alpha r_z/\lambda$

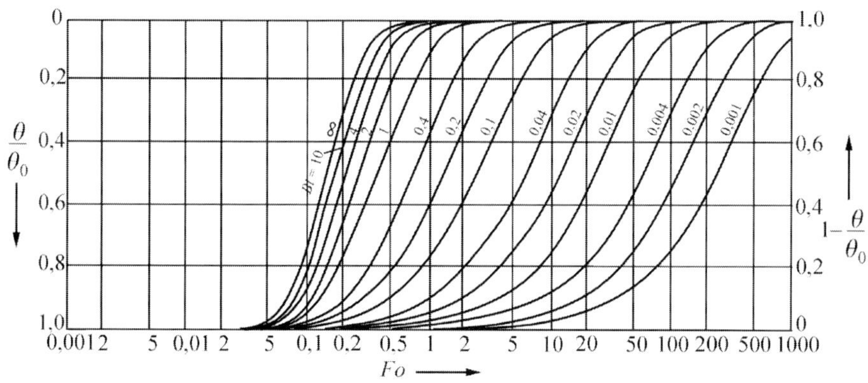

Fig. 15.25. Changes in the dimensionless temperature in the center of a sphere $\theta(0,F_0)/\theta_0 = [T(0,t) - T_{cz}]/(T_0 - T_{cz})$ in the Fourier number function $Fo = at/r_z^2$, $Bi = \alpha r_z/\lambda$

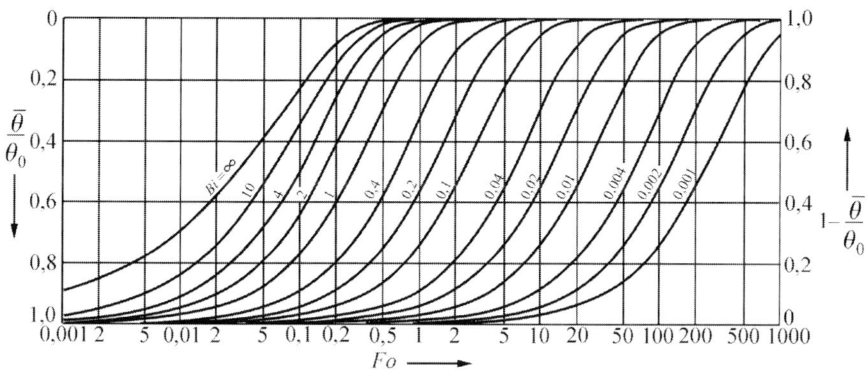

Fig. 15.26. Changes in the average dimensionless temperature $\overline{\theta}/\theta_0 = [\overline{T}(Fo) - T_{cz}]/(T_0 - T_{cz})$ in the Fourier number function $Fo = at/r_z^2$, $Bi = \alpha r_z/\lambda$

$$\bar{T} = \mathscr{L}\left[T(x,t)\right] = \int_0^\infty T(x,t)e^{-st}dt \,. \tag{8}$$

The solution to the boundary problem (5)–(7) is

$$\bar{T} = \frac{T_0}{s} + \frac{\dot{q}_s \cosh\left(\sqrt{\frac{s}{a}}x\right)}{\lambda s \sqrt{\frac{s}{a}} \sinh\left(\sqrt{\frac{s}{a}}L\right)} = \frac{T_0}{s} + \frac{g_1(s)}{g_2(s)} \tag{9}$$

or

$$\bar{T} = \frac{T_0}{s} + \frac{\dot{q}_s \cosh(qx)}{\lambda sq \sinh(qL)} = \frac{T_0}{s} + \frac{g_1(s)}{g_2(s)}, \tag{10}$$

where
$$q = \sqrt{s/a} \,. \tag{11}$$

Denominator is zero, when $s = 0$ or

$$\sinh(qL) = 0 \,. \tag{12}$$

Once notation $\mu = iqL$ is introduced and

$$\sinh(i\alpha) = i\sin\alpha \tag{13}$$

accounted for, (12) can be written in the form

$$\sinh\left(\frac{\mu_n}{i}\right) = 0 \quad \text{lub} \quad \sinh(-i\mu_n) = 0 \,. \tag{14}$$

If we account for the identity of (13), then we have from (14) the characteristic equation

$$\sin(\mu) = 0 \,. \tag{15}$$

The roots of (15) are

$$\mu_n = n\pi, \quad n = 0,\ 1,\ \ldots \tag{16}$$

By accounting that $\mu_n = iq_nL = i\sqrt{\frac{s_n}{a}}L$, one has

$$s_n = -\frac{\mu_n^2 a}{L^2}, \qquad n = 0,\ 1,\ \ldots \tag{17}$$

Therefore, there are a double pole $s = 0$ and single poles (17) for $n = 1$, 2, ... Temperature distribution $T(x,t)$ will be determined from (18) while accounting for (25) (both formulas are listed in Appendix H)

$$T(x,t)=\mathscr{L}^{-1}\left[\bar{T}(s)\right]=\frac{1}{2\pi i}\int_{\delta-i\infty}^{\delta+i\infty}\bar{T}(s)\,e^{st}ds=$$
$$=T_0+\frac{A}{D}t+\frac{B}{D}-\frac{AE}{D^2}+\sum_{n=1}^{\infty}\frac{g_1(s_n)}{g_2'(s_n)}e^{s_n t}. \tag{18}$$

In order to determine coefficients A, B, D and E, the transform

$$F(s)e^{st}=\left[\bar{T}(s)-\frac{T_0}{s}\right]e^{st}=\frac{\dot{q}_s\cosh(qx)}{\lambda sq\sinh(qL)}e^{st} \tag{19}$$

should be transformed by expanding function cosh(qx) and sinh(qL) and e^{st} into Taylor series

$$F(s)e^{st}=\frac{\dot{q}_s\left(1+\frac{q^2x^2}{2!}+\frac{q^4x^4}{4!}+...\right)(1+st+...)}{\lambda sq\left(qL+\frac{q^3L^3}{3!}+...\right)}=$$
$$=\frac{\dot{q}_s\left(1+\frac{sx^2}{2a}+\frac{s^2x^4}{24a^2}+...\right)(1+st+...)}{s^2\left(\frac{\lambda L}{a}+\frac{\lambda L^3}{6a^2}s+...\right)}. \tag{20}$$

The following expressions for constants (Appendix H) are obtained from (20).

$$A=\dot{q}_s,\quad B=\frac{\dot{q}_s x^2}{2a},\quad D=\frac{\lambda L}{a},\quad E=\frac{\lambda L^3}{6a^2}. \tag{21}$$

Next, we must determine the terms, which are present in (18) under the summation symbol. The numerator in (19) can be transformed into a form

$$g_1(s_n)=\dot{q}_s\cosh\left(\sqrt{\frac{s_n}{a}}x\right)=\dot{q}_s\cosh\left(\frac{\mu_n}{i}\frac{x}{L}\right)=\dot{q}_s\cosh\left(-i\mu_n\frac{x}{L}\right)=$$
$$=\dot{q}_s\cos\left(\mu_n\frac{x}{L}\right)=\dot{q}_s\cos\left(n\pi\frac{x}{L}\right). \tag{22}$$

During the transformation of (22), it has been accounted for that cosh($i\alpha$) = $i$cosα and cos($-\alpha$) = cos α. In order to calculate the derivative $g_2'(s_n)$, present in the numerator, one should determine the derivative dq/ds

$$\frac{dq}{ds}=\frac{d}{ds}\sqrt{\frac{s}{a}}=\frac{1}{\sqrt{a}}\frac{1}{2\sqrt{s}}=\frac{1}{2aq}, \tag{23}$$

hence follows that

$$s\frac{dq}{ds}=\frac{q}{2}. \tag{24}$$

Derivative $g_2'(s_n)$ is

$$\begin{aligned} g_2'(s_n)&=\left.\frac{dg_2}{ds}\right|_{s=s_n}=\frac{d}{ds}\left.\left[\lambda sq\sinh(qL)\right]\right|_{s=s_n}= \\ &=\left.\frac{\lambda d(sq)}{ds}\sinh(qL)\right|_{s=s_n}+\left.\lambda sq\cosh(qL)\frac{L}{2aq}\right|_{s=s_n}. \end{aligned} \tag{25}$$

By accounting for (15) and (16), one has

$$\sinh(qL)\big|_{s=s_n}=\sinh\left(\frac{\mu_n}{i}\right)=\sinh(-i\mu_n)=-i\sin(\mu_n)=0 \tag{26}$$

and

$$\begin{aligned} \cosh(qL)\big|_{s=s_n}&=\cosh\left(\frac{\mu_n}{i}\right)=\cosh(-i\mu_n)=\cos(\mu_n)= \\ &=\cos(n\pi)=-(-1)^{n+1}, \qquad n=1,\ 2,\ \ldots, \end{aligned} \tag{27}$$

while (25) assumes the form

$$g_2'(s_n)=-\frac{\lambda s_n}{2a}(-1)^{n+1}, \qquad n=1,\ 2,\ \ldots \tag{28}$$

By accounting for (16) and (17) in (28), one obtains

$$g_2'(s_n)=(-1)^{n+1}\frac{\lambda n^2\pi^2}{2L}, \qquad n=1,\ 2,\ \ldots \tag{29}$$

If (21), (22) and (29) are substituted into (18), then one obtains

$$\begin{aligned} T(x,t)&=T_0+\frac{\dot{q}_s L}{\lambda}\left[Fo+\frac{1}{2}\left(\frac{x}{L}\right)^2-\frac{1}{6}+\right. \\ &\left.+\sum_{n=1}^{\infty}(-1)^{n+1}\frac{2}{n^2\pi^2}\cos\left(n\pi\frac{x}{L}\right)e^{-n^2\pi^2 Fo}\right], \end{aligned} \tag{30}$$

where $Fo = at/L^2$.

Average temperature across the wall thickness is

$$\overline{T}_{sc} = \frac{1}{L}\int_0^L T(x,t)\,dt = T_0 + \frac{\dot{q}_s L}{\lambda} Fo. \qquad (31)$$

for dimensionless time $Fo \geq 0.5$, while (30) is simplified to a form

$$T(x,t) = T_0 + \frac{\dot{q}_s L}{\lambda}\left[Fo + \frac{1}{2}\left(\frac{x}{L}\right)^2 - \frac{1}{6}\right]. \qquad (32)$$

Exercise 15.11 A Program and Calculation Results for Temperature Distribution in a Plate with Boundary Conditions of 2nd Kind

Write a program for the calculation of temperature distribution in an infinite plate with thickness L heated in time $t > 0$ by a heat flow at constant density $\dot{q}_s$ (Fig. 15.29). The back surface of the plate ($x = L$) is thermally insulated. Material properties λ, ρ, c are constant, while initial temperature T_0 is uniform. Temperature distribution in the plate is expressed by (30), derived in Ex. 15.10. Present the calculation results in a tabular and graphical form.

Solution

A program for calculating temperature distribution in an infinite plate heated by a heat flow with density $\dot{q} = \dot{q}_s$

```
C      Calculation of temperature distribution in an infinite
C      plate heated by a heat flow on the front surface and
C      insulated on the back surface
       program p15_11
       open(unit=1,file='p15_11.in')
       open(unit=2,file='p15_11.out')
       read(1,*)ne
       write(2,'(a)')
     & "CALCULATION OF TEMPERATURE DISTRIBUTION IN A PLATE"
       write(2,'(/a)') "INPUT DATA"
       write(2,'(a,i10)')     "ne      =",ne
       write(2,'(/a)')"CALCULATED TEMPERATURE [C]"
       write(2,'(a,a)')
```

```
     &"   Fo       x/L=0         x/L=0,2        x/L=0,4       x/L=0,5",
     &     "       x/L=0,6         x/L=0,8         x/L=1,0"

      Fo=0.
      do while (Fo.lt.0.2)
         write(2,'(f5.2,7(3x,f10.6))')Fo,
     &         temperature_bezw(Fo,0.0,ne),
     &         temperature_bezw(Fo,0.2,ne),
     &         temperature_bezw(Fo,0.4,ne),
     &         temperature_bezw(Fo,0.5,ne),
     &         temperature_bezw(Fo,0.6,ne),
     &         temperature_bezw(Fo,0.8,ne),
     &         temperature_bezw(Fo,1.0,ne)
         Fo=Fo+.01
      enddo

      do while (Fo.le.1.2)
         write(2,'(f5.2,7(3x,f10.6))')Fo,
     &         temperature_bezw(Fo,0.0,ne),
     &         temperature_bezw(Fo,0.2,ne),
     &         temperature_bezw(Fo,0.4,ne),
     &         temperature_bezw(Fo,0.5,ne),
     &         temperature_bezw(Fo,0.6,ne),
     &         temperature_bezw(Fo,0.8,ne),
     &         temperature_bezw(Fo,1.0,ne)
         Fo=Fo+.05
      enddo
      end  program p15_11

c     according to equation (31) in 15.10
      function temperature_bezw(Fo,x_przez_L,ne)
      pi=3.141592654
      xpL=x_przez_L
      teta=0.
      do n=1,ne
      teta=teta+(-1.)**(n+1)*2.*cos(n*pi*xpL)*
     &exp(-n**2*pi**2*Fo)/
     &     n**2/pi**2
      enddo

      temperature_bezw=teta+Fo+0.5*xpL**2-1./6.
      end function
```

Plate temperature distribution for selected coordinates x/L is presented in Fig. 15.29 and Table 15.4.

Table 15.4. Dimensionless temperature $[T(x,t) - T_0]/(\dot{q}_S L/\lambda)$ for the selected values of the dimensionless coordinate $X = x/L$ and the Fourier number $Fo = at/L^2$; the plate is being heated on the front face $x = L$, while the back surface $x = 0$ is thermally insulated

Fo	$x/L = 0$	$x/L = 0.2$	$x/L = 0.4$	$x/L = 0.5$	$x/L = 0.6$	$x/L = 0.8$	$x/L = 1.0$
0.00	0.000000	0.000000	0.000000	0.000000	0.000000	0.000000	0.000203
0.01	0.000000	0.000000	0.000001	0.000014	0.000196	0.010051	0.112838
0.02	0.000000	0.000003	0.000153	0.000802	0.003396	0.033326	0.159577
0.03	0.000005	0.000071	0.001147	0.003722	0.010530	0.057196	0.195441
0.04	0.000057	0.000393	0.003449	0.008754	0.020102	0.079856	0.225676
0.05	0.000269	0.001166	0.007040	0.015366	0.031010	0.101159	0.252313
0.06	0.000786	0.002515	0.011718	0.023074	0.042621	0.121223	0.276395
0.07	0.001735	0.004489	0.017263	0.031528	0.054573	0.140203	0.298541
0.08	0.003207	0.007098	0.023492	0.040486	0.066658	0.158238	0.319154
0.09	0.005251	0.010323	0.030261	0.049784	0.078753	0.175447	0.338514
0.10	0.007885	0.014132	0.037461	0.059311	0.090788	0.191930	0.356826
0.11	0.011104	0.018489	0.045011	0.068992	0.102722	0.207771	0.374245
0.12	0.014887	0.023353	0.052850	0.078777	0.114535	0.223039	0.390892
0.13	0.019205	0.028684	0.060933	0.088632	0.126218	0.237798	0.406863
0.14	0.024023	0.034443	0.069223	0.098535	0.137770	0.252099	0.422240
0.15	0.029306	0.040594	0.077692	0.108469	0.149195	0.265989	0.437089
0.16	0.035017	0.047102	0.086317	0.118425	0.160498	0.279508	0.451466
0.17	0.041121	0.053935	0.095079	0.128395	0.171687	0.292694	0.465422
0.18	0.047584	0.061063	0.103964	0.138375	0.182770	0.305578	0.479000
0.19	0.054375	0.068460	0.112957	0.148361	0.193755	0.318189	0.492236
0.20	0.061464	0.076101	0.122047	0.158352	0.204650	0.330554	0.505165
0.25	0.100516	0.117236	0.168646	0.208336	0.258025	0.389430	0.566146
0.30	0.143824	0.161821	0.216576	0.258334	0.310092	0.444845	0.622842
0.35	0.189738	0.208515	0.265313	0.308333	0.361354	0.498152	0.676928
0.40	0.237244	0.256497	0.314542	0.358333	0.412125	0.550170	0.729423
0.45	0.285721	0.305265	0.364071	0.408333	0.462596	0.601402	0.780946
0.50	0.334791	0.354512	0.413784	0.458333	0.512883	0.652154	0.831876
0.55	0.384223	0.404053	0.463608	0.508333	0.563058	0.702614	0.882444
0.60	0.433877	0.453773	0.513501	0.558333	0.613166	0.752894	0.932790
0.65	0.483665	0.503602	0.563436	0.608333	0.663231	0.803065	0.983002
0.70	0.533536	0.553497	0.613396	0.658333	0.713271	0.853170	1.033131
0.75	0.583457	0.603433	0.663372	0.708333	0.763295	0.903233	1.083210
0.80	0.633409	0.653395	0.713357	0.758333	0.813310	0.953272	1.133258
0.85	0.683380	0.703371	0.763348	0.808333	0.863319	1.003296	1.183287
0.90	0.733362	0.753356	0.813342	0.858333	0.913325	1.053311	1.233305
0.95	0.783351	0.803347	0.863339	0.908333	0.963328	1.103320	1.283316
1.00	0.833344	0.853342	0.913337	0.958333	1.013330	1.153325	1.333323
1.05	0.883340	0.903339	0.963335	1.008333	1.063331	1.203328	1.383327
1.10	0.933337	0.953337	1.013335	1.058333	1.113332	1.253330	1.433329
1.15	0.983336	1.003335	1.063334	1.108333	1.163333	1.303331	1.483331
1.20	1.033335	1.053334	1.113334	1.158333	1.213333	1.353332	1.533332

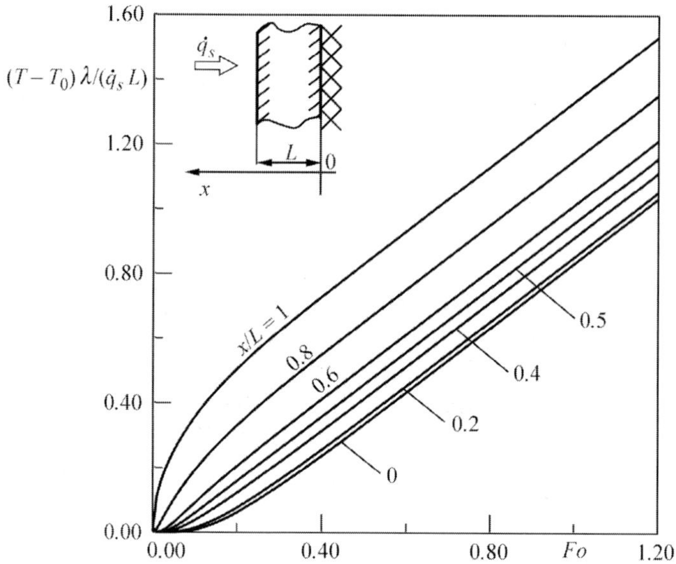

Fig. 15.29. Temperature distribution along the thickness of an infinitely long plate heated on the front face ($x = L$) and thermally insulated on the back surface ($x = 0$)

Exercise 15.12 Formula Derivation for Temperature Distribution in an Infinitely Long Cylinder with Boundary Conditions of 2nd Kind

Determine a formula for temperature distribution in an infinitely long cylinder with an outer surface radius r_z heated by a heat flow at constant density $\dot{q}_s$ (Fig. 15.30). Thermo-physical properties ρ, λ, c are constant. Initial temperature of the plate is uniform and is T_0. Apply the Laplace transform to solve the initial-boundary problem.

Solution

Temperature distribution in the cylinder is governed by the heat conduction equation

$$a\left(\frac{\partial^2 T}{\partial r^2}+\frac{1}{r}\frac{\partial T}{\partial r}\right)=\frac{\partial T}{\partial t}, \tag{1}$$

by boundary conditions

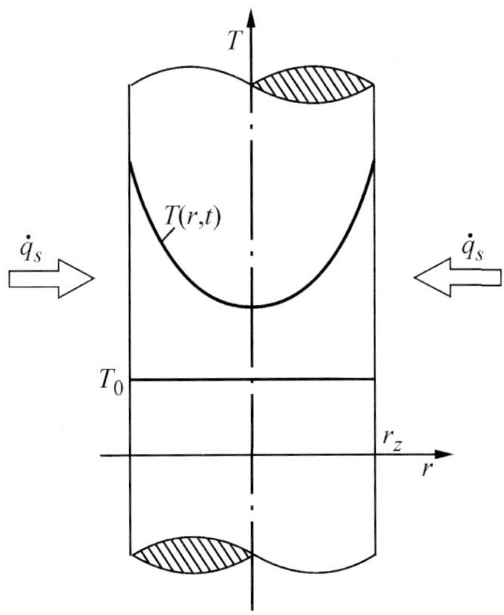

Fig. 15.30. Diagram that depicts heating of an infinitely long cylinder by a heat flow with constant density $\dot{q}_s$

$$\left.\frac{\partial T}{\partial r}\right|_{r=0} = 0, \tag{2}$$

$$\lambda \left.\frac{\partial T}{\partial r}\right|_{r=r_z} = \dot{q}_s \tag{3}$$

and by initial condition

$$T\big|_{t=0} = T_0. \tag{4}$$

Once Laplace transform is applied to (1)–(3) while accounting for the initial condition (4), one has

$$a\left(\frac{d^2\overline{T}}{dr^2} + \frac{1}{r}\frac{d\overline{T}}{dr}\right) - s\overline{T} = T_0, \tag{5}$$

$$\left.\frac{d\overline{T}}{dr}\right|_{r=0} = 0, \tag{6}$$

$$\lambda \left.\frac{d\overline{T}}{dr}\right|_{r=r_z} = \frac{\dot{q}_s}{s}, \tag{7}$$

where

$$\bar{T} = \bar{T}(x,s) = \mathscr{L}[T(x,t)] = \int_0^t T(x,t)e^{-st}dt . \tag{8}$$

The solution of the boundary problem (5)–(7) is

$$\bar{T} - \frac{T_0}{s} = \frac{\dot{q}_s I_0(qr)}{\lambda q s I_1(qr_z)} = \frac{g_1(s)}{g_2(s)}, \tag{9}$$

where $q = \sqrt{s/a}$.

While deriving (9), the following formula for the derivative is used

$$\frac{d}{dr}\left[I_0(qr)\right] = qI_1(qr). \tag{10}$$

Next, accounting that

$$I_0(qr) = J_0(iqr) \qquad \text{and} \qquad I_1(qr_z) = \frac{1}{i}J_1(ir_z) \tag{11}$$

Equation (9) assumes the form

$$\bar{T} - \frac{T_0}{s} = i\frac{\dot{q}_s J_0(iqr)}{\lambda q s J_1(iqr_z)}. \tag{12}$$

By substituting

$$\mu = iqr_z \tag{13}$$

the following characteristic equation is obtained

$$J_1(\mu) = 0 , \tag{14}$$

whose roots are μ_1 = 3.8317; μ_2 = 7.0156; μ_3 = 10.1735; μ_4 = 13.3237; μ_5 = 16.4706; μ_6 = 19.6159. The subsequent zeros of the Bessel function $J_1(\mu)$ can be found in tables for special functions, e.g. in [5]. Once functions $I_0(qr)$ and $I_1(qr_z)$ are expanded into Taylor series, (12) has the form

$$\bar{T} - \frac{T_0}{s} = \frac{\dot{q}_s}{\lambda} \frac{1 + \frac{q^2r^2}{4} + \frac{q^4r^4}{64} + \ldots}{sq\left(\frac{qr_z}{2} + \frac{q^3r_z^3}{16} + \frac{q^5r_z^5}{384} + \ldots\right)}, \tag{15}$$

hence, after transformations

$$\overline{T}-\frac{T_0}{s}=\frac{2\dot{q}_s a}{\lambda r_z}\frac{1+\dfrac{sr^2}{4a}+\dfrac{s^2r^4}{64a^2}+\ldots}{s^2\left(1+\dfrac{sr_z^2}{8a}+\dfrac{s^2r_z^4}{192a^2}+\ldots\right)}. \tag{16}$$

A double pole exists in $s = 0$, while single poles in $s_n = -\mu_n^2 a/r_z^2$, $n = 1, 2 \ldots$

Temperature distribution is obtained using the inverse Laplace tranformation (Appendix H). From the analysis of (16), it follows that

$$A=\frac{2\dot{q}_s a}{\lambda r_z},\quad B=\frac{2\dot{q}_s a}{\lambda r_z}\cdot\frac{r^2}{4a},\quad D=1,\quad E=\frac{r_z^2}{8a}. \tag{17}$$

Temperature distribution $T(r,t)$ is determined form formula (Appendix H)

$$\begin{aligned}T(r,t)&=\mathscr{L}^{-1}\left[\overline{T}(x,s)\right]=\frac{1}{2\pi i}\int_{\delta-i\infty}^{\delta+i\infty}\overline{T}(r,s)e^{st}ds=\\&=T_0+\frac{A}{D}t+\frac{B}{D}-\frac{AE}{D^2}+\sum_{n=1}^{\infty}\frac{g_1(s_n)}{g_2'(s_n)}e^{s_nt}.\end{aligned} \tag{18}$$

Derivative $g_2'(s_n)$ can be calculated from dependence

$$g_2'(s_n)=\left.\frac{dg_2}{ds}\right|_{s=s_n}=\left.\frac{d}{ds}(\lambda qs)I_1(qr_z)\right|_{s=s_n}+\left[\lambda qs\frac{dI_1(qr_z)}{ds}\right]\Bigg|_{s=s_n}. \tag{19}$$

By accounting, next, that the first term on the right-hand-side equals zero, due to the fact that the characteristic equation $I_1(q_n r_z) = J_1(iq_n r_z) = J_1(\mu_n) = 0$, the derivative $g_2'(s_n)$ can be expressed in the following way:

$$\begin{aligned}g_2'(s_n)&=\left[\lambda qs\frac{dI_1(qr_z)}{ds}\right]\Bigg|_{s=s_n}=\lambda qs\left[I_0(qr_z)-\frac{1}{qr_z}I_1(qr_z)\right]\frac{r_z}{2aq}\Bigg|_{s=s_n}=\\&=\frac{\lambda s_n r_z}{2a}I_0(q_n r_z)=-\frac{\lambda\mu_n^2}{2r_z}J_0(iq_n r_z)=-\frac{\lambda\mu_n^2}{2r_z}J_0(\mu_n).\end{aligned} \tag{20}$$

The numerator $g_1(s_n)$ can be transformed in the following way:

$$g_1(s_n)=\dot{q}_s I_0(q_n r)=\dot{q}_s J_0(iq_n r)=\dot{q}_s J_0\left(\mu_n\frac{r}{r_z}\right). \tag{21}$$

By substituting (17), (20) and (21) into (18), one has

$$T(r,t)=T_0+\frac{2\dot{q}_s a}{\lambda r_z}\left(t+\frac{r^2}{4a}-\frac{r_z^2}{8a}\right)-\sum_{n=1}^{\infty}\frac{2\dot{q}_s r_z}{\lambda\mu_n^2}\frac{J_0\left(\mu_n\frac{r}{r_z}\right)}{J_0(\mu_n)}e^{-\mu_n^2\frac{at}{r_z^2}}. \tag{22}$$

If dimensionless quantities are introduced

$$R=\frac{r}{r_z} \quad \text{i} \quad Fo=\frac{at}{r_z^2}, \tag{23}$$

temperature distribution $T(r,t)$ can be expressed in the following way:

$$T(r,t)=T_0+\frac{\dot{q}_s r_z}{\lambda}\left[2Fo+\frac{1}{2}R^2-\frac{1}{4}-2\sum_{n=1}^{\infty}\frac{J_0(\mu_n R)}{\mu_n^2 J_0(\mu_n)}e^{-\mu_n^2 Fo}\right]. \tag{24}$$

For the Fourier number $Fo > 0.5$, one can neglect the infinite series in (24).

Exercise 15.13 Program and Calculation Results for Temperature Distribution in an Infinitely Long Cylinder with Boundary Conditions of 2nd Kind

Write a program for the calculation of temperature distribution in an infinitely long cylinder with an outer surface radius r_z heated by a heat flow with constant density $\dot{q}_s$ (Fig. 15.31). Thermo-physical properties λ, ρ, c are constant. Carry out the calculations by means of (24) derived in Ex. 15.12. Present calculation results in a tabular and graphical form

Solution

A program for the calculation of temperature distribution in a infinitely long cylinder heated by a heat flow with constant density $\dot{q}_s$

```
C     Calculation of temperature distribution in a long
C     cylinder heated by a heat flow with constant q_s
      program p15_13
      dimension zero(50)
      open(unit=1,file='p15_13.in')
      open(unit=2,file='p15_13.out')
      read(1,*)ne
      read(1,*)nc
      read(1,*)(zero(i),i=1,nc)
      write(2,'(a)')"CALCULATION OF TEMPERATURE IN CYLINDER"
      write(2,'(/a)') "INPUT DATA"
```

```
      write(2,'(a,i10)')     "ne      =",ne
      write(2,'(/a)')"CALCULATED TEMPERATURE [C]"
      write(2,'(a,a)')
     &"  Fo     r/r_z=0     r/r_z=0,2    r/r_z=0,4    r/r_z=0,5",
     &"    r/r_z=0,6     r/r_z=0,8     r/r_z=1,0"
      Fo=0.
      do while (Fo.lt.0.2)
         write(2,'(f5.2,7(3x,f10.5))')Fo,
     &      temperature_bezw_cyl(zero,Fo,0.0,ne),
     &      temperature_bezw_cyl(zero,Fo,0.2,ne),
     &      temperature_bezw_cyl(zero,Fo,0.4,ne),
     &      temperature_bezw_cyl(zero,Fo,0.5,ne),
     &      temperature_bezw_cyl(zero,Fo,0.6,ne),
     &      temperature_bezw_cyl(zero,Fo,0.8,ne),
     &      temperature_bezw_cyl(zero,Fo,1.0,ne)
          Fo=Fo+.01
      enddo
      do while (Fo.le.1.2)
         write(2,'(f5.2,7(3x,f10.5))')Fo,
     &      temperature_bezw_cyl(zero,Fo,0.0,ne),
     &      temperature_bezw_cyl(zero,Fo,0.2,ne),
     &      temperature_bezw_cyl(zero,Fo,0.4,ne),
     &      temperature_bezw_cyl(zero,Fo,0.5,ne),
     &      temperature_bezw_cyl(zero,Fo,0.6,ne),
     &      temperature_bezw_cyl(zero,Fo,0.8,ne),
     &      temperature_bezw_cyl(zero,Fo,1.0,ne)
         Fo=Fo+.05
      enddo
      end  program p15_13
c according to equation (24) in 15.12 R=r/r_z, Fo=at/r_z/r_z
      function temperature_bezw_cyl(zero,Fo,R,ne)
      dimension zero(*)
      pi=3.141592654
      teta=0.
      do n=1,ne
         s=zero(n)
       teta=teta+bessj0(s*R)*exp(-s**2*Fo)/( s**2*bessj0(s) )
      enddo
      temperature_bezw_cyl=-2.*teta+2.*Fo+0.5*R**2-1./4.
      end function
      FUNCTION bessj0(x)
      REAL bessj0,x
      REAL ax,xx,z
      DOUBLE PRECISION p1,p2,p3,p4,p5,q1,q2,q3,q4,q5,
     & r1,r2,r3,r4,r5,r6,s1,s2,s3,s4,s5,s6,y
      SAVE p1,p2,p3,p4,p5,q1,q2,q3,q4,q5,r1,r2,
     & r3,r4,r5,r6,s1,s2,s3,s4,s5,s6
```

```
 DATA p1,p2,p3,p4,p5/1.d0,-.1098628627d-2,.2734510407d-
&4,-.2073370639d-5,.2093887211d-6/, q1,q2,q3,q4,q5/
&-.1562499995d-1,.1430488765d-3,-.6911147651d-5,
&.7621095161d-6,-.934945152d-7/
 DATA r1,r2,r3,r4,r5,r6/57568490574.d0,-13362590354.d0,
&651619640.7d0,-11214424.18d0,77392.33017d0,
&-184.9052456d0/,s1,s2,s3,s4,s5,s6/57568490411.d0,
&1029532985.d0,9494680.718d0,59272.64853d0,
&267.8532712d0,1.d0/
 if(abs(x).lt.8.)then
   y=x**2
 bessj0=(r1+y*(r2+y*(r3+y*(r4+y*(r5+y*r6)))))/
&(s1+y*(s2+y*(s3+y*(s4+y*(s5+y*s6)))))
 else
   ax=abs(x)
   z=8./ax
   y=z**2
   xx=ax-.785398164
   bessj0=sqrt(.636619772/ax)*(cos(xx)*
&(p1+y*(p2+y*(p3+y*
&(p4+y*p5))))-z*sin(xx)*(q1+y*(q2+y*(q3+y*(q4+y*q5)))))
 endif
 return
 END
```

Dimensionless temperature distribution for selected dimensionless coordinates r/r_z is presented in Fig. 15.31 and Table 15.5.

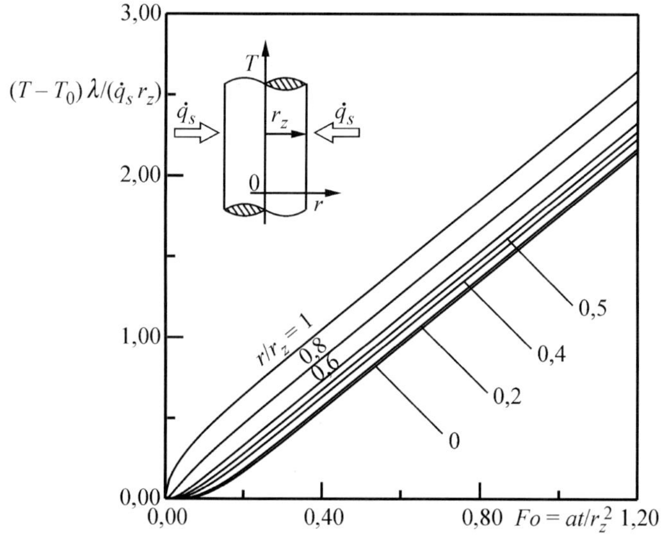

Fig. 15.31. Temperature transient in an infinitely long cylinder heated by a constant density heat flow

Table 15.5. Dimensionless temperature in an infinitely long cylinder $[T(r,t) - T_0]/(\dot{q}_S L/\lambda)$ for the selected values of the dimensionless coordinate $R = r/r_z$ and the Fourier number $Fo = at/r_z^2$

Fo	$r/r_z = 0$	$r/r_z = 0.2$	$r/r_z = 0.4$	$r/r_z = 0.5$	$r/r_z = 0.6$	$r/r_z = 0.8$	$r/r_z = 1.0$
0.00	0.00256	0.00036	0.00024	0.00022	0.00021	0.00027	0.01026
0.01	0.00000	0.00000	0.00000	0.00002	0.00026	0.01159	0.11814
0.02	0.00000	0.00001	0.00025	0.00118	0.00457	0.03918	0.17046
0.03	0.00003	0.00017	0.00192	0.00556	0.01439	0.06830	0.21210
0.04	0.00028	0.00096	0.00585	0.01326	0.02785	0.09669	0.24829
0.05	0.00120	0.00291	0.01211	0.02359	0.04351	0.12402	0.28104
0.06	0.00325	0.00636	0.02044	0.03588	0.06054	0.15036	0.31139
0.07	0.00676	0.01147	0.03052	0.04963	0.07842	0.17582	0.33995
0.08	0.01188	0.01822	0.04206	0.06449	0.09685	0.20051	0.36712
0.09	0.01862	0.02655	0.05483	0.08021	0.11567	0.22456	0.39317
0.10	0.02692	0.03632	0.06863	0.09661	0.13474	0.24805	0.41833
0.11	0.03667	0.04741	0.08331	0.11357	0.15400	0.27105	0.44273
0.12	0.04771	0.05965	0.09873	0.13097	0.17339	0.29364	0.46650
0.13	0.05993	0.07293	0.11479	0.14876	0.19290	0.31588	0.48973
0.14	0.07317	0.08710	0.13140	0.16686	0.21248	0.33781	0.51252
0.15	0.08731	0.10205	0.14847	0.18523	0.23213	0.35948	0.53492
0.16	0.10223	0.11769	0.16594	0.20382	0.25183	0.38091	0.55698
0.17	0.11784	0.13391	0.18376	0.22261	0.27157	0.40215	0.57876
0.18	0.13405	0.15065	0.20188	0.24157	0.29136	0.42323	0.60030
0.19	0.15077	0.16784	0.22026	0.26067	0.31117	0.44415	0.62163
0.20	0.16794	0.18540	0.23886	0.27989	0.33101	0.46495	0.64277
0.25	0.25861	0.27739	0.33425	0.37735	0.43048	0.56758	0.74653
0.30	0.35413	0.37355	0.43204	0.47613	0.53023	0.66884	0.84834
0.35	0.45198	0.47170	0.53098	0.57554	0.63011	0.76944	0.94920
0.40	0.55095	0.57082	0.63047	0.67526	0.73005	0.86973	1.04962
0.45	0.65046	0.67039	0.73023	0.77512	0.83003	0.96987	1.14982
0.50	0.75022	0.77019	0.83011	0.87506	0.93001	1.06994	1.24991
0.55	0.85011	0.87009	0.93005	0.97503	1.03001	1.16997	1.34996
0.60	0.95005	0.97004	1.03003	1.07501	1.13000	1.26999	1.44998
0.65	1.05002	1.07002	1.13001	1.17501	1.23000	1.36999	1.54999
0.70	1.15001	1.17001	1.23001	1.27500	1.33000	1.47000	1.65000
0.75	1.25001	1.27000	1.33000	1.37500	1.43000	1.57000	1.75000
0.80	1.35000	1.37000	1.43000	1.47500	1.53000	1.67000	1.85000
0.85	1.45000	1.47000	1.53000	1.57500	1.63000	1.77000	1.95000
0.90	1.55000	1.57000	1.63000	1.67500	1.73000	1.87000	2.05000
0.95	1.65000	1.67000	1.73000	1.77500	1.83000	1.97000	2.15000
1.00	1.75000	1.77000	1.83000	1.87500	1.93000	2.07000	2.25000
1.05	1.85000	1.87000	1.93000	1.97500	2.03000	2.17000	2.35000
1.10	1.95000	1.97000	2.03000	2.07500	2.13000	2.27000	2.45000
1.15	2.05000	2.07000	2.13000	2.17500	2.23000	2.37000	2.55000
1.20	2.15000	2.17000	2.23000	2.27500	2.33000	2.47000	2.65000

Exercise 15.14 Formula Derivation for Temperature Distribution in a Sphere with Boundary Conditions of 2nd Kind

Determine a formula for temperature distribution in a sphere with an outer surface radius r_z heated by a heat flow with constant density $\dot{q}_s$ (Fig. 15.32). Thermo-physical properties ρ, λ, c are constant. Initial temperature of the sphere is uniform and measures T_0. Apply Laplace transform to solve the initial-boundary problem.

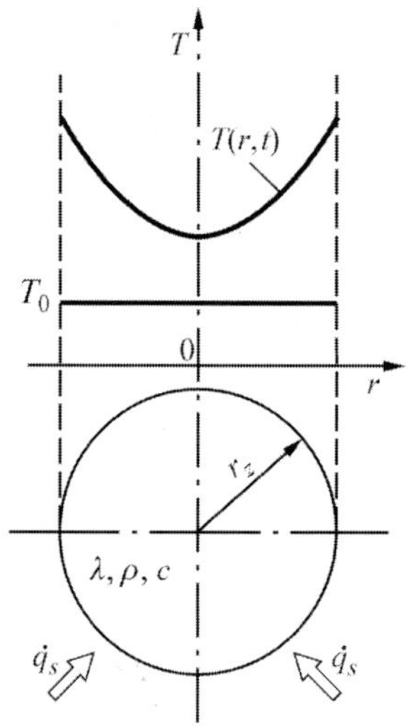

Fig. 15.32. The sphere heating by a heat flow with constant density $\dot{q}_s$

Solution

Temperature distribution in a sphere is governed by the heat conduction equation

$$a\left(\frac{\partial^2 T}{\partial r^2}+\frac{2}{r}\frac{\partial T}{\partial r}\right)=\frac{\partial T}{\partial t}, \tag{1}$$

by boundary conditions

$$\left.\frac{\partial T}{\partial r}\right|_{r=0}=0, \tag{2}$$

$$\lambda\left.\frac{\partial T}{\partial r}\right|_{r=r_z}=\dot{q}_s \tag{3}$$

and by initial-boundary condition

$$T\big|_{t=0} = T_0 . \tag{4}$$

Once Laplace transform is applied to (1)–(3), while accounting for initial condition (4), one has

$$a\left(\frac{d^2\bar{T}}{dr^2} + \frac{2}{r}\frac{d\bar{T}}{dr}\right) - s\bar{T} = T_0 , \tag{5}$$

$$\left.\frac{d\bar{T}}{dr}\right|_{r=0} = 0 , \tag{6}$$

$$\lambda\left.\frac{d\bar{T}}{dr}\right|_{r=r_z} = \frac{\dot{q}_s}{s} , \tag{7}$$

where

$$\bar{T} = \bar{T}(r,s) = \mathscr{L}\left[T(r,t)\right] = \int_0^\infty T(r,t)e^{-st}dt . \tag{8}$$

Solution of the boundary problem (5)–(7) has the form

$$\bar{T} = \frac{T_0}{s} + A\frac{\sinh(qr)}{r} + B\frac{\cosh(qr)}{r} , \tag{9}$$

where $q = \sqrt{s/a}$.

Due to the finite temperature value inside the sphere $r = 0$ constant B equals zero. Constant A determined from condition (7) has the form

$$A = \frac{\dot{q}_s r_z^2}{\lambda s}\frac{1}{qr_z\cosh(qr_z) - \sinh(qr_z)} . \tag{10}$$

By substituting A into (9), while noting that $B = 0$, a formula for a temperature distribution in the image domain is obtained

$$\bar{T} - \frac{T_0}{s} = \frac{\dot{q}_s r_z}{\lambda}\frac{r_z}{r}\frac{\sinh(qr)}{s\left[qr_z\cosh(qr_z) - \sinh(qr_z)\right]} = \frac{g_1(s)}{g_2(s)} . \tag{11}$$

Function (11) has a single pole in $s = 0$ and single poles, which constitute the roots of the following transcendental equation

$$g_2(s) = 0 , \tag{12}$$

hence,

$$qr_z\cosh(qr_z) - \sinh(qr_z) = 0 . \tag{13}$$

By substituting $\mu = iqr_z$ into (13), one has

$$\frac{\mu}{i}\cosh\left(\frac{\mu}{i}\right)-\sinh\left(\frac{\mu}{i}\right)=0,$$
$$\frac{\mu}{i}\cosh(-i\mu)-\sinh(-i\mu)=0, \tag{14}$$

hence, by accounting that

$$\cosh(-i\mu)=\cos\mu \quad \text{oraz} \quad \sinh(-i\mu)=-i\sin\mu \tag{15}$$

one obtains

$$\mu\cos\mu-\sin\mu=0,$$
$$\text{tg}\mu=\mu. \tag{16}$$

If we assume that

$$\mu_n = iq_n r_z, \tag{17}$$

then

$$q_n=\frac{\mu_n}{ir_z}, \qquad s_n=-\frac{\mu_n^2 a}{r_z^2}. \tag{18}$$

The roots of the characteristic equation (16) (the first five) are $\mu_1 = 4.4934$; $\mu_2 = 7.7253$; $\mu_3 = 10.9041$; $\mu_4 = 14.4934$ and $\mu_5 = 17.2208$.

Once the functions, sinh(qr), cosh(qr_z) and sinh(qr_z) in the (11) are expanded into the Taylor series (see Appendix H), one has

$$F(s)e^{st}=\frac{\dot{q}_s r_z^2}{\lambda r}\frac{\left(qr+\frac{q^3r^3}{6}+\ldots\right)(1+st+\ldots)}{s\left[qr_z\left(1+\frac{q^2r_z^2}{2}+\frac{q^4r_z^4}{24}+\ldots\right)-\left(qr_z+\frac{q^3r_z^3}{6}+\frac{q^5r_z^5}{120}+\ldots\right)\right]}=$$
$$=\frac{\dot{q}_s r_z}{\lambda}\frac{\left(1+\frac{sr^2}{6a}+\ldots\right)(1+st+\ldots)}{s^2\left(\frac{r_z^2}{3a}+\frac{sr_z^4}{30a^2}\ldots\right)}. \tag{19}$$

From the analysis of this expression it follows that (see Appendix H)

$$A=\frac{\dot{q}_s r_z}{\lambda}, \qquad B=\frac{\dot{q}_s r_z}{\lambda}\frac{r^2}{6a}, \qquad D=\frac{r_z^2}{3a}, \qquad E=\frac{r_z^4}{30a^2}. \tag{20}$$

By accounting for formula (H.25) from Appendix H, one obtains ($k = 2$)

$$c_{-1} = \frac{1}{D}\left(At + B - \frac{AE}{D} \right) = \frac{\dot{q}_s r_z}{\lambda}\left(\frac{3a}{r_z^2} t + \frac{1}{2}\frac{r^2}{r_z^2} - \frac{3}{10} \right). \tag{21}$$

Temperature distribution $T(r,t)$ is determined from formula (see Appendix H)

$$\begin{aligned} T(r,t) &= \mathscr{L}^{-1}\left[\overline{T}(r,s) \right] = \frac{1}{2\pi i}\int_{\delta - i\infty}^{\delta + i\infty} \overline{T}(r,s) e^{-st} ds = \\ &= T_0 + \frac{1}{D}\left(At + B - \frac{AE}{D} \right) + \sum_{n=1}^{\infty} \frac{g_1(s_n)}{g_2'(s_n)} e^{s_n t}. \end{aligned} \tag{22}$$

For a single pole $s = 0$ ((H.8) in Appendix H), the derivative $g_2'(s_n)$ has the form

$$\begin{aligned} g_2'(s_n) &= \lambda r s_n h'(s_n) = \lambda r s_n \frac{d}{ds}\left[q r_z \cosh(q r_z) - \sinh(q r_z) \right]\Big|_{s=s_n} = \\ &= \lambda r s_n \left\{ \frac{dq}{ds} r_z \cosh(q r_z) + q r_z \sinh(q r_z) \frac{dq}{ds} r_z - \right. \\ &\left. - \cosh(q r_z) \frac{dq}{ds} r_z \right\}\Big|_{s=s_n} = \lambda r r_z \frac{q_n^2}{2} \sinh(q_n r_z) = \\ &= \lambda r \frac{\mu_n^2}{2i^2} \sinh\left(\frac{\mu_n}{i} \right) = -\frac{1}{2}\lambda r \mu_n^2 \sinh(-i\mu_n) = \frac{1}{2} i \lambda r \mu_n^2 \sin \mu_n. \end{aligned} \tag{23}$$

When determining $g_2'(s_n)$, the following is accounted for

$$s\frac{dq}{ds} = \frac{q}{2}. \tag{24}$$

Next, numerator is determined

$$\begin{aligned} g_1(s_n) &= \dot{q}_s r_z^2 \sinh(q_n r) = \dot{q}_s r_z^2 \sinh\left(\frac{\mu_n}{i r_z} r \right) = \\ &= \dot{q}_s r_z^2 \sinh\left(-i\mu_n \frac{r}{r_z} \right) = -i \dot{q}_s r_z^2 \sin\left(\mu_n \frac{r}{r_z} \right). \end{aligned} \tag{25}$$

By substituting dependencies (21), (23) and (25) into (22), one has

$$T(r,t) = T_0 + \frac{\dot{q}_s r_z}{\lambda}\left(\frac{3at}{r_z^2} + \frac{1}{2}\frac{r^2}{r_z^2} - \frac{3}{10} \right) - \frac{2\dot{q}_s r_z^2}{\lambda r} \sum_{n=1}^{\infty} \frac{\sin\left(\mu_n \frac{r}{r_z} \right)}{\mu_n^2 \sin \mu_n} e^{\mu_n^2 \frac{at}{r_z^2}}. \tag{26}$$

Assuming that the dimensionless radius is $R = (r/r_z)$ and the Fourier number is $Fo = at/r_z^2$, (26) can be written in the form

$$\frac{\left[T(r,t)-T_0\right]\lambda}{\dot{q}_s r_z}=3Fo+\frac{1}{2}R^2-\frac{3}{10}-2\sum_{n=1}^{\infty}\frac{\sin\left(\mu_n R\right)}{\mu_n R}\frac{e^{-\mu_n^2 Fo}}{\mu_n \sin \mu_n}. \qquad (27)$$

For $Fo > 0.5$ one can neglect the infinite series in the above formula.

Exercise 15.15 Program and Calculation Results for Temperature Distribution in a Sphere with Boundary Conditions of 2nd Kind

Write a program for the calculation of temperature distribution in an infinitely long cylinder with the outer surface radius r_z heated by a heat flow with constant density $\dot{q}_s$ (Fig. 15.33). Thermo-physical properties λ, ρ, c are constant. Carry out the calculations using (26) derived in Ex. 15.14. Present the calculation results in a tabular and graphical form.

Solution

Program for calculating temperature distribution in a sphere heated by a heat flow with constant heat flux $\dot{q}_s$

```
C      Calculation of temperature distribution in a sphere
C      heated by a heat flow with constant heat flux q_s
       program p15_15
       dimension eigen(50)
       open(unit=1,file='p15_15.in')
       open(unit=2,file='p15_15.out')
       read(1,*)ne
       write(2,'(a)')
     & "CALCULATION OF TEMPERATURE DISTRIBUTION IN SPHERE"
       write(2,'(/a)') "INPUT DATA"
       write(2,'(a,i10)')      "ne       =",ne
       call equation_roots_sph(0.0,ne+1,eigen)
       write(*,*) (eigen(i),i=1,6)
       write(2,'(/a)')"CALCULATED TEMPERATURE [C]"
       write(2,'(a,a)')
     &"    Fo          r/r_z=0,0    r/r_z=0,2     r/r_z=0,4
     &"r/r_z=0,5",          r/r_z=0,6     r/r_z=0,8     r/r_z=1,0"
       Fo=0.
       do while (Fo.lt.0.2)
          write(2,'(f5.2,7(3x,f10.5))')Fo,
     &       temperature_bezw_sph(eigen,Fo,0.,ne),
     &       temperature_bezw_sph(eigen,Fo,0.2,ne),
     &       temperature_bezw_sph(eigen,Fo,0.4,ne),
     &       temperature_bezw_sph(eigen,Fo,0.5,ne),
```

```
      &     temperature_bezw_sph(eigen,Fo,0.6,ne),
      &     temperature_bezw_sph(eigen,Fo,0.8,ne),
      &     temperature_bezw_sph(eigen,Fo,1.0,ne)
       Fo=Fo+.01
       enddo
       do while (Fo.le.1.2)
          write(2,'(f5.2,7(3x,f10.5))')Fo,
      &     temperature_bezw_sph(eigen,Fo,0.,ne),
      &     temperature_bezw_sph(eigen,Fo,0.2,ne),
      &     temperature_bezw_sph(eigen,Fo,0.4,ne),
      &     temperature_bezw_sph(eigen,Fo,0.5,ne),
      &     temperature_bezw_sph(eigen,Fo,0.6,ne),
      &     temperature_bezw_sph(eigen,Fo,0.8,ne),
      &     temperature_bezw_sph(eigen,Fo,1.0,ne)
       Fo=Fo+.05
       enddo
       end  program p15_15
c      according to Eq. (27) in 15.14, R=r/r_z, Fo=at/r_z/r_z
       function temperature_bezw_sph(eigen,Fo,R,ne)
       dimension eigen(*)
       pi=3.141592654
       teta=0.
       do n=1,ne
          s=eigen(n+1)
          if (R.eq.0.) then
             teta=teta+exp(-s**2*Fo)/(s*sin(s))
           else
             teta=teta+sin(s*R)*exp(-s**2*Fo)/(s*R*s*sin(s))
          endif
       enddo
       temperature_bezw_sph=-2.*teta+3.*Fo+0.5*R**2-3./10.
       end function
c      procedure calculates roots of the ch. eq. x*cot(x)=1-Bi
c      where Bi is Biot number, ne is number of calculated
c      roots, eigen is output vector with calculated roots
       subroutine equation_roots_sph(bi,ne,eigen)
       dimension eigen(*)
       pi=3.141592654
       if ((bi.eq.1.).or.(bi.gt.10000.)) then
        do  i=1,ne
           if(bi.eq.1.) eigen(i)=(2.*float(i)-1.)*pi/2.
            if(bi.gt.10000.) eigen(i)=float(i)*pi
        enddo
       else
        h=1.-bi
        if (h.lt.0.) then
           h1=pi/2.
```

```
         h2=pi
     else
        h1=0.
        h2=pi/2.
     endif
     do i=1,ne
        xi=h1+(float(i)-1.)*pi
        xf=h2+(float(i)-1.)*pi
        do while (abs(xf-xi).ge.5.E-06)
           xm=(xi+xf)/2.
           y=xm*cos(xm)/sin(xm)-h
           if (y.lt.0.) then
              xf=xm
           else
              xi=xm
           endif
        enddo
        eigen(i)=xm
     enddo
    endif
    return
    end
```

Dimensionless temperature distribution for the selected coordinates r/r_z is presented in Fig. 15.33 and Table 15.6.

Table 15.6. Dimensionless temperature in a sphere $[T(r,t) - T_0]/(\dot{q}_s L/\lambda)$ for the selected values of the dimensionless coordinate $R=r/r_z$ and the Fourier number $Fo = at/r_z^2$

Fo	$r/r_z = 0$	$r/r_z = 0.2$	$r/r_z = 0.4$	$r/r_z = 0.5$	$r/r_z = 0.6$	$r/r_z = 0.8$	$r/r_z = 1.0$
0.00	–0.02891	–0.00266	–0.00248	0.00234	–0.00213	–0.00124	0.01842
0.01	0.00000	0.00000	0.00000	0.00003	0.00034	0.01331	0.12364
0.02	0.00000	0.00001	0.00040	0.00170	0.00607	0.04575	0.18192
0.03	0.00009	0.00038	0.00310	0.00811	0.01931	0.08084	0.22985
0.04	0.00088	0.00211	0.00950	0.01948	0.03772	0.11576	0.27260
0.05	0.00342	0.00631	0.01976	0.03489	0.05942	0.15002	0.31217
0.06	0.00865	0.01358	0.03346	0.05337	0.08327	0.18360	0.34956
0.07	0.01699	0.02404	0.05008	0.07421	0.10860	0.21656	0.38538
0.08	0.02848	0.03753	0.06912	0.09686	0.13497	0.24900	0.42002
0.09	0.04288	0.05373	0.09015	0.12093	0.16210	0.28100	0.45375
0.10	0.05988	0.07229	0.11281	0.14614	0.18982	0.31264	0.48676
0.11	0.07911	0.09284	0.13682	0.17224	0.21799	0.34398	0.51920
0.12	0.10023	0.11507	0.16192	0.19907	0.24651	0.37508	0.55119
0.13	0.12293	0.13870	0.18791	0.22649	0.27530	0.40598	0.58281
0.14	0.14694	0.16347	0.21464	0.25439	0.30433	0.43671	0.61413
0.15	0.17203	0.18919	0.24196	0.28267	0.33353	0.46731	0.64520
0.16	0.19801	0.21569	0.26977	0.31126	0.36289	0.49780	0.67608

Table 15.6. (cont.)

Fo	$r/r_z = 0$	$r/r_z = 0.2$	$r/r_z = 0.4$	$r/r_z = 0.5$	$r/r_z = 0.6$	$r/r_z = 0.8$	$r/r_z = 1.0$
0.17	0.22472	0.24282	0.29799	0.34012	0.39236	0.52821	0.70680
0.18	0.25203	0.27048	0.32653	0.36918	0.42193	0.55853	0.73738
0.19	0.27983	0.29856	0.35533	0.39842	0.45157	0.58880	0.76786
0.20	0.30804	0.32700	0.38436	0.42779	0.48129	0.61902	0.79825
0.25	0.45293	0.47255	0.53159	0.57602	0.63047	0.76964	0.94936
0.30	0.60107	0.62093	0.68058	0.72537	0.78017	0.91987	1.09977
0.35	0.75039	0.77034	0.83021	0.87514	0.93006	1.06995	1.24992
0.40	0.90014	0.92012	0.98008	1.02505	1.08002	1.21998	1.39997
0.45	1.05005	1.07005	1.13003	1.17502	1.23001	1.36999	1.54999
0.50	1.20002	1.22002	1.28001	1.32501	1.38000	1.52000	1.70000
0.55	1.35001	1.37001	1.43000	1.47500	1.53000	1.67000	1.85000
0.60	1.50000	1.52000	1.58000	1.62500	1.68000	1.82000	2.00000
0.65	1.65000	1.67000	1.73000	1.77500	1.83000	1.97000	2.15000
0.70	1.80000	1.82000	1.88000	1.92500	1.98000	2.12000	2.30000
0.75	1.95000	1.97000	2.03000	2.07500	2.13000	2.27000	2.45000
0.80	2.10000	2.12000	2.18000	2.22500	2.28000	2.42000	2.60000
0.85	2.25000	2.27000	2.33000	2.37500	2.43000	2.57000	2.75000
0.90	2.40000	2.42000	2.48000	2.52500	2.58000	2.72000	2.90000
0.95	2.55000	2.57000	2.63000	2.67500	2.73000	2.87000	3.05000
1.00	2.70000	2.72000	2.78000	2.82500	2.88000	3.02000	3.20000
1.05	2.85000	2.87000	2.93000	2.97500	3.03000	3.17000	3.35000
1.10	3.00000	3.02000	3.08000	3.12500	3.18000	3.32000	3.50000
1.15	3.15000	3.17000	3.23000	3.27500	3.33000	3.47000	3.65000
1.20	3.30000	3.32000	3.38000	3.42500	3.48000	3.62000	3.80000

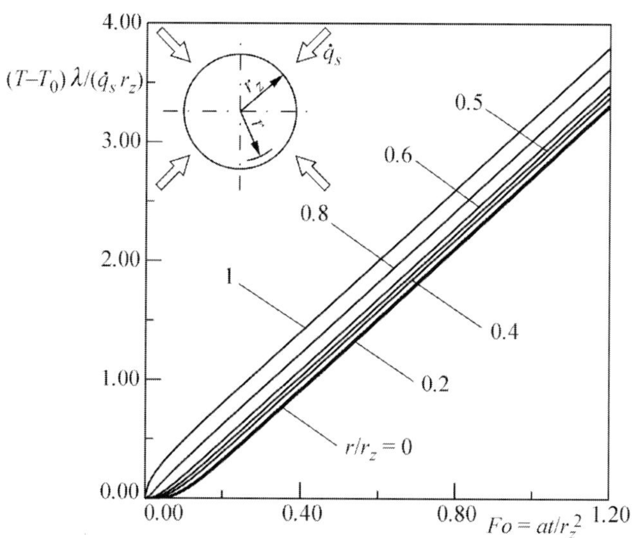

Fig. 15.33. Temperature transient of a sphere heated by a heat flow with constant density $\dot{q}_s$

Exercise 15.16 Heating Rate Calculations for a Thick-Walled Plate

Steel plate, $2L = 300$ mm thick is heated in a furnace, whose chamber temperature is $T_{cz} = 850°C$. Temperature of the plate, before it is placed in the furnace is constant and is $T_0 = 20°C$. Heat transfer coefficient on the plate surface is constant and equals $\alpha = 250\ W/(m^2·K)$. Due to a considerable width and length of the plate with respect to its thickness, assume that the plate is infinite, i.e. that the temperature distribution in the plate is one-dimensional. Calculate the amount of heat (energy) accumulated by 1 m^2 of the plate during 1 hour from the moment the plate is placed in the furnace. Also calculate the temperature in the center and on the plate surface. Thermal conductivity of the steel is $\lambda = 36\ W/(m·K)$, while temperature diffusivity coefficient $a = 9·10^{-6}\ m^2/s$.

Solution

The amount of energy transferred by 1 m^2 of the plate can be calculated from formula

$$\Delta E = E_2 - E_1 = \rho A c\left[\int_{-L}^{L} T(x,t_2)\,dx - \int_{-L}^{L} T(x,t_1)\,dx\right] =$$

$$= \rho A c\left[2\int_{0}^{L} T(x,t_2)\,dx - 2LT_0\right] = \rho A c\left[2L\frac{1}{L}\int_{0}^{L} T(x,t_2)\,dx - 2LT_0\right], \tag{1}$$

$$\Delta E = 2L\rho A c\left[\bar{T}(t_2) - T_0\right], \tag{2}$$

where $A = 1\ m^2$ is the surface area of the heated plate (from one side only), $\rho c = \lambda/a = 36/(9·10^{-6}) = 4·10^6\ J/(m^3·K)$. Carry out the calculations by means of the program developed in Ex. 15.2. Average temperature $\bar{T}(t_2)$ is calculated as follows

$$\bar{T}_{sc}(t)\Big|_{t_2=3600s} = \frac{1}{L}\int_{0}^{L} T(x,t_2)\,dx = 577°C.$$

Plate surface temperature is $T(L,t_2) = 650°C$, while temperature of the plate center is $T(0,t_2) = 539°C$.

The energy amount is determined from (2)

$$\Delta E = 2·1·4·10^6(577 - 20) = 4{,}456·10^9\ J.$$

Calculations can also be done using the diagrams shown in Ex.15.3. Fourier Fo and Biot numbers are

$$Fo = \frac{at_2}{L^2} = \frac{9 \cdot 10^{-6} \cdot 3600}{0.15^2} = 1.44 , \qquad Bi = \frac{\alpha L}{\lambda} = \frac{250 \cdot 0.15}{36} = 1.042 .$$

Average temperature $\bar{\theta}/\theta_0$ taken from the diagram in Fig. 15.8 is

$$\frac{\bar{\theta}}{\theta_0} = \frac{T_{cz} - \bar{T}(t_2)}{T_{cz} - T_0} \approx 0.33 ,$$

hence,

$$\bar{T}(t_2) = T_{cz} - 0.33(T_{cz} - T_0) = 850 - 0.33(850 - 20) = 576.1°\mathrm{C} .$$

Plate surface temperature is (diagram, Fig. 15.6)

$$\frac{\theta(L,t_2)}{\theta_0} = \frac{T_{cz} - T(L,t_2)}{T_{cz} - T_0} \approx 0.24 ,$$

hence,

$$T(L,t_2) = T_{cz} - 0.24(T_{cz} - T_0) = 850 - 0.24(850 - 20) = 650.8°\mathrm{C} .$$

Temperature of the plate center taken from the diagram in Fig. 15.7 is

$$\frac{\theta(0,t_2)}{\theta_0} = \frac{T_{cz} - T(0,t_2)}{T_{cz} - T_0} \approx 0.37 ,$$

hence,

$$T(0,t_2) = T_{cz} - 0.4(T_{cz} - T_0) = 850 - 0.4(850 - 20) = 518°\mathrm{C} .$$

The amount of energy accumulated by the plate, and calculated on the basis of temperatures taken from the diagrams, is

$$\Delta E = 2 \cdot 1 \cdot 4 \cdot 10^6 (576.1 - 20) = 4.45 \cdot 10^9\,\mathrm{J} .$$

Calculation results obtained by means of the program are more accurate.

Exercise 15.17 Calculating the Heating Rate of a Steel Shaft

Long steel shaft with a diameter that is $d_z = 2r_z = 120$ mm was placed in a furnace with a temperature of 800°C. How long should a shaft with an initial temperature of $T_0 = 20$°C be heated so that temperature would reach $T(0,t_n) = 644$°C at the shaft axis? Also determine surface temperature of the shaft at the end of the heating process, i.e. the temperature $T(r_z,t_n)$. Use the following data for the calculation: thermal conductivity $\lambda = 21$

W/(m·K), temperature diffusivity $a = 6\cdot10^{-6}$ m²/s, heat transfer coefficient on the shaft surface α = 140 W/(m²·K). Carry out the calculations by means of the program presented in Ex. 15.5 and by using the diagrams from Ex. 15.6.

Solution

Once the calculations are completed by means of the program developed in Ex.15.5, the following results are obtained:

- time t_n = 1405.31 s,
- shaft surface temperature after time t_n: $T(r_z,t_n)$ = 671.02°C.

In order to make use of the diagrams presented in Ex. 15.6, calculate dimensionless temperature θ/θ_0 and the Biot number Bi first

$$\frac{\theta}{\theta_0} = \frac{T(0,t_n) - T_{cz}}{T_0 - T_{cz}} = \frac{644 - 800}{20 - 800} = 0.2\,,$$

$$Bi = \frac{\alpha r_z}{\lambda} = \frac{140\cdot 0.06}{21} = 0.4\,.$$

From Fig. 15.16, Ex. 15.6, one has

$$Fo = \frac{a t_n}{r_z^2} = 2.4$$

hence

$$t_n = \frac{Fo\, r_z^2}{a} = \frac{2.4\cdot 0.06^2}{6\cdot 10^{-6}} = 1440 \text{ s} = 24\,\text{min}\,.$$

Dimensionless temperature of the surface θ/θ_0 for Fo = 2,4 taken from the diagram in Fig. 15.15, Ex. 15.6 is

$$\frac{\theta}{\theta_0} = \frac{T(r_z,t_n) - T_{cz}}{T_0 - T_{cz}} \approx 0.15\,,$$

hence

$$T(r_z,t_n) = 0.15(20 - 800) + 800 = 683^{\circ}\text{C}\,.$$

Results obtained by means of the computational program are more accurate than the results determined by means of the diagrams.

Exercise 15.18 Determining Transients of Thermal Stresses in a Cylinder and a Sphere

Determine transients of thermal stresses in the centre line of an infinitely long cylinder and inside a sphere. In both cases, assume the following data for the calculation: $r_z = 0.025$ m, $\lambda = 40$ W/(m·K), $a = 8\cdot 10^{-6}$ m^2/s, $\alpha = 200$ W/(m^2·K), $\Phi_w = E\beta/(1-\nu) = 3.7$ MPa/K, $T_{cz} = 20°C$, $T_0 = 750°C$. In order to determine stresses, apply computational programs developed in Ex. 15.5 and 15.8.

Solution

Thermal stresses are calculated from formula [6]

$$\sigma_T = \Phi_w\left[\overline{T}_w(t) - T(r,t)\right], \tag{1}$$

while the material constant Φ_w is given by

$$\Phi_w = \frac{E\beta}{1-\nu}, \tag{2}$$

where E [MPa] is the longitudinal elasticity module, β [1/K] a linear thermal expansion coefficient, ν a Poisson ratio.

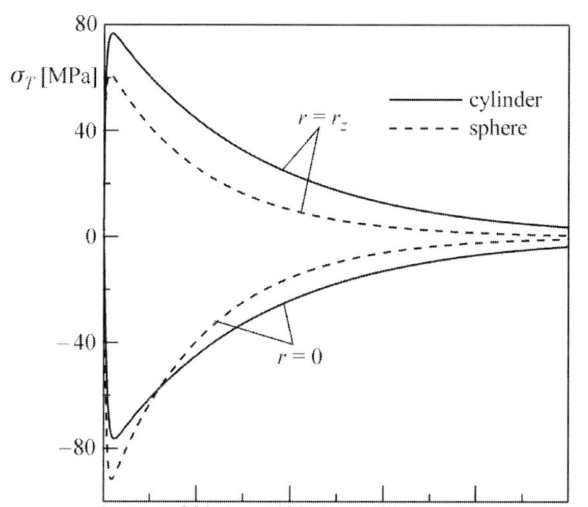

Fig. 15.34. Thermal stresses transient in a cylinder and a sphere caused by a sudden drop in temperature (cooling)

Average temperature is defined as

$$\overline{T}_w(t) = \frac{m+1}{r_z^{m+1}} \int_0^{r_z} r^m T(r,t)\,dr \,, \tag{3}$$

where: $m = 1$ for a cylinder and $m = 2$ for a sphere.

Calculation results are presented in Fig.15.34. One can conclude from them that thermal stresses are extensible on the surface, while compressible in the centre line of the cylinder and inside the sphere. Maximum compressible or extensible stresses occur at the beginning of the cooling process. Despite the large difference in temperature between initial temperature $T_0 = 750°C$ and medium's temperature $T_{cz} = 20°C$, the absolute stress value is not very high, since the heat transfer coefficient is relatively small.

Exercise 15.19 Calculating Temperature and Temperature Change Rate in a Sphere

Calculate temperature and temperature change rate inside a sphere with a radius of $r_z = 30$ mm in time $t = 10$ s. Initial temperature of the copper sphere is $T_0 = 30°C$. The sphere surface is tarnished black and its emissivity ε is of 0.97. The sphere was suddenly placed in a furnace chamber of a power boiler, whose temperature was $T_k = 1350°C$. Assume the following thermo-physical properties of copper for the calculation: $\lambda = 386$ W/(m·K), $a = 11.23 \cdot 10^{-5}$ m²/s.

Solution

First calculate heat flux on the sphere surface

$$\dot{q}_s = \varepsilon\sigma\left[T_k^4 - T^4(r_z,t)\right], \tag{1}$$

where $\sigma = 5.67 \cdot 10^{-8}$ W/(m²·K⁴) is the Stefan-Boltzmann constant.

Taking into account that at an initial heating stage $T_k >> T(r_z,t)$, (1) is simplified to a form

$$\dot{q}_s = \varepsilon\sigma T_k^4 = 0.97 \cdot 5.67 \cdot 10^{-8} \cdot (1350 + 273)^4 = 381\,618 \text{ W/m}^2 \,. \tag{2}$$

Assuming that

$$Fo = \frac{at}{r_z^2} = \frac{11.23 \cdot 10^{-5} \cdot 10}{0.015^2} = 4.99 >> 0.5 \,,$$

Equation (26) in Ex. 15.14 for the calculation of sphere temperature can be simplified to a form

$$T(r,t) = T_0 + \frac{\dot{q}_s r_z}{\lambda}\left(\frac{3at}{r_z^2} + \frac{1}{2}\frac{r^2}{r_z^2} - \frac{3}{10}\right), \tag{3}$$

where from it is easy to determine temperature change rate

$$v_T = \frac{dT}{dt} = \frac{\dot{q}_s r_z}{\lambda} \cdot \frac{3a}{r_z^2}. \tag{4}$$

Temperature inside the sphere after time t = 10 s from the moment the sphere is placed inside the furnace, calculated from (3) is

$$T_c = T(0,\ 10\ \text{s}) = 30 + \frac{381618 \cdot 0.015}{386}\left(\frac{3 \cdot 11.23 \cdot 10^{-5} \cdot 10}{0.015^2} - \frac{3}{10}\right) = 247.6\,^\circ\text{C}.$$

Temperature change rate inside the sphere is

$$v_T = \frac{381618 \cdot 0.015}{386}\,\frac{3 \cdot 11.23 \cdot 10^{-5}}{0.015^2} = 22.2\,\frac{\text{K}}{\text{s}}.$$

Exercise 15.20 Calculating Sensor Thickness for Heat Flux Measuring

A flat sensor (a slug calorimeter), insulated on the back and lateral surfaces (Fig. 15.35) was used for measuring heat flux $\dot{q}_s$ absorbed by a furnace chamber water-wall of a boiler (thermal load). Maximum value $\dot{q}_s$ does not exceed 400 000 W/m^2. Due to the accuracy of the measurement, the temperature change rate v_T measured on the sensor's back surface should not be very high. Assume that the sensor is made of a chrome-nickel steel (20% Cr, 15% Ni) and calculate the thickness of the sensor L, so that the rate of temperature increase would be $v_T \leq 5$ K/s. Assume the following steel properties for the calculation: λ =15.1 W/(m·K), a = 4.2·10^{-6} m^2/s.

Solution

Measurements are taken during a quasi-steady state, when the following condition is satisfied

$$Fo = \frac{at}{L^2} > 0.5, \tag{1}$$

Table 16.1. cont

Entry no.	η	T^*	Entry no.	η	T^*	Entry no.	η	T^*
58	5.8	0.66632	73	7.3	0.69240	88	8.8	0.71271
59	5.9	0.66830	74	7.4	0.69391	89	8.9	0.71391
60	6.0	0.67024	75	7.5	0.69539	90	9.0	0.71510
61	6.1	0.67214	76	7.6	0.69685	91	9.1	0.71626
62	6.2	0.67400	77	7.7	0.69828	92	9.2	0.71742
63	6.3	0.67583	78	7.8	0.69969	93	9.3	0.71856
64	6.4	0.67762	79	7.9	0.70109	94	9.4	0.71968
65	6.5	0.67938	80	8.0	0.70246	95	9.5	0.72079
66	6.6	0.68111	81	8.1	0.70380	96	9.6	0.72188
67	6.7	0.68281	82	8.2	0.70513	97	9.7	0.72296
68	6.8	0.68448	83	8.3	0.70644	98	9.8	0.72403
69	6.9	0.68612	84	8.4	0.70773	99	9.9	0.72508
70	7.0	0.68773	85	8.5	0.70900	100	10.0	0.72612
71	7.1	0.68931	86	8.6	0.71026			
72	7.2	0.69087	87	8.7	0.71149			

One needs to carry out a large number of transformations in order to determine temperature distribution by means of the Duhamel integral. It is much easier to determine the distribution by applying Duhamel integral calculated by means of approximation using the rectangles method. Temperature transient with a unit function can also be in such a case numerically calculated, e.g. by means of the difference method or finite element method.

Exercise 16.3 Derivation of an Approximate Formula for a Half-Space Surface Temperature with an Arbitrary Change in Medium's Temperature in the Function of Time

Determine a formula for surface temperature of a body with a convective boundary condition and time-variable medium's temperature $T_{cz}(t)$ and constant heat transfer coefficient α. Treat the body with an arbitrary shape as a semi-infinite body, while assuming that the heating or cooling processes are short-lasting. Determine the value of heat transfer coefficient on the surface of complex-shape bodies [3–5] by means of a formula for surface temperature of a semi-infinite body. The value of the heat transfer coefficient α is chosen in such way that both, the body's measured surface temperature $T_{s,z}(t_p)$ and calculated surface temperature $T_{s,z}(t_p)$ are equal after time t_p from the beginning of the heating or cooling process. Body surface temperature is frequently measured by means of liquid crystals.

Solution

A diagram of a semi-infinite body is shown in Fig. 16.2.

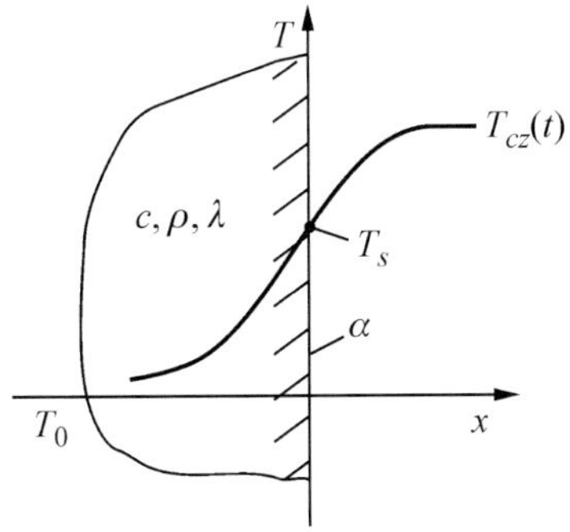

Fig. 16.2. A diagram of a semi-infinite body

To determine body surface temperature, (15) is used; it is derived in Ex. 16.1. Backward approximation of the derivative $df/d\theta$ by a difference quotient yields (14), from which, after transformations, (13), Ex. 16.1, is obtained. Equation (13), in Ex. 16.1, was a starting point for the derivation of the Duhamel integral. Surface temperature $T_s = T(0,t)$ is, therefore, formulated as follows:

$$\begin{aligned} T_s(t_M) &= T(0,t_M) = T_0 + f_1 u(0,t_M-\theta_0) + (f_2 - f_1)u(0,t_M-\theta_1) + \\ &+ \ldots + (f_M - f_{M-1})u(0,t_M-\theta_{M-1}). \end{aligned} \tag{1}$$

Temperature of the medium $T_{cz}(t)$ is approximated by a stepped line (Fig. 16.3).

Function $u\,(0,t)$ is a body surface temperature when the medium's temperature undergoes a unit step-increase.

$$u(0,t) = 1 - \left[1 - \operatorname{erf}\left(\frac{\alpha}{\lambda}\sqrt{at}\right)\right] e^{\frac{\alpha^2 at}{\lambda^2}} \tag{2}$$

The coordinates of temporal points θ_i, in which temperature $T_{cz,i} = T_{cz}(\theta_i)$ is measured, are indicated in Fig. 16.3. Once we account for

$$f_1 = \Delta T_{cz(1,0)} \tag{3}$$

in (1) and

$$f_i - f_{i-1} = \Delta T_{cz(i,i-1)} = T_{cz,i} - T_{cz,i-1}, \qquad i = 2,\ \ldots\ M\,, \tag{4}$$

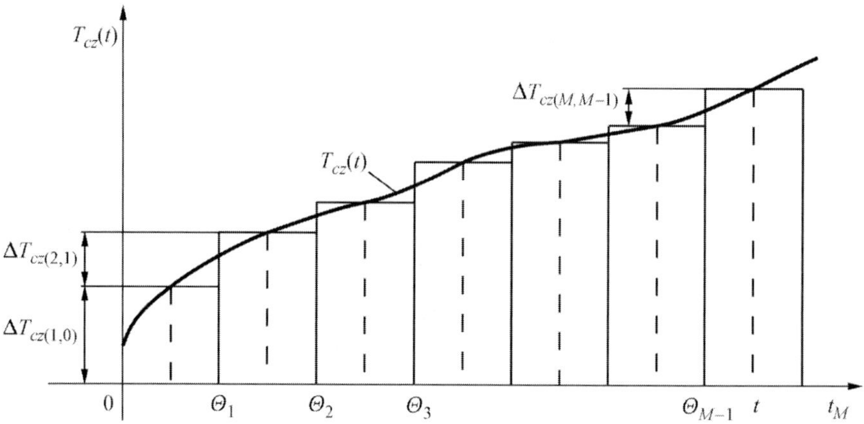

Fig. 16.3. Approximation of temperature changes in a medium $T_{cz}(t)$ by a stepped line

and the influence function (2), we have

$$T_s(t_M) = T_0 + \sum_{i=1}^{M} \left\{ 1 - \exp\left[\frac{\alpha^2 a(t_M - \theta_i)}{\lambda^2} \right] \times \right.$$
$$\left. \times \left[1 - \operatorname{erf}\left(\frac{\alpha\sqrt{a(t_M - \theta_i)}}{\lambda} \right) \right] \right\} \cdot \Delta T_{cz(i,i-1)}. \tag{5}$$

If,

$$\eta_M = \frac{\alpha^2 a t_M}{\lambda^2}, \tag{6}$$

then (5) can be written in the form

$$T_s(\eta_M) - T_0 = \sum_{i=1}^{M} \left\{ 1 - \exp\left[\eta_M \left(1 - \frac{\theta_i}{t_M} \right) \right] \times \right.$$
$$\left. \times \left[1 - \operatorname{erf}\sqrt{\eta_M \left(1 - \frac{\theta_i}{t_M} \right)} \right] \right\} \Delta T_{cz(i,i-1)}. \tag{7}$$

Using (5), one can determine heat transfer coefficient α while accounting for the medium's time-variable temperature $T_{cz}(t)$. By measuring half-space surface temperature in time t_M by means of the liquid crystals and by comparing it to temperature calculated from (5), one is able to determine α from the following non-linear algebraic equation

$$T_{s,o}\left(t_M\right)-T_{s,z}\left(t_M\right)=0\,, \tag{8}$$

where $T_{s,o}(t_M)$ is the temperature given by (5), while $T_{s,z}(t_M)$ a measured half-space surface temperature.

For details about the conducted experiment refer to papers [3÷5].

Exercise 16.4 Definition of an Approximate Formula for a Half-Space Surface Temperature when Temperature of a Medium Undergoes a Linear Change in the Function of Time

Temperature of a medium, which heats up a half-space whose initial temperature is constant and is T_0 increases according to formula $T_{cz}(t) = T_0 + bt$, where b is the constant heating rate. Calculate half-space surface temperature by means of (5) from Ex. 16.3. Present calculation results in the form $T^* = T^*(\eta_M)$, where $T^* = [T(0,t) - T_0]/(bt_M)$. Compare calculation results with the results presented in Table 16.1, which were obtained by means of the analytical (17) (Ex. 16.2).

Solution

Temperature changes in the medium are presented in Fig. 16.4.

From the above diagram, one can see that

$$\Delta T_{cz(i,i-1)}=b\left(\theta_i-\theta_{i-1}\right). \tag{1}$$

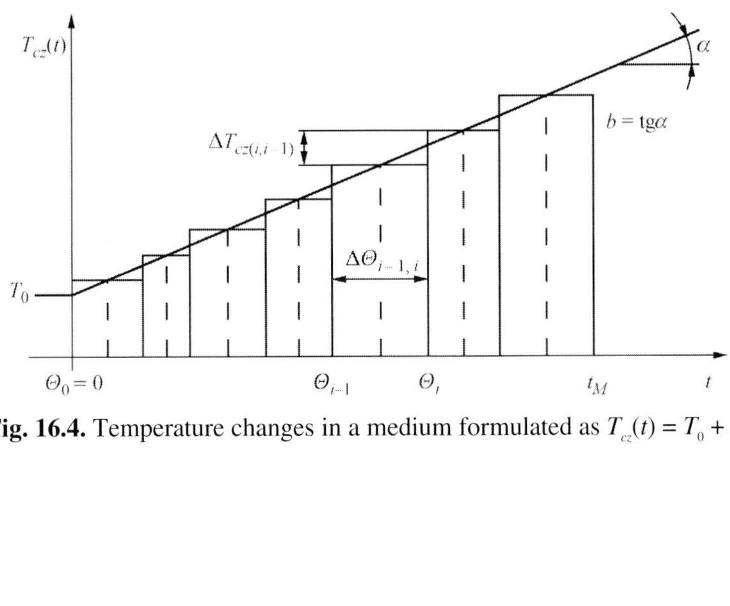

Fig. 16.4. Temperature changes in a medium formulated as $T_{cz}(t) = T_0 + bt$

$$T(\mathbf{x},t) = B_1 u(\mathbf{x}, t - t_1), \qquad t_1 < t \le t_2, \tag{6}$$

$$T(\mathbf{x},t) = B_1 u(\mathbf{x}, t - t_1) - B_1 u(\mathbf{x}, t - t_2), \qquad t_2 < t. \tag{7}$$

Equations (5)-(7) enable us to calculate temperature at a any spatial point $\mathbf{x}$ and time t.

Example 2

Function $B(t)$ can be expressed as follows (Fig. 16.7):

$$B(t) = B_1 \quad \text{for} \quad 0 < t \le t_1, \tag{8}$$

$$B(t) = B_1 + (B_2 - B_1) \quad \text{for} \quad t_1 < t \le t_2, \tag{9}$$

$$B(t) = B_1 + (B_2 - B_1) + (B_3 - B_2) \quad \text{for} \quad t_2 < t. \tag{10}$$

If the solutions of partial problems T_I, T_{II} and T_{III} are known when boundary conditions are as given in Fig. 16.7a, then we can present the solution T as the sum of solutions to partial problems, while accounting for the initial stage of operation and duration of a given boundary problem. For individual partial problems, time is measured from the moment the boundary condition exerts its influence, i.e. as a time variable in problem I (Fig. 16.7b) time t will be assumed, in problem II (Fig. 16.7c) time $(t - t_1)$, while in III (Fig. 16.8d) time $(t - t_2)$.

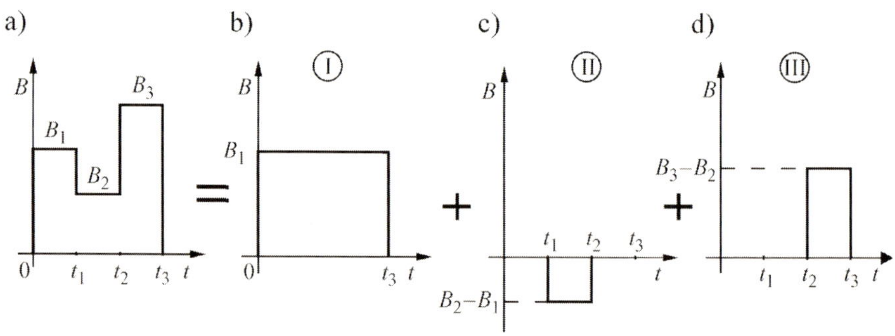

Fig. 16.7. Division of function $B(t)$ into three components

Example 3

The solution of a transient heat conduction problem, with the boundary condition shown in Fig. 16.8a, has the form

$$T(\mathbf{x},t) = T_{\mathrm{I}}(\mathbf{x},t), \qquad 0 \le t \le t_1, \tag{11}$$

$$T(\mathbf{x},t) = T_{\mathrm{I}}(\mathbf{x},t) + T_{\mathrm{II}}(\mathbf{x},t-t_1), \qquad t_1 < t \le t_2, \tag{12}$$

$$T(\mathbf{x},t) = T_{\mathrm{I}}(\mathbf{x},t) + T_{\mathrm{II}}(\mathbf{x},t-t_1) + T_{\mathrm{III}}(\mathbf{x},t-t_2), \quad t_2 < t \le t_3, \tag{13}$$

where T_{I}, T_{II} and T_{III} are partial solutions, for which the changes in function B are presented in Fig. 16.8b, c and d.

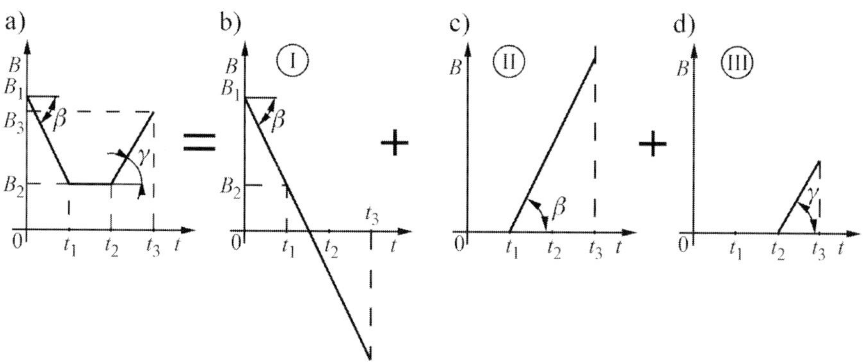

Fig. 16.8. Division of function *B*(*t*) into three components

Example 4

The solution of the transient heat conduction problem with changes in function *B*(*t*), presented in Fig. 16.9a, has the form

$$T(\mathbf{x},t) = T_{\mathrm{I}}(\mathbf{x},t), \qquad 0 < t \le t_1, \tag{14}$$

$$T(\mathbf{x},t) = T_{\mathrm{I}}(\mathbf{x},t) + T_{\mathrm{II}}(\mathbf{x},t-t_1) + T_{\mathrm{III}}(\mathbf{x},t-t_1), \qquad t_1 < t. \tag{15}$$

When using diagrams to determine components T_{I}, T_{II} and T_{III}, one should note that time is measured from the moment a given boundary condition exerts its influence.

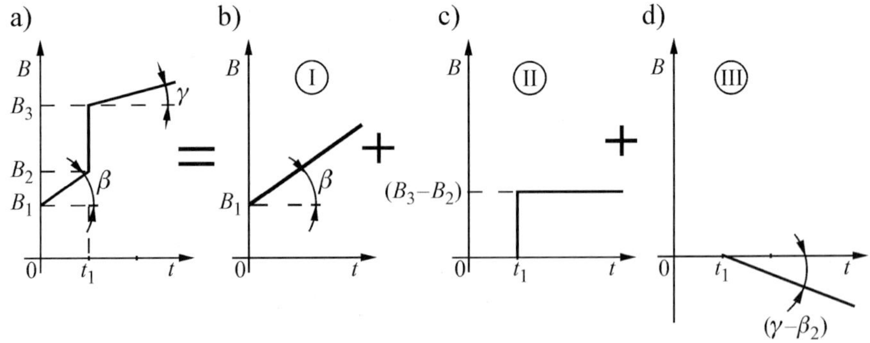

Fig. 16.9. Division function $B(t)$ into three components

Exercise 16.7 Formula Derivation for a Half-Space Surface Temperature with a Change in Surface Heat Flux in the Form of a Triangular Pulse

Using the superposition principle, determine half-space surface temperature for a change in heat flux presented in Fig. 16.10. The initial temperature is uniform and equals T_0. Solve the same problem by means of the Duhamel integral.

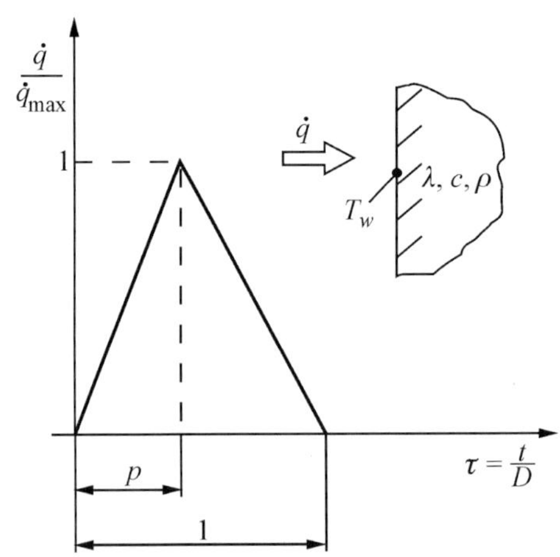

Fig. 16.10. Changes in heat flux on the surface of a half-space in the form of a triangular pulse; $\dot{q}_{max}$ is the maximum heat flux value; D - a pulse duration

Solution

If dimensionless time

$$\tau = \frac{t}{D} \tag{1}$$

and the dimensionless thermal flow $\dot{q}/\dot{q}_{max}$ are introduced, the heat flux changes presented in Fig. 16.10 can be described by means of function

$$\frac{\dot{q}}{\dot{q}_{max}} = \frac{\tau}{p}, \qquad 0 \le \tau \le p, \tag{2}$$

$$\frac{\dot{q}}{\dot{q}_{max}} = \frac{\tau}{p} - \frac{1}{1-p}\frac{\tau - p}{p}, \qquad p \le \tau \le 1, \tag{3}$$

where p is the dimensionless time, in which the heat flux reaches its maximum. Quantity D in (1) is the total pulse duration.

Dimensionless half-space surface temperature

$$\theta = \frac{\sqrt{\pi \lambda c\, \rho / D}}{\dot{q}_{max}} \left(T_w - T_0 \right) \tag{4}$$

is determined for time $0 \le \tau \le p$ using formula [2, 6]

$$\theta(\tau) = \frac{4}{3p}\tau^{3/2}, \qquad 0 \le \tau \le p. \tag{5}$$

Once the form of (3) is analyzed and the superposition principle applied, the temperature in the second stage of the pulse duration $a \le \tau \le 1$ can be expressed as follows:

$$\theta(\tau) = \frac{4}{3p}\tau^{3/2} - \frac{1}{1-p}\frac{4}{3p}(\tau - p)^{3/2},$$

$$\theta(\tau) = \frac{4}{3p}\left[\tau^{3/2} - \frac{(\tau - p)^{3/2}}{1-p} \right], \qquad p \le \tau \le 1. \tag{6}$$

Obtained results can be presented in the dimensional form by means of the equation

$$T_w = T_0 + \frac{4\dot{q}_{max}\sqrt{D}}{3p\sqrt{\pi \lambda c \rho}} \cdot \left(\frac{t}{D} \right)^{3/2}, \qquad 0 \le \frac{t}{D} \le p, \tag{7}$$

$$T_w = T_0 + \frac{4\dot{q}_{\max}\sqrt{D}}{3p\sqrt{\pi\lambda c\rho}}\left[\left(\frac{t}{D}\right)^{3/2} - \frac{\left(\frac{t}{D}-p\right)^{3/2}}{1-p}\right], \qquad p \le \frac{t}{D} \le 1 . \tag{8}$$

Equations (7) and (8) can be obtained using the Duhamel integral ((15), Ex.16.1))

$$T(0,t) = T_0 + \int_0^t \frac{df(\theta)}{d\theta} u(0,t-\theta)\,d\theta , \tag{9}$$

where function $u(0,t)$ is the half-space surface temperature with initial temperature at zero $T_0 = 0$ and with a unit step-increase in heat flux $\dot{q}_s = 1$ W/m^2. From (12) in Ex. 16.3, one has

$$u(0,t) = 2\sqrt{\frac{t}{\pi\lambda c\rho}} \tag{10}$$

Taking into account that

$$\frac{d\dot{q}}{dt} = \dot{q}_{\max}\frac{1}{pD}, \qquad 0 \le \tau \le p \tag{11}$$

and

$$\frac{d\dot{q}}{dt} = -\dot{q}_{\max}\frac{1}{(1-p)D}, \qquad p \le \tau \le 1 \tag{12}$$

from (9) the following expression is obtained

$$T(0,t) = T_0 + \frac{\dot{q}_{\max}}{pD}\frac{2}{\sqrt{\pi\lambda c\rho}}\int_0^t \sqrt{t-\theta}\,d\theta, \qquad 0 \le \tau \le p \tag{13}$$

and

$$T(0,t) = T_0 + \frac{\dot{q}_{\max}}{pD}\frac{2}{\sqrt{\pi\lambda c\rho}}\int_0^{pD} \sqrt{t-\theta}\,d\theta - \\ -\frac{\dot{q}_{\max}}{(1-p)D}\frac{2}{\sqrt{\pi\lambda c\rho}}\int_{pD}^{t} \sqrt{t-\theta}\,d\theta, \qquad p \le \tau \le 1. \tag{14}$$

By introducing a new variable

$$z = t - \theta \tag{15}$$

one can easily calculate the integrals in (13) and (14). Because $T_w = T(0,t)$, the calculation of integrals in (13) and (14) yields, respectively, (7) and (8).

Exercise 16.8 Formula Derivation for a Half-Space Surface Temperature with a Mixed Step-Variable Boundary Condition in Time

Using the superposition principle discussed in Ex. 16.6, derive a formula for a half-space temperature with a constant initial temperature T_0. Half-space surface is heated by a heat flow with constant density $\dot{q}_s$, while the surface is simultaneously cooled by an airflow at temperature T_p. Heat transfer coefficient α is constant. At time point t_u, the heat flux is step-wise decreased to a value $\varepsilon\dot{q}_s$ and maintained at this level.

Solution

A diagram, which illustrates half-space and heat flux changes is presented in Fig. 16.11.

Temperature distribution in the half-space is described by the heat conduction equation

$$\frac{1}{a}\frac{\partial T}{\partial t} = \frac{\partial^2 T}{\partial x^2}, \tag{1}$$

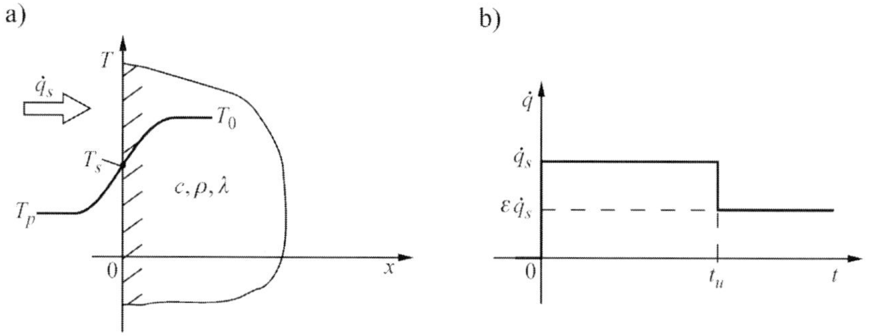

Fig. 16.11. Simultaneous: heating of a half-space by a heat flow with a density shown in the diagram b); cooling by airflow at temperature T_p – a)

by boundary conditions

$$-\lambda \frac{\partial T}{\partial x}\bigg|_{x=0} = \dot{q}_s - \alpha\left(T\big|_{x=0} - T_p\right), \qquad 0 < t \le t_u\,, \tag{2}$$

$$-\lambda \frac{\partial T}{\partial x}\bigg|_{x=0} = \varepsilon\dot{q}_s - \alpha\left(T\big|_{x=0} - T_p\right), \qquad t_u < t \tag{3}$$

and by initial condition

$$T\left(x,t\right)\big|_{t=0} = T_0\,. \tag{4}$$

In accordance with the superposition method, heat flux changes on the half-space surface can be presented as shown in Fig. 16.12.

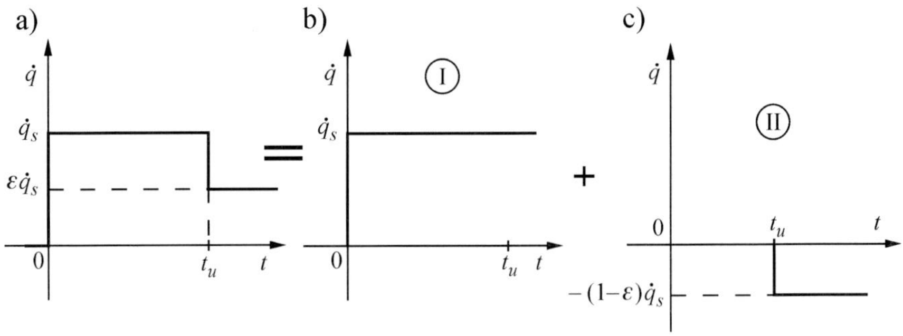

Fig. 16.12. Heat flux division into two components

Temperature distribution $T_I(x,t)$ is determined from the following equation

$$\frac{1}{a}\frac{\partial T_I}{\partial t} = \frac{\partial^2 T_I}{\partial x^2}\,, \tag{5}$$

with a boundary condition

$$-\lambda \frac{\partial T_I}{\partial x}\bigg|_{x=0} = \dot{q}_s - \alpha\left(T_I\big|_{x=0} - T_p\right) \tag{6}$$

and initial condition

$$T_I\left(x,t\right)\big|_{t=0} = T_0\,. \tag{7}$$

Condition (6) can be written in the form

$$\lambda \frac{\partial T_{\mathrm{I}}}{\partial x}\bigg|_{x=0} = \alpha\left(T_{\mathrm{I}}\big|_{x=0} - T_z\right), \tag{8}$$

where equivalent temperature of a medium T_z is formulated as

$$T_z = T_p + \frac{\dot{q}_s}{\alpha}, \tag{9}$$

while surface temperature $T_{\mathrm{I}}(x,t)$ is expressed by a equation

$$\frac{T_{\mathrm{I}}(0,t) - T_0}{T_z - T_0} = 1 - \operatorname{erfc}\left(\frac{\alpha}{\lambda}\sqrt{at}\right) \cdot \exp\left(\frac{\alpha^2 at}{\lambda^2}\right). \tag{10}$$

If temperature of the passing airflow T_p equals the initial temperature T_0, then (10) is simplified to a form

$$\alpha \frac{T_{\mathrm{I}}(0,t) - T_0}{\dot{q}_s} = 1 - \operatorname{erfc}\left(\frac{\alpha}{\lambda}\sqrt{at}\right) \cdot \exp\left(\frac{\alpha^2 at}{\lambda^2}\right). \tag{11}$$

In accordance with the superposition principle, the half-space surface temperature $T_s(t)$ is formulated as

$$T_s(t) = T_{\mathrm{I}}(0,t), \qquad 0 < t \le t_u, \tag{12}$$

$$T_s(t) = T_{\mathrm{I}}(0,t) + T_{\mathrm{II}}(0,t-t_u), \qquad t_u < t. \tag{13}$$

Once we account for (11) and decompose the heat flux into two components (Fig. 16.12), we can express the surface temperature $T_s(t)$ as follows:

$$\alpha \frac{T_s - T_0}{\dot{q}_s} = 1 - \operatorname{erfc}\left(\frac{\alpha}{\lambda}\sqrt{at}\right) \cdot \exp\left(\frac{\alpha^2 at}{\lambda^2}\right), \qquad 0 < t \le t_u, \tag{14}$$

$$\begin{aligned} \alpha \frac{T_s - T_0}{\dot{q}_s} &= 1 - \operatorname{erfc}\left(\frac{\alpha}{\lambda}\sqrt{at}\right) \cdot \exp\left(\frac{\alpha^2 at}{\lambda^2}\right) - \\ &-(1-\varepsilon)\left[1 - \operatorname{erfc}\left(\frac{\alpha}{\lambda}\sqrt{a(t-t_u)}\right) \cdot \exp\left(\frac{\alpha^2 a(t-t_u)}{\lambda^2}\right)\right], \qquad t_u < t. \end{aligned} \tag{15}$$

After transformations, (15) assumes the form

$$\alpha \frac{T_s - T_0}{\dot{q}_s} = \varepsilon - \operatorname{erfc}\left(\frac{\alpha}{\lambda}\sqrt{at}\right)\cdot \exp\left(\frac{\alpha^2 at}{\lambda^2}\right) + \\ +(1-\varepsilon)\operatorname{erfc}\left(\frac{\alpha}{\lambda}\sqrt{a(t-t_u)}\right)\cdot \exp\left(\frac{\alpha^2 a(t-t_u)}{\lambda^2}\right), \qquad t_u < t. \tag{16}$$

Equations (14) and (16) enable us to calculate the half-space surface temperature when step-function is applied, as shown in Fig.16.12a. These formulas are applied when the heat transfer coefficient α [8] is determined experimentally.

The surface of the analyzed model, usually made of plexiglass, is covered by a thin electric resistance heating foil. Initially, for $0 < t \le t_u$, the heat flux transferred by the foil to a base is $\dot{q}_s$. After time t_u the heating power is reduced and the heat flux $\dot{q}_s$ decreases to $\varepsilon\dot{q}_s$ (Fig. 16.11b), where $0 < \varepsilon \le 1$.

A standardized transient of the half-space surface temperature is shown in Fig. 16.13.

The model's initial temperature T_0 equals the temperature of surroundings. The foil surface temperature $T_{s,z}$ is assumed to be a body temperature and is measured by means of liquid crystals. The colour change in crystals occurs at a specific constant temperature; this is why it is easy to register temperature $T_{s,z}$ by means of a video camera. During the first stage $0 < t \le t_u$ temperature $T_{s,z}$ is obtained at the time point t_1, while during the second stage $t_u < t$ at the time point t_2.

From the comparison $T_s(t_1) = T_s(t_2)$, the following non-linear equation is obtained with respect to α once (14) and (16) are accounted for:

$$1 - \operatorname{erfc}\left(\frac{\alpha}{\lambda}\sqrt{at_1}\right)\cdot \exp\left(\frac{\alpha^2 at_1}{\lambda^2}\right) = \varepsilon - \operatorname{erfc}\left(\frac{\alpha}{\lambda}\sqrt{at_2}\right)\cdot \exp\left(\frac{\alpha^2 at_2}{\lambda^2}\right) + \\ +(1-\varepsilon)\operatorname{erfc}\left(\frac{\alpha}{\lambda}\sqrt{a(t_2-t_u)}\right)\cdot \exp\left(\frac{\alpha^2 a(t_2-t_u)}{\lambda^2}\right), \qquad t_u < t. \tag{17}$$

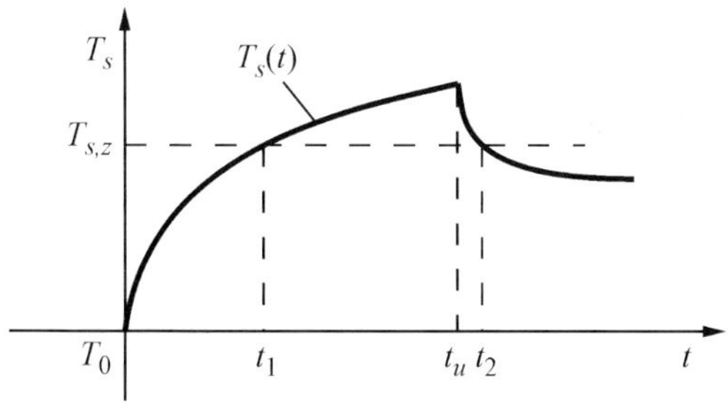

Fig. 16.13. Temperature transient of the model's surface

The above equation contains neither $\dot{q}_s$ nor $T_{s,z}$, which in itself is an attribute of the earlier described method for determining a heat transfer coefficient. Equation (17) can be easily solved using, for e.g., the interval halving method or the interval searching method, or any other widely available methods for solving non-linear algebraic equations.

Exercise 16.9 Formula Derivation for a Plate Surface Temperature with a Surface Heat Flux Change in the Form of a Triangular Pulse and the Calculation of This Temperature

Using the superposition method, determine temperature distribution in a plate heated by a heat flow whose density changes in time as shown on the diagram in Fig. 16.14. Initial plate temperature is uniform and is T_0. Back plate surface is thermally insulated.

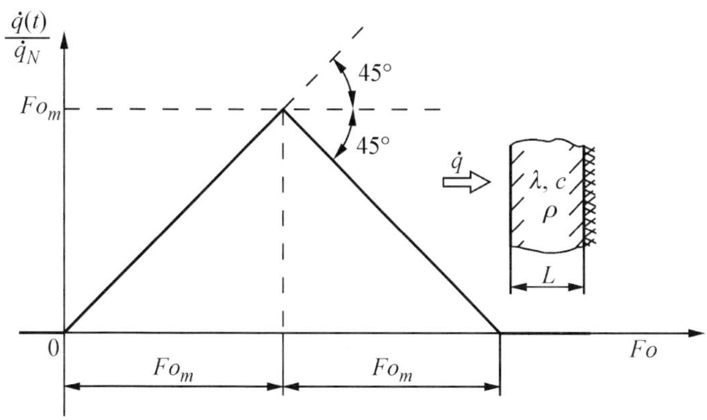

Fig. 16.14. Heat flux changes $\dot{q}(t)/\dot{q}_N$ in time, $Fo = at/L^2$

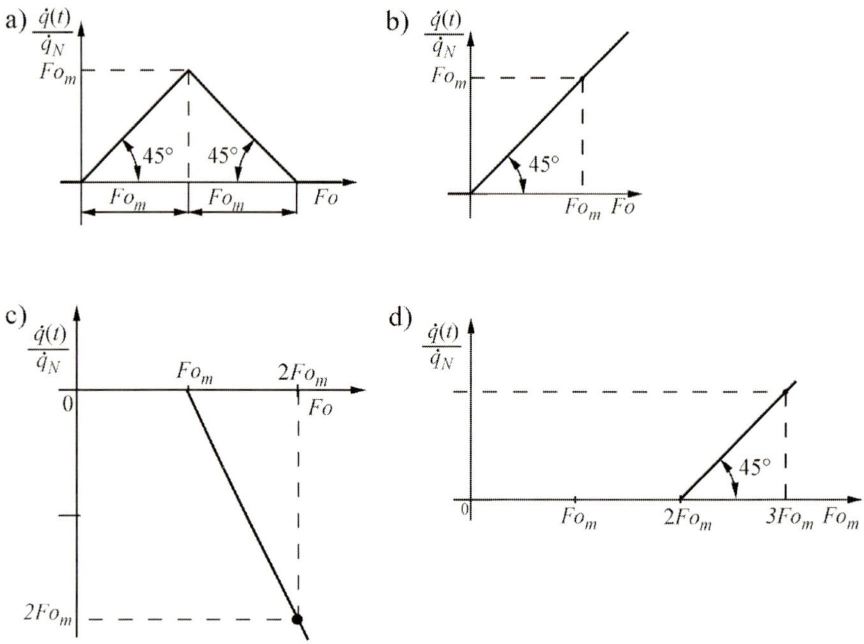

Fig. 16.15. Division of heat flux changes presented in Fig. a) into three components – Fig. b), c) and d)

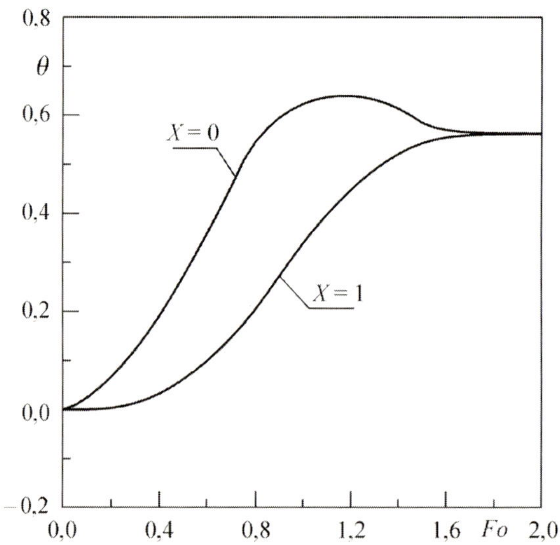

Fig. 16.16. Plate temperature determined on the front face and on the thermally insulated back surface, $Fo_m = 0.75$

The transient of calculated temperatures is shown in Fig.16.16. The print-out of the computational program in the FORTRAN language is presented below.

Computational program in FORTRAN language for calculating plate temperature with time-dependent heat flux (Fig. 16.15a) on the front face and thermally insulated back surface

```
C      Program for determining temperature distribution
C      on the back surface-insulated plate. Heat flux
C      varies in time in the triangle form
       PROGRAM tr
       OPEN(2,FILE='16_9_tr.out',STATUS='OLD')
       DFo=0.05
       write(2,*) 'Fo,     X=0.,    X=1.'
       Fo_m=0.75
       Fo=0.0
       DO WHILE (Fo.LE.Fo_m)
         write(2,300) Fo,TETA(0.,Fo),TETA(1.,Fo)
         Fo=Fo+DFo
       ENDDO
       DO WHILE ((Fo.GT.Fo_m).and.(Fo.LE.2.*Fo_m))
         write(2,300) Fo,TETA(0.,Fo)-2.*TETA(0.,Fo-Fo_m),
      &             TETA(1.,Fo)-2.*TETA(1.,Fo-Fo_m)
         Fo=Fo+DFo
       ENDDO
       DO WHILE (Fo.LE.2.0)
       write(2,300)Fo,TETA(0.,Fo)-2.*TETA(0.,Fo-Fo_m)+
      & TETA(0.,Fo-2.*Fo_m),
      & TETA(1.,Fo)-2.*TETA(1.,Fo-Fo_m)+TETA(1.,Fo-2.*Fo_m)
         Fo=Fo+DFo
       ENDDO
       STOP
300    FORMAT (F4.2,3x,F9.7,3x,F9.7)
       END

       REAL FUNCTION TETA(X,Fo)
       REAL X,Fo,Pi
       Pi=3.14159
       suma=0.0
          DO n=1,10
          s_mi=FLOAT(n)*Pi
             a_n=(-1.)**(n+1)*2./s_mi**4
             suma=suma+a_n*cos(s_mi*(1.-X))*exp(-s_mi**2*Fo)
          ENDDO
          TETA=0.5*Fo**2+Fo/3.+0.5*Fo*X**2-Fo*X+1./24.*X**4
```

```
&     -1./6.*X**3+1./6.*X**2-1./45.-suma
 RETURN
 END
```

Exercise 16.10 Formula Derivation for a Plate Surface Temperature with a Surface Heat Flux Change in the Form of a Rectangular Pulse; Temperature Calculation

Using the superposition principle, determine plate temperature distribution with a time-dependent heat flux $\dot{q}$, as shown on the diagram in Fig. 16.17. Plate temperature is T_0.

Solution

Temperature distribution in a plate is described by the heat conduction equation

$$\frac{\partial \theta}{\partial Fo} = \frac{\partial^2 \theta}{\partial X^2}, \qquad 0 \le X \le 1, \tag{1}$$

with boundary conditions (Fig. 16.17)

$$-\left.\frac{\partial \theta}{\partial X}\right|_{X=0} = \begin{cases} 1, & Fo_1 < Fo < Fo_2, \\ 0, & Fo_2 \le Fo \text{ lub } Fo \le Fo_1, \end{cases} \tag{2}$$

$$\left.\frac{\partial \theta}{\partial X}\right|_{X=1} = 0 \tag{3}$$

and initial condition

$$\theta(X,0) = 0, \qquad Fo \le 0, \tag{4}$$

where

$$Fo = at/L^2, \qquad \theta = (T - T_0)\lambda/\dot{q}_N L, \qquad X = x/L,$$

$$Fo_1 = Fo_c - \Delta Fo/2, \qquad Fo_2 = Fo_c + \Delta Fo/2.$$

In accordance with the superposition principle, the heat flux changes are defined as the sum of two heat flux step-changes at time points Fo_1 and Fo_2 (Fig. 16.18).

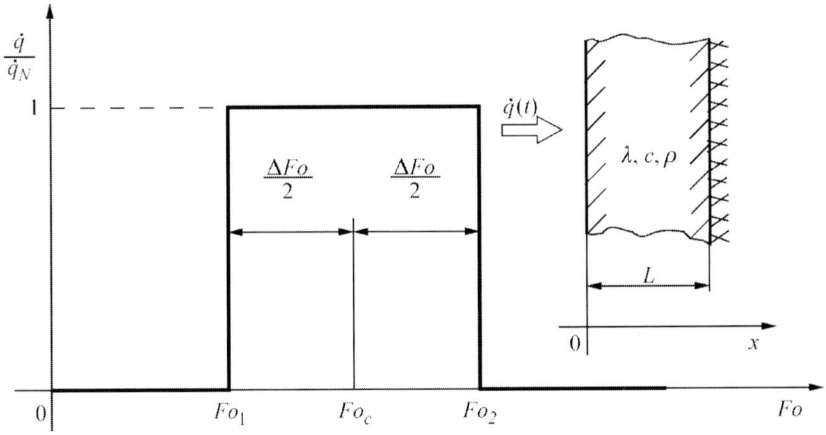

Fig. 16.17. Dimensionless heat flux changes $\dot{q}(t)/\dot{q}_N$ in the Fourier number function, $Fo = at/L^2$

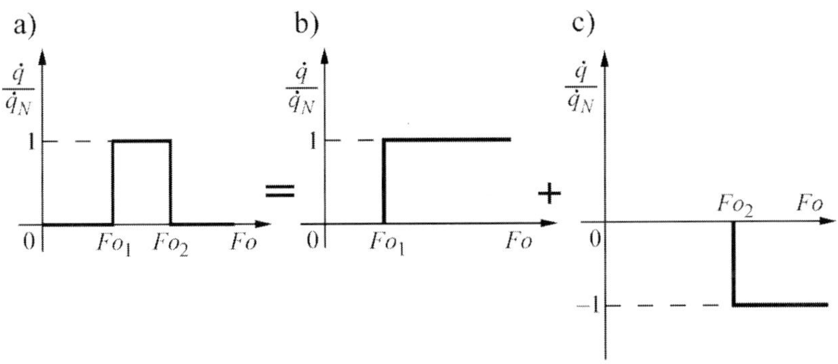

Fig. 16.18. Division of heat flux into two componen ts

Plate temperature distribution with a heat flux step-change is expressed as

$$\theta(X,Fo) = Fo + \frac{X^2}{2} - X + \frac{1}{3} - \frac{2}{\pi^2}\sum_{n=1}^{\infty}\frac{1}{n^2}\cos(n\pi X)\exp\left(-n^2\pi^2 Fo\right). \quad (5)$$

In accordance with the superposition principle, the solution of the problem (1)-(4) has the form

$$\theta = 0, \qquad 0 < Fo \le Fo_1, \quad (6)$$

$$\theta = \theta(X, Fo - Fo_1), \qquad Fo_1 < Fo < Fo_2, \quad (7)$$

$$\theta = \theta(X, Fo - Fo_1) - \theta(X, Fo - Fo_2), \qquad Fo_2 \le Fo. \quad (8)$$

Dimensionless temperature transient $\theta(X, Fo)$ for $X = 0$ and $X = 1$ is presented in Table 16.4.

Also, calculated temperature transient is presented in Fig.16.19.

Table 16.4. Dimensionless plate temperature transient on the butting face $\theta(0,Fo)$ and the back surface $\theta(1,Fo)$ for $Fo_1 = 0.75$ and $Fo_2 = 1.25$

Fo	$\theta(0,Fo)$	$\theta(1,Fo)$	Fo	$\theta(0,Fo)$	$\theta(1,Fo)$	Fo	$\theta(0,Fo)$	$\theta(1,Fo)$
0.00	0.0000000	0.0000000	0.75	0.0192868	–0.0009125	1.50	0.5170643	0.4829410
0.05	0.0000000	0.0000000	0.80	0.2523133	0.0002696	1.55	0.5104165	0.4895842
0.10	0.0000000	0.0000000	0.85	0.3568263	0.0078856	1.60	0.5063590	0.4936410
0.15	0.0000000	0.0000000	0.90	0.4370888	0.0293066	1.65	0.5038822	0.4961178
0.20	0.0000000	0.0000000	0.95	0.5051653	0.0614640	1.70	0.5023701	0.4976300
0.25	0.0000000	0.0000000	1.00	0.5661457	0.1005160	1.75	0.5014470	0.4985531
0.30	0.0000000	0.0000000	1.05	0.6228415	0.1438246	1.80	0.5008833	0.4991167
0.35	0.0000000	0.0000000	1.10	0.6769283	0.1897384	1.85	0.5005393	0.4994607
0.40	0.0000000	0.0000000	1.15	0.7294230	0.2372436	1.90	0.5003293	0.4996708
0.45	0.0000000	0.0000000	1.20	0.7809461	0.2857205	1.95	0.5002011	0.4997990
0.50	0.0000000	0.0000000	1.25	0.8318758	0.3347906	2.00	0.5001227	0.4998773
0.55	0.0000000	0.0000000	1.30	0.6301309	0.3839533	2.05	0.5000749	0.4999251
0.60	0.0000000	0.0000000	1.35	0.5759643	0.4259908	2.10	0.5000458	0.4999543
0.65	0.0000000	0.0000000	1.40	0.5459132	0.4543583			
0.70	0.0000000	0.0000000	1.45	0.5279658	0.4720718			

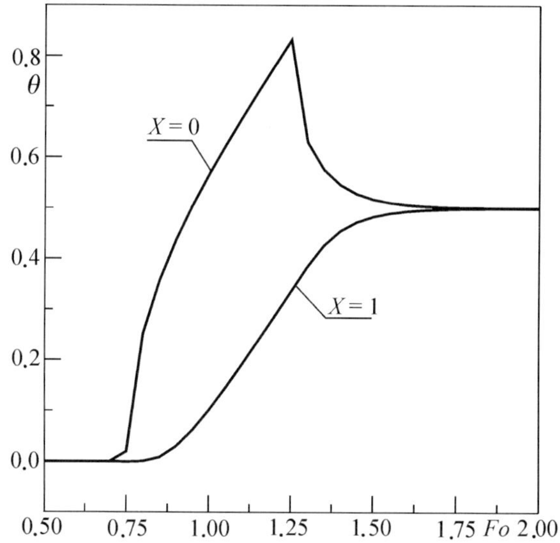

Fig. 16.19. Dimensionless temperature $\theta = (T - T_0)\lambda/(\dot{q}_N L)$ of the front face ($X = 0$) and the back surface ($X = 1$) with heat flux changes shown in Fig. 16.17 for $Fo_1 = 0.75$ and $Fo_2 = 1.25$

Exercise 16.11 A Program and Calculation Results for a Half-Space Surface Temperature with a Change in Surface Heat Flux in the Form of a Triangular Pulse

Calculate a half-space temperature transient in time. Heat flux changes in time (Fig. 16.10, Ex. 16.7). Initial temperature of the copper half-space is $T_0 = 0°C$; maximum heat flux value $\dot{q}_{\max} = 3\,000\,000$ W/m^2; pulse duration $D = 60$ s. Assume for the calculation that $p = 0{,}4$ and $(\lambda c \rho)^{1/2} = 35\,968$ J/(m^2·K·s$^{1/2}$) (see Table 17.1, Ex. 17.1). In order to simulate measurement data recorded at the step $\Delta t = 1$ s, add normally distributed pseudo-random numbers with a mean equal to zero and a variance of 1.0 to calculated temperature. Present calculation results in a tabular and graphical form. For the calculation, use formulas derived in Ex. 16.7. Attach a printout of the computational program.

Solution

Normally distributed pseudo-random numbers ε_i with a mean equal to zero and a variance $\sigma^2 = 1.0$ were added to a temperature transient calculated by means of (7) and (8) from Ex. 16.7. The probability that the errors calculated this way lie in the interval $\pm 3\sigma = \pm 3°C$ is $P = 99.7\%$. 'Measurement data" $T^z_{w,i}$ disturbed by random errors was calculated from formula

$$T^z_{w,i} = T_{w,i} + \varepsilon_i, \qquad i = 0, ..., 60, \tag{1}$$

where ε_i are pseudo-random numbers, while T_w temperatures $T_w(t_i)$ calculated by means of (7) and (8) in Ex. 16.7. Calculation results are listed in Table 16.5 and Fig. 16.20.

A program for determining a half-space temperature transient in time

```
PROGRAM temperature
Pi=3.14159
OPEN(2,FILE='16_11.wyn',STATUS='OLD')
write(2,*) 't,     Tw,    Tw_z'
D=60.
p=0.4
p_l_c_ro=35968.
q_max=3000000.
dt=1.
temp_0=0.
```

Exercise 16.12 Calculation of a Half-Space Temperature with a Mixed Step-Variable Boundary Condition in Time

Calculate temperature of a half-space made of a plexiglass, initially heated by a heat flow at density $\dot{q} = 6000$ W/m^2. A step-decrease in the heat flow density occurs at time $t_u = 30$ s and is $\varepsilon\dot{q} = 0.55 \cdot 6000 = 3300$ W/m^2. Initial temperature T_0 of the half-space and the temperature of the passing airflow T_p are identical and are $T_0 = T_p = 20$°C. The half-space-to-air heat transfer coefficient is $\alpha = 200$ W/(m^2·K). Also, carry out calculations when $\alpha = 150$ W/(m^2·K). Assume the following plexiglass values for the calculation: $\lambda = 0.184$ W/(m·K), $\rho = 1180$ kg/m^3, $c = 1440$ J/(kg·K). Determine time-points t_1 and t_2, in which the half-space temperature measures 38.2°C. At this temperature, the liquid crystals (Hallcrest, BM/R38C5W/C17-10[3] or Hallcrest, BM/R100F2W/C17-10[4]) change colour from colourless to red, which is something that can be easily recorded by means of a camera. Apply to the calculation formulas derived in Ex. 16.8.

Table 16.6. Half-space surface temperatures

t [s]	T [°C] $\alpha = 150$ W/(m^2·K)	T [°C] $\alpha = 200$ W/(m^2·K)	t [s]	T [°C] $\alpha = 150$ W/(m^2·K)	T [°C] $\alpha = 200$ W/(m^2·K)	t [s]	T [°C] $\alpha = 150$ W/(m^2·K)	T [°C] $\alpha = 200$ W/(m^2·K)
0	20.000	20.000	21	45.113	40.994	42	38.599	34.740
1	29.721	29.101	22	45.356	41.158	43	38.509	34.668
2	32.673	31.619	23	45.587	41.313	44	38.432	34.606
3	34.629	33.224	24	45.806	41.460	45	38.364	34.553
4	36.103	34.401	25	46.016	41.599	46	38.306	34.506
5	37.285	35.325	26	46.215	41.732	47	38.255	34.465
6	38.271	36.081	27	46.406	41.859	48	38.211	34.430
7	39.115	36.719	28	46.589	41.979	49	38.173	34.399
8	39.851	37.268	29	46.765	42.095	50	38.140	34.371
9	40.502	37.747	30	46.933	42.205	51	38.111	34.348
10	41.085	38.172	31	42.720	38.215	52	38.087	34.327
11	41.612	38.553	32	41.548	37.184	53	38.066	34.309
12	42.092	38.896	33	40.817	36.559	54	38.047	34.293
13	42.532	39.209	34	40.299	36.123	55	38.032	34.280
14	42.938	39.495	35	39.906	35.798	56	38.019	34.268
15	43.313	39.758	36	39.597	35.544	57	38.008	34.258
16	43.663	40.001	37	39.347	35.341	58	38.000	34.250
17	43.990	40.227	38	39.142	35.175	59	37.993	34.243
18	44.296	40.438	39	38.971	35.038	60	37.987	34.237
19	44.584	40.635	40	38.827	34.922			
20	44.856	40.82	41	38.704	34.824			

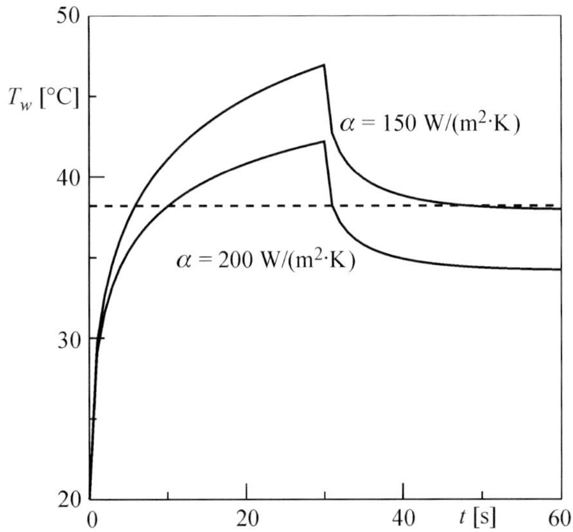

Fig. 16.21. Half-space surface temperature transient

Solution

Time-points t_1 and t_2, at which surface temperature measures 38.2°C, can be determined using surface temperature transients $T_w(t)$ presented in Table 16.6 or Fig. 16.21.

Exercise 16.13 Calculating Plate Temperature by Means of the Superposition Method with Diagrams Provided

Calculate front face temperature of a back-surface-insulated flat wall at a time-point t_2 = 150 s (Fig. 16.22). The flat wall is L = 0.05 m thick, thermal diffusivity coefficient $a = 1 \cdot 10^{-5}$ m²/s, while the thermal conductivity λ = 50 W/(m·K).

Heat flux changes in time as follows:

$$\dot{q}_1 = 2 \cdot 10^5 \text{ W/m}^2, \qquad 0 < t < t_1, \tag{1}$$

$$\dot{q}_2 = 1 \cdot 10^5 \text{ W/m}^2, \qquad t_1 \le t, \tag{2}$$

where t_1 = 50 s. Initial temperature of the plate is T_0 = 20°C.

Use the diagram from Fig 16.23.

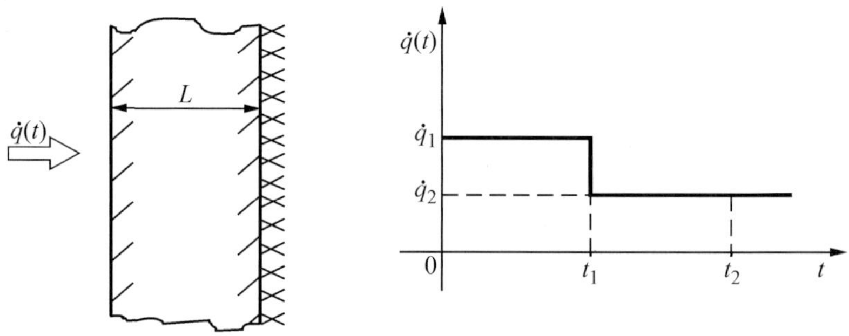

Fig. 16.22. Flat wall heating

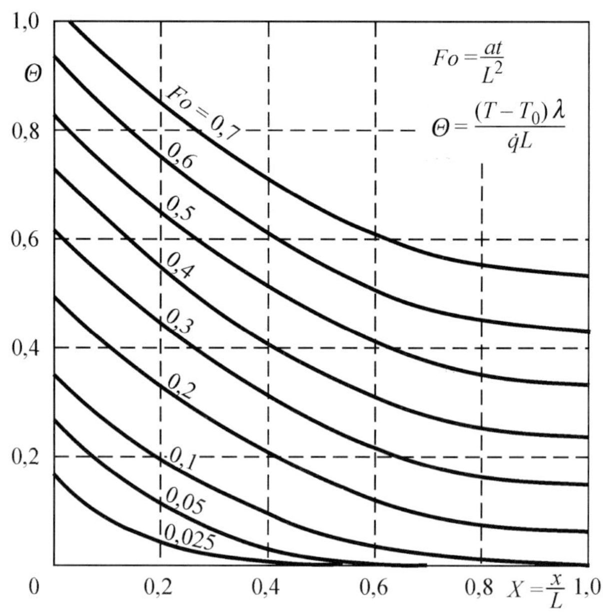

Fig. 16.23. Dimensionless temperature of a flat wall heated by a heat flow at constant density $\dot{q}$ with back surface insulated

Solution

Heat flux changes shown in Fig. 16.22 can be presented as the sum of two components (Fig. 16.24).

The Fourier number for the first problem (Fig. 16.24a) is

$$Fo = \frac{at_2}{L^2} = \frac{1 \cdot 10^{-5} \cdot 150}{(0.05)^2} = 0.6\,, \tag{3}$$

while for the second (Fig. 16.24b)

$$Fo = \frac{a(t_2 - t_1)}{L^2} = \frac{1 \cdot 10^{-5}(150 - 50)}{(0.05)^2} = 0.4. \tag{4}$$

From diagram (Fig. 16.23) for $X = 0$, one obtains:

$$\theta_1 = \theta(0; 0.6) = 0.94, \tag{5}$$

$$\theta_2 = \theta(0; 0.4) = 0.73. \tag{6}$$

Temperature of the plate front surface at a time $t_2 = 150$ s is

$$T(0, t_2) = \theta_1 \frac{\dot{q}_1 L}{\lambda} + \theta_2 \frac{(\dot{q}_2 - \dot{q}_1)L}{\lambda} + T_0 \tag{7}$$

As a result of substitution, one has

$$T(0, t_2) = 0.94 \cdot \frac{2 \cdot 10^5 \cdot 0.05}{50} + 0,73 \frac{(1-2) \cdot 10^5 \cdot 0.05}{50} + 20 = \tag{8}$$

$$= 188 - 73 + 20 = 135^\circ \mathrm{C}.$$

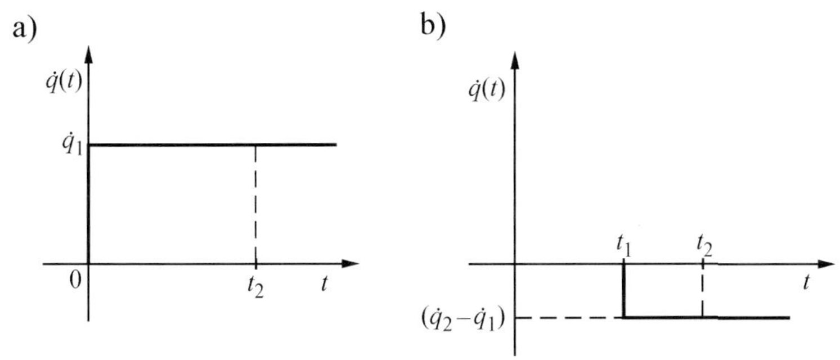

Fig. 16.24. Division of heat flux- the two components

Exercise 16.14 Calculating the Temperature of a Paper in an Electrostatic Photocopier

Electrostatic photocopier operates on the principle that a paper is electrostatically charged on the side, which displays the original imprint. Dark regions are positively charged, while the light regions are negatively charged.

Next, the paper comes into contact with negatively charged carbon molecules encapsulated in a polymer foil (Fig. 16.25) [1]. Carbon molecules attach themselves to a positively charged regions of the paper and due to a rise in the paper surface temperature T_{max} = 115.5°C, get permanently affixed to the paper. If the temperature is too low, stains form on the paper surface.

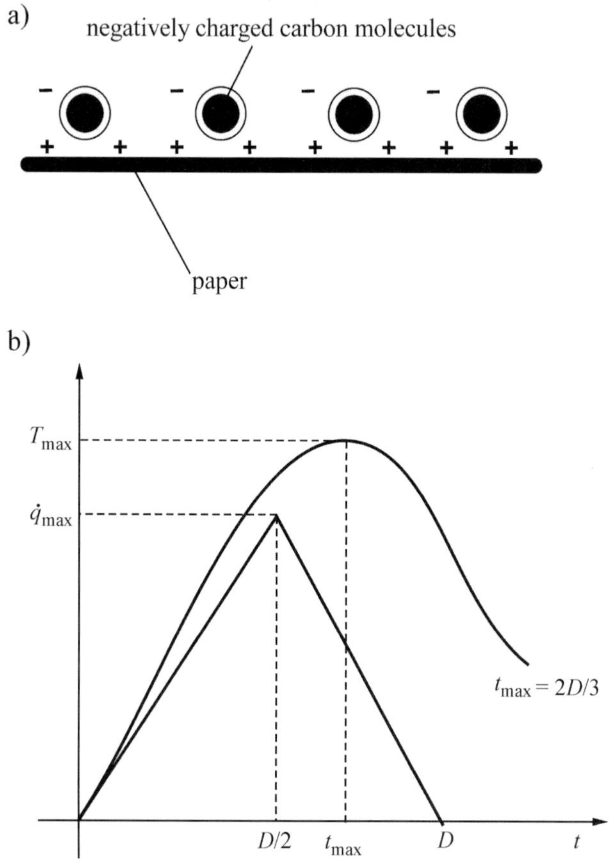

Fig. 16.25. A diagram, which illustrates the heating of a paper in a xerographic printer: (**a**) xerographic paper, with attached negatively charged carbon molecules, (**b**) heat flux transient and surface temperature; D –pulse duration in the form of a triangle

At the initial stages of development of the fast photocopying machines, engineers encountered technical problems; they found it difficult to design a photocopier that would warm up a paper to a temperature T_{max} in a very short period of time. This difficulty, however, was quickly overcome. The idea was to heat up a paper by a very large short-lasting heat flow generated by means of the flash-lamp-condenser system. The heat flux, which falls on the paper surface, changes in time in the form of an isosceles triangle (Fig. 16.25b).

The duration of paper heating is very short and lasts $D = 1\cdot10^{-3}$ s. In order to appropriately design the flash-lamp-condenser system, it is necessary to know what $\dot{q}_{max}$, is, since this ensures that the paper surface is heated to T_{max}. There are delays in surface temperature changes in comparison to heat flux changes that occur on the paper surface. The maximum half-space surface temperature occurs at a time point located in the interval $0.5D \le t \le D$. Determine value $\dot{q}_{max}$ which ensures that T_{max} = 115.5°C is reached, by assuming the following data for the calculation [1]:

- thermal conductivity for the paper is $\lambda = 0.1295$ W/(m·K),
- thermal diffusivity for the paper is $a = 6.65\cdot10^{-8}$ m^2/s,
- paper thickness is $L = 7.63\cdot10^{-5}$ m,
- initial temperature is $T_0 = 20.1$°C.

Account for the fact that only 30% of the incident heat flow is absorbed by the paper, while the paper heating lasts $D = 1\cdot10^{-3}$ s.

Solution

Paper can be regarded as a semi-infinite body, since

$$Fo = \frac{aD}{L^2} = \frac{6.65\cdot10^{-8}\cdot1\cdot10^{-3}}{\left(7.63\cdot10^{-5}\right)^2} = 0.0114$$

and the back surface temperature of the paper equals initial temperature. The depth of heat penetration at time D is smaller than the thickness of paper L. Formulas derived in Ex. 16.7 will be used to solve this problem. Half-space surface temperature in the time interval $pD \le t \le D$ is derived in (8), Ex. 16.7

$$T_w = T(0,t) = T_0 + \frac{4\dot{q}_{max}\sqrt{D}}{3p\sqrt{\pi\lambda c\rho}}\left[\tau^{3/2} - \frac{(\tau - p)^{3/2}}{1-p}\right], \tag{1}$$

where $\tau = t/D$. Coefficient p is the dimensionless time, in which $\dot{q}(t)$ reaches the maximum value, with respect to the whole time interval D. In the given case, $p = 0.5$. The maximum of function $T_w(t)$ occurs under the condition that the first derivative with respect to time is equal zero.

$$\frac{dT_w}{dt} = \frac{dT_w}{d\tau} \cdot \frac{d\tau}{dt} = \frac{2\dot{q}_{\max}\sqrt{D}}{p\sqrt{\pi\lambda c\rho}} \left[\tau^{1/2} - \frac{(\tau - p)^{1/2}}{1-p} \right] \cdot \frac{1}{D} = 0 . \tag{2}$$

Once equation (2) is solved, one has

$$\tau_{\max} = \frac{1}{2-p}, \tag{3}$$

where τ_{max} is the time point, at which $T_w(t)$ reaches its maximum value (Fig. 16.25).

By substituting $p = 0.5$ into (3), one obtains

$$\tau_{\max} = \frac{2}{3} D = \frac{2}{3} \cdot 1 \cdot 10^{-3} \text{s} = 6.667 \cdot 10^{-4} \text{s} .$$

By substituting $\tau = \tau_{max}$ into (1), one has

$$T_{\max} = T_w(\tau_{\max}) = \frac{4\dot{q}_{\max}\sqrt{D}}{3p\sqrt{\pi\lambda c\rho}} \left[\tau_{\max}^{3/2} - \frac{(\tau_{\max} - p)^{3/2}}{1-p} \right] + T_0 . \tag{4}$$

Following that $\dot{q}_{\max}$ is determined from equation (4)

$$\dot{q}_{\max} = \frac{3p}{4\sqrt{D}} \frac{\sqrt{\pi\lambda c\rho}\,(T_{\max} - T_0)}{\left[\tau_{\max}^{3/2} - \frac{(\tau_{\max} - p)^{3/2}}{1-p} \right]}, \tag{5}$$

where $\tau_{max} = 4p/3 = 4{\cdot}0.5/3 = 2/3$.

By substituting data into (5), one obtains (after $\lambda c\rho = \lambda^2/a$ is accounted for)

$$\dot{q}_{\max} = \frac{3 \cdot 0.5}{4\sqrt{0.001}} \frac{\sqrt{3.1416 \frac{0.1295^2}{6.65 \cdot 10^{-8}}}\,(115.5 - 20.1)}{\left[\left(\frac{2}{3}\right)^{3/2} - \frac{(2/3 - 0.5)^{3/2}}{1-0.5} \right]} = 2\,466\,546 \frac{\text{W}}{\text{m}^2} .$$

Lamp-generated heat flux in the xerographic printer should amount to

$$\dot{q}_{k,\max} = \frac{\dot{q}_{\max}}{30} \cdot 100 = 2\,466\,546 \cdot \frac{100}{30} = 8\,221\,820 \frac{\text{W}}{\text{m}^2} .$$

Flash-lamp-condenser system should generate a maximum density heat flow $\dot{q}_{k,\max}$, so that surface temperature of the paper would equal T_{max}.

Literature

1. Basmadjin D (1999) The Art of Modeling in Science and Engineering. Chapman & Hall-CRC. New York
2. Carslaw HS, Jaeger JC (1959) Conduction of Heat in Solids. Oxford University Press, London
3. Hwang JJ, Cheng CS (1999) Augmented heat transfer in a triangular duct by using multiple swirling jets. Transactions of the ASME, J. of Heat Transfer 121: 683-690
4. Kukreja RT, Lau SC (1998) Distributions of local heat transfer coefficient on surfaces with solid and perforated ribs. Enhanced Heat Transfer 5: 9-21
5. Metzger DE, Larson EE (1986) Use of melting point surface coatings for local convective heat transfer measurements in rectangular channel flows with 90-deg turns. Transactions of the ASME, J. of Heat Transfer 108: 48-54
6. Schneider PJ (1973) Conduction. Section 3. In Rohsenow WM, Hartnett JP (eds) Handbook of Heat Transfer. McGraw-Hill. New York: 3.1-3.133
7. Tautz H (1971) Wärmeleitung und Temperaturausgleich. Verlag Chemie, Weinheim
8. Wolfersdorf J, Hoecker R, Sattelmayer T (1993) A hybrid transient step-heating heat transfer measurement technique using heater foils and liquid-crystal thermography. Transactions of the ASME, J. of Heat Transfer 115: 319-324
9. Pekhovitch AI, Dzjidkih WM (1976) Calculations of thermal states of solid bodies (in Russian). Leningrad, Energija
10. Smirnoff MM (1975) Exercises of math physics equation (in Russian). Nauka, Moscow

17 Transient Heat Conduction in a Semi-Infinite Body. The Inverse Problem

In this chapter, inverse transient heat conduction problems in a semi-infinite body are discussed. Particular attention is paid to transient methods for measuring a heat transfer coefficient applied to tests conducted on heat transfer in gas turbines, combustion engines and rocket motors. The authors also derive formulas, which allow to determine both, constant and time-variable heat transfer coefficients on the basis of a measured temperature on the surfaces of construction elements that are considered as semi-infinite bodies and discuss the method for measuring heat flux on the inner surface of the combustion engine cylinder on the basis of a measured cylinder wall temperature in a single point. The authors derive theoretical dependencies, develop computational programs and illustrate how the programs are applied in practice. Overall, the chapter contains eight exercises, which are both, theoretical and computational in character.

Exercise 17.1 Measuring Heat Transfer Coefficient. The Transient Method

Describe a transient method for measuring heat transfer coefficient on the surface of a complex-shape body, based on the step-change in medium's temperature. The solid is regarded as a semi-infinite body due to a short measuring time. Assume that the body surface temperature is measured by means of thermo-chromatic liquid crystals.

Solution

The change in surface temperature of a semi-infinite body with initial temperature T_0 that was suddenly heated or cooled by a medium (usually air) at temperature T_{cz} (Fig. 17.1), is described by (14) in Ex. 14.4.

$$T(0,t) = T_0 + (T_{cz} - T_0)\left\{1 - \left[1 - \operatorname{erf}\left(\frac{\alpha}{\lambda}\sqrt{at}\right)\right] \cdot \exp\left(\frac{\alpha^2 at}{\lambda^2}\right)\right\}. \quad (1)$$

Assuming that surface temperature T_{sm} is known for time t_m from the readings taken (Fig. 17.1), then from the condition of equality of calculated temperature $T(0,t)$ and measured temperature T_{sm} for time $t = t_m$ one obtains the equation that allows to determine α

$$T\left(0,t_m\right)=T_{sm}, \tag{2}$$

hence, once (1) is accounted for, one has

$$T_0+\left(T_{cz}-T_0\right)\left[1-\operatorname{erfc}\left(\frac{\alpha}{\lambda}\sqrt{at_m}\right)\cdot\exp\left(\frac{\alpha^2 at_m}{\lambda^2}\right)\right]=T_{sm}, \tag{3}$$

where $\operatorname{erfc} x = 1-\operatorname{erf} x$.

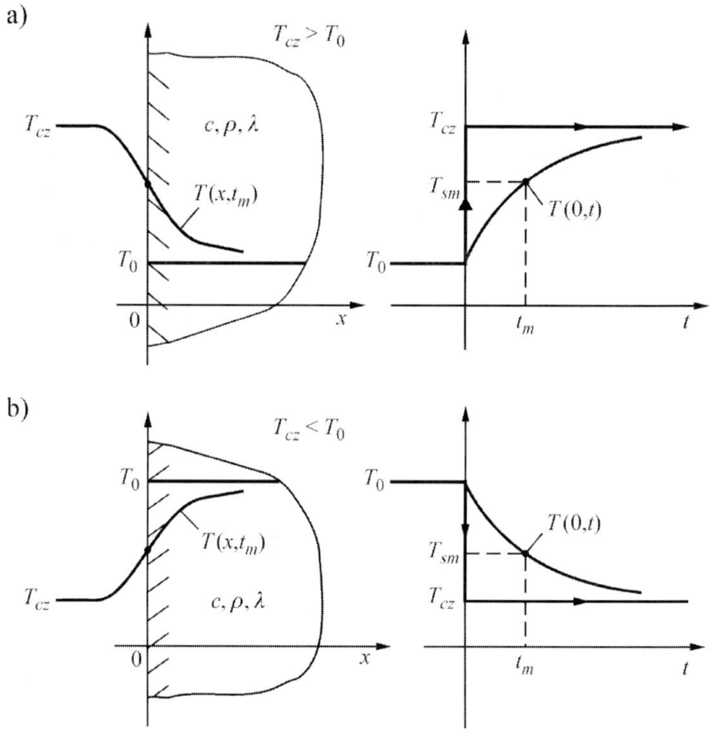

Fig. 17.1. Surface temperature transient of an infinite body $T(0,t)$ with initial temperature T_0 triggered by a step-change in medium's temperature from $T_{cz} = T_0$ for $t \le 0$ to $T_{cz} \neq T_0$ for $t > 0$: a) $T_{cz} > T_0$ (heated body), b) $T_{cz} < T_0$ (cooled body)

The above is a non-linear algebraic equation with respect to α, which can be solved using one of the widely available methods, such as the interval searching method, method of interval bisection, Newton method or secant method.

Coefficient α can also be determined from the graph of (2) or (3), shown in Ex. 14.4. For this purpose, (3) will be written in the form

$$\frac{T_{sm} - T_0}{T_{cz} - T_0} = 1 - \left[1 - \operatorname{erf}\left(\frac{\alpha}{\lambda}\sqrt{at_m}\right)\right] \cdot \exp\left(\frac{\alpha^2 at_m}{\lambda^2}\right). \tag{4}$$

If we first calculate $(T_{sm} - T_0)/(T_{cz} - T_0)$ and then allow for $x/(2\sqrt{at_m} = 0$, we can determine parameter $\alpha\sqrt{at_m}/\lambda$ from the graph of (2) in Ex. 14 or parameter $\alpha^2\sqrt{at_m}/\lambda^2$ from the graph of (3), in the same exercise.

If we take in to account that thermo-physical properties of the body, $a = \lambda/c\rho$ and λ and measurement time t_m are known, we will find it easy to calculate α.

Liquid crystals method is a method for measuring heat transfer coefficient α and is widely applied in practice [1–3, 5, 6, 8, 10, 14, 15], especially when it is necessary to determine the distribution of a heat transfer coefficient on the surface of gas turbine blades. Table 17.1 lists thermo-physical properties of materials, which are frequently used for building models of the tested construction elements.

Table 17.1. Thermo-physical properties of the selected materials at temperature 20°C

Material	λ [W/(m·K)]	ρ [kg/m³]	c [J/(kg·K)]	$(\lambda c\rho)^{1/2}$ [J/(m²·K·s^{1/2})]
Aluminum (pure)	204	2707	896	22244
Copper (commercial)	372	8300	419	35968
Silver (99,90% Ag)	57.5	7839	464	14461.8
Carbon steel (0,1%)	411	10524	236	31949.7
Alloy steel (0,19% C, 11,8% Cr, 1,07% Mo, 0,8% Ni)	24.0	7760	456	9215.5
Heat-resistant glass: Pyrex	1.4	2225	835	1612.8
Plexiglass	0.184	1180	1440	559.2
Polyethylene	0.35	920	2300	860.6
Teflon	0.23	2200	1040	725.4
Polyvinyl chloride (PVC)	0.15	1380	960	445.8
Pyroceram 9606*	3.98	2600	808	2891.6

* Product trade name of the firm Corning Glass, New York USA

The main advantages of this method, in comparison to other standard methods, are high spatial resolution and its adaptability to heat transfer testing in complex-shape bodies. It is a non-expensive and non-invasive method, more frequently used than thermo-element-based measuring method when temperatures are low and resemble the temperature of surroundings.

If we allow that the models of construction elements are made of materials with low thermal conductivity and that the duration of the experiment is very short, then the depth of heat penetration is smaller than the thickness of the model wall. In such a case, the model can be regarded as a semi-infinite body.

In liquid crystals method, the model with a uniform initial temperature is heated or cooled when there is a step-change in the flowing medium's temperature, which remains constant during the duration of the experiment.

The experiment can be conducted in two ways:

1. Model remains in the temperature of surroundings and is suddenly placed into the flow of the passing medium, whose temperature is higher than the temperature of surroundings (Fig. 17.1a).
2. Model is initially heated by a hot air or water and then is suddenly placed into the passing airflow, whose temperature equals the temperature of surroundings (Fig. 17.1b).

The measurement method is thoroughly described in reference [1÷3, 5, 6, 8, 10, 14, 15]. Paper [3] discusses the method for determining the distribution of a heat transfer coefficient on the surface of a turbine blade by means of the liquid crystals (BM/R32C5W/17-10), which change color from colourless to red, from red to green and from green to blue at temperature 31.6°C, 32.7°C and 37.2°C, respectively. The crystal-covered blade model was first heated to an initial temperature T_0, which was higher than 37.2°C (the blue colour of liquid crystals). Next, the model was placed in 0.1 s into the passing airflow, whose temperature equaled the temperature of surroundings, and time t_{sm} was recorded, after which the blade surface reached the temperature of T_{sm} = 32.7°C, at which the liquid crystals changed their colour from green to red. The colour changes in liquid crystals were recorded by a video camera and a computer system.

Exercise 17.2 Deriving a Formula for Heat Flux on the Basis of Measured Half-Space Surface Temperature Transient Interpolated by a Piecewise Linear Function

Derive a formula for the half-space surface heat flux $\dot{q}(t)$ by assuming that measured time transient of the half-space surface temperature is interpolated by a piecewise linear function.

Solution

The starting point is (19) derived in Ex. 14.5, according to which the heat flux $\dot{q}(t)$ on the half-space surface is formulated as follows

$$\dot{q}(t)=\sqrt{\frac{\lambda c \rho}{\pi}}\int_0^t \frac{df(\Theta)}{d\Theta}\frac{1}{\sqrt{t-\Theta}}d\Theta, \tag{1}$$

where $f(t)$ is the measured half-space surface temperature transient.

Approximation by a piecewise linear function of the measured half-space surface temperature transient (Fig. 17.2) enables one to write the integral (1) in the following way:

$$\dot{q}(t_{M+1})=\sqrt{\frac{\lambda c \rho}{\pi}}\sum_{i=1}^{M}\int_{t_i}^{t_{i+1}} \frac{df(\Theta)}{d\Theta}\frac{1}{\sqrt{t_{M+1}-\Theta}}d\Theta. \tag{2}$$

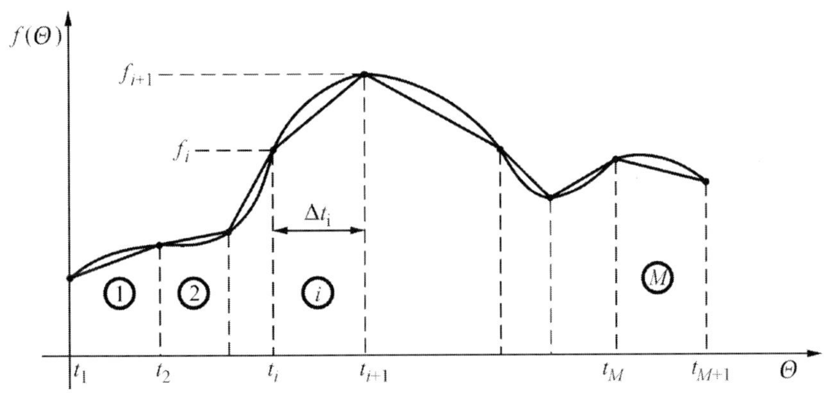

Fig. 17.2. Interpolation of measured half-space temperature transient by a piecewise linear function

Derivative with respect to time $df(\Theta)/d\Theta$ can be approximated by a forward difference quotient

$$\frac{df(\Theta)}{d\Theta}=\frac{f_{i+1}-f_i}{t_{i+1}-t_i}. \tag{3}$$

By accounting for (3) in (2), one has

$$\dot{q}(t_{M+1})=\sqrt{\frac{\lambda c \rho}{\pi}}\sum_{i=1}^{M}\frac{f_{i+1}-f_i}{t_{i+1}-t_i}\int_{t_i}^{t_{i+1}}\frac{1}{\sqrt{t_{M+1}-\Theta}}d\Theta. \tag{4}$$

Next, by substituting $t_{M+1} - \Theta = z^2$, which yields $\sqrt{t_{M+1} - \Theta} = z$, $2z dz = -d\Theta$ the following form of (4) is obtained:

$$\dot{q}\left(t_{M+1}\right) = \sqrt{\frac{\lambda c \rho}{\pi}} \sum_{i=1}^{M} \frac{f_{i+1} - f_i}{t_{i+1} - t_i} \int_{\sqrt{t_{M+1} - t_i}}^{\sqrt{t_{M+1} - t_{i+1}}} \left(-\frac{2z}{z}\right) dz \tag{5}$$

From the integration of the (5), one obtains a formula, which allows to calculate heat flux at time point t_{M+1} [12]

$$\dot{q}\left(t_{M+1}\right) = 2\sqrt{\frac{\lambda c \rho}{\pi}} \sum_{i=1}^{M} \frac{f_{i+1} - f_i}{t_{i+1} - t_i} \left(\sqrt{\left(t_{M+1} - t_i\right)} - \sqrt{\left(t_{M+1} - t_{i+1}\right)}\right) \tag{6}$$

Once we multiply the nominator and denominator under the summation symbol by

$$\left(\sqrt{t_{M+1} - t_i} + \sqrt{t_{M+1} - t_{i+1}}\right) \tag{7}$$

we obtain an alternative form of (6)

$$\dot{q}\left(t_{M+1}\right) = 2\sqrt{\frac{\lambda c \rho}{\pi}} \sum_{i=1}^{M} \frac{f_{i+1} - f_i}{\left(\sqrt{t_{M+1} - t_i} + \sqrt{t_{M+1} - t_{i+1}}\right)}, \qquad M = 1,\ 2,\ 3,\ \ldots \tag{8}$$

Sampling intervals $\Delta t_i = t_{i+1} - t_i$ do not have to be identical. Measurement data f_i, i = 1, 2, ..., (M + 1) is often disturbed by random errors, which cause oscillations in the calculated transient $\dot{q}(t)$. In order to reduce or eliminate the oscillations, the data can be "smoothed out" using digital filters, which are based on the local polynomial approximation [12]. When locally approximating seven subsequent measurement points by means of the third degree polynomial, one has

$$y_1 = \frac{39 f_1 + 8 f_2 - 4\left(f_3 + f_4 - f_5\right) + f_6 - 2 f_7}{42}, \tag{9}$$

$$y_2 = \frac{8 f_1 + 19 f_2 + 16 f_3 + 6 f_4 - 4 f_5 - 7 f_6 + 4 f_7}{42}, \tag{10}$$

$$y_3 = \frac{-4 f_1 + 16 f_2 + 19 f_3 + 12 f_4 + 2 f_5 - 4 f_6 + f_7}{42}, \tag{11}$$

$$y_i = \frac{7 f_i + 6\left(f_{i+1} + f_{i-1}\right) + 3\left(f_{i+2} + f_{i-2}\right) - 2\left(f_{i+3} + f_{i-3}\right)}{21}, \tag{12}$$

$i = 4, \ldots, N - 3.$

For the last three points $(M-1)$, M and $(M+1)$, the expressions that describe the smoothed temperature values have the form

$$y_{N-2} = \frac{f_{N-6} - 4f_{N-5} + 2f_{N-4} + 12f_{N-3} + 19f_{N-2} + 16f_{N-1} - 4f_N}{42}, \quad (13)$$

$$y_{N-1} = \frac{4f_{N-6} - 7f_{N-5} - 4f_{N-4} + 6f_{N-3} + 16f_{N-2} + 19f_{N-1} + 8f_N}{42}, \quad (14)$$

$$y_N = \frac{-2f_{N-6} + 4f_{N-5} + f_{N-4} - 4f_{N-3} - 4f_{N-2} + 8f_{N-1} + 39f_N}{42} \quad (15)$$

where $N = M + 1$ (Fig. 17.2).

"Smoothed" temperature values are used to calculate heat flux $\dot{q}\left(t_{M+1}\right)$

$$\dot{q}\left(t_{M+1}\right) = 2\sqrt{\frac{\lambda c \rho}{\pi}} \sum_{i=1}^{M} \frac{y_{i+1} - y_i}{\left(\sqrt{t_{M+1} - t_i} + \sqrt{t_{M+1} - t_{i+1}}\right)}, \qquad M = 1, 2, 3, \ldots \quad (16)$$

Approximation by means of the spline function can be used to smooth measured temperature transients. Number of programs are available in such software packages as IMSL [7], Table Curve [11] and MATLAB [4, 9]. The Table Curve Program [11] is user friendly, since it allows to make a quick graphical presentation of the obtained results and to evaluate the quality of the approximation.

Exercise 17.3 Deriving Heat Flux Formula on the Basis of a Measured and Polynomial-Approximated Half-Space Surface Temperature Transient

Derive a formula for heat flux $\dot{q}(t)$ on the half-space surface by assuming that measured time transient of the half-space surface temperature is approximated by the polynomial of m degree with respect to time.

Solution

Measured surface temperature transient $f(t_i)$, $i = 1, 2, \ldots, N$ is approximated by the polynomial

$$y\left(t\right) = A_0 + A_1 t + A_2 t^2 + \ldots + A_m t^m, \qquad m \le N - 1, \quad (1)$$

Polynomial coefficients are determined by the least-squares-method

$$\sum_{i=1}^{N}\left[y(t_i)-f_i\right]^2=\min, \tag{2}$$

where f_i is the measured half-space surface temperature at time point t_i. Commercial programs can be used to determine coefficients A_j, $j = 0, \ldots, m$.

Heat flux on the half-space surface is calculated from (1) in Ex. 17.2

$$\dot{q}(t)=\sqrt{\frac{\lambda c\rho}{\pi}}\int_0^t \frac{dy(\Theta)}{d\Theta}\frac{1}{\sqrt{t-\Theta}}d\Theta. \tag{3}$$

Once (1) is substituted into (3), one has

$$\dot{q}(t)=\sqrt{\frac{\lambda c\rho}{\pi}}\int_0^t\left(A_1+2A_2\Theta+\ldots+mA_m\Theta^{m-1}\right)\frac{1}{\sqrt{t-\Theta}}d\Theta. \tag{4}$$

Due to the substitution of $z^2 = t - \Theta$ and the application of an algebraic binomial formula

$$(a+b)^n=\sum_{j=0}^{n}\frac{n!}{j!(n-j)!}a^{n-j}b^j, \qquad n=1,2,\ldots \tag{5}$$

the following formula [12, 13] is obtained from (4) after transformations:

$$\dot{q}(t)=2\sqrt{\frac{\lambda c\rho}{\pi}}\left\{A_1\sqrt{t}+\sum_{i=2}^{m}iA_i t^{\frac{2i-1}{2}}\times\right.$$
$$\left.\times\left[1+(i-1)!\sum_{k=1}^{i-1}\frac{(-1)^k}{(2k+1)k!\,(i-1-k)!}\right]\right\}. \tag{6}$$

In a case when $m = 7$, the following expression for $\dot{q}(t)$ is obtained from (6):

$$\dot{q}(t)=2\sqrt{\frac{\lambda c\rho}{\pi}}\left(A_1\sqrt{t}+\frac{4}{3}A_2\sqrt{t^3}+\frac{8}{5}A_3\sqrt{t^5}+\frac{64}{35}A_4\sqrt{t^7}+\frac{128}{63}A_5\sqrt{t^9}+\right.$$
$$\left.+\frac{512}{231}A_6\sqrt{t^{11}}+\frac{1024}{429}A_7\sqrt{t^{13}}\right). \tag{7}$$

Equation (7) can also be applied to polynomials of a lower degrees. When $m=3$, one should assume then that $A_4 = A_5 = A_6 = A_7 = 0$, while when $m=5$, it is assumed that $A_6 = A_7 = 0$. In the case of higher degree polynomials ($m > 5$), one should take into consideration the fact that oscillations may occur.

Exercise 17.4 Formula Derivation for a Heat Flux Periodically Changing in Time on the Basis of a Measured Temperature Transient at a Point Located under the Semi-Space Surface

Determine a formula for periodically changing heat flux in time under the assumption that half-space surface temperature $y(t)$ is known from the readings taken as well as the temperature f_E inside the half-space, within distance $E \geq \delta$, where δ is the depth of heat flow penetration (Ex. 14.8) into the half-space interior. Thermo-physical properties of the semi-infinite body: λ, c and ρ are constant.

Solution

From the surfaces of combustion chambers in combustion engines and from the surfaces of pipes submersed in a fluid bed, heat flux $\dot{q}(t)$ (Fig. 17.3) can be divided into two components: $\dot{q}$ [steady-state; independent of time] and $\dot{q}_n(t)$ [transient; changes periodically in time].

$$\dot{q}(t) = \bar{\dot{q}} + \dot{q}_n(t) \tag{1}$$

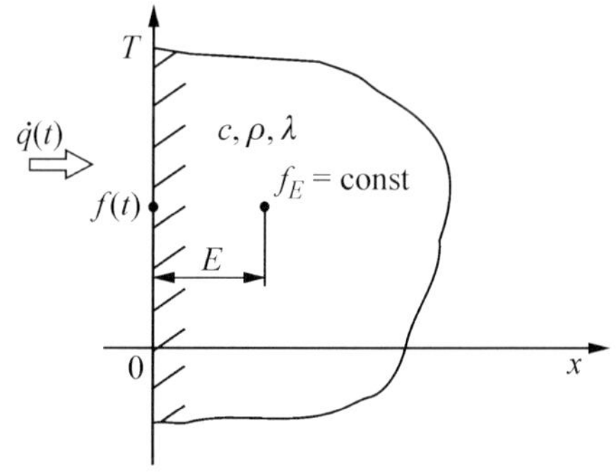

Fig. 17.3. Diagram of the location of temperature measurement points

If the frequency of heat flux changes $\dot{q}_n(t)$ is high, then the temperature and heat flux change in time only in the vicinity of the element's surface, at a depth of about 1÷2 mm, while the temperature inside the element remains constant with time. In order to determine components $\bar{\dot{q}}$ and $\dot{q}_n(t)$, temperature is measured at two points (Fig. 17.3): on the heat-transferring surface (x = 0) and within distance $x = E$ from this surface. Distance E

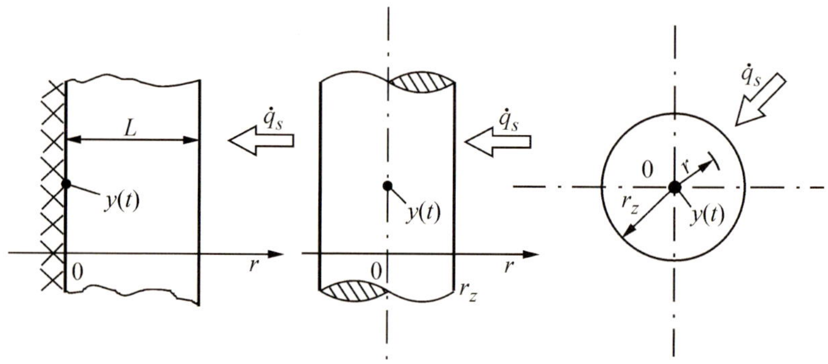

Fig. 18.1. Diagram of an inverse transient heat conduction problem

Solution

Temperature distribution is defined by the heat conduction equation

$$\frac{1}{r^m}\frac{\partial}{\partial r}\left(r^m\frac{\partial T}{\partial r}\right)=\frac{1}{a}\frac{\partial T}{\partial t} \tag{1}$$

and conditions

$$T\big|_{r=0}=y(t), \tag{2}$$

$$\left.\frac{\partial T}{\partial r}\right|_{r=0}=0, \tag{3}$$

where: $m = 0$ for the plate, $m = 1$ for the cylinder and $m = 2$ for the sphere.

In contrast to direct heat conduction problems, conditions (2) and (3) are assigned in an inverse problem on the same boundary $r = 0$. Temperature $T_s(t)$ and heat flux $\dot{q}_s(t)$ are determined on the boundary $r = L$ or $r = r_z$. The unknown temperature distribution has the following form [1, 9]:

$$T(r,t)=\sum_{n=0}^{\infty}h_n(r)\frac{d^n y}{dt^n}. \tag{4}$$

By substituting (4) into (1), one has

$$\frac{1}{r^m}\frac{d}{dr}\left(r^m\frac{dh_0}{dr}\right)=0 \quad \text{dla} \quad n=0 \tag{5}$$

and

$$\frac{1}{r^m}\frac{d}{dr}\left(r^m\frac{dh_n}{dr}\right)=\frac{1}{a}h_{n-1} \quad \text{dla} \quad n=1,\,2,\,3,\,... \tag{6}$$

By substituting (4) into conditions (2) and (3), one obtains

$$h_0(0)=1, \quad h_n(0)=0, \quad n=1, 2, 3, \ldots \tag{7}$$

and

$$\left.\frac{dh_n}{dr}\right|_{r=0}=0, \quad n=0, 1, 2, 3, \ldots \tag{8}$$

The solutions of (5) and (6) with conditions (7) and (8) will be discussed for the plate, cylinder and sphere below.

Plate

The solution of (5) and (6) with conditions (7) and (8) has the following form for $m = 0$:

$$h_n(r)=\frac{1}{(2n)!}\frac{r^{2n}}{a^n}, \qquad n=0, 1, \ldots \tag{9}$$

By substituting (9) into (4), a formula for the plate temperature distribution is obtained:

$$T(r,t)=y(t)+\sum_{n=1}^{\infty}\frac{1}{(2n)!}\left(\frac{r^2}{a}\right)^n\frac{d^n y}{dt^n}. \tag{10}$$

Heat flux is determined from equation

$$\dot{q}(r)=\lambda\frac{\partial T}{\partial r}=c\rho r\sum_{n=1}^{\infty}\frac{1}{(2n-1)!}\left(\frac{r^2}{a}\right)^{n-1}\frac{d^n y}{dt^n}. \tag{11}$$

By substituting $r = L$, the formula for heat flux on the surface of the plate front face is obtained

$$\dot{q}_s=\left.\dot{q}(r)\right|_{r=L}=c\rho L\sum_{n=1}^{\infty}\frac{1}{(2n-1)!}\left(\frac{L^2}{a}\right)^{n-1}\frac{d^n y}{dt^n}=\lambda\sum_{n=1}^{\infty}\frac{L^{2n-1}}{(2n-1)!}\frac{1}{a^n}\frac{d^n y}{dt^n}. \tag{12}$$

Cylinder

The solution of (5) and (6) has the following form for the sphere and cylinder:

$$h_n(r)=\frac{r^{2n}}{2^n(n!)(m+1)(m+3)\ldots(m+2n-1)a^n}. \tag{13}$$

By substituting (13) into (4), one obtains for the cylinder (m = 1)

$$T(r,t)=y(t)+\sum_{n=1}^{\infty}\frac{r^{2n}}{2^{2n}(n!)^2 a^n}\frac{d^n y}{dt^n}. \tag{14}$$

Heat flux is determined from Fourier Law

$$\dot{q}(r,t)=\lambda\frac{\partial T}{\partial r}=\lambda\sum_{n=1}^{\infty}\frac{nr^{2n-1}}{2^{2n-1}(n!)^2 a^n}\frac{d^n y}{dt^n}. \tag{15}$$

Heat flux on the cylinder surface $\dot{q}_s(t)$ is obtained by substituting $r = r_z$ into (15).

Sphere

Since for the sphere m = 2, then the expression for temperature distribution is obtained by substituting (13) into (4)

$$T(r,t)=y(t)+\sum_{n=1}^{\infty}\frac{r^{2n}}{(2n+1)!a^n}\frac{d^n y}{dt^n}. \tag{16}$$

Heat flux $\dot{q}(r,t)$ is determined from Fourier Law

$$\dot{q}(r,t)=\lambda\frac{\partial T}{\partial r}=\lambda\sum_{n=1}^{\infty}\frac{2nr^{2n-1}}{(2n+1)!a^n}\frac{d^n y}{dt^n}. \tag{17}$$

Heat flux on the sphere surface $\dot{q}_s(t)$ is obtained by substituting $r = r_z$ into expression (17).

The solutions for temperature distribution $T(r,t)$ and heat flux $\dot{q}(r,t)$ presented above are very sensitive to the small temperature measurement random errors $f(t)$. This is the result of temperature and heat flux extrapolation, measured or known (e.g. from symmetry condition of temperature field) at a point that lies inside the body in the direction of the body surface.

In terms of the direct problems, fast changes in heat flux or temperature, which occur on the body surface, are quickly suppressed inside it when boundary conditions are known.

In the case of inverse problems, small measurement errors are amplified even dozens of times, causing large oscillations in the determined heat flux or surface temperature.

Additional information regarding the inverse problems can be found in references [1, 9–12].

Exercise 18.2 Formula Derivation for a Temperature of a Medium when Linear Time Change in Plate Surface Temperature is Assigned

Determine temperature changes of a fluid, which warms up an infinite plate with thickness L. The surface temperature should change with time at constant rate v_T. The back plate surface is thermally insulated (Fig. 18.2).

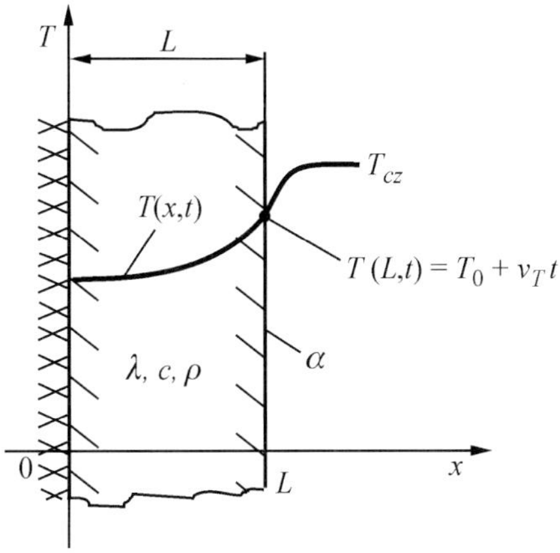

Fig. 18.2. Plate heating; surface temperature $T(L,t)$ increases at constant rate v_T

Solution

Plate temperature distribution is defined by the heat conduction equation

$$\frac{\partial^2 T}{\partial x^2} = \frac{1}{a}\frac{\partial T}{\partial t}, \tag{1}$$

by boundary conditions

$$\lambda \left.\frac{\partial T}{\partial x}\right|_{x=0} = 0, \tag{2}$$

$$T\big|_{x=L} = T_0 + v_T t \tag{3}$$

and by initial condition

$$T\big|_{t=0} = T_0. \tag{4}$$

The solution for initial-boundary problems (1)–(4) has the form [2, 15, 17]

$$T(x,t) = T_0 + \frac{v_T L^2}{a}\left\{\frac{at}{L^2} + \frac{1}{2}\left(\frac{x}{L}\right)^2 - \frac{1}{2} + \right.$$

$$\left. +2\sum_{n=1}^{\infty}\frac{(-1)^{n-1}}{\left(\frac{2n-1}{2}\pi\right)^3}\cos\left[\frac{(2n-1)\pi}{2}\frac{x}{L}\right]\exp\left[-\left(\frac{2n-1}{2}\pi\right)^2\frac{at}{L^2}\right]\right\}. \tag{5}$$

Heat flux is expressed as

$$\dot{q}(x,t) = \lambda\frac{\partial T}{\partial x} = \frac{\lambda v_T L}{a}\left\{\frac{x}{L} - \right.$$

$$\left. -2\sum_{n=1}^{\infty}\frac{(-1)^{n-1}}{\left(\frac{2n-1}{2}\pi\right)^2}\sin\left(\frac{2n-1}{2}\pi\frac{x}{L}\right)\exp\left[-\left(\frac{2n-1}{2}\pi\right)^2\frac{at}{L^2}\right]\right\}. \tag{6}$$

Heat flux on the plate surface $x = L$ is

$$\dot{q}_s = \dot{q}(L,t) = \frac{\lambda v_T L}{a}\left\{1 - 2\sum_{n=1}^{\infty}\left(\frac{2}{(2n-1)\pi}\right)^2\exp\left[-\left(\frac{2n-1}{2}\pi\right)^2\frac{at}{L^2}\right]\right\}. \tag{7}$$

Temperature of the medium $T_{cz}(t)$ can be determined from the boundary condition of 3rd kind

$$\lambda\left.\frac{\partial T}{\partial x}\right|_{x=L} = \alpha\left(T_{cz} - T|_{x=L}\right)$$

$$\dot{q}_s = \alpha\left(T_{cz} - T_0 - v_T t\right), \tag{8}$$

and is

$$T_{cz}(t) = T_0 + v_T t + \frac{\dot{q}_s}{\alpha}. \tag{9}$$

Once (7) is substituted into (9) and transformations are carried out, one has

$$T_{cz}(t) = T_0 + v_T\left\{t + \frac{1}{Bi}\frac{L^2}{a} - \right.$$
$$\left. - \frac{2}{Bi}\frac{L^2}{a}\sum_{n=1}^{\infty}\left[\frac{2}{(2n-1)\pi}\right]^2 \exp\left[-\left(\frac{2n-1}{2}\pi\right)^2 Fo\right]\right\}, \tag{10}$$

where $Bi = \alpha L/\lambda$, $Fo = at/L^2$.

Exercise 18.3 Determining Temperature Transient of a Medium for Which Plate Temperature at a Point with a Given Coordinate Changes According to the Prescribed Function

Determine temperature transient of a medium $T_{cz}(t)$ so that plate temperature at a point with coordinate x_T would change according to the given time function $y(t)$. Back plate surface is thermally insulated (Fig. 18.3). Initial plate temperature is T_0.

Solution

If the medium's temperature $T_{cz}(t)$ is a function of time, then plate temperature at point x_T can be determined by means of the Duhamel integral [2, 13]

$$T(x_T, t) = T_0 + \int_0^t \frac{\partial u(x_T, t-\theta)}{\partial t}\left[T_{cz}(\theta) - T_0\right] d\theta =$$
$$= T_0 + \int_0^t g(x, t-\theta)\left[T_{cz}(\theta) - T_0\right] d\theta, \tag{1}$$

where $u(x,t)$ is the temperature transient of the plate with initial temperature at zero and with a unit step-increase in medium's temperature T_{cz} = 1°C for $t > 0$ s, while $g(x,t)$ is the plate temperature transient an impulse function, i.e. Dirac's function $\delta(t)$ [8, 16]. By applying Laplace transform to the first integral from (1), one has

$$\bar{T}(x_T, s) = \frac{T_0}{s} + \left[s\bar{u}(x_T, s) - u(x_T, 0)\right]\left[\bar{T}_{cz}(s) - \frac{T_0}{s}\right]. \tag{2}$$

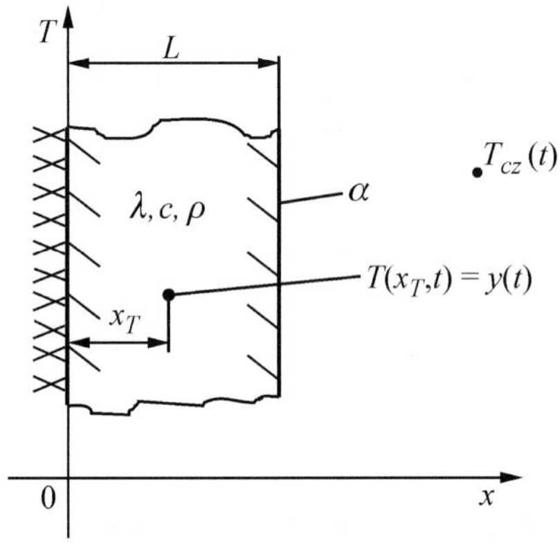

Fig. 18.3. Plate heating or cooling in time; the plate temperature at the point with coordinate x_T changes according to the prescribed function set, $y(t)$, i.e. $T(x,t)=y(t)$

If $u(x_T,0) = 0$, then the medium's temperature transform $\overline{T}_{cz}(s)$ determined from (2) has the form

$$\overline{T}_{cz}(s) = \frac{T_0}{s} + \frac{\overline{T}(x_T,s) - \frac{T_0}{s}}{s\overline{u}(x_T,s)} = \frac{T_0}{s} + \frac{\overline{y}(s) - \frac{T_0}{s}}{s\overline{u}(x_T,s)} . \tag{3}$$

One can determine transform $\overline{u}(x_T,s)$ by applying Laplace transform to the following initial-boundary problem:

$$\frac{\partial^2 u}{\partial x^2} = \frac{1}{a}\frac{\partial u}{\partial t}, \tag{4}$$

$$\left.\frac{\partial u}{\partial x}\right|_{x=0} = 0, \tag{5}$$

$$\lambda \left.\frac{\partial u}{\partial x}\right|_{x=L} = \alpha\left[1 - u(L,t)\right], \tag{6}$$

$$u|_{t=0} = 0, \tag{7}$$

hence

$$\frac{d^2\overline{u}}{dx^2} - \frac{s}{a}\overline{u} = 0, \tag{8}$$

$$\left.\frac{d\overline{u}}{dx}\right|_{x=0} = 0, \tag{9}$$

$$\lambda \left.\frac{d\overline{u}}{dx}\right|_{x=L} = \alpha\left[\frac{1}{s} - \overline{u}(L,s)\right]. \tag{10}$$

The solution of (8) with boundary conditions (9) and (10) is

$$\overline{u}(x,s) = \frac{\alpha}{\lambda}\frac{\cosh(qx)}{s\left[q\sinh(qL) + \frac{\alpha}{\lambda}\cosh(qL)\right]}, \tag{11}$$

where $q = \sqrt{s/a}$.

By substituting (11) into (3), one

$$\overline{T}_{cz}(s) = \frac{T_0}{s} + \left[\overline{y}(s) - \frac{T_0}{s}\right]\frac{\frac{\lambda}{\alpha}q\sinh(qL) + \cosh(qL)}{\cosh(qx_T)}. \tag{12}$$

In order to determine the medium's temperature transient $T_{cz}(t)$, one should either find the inverse Laplace transform of function (12) numerically or by using analytical [3–6].

In the case of more complex time transients of function $y(t)$, it is rather difficult to find analytically the inverse Laplace transform of function (12). This is why the transient $T_{cz}(t)$ can be determined numerically by inverting (12) by means of, for instance, the INLAP program from IMSL library [9].

Exercise 18.4 Formula Derivation for a Temperature of a Medium, which is Warming an Infinite Plate; Plate Temperature at a Point with a Given Coordinate Changes at Constant Rate

Determine temperature transient of a medium, which is warming up an infinite plate with thickness L; during the heating process, the plate temperature at the point with coordinate x_T changes at constant rate v_T. Back plate surface is thermally insulated (Fig. 18.4). Determine temperature transient of the medium $T_{cz}(t)$ using Laplace transformation.

Solution

Temperature distribution in the plate is defined by the heat conduction equation

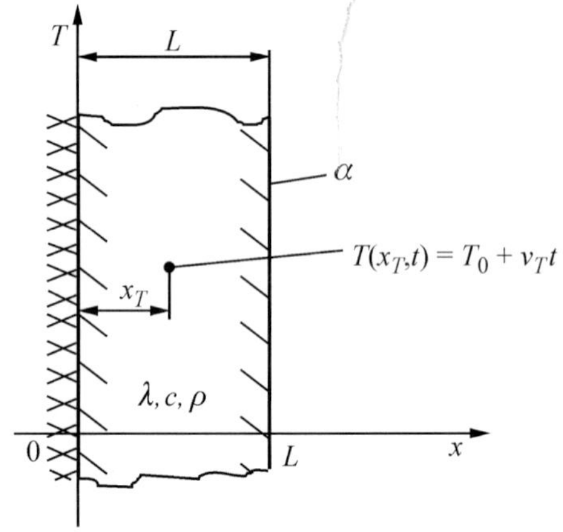

Fig. 18.4. Plate heating; during the heating process, the plate temperature at the point x_T increases at constant rate v_T

$$\frac{\partial^2 T}{\partial x^2} = \frac{1}{a}\frac{\partial T}{\partial t}, \tag{1}$$

by conditions

$$\lambda \left.\frac{\partial T}{\partial x}\right|_{x=0} = 0, \tag{2}$$

$$T\big|_{x=x_T} = T_0 + v_T t \tag{3}$$

and by initial condition

$$T\big|_{t=0} = T_0. \tag{4}$$

One can determine medium's temperature $T_{cz}(t)$ by applying the inverse Laplace transform to (12) from Ex. 18.3

$$\overline{T}_{cz}(s) = \frac{T_0}{s} + \left[\overline{y}(s) - \frac{T_0}{s}\right] \frac{\frac{\lambda}{\alpha} q \sinh(qL) + \cosh(qL)}{\cosh(q x_T)} \tag{5}$$

Because in the given case, transform $\overline{y}(s)$ is formulated as

$$\overline{y}(s) = \frac{T_0}{s} + \frac{v_T}{s^2}, \tag{6}$$

Equation (5) assumes the form

$$\overline{T}_{cz}(s)=\frac{T_0}{s}+\frac{v_T}{s^2}\frac{\frac{\lambda}{\alpha}q\sinh(qL)+\cosh(qL)}{\cosh(qx_T)}. \tag{7}$$

According to Appendix H, first we will determine this part of the inverse c_{-1} ((H.17), Appendix H), which corresponds to the double pole s = 0. Once we expand function

$$F(s)e^{st}=\left[\overline{T}_{cz}(s)-\frac{T_0}{s}\right]e^{st}=$$
$$=v_T\frac{\frac{\lambda}{\alpha}q\sinh(qL)+\cosh(qL)}{s^2\cosh(qx_T)}e^{st} \tag{8}$$

into Taylor series, we have

$$\left[\overline{T}_{cz}(s)-\frac{T_0}{s}\right]e^{st}=$$
$$=v_T\frac{\frac{\lambda}{\alpha}q\left(qL+\frac{q^3L^3}{6}+\ldots\right)+\left(1+\frac{q^2L^2}{2}+\frac{q^4L^4}{24}+\ldots\right)}{s^2\left(1+\frac{q^2x_T^2}{2}+\frac{q^4x_T^4}{24}\right)}\times$$
$$\times\left(1+st+\frac{s^2t^2}{2}+\ldots\right)=$$
$$=v_T\frac{1+\frac{sL^2}{a}\left(\frac{\lambda}{\alpha L}+\frac{1}{2}\right)+\frac{s^2L^4}{a^2}\left(\frac{1}{6}\frac{\lambda}{\alpha L}+\frac{1}{24}\right)+\ldots}{s^2\left(1+\frac{sx_T^2}{2a}+\frac{s^2x_T^4}{24a^2}\right)}\times$$
$$\times\left(1+st+\frac{1}{2}s^2t^2+\ldots\right). \tag{9}$$

Taking into account that (Appendix H)

$$A = v_T, \qquad B = \frac{v_T L^2}{a}\left(\frac{\lambda}{\alpha L} + \frac{1}{2}\right), \qquad C = \frac{v_T L^4}{a^2}\left(\frac{1}{6}\frac{\lambda}{\alpha L} + \frac{1}{24}\right),$$
$$D = 1, \qquad E = \frac{x_T^2}{2a}, \qquad F = \frac{x_T^4}{24a^2}, \tag{10}$$

the quasi-steady part of the solution will have the form

$$\frac{A}{D}t + \frac{B}{D} - \frac{AE}{D^2} = v_T\left\{t + \frac{L^2}{a}\left[\frac{1}{Bi} + \frac{1}{2}\left(1 - \frac{x_T^2}{L^2}\right)\right]\right\}. \tag{11}$$

Next, we will determine this part of the inverse transform, which results from the existence of multiple poles s_n

$$\sum_{n=1}^{\infty} \frac{g_1(s_n)}{g_2'(s_n)} e^{s_n t}. \tag{12}$$

Denominator in (12) is

$$g_2(s) = s^2 \cosh(q x_T). \tag{13}$$

In order to find the poles s_n, the characteristic equation has to be solved, i.e.

$$\cosh(q x_T) = 0. \tag{14}$$

By substituting $qx_T = \mu/i$, the characteristic (14) has the form

$$\cosh\left(\frac{\mu}{i}\right) = \cosh(-i\mu) = \cos\mu = 0. \tag{15}$$

Roots of (15) are

$$\mu_n = i q_n x_T = \frac{(2n-1)\pi}{2}, \qquad n = 1, 2, 3, \ldots \tag{16}$$

From (16), one obtains

$$\mu_n^2 = -q_n^2 x_T^2, \qquad \mu_n^2 = -\frac{s_n}{a} x_T^2,$$

$$s_n = -\frac{a\mu_n^2}{x_T^2} = -\left(\frac{2n-1}{2}\right)^2 \frac{\pi^2}{x_T^2} a. \tag{17}$$

Derivative $g_2'(s_n)$ is determined as

$$g_2'(s_n)=\left.\frac{dg_2(s)}{ds}\right|_{s=s_n}=\left.\left[2s\cosh(qx_T)+\frac{s^2x_T}{2aq}\sinh(qx_T)\right]\right|_{s=s_n}=$$

$$=\left.\frac{s^2x_T}{2aq}\sinh(qx_T)\right|_{s=s_n}=\frac{is_n^2x_T^2}{2a\left(\frac{2n-1}{2}\right)\pi}\sinh\left(\frac{\mu_n}{i}\right)= \tag{18}$$

$$=\frac{-s_n^2x_T^2i^2}{2a\left(\frac{2n-1}{2}\right)\pi}\sin\mu_n=\frac{1}{2}\frac{a}{x_T^2}\left(\frac{2n-1}{2}\pi\right)(-1)^{n-1},\quad n=1,2,3,\ldots$$

Next, the numerator is determined

$$g_1(s_n)=v_T\left.\left[\frac{\lambda}{\alpha}q\sinh(qL)+\cosh(qL)\right]\right|_{s=s_n}. \tag{19}$$

If we take into account that

$$\frac{\lambda}{\alpha}q_n\sinh(q_nL)=\frac{\lambda}{\alpha}\frac{\mu_n}{ix_T}\sinh\left(\frac{\mu_n}{i}\frac{L}{x_T}\right)=\frac{\lambda}{\alpha}\frac{\mu_n}{ix_T}\sinh\left(-i\mu_n\frac{L}{x_T}\right)=$$

$$=\frac{\lambda}{\alpha}\frac{\mu_n}{ix_T}\left[-i\sin\left(\mu_n\frac{L}{x_T}\right)\right]=-\frac{\lambda}{\alpha L}\frac{L}{x_T}\mu_n\sin\left(\mu_n\frac{L}{x_T}\right) \tag{20}$$

and

$$\left.\cosh(qL)\right|_{s=s_n}=\cosh\left(\frac{\mu_n}{i}\frac{L}{x_T}\right)=\cos\left(\mu_n\frac{L}{x_T}\right), \tag{21}$$

then (19) assumes the form

$$g_1(s_n)=v_T\left\{-\frac{1}{Bi}\left(\frac{L}{x_T}\right)\frac{(2n-1)}{2}\pi\sin\left(\frac{2n-1}{2}\pi\frac{L}{x_T}\right)+\right.$$

$$\left.+\cos\left(\frac{2n-1}{2}\pi\frac{L}{x_T}\right)\right\}, \tag{22}$$

where $Bi=\alpha L/\lambda$.

Once we account for (18) and (22), then (12) has the form

$$\sum_{n=1}^{\infty}\frac{g_1(s_n)}{g_2'(s_n)}e^{s_n t}=$$

$$=v_T\sum_{n=1}^{\infty}\frac{\cos\left(\frac{2n-1}{2}\pi\frac{L}{x_T}\right)-\frac{1}{Bi}\left(\frac{L}{x_T}\right)\left(\frac{2n-1}{2}\pi\right)\sin\left(\frac{2n-1}{2}\pi\frac{L}{x_T}\right)}{\frac{a}{2x_T^2}(-1)^{n-1}\left(\frac{2n-1}{2}\right)^3\pi^3}\times \tag{23}$$

$$\times\exp\left[-\frac{at}{x_T^2}\left(\frac{2n-1}{2}\pi\right)^2\right].$$

The solution to the problem (1)–(4) is given by

$$T_{cz}(t)=T_0+\frac{1}{D}\left(At+B-\frac{AE}{D}\right)+\sum_{n=1}^{\infty}\frac{g_1(s_n)}{g_2'(s_n)}e^{s_n t}. \tag{24}$$

Substituting (11) and (23) into (24), yields

$$T_{cz}(t)=T_0+v_T\left\{t+\frac{L^2}{a}\left[\frac{1}{Bi}+\frac{1}{2}\left(1-\frac{x_T^2}{L^2}\right)\right]\right\}+$$

$$+\frac{2x_T^2v_T}{a}\sum_{n=1}^{\infty}\frac{\cos\left(\frac{2n-1}{2}\pi\frac{L}{x_T}\right)-\frac{1}{Bi}\left(\frac{L}{x_T}\right)\left(\frac{2n-1}{2}\pi\right)\sin\left(\frac{2n-1}{2}\pi\frac{L}{x_T}\right)}{(-1)^{n-1}\left(\frac{2n-1}{2}\right)^3\pi^3}\times \tag{25}$$

$$\times\exp\left[-\frac{at}{x_T^2}\left(\frac{2n-1}{2}\pi\right)^2\right],$$

where $Bi = \alpha L/\lambda$.

When $x_T = L$, then, noting that $\cos\left(\frac{2n-1}{2}\pi\right)=0$, from (25), we have

$$T_{cz}(t) = T_0 + v_T\left(t + \frac{L^2}{a}\frac{1}{Bi}\right) -$$

$$-\frac{2}{Bi}\frac{v_T L^2}{a}\sum_{n=1}^{\infty}\left(\frac{2}{(2n-1)\pi}\right)^2 \exp\left[-\left(\frac{2n-1}{2}\pi\right)^2 Fo\right], \tag{26}$$

where $Bi = \alpha L/\lambda$ and $Fo = at/L^2$.

Equation (26) and (10) derived in Ex. 18.2 are identical.

Exercise 18.5 Determining Temperature and Heat Flux on the Plate Front Face on the Basis of a measured Temperature Transient on an Insulated Back Surface; Heat Flow on the Plate Surface is in the Form of a Triangular Pulse

Determine heat flux and temperature on the plate front face on the basis of measured temperature transient on the insulated back surface. Generate measurement data, using dimensionless temperature $\theta\,(1,Fo)$ (Table 16.3 in Ex. 16.9). Assume the following data for the calculation, L = 0.1 m, $\dot{q}_N$ = 100000 W/m^2, λ = 50 W/(m·K), a = $1\cdot10^{-5}$ m^2/s, T_0 = 20°C, Fo_m = at_m/L^2 = 0.75 and apply (11) and (12) derived in Ex. 18.1. Calculate time derivatives $d^n y/dt^n$ using the local polynomial approximation of the measured temperature transient $f_i = f(t_i)$, i = 1, 2, ..., 41. Develop a computational program in the FORTRAN language.

Solution

Measurement data contained in Table 16.3 (Ex. 16.9) was generated using step $\Delta Fo = 0.05$. Therefore, time step Δt is

$$\Delta t = \frac{\Delta Fo L^2}{a} = \frac{0.05\cdot 0{,}1^2}{1\cdot 10^{-5}} = 50 \text{ s}. \tag{1}$$

"Measured" temperature values were calculated from formula

$$f_i = T_i = T_0 + \frac{\dot{q}_N L}{\lambda}\theta_i = 20 + \frac{100000\cdot 0.1}{50}\theta_i = 20 + 200\theta_i. \tag{2}$$

Formulas given in paper [14] will be used to calculate derivatives dy/dt, d^2y/dt^2 and d^3y/dt^3 that one can determine by differentiating a local polynomial $y(t)$ of the third degree, which approximates nine subsequent measurement points f_i.

$$\begin{aligned}
y_i &= y(t_i) = \frac{1}{693}(-63f_{i-4} + 42f_{i-3} + 117f_{i-2} + 162f_{i-1} + 177f_i + 162f_{i+1} + \\
&\quad +117f_{i+2} + 42f_{i+3} - 63f_{i+4}), \\
y_i' &= \left.\frac{dy}{dt}\right|_{t=t_i} = \frac{1}{1188\Delta t}(86f_{i-4} - 142f_{i-3} - 193f_{i-2} - 126f_{i-1} + 126f_{i+1} + \\
&\quad +193f_{i+2} + 142f_{i+3} - 86f_{i+4}), \\
y_i'' &= \left.\frac{d^2y}{dt^2}\right|_{t=t_i} = \frac{1}{462(\Delta t)^2}(28f_{i-4} + 7f_{i-3} - 8f_{i-2} - 17f_{i-1} - 20f_i - \\
&\quad -17f_{i+1} - 8f_{i+2} + 7f_{i+3} + 28f_{i+4}), \\
y_i''' &= \left.\frac{d^3y}{dt^3}\right|_{t=t_i} = \frac{1}{198(\Delta t)^3}(-14f_{i-4} + 7f_{i-3} + 13f_{i-2} + 9f_{i-1} - 9f_{i+1} - \\
&\quad -13f_{i+2} - 7f_{i+3} + 14f_{i+4}),
\end{aligned} \tag{3}$$

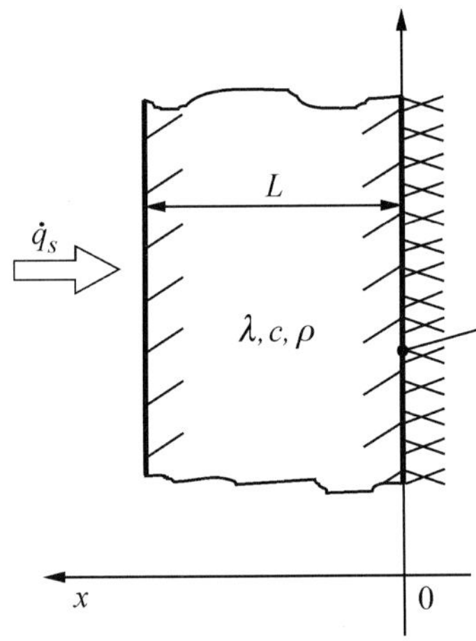

Fig. 18.5. Diagram that illustrates how heat flux $\dot{q}_s$ and plate surface temperature $x = L$ is determined on the basis of temperature $f(t)$ measured on the back plate surface

In (10) and (12), Ex. 18.1, only the first three terms are accounted for in the infinite series.

Calculation results are presented in Table 18.2.

Table 18.1. Measurement data : temperature of the insulated back plate surface

Ent. no.	t [s]	f [°C]	Lp.	t [s]	f [°C]	Lp.	t [s]	f [°C]
1	–200	20.00000	16	550	35.78754	31	1300	117.98048
2	–150	20.00000	17	600	39.87788	32	1350	121.38310
3	–100	20.00000	18	650	44.46552	33	1400	124.29120
4	–50	20.00000	19	700	49.55146	34	1450	126.70264
5	0	20.00016	20	750	55.13610	35	1500	128.61612
6	50	20.00032	21	800	61.22006	36	1550	130.03116
7	100	20.03000	22	850	67.74462	37	1600	130.97630
8	150	20.20522	23	900	74.47792	38	1650	131.56744
9	200	20.65190	24	950	81.16812	39	1700	131.93036
10	250	21.45732	25	1000	87.64070	40	1750	132.15220
11	300	22.67628	26	1050	93.78620	41	1800	132.28766
12	350	24.34242	27	1100	99.53728	42	1850	132.37036
13	400	26.47632	28	1150	104.85286	43	1900	132.42086
14	450	29.09052	29	1200	109.70782	44	1950	132.45168
15	500	32.19270	30	1250	114.08682	45	2000	132.47048

Table 18.2. Results of temperature and heat flux calculations on the plate front face

t [s]	$T_s = T(L,t)$ [°C]	$\dot{q}_s$ [W/m^2]	t [s]	$T_s = T(L,t)$ [°C]	$\dot{q}_s$ [W/m^2]
0	20.89	2454.00	950	143.10	56350.00
50	22.32	5379.00	1000	145.30	50920.00
100	24.78	9405.00	1050	146.70	45580.00
150	28.37	14180.00	1100	147.60	40350.00
200	33.00	19320.00	1150	148.00	35220.00
250	38.47	24540.00	1200	148.00	30130.00
300	44.54	29710.00	1250	147.40	25080.00
350	51.13	34820.00	1300	146.30	20050.00
400	58.22	39890.00	1350	144.70	15030.00
450	65.80	44930.00	1400	142.60	10130.00
500	73.89	49960.00	1450	140.30	5776.00
550	82.47	54970.00	1500	137.80	2461.00
600	91.55	59980.00	1550	135.70	383.30
650	101.10	64770.00	1600	134.10	–592.30
700	110.80	68470.00	1650	133.10	–816.20
750	120.00	70090.00	1700	132.60	–676.90
800	128.20	69240.00	1750	132.50	–458.80
850	134.90	66190.00	1800	132.50	–287.70
900	139.80	61630.00			

From the analysis of results presented in Table 18.2, Fig. 18.6 and 18.7, it follows that the accuracy of the obtained results is very good.

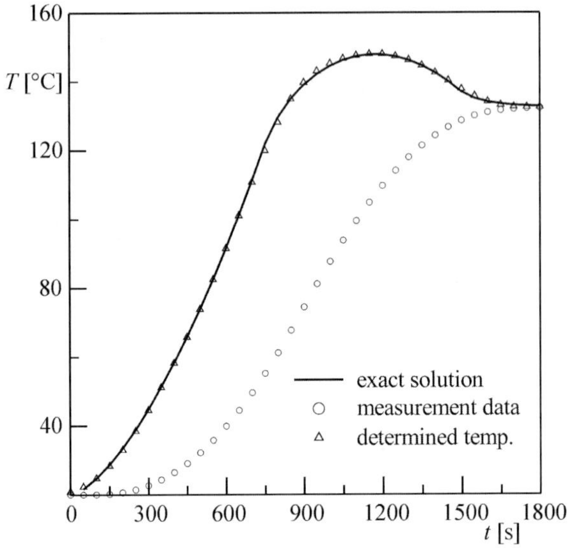

Fig. 18.6. Measurement data and determined temperature of the plate front face

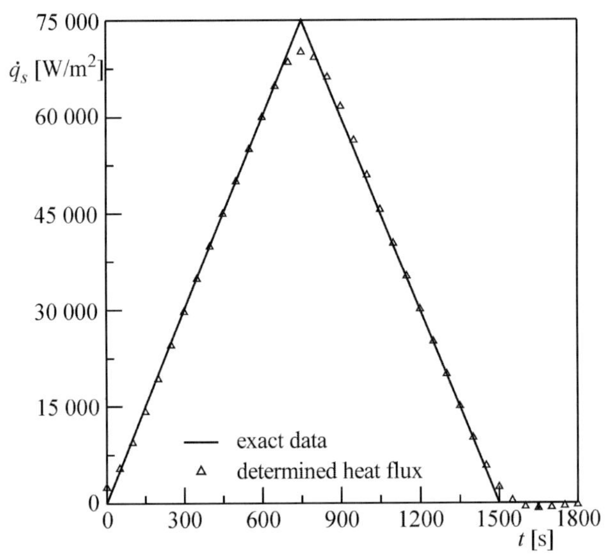

Fig. 18.7. Comparison of heat flux $\dot{q}_s$ determined on the surface of the plate front face with exact data

Data for the program used to calculate heat flux and temperature of the plate front face on the basis of measured temperature transient on the thermally insulated back surface

```
p18_5.in
45 4  50.
20. 0.1 1.E-5
50. 100000.
0      20
0      20
0      20
0      20
0      20.00016
50     20.00032
100    20.03
150    20.20522
200    20.6519
250    21.45732
300    22.67628
350    24.34242
400    26.47632
450    29.09052
500    32.1927
550    35.78754
600    39.87788
650    44.46552
700    49.55146
750    55.1361
800    61.22006
850    67.74462
900    74.47792
950    81.16812
1000   87.6407
1050   93.7862
1100   99.53728
1150   104.85286
1200   109.70782
1250   114.08682
1300   117.98048
1350   121.3831
1400   124.2912
1450   126.70264
1500   128.61612
1550   130.03116
1600   130.9763
1650   131.56744
1700   131.93036
1750   132.1522
```

```
1800   132.28766
1850   132.37036
1900   132.42086
1950   132.45168
2000   132.47048
```

Program for the calculation of heat flux and temperature of the plate front face on the basis of measured temperature transient on the thermally insulated back surface

```
      program p18_5
      dimension w_f(50),w_y(50)
      open(unit=1,file='p18_5.in')
      open(unit=2,file='p18_5.out')
      read(1,*)n_max,n_start,dt
      read(1,*)t_o,s_l,s_a
      read(1,*)s_lam, s_q
      do i=1,n_max
         read(1,*)t, w_f(i)
      enddo
      write(2,'(a)')
     &"CALCULATION OF PLATE TEMPERATURE DISTRIBUTION"
      write(2,'(/a)') "INPUT DATA"
      write(2,'(a,i10)')      "n_max   =",n_max
      write(2,'(a,i10)')      "n_start  =",n_start
      write(2,'(a,e10.5,a)') "t_o     =",t_o, " [C]"
      write(2,'(a,e10.5,a)') "L       =",s_l," [m]"
      write(2,'(a,e10.5,a)') "a       =",s_a, " [m^2/s]"
      write(2,'(a,e10.5,a)') "lambda =",s_lam," [W/mK]"
      write(2,'(a,e10.5,a)') "q_N    =",s_q," [W/m2]"
      s_cro=s_lam/s_a
      write(2,'(/a)')"CALCULATED TEMPERATURE[C] AND q[W/m2]"
      write(2,'(a,a)')"  t[s]     T(L,t)       q_s"
      t=0.
      t_max=(n_max-2*n_start-1)*50.
      do while (t.le.t_max)
         write(2,'(f5.0,5(3x,e10.4))')t,
     &      temperature(s_l,t,s_a,w_f,w_y,n_max,n_start,dt),
     &      flux_q(s_l,t,s_a,w_f,w_y,n_max,n_start,s_cro,dt)
         t=t+dt
      enddo
      end program p18_5
c     according to equation (11)
      function flux_q(r,t,s_a,w_f,w_y,n_max,n_start,s_cro,dt)
      dimension w_f(*),w_y(*)
      i=1+t/50
      call filtr_9p1(w_f,w_y,n_max,n_start,dt)
      str=w_y(i)*s_cro*r
```

```
      call filtr_9p2(w_f,w_y,n_max,n_start,dt)
      str=str+w_y(i)*s_cro*r*(r**2/s_a)/6
      call filtr_9p3(w_f,w_y,n_max,n_start,dt)
      str=str+w_y(i)*s_cro*r*(r**2/s_a)**2/120
      flux_q=str
      end function
c     according to equation (10)
      function temperature(r,t,s_a,w_f,w_y,n_max,n_start,dt)
      dimension w_f(*),w_y(*)
      i=1+t/50
      call filtr_9(w_f,w_y,n_max,n_start)
      temp=w_y(i)
      call filtr_9p1(w_f,w_y,n_max,n_start,dt)
      temp=temp+w_y(i)*(r**2/s_a)/2
      call filtr_9p2(w_f,w_y,n_max,n_start,dt)
      temp=temp+w_y(i)*(r**2/s_a)**2/24
      call filtr_9p3(w_f,w_y,n_max,n_start,dt)
      temp=temp+w_y(i)*(r**2/s_a)**3/720
      temperature=temp
      end function
c     according to equation (3)
      subroutine filtr_9(w_f,w_y,n_max,n_start)
      dimension w_f(*),w_y(*)
      do i=1+n_start, n_max-n_start
         j=i-n_start
         w_y(j)=1./693.*
     &(-63.*w_f(i-4)+42.*w_f(i-3)+117.*w_f(i-2)+162.*w_f(i-1)
     & +177.*w_f(i)
     & +162.*w_f(i+1)+117.*w_f(i+2)+42.*
     $ w_f(i+3)-63.*w_f(i+4))
      enddo
      end
      subroutine filtr_9p1(w_f,w_y,n_max,n_start,dt)
      dimension w_f(*),w_y(*)
      do i=1+n_start, n_max-n_start
         j=i-n_start
         w_y(j)=1./(1188.*dt)*
     &(86.*w_f(i-4)-142.*w_f(i-3)-193.*w_f(i-2)-126.*w_f(i-1)
     & +126.*w_f(i+1)+193.*w_f(i+2)+142.*
     & w_f(i+3)-86.*w_f(i+4) )
      enddo
      end
      subroutine filtr_9p2(w_f,w_y,n_max,n_start,dt)
      dimension w_f(*),w_y(*)
      do i=1+n_start, n_max-n_start
         j=i-n_start
```

```
        w_y(j)=1./(462.*dt*dt)*
&        ( 28.*w_f(i-4)+7.*w_f(i-3)-8.*w_f(i-2)-17.*w_f(i-1)
&        -20*w_f(i)
&        -17.*w_f(i+1)-8.*w_f(i+2)+7.*w_f(i+3)+28.*w_f(i+4))
      enddo
      end
      subroutine filtr_9p3(w_f,w_y,n_max,n_start,dt)
      dimension w_f(*),w_y(*)
      do i=1+n_start, n_max-n_start
         j=i-n_start
         w_y(j)=1./(198.*dt*dt*dt)*
&( -14.*w_f(i-4)+7.*w_f(i-3)+13.*w_f(i-2)+9.*w_f(i-1)
&  -9.*w_f(i+1)-13.*w_f(i+2)-7.*w_f(i+3)+14.*w_f(i+4) )
      enddo
      end
```

Exercise 18.6 Determining Temperature and Heat Flux on the Surface of a Plate Front Face on the Basis of a Measured Temperature Transient on an Insulated Back Surface; Heat Flow on the Plate Surface is in the Form of a Rectangular Pulse

Determine heat flux and temperature of the plate butting face on the basis of a measured temperature transient on an insulated back surface. Generate measurement data using dimensionless temperatue θ (1,*Fo*) from Table 16.4 in Ex.16.10. Assume the following data for the calculation, $L = 0.1$ m, $\dot{q}_N = 100\,000$ W/m^2, $\lambda = 50$ W/(m·K), $a = 1 \cdot 10^{-5}$ m^2/s, $T_0 = 20$°C and apply (10) and (12) derived in Ex. 18.1. Calculate time derivatives $d^n y/dt^n$ using local polynomial approximation of the measured temperature transient $f_i = f(t_i)$, $i = 1, 2, ..., 43$. Carry out the calculations using the program developed in Ex. 18.5.

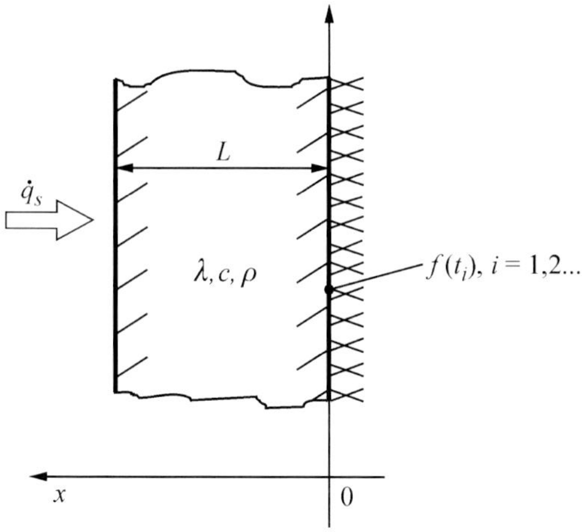

Fig. 18.8. Diagram that illustrates how heat flux $\dot{q}_s$ and temperature are determined on the plate surface $x = L$ on the basis of a measured temperature $f(t)$ on the back plate surface

Table 18.3. Measurement data: temperature of an insulated back plate surface

Entry no.	t [s]	f [°C]	Lp.	t [s]	f [°C]	Entry no.	t [s]	f [°C]
1	–200	20.00000	16	550	20.00000	31	1300	96.79066
2	–150	20.00000	17	600	20.00000	32	1350	105.19816
3	–100	20.00000	18	650	20.00000	33	1400	110.87166
4	–50	20.00000	19	700	20.00000	34	1450	114.41436
5	0	20.00000	20	750	19.81750	35	1500	116.58820
6	50	20.00000	21	800	20.05392	36	1550	117.91684
7	100	20.00000	22	850	21.57712	37	1600	118.72820
8	150	20.00000	23	900	25.86132	38	1650	119.22356
9	200	20.00000	24	950	32.29280	39	1700	119.52600
10	250	20.00000	25	1000	40.10320	40	1750	119.71062
11	300	20.00000	26	1050	48.76492	41	1800	119.82334
12	350	20.00000	27	1100	57.94768	42	1850	119.89214
13	400	20.00000	28	1150	67.44872	43	1900	119.93416
14	450	20.00000	29	1200	77.14410	44	1950	119.95980
15	500	20.00000	30	1250	86.95812	45	2000	119.97546

Solution

Measurement data from Table 16.4 in Ex. 16.10 was generated with a step $\Delta Fo = 0.05$. Therefore, time step Δt measures

Solution

INLAP program from IMSL library [9] will be used to find numerically inverse transform (7) from Ex. 18.4

$$T_{cz}(s)-\frac{T_0}{s}=\frac{v_T}{s^2}\frac{\frac{\lambda}{\alpha}q\sinh(qL)+\cosh(qL)}{\cosh(qx_T)}, \tag{2}$$

where $q = (s/a)^{1/2}$ is based on the application of epsilon algorithm to complex Fourier series. Calculations for $x_T = 0.3$, $L = 0.03$ m were carried out for time step $\Delta t = t_{i+1}-t_i = 30$ s due to the fact that the inverse problem is badly conditioned, since point x_T is at a distance $(L-x_T) = 0.1 - 0.03 = 0.07$ m from the plate front face, which is heated. When $x_T = L$, i.e. the point at which the plate temperature should be increasing at a constant rate, lies on the plate front face, the problem becomes well conditioned and very good results are obtained when the time step is significantly reduced to $\Delta t = 3$ s. Calculation results $T_{cz}(t)$ for $x_T = L$ are presented in Fig. 18.11, while for $x_T = 0.3L$ in Fig. 18.12.

From the analysis of Fig. 18.11 and Fig. 18.12 it follows that both, the numerical inversion of transform (2) and the analytical (25) from Ex. 18.4 yield the same results. When $x_T = L$ (Fig. 18.11), both methods for determining $T_{cz}(t)$ yield results that show slight divergence only in the second place after the decimal point.

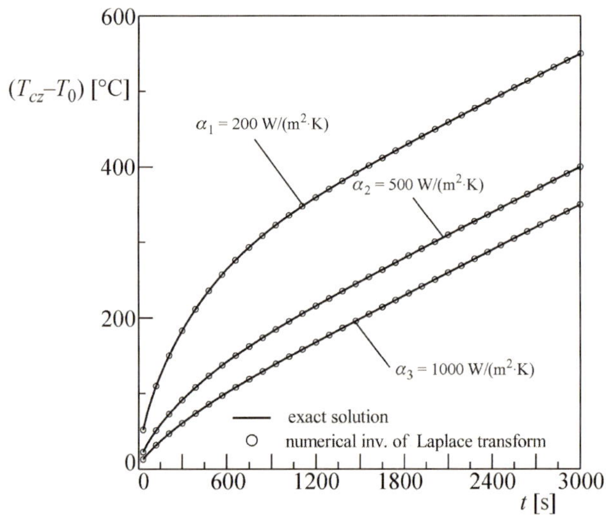

Fig. 18.11. Transient of a medium's temperature excess $[T_{cz}(t) - T_0]$; the plate surface temperature $T(L,t)$ increases at a constant rate of $v_T = 0.1$°C/s, $\Delta t = 3$ s

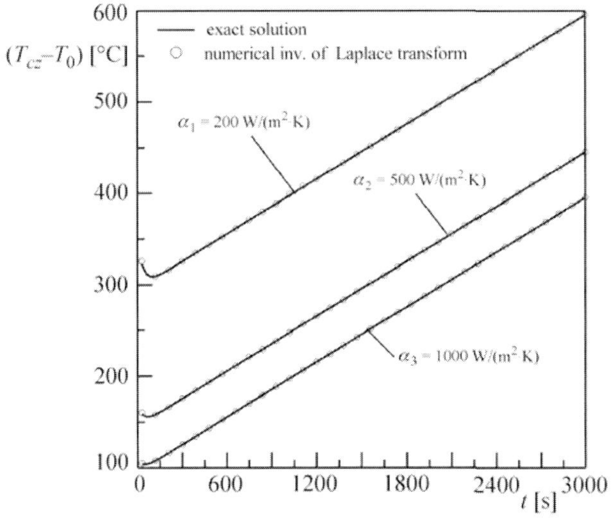

Fig. 18.12. Transient of a medium's temperature excess $[T_{cz}(t) - T_0]$; the plate temperature at point $x_T = 0.03$ m, which lies at a distance $(L - x_T) = 0.07$ m from the front face, increases at a constant rate of $v_T = 0.1$°C, $\Delta t = 30$ s

A program for calculating medium's temperature when plate temperature is known at a point located at a distance x_T from the insulated back plate surface by means of the analytical formula and with the use of the numerical inversion of the Laplace transform

```
      program p18_7
      INTEGER I, KMAX, N
      REAL    ALPHA,EXP,FINV(10000),FLOAT, RELERR, T(10000),
     &        TRUE(10000),FO(10000)
      COMPLEX    F
      INTRINSIC  EXP, FLOAT
      EXTERNAL   F, INLAP, UMACH
      common a,vt,al,xt,slam,alfa
      OPEN (2,FILE='p18_7.out',STATUS='OLD')
c     a - thermal diffusivity m2/s
c     vt - temp change rate K/s
c     slam - lambda W/mK
c     points calculated from the insulated back surface
c     (where temperature is set)
c     al- plate thickness m
c     alfa- heat transfer coefficient W/m2K
c     dt - time step s
      a=1.E-05
      vt=0.1
```

```
      al=0.1
c     for next solution xt=al, relerr=0.1E-01, dt=0.3 s
c     xt=al*0.3
          xt=al
          slam=50.
          ALFA=1000.
          dt=30.
      DO 10  I=1,1000
         T(I) = dt*FLOAT(I)
   10 CONTINUE
c     goto 40
      N      =1000
      ALPHA  = 0
      KMAX   = 1000
      RELERR = 0.2E-01
      CALL INLAP (F, N, T, ALPHA, RELERR, KMAX, FINV)
c     exact solution
      pi=3.141592653
      BI=ALFA*AL/SLAM
      do  i=1,1000
      s1=0.
      do j=1,20
         beta=float(2*j-1)/2.*pi
         fo(i)=a*t(i)/xt/xt
         s1=s1 +float((-1)**(j-1))*(cos(beta*al/xt)-
     & (1./bi)*(al/xt)
     1   *beta*sin(beta*al/xt))
     2     /(beta**3)*exp(-(beta**2)*fo(i))
      enddo
         TRUE(i) =vt*(t(i)+al*al/a*
     $     (0.5-0.5*xt*xt/al/al+1./bi))
     1   +(vt/a)*2.*xt*xt*s1
      enddo
      write(2,'(a)')"      ALFA          RELERR          XT"
      write(2,*)ALFA,RELERR,XT
      write(2,'(/a)')"CALCULATED MEDIUM TEMPERATURE [C]"
      write(2,'(a,a)')"   t[s]       T(t)      T_ exact (t) "
      write(2,'(3(3x,e11.5))')(T(I),FINV(I),TRUE(I),I=1,100)
      end
      COMPLEX FUNCTION F (S)
      COMPLEX    S,q,csinh,ccosh
      common a,vt,al,xt,slam,alfa
      q=csqrt(s/a)
      F=(vt/s**2)*((slam/alfa)*q*csinh(q*al)+ccosh(q*al))
     & /ccosh(q*xt)
      RETURN
      END
      COMPLEX FUNCTION ccosh(s)
      COMPLEX s
      ccosh=(cexp(s)+cexp(-s))/2.
      RETURN
      COMPLEX FUNCTION csinh(s)
```

```
COMPLEX s
END
csinh=(cexp(s)-cexp(-s))/2.
RETURN
END
```

Exercise 18.8 Determining Time-Temperature Transient of a Medium, for which the Plate Temperature at a Point with a Given Coordinate Changes According to the Square Function Assigned

Plate temperature $y(t)$ in a point located at a distance x_T from thermally insulated back surface should change in time according to the following formula:

$$y(t) = T(x_T, t) = T_0 + C_1 t + C_2 t^2 , \tag{1}$$

where C_1 and C_2 are constants. For the calculation assume that T_0 = 20°C, C_1 = 297/1620°C/s and C_2 = –33/1296000°C/s^2. Plate thickness is L = 0.1 m. Thermo - physical properties are as follow: λ = 50 W/(m·K), $a = \lambda/(c\rho)$ = $1\cdot 10^{-5}$ m^2/s. Carry out calculations for the two values x_T: $x_T = 0.3L = 0.03$ m and $x_T = L = 0.1$ m and for the two values of the heat transfer coefficient: α_1 = 200 W/(m^2·K) and α_2 = 1000 W/(m^2·K) Determine temperature transient $T_{cz}(t)$ by numerically inversing Laplace transform defined by (12) in Ex. 18.3. Carry out calculations by means of the INLAP program from IMSL library, assuming that time step is Δt = 60 s for x_T = 0.03 m and Δt = 3 s for x_T = 0.1 m.

Solution

Wall temperature transient $y(t)$ at the point x_T is presented in Fig. 18.13. Laplace transform of function $y(t)$ formulated in (1) has the form

$$\bar{y}(s) = \frac{T_0}{s} + \frac{C_1}{s^2} + \frac{C_2}{s^3} . \tag{2}$$

By accounting for (2) in (12), Ex. 18.3, one obtains

$$\bar{T}_{cz}(s) - \frac{T_0}{s} = \left(\frac{C_1}{s^2} + \frac{C_2}{s^3} \right) \frac{\frac{\lambda}{\alpha} q \sinh(qL) + \cosh(qL)}{\cosh(q x_T)} . \tag{3}$$

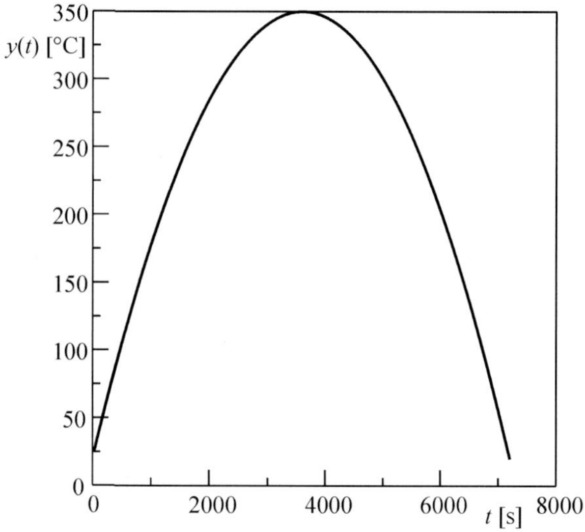

Fig. 18.13. Temperature transient $y(t)$ at a point with coordinate x_T

Determined temperature transients $T_{cz}(t)$ that ensure the required wall temperature transient $y(t)$ at the point with coordinate x_T, are presented, respectively, in Fig. 18.14 and Fig. 18.15 for $x_T = 0.1$ m and $x_T = 0.03$ m.

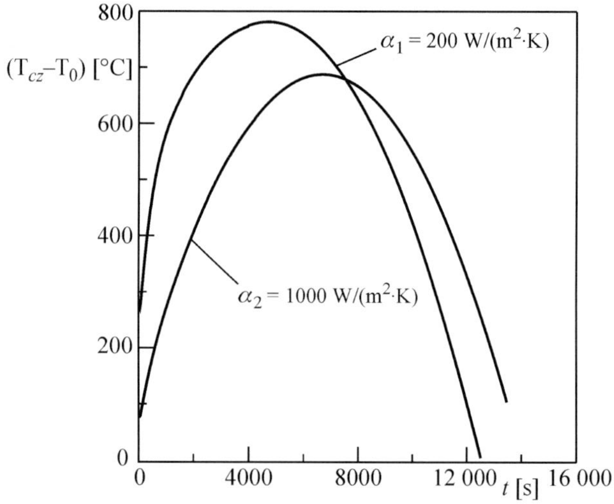

Fig. 18.14. Temperature transient of a medium $T_{cz}(t)$, which ensures assumed wall's front face temperature transient presented in Fig. 18.13, $\Delta t = 3$ s

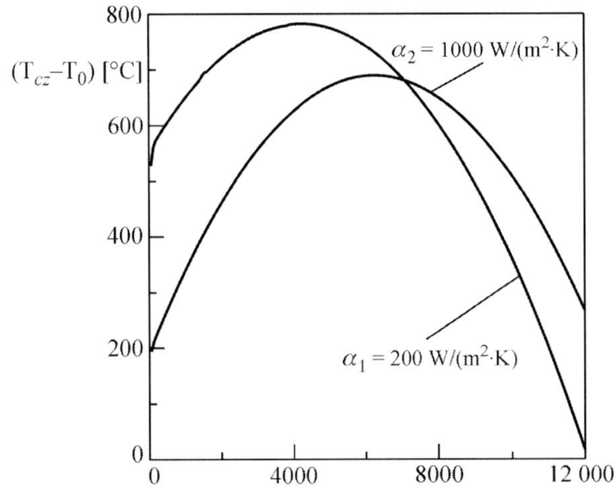

Fig. 18.15. Temperature transient of a medium $T_{cz}(t)$, which ensures assumed wall′s butting face temperature transient $y(t)$ presented in Fig. 18.13 in a point located at a distance $x_T = 0.03$ m from thermally insulated back surface, $\Delta t = 60$ s

Literature

1. Burggraf OR (1964) An exact solution of the inverse problem in heat conduction theory and applications. Transactions of the ASME, J. of Heat Transfer: 373-382
2. Carslaw HS, Jaeger JC (1959) Conduction of Heat in Solids. Clarendon Press, Oxford
3. Ciałkowski M (1996) Selected methods and algorithms of solving inverse problem for heat conduction equation (in Polish). Poznań Univ. of Technology Press, Poznań
4. Ciałkowski M, Frąckowiak A (2000) Thermal function and its applying for solving problems of heat transfer and mechanic (in Polish). Poznań Univ. of Technology Press, Poznań
5. Crump KS (1976) Numerical inversion of Laplace transforms using a Fourier series approximation. Journal of the Association for Computing Machinery 23: 89-96
6. Durbin F (1973) Numerical inversion of Laplace transform: efficient improvement to Dubner and Abate`s method. Comp. J 17: 371-376
7. Honig G, Hirdes U (1984) A method for the numerical inversion of Laplace transforms. J. Comp. Appl. Math. 9: 113-132
8. Hoog FR, Knight JH, Stokes AN (1982) An improved method for the numerical inversion of Laplace transforms. SIAM J. on Scientific and Statistical Com. 3: 357-366

9. IMSL. Fortran Subroutines for Mathematical Applications (1994). Visual Numerics, Houston
10. Naresh Sinha K (1991) Linear Systems. Wiley, New York
11. Taler J (1996) A semi-numerical method for solving inverse heat conduction problems. Heat and Mass Transfer 31: 105-111
12. Taler J (1997) Überwachung von instationären Wärmespannungen in dickwandigen Bauteilen. Forschung im Ingenieurwesen 63: 127-135
13. Taler J (1997) Analytical solution of the overdetermined inverse heat conduction problem with an application to monitoring thermal stresses. Heat and Mass Transfer 33: 209-218
14. Taler J, Zima W (1999) Solution of inverse heat conduction problems using control volume approach. Int. J. of Heat and Mass Transfer 42: 1123-1140
15. Taler J, Zborowski M (1998) Solution of the inverse problems in heat transfer and thermal stress analysis. Journal of Thermal Stresses 21: 563-579
16. Taler J (1995) Theory and practice of heat transfer (in Polish). Ossolineum, Wrocław
17. Tautz H (1971) Wärmeleitung und Temperaturausgleich. Verlag Chemie, Weinheim
18. Zill DG (1986) Differential Equations with Boundary-Value Problems. PWS, Boston
19. Pekhovitch AI, Dzjidkhin WM (1976) Calculations of thermal states of solid bodies (in Russian). Leningrad, Energija

19 Multidimensional Problems. The Superposition Method

In Chap. 19, the authors discuss how one-dimensional solutions can be employed to determine two or three-dimensional temperature distributions, using the superposition method. They also compile formulas for calculating temperature distribution with boundary conditions of 1st, 2nd and 3rd kind. Formulas are derived for a transient temperature distribution in a plate with convective boundary conditions or a surface-set heat flux. The authors also present two computational examples that illustrate how temperature can be determined in a cylinder of a finite length and in a cuboidal steel block.

Exercise 19.1 The Application of the Superposition Method to Multidimensional Problems

By knowing the solutions to transient heat conduction problems in a half-space, plate and cylinder, one can determine formulas that describe temperature distribution in the elements formed as a result of a mutual intersection of the aforementioned simple-shape bodies. Axes of these bodies are perpendicular. The bodies can be anisotropic, but the thermo-physical properties cannot be temperature-dependent. Initial temperature of a body (T_0) is constant.

Solution

Due to the fact that different procedures must be applied, boundary conditions of 1st and 3rd kind will be analyzed separately from boundary conditions of 2nd kind.

Boundary Conditions of 1st and 3rd Kind

If we know the solutions to transient problems in a half-space (Ex. 14.4), plate (Ex. 15.1) and cylinder (Ex. 15.4), then temperature distribution in the body formed from the intersection of simple bodies is the product of dimensionless solutions for these bodies. By denoting dimensionless temperature of the plate as

$$\Theta^* = \frac{T(x,t) - T_{cz}}{T_0 - T_{cz}}, \tag{1}$$

the temperature of the cuboid is calculated from formula

$$\Theta^*(x,y,z,t) = \Theta^*(x,t) \cdot \Theta^*(y,t) \cdot \Theta^*(z,t). \tag{2}$$

By substituting (2) into a heat conduction equation and boundary conditions, one obtains three one-dimensional problems for the plate. If, subsequently, absolute temperature (1) is denoted for the half-space as $S(x,t)$, for the plate as $P(x,t)$ and for the cylinder as $C(r,t)$, then temperature of body Θ^* can be expressed as the product of the listed solutions. The form of the solution for the nine different geometric figures is presented in Table 19.1. It should be noted that temperature $(T(x,t) - T_{cz})/(T_{cz} - T_0)$ (Ex. 14.4) was determined for the semi-infinite body; therefore, temperature $\Theta^*(x,t)$ is formulated as

$$S(x,t) = \Theta^*(x,t) = \frac{T(x,t) - T_{cz}}{T_0 - T_{cz}} = 1 - \frac{T(x,t) - T_0}{T_{cz} - T_0}. \tag{3}$$

In the case of the boundary condition of the 1st kind, the body surface temperature undergoes a step-increase from initial temperature T_0 to temperature T_{cz}. Basic geometric forms are presented in Fig. 19.1.

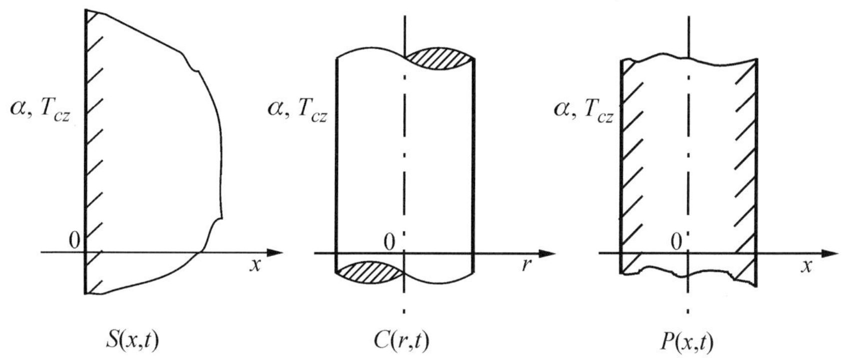

Fig. 19.1. Diagram of a simple-shape body with an assigned boundary conditions of 1st or 3rd kind

Boundary Conditions of 2nd Kind

For boundary conditions of 2nd kind, temperature distribution in a body, which is a combination of simple-shape bodies, is the sum of the initial temperature and temperatures that describe temperature distribution in these bodies. In the case of a cuboid (Table 19.1), the temperature field is formulated as

$$T(x,y,z,t) = T_0 + T_p(x,t) + T_p(y,t) + T_p(z,t), \tag{4}$$

where T_p is the solution of a one-dimensional problem for the plate, when the boundary condition of 2nd kind (Ex. 15.10) and the zero initial condition are assigned. It should be emphasized, however, that body surface heat flux $\dot{q}_s$ can be time-dependent. In Table 19.1, the authors list analytical formulas for boundary conditions of 1st or 3rd kind and boundary conditions of 2nd kind.

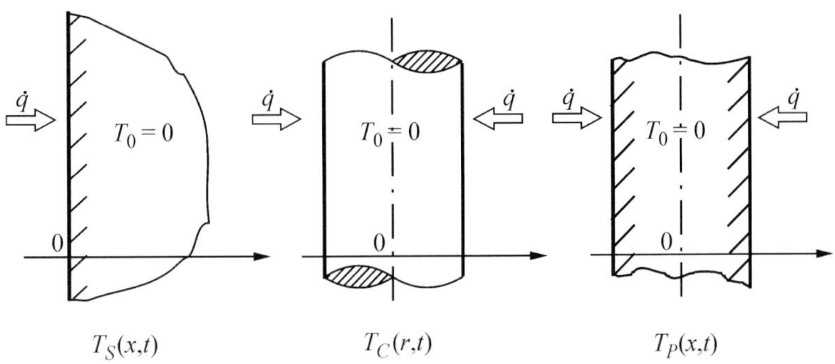

Fig. 19.2. Diagram of a simple-shape body with an assigned boundary conditions of 2nd kind

Table 19.1. Formulas for the calculation of temperature by means of the superposition method

Body shape	Diagram	Analytical formula: Boundary condition of 1st or 3rd kind	Analytical formula: Boundary condition of 2nd kind
Two-dimensional corner	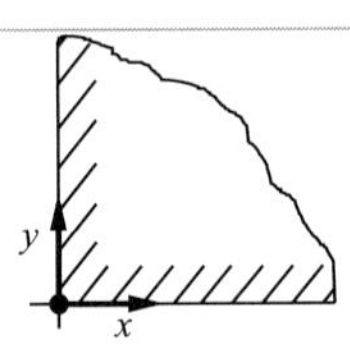	z	$T(x,y,t) = T_0 + T_S(x,t) + T_S(y,t)$
Three-dimensional corner	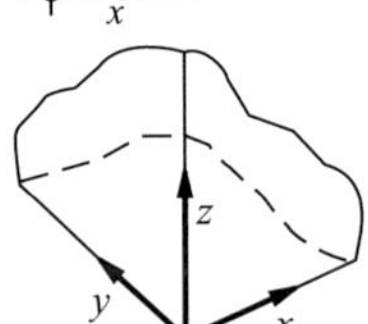	$\Theta^*(x,y,z,t) = S(x,t) \cdot S(y,t) \cdot S(z,t)$	$T(x,y,z,t) = T_0 + T_S(x,t) + T_S(y,t) + T_S(z,t)$

Table 19.1. (cont.)

Body shape	Diagram	Analytical formula Boundary cond. of 1st or 3rd kind	 Boundary condition of 2nd kind
Infinite plate	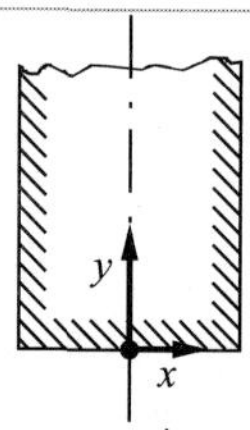	$\Theta^*(x,y,t) = P(x,t) \cdot S(y,t)$	$T(x,y,t) = T_0 + T_P(x,t) + T_S(y,t)$
Infinitely long rod with a rectangular cross-section	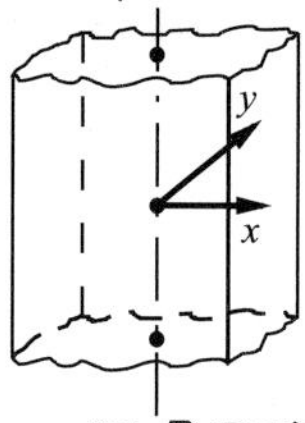	$\Theta^*(x,y,t) = P(x,t) \cdot P(y,t)$	$T(x,y,t) = T_0 + T_P(x,t) + T_P(y,t)$
Semi-finite rod with a rectangular cross-section	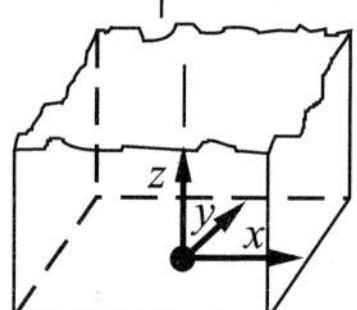	$\Theta^*(x,y,z,t) = P(x,t) \cdot P(y,t) \cdot P(z,t)$	$T(x,y,z,t) = T_0 + T_P(x,t) + T_S(y,t) + T_P(z,t)$
A corner with a finite width	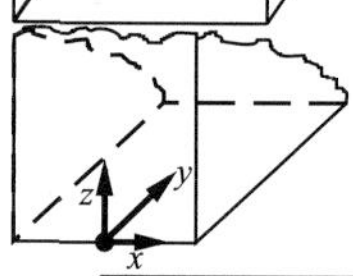	$\Theta^*(x,y,z,t) = P(x,t) \cdot S(y,t) \cdot S(z,t)$	$T(x,y,z,t) = T_0 + T_P(x,t) + T_S(y,t) + T_P(z,t)$
A cuboid with a coordinate system that begins in the center of the object	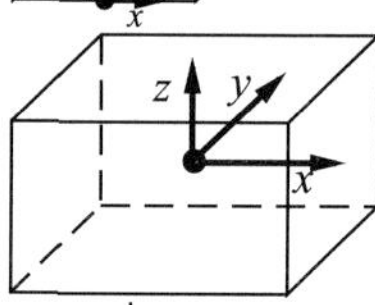	$\Theta^*(x,y,z,t) = P(x,t) \cdot P(y,t) \cdot P(z,t)$	$T(x,y,z,t) = T_0 + T_P(x,t) + T_P(y,t) + T_P(z,t)$
Semi-infinite cylinder	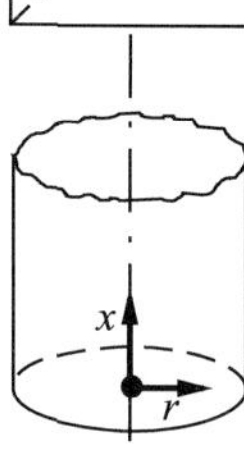	$\Theta^*(x,r,t) = S(x,t) \cdot C(r,t) \cdot P(z,t)$	$T(x,r,t) = T_0 + T_S(x,t) + T_C(r,t)$
A cylinder with a finite length (the coordinate system has a point of origin in the cylinder center)	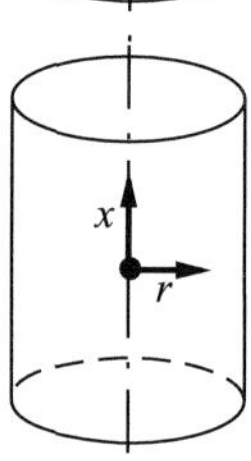	$\Theta^*(x,r,t) = P(x,t) \cdot C(r,t)$	$T(x,r,t) = T_0 + T_P(x,t) + T_C(r,t)$

Exercise 19.2 Formula Derivation for Temperature Distribution in a Rectangular Region with a Boundary Condition of 3rd Kind

Using the superposition method, derive a formula for temperature distribution in an infinitely long rod with a rectangular cross-section $2b \times 2d$, and is heated or cooled by a medium at constant temperature T_{cz} (Fig. 19.3). Heat transfer coefficient on the side, which measures $2b$ in length is α_b, while on the side $2d$ equals α_d. Initial temperature of the rod is constant and equals T_0, while thermo-physical properties are temperature-invariant.

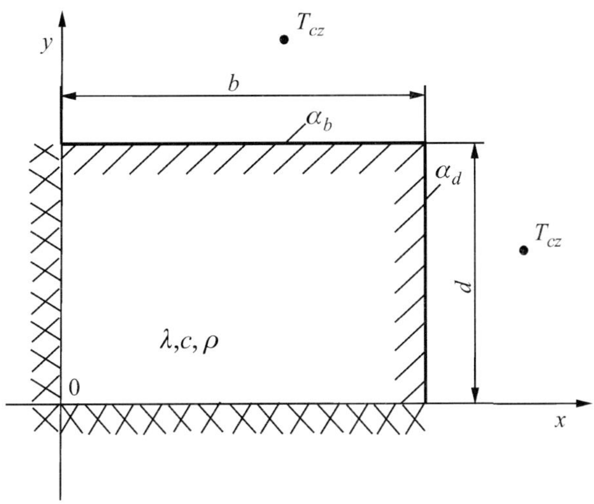

Fig. 19.3. The quarter of a rectangular cross-section of an infinitely long rod

Solution

Due to a symmetry of the temperature field, 1/4 of the rod's rectangular cross-section will be analyzed here. Temperature distribution within the rectangular cross-section is described by a differential heat conduction equation

$$\frac{\partial^2 T}{\partial x^2} + \frac{\partial^2 T}{\partial y^2} = \frac{1}{a}\frac{\partial T}{\partial t}, \tag{1}$$

by boundary conditions

$$\lambda \frac{\partial T}{\partial x}\bigg|_{x=0} = 0 , \tag{2}$$

$$-\lambda \frac{\partial T}{\partial x}\bigg|_{x=b} = \alpha_d \left(T\big|_{x=b} - T_{cz} \right), \tag{3}$$

$$\lambda \frac{\partial T}{\partial y}\bigg|_{y=0} = 0 , \tag{4}$$

$$-\lambda \frac{\partial T}{\partial y}\bigg|_{y=d} = \alpha_b \left(T\big|_{y=d} - T_{cz} \right) \tag{5}$$

and by initial condition

$$T\big|_{t=0} = T_0 . \tag{6}$$

According to with the superposition method, the solution of an initial-boundary problem (1)–(6) will be searched for in the form

$$\theta^* (x,y,t) = \theta_1^* (x,t) \cdot \theta_2^* (y,t), \tag{7}$$

where,

$$\theta^* (x,y,t) = \frac{T(x,y,t) - T_{cz}}{T_0 - T_{cz}}, \tag{8}$$

$$\theta_1^* (x,t) = \frac{T_1(x,t) - T_{cz}}{T_0 - T_{cz}}, \tag{9}$$

$$\theta_2^* (y,t) = \frac{T_2(y,t) - T_{cz}}{T_0 - T_{cz}}. \tag{10}$$

By substituting (7) into (1)–(6), two one-dimensional problems are obtained

$$\frac{\partial^2 \theta_1^*}{\partial x^2} = \frac{1}{a} \frac{\partial \theta_1^*}{\partial t}, \tag{11}$$

$$\frac{\partial \theta_1^*}{\partial x}\bigg|_{x=0} = 0 , \tag{12}$$

$$-\lambda \frac{\partial \theta_1^*}{\partial x}\bigg|_{x=b} = \alpha_d \theta_1^*\big|_{x=b}, \tag{13}$$

$$\theta_1^*\big|_{t=0} = 0 \tag{14}$$

and

$$\frac{\partial^2 \theta_2^*}{\partial y^2} = \frac{1}{a}\frac{\partial \theta_2^*}{\partial t} \tag{15}$$

$$\frac{\partial \theta_2^*}{\partial y}\bigg|_{y=0} = 0, \tag{16}$$

$$-\lambda \frac{\partial \theta_2^*}{\partial y}\bigg|_{y=d} = \alpha_b \theta_2^*\big|_{y=d}, \tag{17}$$

$$\theta_2^*\big|_{t=0} = 0. \tag{18}$$

The solution of a one-dimensional problem was obtained in Ex. 15.1. The appropriate solutions of the initial-boundary problems (11)–(14) and (15)–(18) have the form

$$\theta_1^*(x,t) = \sum_{n=1}^{\infty} \frac{2\sin\mu_n}{\mu_n + \sin\mu_n \cos\mu_n} \cos\left(\mu_n \frac{x}{b}\right) \exp\left(-\mu_n^2 \frac{at}{b^2}\right), \tag{19}$$

where μ_n are the roots of the characteristic equation

$$\mathrm{ctg}\mu = \frac{\mu\lambda}{\alpha_d b}, \tag{20}$$

and

$$\theta_2^*(y,t) = \sum_{n=1}^{\infty} \frac{2\sin\beta_n}{\beta_n + \sin\beta_n \cos\beta_n} \cos\left(\beta_n \frac{y}{d}\right) \exp\left(-\beta_n^2 \frac{at}{d^2}\right), \tag{21}$$

where β_n are the roots of the following characteristic equation

$$\mathrm{ctg}\beta = \frac{\beta\lambda}{\alpha_b d}. \tag{22}$$

Temperature $T(x,y,t)$ can be determined from (8) once temperatures $\theta_1^*(x,t)$ and $\theta_2^*(y,t)$ are calculated by means of the program developed in Ex. 15.1 or by using the diagrams annexed to Ex. 15.3.

Exercise 19.3 Formula Derivation for Temperature Distribution in a Rectangular Region with Boundary Conditions of 2nd Kind

By means of the superposition method, derive a formula for temperature distribution in an infinitely long rod with a rectangular cross-section $2b \times 2d$ and is heated by a heat flux $\dot{q}_1$ on the side, which is in length $2d$ and by a heat flux $\dot{q}_2$ on the side, which is in length $2b$ (Fig. 19.4). Initial temperature of the rod is constant and equals T_0, while thermo-physical properties are temperature-independent.

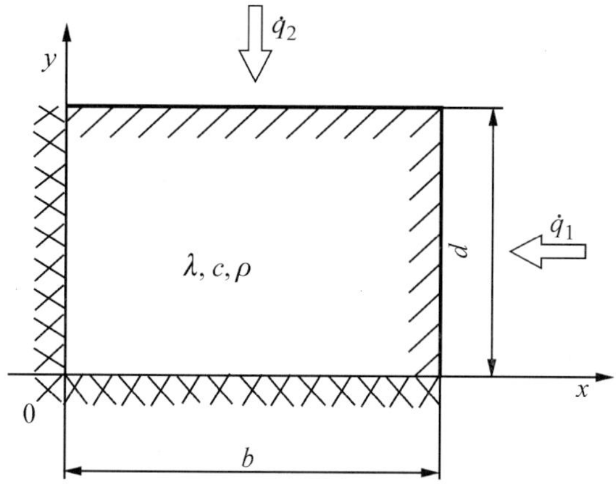

Fig. 19.4. Quarter of a rectangular cross-section of an infinitely long rod

Solution

Due to a symmetry of the temperature field, 1/4 of the entire cross-section of the rod will be analyzed below. Temperature distribution within the cross-section is described by a differential heat conduction equation

$$\frac{\partial^2 T}{\partial x^2} + \frac{\partial^2 T}{\partial y^2} = \frac{1}{a}\frac{\partial T}{\partial t}, \tag{1}$$

by boundary conditions

$$\lambda \left. \frac{\partial T}{\partial x} \right|_{x=0} = 0, \tag{2}$$

$$\lambda \frac{\partial T}{\partial x}\bigg|_{x=b} = \dot{q}_1 , \tag{3}$$

$$\lambda \frac{\partial T}{\partial y}\bigg|_{y=0} = 0 , \tag{4}$$

$$\lambda \frac{\partial T}{\partial y}\bigg|_{y=d} = \dot{q}_2 \tag{5}$$

and by initial condition

$$T\big|_{t=0} = T_0 . \tag{6}$$

According to with the superposition method, the solution of the initial-boundary problems (1)÷(6) will be searched for in the form

$$T - T_0 = T_1\left(x,t\right) + T_2\left(y,t\right). \tag{7}$$

Once (7) is substituted into (1)–(6), two one-dimensional problems are obtained:

- in the direction of x axis

$$\frac{\partial^2 T_1}{\partial x^2} = \frac{1}{a}\frac{\partial T_1}{\partial t} , \tag{8}$$

$$\lambda \frac{\partial T_1}{\partial x}\bigg|_{x=0} = 0 , \tag{9}$$

$$\lambda \frac{\partial T_1}{\partial x}\bigg|_{x=b} = \dot{q}_1 , \tag{10}$$

$$T_1\big|_{t=0} = 0 , \tag{11}$$

- in the direction of y axis

$$\frac{\partial^2 T_2}{\partial y^2} = \frac{1}{a}\frac{\partial T_2}{\partial t} , \tag{12}$$

$$\lambda \frac{\partial T_2}{\partial y}\bigg|_{y=0} = 0 , \tag{13}$$

$$\lambda \frac{\partial T_2}{\partial y}\bigg|_{y=d} = \dot{q}_2 , \tag{14}$$

$$T_2\big|_{t=0} = 0 \,. \tag{15}$$

The solution to (8)–(11) and (12)–(15) is presented in Ex. 15.1.

Using solution (31) from Ex. 15.10, temperature distribution given by (7) can be presented in the form

$$T(x,y,z,t) - T_0 =$$

$$= \frac{\dot{q}_1 b}{\lambda}\left[\frac{at}{b^2} + \frac{1}{2}\left(\frac{x}{b}\right)^2 - \frac{1}{6} - \frac{2}{\pi^2}\sum_{n=1}^{\infty}\frac{(-1)^n}{n^2}\cos\left(n\pi\frac{x}{b}\right)\exp\left(-n^2\pi^2\frac{at}{b^2}\right)\right] + \tag{16}$$

$$+\frac{\dot{q}_2 d}{\lambda}\left[\frac{at}{d^2} + \frac{1}{2}\left(\frac{y}{d}\right)^2 - \frac{1}{6} - \frac{2}{\pi^2}\sum_{n=1}^{\infty}\frac{(-1)^n}{n^2}\cos\left(n\pi\frac{y}{d}\right)\exp\left(-n^2\pi^2\frac{at}{d^2}\right)\right].$$

Numerical results obtained for a one-dimensional problem, presented in Table 15.4, Chapt. 15, enable one to determine temperature distribution, directly from (7).

Exercise 19.4 Calculating Temperature in a Steel Cylinder of a Finite Height

A steel cylinder with a diameter of d_z = 50 mm, a height of $2h$ = 100 mm and initial temperature of T_0 = 650°C is abruptly submersed in a liquid, whose temperature is T_{cz} = 95°C. Heat transfer coefficient from the cylinder surface to a liquid is α = 344 W/(m^2·K).

The cylinder is made of a carbon steel with a carbon content of C = 0.5%. Thermo-physical properties of the steel at average temperature of $T_{\acute{s}r}$ = 0,5(T_0 + T_{cz}) = 372.5°C are λ = 43 W/(m·K) and a = 1.2·10^{-5} m^2/s. Calculate temperature inside the cylinder at the point O after t = 52.1 s, from the moment the cylinder is submersed in the liquid.

Solution

Temperature inside the cylinder at the point O (Fig. 19.5) will be determined according the superposition method

$$\theta^*(0,0,t) = P(0,t)\cdot C(0,t), \tag{1}$$

$$\frac{T(0,0,t) - T_{cz}}{T_0 - T_{cz}} = \frac{T_p(0,t) - T_{cz}}{T_0 - T_{cz}}\cdot\frac{T_c(0,t) - T_{cz}}{T_0 - T_{cz}}. \tag{2}$$

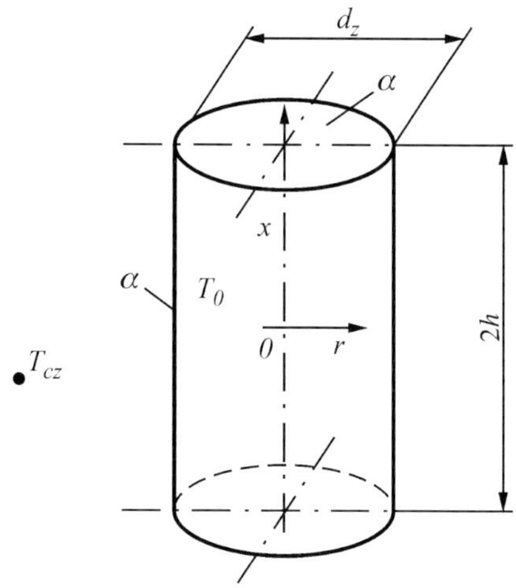

Fig. 19.5. Diagram of a cylinder with a finite height

First we will determine the Biot number Bi and the Fourier number Fo for the plate

$$Bi = \frac{\alpha h}{\lambda} = \frac{344 \cdot 0.05}{43} = 0.4 ,$$

$$Fo = \frac{at}{h^2} = \frac{1.2 \cdot 10^{-5} \cdot 52.1}{0.05^2} = 0.25 .$$

From the diagram in Ex. 15.3, one has

$$P(0, Fo) = \frac{T_p(0, Fo) - T_{cz}}{T_0 - T_{cz}} = 0.96 .$$

The Biot and Fourier numbers for an infinitely long cylinder are

$$Bi = \frac{\alpha r_z}{\lambda} = \frac{344 \cdot 0.025}{43} = 0.2 ,$$

$$Fo = \frac{at}{r_z^2} = \frac{1.2 \cdot 10^{-5} \cdot 52.1}{0.025^2} = 1.0 .$$

From the Fig. 15.15 in Ex. 15.6, one obtains

$$C(0, Fo) = \frac{T_c(0, Fo) - T_{cz}}{T_0 - T_{cz}} = 0.72 .$$

Dimensionless temperature of the cylinder center determined from (2) is

$$\frac{\lambda T_P\left(x, Fo_x\right)\Big|_{x=0}}{\dot{q}_s L_x} = 0.035\,,$$

hence, one has

$$T_P\left(x, 120\ \text{s}\right)\Big|_{x=0} = 0.035\frac{\dot{q}_s L_x}{\lambda} = 0.035\cdot\frac{355143\cdot 0.075}{37} = 25.2^{\circ}\text{C}\,,$$

$$\frac{\lambda T_P\left(y, Fo_y\right)\Big|_{y=0}}{\dot{q}_s L_y} = 0.00$$

and

$$T_P\left(y, 120\ \text{s}\right)\Big|_{y=0} = 0.0\cdot\frac{\dot{q}_s L_y}{\lambda} = 0.0\cdot\frac{355143\cdot 0.3}{37} = 0^{\circ}\text{C}\,,$$

and

$$\frac{\lambda T_P\left(z, Fo_z\right)\Big|_{z=0}}{\dot{q}_s L_z} = 5.7\cdot 10^{-6}\,,$$

hence,

$$T_P\left(z, 120\ \text{s}\right)\Big|_{z=0} = 5.7\cdot 10^{-6}\cdot\frac{\dot{q}_s L_z}{\lambda} = 5.7\cdot 10^{-6}\cdot\frac{355143\cdot 0.15}{37} = 8.2\cdot 10^{-3}\,^{\circ}\text{C}$$

Temperature inside the block is then

$$T\left(0,0,0,t\right)\Big|_{t=120\text{s}} = T_0 + T_P\left(x, 120\ \text{s}\right)\Big|_{x=0} + T_P\left(y, 120\ \text{s}\right)\Big|_{y=0} +$$

$$+ T_P\left(z, 120\ \text{s}\right)\Big|_{z=0} = 20 + 25.2 + 0 + 0.008 \approx 45.2^{\circ}\text{C}.$$

20 Approximate Analytical Methods for Solving Transient Heat Conduction Problems

One of the most popular approximate analytical methods are *thermal balance method* [3, 4, 6, 7, 12–19, 22÷25] and the *Biot method* [1, 2, 9–11, 20]. Also the *Gauss' principle of least constraint,* known from the analytical mechanics, can also be applied when approximately solving the differential heat conduction equation [20]. In this chapter, the first of the aforementioned methods will be discussed in greater detail.

Paper [21] lists number of examples in which the approximate methods are applied when solving one and two-dimensional heat conduction problems. Approximate analytical methods are especially useful when carrying out different engineering analyses, since it is important to obtain the results quickly and easily; the accuracy of results, however, is of less importance.

Exercise 20.1 Description of an Integral Heat Balance Method by Means of a One-Dimensional Transient Heat Conduction Example

Characterize the integral heat balance method using the example of a one-dimensional transient heat conduction.

Solution

Integral heat balance method was for the first time presented by *Karman* [5], and following that extended by *Pohlhausen* [8]. Both papers are on the integration of laminar boundary layer equations. In the integral heat balance method, the concept called the *depth of heat penetration* $\delta(t)$ is used; it can be defined as the largest distance covered in a given time by a penetrating heat. This means that within distance $x \geq \delta$ from the surface, one can assume the body temperature to be approximately equal to the initial temperature and the heat is transferred over a distance no larger than $\delta(t)$. The depth of heat penetration corresponds to the thermal thickness of a boundary layer in a convective heat transfer. In heat conduction calculations, the concept of depth of heat penetration $\delta(t)$ was introduced by *Biot* [2]. In the integral heat balance method, the differential equation of the transient heat conduction

$$c\rho\frac{\partial T}{\partial t}=\lambda\frac{\partial^2 T}{\partial x^2} \tag{1}$$

is integrated over variable x within the limits from $w(t)$ to $v(t)$

$$c\rho\int_{w(t)}^{v(t)}\frac{\partial T}{\partial t}dx=\lambda\left[\left.\frac{\partial T}{\partial x}\right|_{x=v(t)}-\left.\frac{\partial T}{\partial x}\right|_{x=w(t)}\right]. \tag{2}$$

If we account for integral differentiation rules over parameter (Leibniz rule), then (2) will have the form

$$\begin{aligned}&c\rho\left\{\frac{d}{dt}\int_{w(t)}^{v(t)}T(x,t)dx-T\left[v(t),t\right]\frac{dv}{dt}+T\left[w(t),t\right]\frac{dw}{dt}\right\}=\\&=\lambda\left[\left.\frac{\partial T}{\partial x}\right|_{x=v(t)}-\left.\frac{\partial T}{\partial x}\right|_{x=w(t)}\right].\end{aligned} \tag{3}$$

When body dimensions do not change in time, for example, due to a change of state, one can assume then that $w(t) = 0$. It is assumed, moreover, that heat transfer conditions only change on the plate front face $x = 0$.

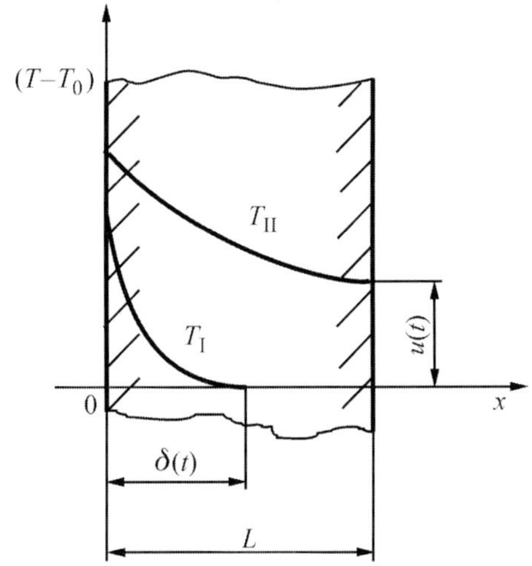

Fig. 20.1. Diagram of heat penetration in the first and second phase; $\delta(t)$ is the depth of heat penetration

This can be either the warming-up by a medium at a step-increasing temperature or the warming-up by a heat flow at a density assigned for $t > 0$. The upper integration limit equals $\delta(t)$ in the case of a semi-infinite space. If temperature field is determined in an element with finite dimensions, for example in a plate (Fig. 20.1), then one can distinguish two heat transfer phases: the first one, when the depth of heat penetration is smaller than the wall thickness, i.e. $\delta \le L$; the second, when $\delta = L$. For the first heat transfer phase, once we account for the conditions that arise from the depth of the heat penetration concept $v(t) = \delta(t)$, we have

$$T\left[v(t),t\right] = T_0\,, \tag{4}$$

$$\left.\frac{\partial T}{\partial x}\right|_{x=v(t)} = 0\,, \tag{5}$$

and once we account that $w(t) = 0$ and $dw/dt = 0$, (3) assumes the form

$$\frac{d}{dt}\left[c\rho\delta\left(\bar{T}_{\mathrm{I}} - T_0\right)\right] = -\lambda\left.\frac{\partial T}{\partial x}\right|_{x=0}\,, \tag{6}$$

where average temperature $\bar{T}_{\mathrm{I}}$ is defined as

$$\bar{T}_{\mathrm{I}} = \frac{1}{\delta}\int_0^{\delta} T(x,t)\,dx, \qquad 0 \le \delta \le L\,. \tag{7}$$

In the second heat transfer phase $\delta(t) = L$, the temperature of the rear surface changes, i.e. $T|_{x=L} = u(t)$ (Fig. 20.1). Equation (3) assumes the form

$$c\rho L\frac{d\bar{T}_{\mathrm{II}}}{dt} = \lambda\left(\left.\frac{\partial T}{\partial x}\right|_{x=L} - \left.\frac{\partial T}{\partial x}\right|_{x=0}\right), \tag{8}$$

where $\bar{T}_{\mathrm{II}}$ is the average temperature equal to

$$\bar{T}_{\mathrm{II}} = \frac{1}{L}\int_0^L T\,dx\,. \tag{9}$$

In order to determine temperature $T(x,t)$, one should assume a function form that approximates temperature distribution in both, the first and second heat transfer phase. Usually these are polynomials of 2nd or 3rd degree. Integral heat balance method can be also applied to non-linear heat conduction equations and non-linear boundary conditions [1]. When $c\rho = f_1(T)$ and $\lambda = f_2(T)$, then once integral transformation is applied

$$\vartheta(T) = \int_0^T c\rho dT \tag{10}$$

calculation method resembles the method used when constant thermal properties c, ρ and λ are applied. Also in the case of heterogeneous materials, i.e. when $c\rho = f_1(x)$ and $\lambda = f_2(x)$, once integral transformation is applied

$$\omega = \int_0^x \frac{dx}{\lambda(x)} \tag{11}$$

it is possible to apply the heat balance method.

Exercise 20.2 Determining Transient Temperature Distribution in a Flat Wall with Assigned Conditions of 1st, 2nd and 3rd Kind

Determine transient temperature field in a flat wall (plate) with boundary conditions of 1st, 2nd and 3rd kind. Assume that heat transfer conditions only change on the plate front face. Use the integral heat balance method to solve the problem.

Solution

Plate temperature distribution is defined by the heat conduction equation

$$\frac{1}{a}\frac{\partial T}{\partial t} = \frac{\partial^2 T}{\partial x^2}, \tag{1}$$

by boundary conditions

$$-\lambda \left.\frac{\partial T}{\partial x}\right|_{x=0} + \beta_1 T\big|_{x=0} = \gamma_1 T_{01}(t), \tag{2}$$

$$-\lambda \left.\frac{\partial T}{\partial x}\right|_{x=L} + \beta_2 T\big|_{x=L} = \gamma_2 T_{02}(t) \tag{3}$$

and by initial condition

$$T(x,t)\big|_{t=0} = T_{02}\,. \tag{4}$$

Such formulation of boundary conditions enables one, by appropriately selecting coefficients, to assign temperature, heat flux or convective heat transfer on the plate surfaces, therefore, the boundary conditions of 1st, 2nd or 3rd kind respectively:

a) boundary conditions of 1st kind

$$\begin{aligned} &\beta_1 = \gamma_1 \to \infty, \qquad T\big|_{x=0} = T_{01}(t), \\ &\beta_2 = \gamma_2 \to \infty, \qquad T\big|_{x=L} = T_{02}; \end{aligned} \tag{5}$$

b) boundary conditions of 2nd kind

$$\begin{aligned} &\beta_1 = 0, \qquad -\lambda \frac{\partial T}{\partial x}\bigg|_{x=0} = \gamma_1 T_{01}(t) = \dot{q}_{01}(t), \\ &\beta_2 = 0, \qquad -\lambda \frac{\partial T}{\partial x}\bigg|_{x=L} = \gamma_2 T_{02} = \dot{q}_{02}; \end{aligned} \tag{6}$$

c) boundary conditions of 3rd kind

$$\begin{aligned} &\beta_1 = \gamma_1 = \alpha_1, \qquad -\lambda \frac{\partial T}{\partial x}\bigg|_{x=0} = \alpha_1\left[T_{01}(t) - T\big|_{x=0}\right], \\ &\beta_2 = \gamma_2 = -\alpha_2, \qquad -\lambda \frac{\partial T}{\partial x}\bigg|_{x=L} = \alpha_2\left[T\big|_{x=L} - T_{02}\right]. \end{aligned} \tag{7}$$

According to the integral heat balance method, temperature distribution is determined in two phases.
In the first heat transfer phase, when $0 \le \delta(t) \le L$, temperature distribution should be approximated by a polynomial of the second degree

$$T_1 = a_0 + a_1 x + a_2 x^2\,. \tag{8}$$

Constants a_0, a_1 and a_2 are determined from the boundary condition (2) and from conditions

$$\begin{aligned} &T\big|_{x=\delta} = T_{02}, \\ &\lambda \frac{\partial T}{\partial x}\bigg|_{x=\delta} = 0 \end{aligned} \;. \tag{9}$$

By substituting the determined constants into (8), temperature distribution in the first heat transfer phase has the form

$$T_{\mathrm{I}} = T_{02} - \frac{1}{2}\delta \frac{\beta_1 T_{02} - \gamma_1 T_{01}(t)}{\frac{1}{2}\beta_1\delta + \lambda} + \frac{\beta_1 T_{02} - \gamma_1 T_{01}(t)}{\frac{1}{2}\beta_1\delta + \lambda} x - \frac{1}{2\delta}\frac{\beta_1 T_{02} - \gamma_1 T_{01}(t)}{\frac{1}{2}\beta_1\delta + \lambda} x^2, \qquad 0 \le x \le \delta \tag{10}$$

and

$$T_{\mathrm{I}} = T_{02}, \qquad \delta \le x \le L. \tag{11}$$

If (10) is substituted into (7) (Ex. 20.1) for average temperature $\overline{T}_{\mathrm{I}}$, then the following equation is obtained

$$\overline{T}_{\mathrm{I}} = T_{02} - \frac{1}{3}\frac{\frac{\beta_1\delta}{\lambda}T_{02} - \frac{\gamma_1\delta}{\lambda}T_{01}(t)}{\frac{1}{2}\beta_1\delta + \lambda}. \tag{12}$$

However, when (12) is substituted into (6) (Ex. 20.1) and boundary condition (2) is accounted for, one obtains, after transformation, the following differential equation to describe $\delta(t)$:

$$c\rho\left\{\left[\frac{\frac{2\beta_1\delta}{\lambda}T_{02}\left(\frac{\beta_1\delta}{\lambda}+2\right) - \frac{\beta_1^2\delta^2}{\lambda^2}T_{02} - \frac{2\beta_1\delta}{\lambda}\frac{\gamma_1\delta}{\lambda}T_{01}(t)}{\left(\frac{\beta_1\delta}{\lambda}+2\right)^2} - \frac{\frac{4\gamma_1\delta}{\lambda}T_{01}(t) - \frac{\beta_1\gamma_1\delta^2}{\lambda^2}T_{01}(t)}{\left(\frac{\beta_1\delta}{\lambda}+2\right)^2}\right]\frac{d\delta}{dt} - \frac{\frac{\gamma_1\delta^2}{\lambda}\frac{\beta_1\delta}{\lambda}}{\left(\frac{\beta_1\delta}{\lambda}+2\right)^2}\frac{dT_{01}(t)}{dt} - \frac{2\frac{\gamma_1\delta^2}{\lambda}}{\left(\frac{\beta_1\delta}{\lambda}+2\right)^2}\frac{dT_{01}(t)}{dt}\right\} = -6\frac{\gamma_1 T_{01}(t) - \beta_1 T_{02}}{\frac{\beta_1\delta}{\lambda}+2}. \tag{13}$$

In the second heat transfer phase, temperature distribution should also be approximated by a polynomial of the second degree

$$T_{\mathrm{II}} = a_3 + a_4 x + a_5 x^2 . \tag{14}$$

Once constants a_3, a_4 and a_5 are determined from boundary conditions (2) and (3) and from condition

$$T_{\mathrm{II}}\big|_{x=L} = u(t) \tag{15}$$

and again substituted into (14), one has

$$\begin{aligned} T_{\mathrm{II}} = {} & \frac{\frac{\gamma_1 L}{\lambda} T_{01}(t) - \frac{\beta_2 L}{\lambda} u(t) + 2u(t) + \frac{\gamma_2 L}{\lambda} T_{02}}{\frac{\beta_1 L}{\lambda} + 2} + \\ & + \frac{\frac{-\beta_1 \beta_2 L}{\lambda^2} u(t) - 2\frac{\gamma_1}{\lambda} T_{01}(t) + \frac{\beta_1 \gamma_2 L}{\lambda^2} T_{02} + 2\frac{\beta_1}{\lambda} u(t)}{\frac{\beta_1 L}{\lambda} + 2} x + \\ & + \frac{\frac{\beta_1 \gamma_2}{\lambda^2} T_{02} + \frac{\beta_1 \beta_2}{\lambda^2} u(t) + \frac{\beta_2}{\lambda L} u(t) - \frac{\beta_1}{\lambda L} u(t) - \frac{\gamma_2}{\lambda L} T_{02} + \frac{\gamma_1}{\lambda L} T_{01}(t)}{\frac{\beta_1 L}{\lambda} + 2} x^2 . \end{aligned} \tag{16}$$

Accounting for (16) in (9) (Ex. 20.1), one obtains, after transformation

$$\begin{aligned} \bar{T}_{\mathrm{II}} = {} & \frac{2u(t) - \frac{2}{3}\frac{\beta_2 L}{\lambda} u(t) - \frac{4}{3}\frac{\gamma_2 L}{\lambda} T_{02} - \frac{1}{6}\frac{\beta_1 \beta_2 L^2}{\lambda^2} u(t)}{\frac{\beta_1 L}{\lambda} + 2} + \\ & + \frac{\frac{2}{3}\frac{\beta_1 L}{\lambda} u(t) + \frac{1}{6}\frac{\beta_1 \gamma_2 L^2}{\lambda^2} T_{02} + \frac{1}{3}\frac{\gamma_1 L}{\lambda} T_{01}(t)}{\frac{\beta_1 L}{\lambda} + 2} . \end{aligned} \tag{17}$$

If (17) is substituted into heat balance (8) (Ex. 20.1), then differential equation is obtained that serves as a tool for determining $u(t)$

$$c\rho L\left(2-\frac{2}{3}\frac{\beta_2 L}{\lambda}-\frac{1}{6}\frac{\beta_1\beta_2 L^2}{\lambda^2}+\frac{2}{3}\frac{\beta_1 L}{\lambda}\right)\frac{du}{dt}+\frac{1}{3}\frac{\gamma_1 L}{\lambda}\frac{dT_{01}(t)}{dt}=$$
$$=2u(t)(\beta_2-\beta_1)+2\frac{\beta_1\beta_2 L}{\lambda}u(t)-2\frac{\beta_1\gamma_2 L}{\lambda}T_{02}+2\gamma_1 T_{01}(t)-2\gamma_2 T_{02}. \tag{18}$$

In the examples given, (13) and (18) will be solved for the few cases that are of particular importance in the field of engineering.

Example 1

$$\beta_1=\gamma_1=\alpha_1, \qquad \beta_2=\gamma_2=0, \qquad T_{01}(t)=T_{01}=\text{const}, \qquad T_{02}=0. \tag{19}$$

This case corresponds to a so called thermal shock in tanks or other constructions, when a wall, insulated on one side, is suddenly warmed-up or convectively cooled down on the other side.

Temperature distribution in the first heat transfer phase is obtained once (19) is substituted into (10)

$$T_{\mathrm{I}}=\begin{cases}\dfrac{Bi(\delta)}{Bi(\delta)+2}T_{01}\left(1-\dfrac{x}{\delta}\right)^2, & 0\le x\le\delta,\\[2ex] 0, & \delta\le x\le L,\end{cases} \tag{20}$$

where $Bi(\delta) = \alpha_1\delta/\lambda$.

In the second phase, temperature distribution is obtained from (16) once (19) is allowed for

$$T_{\mathrm{II}}=\frac{Bi(L)}{Bi(L)+2}(T_{01}-u)\left(1-\frac{x}{L}\right)^2+u, \qquad Fo\ge Fo_1, \tag{21}$$

where $Fo_1 = at_1/L^2$; time t_1 is a moment when $\delta = L$. The solution of (13), when initial condition is $\delta|_{t=0} = 0$ and (19) accounted for, is

$$\frac{1}{12}+\frac{1}{3Bi(\delta)}-\frac{2}{3\left[Bi(\delta)\right]^2}\ln\frac{Bi(\delta)+2}{2}=Fo(\delta), \tag{22}$$

where $Fo(\delta) = at/\delta^2$.

Fourier number Fo_1 is obtained once $\delta = L$ is substituted into (22). Calculation of temperature distribution in the first phase is done in the following way. First we assume that δ, for example, $\delta = 0.5L$, then we determine time t from (22) that corresponds to the assumed value δ. (20) enables one to determine spatial temperature distribution. Following that (18) must be solved when initial condition is $u(Fo_1) = 0$. Once we account for (19), we have

$$u(t) = T_{01}\left[1 - e^{-\mu^2(Fo - Fo_1)}\right], \qquad Fo \geq Fo_1, \tag{23}$$

where

$$\mu^2 = \frac{3Bi(L)}{3 + Bi(L)}. \tag{24}$$

One can easily observe, by comparing (23) to the analytical solution (Ex. 15.1), that μ is an approximation of the first root of the characteristic equation

$$\text{ctg}\mu = \frac{\mu}{Bi}. \tag{25}$$

In order to evaluate the accuracy of the approximate solution (24), one should compare the first root of (25), determined from (24) with the exact values μ_1^* calculated in Ex. 15.2.

Table 20.1. Comparison between the exact value of the first root μ_1^* of the characteristic (25) and approximate values μ_1 calculated from (24)

Bi	0.001	0.002	0.004	0.01	0.2	0.8
μ_1	0.0316	0.0447	0.0632	0.0998	0.4330	0.7947
μ_1^*	0.0316	0.0447	0.0632	0.0998	0.4328	0.7910
Bi	2.0	4.0	10.0	30.0	60.0	100.0
μ_1	1.0954	1.30993	1.5191	1.6514	1.6902	1.7066
μ_1^*	1.0769	1.2646	1.4289	1.5202	1.5451	1.5552

From Table 20.1 it follows that the accuracy of the approximate (24) is higher for the smaller values of the Biot number Bi. Temperature distribution in the second heat transfer phase is calculated from (21) by accounting for (23).

Example 2

$$\beta_1 = \gamma_1 = \alpha_1 \to \infty, \qquad \beta_2 = \gamma_2 = 0, \qquad T\big|_{x=0} = T_{01}(t), \qquad T_{02} = 0. \tag{26}$$

This case is a good example of an ideal thermal shock on a unilaterally insulated flat wall surface.

Temperature distribution in the first heat transfer phase is obtained from (10) by accounting for (26)

$$T_1 = \begin{cases} T_{01}(t)\left(1-\dfrac{x}{\delta}\right)^2, & 0 \le x \le \delta, \quad (27) \\ 0, & \delta \le x \le L. \quad (28) \end{cases}$$

From the heat balance (13), the following differential equation is obtained

$$T_{01}(t)\delta\frac{d\delta}{dt} + \delta^2\frac{dT_{01}}{dt} = 6aT_{01}(t). \tag{29}$$

By substituting $z = \delta^2$ into (29), one reduces it to a linear differential equation

$$T_{01}(t)\frac{dz}{dt} + 2z\frac{dT_{01}}{dt} = 12T_{01}(t). \tag{30}$$

It is easy to solve this equation by means of the integrating factor. The solution of (30), when initial conditions are $\delta|_{t=0} = 0$, has the form

$$z = \delta^2 = 12a\frac{\int_0^t \exp\left(\int_0^t p\,dt\right)dt}{\exp\left(\int_0^t p\,dt\right)}, \tag{31}$$

where

$$p = \frac{2}{T_{01}(t)}\frac{dT_{01}(t)}{dt}. \tag{32}$$

When surface temperature is defined by function

$$T_{01} = Ct^m, \tag{33}$$

then, by substituting (33) into (32), the following is obtained from (31):

$$z = \delta^2 = \frac{12at}{2m+1}. \tag{34}$$

For different values of m, we have

$$\delta = \begin{cases} 3.464\sqrt{at}, & m=0, \\ 2\sqrt{at}, & m=1, \\ 1.549\sqrt{at}, & m=2, \\ 1.309\sqrt{at}, & m=3. \end{cases} \tag{35}$$

Due to the application of the Biot method, one has $\delta = 3.36\sqrt{at}$ for $m = 0$ and $\delta = 2.29\sqrt{at}$ for $m = 1$. In the second heat transfer phase, the temperature field assumes the following form, once we account for conditions (26) in (18)

$$T_{\mathrm{II}} = \left[T_{01}(t) - u\right]\left(1 - \frac{x}{L}\right)^2 + u, \qquad Fo_1 \le Fo. \tag{36}$$

When all mathematical operations, which resemble those in example 1, are carried out, the following is obtained for $m = 1$

$$u = \frac{CL^2}{a}\left[Fo - \frac{1}{2} - \left(Fo_1 - \frac{1}{2}\right)e^{-3(Fo - Fo_1)}\right], \tag{37}$$

where $Fo_1 = at_1/L^2$.

Time t_1 is determined from (34) for $m = 1$ once we substitute $\delta = L$.

The Fourier number Fo_1 is of 0.25. Temperature distribution in the second heat transfer phase can be also easily determined for $m = 0, 2, ..., n$. Plate temperature distribution for the first and second heat transfer phase, for $m = 1$ is presented in Fig. 20.2 and 20.3.

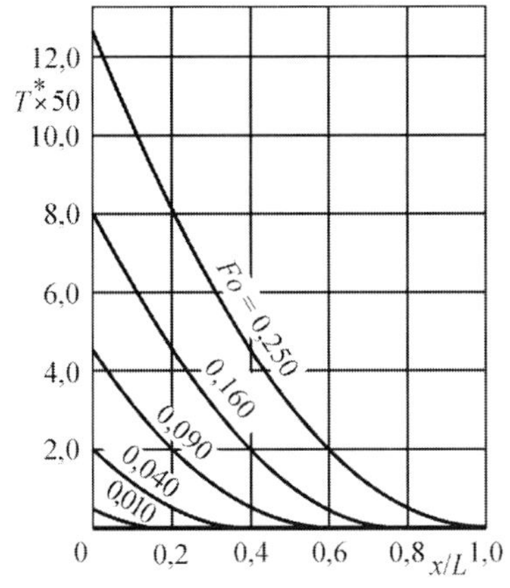

Fig. 20.2. Temperature distribution in a flat wall; dependence on the Fourier number $Fo = at/L^2$ in the first heat transfer phase (Example 2), $T^* = T_1 a/(CL^2)$

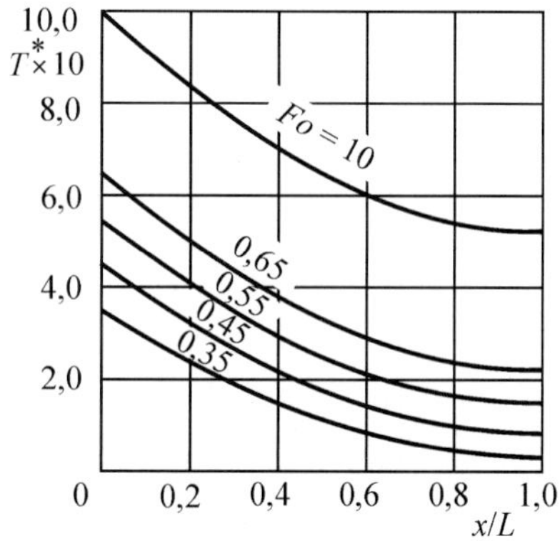

Fig. 20.3. Temperature distribution in a flat wall; dependence on the Fourier number $Fo = at/L^2$, $T^* = T_{II}a/(CL^2)$ in the second heat transfer phase.

Example 3

$$\beta_1 = 0, \quad \beta_2 = \gamma_2 = 0, \quad T_{02} = 0, \quad \gamma_1 T_{01}(t) = \dot{q}_0(t), \quad \dot{q}_0(t) = C_1 t^n. \quad (38)$$

Temperature distribution in the first heat transfer phase is obtained from (10) by accounting for (38) first

$$T_I = \begin{cases} \dfrac{\dot{q}_0(t)\delta}{2}\left(1 - \dfrac{x}{\delta}\right)^2, & 0 \le x \le \delta, \\ 0, & \delta \le x \le L. \end{cases} \quad (39)$$

Equation (13), once (38) is accounted for, assumes the following form:

$$\frac{d}{dt}\left[\dot{q}_0(t)\delta^2\right] = 6a\dot{q}_0(t). \quad (40)$$

The solution of (40), when initial condition is $\delta|_{t=0} = 0$, is function

$$\delta = \sqrt{6a}\left[\frac{1}{\dot{q}_0(t)}\int_0^t \dot{q}_0(t)\,dt\right]^{1/2}, \quad (41)$$

which yields, once we account for (38)

$$\delta = \sqrt{\frac{6at}{n+1}}. \quad (42)$$

In the second heat transfer phase, temperature distribution is expressed as.

$$T_{\mathrm{II}} = \frac{\dot{q}_0(t)L}{2\lambda}\left(1-\frac{x}{L}\right)^2 + u, \qquad Fo_1 \le Fo. \tag{43}$$

Once we account for conditions (38) in (18) and integrate the equation when the initial condition is $u|_{t=t1} = 0$, we have

$$u(t) = \frac{\dot{q}_0(t)L}{2\lambda}\left(\frac{2Fo}{n+1}-1\right) + \frac{\dot{q}_0(t_1)L}{2\lambda}\left(1-\frac{2Fo_1}{n+1}\right), \tag{44}$$

where time t_1, determined from (42) once $\delta = L$ is substituted, is $t_1 = (n+1)L^2/6a$.

Example 4

$$\beta_1 = \gamma_1 = \alpha_1, \qquad T_{02} = 0, \qquad \beta_2 = \gamma_2 = -\alpha_2. \tag{45}$$

The first heat transfer phase is described by formulas that are identical to the ones used in Example 1. In the second phase, temperature distribution is expressed by (16), which, once we account for (45), assumes the form

$$T_{\mathrm{II}} = \frac{Bi_1 T_{01} + Bi_2 u + 2u}{Bi + 2} + \frac{2Bi_1 u + Bi_1 Bi_2 u - 2Bi_1 T_{01}}{Bi_1 + 2}\frac{x}{L} + \frac{Bi_1 T_{01} - Bi_1 Bi_2 u - Bi_2 u - Bi_1 u}{Bi_1 + 2}\left(\frac{x}{L}\right)^2, \qquad Fo_1 \le Fo. \tag{46}$$

When (18) is integrated, while (45) is accounted for when initial condition is $u|_{t=t_1} = 0$, one has

$$u = \frac{Bi_1 T_{01}}{Bi_1 + Bi_1 Bi_2 + Bi_2}\left[1 - e^{-\mu^2 (Fo - Fo_1)}\right], \tag{47}$$

where

$$\mu^2 = \frac{12\left(Bi_1 + Bi_2 + Bi_1 Bi_2\right)}{12 + 4Bi_1 + Bi_1 Bi_2 + 4Bi_2}. \tag{48}$$

Equation (48) is the approximation of the first root of characteristic equation

$$\sigma = \sigma^* \frac{E\beta T_{od}}{1-\nu} = -0.9 \frac{19.13 \cdot 1.325 \cdot 10^{-5} \cdot 100}{1-0.3} = -296.65 \text{ MPa} . \qquad (9)$$

These stresses occur in time

$$t = \frac{Fo\,L^2}{a} = \frac{0.0075 \cdot 0.06^2}{1.325 \cdot 10^{-6}} = 2.04\text{s} .$$

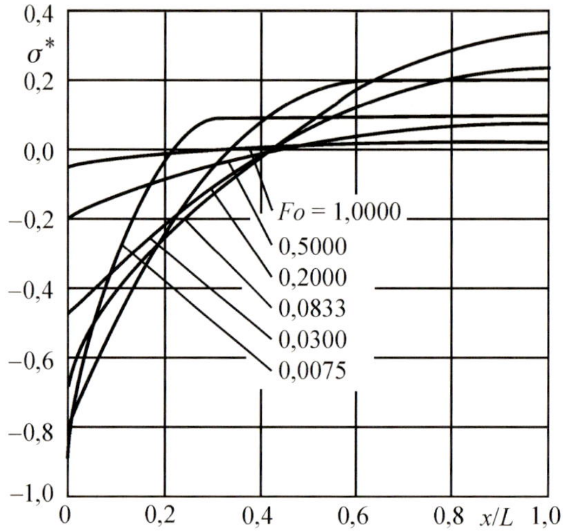

Fig. 20.4. Distribution of thermal stresses in a back-wall-insulated flat wall with a step-change in the front face temperature . $\sigma^* = \sigma\,(1-\nu)/(E\beta T_{01})$

Literature

1. Barry GW, Goodling JS (1987) A Stefan problem with contact resistance. Transactions of the ASME, J. of Heat Transfer 109: 820-825
2. Biot MA(1970) Variational Principles in Heat Transfer. Clarendon Press, Oxford
3. Chung BTF, Hsiao JS (1985) Heat transfer with ablation in a finite slab subjected to time-variant heat fluxes. AIAA Journal 23: 145-150
4. Goodman TR (1964) Application of Integral Methods to Transient Non-linear Heat Transfer. Advances in Heat Transfer 1: 51-122
5. Karman T (1921) Über laminare und turbulente Reibung. Zeitschrift für angewandte Mathematik und Mechanik 1: 233-252
6. Młynarski F, Taler J (1976) The effect of thermal-shock-caused stresses on the durability of radiant tubes of converter waste-heat boilers. Power Engineering Archives 3: 115-127

7. Młynarski F, Taler J ((1982) Temperature and stress field triggered by a thermal shock. Chemical and Process Engineering 3(1): 149-162
8. Pohlhausen K (1921) Zur näherungsweisen der Differentialg-leichung der laminaren Reibungsschicht. Zeitschrift für angewandte Mathematik und Mechanik 1: 252-268
9. Prasad A, Agrawal HC (1972) Biot's variational principle for a Stefan problem. AIAA Journal 10 (3): 325-327
10. Prasad A, Sinha SN (1975) Radiative ablation of melting solids. AIAA Journal 14(10): 1494-1497
11. Rao VD, Sarma PK, Raju GJVJ (1985) Biot's variational method to fluidized-bed coating on thin plates. Transactions of the ASME, J. of Heat Transfer 107: p. 258-260
12. Razelos P (1973) Methods of Obtaining Approximate Solution. Section 4 *In*: Handbook of Heat Transfer, Ed. by WM Rohsenow, JP Hartnett. McGraw-Hill, New York
13. Rup K, Taler J (1977) Defining transient temperature field in a flat wall with a variable heat conduction coefficient. Theoretical and Applied Mechanics 15 (1): 21-28
14. Rup K, Taler J (1979) Approximate analysis of transient temperature field in simple fins. Engineers' Thesis 1: 145-153
15. Taler J, Rup K (1976) Applying heat balance method when determining transient temperature and stress field in a flat wall. Chemical Engineering 6 (3): 657-672
16. Taler J (1978) Approximation of transient temperature field in cylindrical and spherical bodies. Theoretical and Applied Mechanics 16 (2): 247-263
17. Taler J (1979) Approximation of transient temperature fields by one-dimensional polynomials. Chemical Engineering 9(1): 243-258
18. Taler J, Rup K (1997) Calculating fins by means of 'averaging functional corrections' method. Machinery Construction Archives 26(1): 143-153
19. Taler J (1980) 'Averaging functional corrections' method and its relation to heat balance method. Chemical and Process Engineering 3 (1): 609-626
20. Taler J (1977) Application of Gauss method to an approximate solving of heat conduction differential equations. Engineers' Thesis 25 (2): 349-368
21. Taler J (1980) Weighted residuums method and its application to the calculation of temperature fields in boilers' elements. Monography (14), Pub. Krakow University of Technology, Kraków
22. Venkateshan SP, Solaiappan O (1988) Approximate solution of non-linear transient heat conduction in one dimension. Wärme- und Stoffübertragung 23:229-233
23. Venkateshan SP, Rao VR (1991) Approximate solution of non-linear transient heat conduction in cylindrical geometry. Wärme- und Stoffübertragung 26: 97-102
24. Yang JW, Bankoff SG (1987) Solidification effects on the fragmentation of molten metal drops behind a pressure shock wave. Transactions of the ASME, J. of Heat Transfer 109: 226-231

25. Yang KT, Szewczyk A (1959) An approximate treatment of unsteady heat conduction in semi-infinite solids with variable thermal properties. Transactions of the ASME, J. of Heat Transfer 81: 251-252

21 Finite Difference Method

Analytical methods allow to determine the temperature field in a simple shape bodies, such as the plate, cylinder or sphere. It is possible to find the analytical solution to two-dimensional problems only for regular-shape-regions, e.g. for a rectangle or spherical regions. Another problem, which stands as a barrier in the application of analytical methods is the complex mathematical apparatus, which everyone who intends to determine analytical solutions should be thoroughly familiar with. It should be emphasized, however, that analytical solutions are frequently used to help evaluate the accuracy of numerical solutions.

Finite difference method belongs to one of the earliest developed methods that allow to determine temperature values at discrete spatial points (nodes) and temporal points [1–9, 11, 13]. In the finite difference method, one can approximate partial derivatives in the heat conduction equation by difference quotients, without the need of going into the physics of the problem.

Similar method to finite difference method was simultaneously being developed; it was called the *elementary balance method,* also known as the *control volume method* [9] or, lately, *finite volume method* [1, 2, 5, 12].

Finite volume method was developed mainly by engineers, who were attempting to solve complex heat-flow problems, which they came across in practice, by means of the simplest methods available. This method is based on an energy (heat) balance for an isolated finite volume. The condition that satisfies the heat balance for individual control volumes enables one to determine, with high accuracy, the transient heat flux and temperature distribution.

In the case of the classical finite element method (FEM), the energy preservation condition for fictious control volumes around the nodes of finite element mesh was not satisfied. In order to ensure appropriate calculation accuracy, the analyzed region must be divided into a very large number of finite elements. For this reason, the finite element balance method or finite volume method are utilized in the heat transfer and fluid mechanics. The differences between these two methods dissipate. The main difference lies in the fact that in FEM the functions are used to interpolate temperature distribution (or other quantities) inside the element, while they are not used in the finite volume method. Furthermore, in the classical FEM, thermal capacity of an element is distributed at different weight among individual element nodes. In control volume method, thermal capacity of a control volume is concentrated in a single point. In comparison to finite difference method, it is easier to apply the control volume method in practice, especially when the analyzed region has an irregular boundary and the thermo-physical properties are temperature-dependent.

In this chapter, three methods will be evaluated in greater detail: *explicit*, *implicit* and *Crank-Nicolson method.* They are preceded with a basic information on the boundary condition approximation methods.

Exercise 21.1 Methods of Heat Flux Approximation on the Plate Surface

Discuss methods of heat flux approximation on the plate surfaces with different degrees of accuracy by assuming that $T = T(x, t)$.

Solution

Plate discretisation together with the marked nodes are shown in Fig. 21.1.

Heat flux on the front face and the back plate surface (Fig. 21.1) is formulated as

$$\dot{q}\big|_{x=0} = -\lambda \frac{\partial T}{\partial x}\bigg|_{x=0}, \tag{1}$$

$$\dot{q}\big|_{x=L} = -\lambda \frac{\partial T}{\partial x}\bigg|_{x=L}. \tag{2}$$

Derivatives in (1) and (2) should be approximated by difference quotients, i.e. by means of the finite difference method. In order to derive the difference formulas, a function expanded into a Taylor series was applied around point $x_i = (i - 1)\Delta x$, $\Delta x = L/(N - 1)$. Using the notation $T_i = T(x_i,t)$, the value of temperature $T_{i\ 1} = T(x_{i+1},t)$ can be calculated by means of the Taylor series

$$T_{i+1} = T_i + \left(\frac{\partial T}{\partial x}\right)_i \Delta x + \left(\frac{\partial^2 T}{\partial x^2}\right)_i \frac{(\Delta x)^2}{2} + \left(\frac{\partial^3 T}{\partial x^3}\right)_i \frac{(\Delta x)^6}{6} + \ldots \tag{3}$$

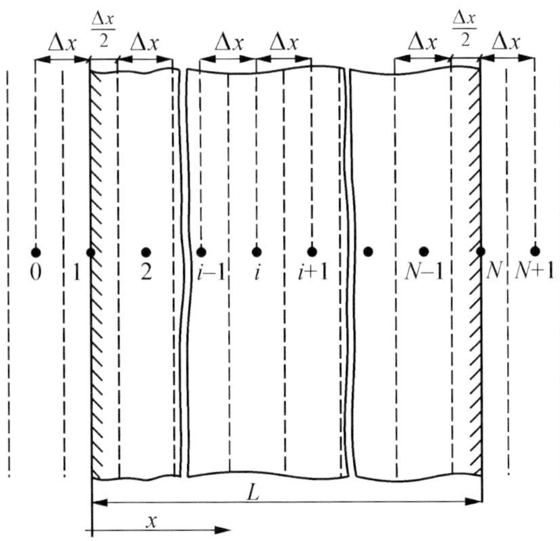

Fig. 21.1. Difference mesh (control volumes) together with the marked nodes

The first derivative $(\partial T/\partial x)_i$ determined from (3) is

$$\left(\frac{\partial T}{\partial x}\right)_i = \frac{T_{i+1}-T_i}{\Delta x} - \left(\frac{\partial^2 T}{\partial x^2}\right)_i \frac{(\Delta x)}{2} - \left(\frac{\partial^3 T}{\partial x^3}\right)_i \frac{(\Delta x)^2}{6} + \ldots \tag{4}$$

The first term on the right-hand-side is the forward difference quotient

$$\left(\frac{\partial T}{\partial x}\right)_i \approx \frac{T_{i+1}-T_i}{\Delta x}. \tag{5}$$

The remaining terms

$$O(\Delta x) = -\left(\frac{\partial^2 T}{\partial x^2}\right)_i \frac{(\Delta x)}{2} - \left(\frac{\partial^3 T}{\partial x^3}\right)_i \frac{(\Delta x)^2}{6} + \ldots \tag{6}$$

are the Taylor series truncation error, while the first term contains Δx to the first power. One can state, therefore, that the approximation accuracy of derivative (5) is of the first order. Symbol $O(\Delta x)$ stands for the first approximation order and, at the same time, means that the lowest power with Δx equals 1 in the omitted part of the Taylor series.

Backward difference quotient can be determined in a similar way. Expansion of T_{i-1} into Taylor series around T_i has the form

$$T_{i-1} = T_i + \left(\frac{\partial T}{\partial x}\right)_i (-\Delta x) + \left(\frac{\partial^2 T}{\partial x^2}\right)_i \frac{(-\Delta x)^2}{2} + \left(\frac{\partial^3 T}{\partial x^3}\right)_i \frac{(-\Delta x)^3}{6} + \ldots, \tag{7}$$

Heat flux on the boundary $x = 0$ can be, therefore, approximated by means of the following formulas:

$$\dot{q}_1 = -\lambda \left.\frac{\partial T}{\partial x}\right|_{x=0} = -\lambda \frac{T_2 - T_1}{\Delta x} + \mathrm{O}(\Delta x), \tag{22}$$

$$\dot{q}_1 = -\lambda \frac{T_2 - T_0}{2(\Delta x)} + \mathrm{O}\left[(\Delta x)^2\right], \tag{23}$$

$$\dot{q}_1 = -\lambda \frac{-3T_1 + 4T_2 - T_3}{2(\Delta x)} + \mathrm{O}\left[(\Delta x)^2\right]. \tag{24}$$

Temperature at an apparent (fictious) point T_0, which lies beyond the analyzed region, is eliminated by means of the discrete form of the differential equation at point $x_1 = 0$. Heat flux on the second boundary $x = x_N$ can be determined in a similar way

$$\dot{q}_N = -\lambda \left.\frac{\partial T}{\partial x}\right|_{x=L} = -\lambda \frac{T_N - T_{N-1}}{\Delta x} + \mathrm{O}(\Delta x), \tag{25}$$

$$\dot{q}_N = -\lambda \frac{T_{N+1} - T_{N-1}}{2(\Delta x)} + \mathrm{O}\left[(\Delta x)^2\right], \tag{26}$$

$$\dot{q}_N = -\lambda \frac{3T_N - 4T_{N-1} + T_{N-2}}{2(\Delta x)} + \mathrm{O}\left[(\Delta x)^2\right]. \tag{27}$$

It should be added that all types of heat flux approximation, formulated in (22)–(27), are applied in practice to transient temperature field calculations. When applying (22) and (25) with the accuracy of the first order, the node number N should be large , so that Δx and the approximation error will be small.

Exercise 21.2 Explicit Finite Difference Method with Boundary Conditions of 1st, 2nd and 3rd Kind

Describe how explicit finite difference method is applied to a one-dimensional transient heat conduction problem. Derive appropriate equations, while allowing for boundary conditions of 1st, 2nd and 3rd kind.

Solution

In order to solve the heat conduction equation

$$\frac{1}{a}\frac{\partial T}{\partial t}=\frac{\partial^2 T}{\partial x^2} \tag{1}$$

a flat wall should be divided into a finite number of control volumes with a width of $\Delta x = L/(N-1)$ (Fig. 21.2).

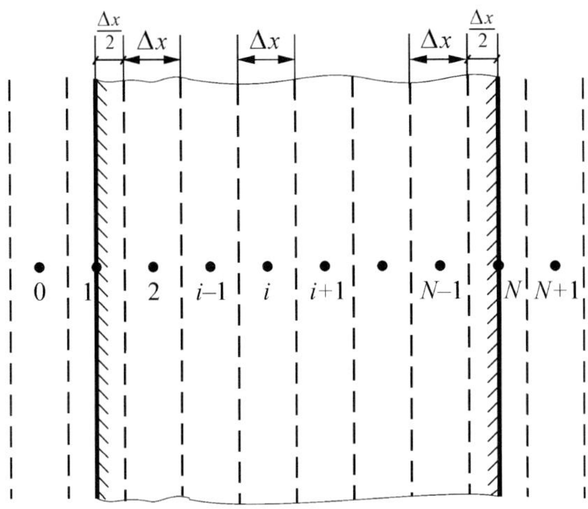

Fig. 21.2. Flat wall division into control volumes (finite)

Temperature will be calculated in nodes $x_i = (i-1)\Delta x$, $i = 1, ..., N$ at time points $t_n = n\Delta t$, $n = 0, 1, ...$. Derivatives after time and space will be approximated by means of the forward and central difference quotient

$$\left(\frac{\partial T}{\partial t}\right)_i^n=\frac{T_i^{n+1}-T_i^n}{\Delta t}+\mathrm{O}(\Delta t)\ , \tag{2}$$

$$\left(\frac{\partial^2 T}{\partial x^2}\right)_i^n=\frac{T_{i+1}^n-2T_i^n+T_{i-1}^n}{(\Delta x)^2}+\mathrm{O}\left[(\Delta x)^2\right],\quad \text{gdzie}\quad T(x_i,t_n)=T_i^n . \tag{3}$$

By substituting (2) and (3) into (1), one has

$$\frac{1}{a}\frac{T_i^{n+1}-T_i^n}{\Delta t}=\frac{T_{i+1}^n-2T_i^n+T_{i-1}^n}{(\Delta x)^2} . \tag{4}$$

From (4), the unknown T_i^{n+1} is determined

$$T_i^{n+1} = (\Delta Fo) T_{i-1}^n + [1 - 2(\Delta Fo)] T_i^n + (\Delta Fo) T_{i+1}^n, \tag{5}$$

where $\Delta Fo = a(\Delta t)/(\Delta x)^2$.

A diagram of space-marching in time created on the basis of (5) is presented in Fig. 21.3. The order of accuracy of the explicit method formulated in (5) is $O(\Delta t) + O[(\Delta x)^2]$. To obtain a good calculation accuracy, both time step Δt and spatial step Δx should be very small. If dimensionless time step ΔFo is too big, the explicit method becomes unstable; this is manifested by very big temperature oscillations at individual time steps and by the inability to reach the steady-state when $t \to \infty$.

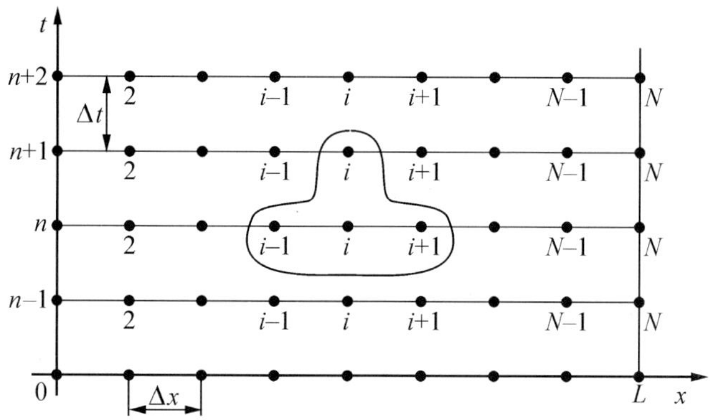

Fig. 21.3. Diagram of a computational molecule that illustrates space-marching in time in the finite difference method

From the stability analysis of the explicit method carried out by means of the von Neumann method or matrix method, it is evident that stable solution is obtained when the coefficient, which stands next to a temperature value in a given node in the previous time step, as for e.g. in T_i^n, is larger than or equal to zero. From (5) it follows, therefore, that the condition of solution stability for internal nodes $1 < i < (N-1)$ has the form

$$1 - 2(\Delta Fo) \geq 0, \tag{6}$$

hence

$$\Delta Fo = a \frac{(\Delta t)}{(\Delta x)^2} \leq \frac{1}{2}. \tag{7}$$

In the case of the boundary condition of 3rd kind, time step Δt should be even smaller than the computational step from (7).

Equations for boundary nodes 1 and N have the form:

- for boundary conditions of 1st kind

$$T\big|_{x=0} = f_1(t), \tag{8}$$

$$T\big|_{x=L} = f_N(t), \tag{9}$$

where $f_1(t)$ and $f_N(t)$ are the assigned functions.

Temperature T_i^{n+1} is calculated for nodes $2 \le i \le (N-1)$, since temperatures in nodes 1 and N are known:

$$T_1^n = f_1(t_n), \tag{10}$$

$$T_N^n = f_N(t_n); \tag{11}$$

- for boundary conditions of 2nd kind

$$-\lambda \frac{\partial T}{\partial x}\bigg|_{x=0} = \dot{q}_1(t), \tag{12}$$

$$-\lambda \frac{\partial T}{\partial x}\bigg|_{x=L} = \dot{q}_N(t), \tag{13}$$

where $\dot{q}_1(t)$ is the flow density, which flows into the plate through the surface $x = 0$; $\dot{q}_N(t)$ is the heat flux, which recedes from the plate through the surface $x = L$.

In boundary condition (12), the derivative at point $x = 0$ is approximated by the central difference quotient ((11), Ex. 21.1)

$$-\lambda \frac{T_2^n - T_0^n}{2\Delta x}\bigg|_{x=0} = \dot{q}_1^n, \tag{14}$$

where $\dot{q}_1^n = \dot{q}_1(t_n)$.

Equation (5) assumes for node $i = 1$ the following form:

$$T_1^{n+1} = (\Delta Fo)T_0^n + \left[1 - 2(\Delta Fo)\right]T_1^n + (\Delta Fo)T_2^n. \tag{15}$$

Temperature is eliminated from (14) and (15) in an apparent node T_0. From (14) one has

$$T_0^n = T_2^n + 2\frac{\dot{q}_1^n \Delta x}{\lambda}. \tag{16}$$

By substituting (16) into (15), one has

$$T_1^{n+1} = \left[1 - 2(\Delta Fo)\right]T_1^n + 2(\Delta Fo)T_2^n + 2\frac{\dot{q}_1^n \Delta x}{\lambda}\Delta Fo, \qquad n = 0, 1, \ldots \quad (17)$$

Approximation accuracy of the differential heat conduction equation at node 1 achieved by means of (17) is of the order O(Δt) and O[$(\Delta x)^2$].

One should emphasize that (17) can be easily obtained from the energy balance equation for node 1, if we account that the width of control volume is $\Delta x/2$. Energy balance equation for node 1 has the following form:

$$c\rho\frac{\Delta x}{2}\frac{dT_1}{dt} = \lambda\left.\frac{\partial T}{\partial x}\right|_{i+1/2} + \dot{q}_1, \quad (18)$$

hence, we have

$$c\rho\frac{\Delta x}{2}\frac{T_1^{n+1} - T_1^n}{\Delta t} = \lambda\frac{T_2^n - T_1^n}{\Delta x} + \dot{q}_1^n, \quad (19)$$

$$T_1^{n+1} = \left[1 - 2(\Delta Fo)\right]T_1^n + 2(\Delta Fo)T_2^n + 2\frac{\dot{q}_1^n \Delta x}{\lambda}\Delta Fo. \quad (20)$$

Equations (17) and (20) are identical. Boundary condition (13) is approximated in a similar way, by simply approximating the derivative by means of the central difference quotient

$$-\lambda\left.\frac{T_{N+1}^n - T_{N-1}^n}{2\Delta x}\right|_{x=L} = \dot{q}_N^n, \quad (21)$$

where $\dot{q}_N^n = \dot{q}_N(t_n)$.

Equation (5) assumes for node N the following form:

$$T_N^{n+1} = (\Delta Fo)T_{N-1}^n + \left[1 - 2(\Delta Fo)\right]T_N^n + (\Delta Fo)T_{N+1}^n. \quad (22)$$

By eliminating temperature T_{N+1} from (21), one has

$$T_{N+1}^n = T_{N-1}^n - 2\frac{\dot{q}_N^n \Delta x}{\lambda}. \quad (23)$$

Once (23) is substituted into (22) and transformations are carried out

$$T_N^{n+1} = 2(\Delta Fo)T_{N-1}^n + \left[1 - 2(\Delta Fo)\right]T_N^n - 2\frac{\dot{q}_N^n \Delta x}{\lambda}\Delta Fo. \quad (24)$$

Temperature in nodes i = 1, ..., N is calculated by means of the explicit method from the following formulas:

$$T_1^{n+1} = \left[1-2(\Delta Fo)\right]T_1^n + 2(\Delta Fo)T_2^n + 2\frac{\dot{q}_1^n \Delta x}{\lambda}\Delta Fo\,, \tag{25}$$

$$T_i^{n+1} = (\Delta Fo)T_{i-1}^n + \left[1-2(\Delta Fo)\right]T_i^n + (\Delta Fo)T_{i+1}^n, \qquad n = 2, ..., N-1\,, \tag{26}$$

$$T_N^{n+1} = 2(\Delta Fo)T_{N-1}^n + \left[1-2(\Delta Fo)\right]T_N^n - 2\frac{\dot{q}_N^n \Delta x}{\lambda}\Delta Fo\,. \tag{27}$$

Calculations begin from $n = 0$. Temperatures in nodes are present on the right-hand-side of (25)–(27) at an initial moment $t = 0$. In terms of the explicit method, therefore, there is no need to solve the algebraic equation system. It is easy to program (25)–(27):

- for boundary conditions of 3rd kind

$$-\lambda\left.\frac{\partial T}{\partial x}\right|_{x=0} = \alpha_1(-T|_{x=0} + T_{cz,1})\,, \tag{28}$$

$$-\lambda\left.\frac{\partial T}{\partial x}\right|_{x=L} = \alpha_N(T|_{x=L} - T_{cz,2})\,, \tag{29}$$

where $T_{cz,1}$ is the temperature of the medium that is in contact with surface $x = 0$; $T_{cz,2}$ the temperature of the medium that flows over surface $x = L$. Heat transfer coefficients α_1 and α_N are assigned on surfaces $x = 0$ and $x = L$, respectively.

Boundary conditions (28) and (29) can be approximated by means of the central difference quotient, using apparent points. The same results, as shown in the case of the boundary conditions of 2nd kind, can be obtained by writing heat balance equations for nodes 1 and 2 (Fig. 21.2)

$$c\rho\frac{\Delta x}{2}\frac{dT_1}{dt} = \alpha_1\left(T_{cz,1} - T_1\right) + \lambda\left.\frac{\partial T}{\partial x}\right|_{i+1/2}\,, \tag{30}$$

$$c\rho\frac{\Delta x}{2}\frac{dT_N}{dt} = \alpha_N\left(T_{cz,2} - T_N\right) + \lambda\left.\frac{\partial T}{\partial x}\right|_{N-1/2}\,. \tag{31}$$

As a result of the derivatives approximation after time with forward difference quotients, one obtains, respectively

$$c\rho\frac{\Delta x}{2}\frac{T_1^{n+1} - T_1^n}{\Delta t} = \alpha_1\left(T_{cz,1}^n - T_1^n\right) + \lambda\frac{T_2^n - T_1^n}{\Delta x}\,, \tag{32}$$

$$T_{i,j}^{n+1} = T_{i,j}^{n} + \left(\Delta Fo_x\right)\left(T_{i-1,j}^{n} - 2T_{i,j}^{n} + T_{i+1,j}^{n}\right) + \Delta Fo_y\left(T_{i,j-1}^{n} - 2T_{i,j}^{n} + T_{i,j+1}^{n}\right), \tag{3}$$

where $\Delta Fo_x = a(\Delta t)/(\Delta x)^2$ and $\Delta Fo_y = a(\Delta t)/(\Delta y)^2$.

In terms of the uniform mesh $\Delta x = \Delta y$, (3) assumes the form

$$T_{i,j}^{n+1} = \left(\Delta Fo\right)\left(T_{i-1,j}^{n} + T_{i+1,j}^{n} + T_{i,j-1}^{n} + T_{i,j+1}^{n}\right) + \left(1 - 4\Delta Fo\right)T_{i,j}^{n}. \tag{4}$$

Next, one should derive the heat balance equation for a node that lies on the boundary on which the boundary conditions of 3rd kind are assigned.

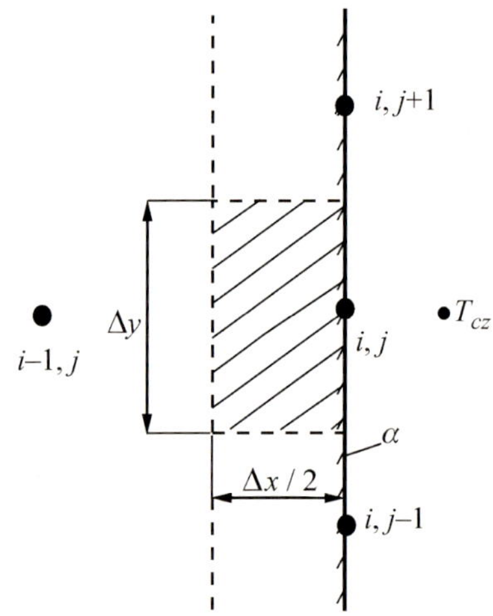

Fig. 21.5. Diagram of a control volume attached to a body edge

Heat balance equation for node (i, j) that lies on the boundary (Fig. 21.5) has the form

$$c\rho\frac{\Delta x}{2}\Delta y\frac{dT_{i,j}}{dt} = \lambda\Delta y\frac{T_{i-1,j} - T_{i,j}}{\Delta x} + \alpha\Delta y\left(T_{cz} - T_{i,j}\right) + \tag{5}$$

$$+\lambda\frac{\Delta x}{2}\frac{T_{i,j+1} - T_{i,j}}{\Delta y} + \lambda\frac{\Delta x}{2}\frac{T_{i,j-1} - T_{i,j}}{\Delta y},$$

hence, once the derivative after time is approximated by the forward difference quotient and the right-hand-side of equation calculated at the time point $t_n = t_{n+1} - \Delta t$, one has

$$\frac{T_{i,j}^{n+1}-T_{i,j}^{n}}{\Delta t}=a\left[2\frac{T_{i-1,j}^{n}-T_{i,j}^{n}}{(\Delta x)^{2}}+2\frac{\alpha}{\lambda(\Delta x)}\left(T_{cz}^{n}-T_{i,j}^{n}\right)+\right.$$
$$\left.+\frac{T_{i,j+1}^{n}-T_{i,j}^{n}}{(\Delta y)^{2}}+\frac{T_{i,j-1}^{n}-T_{i,j}^{n}}{(\Delta y)^{2}}\right]. \quad (6)$$

Next, temperature $T_{i,j}^{n+1}$ is determined from (6)

$$T_{i,j}^{n+1}=(\Delta Fo_{x})\left[2T_{i-1,j}^{n}+2(\Delta Bi_{x})T_{cz}^{n}\right]+(\Delta Fo_{y})\left(T_{i,j-1}^{n}+T_{i,j+1}^{n}\right)+$$
$$+\left[1-2(\Delta Fo_{x})-2(\Delta Fo_{y})-2(\Delta Bi_{x})(\Delta Fo_{x})\right]T_{i,j}^{n}, \quad (7)$$

where $\Delta Bi_x = \alpha(\Delta x)/\lambda$.

In the case of the uniform mesh of control volumes $\Delta x = \Delta y$, (7) is simplified to a form

$$T_{i,j}^{n+1}=(\Delta Fo)\left[2T_{i-1,j}^{n}+T_{i,j-1}^{n}+T_{i,j+1}^{n}+2(\Delta Bi)T_{cz}^{n}\right]+$$
$$+\left[1-4(\Delta Fo)-2(\Delta Bi)(\Delta Fo)\right]T_{i,j}^{n}. \quad (8)$$

From condition

$$1-4(\Delta Fo)-2(\Delta Bi)(\Delta Fo)\geq 1 \quad (9)$$

one obtains the condition of calculation stability

$$(\Delta Fo)(2+\Delta Bi)\leq\frac{1}{2}. \quad (10)$$

Heat balance equations for corner nodes are derived in a similar way; one should analyze the outer corner first (Fig. 21.6).

Heat balance equation for node (i, j) has the form

$$c\rho\frac{\Delta x}{2}\frac{\Delta y}{2}\frac{dT_{i,j}}{dt}=\lambda\frac{\Delta y}{2}\frac{T_{i-1,j}-T_{i,j}}{\Delta x}+\alpha\frac{\Delta y}{2}\left(T_{cz}-T_{i,j}\right)+$$
$$+\lambda\frac{\Delta x}{2}\frac{T_{i,j-1}-T_{i,j}}{\Delta y}+\alpha\frac{\Delta x}{2}\left(T_{cz}-T_{i,j}\right). \quad (11)$$

By carrying out the same operations as before, from (11) one has

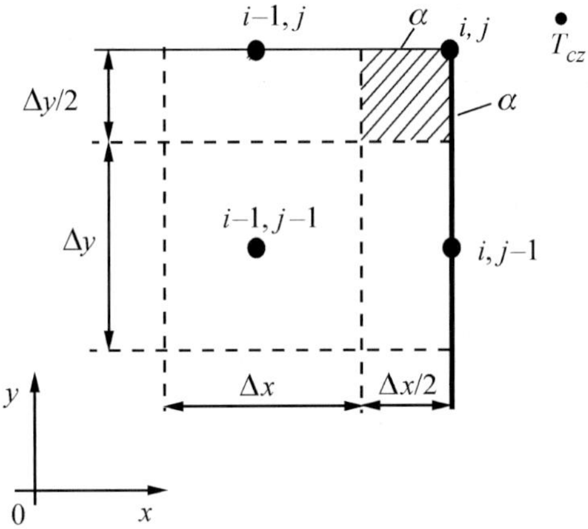

Fig. 21.6. Diagram of a corner control volume

$$\frac{T_{i,j}^{n+1}-T_{i,j}^{n}}{\Delta t}=a\left[2\frac{T_{i-1,j}^{n}-T_{i,j}^{n}}{(\Delta x)^2}+2\frac{\alpha}{\lambda(\Delta x)}\left(T_{cz}^{n}-T_{i,j}^{n}\right)+\right.$$
$$\left.+2\frac{T_{i,j-1}^{n}-T_{i,j}^{n}}{(\Delta y)^2}+2\frac{\alpha}{\lambda(\Delta y)}\left(T_{cz}^{n}-T_{i,j}^{n}\right)\right], \tag{12}$$

hence, one can determine temperature $T_{i,j}^{n+1}$

$$T_{i,j}^{n+1}=2(\Delta Fo_x)\left[T_{i-1,j}^{n}+(\Delta Bi_x)T_{cz}^{n}\right]+2(\Delta Fo_y)\left[T_{i,j-1}^{n}+(\Delta Bi_y)T_{cz}^{n}\right]+$$
$$+\left[1-2(\Delta Fo_x)-2(\Delta Fo_y)-2(\Delta Bi_x)(\Delta Fo_x)-2(\Delta Bi_y)(\Delta Fo_y)\right]T_{i,j}^{n}, \tag{13}$$

where $\Delta Bi_x = \alpha(\Delta x)/\lambda$ and $\Delta Bi_y = \alpha(\Delta y)/\lambda$.

When $\Delta x = \Delta y$, then (13) assumes the form

$$T_{i,j}^{n+1}=2(\Delta Fo)\left[T_{i-1,j}^{n}+T_{i,j-1}^{n}+2(\Delta Bi)T_{cz}^{n}\right]+$$
$$+\left[1-4(\Delta Fo)-4(\Delta Bi)(\Delta Fo)\right]T_{i,j}^{n}. \tag{14}$$

From condition

$$\left[1-4(\Delta Fo)-4(\Delta Bi)(\Delta Fo)\right]\geq 0 \tag{15}$$

one obtains the condition of calculation stability

$$(\Delta Fo)\left[1+(\Delta Bi)\right]\leq\frac{1}{4}. \qquad (16)$$

In terms of the inner corner, operations are carried out in a similar way. Heat balance equation for node (i, j) has the form

$$\frac{3}{4}c\rho\Delta x\Delta y\frac{dT_{i,j}}{dt}=\lambda\Delta y\frac{T_{i-1,j}-T_{i,j}}{\Delta x}+\lambda\frac{\Delta y}{2}\frac{T_{i+1,j}-T_{i,j}}{\Delta x}+\lambda\Delta x\frac{T_{i,j+1}-T_{i,j}}{\Delta y}+$$
$$+\lambda\frac{\Delta x}{2}\frac{T_{i,j-1}-T_{i,j}}{\Delta y}+\alpha\frac{\Delta y}{2}\left(T_{cz}-T_{i,j}\right)+\alpha\frac{\Delta x}{2}\left(T_{cz}-T_{i,j}\right). \qquad (17)$$

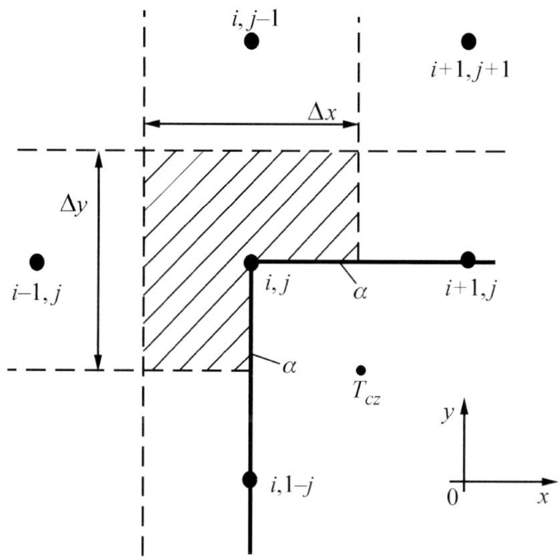

Fig. 21.7. Diagram of inner corner

By carrying out similar operations as before, one has

$$\frac{T_{i,j}^{n+1}-T_{i,j}^{n}}{\Delta t}=a\left[\frac{4}{3}\frac{T_{i-1,j}^{n}-T_{i,j}^{n}}{(\Delta x)^2}+\frac{2}{3}\frac{T_{i+1,j}^{n}-T_{i,j}^{n}}{(\Delta x)^2}+\frac{4}{3}\frac{T_{i,j+1}^{n}-T_{i,j}^{n}}{(\Delta y)^2}+\right.$$
$$\left.+\frac{2}{3}\frac{T_{i,j-1}^{n}-T_{i,j}^{n}}{(\Delta y)^2}+\frac{2}{3}\frac{\alpha}{\lambda(\Delta x)}\left(T_{cz}^{n}-T_{i,j}^{n}\right)+\frac{2}{3}\frac{\alpha}{\lambda(\Delta y)}\left(T_{cz}^{n}-T_{i,j}^{n}\right)\right], \qquad (18)$$

hence, one can determine temperature $T_{i,j}^{n+1}$

$$T_{i,j}^{n+1} = \frac{2}{3}(\Delta Fo_x)\left[2T_{i-1,j}^{n} + T_{i+1,j}^{n} + (\Delta Bi_x)T_{cz}^{n}\right] + \\ + \frac{2}{3}(\Delta Fo_y)\left[2T_{i,j+1}^{n} + T_{i,j-1}^{n} + (\Delta Bi_y)T_{cz}^{n}\right] + \\ + \left[1 - 2(\Delta Fo_x) - 2(\Delta Fo_y) - \frac{2}{3}(\Delta Bi_x)(\Delta Fo_x)\right]T_{i,j}^{n}. \tag{19}$$

For uniform mesh, when $\Delta x = \Delta y$ (19) is simplified to a form

$$T_{i,j}^{n+1} = \frac{2}{3}\Delta Fo\left[2T_{i-1,j}^{n} + T_{i+1,j}^{n} + 2T_{i,j+1}^{n} + T_{i,j-1}^{n} + 2(\Delta Bi)T_{cz}^{n}\right] + \\ + \left[1 - 4(\Delta Fo) - \frac{4}{3}(\Delta Bi)(\Delta Fo)\right]T_{i,j}^{n}. \tag{20}$$

The condition of calculation stability has the form

$$(\Delta Fo)(3 + \Delta Bi) \le \frac{3}{4}. \tag{21}$$

Exercise 21.4 Solving Two-Dimensional Problems by Means of the Implicit Difference Method

Derive formulas for the calculation of transient temperature fields by means of the implicit difference method. To derive the difference equations, apply the control volume method. Compile computational formulas in a table by assuming that $\Delta x = \Delta y$.

Solution

Implicit method will be presented on the basis of an example in which the transient temperature distribution is determined in a thick-walled plate, which is convectively heated or cooled on its front face or thermally insulated on its back surface. Control volume division and node location are presented in Fig. 21.2. Once the derivative after time is approximated by the forward difference quotient in the heat conduction equation ((2), Ex. 21.2) and the right-hand-side of equation is expressed in the node temperature function at an instant $t + \Delta t$, one gets

$$\frac{1}{a}\frac{T_i^{n+1}-T_i^n}{\Delta t}=\frac{T_{i+1}^{n+1}-2T_i^{n+1}+T_{i-1}^{n+1}}{(\Delta x)^2}, \tag{1}$$

where from, after transformation, the following algebraic equation is obtained

$$-(\Delta Fo)T_{i-1}^{n+1}+\left[1+2(\Delta Fo)\right]T_i^{n+1}-(\Delta Fo)T_{i+1}^{n+1}=T_i^n, \quad i=2, \ldots, N, \tag{2}$$

where $\Delta Fo = a\Delta t/(\Delta x)^2$. For node 1, which lies on the boundary, the difference equation has a similar form to the form of (32) from Ex. 21.2. The right-hand-side of this equation in the implicit method is expressed by temperatures at time point $t + \Delta t$

$$c\rho\frac{\Delta x}{2}\frac{T_1^{n+1}-T_1^n}{\Delta t}=\alpha_1\left(T_{cz}^{n+1}-T_1^{n+1}\right)+\lambda\frac{T_2^{n+1}-T_1^{n+1}}{\Delta x}, \tag{3}$$

After transformations, the algebraic (3) has the form

$$\left[1+2(\Delta Fo)(1+\Delta Bi)\right]T_1^{n+1}-2(\Delta Fo)T_2^{n+1}=2(\Delta Fo)(\Delta Bi)T_{cz}^{n+1}+T_1^n, \tag{4}$$

where $\Delta Bi = \alpha(\Delta x)/\lambda$.

For node $i = N$, one has

$$-2(\Delta Fo)T_{N-1}^{n+1}+\left[1+2(\Delta Fo)\right]T_N^{n+1}=T_N^n. \tag{5}$$

The implicit method is unconditionally stable. The accuracy order of the method is $O(\Delta t)$ and $O[(\Delta x)^2]$. In order to obtain a good calculation accuracy, both Δt and Δx should be, respectively, small. It is best to carry out the calculations for a few decreasing values, Δt and Δx. If calculation results do not change as Δt and Δx decreases, then one can assume that the accuracy of results is high.

Basic difference equations, which result from the application of the implicit finite difference method to the solution of two-dimensional transient heat conduction problems, are compiled in Table 21.1.

In the implicit method, the algebraic equation system must be solved at every time step $n + 1$, i.e. at time point $t + \Delta t$. For one-dimensional problems, the system is tri-diagonal, while for two-dimensional, penta-diagonal. To solve these systems, one can use the direct methods from linear algebra like, for example, Gauss elimination method, Gauss-Jordan method or iterative methods, such as Gauss-Seidel method or over-relaxation method. An algebraic equation system, whose unknowns are temperatures T_i^{n+1}, $i = 1, \ldots, N$ can be written in the form

Table 21.1. Implicit difference equations in two-dimensional transient heat conduction problems

Node type	Diagram and the difference equation for a node
Internal node	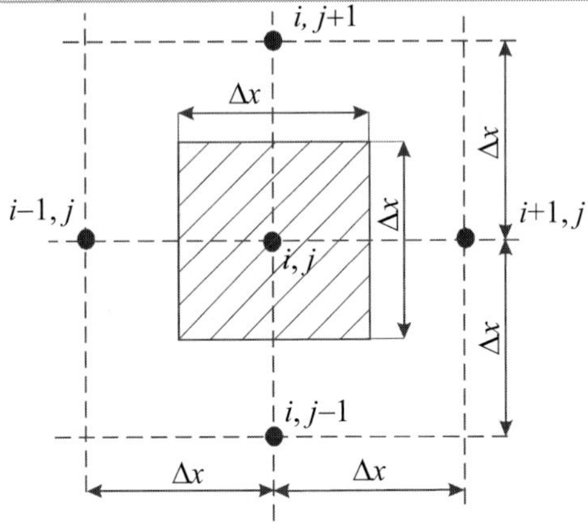 $$\left[1+4(\Delta Fo)\right]T_{i,j}^{n+1}-(\Delta Fo)\left[T_{i-1,j}^{n+1}+T_{i+1,j}^{n+1}+T_{i,j-1}^{n+1}+T_{i,j+1}^{n+1}\right]=T_{i,j}^{n} \quad (1)$$
Body surface node	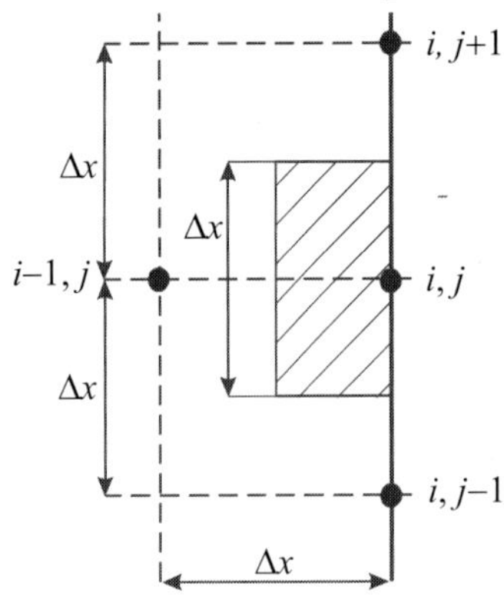 $$\left[1+2(\Delta Fo)(2+\Delta Bi)\right]T_{i,j}^{n+1}-(\Delta Fo)\left[2T_{i-1,j}^{n+1}+T_{i,j-1}^{n+1}+T_{i,j+1}^{n+1}+(\Delta Bi)T_{cz}^{n+1}\right]=T_{i,j}^{n} \quad (2)$$
Outer corner node	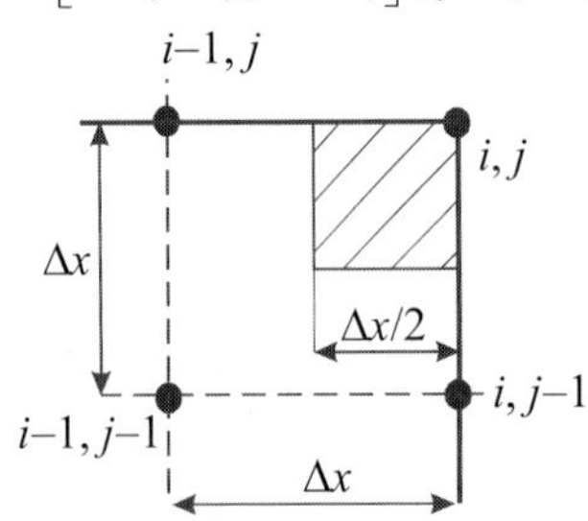 $$\left[1+4(\Delta Fo)(1+\Delta Bi)\right]T_{i,j}^{n+1}-2(\Delta Fo)\left[2T_{i-1,j}^{n+1}+T_{i,j-1}^{n+1}+2(\Delta Bi)T_{cz}^{n+1}\right]=T_{i,j}^{n} \quad (3)$$

Table 21.1. (cont.)

Node type	Diagram and the difference equation for a node
Internal node	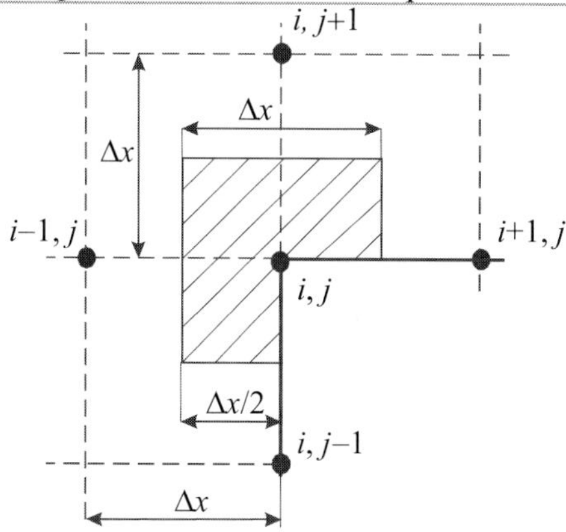 $\left[1+4(\Delta Fo)\left(1+\frac{\Delta Bi}{3}\right)\right]T_{i,j}^{n+1}-\frac{2}{3}(\Delta Fo)\left[2T_{i-1,j}^{n+1}+T_{i+1,j}^{n+1}+2T_{i,j-1}^{n+1}+2T_{i,j+1}^{n+1}+2(\Delta Bi)T_{cz}^{n+1}\right]=T_{i,j}^{n}$ (4)

$$\mathbf{AT} = \mathbf{b} \tag{6}$$

where $\mathbf{A}$ is the coefficients matrix whose dimensions are $N \cdot N$, $\mathbf{T} = (T_1^{n+1}, T_2^{n+1}, ..., T_N^{n+1})^T$ a column vector, while $\mathbf{b}$ a column vector of the equations' right-hand-side.

It is easy to find the solution once the inverse matrix $\mathbf{A}^{-1}$ is determined first:

$$\mathbf{T} = \mathbf{A}^{-1}\mathbf{b}. \tag{7}$$

In simple (direct) transient heat conduction problems, matrix $\mathbf{A}$ is not singular when the boundary conditions and the initial condition are known, while the inverse matrix $\mathbf{A}^{-1}$ can be determined with a great accuracy. In the case of one-dimensional problems, Thomas algorithm is used to solve the equation system.

Finally, to culminate the discussion on the implicit finite difference method, one should mention here about the Crank-Nicolson method in which the right-hand-side of the heat conduction equation is in one-half approximated by the temperature function T_i^n, while in the second-half by the temperature function T_i^{n+1}

$$\frac{T_i^{n+1}-T_i^n}{\Delta t} = a\left[\frac{1}{2}\frac{T_{i-1}^n - 2T_i^n + T_{i+1}^n}{(\Delta x)^2} + \frac{1}{2}\frac{T_{i-1}^{n+1} - 2T_i^{n+1} + T_{i+j}^{n+1}}{(\Delta x)^2}\right]. \tag{8}$$

The order of accuracy of the method is $O[(\Delta t)^2] + O[(\Delta x)^2]$; it is higher, therefore, than it is in the case of the explicit and implicit method. Due to the ongoing advances in computer technology, namely in the computer processing speed, such advantage no longer plays the kind of role it did few decades ago. By carrying out calculations by means of, for e.g., the implicit method with a smaller time step Δt, one can obtain the same accuracy as one does in the case of the Crank–Nicolson method.

Exercise 21.5 Algorithm and a Program for Solving a Tridiagonal Equation System by Thomas Method

Describe Thomas algorithm for solving tridiagonal algebraic equations system. Write an appropriate program in FORTRAN language.

Solution

The system in question, i.e. N equations system with N temperature unknowns in nodes has the form

$$d_1 T_1 + b_1 T_2 = c_1 \tag{1}$$

$$a_2 T_1 + d_2 T_2 + b_2 T_3 = c_2 \tag{2}$$

$$a_3 T_2 + d_3 T_3 + b_3 T_4 = c_3 \tag{3}$$

$$\cdots\cdots\cdots\cdots$$

$$a_{N-1} T_{N-2} + d_{N-1} T_{N-1} + b_{N-1} T_N = c_{N-1} \tag{4}$$

$$a_N T_{N-1} + d_N T_N = c_N . \tag{5}$$

The system is tri-diagonal with other-than-zero coefficients only on the main diagonal (coefficients d_i), lower diagonal (coefficients a_i) and upper diagonal (coefficients b_i). The above equation system can be written in the form

$$\mathbf{AT} = \mathbf{c} \tag{6}$$

$$\mathbf{A} = \begin{bmatrix} d_1 & b_1 & 0 & 0 & 0 & \cdots & 0 & 0 \\ a_2 & d_2 & b_2 & 0 & 0 & \cdots & 0 & 0 \\ 0 & a_3 & d_3 & b_3 & 0 & \cdots & 0 & 0 \\ \cdot & \cdot & \cdot & \cdot & \cdot & \cdot & \cdot & \cdot \\ 0 & 0 & \cdot & \cdot & \cdot & a_{N-1} & d_{N-1} & b_{N-1} \\ 0 & 0 & \cdot & \cdot & \cdot & & a_N & d_N \end{bmatrix}. \tag{7}$$

Thomas algorithm is the result of an application of the Gaussian elimination method to a tri-diagonal equation system. In order to eliminate coefficients from the lower diagonal, (1) is multiplied by a_2

$$a_2 d_1 T_1 + a_2 b_1 T_2 = c_1 a_2 . \tag{8}$$

By multiplying (2) by d_1 , one has

$$a_2 d_1 T_1 + d_1 d_2 T_2 + d_1 b_2 T_3 = d_1 c_2 . \tag{9}$$

Subtracting (8) from (9) results in

$$\left(d_1 d_2 - b_1 a_2\right) T_2 + d_1 b_2 T_3 = d_1 c_2 - c_1 a_2 , \tag{10}$$

hence, once both sides are divided by $d_{1,}$ one obtains

$$\left(d_2 - \frac{a_2\, b_1}{d_1}\right) T_2 + b_2 T_3 = c_2 - \frac{c_1 a_2}{d_1} \tag{11}$$

If (11) coefficients are denoted by

$$d_2' = d_2 - \frac{b_1\, a_2}{d_1}, \tag{12}$$

$$c_2' = c_2 - \frac{c_1 a_2}{d_1}, \tag{13}$$

then (11) can be written in a simpler form

$$d_2' T_2 + b_2 T_3 = c_2' . \tag{14}$$

By continuing the elimination process, (14) should be multiplied by a_3

$$a_3 d_2' T_2 + a_3 b_2 T_3 = a_3 c_2' . \tag{15}$$

By multiplying (3) by d'_2, one has

$$a_3 d_2' T_2 + d_2' d_3 T_3 + d_2' b_3 T_4 = c_3 d_2' . \tag{16}$$

Subtracting (15) from (16) yields

```
      t(k)=(c(k)-b(k)*t(k+1))/d(k)
   enddo
c  write(2,'(3ht =,10(3x,e11.4))') (t(i),i=1,n)
   return
   end
```

Exercise 21.6 Stability Analysis of the Explicit Finite Difference Method by Means of the von Neumann Method

Conduct the stability analysis for the explicit finite difference method by means of the von Neumann method.

Solution

In numerical methods, calculations are done by marching in time with an assigned time-step Δt. If calculation error increases when moving from a given time point t to the next $t+\Delta t$, the error is reinforced and the calculations become unstable. Calculation instability is manifested by large oscillations in calculated temperature in time, whose amplitude tends towards infinity as it marches in time. If calculation error decreases when moving to the next step, calculations are usually stable [1, 2, 5, 10].

Numerical solution of the partial equation is burdened with a discretization error d_i

$$d_i = T_{a,i} - T_{d,i}, \qquad i = 1, \ldots, N \tag{1}$$

and rounding error ε_i in computer calculations

$$d_i = T_i - T_{d,i}, \tag{2}$$

where $T_{a,i}$ is the analytical solution of the partial equation in node i; $T_{d,i}$, an accurate solution of the difference equation in node i; T_i a temperature in node i is computer-calculated with finite accuracy by means of the numerical method.

Numerical solution T_i^n formulated as

$$T_i^n = T_{d,i}^n + \varepsilon_i^n \tag{3}$$

satisfies (4) from Ex. 21.2

$$\frac{1}{a}\frac{T_i^{n+1} - T_i^n}{\Delta t} = \frac{T_{i+1}^n - 2T_i^n + T_{i-1}^n}{(\Delta x)^2}. \tag{4}$$

By substituting (3) into (4), one has

$$\frac{1}{a}\frac{T_{d,i}^{n+1}+\varepsilon_i^{n+1}-T_{d,i}^{n}-\varepsilon_i^{n}}{\Delta t}=$$
$$=\frac{T_{d,i+1}^{n+1}+\varepsilon_{i+1}^{n+1}-2T_{d,i}^{n}-2\varepsilon_i^{n}+T_{d,i-1}^{n}-2\varepsilon_{i-1}^{n}}{\left(\Delta x\right)^2}. \tag{5}$$

Equation (5) is simplified to a form

$$\frac{1}{a}\frac{\varepsilon_i^{n+1}-\varepsilon_i^{n}}{\Delta t}=\frac{\varepsilon_{i+1}^{n+1}-2\varepsilon_i^{n}-2\varepsilon_{i-1}^{n}}{(\Delta x)^2}, \tag{6}$$

since the numerical solution T_d fully satisfies, in accordance with the definition, the difference (4). The solution of (4) given by (5), Ex. 15.2 is stable, when the ratio of the error's absolute values at time step n+1 and n satisfies the condition

$$G=\left|\frac{\varepsilon_i^{n+1}}{\varepsilon_i^{n}}\right|\le 1. \tag{7}$$

If the boundary conditions are periodic, then error ε_i^n can be expanded into a Fourier series. Assuming that transient temperature distribution is searched for in the region $-L \le x \le L$, then the error can be approximated by function

$$\varepsilon\left(x,t\right)=\sum_{m=-N}^{N}A_m\left(t\right)e^{ik_m x}, \tag{8}$$

where $A_m(t)$ is the time-dependent amplitude, $i=\sqrt{-1}$, k_m a wave number, N the intervals number ($N = L/\Delta x$).

Figure 21.8 shows the example of an error transient $\varepsilon^n = \varepsilon(t_n,x)$ and the accepted notations for the stability analysis. Maximum wave length equal to $\lambda_{max} = 2L$ is the basic frequency. The corresponding wave number is at its minimum then and is equal to $k_{min} = 2\pi/\lambda_{max} = 2\pi/2L = \pi/L$. The shortest wave length on the spatial grid with a step Δx is $\lambda_{min} = 2\Delta x$ (Fig. 21.8b). The largest is the wave number that corresponds to the minimum wave length; it is

$$k_{\max}=\frac{2\pi}{\lambda_{\min}}=\frac{2\pi}{2\Delta x}=\frac{N\pi}{L}. \tag{9}$$

Wave number k_m that appears in (8) can be presented in the following way:

$$k_m = mk_{\min} = m\frac{\pi}{L}, \qquad m = 1, 2, \ldots \tag{10}$$

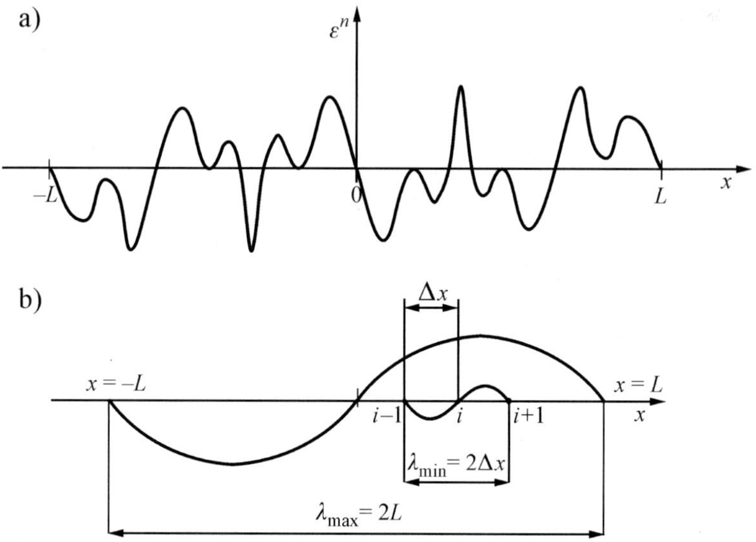

Fig. 21.8. Approximation of the error $\varepsilon^n(x)$ by means of the Fourier series within region $-L \le x \le L$

Therefore, error ε_i^n can be written in the form

$$\varepsilon_i^n = \varepsilon(x_i, t_n) = \sum_{m=-N}^{N} A_m(t_n) e^{ik_m x_i}, \tag{11}$$

where x_i is the coordinate of the spatial i-node. If the amplitude $A_m(t_n)$ is expressed as

$$A_m(t_n) = e^{\beta_m t_n}, \tag{12}$$

where β_m is a constant, then (11) can be written as follows:

$$\varepsilon_i^n = \sum_{m=-N}^{N} e^{\beta_m t_n} e^{ik_m x_i}. \tag{13}$$

By accounting that the difference (6) is linear and by applying the superposition principle, it is sufficient to analyze only one term in the series (13)

$$\varepsilon_m(x_i, t_n) = e^{\beta_m t_n} e^{ik_m x_i}. \tag{14}$$

Amplification factor G formulated in (7) can be calculated using (14)

$$G=\left|\frac{\varepsilon_i^{n+1}}{\varepsilon_i^n}\right|=\left|\frac{e^{\beta_m(t_n+\Delta t)}e^{ik_m x_i}}{e^{\beta_m t_n}e^{ik_m x_i}}\right|=\left|e^{\beta_m \Delta t}\right|. \tag{15}$$

In order to determine $e^{\beta_m \Delta t}$, (14) should be substituted into (6)

$$\frac{1}{a}\frac{e^{\beta_m(t_n+\Delta t)}e^{ik_m x_i}-e^{\beta_m t_n}e^{ik_m x_i}}{\Delta t}= \\ =\frac{e^{\beta_m t_n}e^{ik_m(x_i+\Delta x)}-2e^{\beta_m t_n}e^{ik_m x_i}+e^{\beta_m t_n(x_i-\Delta x)}}{(\Delta x)^2}. \tag{16}$$

Once (16) is divided by $e^{\beta_m t_n}e^{ik_m x_i}$, one has

$$\frac{1}{a}\frac{e^{\beta_m \Delta t}-1}{\Delta t}=\frac{e^{ik_m \Delta x}-2+e^{-ik_m \Delta x}}{(\Delta x)^2}, \tag{17}$$

hence, after transformation

$$e^{\beta_m \Delta t}=1+\frac{a\Delta t}{(\Delta x)^2}(e^{ik_m \Delta x}+e^{-ik_m \Delta x}-2). \tag{18}$$

If,

$$\cos\left(k_m \Delta x\right)=\frac{e^{ik_m \Delta x}+e^{-ik_m \Delta x}}{2}, \tag{19}$$

then (18) can be transformed into a form

$$e^{\beta_m \Delta t}=1+\frac{2a\Delta t}{\left(\Delta x\right)^2}\left[\cos\left(k_m \Delta x\right)-1\right]. \tag{20}$$

By accounting for condition (7) and (15) and (20), a formula for amplification factors is obtained

$$G=\left|1+2\Delta Fo\left[\cos\left(k_m \Delta x\right)-1\right]\right|\le 1. \tag{21}$$

Condition (21) can be transformed using identity

$$\sin^2\frac{k_m \Delta x}{2}=\frac{1-\cos\left(k_m \Delta x\right)}{2}. \tag{22}$$

One has then

$$\left|1-4\Delta Fo\sin^2\left(\frac{k_m \Delta x}{2}\right)\right|\le 1, \tag{23}$$

where $\Delta Fo = a\Delta t/(\Delta x)^2$.

If,

$$1-4\Delta Fo\sin^2\left(\frac{k_m\Delta x}{2}\right)\geq 0, \tag{24}$$

then (23) is always satisfied, since $\Delta Fo > 0$. When

$$1-4\Delta Fo\sin^2\left(\frac{k_m\Delta x}{2}\right)< 0, \tag{25}$$

then (23) assumes the form

$$1-4\Delta Fo\sin^2\left(\frac{k_m\Delta x}{2}\right)\geq -1. \tag{26}$$

Inequality (26) is satisfied when

$$\Delta Fo \leq \frac{1}{2}. \tag{27}$$

Condition (27) must be satisfied in order to make the calculations, carried out by means of the explicit finite difference method, stable. If we assume that the region is divided into control volumes with a width Δx, then time step Δt should satisfy the condition that follows from (27)

$$\Delta t \leq \frac{(\Delta x)^2}{2a}. \tag{28}$$

Exercise 21.7 Calculating One-Dimensional Transient Temperature Field by Means of the Explicit Method and a Computational Program

A thick infinite plate with the initial temperature of $T_0 = 300°C$ is suddenly cooled on one of its surfaces by a jet of flowing water at a temperature of $T_{cz} = 20°C$. Heat transfer coefficient on the plate surface is $\alpha = 2000$ W/(m^2·K).

Assume that the thermo-physical properties of the plate material (steel) are taken for the calculation as follow: $\lambda = 50$ W/(m·K), $a = 1\cdot10^{-5}$ m^2/s. Determine temperature distribution in the plate for the first 60 s, while treating the plate as a semi-infinite body. Use explicit finite difference method to determine temperature distribution. Compare the obtained results with temperature values calculated by means of the analytical solution presented in Ex. 14.4.

Solution

The division of the plate into control volumes is shown in Fig. 21.9.

The solving process of one-dimensional transient heat conduction problems by means of the explicit finite difference method is discussed in Ex. 21.2.

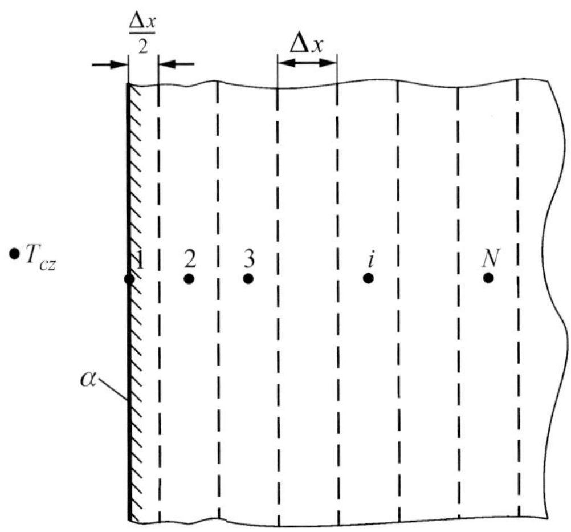

Fig. 21.9. Division of a half-space into control volumes

Temperature in node 1 (Fig. 21.9) is defined by (34), which in the given example assumes the form

$$T_1^{n+1} = 2(\Delta Fo)\left[T_2^n + (\Delta Bi)T_{cz}\right] + \left[1 - 2(\Delta Fo) - 2(\Delta Bi)(\Delta Fo)\right]T_1^n, \tag{1}$$

where $\Delta Fo = a\Delta t/(\Delta x)^2$ and $\Delta Bi = \alpha(\Delta x)/\lambda$.

In node 2 and in subsequent nodes, temperature is calculated by means of (5) from Ex. 21.2

$$T_i^{n+1} = (\Delta Fo)\left(T_{i-1}^n + T_{i+1}^n\right) + \left[1 - 2(\Delta Fo)\right]T_i^n, \qquad i = 2, \ldots, N. \tag{2}$$

The condition of calculation stability (1) has the form

$$(\Delta Fo)(1 + \Delta Bi) \le \frac{1}{2}. \tag{3}$$

For internal nodes, calculations are stable, when

$$\Delta Fo \le \frac{1}{2}. \tag{4}$$

If one assumes for the calculation that $\Delta x = 0.025$ m, then one obtains

$$\Delta Bi = \frac{\alpha(\Delta x)}{\lambda} = \frac{2000 \cdot 0.025}{50} = 1.0\,.$$

From (3) it follows that

$$(\Delta Fo)(1+1) \le \frac{1}{2}, \qquad \Delta Fo \le \frac{1}{4},$$

$$\Delta t \le \frac{1}{4}\frac{(\Delta x)^2}{a} = \frac{1}{4}\frac{(0.025)^2}{1 \cdot 10^{-5}} = 15.625 \text{ s.}$$

In order to increase calculation accuracy, a smaller time step is assumed, i.e. $\Delta t = 5$ s ($\Delta Fo = 0.08$). Equations (1) and (2) have then the form

$$T_1^{n+1} = 0.16\left(T_2^n + 20\right) + 0.68 T_1^n, \tag{5}$$

$$T_i^{n+1} = 0.08\left(T_{i-1}^n + T_{i+1}^n\right) + 0.84 T_i^n, \qquad i = 2, \ldots, N. \tag{6}$$

At an initial moment $t = 0$, all node temperatures equal initial temperature

$$T_i^0 = 300^\circ\text{C}, \;\; i = 1, \ldots, N+1. \tag{7}$$

It is assumed, moreover, that the depth of heat penetration in time $0 \le t \le 60$ s is smaller than $x_{N+1} = (N-1)\Delta x$, i.e. $T_{N+1}^n = T_0$, $n = 0, 1, \ldots$. For the subsequent calculations, one assumes that $N = 10$, therefore $T_{N+1}^n = 300^\circ\text{C}$, $n = 0, 1, \ldots$.

Table 21.2 gives calculation results for nodes no. 1 ($x = 0$ m) and 5 ($x = 0.1$ m) and shows how they compare with temperatures calculated by means of the analytical (9), derived in Ex. 14.4.

Table 21.2. Comparison of temperature values in nodes no. 1 ($x = 0$ m) and no. 5 ($x = 0.1$ m) calculated by means of the explicit difference method and analytical method

t [s]	Node no. 1 (x = 0 m) Solution		Node no. 5 (x = 0.1 m) Solution	
	Numerical, T [°C]	Analytical, T [°C]	Numerical, T [°C]	Analytical, T [°C]
0	300.00	300.00	300.00	300.00
5	255.20	229.00	300.00	300.00
10	224.74	207.80	300.00	300.00
15	203.45	193.90	300.00	300.00
20	188.10	183.40	300.00	300.00
25	176.65	175.00	300.00	300.00
30	167.82	168.10	299.99	300.00

Table 21.2. (cont.)

t [s]	Node no. 1 (x = 0 m) Solution Numerical, T [°C]	Node no. 1 (x = 0 m) Solution Analytical, T [°C]	Node no. 5 (x = 0.1 m) Solution Numerical, T [°C]	Node no. 5 (x = 0.1 m) Solution Analytical, T [°C]
35	160.77	162.10	299.97	300.00
40	154.96	156.90	299.94	300.00
45	150.06	152.40	299.88	299.90
50	145.82	148.30	299.80	299.90
55	142.08	144.60	299.68	299.80
60	138.74	141.30	299.53	299.70

Table 21.3. Node temperatures calculated by means of the explicit difference method and the exact analytical method at two time points $t = 20$ s and $t = 40$ s

x [m]		0	0,025	0,05	0,075	0,1	0,125	0,15	0,175	0,2	0,225
Node no.		1	2	3	4	5	6	7	8	9	10
t = 20 s	Numerical method	188.1	284.7	299.0	299.9	300.0	300.0	300.0	300.0	300.0	300.0
	Analytical method	183.4	283.1	299.3	299.9	300.0	300.0	300.0	300.0	300.0	300.0
t = 40 s	Numerical method	154.9	259.3	292.8	299.2	299.9	300.0	300.0	300.0	300.0	300.0
	Analytical method	156.9	257.8	293.1	299.4	299.9	300.0	300.0	300.0	300.0	300.0

Table 21.3 presents the results of node temperature calculations in time $t = 20$ s i $t = 40$ s and compares them to the analytical solution.

From the comparisons shown in Tables 21.2 and 21.3 one can conclude that the accuracy of the obtained results is very good, despite the presence of a large spatial step Δx. Moreover, a computational algorithm, which is simple and easy to programme, is an additional advantage of the explicit method. The program written in FORTRAN language, by means of which the results given in Tables 21.2 and 21.3 were obtained, is presented below.

Program in FORTRAN language for calculating one-dimensional transient temperature distribution by means of the explicit method and the exact analytical solution

```
p_21_07.in
10 12 300. 20.
1.E-5 2000. 50. 0.025 5.
```

```
p_21_07.for
c     Calculation of one-dimensional transient temperature
c     distribution by means of explicit method and
c     exact analytical solution
      program p_21_07
      dimension t(50),tt(50)
      open(unit=1,file='p_21_07.in')
      open(unit=2,file='p_21_07.out')
      read(1,*)n,n_time,t_pocz,t_cz
      read(1,*)a,alfa,s_lambda,dx,dt

      write(2,'(a)')
     &"CALCULATION OF ONE-DIMENSIONAL TRANSIENT TEMP.
     &   DISTRIB."
      write(2,'(/a)') "INPUT DATA"
      write(2,'(a,i10)')   "equation number n=",n
      write(2,'(a,i10)')   "max number of time steps=",n_time
      write(2,'(a,e10.5,a)')"initial temp. t_pocz=",
     & t_pocz,"[C]"
      write(2,'(a,e10.5,a)')"medium temp. t_cz=",t_cz,"[C]"
      write(2,'(a,e10.5,a)')"coefficient a=",a," [m2/s]"
      write(2,'(a,e10.5,a)')"coefficient alfa=",alfa,"
     &  [W/m2/K]"
      write(2,'(a,e10.5,a)')"coeff. lambda=",s_lambda,"
     &  [W/m/K]"
      write(2,'(a,e10.5,a)')"space step dx=",dx," [m]"
      write(2,'(a,e10.5,a)')"time step dt=",dt," [s]"

      DFo=a*dt/dx**2
      DBi=alfa*dx/s_lambda

      do i=1,n+1
         t(i)=t_pocz
      enddo
      write(2,'(/a)')"CALCULATED TEMPERATURE"
      write(2,'(a)')"   Time       T1[C]       T5[C]"
      i=0
      time=i*dt
      write(2,'(f5.2,3x,e11.6,3x,e11.6)')time,t(1),t(5)
      i=i+1
      do while (i.le.n_time)
         tt(1)=2.*DFo*(t(2)+DBi*t_cz)+(1.-2.*DFo-
     &      2.*DBi*DFo)*t(1)
         do j=2,n
            tt(j)=DFo*(t(j-1)+t(j+1))+(1.-2.*DFo)*t(j)
         enddo
         do j=1,n
            t(j)=tt(j)
         enddo
         time=i*dt
```

```
      write(2,'(f5.2,3x,e11.6,3x,e11.6)')time,t(1),t(5)
      i=i+1
   enddo
   end  program p_21_07
```

Exercise 21.8 Calculating One-Dimensional Transient Temperature Field by Means of the Implicit Method and a Computational Program

Solve Ex. 21.7 by means of the implicit finite difference method. Assume the following data for the calculation: T_0 = 300°C, T_{cz} = 20°C, α = 2000 W/(m^2·K), λ = 50 W/(m·K), $a = 1\cdot10^{-5}$ m^2/s. The plate division into control volumes is presented in fig.21.10.

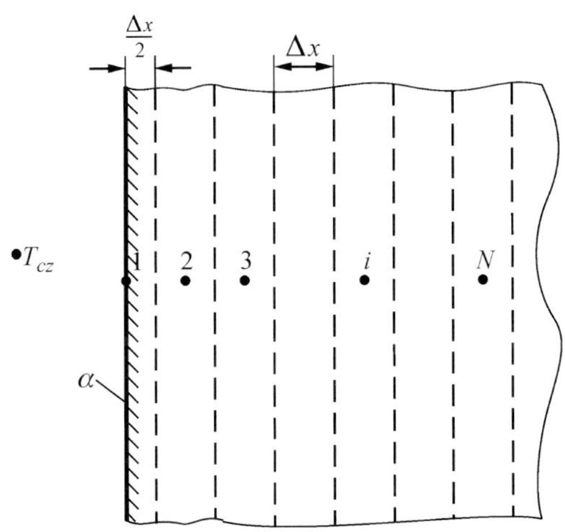

Fig. 21.10. Division of a half-space into control volumes

Solution

Energy balance equation for node 1, which lies on the surface, has the form ((4), Ex. 21.4)

$$\left[1+2(\Delta Fo)(1+\Delta Bi)\right]T_1^{n+1}-2(\Delta Fo)T_2^{n+1}=2(\Delta Fo)(\Delta Bi)T_{cz}^{n+1}+T_1^n. \qquad (1)$$

For internal nodes, energy balance equations have the form ((2), Ex. 21.4)

$$-(\Delta Fo)T_{i-1}^{n+1}+\left[1+2(\Delta Fo)\right]T_i^{n+1}-(\Delta Fo)T_{i+1}^{n+1}=T_i^n, \qquad i=2,...,N-1, \quad (2)$$

where $\Delta Fo = a\Delta t/(\Delta x)^2$, $N = 10$.

It is assumed, moreover, that the temperature in node $T_{N+1}^n = T_{11}^n = T_0 = 300\,^\circ\text{C}$, i.e. at time $0 \le t \le 60$ s when the heat does not penetrate into a depth of $x_{11,}$ and the temperature in node $N + 1 = 11$ equal initial temperature T_0. A system of 10 linear algebraic equations must be solved at every time step $t + \Delta t$. Thomas method, already discussed in Ex. 21.5, will be applied to solve the (1)–(2). Coefficients of the tri-diagonal equation system are formulated as follow:

$$\begin{aligned} &a_1=0, \qquad d_1=1+2(\Delta Fo)(1+\Delta Bi), \\ &b_1=-2(\Delta Fo), \qquad c_1=2(\Delta Fo)(\Delta Bi)T_{cz}^{n+1}+T_1^n, \end{aligned} \quad (3)$$

$$a_i=-\Delta Fo, \quad i=2,...,N, \quad (4)$$

$$d_i=1+2(\Delta Fo), \quad i=2,...,N, \quad (5)$$

$$b_i=-\Delta Fo, \quad i=2,...,N-1, \quad b_N=0, \quad (6)$$

$$c_i=T_i^n, \quad i=2,...,N-1, \quad c_N=T_N^n+(\Delta Fo)T_{N+1}^{n+1}=T_N^n+300(\Delta Fo). \quad (7)$$

Calculations will be made for $\Delta x = 0.025$ m and $\Delta t = 5$s [$\Delta Bi = \alpha(\Delta x)/\lambda = 2000 - 0.025/50 = 1$ and $\Delta \text{Fo} = a\Delta t/(\Delta x)^2 = 1\cdot 10^{-5}\cdot 60/0.025^2 = 0.08$]. The system of (1) and (2) assumes the form

$$\begin{aligned} &1.32T_1^{n+1}-0.16T_2^{n+1}=3.2+T_1^n, \\ &-0.08T_{i-1}^{n+1}+1.16T_i^{n+1}-0.08T_{i+1}^{n+1}=T_i^n, \qquad i=2,...,N-1, \\ &-0.08T_{N-1}^{n+1}+1.16T_N^{n+1}=24+T_N^n, \end{aligned} \quad (8)$$

where $N = 10$.

Table 21.4 presents the results of temperature calculations in nodes no. 1 ($x = 0$ m) and 5 ($x = 0.1$ m) and compares them to temperatures calculated by means of the analytical (9) derived in Ex. 14.4.

Table 21.5 presents the results of node temperature calculations at time $t = 20$ s and $t = 40$ s and compares them with the analytical solution.

Table 21.4. Comparison of temperature values in nodes no. 1 ($x = 0$ m) and no. 5 ($x = 0.1$ m) calculated by means of the implicit difference method and analytical method

t [s]	Node no. 1 (x = 0 m) Solution		Node no. 5 (x = 0.1 m) Solution	
	Numerical, T [°C]	Analytical, T [°C]	Numerical, T [°C]	Analytical, T [°C]
0	300.00	300.00	300.00	300.00
5	265.77	229.00	300.00	300.00
10	239.37	207.80	300.00	300.00
15	218.79	193.80	299.99	300.00
20	202.54	183.40	299.97	300.00
25	189.56	175.00	299.94	300.00
30	179.04	168.10	299.90	300.00
35	170.39	162.10	299.83	300.00
40	163.18	156.90	299.74	300.00
45	157.09	152.40	299.63	299.90
50	151.86	148.30	299.49	299.90
55	147.32	144.60	299.31	299.80
60	143.32	141.30	299.11	299.70

Table 21.5. Node temperatures calculated by means of the implicit difference method and exact analytical method at two time points $t = 20$ s and $t = 40$ s

x [m]		0	0.025	0.05	0.075	0.1	0.125	0.15	0.175	0.2	0.225
Node no.		1	2	3	4	5	6	7	8	9	10
t = 20 s	Numerical method	202.54	283.56	297.75	299.73	299.97	300.0	300.0	300.0	300.0	300.0
	Analytical method	183.40	283.10	299.30	299.90	300.00	300.0	300.0	300.0	300.0	300.0
t = 40 s	Numerical method	163.18	261.31	291.49	298.42	299.74	299.96	300.0	300.0	300.0	300.0
	Analytical method	156.90	257.80	293.10	299.40	299.88	299.99	300.0	300.0	300.0	300.0

From the comparisons presented in Tables 21.4 and 21.5, one can conclude that the accuracy of the obtained results is very good. When comparing, however, the results from Tables 21.2 and 21.3 and 21.4 and 21.5 by means of the explicit and implicit method, respectively, it is evident that both methods are almost equally accurate, when the spatial step and time step are the same. Explicit method is slightly more accurate. However, the order of accuracy in both methods is identical: $O(\Delta t) + O[(\Delta x)^2]$. In order to increase the accuracy of the numerical calculations, one can reduce the spatial step Δx and time step Δt or apply a different method, which demonstrates a higher order of accuracy, for example, the implicit

Crank–Nicolson method with an accuracy of $O[(\Delta t)^2]$ and $O[(\Delta x)^2]$. By reducing Δx and Δt, one can practically obtain the same results as one does in the case of the exact analytical method. A simple computational algorithm and the ease, which the calculations are programmed with are the main advantages of the explicit method. The program in FORTRAN language for solving the equation systems (1) and (2) is presented below.

Program for calculating one-dimensional half-space temperature distribution with an assigned convective boundary condition

```
p_21_08.in
10 12 300. 20.
1.E-5 2000. 50. 0.025 5
```

```
p_21_08.for
c  Calculation of one-dimensional transient temp.
distribution
c  by means of implicit method and exact analytical solution
      program p_21_08
      dimension ta(200),td(200),tb(200),tc(200),tt(200)
      dimension t(200)
      open(unit=1,file='p_21_08.in')
      open(unit=2,file='p_21_08.out')
      read(1,*)n,n_time,t_pocz,t_cz
      read(1,*)a,alfa,s_lambda,dx,dt
      write(2,'(a)')
     &"CALCULATION OF ONE-DIMENSIONAL TRANSIENT TEMP.
     &  DISTRIB."
      write(2,'(/a)') "INPUT DATA"
      write(2,'(a,i10)')  "equation number n=",n
      write(2,'(a,i10)')  "max number of time steps =",n_time
      write(2,'(a,e10.5,a)')" initial temp t_pocz=",
     &t_pocz,"[C]"
      write(2,'(a,e10.5,a)')"medium temp. t_cz=",t_cz,"[C]"
      write(2,'(a,e10.5,a)')"coefficient a=",a," [m2/s]"
      write(2,'(a,e10.5,a)')"coefficient alfa=",alfa,"
     & [W/m2/K]"
      write(2,'(a,e10.5,a)')" coeff. lambda=",s_lambda,"
     & [W/m/K]"
      write(2,'(a,e10.5,a)')"space step dx=",dx," [m]"
      write(2,'(a,e10.5,a)')"time step dt=",dt," [s]"
      DFo=a*dt/dx**2
      DBi=alfa*dx/s_lambda
      do i=1,n
         t(i)=t_pocz
      enddo
      write(2,'(/a)')"CALCULATED TEMPERATURE"
      write(2,'(a)')"Time      T1[C]      T5[C]"
```

```
      j=0
      time=j*dt
      write(2,'(f5.2,3x,e11.6,3x,e11.6)')time,t(1),t(5)
      j=j+1
      do while (j.le.n_time)
         td(1)=1.+2.*DFo*(1.+DBi)
         tb(1)=-2.*DFo
         tc(1)=2.*DFo*DBi*t_cz+t(1)
         do i=2,n
            ta(i)=-DFo
            td(i)=1.+2.*DFo
         enddo
         do i=2,n-1
            tb(i)=-DFo
            tc(i)=t(i)
         enddo
         tb(n)=0.
         tc(n)=t(n)+T_pocz*DFo
         call triada (ta,td,tb,tc,n,tt)
         do i=1,n
            t(i)=tt(i)
         enddo
         time=j*dt
c     write(2,'(f5.2,3x,e11.6,3x,e11.6)')time,t(1),t(5)
      write(2,'(f5.2,10(3x,e11.6))')time,(t(i),i=1,n)
         j=j+1
      enddo
      end  program p_21_08

      subroutine triada (a,d,b,c,n,t)
c     elimination of coefficients ai & calculation new coeff.
      dimension a(*),d(*),b(*),c(*),t(*)
      do i=2,n
         f=a(i)/d(i-1)
         d(i)=d(i)-b(i-1)*f
         c(i)=c(i)-c(i-1)*f
      enddo
c     calculation of ti
      t(n)=c(n)/d(n)
      do i=1,n-1
         k=n-i
         t(k)=(c(k)-b(k)*t(k+1))/d(k)
      enddo
c     write(2,'(3ht =,10(3x,e11.4))') (t(i),i=1,n)
      return
      end
```

Exercise 21.9 Calculating Two-Dimensional Transient Temperature Field by Means of the Implicit Method and a Computational Program; Algebraic Equation System is Solved by Gaussian Elimination Method

A long chamotte element with a rectangular cross-section is unilaterally warmed-up by a superficial electric heater at a temperature T_s of 250°C. Once the whole element is warmed-up to a temperature of $T_o = T_s = 250$°C, it is subjected to an air cooling phase at the temperature of $T_{cz} = 30$°C; the cooling airflow moves along the three sides of the element (Fig. 21.11). During the cooling process, the temperature of the electrically heated surface remains constant and equals initial temperature. Heat transfer coefficient is $\alpha = 50$ W/(m^2·K).

Assume for the calculation that the chamotte has the following thermophysical properties: $\rho = 1851.85$ kg/m^3, $c = 900$ J/(kg·K), $\lambda = 1.0$ W/(m·K). The determined temperature distribution should be compared with the results obtained by means of FEM while using the ANSYS program. Use implicit difference method to determine temperature distribution.

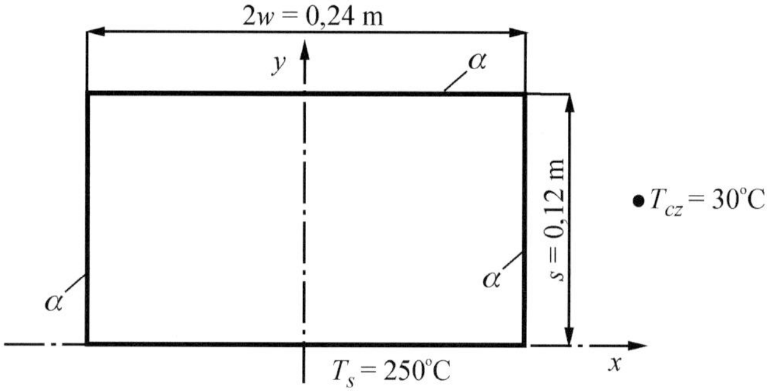

Fig. 21.11. Diagram of an analyzed region and boundary conditions

Solution

Due to the symmetry of the temperature field with respect to y axis, temperature field will be determined only for the one half of the element's cross-section (Fig. 21.12).

Appropriate difference equations were derived in Ex. 21.4.

Equation (3) from Table 21.1 assumes the following form for node1:

$$\left[1+4(\Delta Fo)(1+\Delta Bi)\right]T_1^{n+1} - 2(\Delta Fo)\left[T_4^{n+1} + T_2^{n+1} + 2(\Delta Bi)T_{cz}\right] = T_1^n . \qquad (1)$$

By accounting for (2) from Table 21.1, the difference equation for node 2 can be written in the form

$$\left[1+2(\Delta Fo)(2+\Delta Bi)\right]T_2^{n+1}-(\Delta Fo)\left[2T_3^{n+1}+T_s+T_1^{n+1}+2(\Delta Bi)T_{cz}\right]=T_2^n. \quad (2)$$

The difference equation for node 3 is obtained from (1), Table 21.1

$$\left[1+4(\Delta Fo)\right]T_3^{n+1}-\Delta Fo\left(T_6^{n+1}+T_2^{n+1}+T_s+T_4^{n+1}\right)=T_3^n. \quad (3)$$

By accounting for (2) from Table 21.1, the difference equation for node 4 assumes the form

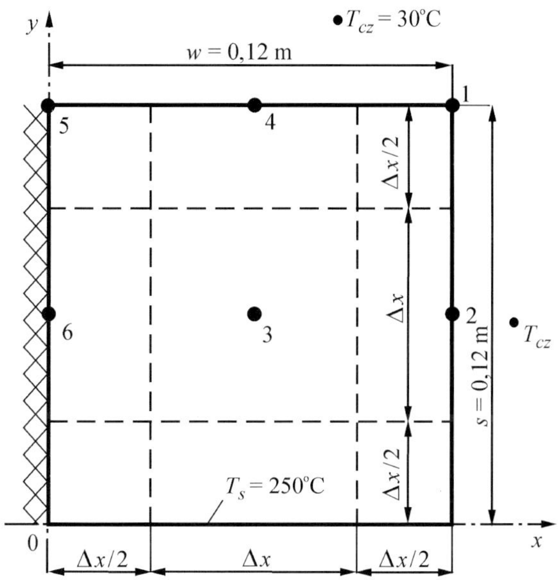

Fig. 21.12. Division of an analyzed region into control volumes

$$\left[1+2(\Delta Fo)(2+\Delta Bi)\right]T_4^{n+1}- \\ -(\Delta Fo)\left[2T_3^{n+1}+T_5^{n+1}+T_1^{n+1}+2(\Delta Bi)T_{cz}\right]=T_4^n. \quad (4)$$

Heat balance equation for node 5 has the following form:

$$c\rho\frac{\Delta x}{2}\cdot\frac{\Delta x}{2}\frac{dT_5}{dt}=\lambda\frac{\Delta x}{2(\Delta x)}\left(T_4-T_5\right)+\lambda\frac{\Delta x}{2(\Delta x)}\left(T_6-T_5\right)+ \\ +\alpha\frac{\Delta x}{2}\left(T_{cz}-T_5\right). \tag{5}$$

As prescribed by the implicit finite difference method, (5) is approximated as follows:

$$c\rho\frac{(\Delta x)^2}{4}\frac{T_5^{n+1}-T_5^n}{\Delta t}=\frac{\lambda}{2}(T_4^{n+1}-T_5^{n+1})+\frac{\lambda}{2}(T_6^{n+1}-T_5^{n+1})+ \\ +\alpha\frac{\Delta x}{2}(T_{cz}-T_5^{n+1}), \tag{6}$$

where from, after transformations, one has

$$-2(\Delta Fo)T_4^{n+1}+\left[1+4(\Delta Fo)+2(\Delta Fo)(\Delta Bi)\right]T_5^{n+1}-2(\Delta Fo)T_6^{n+1}= \\ =2(\Delta Fo)(\Delta Bi)T_{cz}+T_5^n. \tag{7}$$

The equation for node 6 is obtained from (2), Table 21.1 by assuming that $\Delta Bi = 0$

$$\left[1+4(\Delta Fo)\right]T_6^{n+1}-\Delta Fo\left(2T_3^{n+1}+T_s+T_5^{n+1}\right)=T_6^n. \tag{8}$$

In (1)–(8), the following notations were assumed: $\Delta Bi = \alpha(\Delta x)/\lambda$ and $\Delta Fo = a(\Delta t)/(\Delta x)^2$. By allowing that

$$\Delta x=\frac{s}{2}=\frac{w}{2}=\frac{0{,}12}{2}=0.06\text{m},$$

$$a=\frac{\lambda}{c\rho}=\frac{1}{900\cdot 1851.85}=6.0\cdot 10^{-7}\ \frac{\text{m}^2}{\text{s}}$$

and assuming that time step equals $\Delta t = 60$ s, one has

$$\Delta Bi=\frac{\alpha(\Delta x)}{\lambda}=\frac{50\cdot 0.060}{1}=3, \\ \Delta Fo=\frac{a(\Delta t)}{(\Delta x)^2}=\frac{6\cdot 10^{-7}\cdot 60}{0.06^2}=0.01. \tag{9}$$

Once data is ordered and accounted for together with (9) in (1)–(8), one has

$$
\begin{aligned}
&1.16T_1^{n+1} - 0.02T_2^{n+1} - 0.02T_4^{n+1} = 3.6 + T_1^n, \\
&-0.01T_1^{n+1} + 1.1T_2^{n+1} - 0.02T_3^{n+1} = 4.3 + T_2^n, \\
&-0.01T_2^{n+1} + 1.04T_3^{n+1} - 0.01T_4^{n+1} - 0.01T_6^{n+1} = 2.5 + T_3^n, \\
&-0.01T_1^{n+1} - 0.02T_3^{n+1} + 1.1T_4^{n+1} - 0.01T_5^{n+1} = 1.8 + T_4^n, \\
&-0.02T_4^{n+1} + 1.1T_5^{n+1} - 0.02T_6^{n+1} = 1.8 + T_5^n, \\
&-0.02T_3^{n+1} - 0.01T_5^{n+1} + 1.04T_6^{n+1} = 2.5 + T_6^n.
\end{aligned}
\tag{10}
$$

The solution of the equation system (10) has the form

$$\mathbf{T} = \mathbf{A}^{-1}\mathbf{b}, \tag{11}$$

where $\mathbf{A}^{-1}$ is the inverse coefficients matrix, $\mathbf{b}$ the right side vector.

Subprogram MATINV is used to determine the inverse matrix $\mathbf{A}^{-1}$. Transient temperature distribution is also calculated by FEM using ANSYS program.

From the comparison presented in Table 21.6, one can conclude that the accuracy of the implicit difference method is very good, despite the fact that the analysed region is divided into a rather small number of control volumes. Calculations done by means of FEM should be regarded as more accurate, since the analyzed region, presented in Fig. 21.12 was divided into1600 elements (grid 40 × 40). Therefore, the results obtained by means of FEM can be used to evaluate the accuracy of the results obtained by means of the implicit finite difference method. In order to increase the accuracy of calculations carried out by means of the implicit difference method, the spatial step Δx and time step Δt should be reduced.

Table 21.6. Comparison of temperature values in nodes no. 1, 2 and 5 calculated by means of implicit difference method and finite element method

Node no. 1				
	Difference method T [°C]	FEM T [°C]		
t [min]	Δt = 60 s	Δt = 60 s	Δt = 30 s	Δt = 10 s
0	250.00	250.00	250.00	250.00
10	109.87	75.31	74.36	73.74
20	67.35	57.96	57.57	57.31
30	52.55	50.49	50.27	50.12
40	46.65	46.33	46.19	46.09
50	43.73	43.73	43.63	43.56
60	42.18	41.97	41.90	41.85
70	41.22	40.73	40.67	40.64
80	40.56	39.82	39.78	39.75
90	40.08	39.14	39.10	39.08
100	39.71	38.62	38.59	38.57
110	39.42	38.21	38.19	38.17
120	39.20	37.90	37.88	37.87

Node no. 2				
	Difference method T [°C]	FEM T [°C]		
t [min]	Δt = 60 s	Δt = 60 s	Δt = 30 s	Δt = 10 s
0	250.00	250.00	250.00	250.00
10	162.08	128.16	127.40	126.89
20	122.08	105.23	104.78	104.48
30	103.15	92.69	92.36	92.15
40	93.40	84.77	84.53	84.37
50	87.79	79.41	79.23	79.10
60	84.19	75.62	75.47	75.38
70	81.67	72.86	72.74	72.66
80	79.80	70.79	70.70	70.63
90	78.37	69.22	69.14	69.09
100	77.25	68.02	67.95	67.91
110	76.36	67.08	67.02	66.99
120	75.65	66.34	66.30	66.27

Node no. 5				
	Difference method T [°C]	FEM T [°C]		
t [min]	Δt = 60 s	Δt = 60 s	Δt = 30 s	Δt = 10 s
0	250.00	250.00	250.00	250.00
10	160.88	128.83	128.03	127.50
20	118.04	107.58	107.19	106.93
30	96.40	95.83	95.59	95.43
40	84.64	87.65	87.47	87.35

Table 21.6. (cont.)

	Node no. 5			
	Difference method	FEM		
	T [°C]	T [°C]		
t [min]	Δt = 60 s	Δt = 60 s	Δt = 30 s	Δt = 10 s
50	77.62	81.40	81.24	81.13
60	73.00	76.43	76.29	76.19
70	69.71	72.44	72.31	72.22
80	67.21	69.23	69.11	69.03
90	65.25	66.65	66.53	66.46
100	63.68	64.57	64.47	64.40
110	62.42	62.90	62.81	62.75
120	61.39	61.57	61.49	61.44

Computational program in FORTRAN language used for determining node temperatures is presented below.

Program for calculating two-dimensional temperature distribution in the region presented in Fig. 21.12 by means of the implicit difference method

```
p_21_09.in
6 120 60. 250.
 1.16 -0.02  0.00 -0.02  0.00  0.00
-0.01  1.10 -0.02  0.00  0.00  0.00
 0.00 -0.01  1.04 -0.01  0.00 -0.01
-0.01  0.00 -0.02  1.10 -0.01  0.00
 0.00  0.00  0.00 -0.02  1.10 -0.02
 0.00  0.00 -0.02  0.00 -0.01  1.04
 3.60  4.30  2.50  1.80  1.80  2.50
```

```
      p_21_09.for
c     Calculation of two-dimensional temperature
c     distribution in the region presented in Fig. 21.12
c     by means of the implicit difference method
      program p_21_09
      dimension ta(50,50),tbo(50),tb(50),tc(50,50),tt(50)
      dimension t(50)
      open(unit=1,file='p_21_09.in')
      open(unit=2,file='p_21_09.out')
      read(1,*)n,n_time,dt,t_pocz
      write(2,'(a)')
     & "CALCULATION OF TWO-DIMENSIONAL TRANSIENT TEMP.
     &  DISTRIB."
      write(2,'(/a)') "INPUT DATA"
      write(2,'(a,i10)')    "equation number n=",n
```

For a one-dimensional linear element, the thermal capacity matrix $[M^e]$ has the form

$$\left[M^e\right]=\int_0^L c\rho\begin{bmatrix}\left(N_1^e\right)^2 & N_1^e N_2^e\\ N_1^e N_2^e & \left(N_2^e\right)^2\end{bmatrix}dx=\frac{c\rho L}{6}\begin{bmatrix}2 & 1\\ 1 & 2\end{bmatrix}, \tag{15}$$

where L is the length of the element.

The equation system, which one can use to determine node temperature in a single element e, can be written as follows

$$\left[M^e\right]\left\{\dot{T}^e\right\}+\left(\left[K_c^e\right]+\left[K_c^\alpha\right]\right)\left\{T^e\right\}=\left\{f_Q^e\right\}+\left\{f_q^e\right\}+\left\{f_\alpha^e\right\}, \tag{16}$$

where matrixes and vectors from (16) are defined in Ex. 11.10. Symbol $\dot{T}$ stands for $\partial T/\partial t$.

Next, one should create global equation system by summing up along the sides of (16) for all N elements, which the analyzed region was divided to. This problem is discussed in Ex. 11.15. Details regarding the formation of global equation system are also presented in Ex. 11.16–11.19.

References [1, 3–5, 7, 8, 10–12, 14, 15, 21] discuss how FEM is applied to solve transient heat conduction problems; the references also give a detailed description of the FEM method.

Exercise 22.2 Concentrated (Lumped) Thermal Finite Element Capacity in FEM

Discuss how thermal capacity is concentrated in finite elements in one, two and three-dimensional problems by assuming that linear functions are used to interpolate temperature distribution inside an element.

Solution

Ex. 11.15 presents the methods for creating global equation system in FEM. According to the second method for creating such a system, the differential equations for node i shared by element $(i–1)$ and element i (Fig. 22.1) will be written out first.

Temperature distributions in elements $(i–1)$ and i are described by functions

$$T^{i-1} = T_{i-1}N_1^{i-1} + T_i N_2^{i-1} = \frac{x_i - x}{x_i - x_{i-1}} T_{i-1} + \frac{x - x_{i-1}}{x_i - x_{i-1}} T_i, \qquad x_{i-1} \le x \le x_i, \tag{1}$$

$$T^{i} = T_i N_1^{i} + T_{i+1} N_2^{i} = \frac{x_{i+1} - x}{x_{i+1} - x_i} T_i + \frac{x - x_i}{x_{i+1} - x_i} T_{i+1}, \qquad x_i \le x \le x_{i+1}. \tag{2}$$

Differential equation for node i in the global equation system has the following form:

$$c\rho \int_{x_{i-1}}^{x_i} \frac{\partial T^{i-1}}{\partial t} N_2^{i-1} dx + c\rho \int_{x_i}^{x_{i+1}} \frac{\partial T^{i}}{\partial t} N_1^{i} dx - \lambda \int_{x_{i-1}}^{x_i} \frac{\partial^2 T^{i-1}}{\partial x^2} N_2^{i-1} dx -$$
$$-\lambda \int_{x_i}^{x_{i+1}} \frac{\partial^2 T^{i}}{\partial x^2} N_1^{i} dx = \int_{x_{i-1}}^{x_i} \dot{q}_v N_2^{i-1} dx + \int_{x_i}^{x_{i+1}} \dot{q}_v N_1^{i} dx. \tag{3}$$

By substituting (1) and (2) into (3) and carrying out calculations, the equation for node i is obtained. Such equation can be also determined using the capacity matrix $[M^e]$, stiffness matrix $[K^e]$ and vector $\{f_Q^e\}$.

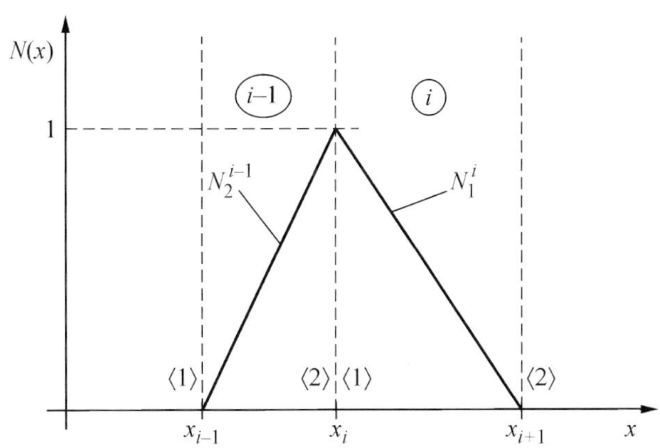

Fig. 22.1. Shape functions N_2^{i-1} i N_1^i for node i that lies on the boundary of elements $(i-1)$ and i

The local numeration of node i in the global coordinate system is number ⟨2⟩. The first term on the left-hand-side is obtained by multiplying the second row in the capacity matrix (15) by the vector of derivatives with respect to time from node temperatures, i.e.

$$c\rho\int\limits_{x_{i-1}}^{x_i}\frac{\partial T^{i-1}}{\partial t}N_2^{i-1}dx = c\rho\int\limits_{x_{i-1}}^{x_i}\left[N_1^{i-1}N_2^{i-1},\left(N_2^{i-1}\right)^2\right]dx\begin{Bmatrix}\dot{T}_1^{i-1}\\ \dot{T}_2^{i-1}\end{Bmatrix}=$$
$$=\frac{c\rho\left(x_i - x_{i-1}\right)}{6}[1,\ 2]\begin{Bmatrix}\dot{T}_{i-1}\\ \dot{T}_i\end{Bmatrix}=\frac{c\rho\Delta x_{i-1}}{6}\left(\dot{T}_{i-1}+2\dot{T}_i\right). \tag{4}$$

The second component on the left-hand-side of (3) can be calculated in a similar way:

$$c\rho\int\limits_{x_i}^{x_{i+1}}\frac{\partial T^{i}}{\partial t}N_1^{i}dx = c\rho\int\limits_{x_i}^{x_{i+1}}\left[\left(N_1^{i}\right)^2,N_1^{i}N_2^{i}\right]dx\begin{Bmatrix}\dot{T}_1^{i}\\ \dot{T}_2^{i}\end{Bmatrix}=$$
$$=\frac{c\rho\left(x_{i+1} - x_i\right)}{6}[2,\ 1]\begin{Bmatrix}\dot{T}_i\\ \dot{T}_{i+1}\end{Bmatrix}=\frac{c\rho\Delta x_i}{6}\left(2\dot{T}_i+\dot{T}_{i+1}\right). \tag{5}$$

The third term on the left-hand-side is obtained by multiplying the second row in the rigidity matrix

$$\left[K^e\right]=\frac{\lambda}{L}\begin{bmatrix}1 & -1\\ -1 & 1\end{bmatrix} \tag{6}$$

for a one-dimensional element by the temperature vector in the nodes of element $(i–1)$

$$-\lambda\int\limits_{x_{i-1}}^{x_i}\frac{\partial^2 T^{i-1}}{\partial x^2}N_2^{i-1}dx=\frac{\lambda}{L^{i-1}}[-1,\ 1]\begin{Bmatrix}T_{i-1}\\ T_i\end{Bmatrix}=\frac{\lambda}{\Delta x_{i-1}}\left(-T_{i-1}+T_i\right). \tag{7}$$

The fourth term on the left-hand-side is obtained as a result of multiplying the first row vector in the stiffness matrix $[K^e]$ formulated in (6) by the temperature vector in nodes of element i

$$-\lambda\int\limits_{x_i}^{x_{i+1}}\frac{\partial^2 T^{i}}{\partial x^2}N_1^{i}dx=\frac{\lambda}{L^{i}}[1,\ -1]\begin{Bmatrix}T_1^i\\ T_2^i\end{Bmatrix}=\frac{\lambda}{\Delta x_i}\left(T_i-T_{i+1}\right), \tag{8}$$

where $\Delta x_{i-1} = x_i - x_{i-1}$, $\Delta x_i = x_{i+1} - x_i$. Accounting that vector $\{f_Q^e\}$ for a one-dimensional element and linear shape functions has the form

$$\left\{f_Q^e\right\}=\frac{\dot{q}_v L}{2}\begin{Bmatrix}1\\ 1\end{Bmatrix} \tag{9}$$

the terms on the right-hand-side can be expressed as follow:

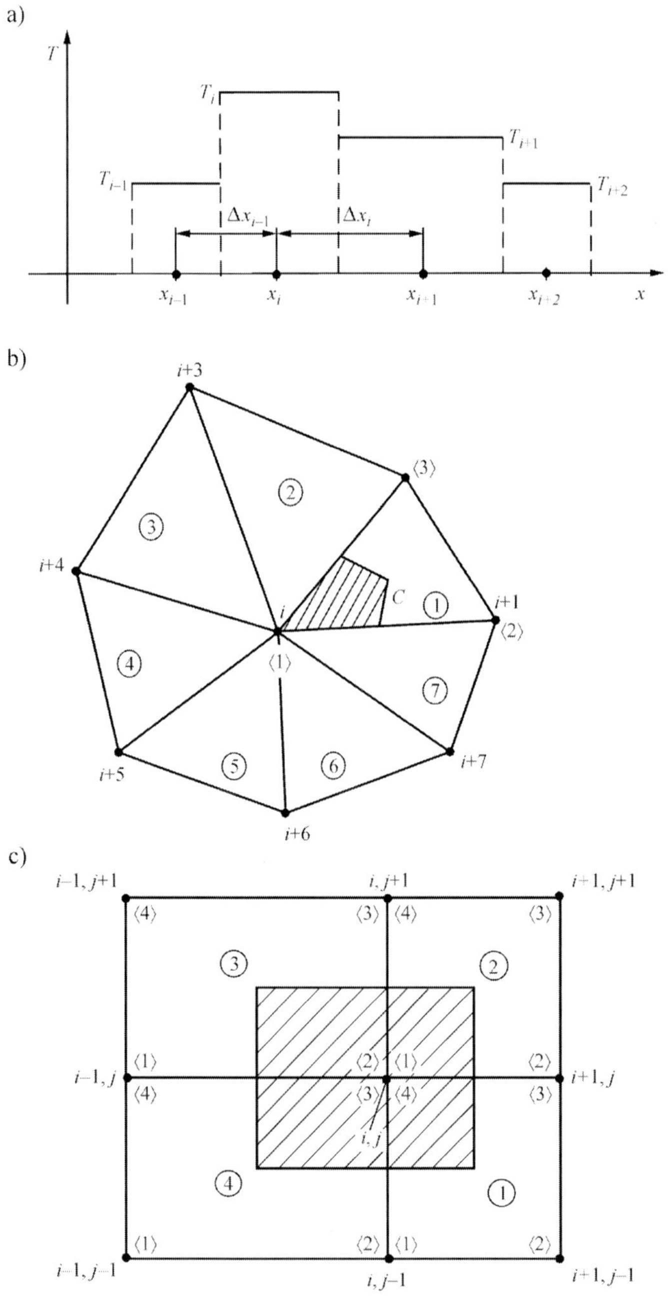

Fig. 22.2. Diagrams that illustrate the concentration of thermal capacity in node i of a) one-dimensional elements, b) two-dimensional triangular elements, c) two-dimensional tetragonal elements

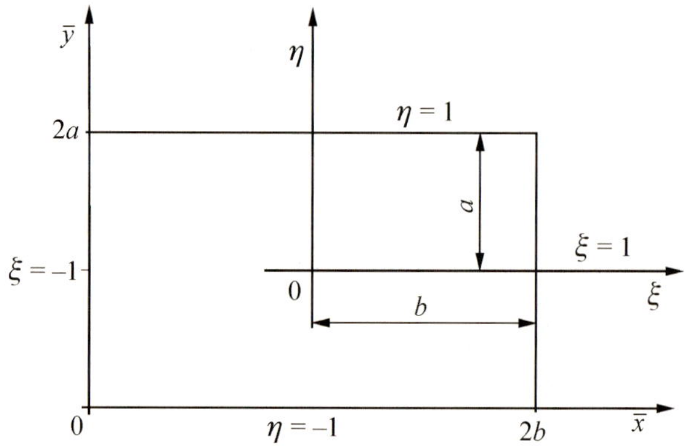

Fig. 22.6. Local $(\bar{x}, \bar{y})$ and natural (ξ, η) coordinate systems

$$
\begin{aligned}
N_1^e &= \frac{1}{4}(1-\xi)(1-\eta), \\
N_2^e &= \frac{1}{4}(1+\xi)(1-\eta), \\
N_3^e &= \frac{1}{4}(1+\xi)(1+\eta), \\
N_4^e &= \frac{1}{4}(1-\xi)(1+\eta).
\end{aligned}
\tag{8}
$$

c. *Two-dimensional triangular elements*

Linear shape functions for a triangular element are defined by (7) in Ex. 11.9. Natural (area) coordinate system is presented in Fig. 22.7.

By connecting point $P = P(x, y)$ with the vertices of triangle 123, three smaller triangles are obtained whose surface areas are A_1, A_2 and A_3 (Fig. 22.7). Natural coordinate system (ξ, η, ζ) is defined as follows:

$$
\begin{aligned}
L_1 &= \xi = \frac{A_1}{A^e}, \\
L_2 &= \eta = \frac{A_2}{A^e}, \\
L_3 &= \zeta = \frac{A_3}{A^e},
\end{aligned}
\tag{9}
$$

where $A^e = A_1 + A_2 + A_3$ is the area of the whole triangle 123 (Fig. 22.7).

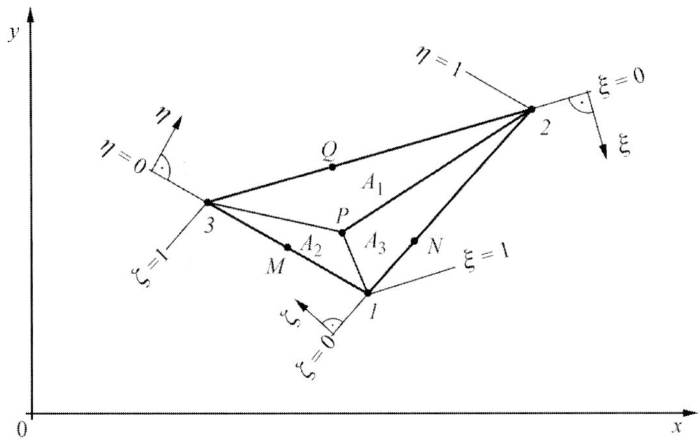

Fig. 22.7. Natural (area) coordinate system

Only two of the natural coordinates ξ, η, ζ are linearly independent, since

$$\frac{A_1}{A^e} + \frac{A_2}{A^e} + \frac{A_3}{A^e} = \frac{A^e}{A^e} = 1 = \xi + \eta + \zeta . \tag{10}$$

From (10) it follows that coordinate ζ, for example, can be expressed by functions ξ and η

$$\zeta = 1 - \xi - \eta . \tag{11}$$

All three natural coordinates change at interval [0,1]. If point *P*, which lies inside the element, is moved to node *1*, then $A_1 = A^e_{123}$ (Fig. 22.7) and natural coordinates equal $\xi = 1$, $\eta = 0$ and $\zeta = 0$.

If point *P* becomes identical with point *Q*, which shifts along side 23, then $A_1 = 0$, $A_2 \neq 0$, $A_3 \neq 0$ and the respective natural coordinates are $\xi = 0$, $\eta \neq 0$ and $\zeta \neq 0$.

Similarly, when point *P* becomes identical with point *M*, which shifts along side 13, then the natural coordinates are $\xi \neq 0$, $\eta = 0$ and $\zeta \neq 0$.

If shape functions that interpolate temperature distribution inside the element are linear, then

$$\begin{aligned} N_1^e &= L_1 = \xi, \\ N_2^e &= L_2 = \eta, \\ N_3^e &= L_3 = \zeta . \end{aligned} \tag{12}$$

Between the natural coordinates ξ, η and ζ and global coordinates *x, y* and *z* the following relationships exist

$$
\begin{aligned}
x &= \xi x_1 + \eta x_2 + \zeta x_3, \\
y &= \xi y_1 + \eta y_2 + \zeta y_3, \\
1 &= \xi + \eta + \zeta.
\end{aligned}
\tag{13}
$$

Once the equation system (13) is solved with respect to ξ, η and ζ, one gets

$$
\begin{aligned}
\xi = N_1^e &= \frac{1}{2A_{123}^e}\left(a_1^e + b_1^e x + c_1^e y\right), \\
\eta = N_2^e &= \frac{1}{2A_{123}^e}\left(a_2^e + b_2^e x + c_2^e y\right), \\
\zeta = N_3^e &= \frac{1}{2A_{123}^e}\left(a_3^e + b_3^e x + c_3^e y\right),
\end{aligned}
\tag{14}
$$

where $2\,A_{123}^e$ is defined by (5) in Chap. 11. Coefficients a_1^e, a_2^e, a_3^e, b_1^e, b_2^e, b_3^e, c_1^e, c_2^e, c_3^e are described by (8) in Chap. 11. The superscript e means that the quantities refer to a single element e.

Natural coordinates are introduced with an aim to simplify the calculation of surface integrals by means of the Gauss-Legendre quadratures method.

Only linear shape functions were analyzed. Analogically, the same procedure applies in the case of higher degree shape functions, for example quadratic or cubic [12, 21].

Exercise 22.6 Coordinate System Transformations and Integral Calculations by Means of the Gauss-Legendre Quadratures

Discuss the transformation of coordinate systems for arbitrarily-shaped tetragonal and triangular elements and the calculation of integrals by means of the Gauss-Legendre quadratures.

Solution

In Chap. 11, the integrals that occur in the coefficients of conduction matrixes and also other integrals of an algebraic equation for i-node are analytically calculated. This is possible for rectangular or triangular elements. If a region is divided into arbitrary quadrilaterals, then the analytical calculation of the integrals becomes rather problematic. In large commercial programs, integrals are usually calculated numerically by means of the Gauss-Legendre quadratures. For this purpose, an arbitrarily-chosen quadrilateral is transformed in the coordinate system (x,y) into the, so called, model element whose dimensions are 2×2 in the new coordinate system (ξ,η). The model is a square: $1 \leq \xi \leq 1$, $1 \leq \eta \leq 1$.

Coordinate transformations are only applied in order to calculate the integrals. The transformation of element B is shown in Fig. 22.8. After the transformation of coordinates (x,y), the elements of integration in the new coordinate system (ξ,η) become more complex; in effect, therefore, a numerical method, usually the Gauss-Legendre method, is applied to calculate these integrals.

The real element B in the system (x,y) is transformed into the model element in the system (ξ,η) by means of the transformation

$$x = \sum_{j=1}^{m} x_j^e N_j^e(\xi,\eta),$$
$$y = \sum_{j=1}^{m} y_j^e N_j^e(\xi,\eta), \tag{1}$$

where $N_j^e(\xi,\eta)$ is the shape function in the model element.

The examples of such transformation are (6) and (13) in Ex. 22.5.

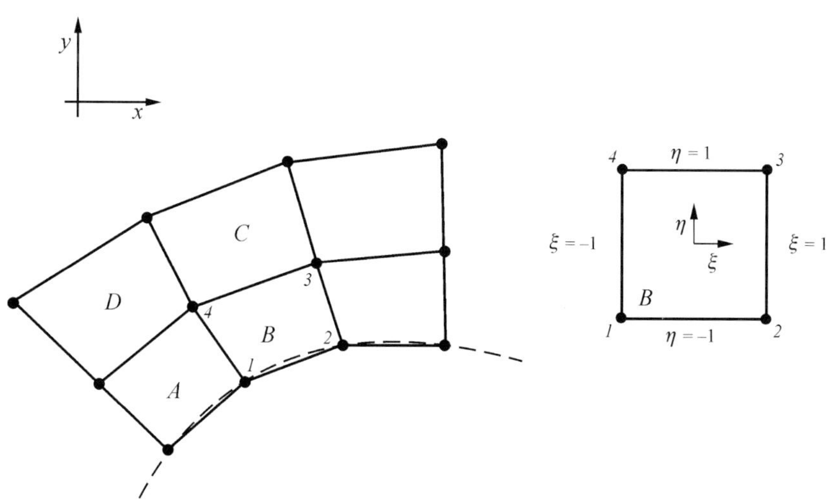

Fig. 22.8. Transformation of an arbitrary tetragonal element B in the Cartesian coordinate system (x,y) into a quadratic model element in the coordinate system (ξ,η)

Temperature distribution in element e is expressed as

$$T^e \equiv T^e(x,y) = \sum_{j=1}^{n} T_j^e N_j^e(x,y), \tag{2}$$

where n is the number of nodes in element e.

Natural number m , which occurs in (1) does not have to be equal to number n in (2). Depending on the relations between m and n, the elements can be divided into

- *subparametric* ($m < n$), approximation order of coordinates x and y is lower than the approximation order of temperature (in the general case of a dependent variable),
- *isoparametric* ($m = n$), approximation orders of coordinates x and y and temperature are identical,
- *superparametric* ($m > n$), approximation order of coordinates x and y is higher than the approximation order of temperature.

Most frequently, the isoparametric elements are used, for which $m = n$.

Next, the transformation of coordinates will be discussed in greater detail. The quantities, which should be transformed are

$$N_j^e(x,y), \quad \frac{\partial N_j^e}{\partial x}, \quad \frac{\partial N_j^e}{\partial y} \quad \text{and} \quad dA = dxdy\,. \tag{3}$$

They occur in integrals, which result from the application of FEM. For instance, the coefficients of conduction matrixes $K_{c,ij}^e$ ((26), Chap. 11) are formulated as follow:

$$K_{c,ij}^e = \int_{\Omega^e} \left(\lambda_x \frac{\partial N_i^e}{\partial x}\frac{\partial N_j^e}{\partial x} + \lambda_y \frac{\partial N_i^e}{\partial y}\frac{\partial N_j^e}{\partial y} \right) dxdy\,. \tag{4}$$

It is easy to express quantity $N_j^e(x,y)$ in functions ξ and η once the (1) is allowed for. The derivatives from the shape function are calculated in the following way:

$$\frac{\partial N_i^e}{\partial \xi} = \frac{\partial N_i^e}{\partial x}\frac{\partial x}{\partial \xi} + \frac{\partial N_i^e}{\partial y}\frac{\partial y}{\partial \xi}, \tag{5}$$

$$\frac{\partial N_i^e}{\partial \eta} = \frac{\partial N_i^e}{\partial x}\frac{\partial x}{\partial \eta} + \frac{\partial N_i^e}{\partial y}\frac{\partial y}{\partial \eta}. \tag{6}$$

Equations (5) and (6) can be written in the matrix form

$$\begin{Bmatrix} \dfrac{\partial N_i^e}{\partial \xi} \\ \dfrac{\partial N_i^e}{\partial \eta} \end{Bmatrix} = \begin{bmatrix} \dfrac{\partial x}{\partial \xi} & \dfrac{\partial y}{\partial \xi} \\ \dfrac{\partial x}{\partial \eta} & \dfrac{\partial y}{\partial \eta} \end{bmatrix} \begin{Bmatrix} \dfrac{\partial N_i^e}{\partial x} \\ \dfrac{\partial N_i^e}{\partial y} \end{Bmatrix}, \tag{7}$$

where the square matrix with dimensions 2×2 is a Jacobian determinant

$$\mathbf{J} = \begin{bmatrix} \dfrac{\partial x}{\partial \xi} & \dfrac{\partial y}{\partial \xi} \\ \dfrac{\partial x}{\partial \eta} & \dfrac{\partial y}{\partial \eta} \end{bmatrix}. \tag{8}$$

The transformation of coordinates is unique when the Jacobian $\mathbf{J}$ is not singular, i.e. when Jacobian determinant is other than zero at every point (ξ, η)

$$J \equiv \det \mathbf{J} = \frac{\partial x}{\partial \xi}\frac{\partial y}{\partial \eta} - \frac{\partial x}{\partial \eta}\frac{\partial y}{\partial \xi} \neq 0 . \tag{9}$$

Derivatives from integral (4) are determined from the transformation of (7)

$$\begin{Bmatrix} \dfrac{\partial N_i^e}{\partial x} \\ \dfrac{\partial N_i^e}{\partial y} \end{Bmatrix} = [J]^{-1} \begin{Bmatrix} \dfrac{\partial N_i^e}{\partial \xi} \\ \dfrac{\partial N_i^e}{\partial \eta} \end{Bmatrix}. \tag{10}$$

The required derivatives for the calculation of the Jacobian determinant (9) are obtained after the differentiation of

$$\begin{aligned} \frac{\partial x}{\partial \xi} &= \sum_{j=1}^{m} x_j \frac{\partial N_j^e}{\partial \xi}, \\ \frac{\partial y}{\partial \xi} &= \sum_{j=1}^{m} y_j \frac{\partial N_j^e}{\partial \xi}, \end{aligned} \tag{11}$$

$$\begin{aligned} \frac{\partial x}{\partial \eta} &= \sum_{j=1}^{m} x_j \frac{\partial N_j^e}{\partial \eta}, \\ \frac{\partial y}{\partial \eta} &= \sum_{j=1}^{m} y_j \frac{\partial N_j^e}{\partial \eta}. \end{aligned} \tag{12}$$

The element of surface area $dxdy$ equals

$$dxdy = \det \mathbf{J} d\xi d\eta \,. \tag{13}$$

From (10) or directly from the solution of the equation systems (5) and (6), one has

$$\frac{\partial N_i^e}{\partial x} = \frac{1}{\det \mathbf{J}} \left[\frac{\partial y}{\partial \eta} \frac{\partial N_j^e}{\partial \xi} - \frac{\partial y}{\partial \xi} \frac{\partial N_j^e}{\partial \eta} \right], \tag{14}$$

$$\frac{\partial N_i^e}{\partial y} = \frac{1}{\det \mathbf{J}} \left[\frac{\partial x}{\partial \xi} \frac{\partial N_i^e}{\partial \eta} - \frac{\partial x}{\partial \eta} \frac{\partial N_i^e}{\partial \xi} \right]. \tag{15}$$

Integral (4) can be transformed into a new coordinate system when the derived relationships are applied. The first component of this integral can be transformed by means of the (13)–(15) into a form

$$\begin{aligned} I^e &= \int_{\Omega^e} \left(\lambda_x \frac{\partial N_i^e}{\partial x} \frac{\partial N_j^e}{\partial x} \right) dxdy = \\ &= \int_{-1}^{1} \int_{-1}^{1} \frac{\lambda_x}{\det \mathbf{J}} \left(\frac{\partial y}{\partial \eta} \frac{\partial N_i^e}{\partial \xi} - \frac{\partial y}{\partial \xi} \frac{\partial N_i^e}{\partial \eta} \right) \left(\frac{\partial y}{\partial \eta} \frac{\partial N_j^e}{\partial \xi} - \frac{\partial y}{\partial \xi} \frac{\partial N_j^e}{\partial \eta} \right) d\xi d\eta. \end{aligned} \tag{16}$$

One can see, therefore, that after the transformation of coordinates, the subintegral expressions are more complex in the new coordinate system (ξ, η) than they are in the coordinate system (x, y). These integrals are usually numerically calculated using the Gauss-Legendre quadrature [1, 3, 11–17, 21]. If the subintegral function is denoted by $F(\xi, \eta)$, one can calculate the integrals in the new coordinate system by means of a relatively simple formulas.

a. *One-dimensional elements*

The integral is calculated by means of formula

$$I = \int_{-1}^{1} F(\xi) d\xi = \sum_{i=1}^{n} w_i F(\xi_i), \tag{17}$$

where ξ_i are the Gaussian point coordinates (Fig. 22.9).

Coordinates ξ_i are the zeros of Legendre polynomials [13]. Coordinates ξ_i and weights w_i, which occur in (17) are compiled in Table 22.1.

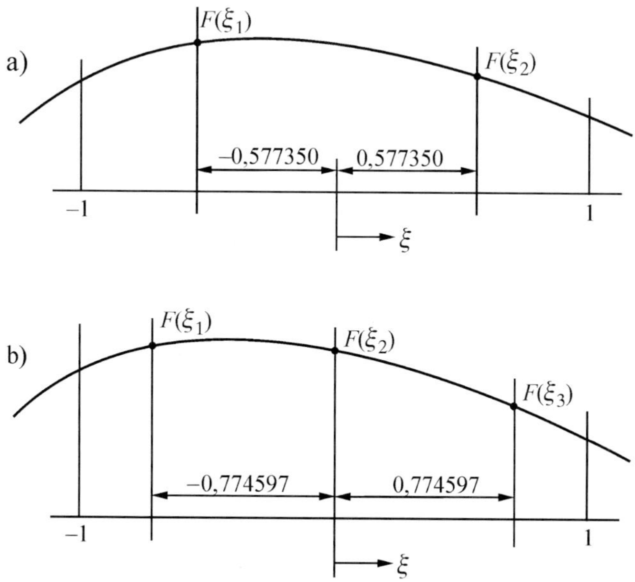

Fig. 22.9. The location of Gaussian points during the calculation of a one-dimensional integral: a) $n = 2$, b) $n = 3$

Table 22.1. Legendre polynomials and the coefficients of the Gauss-Legendre quadratures

n	Nodes ξ_i	Weight coefficients w_i
1	0,0	2,0
2	±0,577350	1,0
3	0,0	8/9 = 0,888 ...
	±0,774597	5/9 = 0,555 ...
4	±0,339982	0,652145
	±0,861136	0,347855
5	0,0	0,568889
	±0,538469	0,478629
	±0,906180	0,236927

The integration by means of the Gaussian quadratures yields accurate results for n integration points, when $F(\xi)$ is the polynomial of degree $2n-1$ or lower. In general, the larger the number of Gaussian points n, the more accurate the calculation of integral (17) is. Now we will discuss the approximate calculation of two-dimensional integrals for tetragonal and triangular elements.

b. *Tetragonal elements*

Formulas for the calculation of an integral in a two-dimensional region are used to determine integrals for quadrilateral elements.

$$I=\int_{-1}^{1}\int_{-1}^{1}F(\xi,\eta)\,d\xi\,d\eta=\int_{-1}^{1}\left[\sum_{j=1}^{n}w_jF(\xi_j,\eta)\right]d\eta=$$
$$=\sum_{k=1}^{n}w_k\left[\sum_{j=1}^{n}w_jF(\xi_j,\eta_k)\right]=\sum_{j=1}^{n}\sum_{k=1}^{n}w_jw_kF(\xi_j,\eta_k),\tag{18}$$

where n is the Gaussian point number in a single direction. For $n = 2$, the integral (18) can be written as follows:

$$I=w_1w_1F(\xi_1,\eta_1)+w_1w_2F(\xi_1,\eta_2)+w_2w_1F(\xi_2,\eta_1)+w_2w_2F(\xi_2,\eta_2),\tag{19}$$

where the coordinates of Gaussian point are ξ_i, $\eta_i = \pm 1/3^{1/2} = \pm 0.57735$ and $w_1^2 = w_1w_2 = w_2^2 = 1$.

Table 22.2. Calculating integrals in a quadrilateral region by means of (20)

m	i	ξ_i	η_i	w_i
1	1	0	0	4
4	1	$-1/3^{1/2}$	$-1/3^{1/2}$	1
	2	$+1/3^{1/2}$	$-1/3^{1/2}$	1
	3	$-1/3^{1/2}$	$+1/3^{1/2}$	1
	4	$+1/3^{1/2}$	$+1/3^{1/2}$	1
9	1	$-(3/5)^{1/2}$	$-(3/5)^{1/2}$	25/81
	2	0	$-(3/5)^{1/2}$	40/81
	3	$+(3/5)^{1/2}$	$-(3/5)^{1/2}$	25/81
	4	$-(3/5)^{1/2}$	0	40/81
	5	0	0	64/81
	6	$+(3/5)^{1/2}$	0	40/81
	7	$-(3/5)^{1/2}$	$+(3/5)^{1/2}$	25/81
	8	0	$+(3/5)^{1/2}$	40/81
	9	$+(3/5)^{1/2}$	$+(3/5)^{1/2}$	25/81

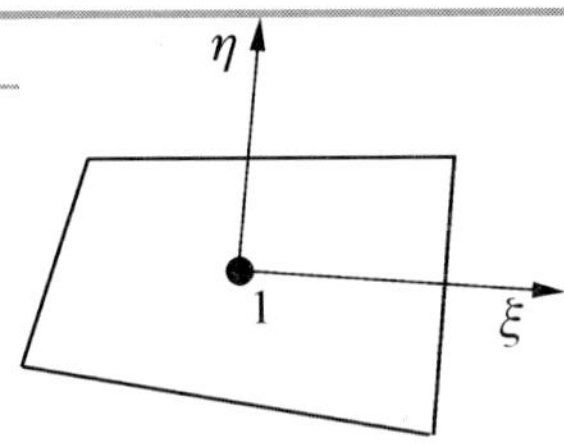

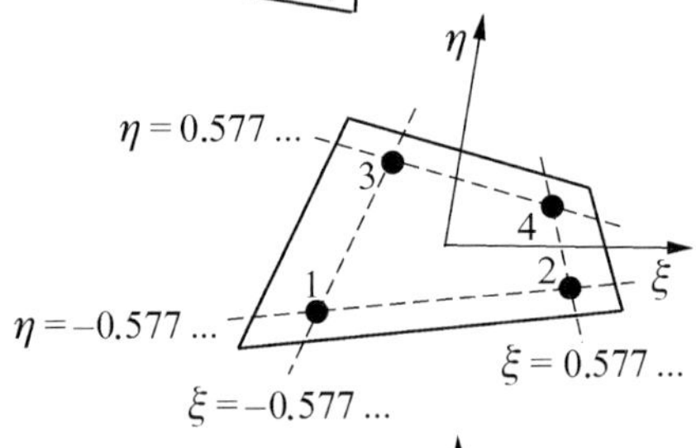

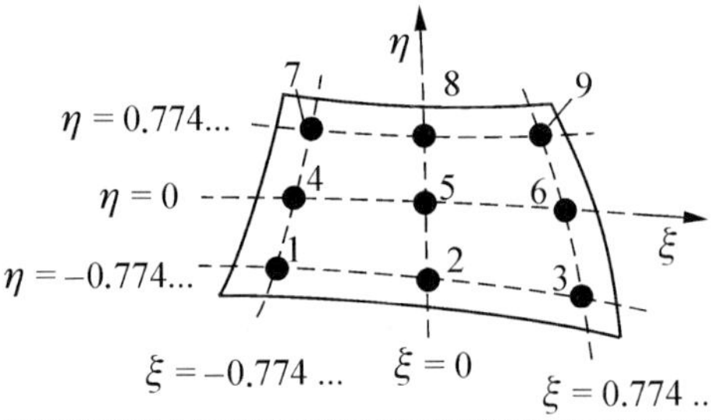

One can see, therefore, that once the numeration of nodes $i = j + (k - 1)n$ is introduced, the integral (18) can be written in the form

$$I = \sum_{i=1}^{m} F(\xi_i, \eta_i) w_i \,, \tag{20}$$

where $m = n^2$, $\xi_i = \alpha_j$, $\eta_i = \alpha_k$ and $w_i = w_j w_k$. Quantities α_j and w_j denote coordinates ξ_i and weights w_i, respectively, compiled in Table 22.2 and are used to calculate two-dimensional integrals.

c. *Triangular elements*

To calculate the integral, coordinates (x,y) are transformed into area coordinates L_1 and L_2, which are linearly independent, since once we account for the (11) from Ex. 22.5, coordinate L_3 can be calculated from formula

$$L_3 = 1 - L_1 - L_2 \,.$$

By allowing for equation

$$\begin{Bmatrix} \dfrac{\partial N_i^e}{\partial x} \\ \dfrac{\partial N_i^e}{\partial y} \end{Bmatrix} = [J]^{-1} \begin{Bmatrix} \dfrac{\partial N_i^e}{\partial L_1} \\ \dfrac{\partial N_i^e}{\partial L_2} \end{Bmatrix}, \tag{21}$$

where

$$[J] = \begin{bmatrix} \dfrac{\partial x}{\partial L_1} & \dfrac{\partial y}{\partial L_1} \\ \dfrac{\partial x}{\partial L_2} & \dfrac{\partial y}{\partial L_2} \end{bmatrix}, \tag{22}$$

surface integral I for a triangular element is calculated from formula

$$I = \int_0^1 \int_0^{1-L_1} F(L_1, L_2, L_3)\, dL_2 dL_1 \approx \sum_{i=1}^{m} F(L_{1,i}, L_{2,i}, L_{3,i}) w_i \tag{23}$$

The locations of integration points and weight coefficients for the triangular element [1, 3, 11, 12, 14, 15, 21] are compiled in Table 22.3.

In both, Chap. 11 and this chapter, the discussion was limited to one and two-dimensional steady-state and transient heat conduction problems. Three-dimensional problems are solved analogically, in keeping with the rules that were applied in the examples of two-dimensional problems, which were discussed above.

Table 22.3. The location of integration and weight points for a triangular element

Integration points number	Point location	Error	Point	Coordinates L_1, L_2, L_3	Weights w_i
1	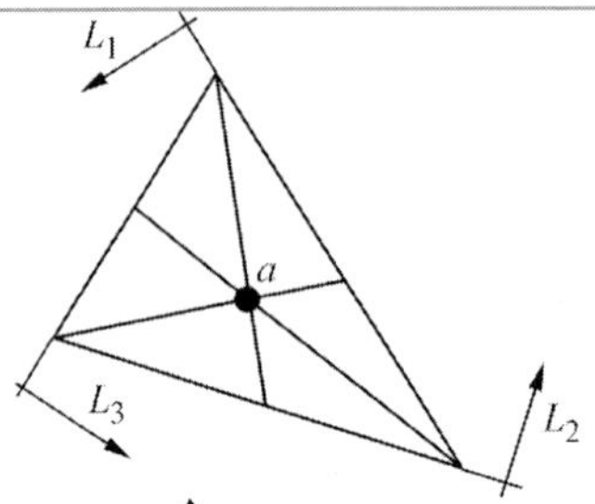	$R = O(h^2)$	a	$\frac{1}{3}, \frac{1}{3}, \frac{1}{3}$	1
3	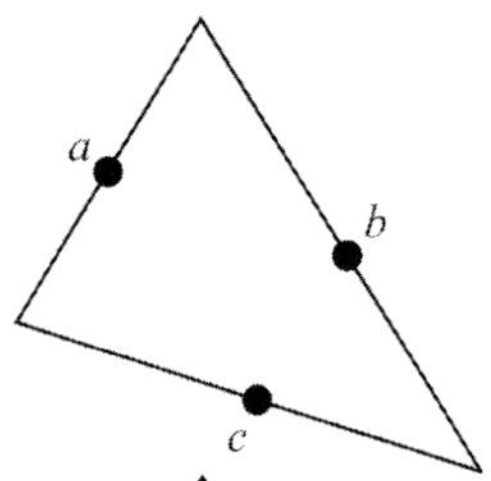	$R = O(h^3)$	a	$\frac{1}{2}, \frac{1}{2}, 0$	$\frac{1}{3}$
			b	$0, \frac{1}{2}, \frac{1}{2}$	$\frac{1}{3}$
			c	$\frac{1}{2}, 0, \frac{1}{2}$	$\frac{1}{3}$
4	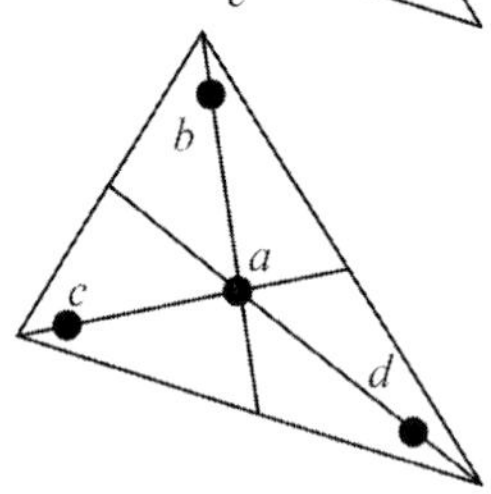	$R = O(h^4)$	a	$\frac{1}{3}, \frac{1}{3}, \frac{1}{3}$	$-\frac{27}{48}$
			b	0,6; 0,2; 0,2	$\frac{25}{48}$ (b, c, d)
			c	0,2; 0,6; 0,2	
			d	0,2; 0,2; 0,6	
7	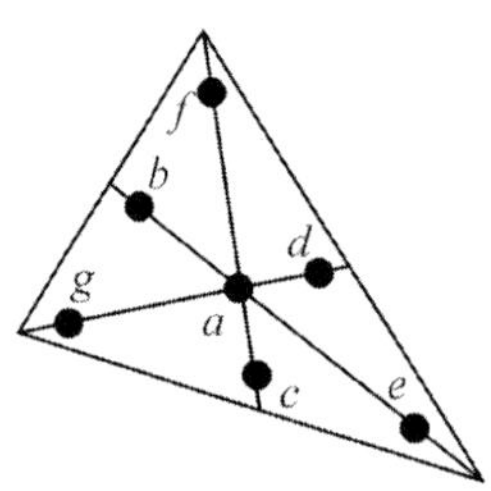	$R = O(h^6)$	a	$\frac{1}{3}, \frac{1}{3}, \frac{1}{3}$	0,2250000000
			b	$\alpha_1 \quad \beta_1 \quad \beta_1$	0,1323941527 (b, c, d)
			c	$\beta_1 \quad \alpha_1 \quad \beta_1$	
			d	$\beta_1 \quad \beta_1 \quad \alpha_1$	
			e	$\alpha_2, \beta_2, \beta_2$	0,1259391805 (e, f, g)
			f	$\beta_2, \alpha_2, \beta_2$	
			g	$\beta_2, \beta_2, \alpha_2$	

Constants α_1, α_2, β_1 and β_2 are $\alpha_1 = 0.059\,715\,871\,7$, $\beta_1 = 0.470\,142\,064\,1$, $\alpha_2 = 0.797\,426\,985\,3$, $\beta_2 = 0.101\,286\,507\,3$.

Exercise 22.7 Calculating Temperature in a Complex-Shape Fin by Means of the ANSYS Program

Oval pipes with an attached aluminum plate-fins, $\delta_{\dot{z}} = 0.08$ mm thick are used in car radiators [19]. Due to the division of plate-fin into fictions (equivalent) fins, temperature distribution in the entire lamella can be determined when temperature distribution only in half of the fin is analyzed [19].

The maximum length of the pipe diameter is $d_{max} = 11.82$ mm, while the minimum $d_{min} = 6.35$ mm (Fig. 22.10). The height of the fin half section is $h = 9.25$ mm, while the width $s = 17.0$ mm. Initial temperature of the fin is $T_0 = 20°\text{C}$. The temperature of surroundings is also $T_{cz} = 20°\text{C}$. At an instant $t = 0$ s, the fin base temperature increases by a step from temperature $T_0 = 20°\text{C}$ to $T_p = 95°\text{C}$.

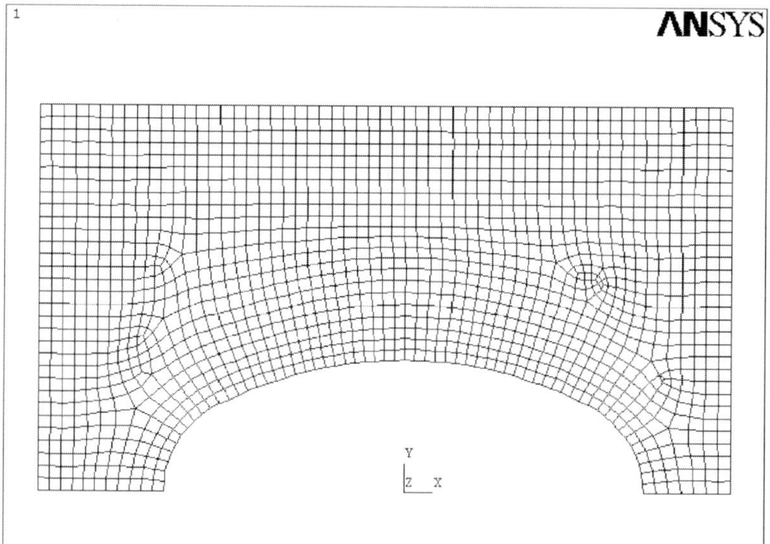

Fig. 22.10. The division of 1/4 of a plate-fin into finite elements

Determine transient fin temperature distribution at time $t_1 = 1.0$ s and $t_1 = 0.2$ s and temperature transient in the upper-left fin corner in the function of time. Heat transfer coefficient on the fin surface is $\alpha = 75$ W/(m^2·K). Heat is given off by the fin to surroundings through the lateral surfaces and through the left lateral front face. The upper and right-hand-side surfaces are thermally insulated. Straight sections of the fin base are also thermally insulated due to the symmetry of the temperature field with respect to the

Explicit Method

Derivative after time is approximated by the forward difference quotient

$$\frac{dT_i}{dt} \approx \frac{T_i(t+\Delta t) - T_i(t)}{\Delta t}. \tag{3}$$

By accounting for (3) in (1), one obtains (assuming that thermo-physical properties and $\dot{q}_v$ are constant) the following expression, which makes it possible to determine temperature T_i in all nodes at the new time point $t + \Delta t$

$$T_i(t+\Delta t) = \frac{\Delta t}{\rho c V_i}\left[\sum_{j=1}^{n} K_{ij} T_j(t) + \left(\frac{\rho c V_i}{\Delta t} - \sum_{j=1}^{n} K_{ij}\right) T_i(t) + \dot{q}_v V_i\right]. \tag{4}$$

Implicit Method

Derivative after time is approximated by a difference quotient (3), while the right side of (1) is calculated in the temperature function at time $t + \Delta t$

$$\rho c V_i \frac{T_i(t+\Delta t) - T_i(t)}{\Delta t} = \sum_{j=1}^{n} K_{ij}\left[T_j(t+\Delta t) - T_i(t+\Delta t)\right] + \dot{q}_v V_i. \tag{5}$$

By transforming (5), one obtains

$$\frac{\Delta t}{\rho c V_i}\left[\left(\frac{\rho c V_i}{\Delta t} + \sum_{j=1}^{n} K_{ij}\right) T_i(t+\Delta t) - \sum_{j=1}^{n} K_{ij} T_j(t+\Delta t)\right] = T_i(t) + \frac{\dot{q}_v \Delta t}{\rho c}. \tag{6}$$

Algebraic equation system (6) must be solved at every time step.

Crank-Nicolson Method

Crank and Nicolson proposed [13] to approximate the right side of (1) using the arithmetic mean taken from the values calculated for t and $t + \Delta t$. Equation (1) is approximated as follows:

$$\rho c V_i \frac{T_i(t+\Delta t)-T_i(t)}{\Delta t} = \frac{1}{2}\left\{\sum_{j=1}^{n} K_{ij}\left[T_j(t)-T_i(t)\right]+ \right.$$
$$\left. +\sum_{j=1}^{n} K_{ij}\left[T_j(t+\Delta t)-T_i(t+\Delta t)\right]\right\}+\dot{q}_v V_i. \tag{7}$$

Once (7) is transformed, the algebraic equation system is obtained; its form is as follows:

$$\left(\frac{\rho c V_i}{\Delta t}+\frac{1}{2}\sum_{j=1}^{n} K_{ij}\right)T_i(t+\Delta t)-\frac{1}{2}\sum_{j=1}^{n} K_{ij}T_j(t+\Delta t)=$$
$$=\left(\frac{\rho c V_i}{\Delta t}-\frac{1}{2}\sum_{j=1}^{n} K_{ij}\right)T_i(t)+\frac{1}{2}\sum_{j=1}^{n} K_{ij}T_j(t)+\frac{\dot{q}_v}{\rho c V_i}. \tag{8}$$

In (1)–(8) V_i is the volume of a control area (control volume), while n is the number of control volumes j, which are adjacent to the analyzed volume i. Nodes are located in the control volumes' centers of gravity.

The procedure discussed above is typical of the finite difference method. In numerical-analytical methods, the system of ordinary differential equations (1) can be directly integrated by means of one of the numerous methods used for solving ordinary differential equations [16, 31].

Runge-Kutta Method of 4th order ensures high solution accuracy. If the equation system (1) is linear, the matrix method can be applied; it requires the calculation of the exponential matrix $e^{\mathbf{A}t}$, where $\mathbf{A}$ is the coefficient matrix of the equation system (1). This method will be discussed in greater detail in Ex. 23.2.

Exercise 23.1 Integration of the Ordinary Differential Equation System by Means of the Runge-Kutta Method

By applying the method of lines, which is used to discretisize only spatial derivatives in the transient heat conduction equation, one obtains the ordinary differential equation system

$$\dot{\mathbf{T}}=\mathbf{A}(\mathbf{T})\mathbf{T}+\mathbf{F}(\mathbf{T},t), \tag{1}$$

where $\dot{\mathbf{T}}$ is the vector of the first derivatives after time of the node temperatures, while the dimensions of the coefficient matrix $\mathbf{A}(\mathbf{T})$ are $N_E \times N_E$. Coefficients of this matrix can be temperature dependent for a variable thermo-physical properties. Vector $\mathbf{F}(\mathbf{T})$, which in general depends on the determined node temperatures, measures N_E and is the boundary condition function.

Discuss Runge-Kutta method of 4th order used for solving (1).

Solution

Runge-Kutta method of 4th order is frequently used in practice, because it is highly accurate [6, 8, 16, 23, 51]. The accuracy order of the method is $O[(\Delta t)^4]$.

Equation (1) will be written in the form

$$\dot{\mathbf{T}} = \mathbf{Q}(\mathbf{T},t), \tag{2}$$

where $\mathbf{Q}(\mathbf{T},t) = \mathbf{A}(\mathbf{T})\,\mathbf{T} + \mathbf{F}(\mathbf{T},t)$.

Initial conditions have the form

$$\mathbf{T}|_{t=0} = \mathbf{T}_0, \tag{3}$$

where vector $\mathbf{T}_0$ contains components, which are the temperatures in nodes at time $t = 0$. Once the integration step Δt is assumed to be constant, temperatures $\mathbf{T}$ are calculated at time points $t_n = n\Delta t$, $n = 1, 2, \ldots$ according to the following algorithm:

$$\mathbf{T}_{n+1} = \mathbf{T}_n + \frac{\Delta t}{6}\left(\mathbf{k}_1 + 2\mathbf{k}_2 + 2\mathbf{k}_3 + \mathbf{k}_4\right), \qquad n = 0,\ 1,\ \ldots, \tag{4}$$

where

$$\begin{aligned}
\mathbf{k}_1 &= \mathbf{Q}\left(\mathbf{T}_n, t_n\right), \\
\mathbf{k}_2 &= \mathbf{Q}\left(\mathbf{T}_n + \frac{\Delta t \mathbf{k}_1}{2}, t_n + \frac{\Delta t}{2}\right), \\
\mathbf{k}_3 &= \mathbf{Q}\left(\mathbf{T}_n + \frac{\Delta t \mathbf{k}_2}{2}, t_n + \frac{\Delta t}{2}\right), \\
\mathbf{k}_4 &= \mathbf{Q}\left(\mathbf{T}_n + \Delta t \mathbf{k}_3, t_n + \Delta t\right).
\end{aligned} \tag{5}$$

In the case of a single differential equation

$$\frac{dT}{dt} = Q(T,t), \qquad T|_{t=0} = T_0 \tag{6}$$

algorithm (5) is simplified to a form

$$T_{n+1} = T_n + \frac{\Delta t}{6}\left(k_1 + 2k_2 + 2k_3 + k_4\right), \qquad n = 0,\ 1,\ \ldots, \tag{7}$$

where

$$\begin{aligned} k_1 &= Q\left(T_n, t_n\right), \\ k_2 &= Q\left(T_n + \frac{\Delta t k_1}{2}, t_n + \frac{\Delta t}{2}\right), \\ k_3 &= Q\left(T_n + \frac{\Delta t k_2}{2}, t_n + \frac{\Delta t}{2}\right), \\ k_4 &= Q\left(T_n + \Delta t k_3, t_n + \Delta t\right). \end{aligned} \tag{8}$$

In order to demonstrate the accuracy of the Runge-Kutta method of 4th order, [51] will be solved

$$\frac{dT}{dt} = \left(t + T - 1\right)^2 \tag{9}$$

with an initial condition

$$T\left(0\right) = 2\,. \tag{10}$$

The obtained solution will be compared with Euler method, according to which the (6) is solved as follows:

$$T_{n+1} = T_n + \Delta t Q\left(T_n, t_n\right), \qquad n = 0,\ 1,\ \ldots \tag{11}$$

The comparison results for Δt = 0,1 s and Δt = 0,05 s are presented in Table 23.1. The Table shows that the results obtained by means of the Runge-Kutta method are highly accurate.

Table 23.1. The solutions of (1) obtained by means of the explicit Euler method and Runge-Kutta method of 4th order for the two integration steps: Δt = 0.1 s and Δt = 0.05 s

t	Δt = 0.1 s		Δt = 0.05 s		Analytical solution
	Eulera Method	Rune-Kutta Method	Euler Method	Runge-Kutta Method	
0.00	2.0000	2.0000	2.0000	2.0000	2.0000
0.10	2.1000	2.1230	2.1105	2.1230	2.1230
0.20	2.2440	2.3085	2.2727	2.3085	2.3085
0.30	2.4525	2.5958	2.5142	2.5958	2.5958
0.40	2.7596	3.0649	2.8845	3.0650	3.0650
0.50	3.2261	3.9078	3.4823	3.9082	3.9082

$$\mathbf{T}_{k+1} = e^{\mathbf{A}(\Delta t)}\mathbf{T}_k +$$
$$+ e^{\mathbf{A}(k+1)(\Delta t)} \int\limits_{k(\Delta t)}^{(k+1)(\Delta t)} e^{-\mathbf{A}s}\mathbf{B}\left\{\mathbf{u}_{k+1} + \left[t_k + (\Delta t) - s\right]\frac{\mathbf{u}_k - \mathbf{u}_{k+1}}{(\Delta t)}\right\} ds = \tag{18}$$
$$= \mathbf{F}\mathbf{T}_k + \mathbf{G}\mathbf{u}_{k+1} + \mathbf{H}\left(\mathbf{u}_k - \mathbf{u}_{k+1}\right),$$

where **F** and **G** are formulated in (14) and (15), respectively and

$$\mathbf{H} = \int\limits_{k(\Delta t)}^{(k+1)(\Delta t)} e^{\mathbf{A}\left[k(\Delta t)+\Delta t - s\right]} \frac{t_k + (\Delta t) - s}{(\Delta t)}\mathbf{B}ds\,. \tag{19}$$

If a new variable is introduced

$$\tau = k(\Delta t) + (\Delta t) - s\,,$$

from where it follows that

$$s = k(\Delta t) + (\Delta t) - \tau, \qquad d\tau = -ds\,, \tag{20}$$

the integral **H** can be written as follows:

$$\begin{aligned}
\mathbf{H} &= -\int\limits_{\Delta t}^{0} e^{\mathbf{A}\tau} \frac{\tau}{(\Delta t)}\mathbf{B}d\tau = \\
&= -\int\limits_{\Delta t}^{0}\left(\mathbf{I} + \mathbf{A}\tau + \frac{\mathbf{A}^2\tau^2}{2!} + \frac{\mathbf{A}^3\tau^3}{3!} + \ldots\right)\frac{\tau}{(\Delta t)}\mathbf{B}d\tau = \\
&= \left\{\frac{\mathbf{I}}{2} + \frac{\mathbf{A}(\Delta t)}{3} + \frac{\left[\mathbf{A}(\Delta t)\right]^2}{2\cdot 4} + \frac{\left[\mathbf{A}(\Delta t)\right]^3}{2\cdot 3\cdot 5} + \ldots + \right. \\
&\left. + \frac{\left[\mathbf{A}(\Delta t)\right]^n (n+1)}{(n+2)!}\right\}(\Delta t)\mathbf{B}.
\end{aligned} \tag{21}$$

One can sequentially determine temperature distribution in the subsequent time points by means of (18). The methods discussed above for integrating the ordinary differential equation system refer to linear systems when thermo-physical properties of a material are temperature-independent. Equations (13) and (18) can be also applied to systems in which thermo-physical properties of a material are temperature-dependent. Once a non-linear equation system (1) is exchanged for a linear system when matrices **A** and **B** are temperature-dependent

$$\dot{\mathbf{T}}^{(n+1)} = \mathbf{A}\left(\mathbf{T}^{(n)}\right)\mathbf{T}^{(n+1)} + \mathbf{B}\left(\mathbf{T}^{(n)}\right)\mathbf{u}^{(n+1)} \tag{22}$$

Equations (13) and (18) can be applied to determine temperature distribution $T_{k+1}^{(n+1)}$. The proper solution is already obtained after few iterations.

The methods described above can be used for solving both, simple [41] and inverse heat conduction problems [17–19, 46, 47].

Exercise 23.3 Determining Steel Plate Temperature by Means of the Method of Lines, while the Plate is Cooled by Air and Boiling Water

A plate made of a low-alloyed steel, which contains 1% Cr and is $2L$ = 0.04 m thick, is submersed in water at the temperature of $T_{cz} = T_n$ = 100°C = 373.15 K and the pressure of p = 1.0133 bar. Initial plate temperature is T_0 = 500°C. Bubble boiling takes place on the plate surface; during the process, the heat transfer coefficient is high and expressed as $\alpha_w = 137.5(T|_{x=L} - T_{cz})^2$ [W/(m^2·K)], where $T|_{x=L}$ is the plate surface temperature.

Determine surface temperature transient of the plate $T|_{x=L}$ and temperature differences $\Delta T = T|_{x=0} - T|_{x=L}$ in the function of time. Compare the obtained results with temperature transient of the plate cooled in an open air. Heat transfer coefficient on the plate surface, which allows for natural convection and radiation, is defined by formula $\alpha_p = 9.7 + 0.04(T|_{x=L} - T_{cz})$ [W/(m^2·K)], where T_{cz} = 20°C = 293.15 K. Assume that thermo-physical properties of the steel are as follow: λ = 52 W/(m·K), c = 460 J/(kg·K), ρ = 7865 kg/m^3.

Determine temperature distribution over the plate thickness for selected time points. Continue calculations until the plate surface temperature reaches 110°C.

Solution

Due to the symmetry of the problem, only one-half of the plate's thickness will be analyzed below (Fig. 23.3).

Heat balance equations written for individual control volumes are as follow:

$$\Delta t_{\max} \le \frac{(\Delta x)^2}{2a(1+\Delta Bi_{\max})} = \frac{0.001^2}{2 \cdot 1.4373 \cdot 10^{-5}(1+42.548)} = \tag{13}$$

$$= 7.988 \cdot 10^{-4}\ \text{s}.$$

Once the plate surface temperature is lowered $T|_{x=L}$ to 400 K, the heat transfer coefficient is

$$\alpha_w = 137.5(400-373.15)^2 = 99127\,\text{W/(m}^2\cdot\text{K)},$$

$$\Delta Bi = \frac{\alpha_w \Delta x}{\lambda} = 1.906\,.$$

From condition (11), one obtains

$$\Delta t \le \frac{(\Delta x)^2}{2a(1+\Delta Bi)} = \frac{0.001^2}{2 \cdot 1.4373 \cdot 10^{-5}(1+1.9063)} = 0.012\ \text{s}\,.$$

Due to the fact that criterion (11) was approximated, the integration step in the Runge-Kutta method will be selected in the following way:

$$\Delta t = 0.0001\text{s} \quad \text{for} \quad 400\ \text{K} \le T|_{x=L}\,,$$

and

$$\Delta t = 0.01\text{s} \quad \text{for} \quad T|_{x=L} < 400\ \text{K}\,.$$

In the case of air cooling, time step Δt can be much larger, since the maximum heat transfer coefficient is rather small

$$\alpha_{p,\max} = 9.7 + 0.04(500-373.15) = 14.78\ \text{W/(m}^2\cdot\text{K)},$$

$$\Delta Bi = \frac{\alpha_{p,\max}\Delta x}{\lambda} = \frac{14.78 \cdot 0.001}{52} = 2.84 \cdot 10^{-4}\,.$$

From (9), one has

$$\Delta t = \frac{(\Delta x)^2}{2a(1+\Delta Bi)} = \frac{0.001^2}{2 \cdot 1.4373 \cdot 10^{-5}(1+2.84 \cdot 10^{-4})} = 0.0347\ \text{s}\,.$$

It is assumed that the integration step during the air-cooling of the plate is $\Delta t = 0.025$ s.

Temperature calculation results for the water-submersed plate are presented in Fig. 23.3. The transient of the plate surface temperature during air-cooling is also presented in Fig.23.3.

Tables 23.2 and 23.3 present spatial temperature distribution across the plate thickness when the plate is air-cooled and boiling-water-cooled. It is evident that it takes longer to cool the plate with air than it does with boiling water.

Table 23.2. Temperature distribution in [K] across the plate thickness during water cooling

Time t [s]	Node no.				
	1	6	11	16	21
	Coordinate x [m]				
	0,020	0,015	0,010	0,005	0,000
0	500,00	500,00	500,00	500,00	500,00
1	391,59	462,65	493,32	499,40	499,95
2	389,69	444,65	479,85	494,73	498,17
5	387,40	424,18	453,97	472,82	479,20
10	385,46	409,29	429,18	442,35	446,95
20	382,62	393,39	402,32	408,21	410,27
30	380,56	385,70	389,91	392,66	393,62

Table 23.3. Temperature distribution in [K] across the plate thickness during air cooling

Time t [s]	Node no.				
	1	6	11	16	21
	Coordinate x [m]				
	0,020	0,015	0,010	0,005	0,000
0	500,00	500,00	500,00	500,00	500,00
10	484,02	488,75	492,18	494,25	494,94
30	472,83	476,30	478,84	480,39	480,91
60	460,68	463,03	464,75	465,80	466,15
120	445,10	446,42	447,38	447,96	448,15
240	428,55	429,17	429,62	429,89	429,98
360	419,51	419,89	420,16	420,32	420,38
480	413,63	413,88	414,07	414,18	414,22
600	409,40	409,59	409,73	409,81	409,84

The transients of the plate surface temperature $T|_{x=L} = T_1$ and temperature difference $\Delta T = T|_{x=L} - T|_{x=0} = T_1 - T_{21}$ are presented in Fig. 23.4 and Fig. 23.5.

From the analysis of Fig.23.4 and 23.5, it is evident that the process of plate cooling by means of water is much faster. Temperature difference across the wall thickness during water-cooling is significantly larger than it is during air-cooling.

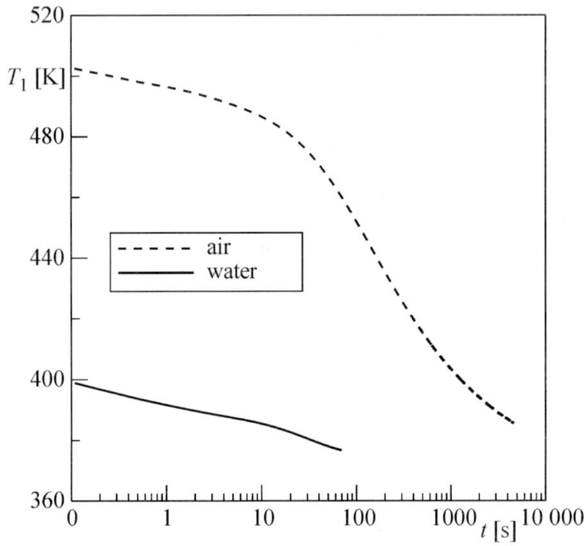

Fig. 23.4. Plate surface temperature transient $T|_{x=L} = T_1$ during water and air cooling

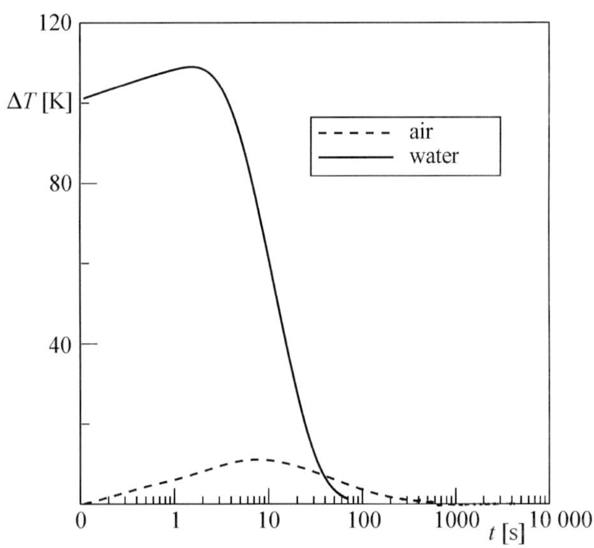

Fig. 23.5. Temperature difference transient across the plate thickness $\Delta T = T|_{x=L} - T|_{x=0} = T_1 - T_{21}$ during water or air cooling

Exercise 23.4 Using the Exact Analytical Method and the Method of Lines to Determine Temperature of a Cylindrical Chamber

Using the method of lines, determine temperature distribution in a cylindrical chamber of a steam attemperator in the 2nd superheater of a steam boiler. A chamber whose outer diameter is d_z = 355 mm and the wall thickness g = 70 mm is made of a steel of type 10CrMo910 (C = 0.15%, Cr= 2÷2.5%, Mo = 0.9÷1.1%, Mn = 0.4÷0.6%, Si = 0.15÷0.5%). Initial temperature of the chamber is T_p = 20°C. Temperature of the medium T_{cz} undergoes a step-increase from the temperature T_p to 100°C. The outer surface of the chamber is thermally insulated. Heat transfer coefficient on the inner surface is α = 5000 W/(m^2·K). For the calculation, assume that the thermo-physical properties of the steel 10CrMo910 with temperature $T_{śr}$ = (20 + 100)/2 = 60°C are, according to Table [48], as follow: λ = 37 W/(m·K), c = 469 J/(kg·K), ρ = 7823 kg/(m^3). Compare the obtained temperature distribution with the results obtained by means of the exact formula for the following time points: t_1 = 60 s, t_2 = 120 s, t_3 = 240 s, t_4 = 480 s, t_5 = 720 s, t_6 = 900 s and t_7 = 1800 s. Furthermore, present temperature distribution determined by means of the method of lines in the graphical form.

Solution

Temperature distribution will be determined by means of the control volume method. Heat balance equation for node i (Fig. 23.6) has the following form, assuming that thermo-physical properties are constant:

$$\pi\left[\left(r_i+\frac{\Delta r}{2}\right)^2-\left(r_i-\frac{\Delta r}{2}\right)^2\right]c\rho\frac{dT_i}{dt}= \\ =2\pi\left(r_i-\frac{\Delta r}{2}\right)\lambda\frac{T_{i-1}-T_i}{\Delta r}+2\pi\left(r_i+\frac{\Delta r}{2}\right)\lambda\frac{T_{i+1}-T_i}{\Delta r}. \tag{1}$$

The above equation can be transformed into the following form:

$$\frac{dT_i}{dt}=\frac{a}{\left(\Delta r\right)^2}\left[\left(1-\frac{\Delta R}{2R_i}\right)\left(T_{i-1}-T_i\right)+\left(1+\frac{\Delta R}{2R_i}\right)\left(T_{i+1}-T_i\right)\right],\quad i=1,\ \ldots,\ N, \tag{2}$$

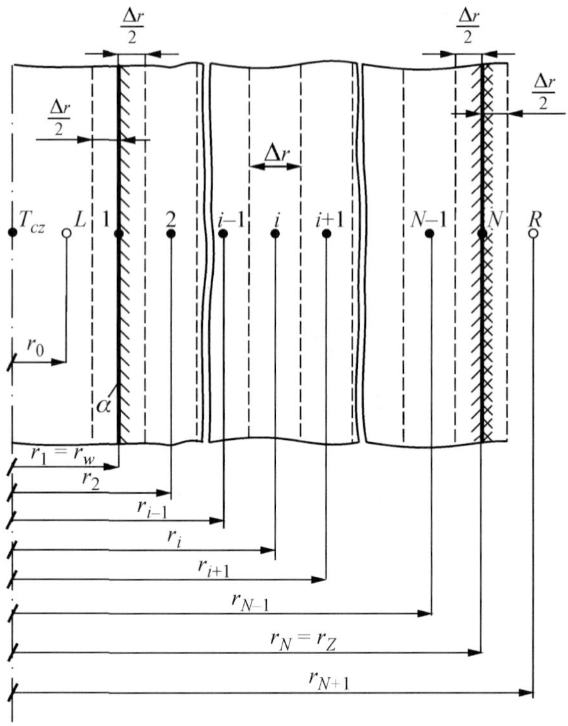

Fig. 23.6. Wall division into control volumes $\Delta r = (r_z - r_w)/(N-1)$; o are apparent nodes

where,

$$a = \frac{\lambda}{c\rho}, \quad R = \frac{r}{r_z}, \quad \Delta R = \frac{\Delta r_z}{r_z}, \quad \Delta r = \frac{r_z - r_w}{N-1}. \tag{3}$$

Temperature is determined in N evenly spaced points (nodes). Apparent nodes L and R (Fig. 23.6) lie outside the cylinder wall and will be used for the approximation of boundary conditions with the accuracy of $O[(\Delta r)^2]$. Node location is expressed as

$$r_i = r_w + (i-1)\Delta r, \qquad i = 0, \ldots, N, \tag{4}$$

$$R_i = R_w + (i-1)\Delta R, \qquad i = 0, \ldots, N+1, \tag{5}$$

where $R_w = r_w/r_z$.

In such equations, temperatures $T_0 = T_L$ and $T_{N+1} = T_R$ appear at the apparent points L and R for $i = 1$ and $i = N$. These temperatures can be eliminated from the equation system (2), using the boundary conditions on the inner and outer surface of the chamber.

From the inner surface boundary condition

$$-\lambda \left.\frac{\partial T}{\partial r}\right|_{r=r_w} = \alpha\left(T_{cz} - T\big|_{r=r_w}\right) \tag{6}$$

and after the approximation of a derivative with a central difference quotient

$$\left.\frac{\partial T}{\partial r}\right|_{r=r_w} \approx \frac{T_2 - T_L}{2\Delta r} \tag{7}$$

one has

$$-\lambda \frac{T_2 - T_L}{2\Delta r} = \alpha\left(T_{cz} - T_1\right). \tag{8}$$

Once (8) is transformed, one obtains

$$T_0 = T_L = T_2 + 2k\Delta R Bi\left(T_{cz} - T_1\right), \tag{9}$$

where

$$\Delta R = \Delta r / r_z, \qquad k = r_z / r_w, \qquad Bi = \alpha\, r_w / \lambda. \tag{10}$$

The outer surface of the chamber is thermally insulated

$$\lambda \left.\frac{\partial T}{\partial r}\right|_{r=r_z} = 0. \tag{11}$$

As a result of the approximation with the central difference quotient, one has

$$\lambda \frac{T_R - T_{N-1}}{2\Delta r} = 0, \tag{12}$$

$$T_{N+1} = T_R = T_{N-1}. \tag{13}$$

Differential equation system (2) and conditions (9) and (13) will be solved by the Runge-Kutta method of the 4th order. Exact analytical solution can be found in paper [9] and also in papers [1, 21]. The graphs, which one can use to determine temperature and heat flux, can be found in paper [52].

Temperature distribution in the chamber wall is formulated as [9, 21]

$$\theta = \frac{T_{cz} - T(r,t)}{T_{cz} - T_p} = \sum_{n=1}^{\infty} A_n M_n \exp\left(-\mu_n^2 Fo\right), \tag{14}$$

where

$$r_i = r_w + (i-1)\Delta r, \qquad i = 0, \ldots, N+1, \tag{10}$$

$$\Delta r = \frac{r_z - r_w}{N-1} \qquad \text{(Fig. 23.6, Ex. 23.4)}. \tag{11}$$

Equations (7), (8) and (9) were obtained from (4) and (5), respectively, once the integrals were calculated with the trapeze method. In terms of the exact solution ((14) from Ex. 23.4), the mean temperature is calculated first

$$\theta_m(t) = \frac{T_{cz} - T_m(t)}{T_{cz} - T_p} \tag{12}$$

by means of (4), which, once the dimensionless quantities are introduced, assumes the form

$$\theta_m(t) = \frac{2}{k^2-1}\int_1^k R\theta \, dR. \tag{13}$$

Substituting (14) from Ex. 23.4 into (13) and accounting that

$$\int RY_0(R\mu_n)dR = \frac{1}{\mu_n} RY_1(R\mu_n), \tag{14}$$

$$\int RJ_0(R\mu_n)dR = \frac{1}{\mu_n} RJ_1(R\mu_n) \tag{15}$$

one obtains, after transformations

$$\theta_m = \frac{2}{k^2-1}\sum_{n=1}^{\infty} A_n \frac{1}{\mu_n}\left[J_1(\mu_n)Y_1(k\mu_n) - J_1(k\mu_n)Y_1(\mu_n)\right]\exp\left(-\mu_n^2 Fo\right), \tag{16}$$

where $k = r_z/r_w$, $Fo = at/r_w^2$.

Constant A_n is formulated in (16), Ex. 23.4. Once θ_m is calculated first, according to (16), $T_m(t)$ is determined next from the (12)

$$T_m(t) = T_{cz} - (T_{cz} - T_p)\theta_m(t) \tag{17}$$

Temperature $T_{r,m}(t)$ defined in a similar way is calculated from formula

$$T_{r,m}(r,t) = T_{cz} - (T_{cz} - T_p)\theta_{r,m}(R,t), \tag{18}$$

where T_p is the initial temperature,

$$\theta_{r,m}(r,t) = \frac{2}{R^2-1}\sum_{n=1}^{\infty} A_n \frac{1}{\mu_n}\left[RJ_1(k\mu_n)Y_1(R\mu_n) - J_1(k\mu_n)Y_1(\mu_n) - \right. \tag{19}$$
$$\left. -RJ_1(R\mu_n)Y_1(k\mu_n) + J_1(\mu_n)Y_1(k\mu_n)\right]\exp\left(-\mu_n^2 Fo\right).$$

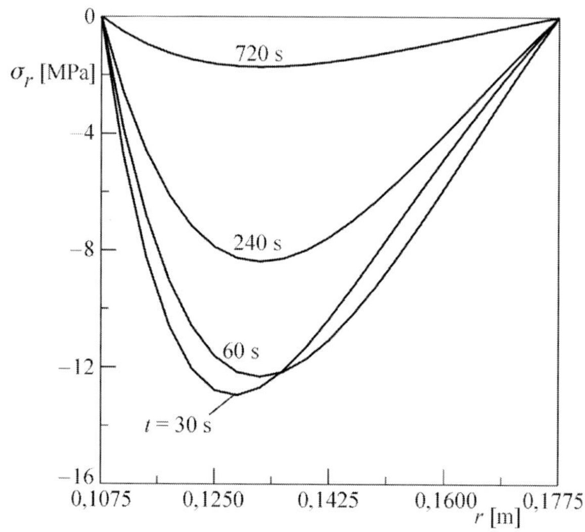

Fig. 23.8. Distribution of radial thermal stresses σ_r in a chamber wall of an attemperator at different moments

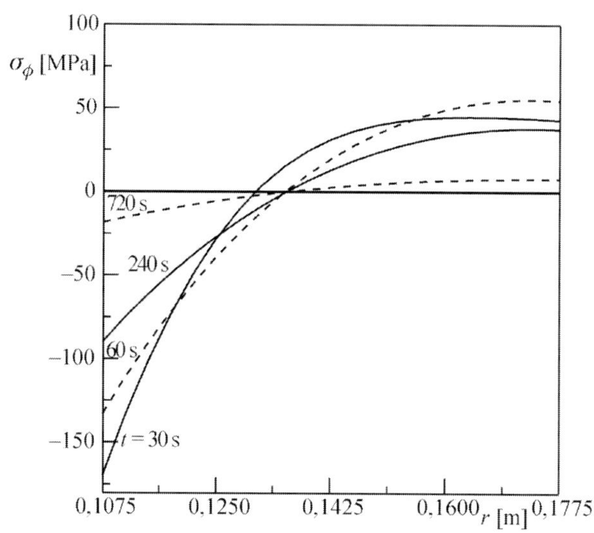

Fig. 23.9. Distribution of circumferential stresses σ_φ in a chamber wall of an attemperator at different moments

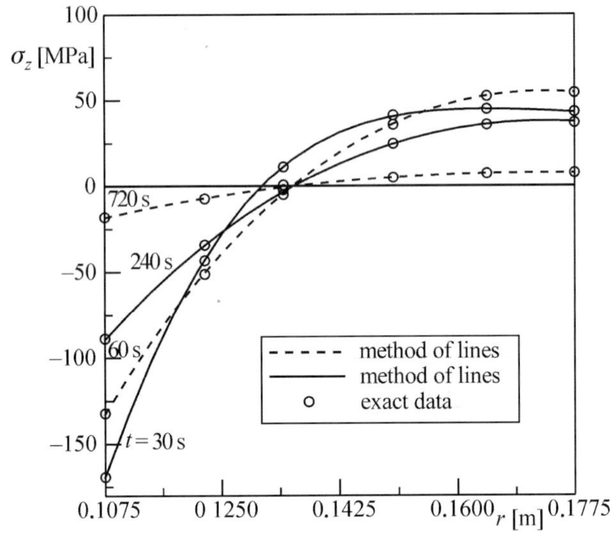

Fig. 23.10. Distribution of axial stresses σ_z in a chamber wall of an attemperator at different moments

By substituting (17) and (18) into (1)–(3), one can calculate thermal stress components.

Figs. 23.8–23.10 show thermal stresses calculated with the method of lines. In Fig. 23.10, the axial stresses calculated by means of the approximate formula are compared with the stresses calculated by means of the exact formula.

The chamber wall was divided into twenty control volumes ($N = 21$).

One can see from the comparison presented in Fig. 23.10 that the accuracy in determining thermal stresses with the method of lines is very good.

Exercise 23.6 Determining Temperature Distribution in a Cylindrical Chamber with Constant and Temperature Dependent Thermo-Physical Properties by Means of the Method of Lines

By means of the method of lines, determine temperature distribution in a cylindrical chamber of a steam attemperator, discussed in Ex. 23.4. Carry out two separate calculations, for the following two cases:

- when thermo-physical properties of a steel 10CrMo910 are constant, temperature-independent and defined at average temperature;

- when thermo-physical properties are temperature-dependent and defined by [35]

$$\lambda(T) = 35.3 + 2.14 \cdot 10^{-2} T - 4.30 \cdot 10^{-5} T^2, \tag{1}$$

$$a(T) = \left(10.2 - 0.48 \cdot 10^{-2} T - 5.06 \cdot 10^{-6} T^2\right) \cdot 10^{-6}, \tag{2}$$

when thermal conductivity λ is expressed in W/(m·K), temperature diffusivity in a in m^2/s, while temperature T in °C. Temperature of a medium is rising at a constant rate of v_T = 3 K/min from the initial temperature T_p = 100°C to finishing temperature T_k = 545°C (Fig. 23.11). Heat transfer coefficient on the inner surface of the chamber is α = 700 W/(m^2·K).

Determine inner and outer surface temperature transient in the cylinder and average temperature across the wall thickness of the chamber in the function of time. Also determine temperature difference transient of the inner surface and average temperature transient across the wall thickness.

Provided that the thermo-physical properties are constant, assume for the calculation that $\lambda = \lambda(T_{śr})$ and $a = a(T_{śr})$, where $T_{śr} = 0.5(T_p + T_k)$ = 0.5(100 + 545) = 322.5°C.

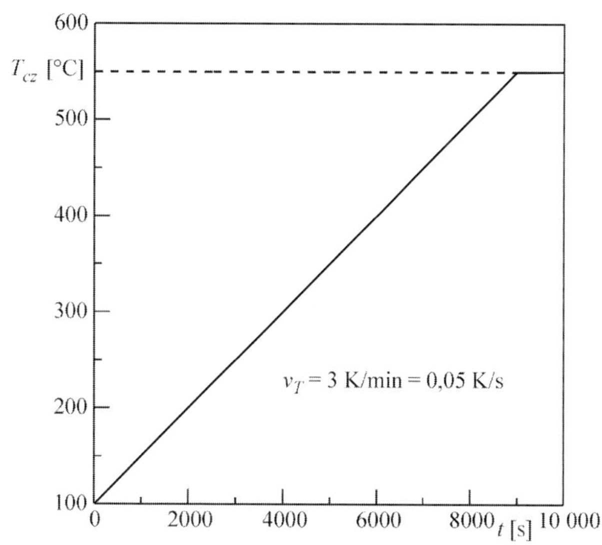

Fig. 23.11. Time changes of fluid temperature $T_{cz}(t)$

Solution

Changes in the thermal conductivity $\lambda(T)$ and temperature diffusivity $a(T)$ expressed, respectively, in (1) and (2) are presented in Fig. 23.12.

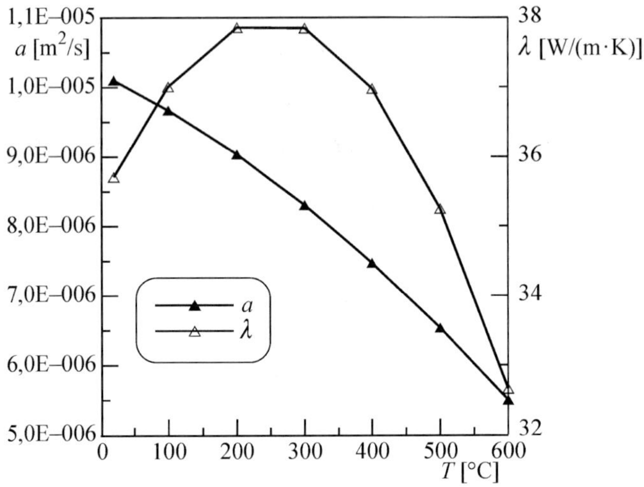

Fig. 23.12. Thermal conductivity λ changes and thermal diffusivity a changes in function of temperature

If we assume that properties are constant, then

$$\lambda\left(T_{sr}\right)=\lambda\left(322.5^{\circ}\mathrm{C}\right)=37.73\ \mathrm{W/(m\cdot K)},$$

$$a\left(T_{sr}\right)=a\left(322.5^{\circ}\mathrm{C}\right)=8.126\cdot 10^{-6}\ \mathrm{m^2/s}.$$

The heat conduction equation, which describes the temperature field in the chamber wall, has the form

$$c(T)\rho(T)\frac{\partial T}{\partial t}=\frac{1}{r}\frac{\partial}{\partial r}\left[\lambda(T)\,r\frac{\partial T}{\partial r}\right]. \tag{3}$$

Boundary condition of 3rd kind is assigned on the inner surface

$$-\left[\lambda(T)\frac{\partial T}{\partial r}\right]\Bigg|_{r=r_w}=\alpha\left(T_{cz}-T\big|_{r=r_w}\right), \tag{4}$$

while the outer surface is thermally insulated

$$-\left[\lambda(T)\frac{\partial T}{\partial r}\right]\Bigg|_{r=r_z}=0\,. \tag{5}$$

Initial temperature T_p is uniform:

$$T\big|_{t=0}=T_p\,. \tag{6}$$

The division of the chamber wall into control volumes is shown in Fig. 23.6, Ex. 23.4.

When nodes Δr are spaced at equal intervals, the lower and upper control volume limit is, respectively

$$\begin{aligned}&\frac{r_{i-1}+r_i}{2}=r_i-\frac{\Delta r}{2} \quad \text{i}\\ &\frac{r_i+r_{i+1}}{2}=r_i+\frac{\Delta r}{2},\end{aligned} \tag{7}$$

where

$\Delta r=\dfrac{r_z-r_w}{N-1}$, N is the node number.

Heat balance equation for the i-control volume has the form [48]

$$\int_{r_i-\Delta r/2}^{r_i+\Delta r/2} c(T)\rho(T)\,rdr=\left[\lambda(T)r\frac{\partial T}{\partial r}\right]\Bigg|_{r_i+\Delta r/2}-\left[\lambda(T)r\frac{\partial T}{\partial r}\right]\Bigg|_{r_i-\Delta r/2}\,. \tag{8}$$

Left side of the equation can be approximated as follows:

$$\begin{aligned}&\int_{r_i-\Delta r/2}^{r_i+\Delta r/2} c(T)\rho(T)\frac{\partial T}{\partial t}rdr\approx\\ &\approx\frac{\left(r_i+\Delta r/2\right)^2-\left(r_i-\Delta r/2\right)^2}{2}c(T_i)\rho(T_i)\frac{dT_i}{dt}.\end{aligned} \tag{9}$$

Derivative $\partial T/\partial r$, which occurs on the right-hand-side of (8), is approximated by the difference quotient, whereby one has

$$\left[\lambda(T)r\frac{\partial T}{\partial r}\right]\Bigg|_{r_i+\Delta r/2}\approx\frac{\lambda(T_i)+\lambda(T_{i+1})}{2}\left(r_i+\frac{\Delta r}{2}\right)\frac{T_{i+1}-T_i}{\Delta r} \tag{10}$$

and

$$\left[\lambda(T)r\frac{\partial T}{\partial r}\right]\Bigg|_{r_i-\Delta r/2} \approx \frac{\lambda(T_{i-1})+\lambda(T_i)}{2}\left(r_i-\frac{\Delta r}{2}\right)\frac{T_i-T_{i-1}}{\Delta r}. \tag{11}$$

Once (9), (10) and (11) are substituted into (8) and transformations carried out, one obtains

$$\begin{aligned}\frac{dT_i}{dt} &= \frac{a_i}{(\Delta r)^2}\left[\frac{\lambda_{i-1}+\lambda_i}{2\lambda_i}\left(1-\frac{\Delta r}{2r_i}\right)(T_{i-1}-T_i)+\right.\\ &\left.+\frac{\lambda_i+\lambda_{i+1}}{2\lambda_i}\left(1+\frac{\Delta r}{2r_i}\right)(T_{i+1}-T_i)\right], \qquad i=1,\ \ldots,\ N,\end{aligned} \tag{12}$$

where $\lambda_i = \lambda(T_i)$, $a_i = a(T_i)$.

In equations (12) for $i = 1$ and $i = N$, temperatures $T_0 = T_L$ and $T_{N+1} = T_R$ in the apparent nodes L and R appear. To eliminate temperatures T_0 and T_{N+1} from the balance (12), the boundary conditions (4) and (5) will be used. Once the derivative $\partial T/\partial r$ is approximated in the condition (4) by the central difference quotient, one obtains

$$-\lambda(T_1)\frac{T_2-T_0}{2\Delta r} = \alpha(T_{cz}-T_1), \tag{13}$$

where from T_0 is determined

$$T_0 = T_2 + 2\frac{\alpha\ \Delta r}{\lambda(T_1)}(T_{cz}-T_1). \tag{14}$$

By following the same procedure in the case of the boundary condition (5), one has

$$\lambda(T_N)\frac{T_{N+1}-T_{N-1}}{2\Delta r} = 0, \tag{15}$$

hence the equality

$$T_{N+1} = T_{N-1}. \tag{16}$$

Ordinary differential equation system (12) together with the condition (14) for $i = 1$ and condition (15) for $i = N$ is solved when the initial conditions, which follow from (6), are

$$T_i\big|_{t=0} = T_p, \qquad i=1,\ \ldots,\ N. \tag{17}$$

Average temperature across the wall thickness is calculated from the expression

$$T_m(t) = \frac{2}{r_z^2 - r_w^2} \int_{r_w}^{r_z} rT(r,t)\,dr\,. \tag{18}$$

Once the integral is calculated using the trapeze method, one has

$$T_m(t) \approx \frac{\Delta r}{r_z^2 - r_w^2} \sum_{i=2}^{N} \left[r_{i-1} T_{i-1}(t) + r_i T_i(t) \right]. \tag{19}$$

Calculation results for N = 21 and Δt = 0.5 s are presented in Fig. 23.13 and 23.14. From the analysis of Fig. 23.13 and 23.14, it is clear that the assumption that thermal conductivity and diffusivity are constant can lead to significant errors in the determined transients of temperature difference $\Delta T = T(r_w,t) - T_m(t)$, which are used for calculating axial thermal stresses (Ex. 23.5).

The method of lines makes it possible to allow for the dependence of the material's thermo-physical properties on temperature, without complicating the computational algorithm.

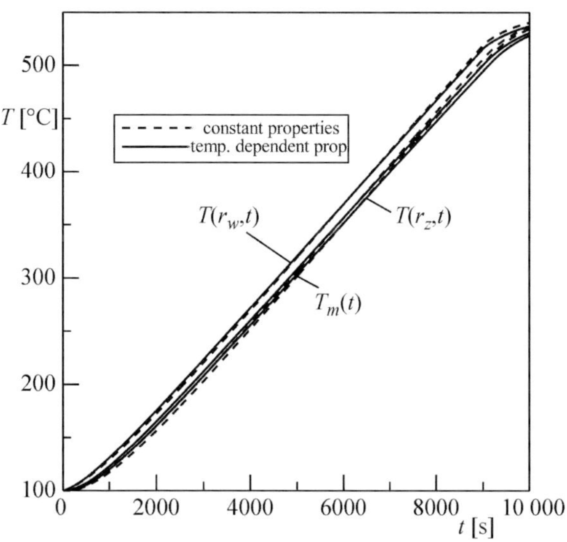

Fig. 23.13. Temperature transients of the inner surface $T(r_w, t)$, outer surface $T(r_z, t)$ and average temperature across the wall thickness $T_m(t)$ for constant and temperature dependent thermo-physical properties

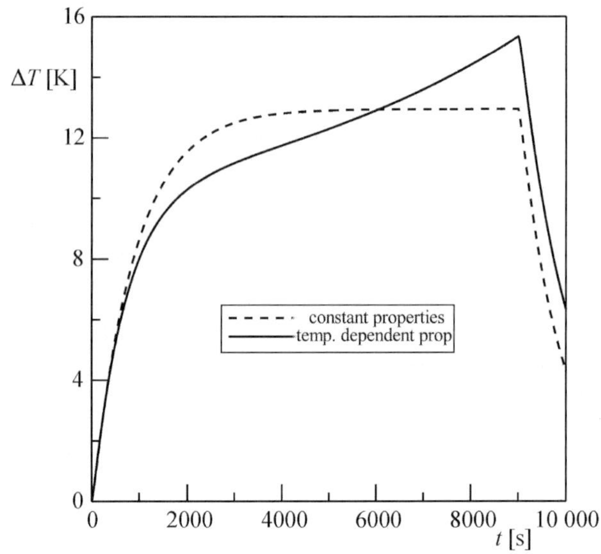

Fig. 23.14. Temperature difference transient $\Delta T = T\,|_{r=r_w} - T_m$ for constant and temperature dependent thermo-physical properties

Exercise 23.7 Determining Transient Temperature Distribution in an Infinitely Long Rod with a Rectangular Cross-Section by Means of the Method of Lines

Determine transient temperature distribution in an infinitely long rod with a rectangular cross-section, with $2d$ = 0.1 m and $2b$ = 0.2 m. Time-dependent heat flux $\dot{q}_2 = 15t$ is assigned on the two opposing surfaces, which are in width $2b$ while heat flux $\dot{q}_1 = 45t$ is assigned on the two remaining surfaces with a width $2d$ = 0.1 m. The rod is made of a steel with the following thermo-physical properties: λ = 40 W/(m·K) and $c\rho = 4 \cdot 10^6$ J/(m^3·K).

Carry out the calculations using the exact analytical method and the method of lines. Apply the Runge-Kutta method of 4th order and the analytical (matrix) method, in which the heat flux changes are approximated by a step line and piecewise linear function, to integrate the ordinary differential equation system. Depict temporal temperature transients at three selected cross-section points calculated by means of the exact analytical method and compare all four calculation methods on the basis of an example of a temperature transient in one of the three previously selected points. Present the comparison results in a tabular form.

Solution

Due to the symmetry of the problem, only a quarter of the rod's cross-section will be analyzed here (Fig. 23.15).

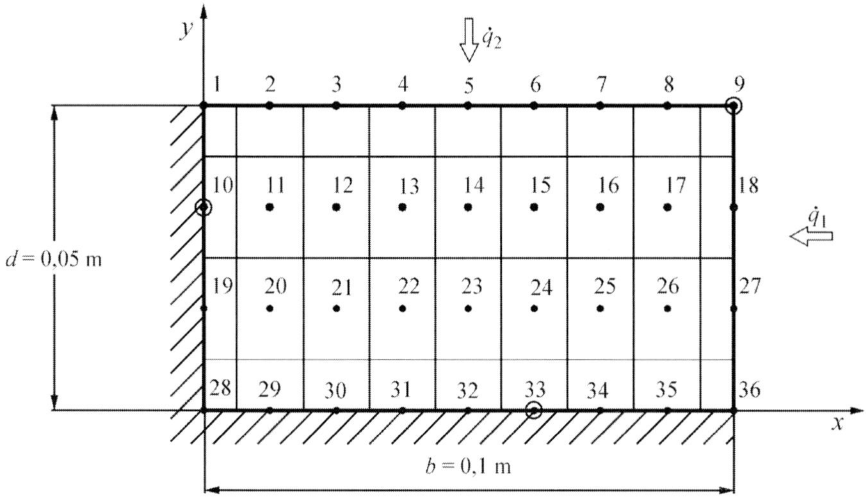

Fig. 23.15. Division of a quarter of a cross-section into control volumes; • – nodes, o – nodes in which temperature transients were determined and presented in Fig. 23.16

Temperature field in the quarter of the rod's cross-section is defined by the heat conduction equation

$$\frac{1}{a}\frac{\partial T}{\partial t}=\frac{\partial^2 T}{\partial x^2}+\frac{\partial^2 T}{\partial y^2} \tag{1}$$

and by boundary conditions

$$\lambda \left.\frac{\partial T}{\partial x}\right|_{x=0}=0\,, \tag{2}$$

$$\lambda \left.\frac{\partial T}{\partial x}\right|_{x=b}=\dot{q}_1\,, \tag{3}$$

$$\lambda \left.\frac{\partial T}{\partial y}\right|_{y=0}=0\,, \tag{4}$$

$$\lambda \frac{\partial T}{\partial y}\bigg|_{y=d} = \dot{q}_2 , \tag{5}$$

where,

$$\dot{q}_1 = p_1 t, \qquad \dot{q}_2 = p_2 t , \tag{6}$$

while p_1 and p_2 are constant heating rates.

If the superposition method is applied and the solution of one-dimensional problem is used for the plate, which is back-surface-insulated and heated on its front face by a heat flux that changes at constant rate [49], then the solution of the problem (1)–(5) can be presented in the form

$$\begin{aligned} T(x,y,z,t) &= T_1(x,t) + T_2(y,t) = \\ &= \frac{1}{12}\frac{p_1 b^3}{\lambda a}\left[6\left(\frac{at}{b^2}+\frac{1}{2}\frac{x^2}{b^2}\right)^2 - \left(\frac{x^2}{b^2}+\frac{1}{2}\right)^2 - \right. \\ &\left. -2\frac{at}{b^2}+\frac{29}{60}-\frac{24}{\pi^4}\sum_{n=1}^{\infty}\frac{(-1)^{n+1}}{n^4}\cos\left(n\pi\frac{x}{b}\right)\exp\left(-n^2\pi^2\frac{at}{b^2}\right)\right] + \\ &+\frac{1}{12}\frac{p_2 d^3}{\lambda a}\left[6\left(\frac{at}{d^2}+\frac{1}{2}\frac{y^2}{d^2}\right)^2 - \left(\frac{y^2}{d^2}+\frac{1}{2}\right)^2 - \right. \\ &\left. -2\frac{at}{d^2}+\frac{29}{60}-\frac{24}{\pi^4}\sum_{n=1}^{\infty}\frac{(-1)^{n+1}}{n^4}\cos\left(n\pi\frac{y}{d}\right)\exp\left(-n^2\pi^2\frac{at}{d^2}\right)\right]. \end{aligned} \tag{7}$$

Heat flux change rates are p_1 = 45 W/(m^2·s) and p_2 = 15 W/(m^2·s). While temperature $T(x,y,z,t)$ was determined, the ten terms in the infinite series were accounted for. Following that, the same problem was solved using the method of lines. Fig. 23.15 shows the division of the analyzed region into control volumes. Energy balance equations for control volumes have the following forms:

$i = 1,$

$$c\rho\frac{\Delta x}{2}\frac{\Delta y}{2}\frac{dT_i}{dt} = \lambda\frac{T_{i+1}-T_i}{\Delta x}\frac{\Delta y}{2} + \lambda\frac{T_{i+N}-T_i}{\Delta y}\frac{\Delta x}{2} + \dot{q}_2\frac{\Delta x}{2};$$

$i = 2,\ \ldots,\ N_x - 1,$

$$c\rho\Delta x\frac{\Delta y}{2}\frac{dT_i}{dt} = \lambda\frac{T_{i-1}-T_i}{\Delta x}\frac{\Delta y}{2} + \lambda\frac{T_{i+1}-T_i}{\Delta x}\frac{\Delta y}{2} + \lambda\frac{T_{i+N}-T_i}{\Delta y}\Delta x + \dot{q}_2\frac{\Delta x}{2};$$

$i = N_x,$

$$c\rho\frac{\Delta x}{2}\frac{\Delta y}{2}\frac{dT_i}{dt} = \lambda\frac{T_{i-1}-T_i}{\Delta x}\frac{\Delta y}{2} + \lambda\frac{T_{i+N}-T_i}{\Delta y}\frac{\Delta x}{2} + \dot{q}_2\frac{\Delta x}{2} + \dot{q}_1\frac{\Delta y}{2};$$

$i = N_x + 1 \quad \text{oraz} \quad i = 2N_x + 1,$

$$c\rho\frac{\Delta x}{2}\Delta y\frac{dT_i}{dt} = \lambda\frac{T_{i+1}-T_i}{\Delta x}\Delta y + \lambda\frac{T_{i-N}-T_i}{\Delta y}\frac{\Delta x}{2} + \lambda\frac{T_{i+N}-T_i}{\Delta y}\frac{\Delta x}{2};$$

$i = N_x + 2,\ \ldots,\ 2N_x - 1 \quad \text{oraz} \quad i = 2N_x + 2,\ \ldots,\ 3N_x - 1,$

$$c\rho\Delta x\Delta y\frac{dT_i}{dt} = \lambda\frac{T_{i-1}-T_i}{\Delta x}\Delta y + \lambda\frac{T_{i+1}-T_i}{\Delta x}\Delta y + \lambda\frac{T_{i-N}-T_i}{\Delta y}\Delta x + \tag{8}$$

$$+\lambda\frac{T_{i+N}-T_i}{\Delta y}\Delta x;$$

$i = 2N_x \quad \text{oraz} \quad i = 3N_x,$

$$c\rho\frac{\Delta x}{2}\Delta y\frac{dT_i}{dt} = \lambda\frac{T_{i-N}-T_i}{\Delta y}\frac{\Delta x}{2} + \lambda\frac{T_{i+N}-T_i}{\Delta y}\frac{\Delta x}{2} + \lambda\frac{T_{i-1}-T_i}{\Delta x}\Delta y + \dot{q}_1\Delta y;$$

$i = 3N_x + 1,$

$$c\rho\frac{\Delta x}{2}\frac{\Delta y}{2}\frac{dT_i}{dt} = \lambda\frac{T_{i-N}-T_i}{\Delta y}\frac{\Delta x}{2} + \lambda\frac{T_{i+1}-T_i}{\Delta x}\frac{\Delta y}{2};$$

$i = 3N_x + 2,\ \ldots,\ 4N_x - 1,$

$$c\rho\Delta x\frac{\Delta y}{2}\frac{dT_i}{dt} = \lambda\frac{T_{i-1}-T_i}{\Delta x}\frac{\Delta y}{2} + \lambda\frac{T_{i+1}-T_i}{\Delta x}\frac{\Delta y}{2} + \lambda\frac{T_{i-N}-T_i}{\Delta y}\Delta x;$$

$i = 4N_x,$

$$c\rho\frac{\Delta x}{2}\frac{\Delta y}{2}\frac{dT_i}{dt} = \lambda\frac{T_{i-N}-T_i}{\Delta y}\frac{\Delta x}{2} + \lambda\frac{T_{i-1}-T_i}{\Delta x}\frac{\Delta y}{2} + \dot{q}_1\frac{\Delta y}{2},$$

where $\Delta x = b/(N-1)$, $\Delta y = d/3$, $N = 9$ is the number of nodes in x axis direction (Fig. 23.15).

System of 36 ordinary differential equations was solved using the Runge-Kutta method of 4th order, by assuming time step $\Delta t = 0.5$ s for the calculation, and with the analytical method using the exponential matrix (Ex. 23.3).

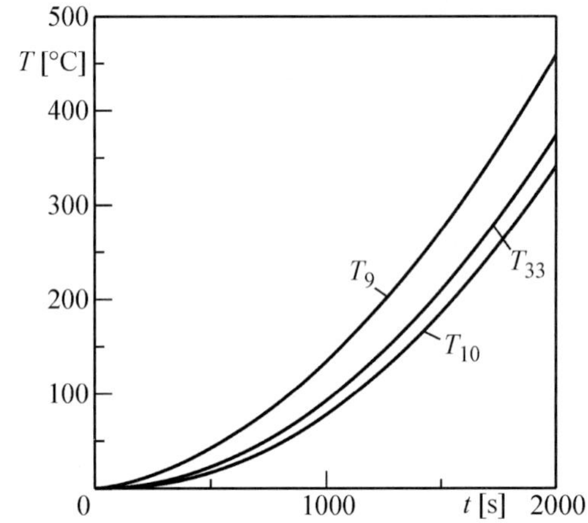

Fig. 23.16. Temperature transients in nodes 9, 10 and 33 determined by means of (7)

Table 23.5. Comparison of temperature transients in node no. 9 in [°C], calculated by means of the exact analytical method and the method of lines; different methods for integrating the ordinary differential equation system were applied

	Method			
		of lines		
t [s]	exact analytical	Runge-Kutta	in matrix form (step line)	in matrix form (piecewise linear)
1.0	–0.18	0.00	0.00	0.00
10.0	0.00	0.08	0.09	0.10
50.0	1.24	1.18	1.20	1.21
100.0	3.57	3.45	3.48	3.51
200.0	10.22	10.04	10.08	10.12
300.0	19.08	18.86	18.91	18.96
500.0	42.73	42.44	42.51	42.58
700.0	73.96	73.61	73.70	73.79
1000.0	134.90	134.45	134.57	134.68
1500.0	273.96	273.35	273.52	273.68
1900.0	418.96	418.23	418.42	418.63

Heat flux changes $\dot{q}_1$ and $\dot{q}_2$ were approximated by a step line and a piecewise linear function. In both cases, the integration step was $\Delta t = 1$ s.

The results of temperature calculations carried out in nodes no. 9, 10 and 33 by means of the exact analytical method (7) are presented in Fig. 23.16.

Results obtained by means of the exact analytical method (7) are compared in Table 23.5 with the results obtained by means of the method of lines, which the three different methods were applied to in order to integrate the ordinary differential equation system. One can see that the accuracy of the finite volume method is very good in all three cases. In comparison to the approximation carried out by means of the step line, the application of the piecewise linear function, in the matrix method, to the approximation of the heat flux changes in time only negligibly improves the accuracy of the obtained results.

Furthermore, one should emphasize that finite volume method is highly accurate in spite of the small number of volumes the analyzed region was divided to.

Literature

1. Albrecht W (1966) Instationäre Wärmespannungen in Hohlzylindern. Konstruktion 18(6): 224-231
2. Arpaci VS, Kao SH, Selamet A (1999) Introduction to Heat Transfer. Prentice Hall, Upper Saddle River
3. Baron B, Marcol A, Pawlikowski S (1999) Numerical methods in Delphi 4 (in Polish). Helion, Gliwice
4. Becker M (1986) Heat Transfer. A Modern Approach. Plenum Press, New York
5. Bejan A (1993) Heat Transfer. Wiley, New York
6. Birkhoff G, Rota GC (1989) Ordinary Differential Equations. Wiley, New York
7. Bronson R (1989) Matrix Operations. Schaum's Outline Series. Mc-Graw-Hill, New York
8. Burden RL, Faires JD (1985) Numerical Analysis. Prindle Weber&Schmidt, Boston
9. Carslaw HS, Jaeger JC (1959) Conduction of Heat in Solids. Ed. 2. Oxford University Press, London
10. Carver MB (1980) Pseudo characteristic method of lines solution of the conservation equations. J. of Computational Physics 35: 57-76
11. Carver MB, Hinds HW (1978) The method of lines and the advective equation. Simulation, August: 59-69
12. Chua LO, Pen-Min L (1981) Computer analysis of electronic systems. Algorithms and computational methods (in Polish). WNT, Warsaw
13. Crank J, Nicolson P (1947) A practical method for numerical integration of solutions of partial differential equations of heat-conduction type. Proc. Cambr. Philos. Soc. 43: 50-67
14. Cutlip MB, Shacham M (1999) Problem Solving in Chemical Engineering with Numerical Methods. Prentice Hall, Upper Saddle River
15. Davidenko K, Ruszczinskij WM (1974) Algorithm for numerical determining transient processes in boilers with supercritical parameters (in Russian). Izvestija Akademii Nauk SSSR, Energetics and Transport 3: 150-161

16. Dormand JR (1996) Numerical Methods for Differential Equations. CRC Press, Boca Raton
17. Duda P, Taler J (2001) Identification of transient temperature and thermal stress distributions using the least squares method. In: European Conference on Computational Mechanics, ECCM
18. Duda P, Taler J (2001) Solution of inverse heat conduction problems using the least squares method. In: Fourth International Congress on Thermal Stresses, Thermal Stressessaka, pp. 523-526
19. Duda P, Taler J (2000) Method for solving inverse problems encountered in monitoring of thermal stresses. Archives of Thermodynamics 21 (3-4): 25-39
20. Elden L (1997) Solving an inverse heat conduction problem by „a method of lines". Transactions of the ASME, J. Heat Transfer 119: 406-412
21. Elizarov DP (1971) Thermal shock in steam lines of thermal power stations (in Russian). Tieploenergetika 18 (2): 78-82
22. Gary J (1976) The method of lines applied to a simple hyperbolic equation. J. of Computational Physics 22: 131-149
23. Gerald CF, Wheatley PO (1994) Applied Numerical Analysis. Addison-Wesley, Reading
24. Goode SW (2000) Differential Equations and Linear Algebra. Prentice Hall, Upper Saddle River
25. Hensel E, Hills RG (1986) An initial value approach to the inverse heat conduction problem. Transactions of the ASME, J. Heat Transfer 108: 248-256
26. Heydweiller JC, Sincovec RF (1976) A stable difference scheme for the solution of hyperbolic equations using the method of lines. J. of Computational Physics 22: 377-388
27. IMSL. FORTRAN subroutines for mathematical applications. Vol. 2 (1994) Visual Numerics Inc.
28. Madsen NK, Sincovec RF (1974) The numerical method of lines for the solution of nonlinear partial differential equations. In: Oden JT et al. (eds) Computational Methods in Nonlinear Mechanics. Texas Inst. for Computational Mechanics, Texas, pp. 371-380
29. Matviejev NM (1986) Integration methods for ordinary differential equations (in Polish). PWN, Warsaw
30. Mills AF (1999) Basic Heat & Mass Transfer. Prentice Hall, Upper Saddle River
31. Ortega JM, Poole WG (1981) An Introduction to Numerical Methods for Differential Equations. Pitman Publishing Inc., London
32. Parkus H (1959) Instationäre Wärmespannungen. Springer, Berlin
33. Press WH, Teukolsky SA, Vetterling WT, Flannery BP (1996) Numerical Recipes in Fortran 77. Ed. 2. Cambridge University Press, Cambridge
34. Recknagel H, Sprenger E, Hönmann W, Schramek E (1994) Heating & Air conditioning. EWFE, Gdańsk
35. Richter F (1983) Physikalische Eigenschaften von Stählen und ihre Temperaturabhängigkeit. Polynome und graphische Darstellungen. Sonderdruck Stahleisen-Sonderberichte 10. Verlag Stahleisen, Düsseldorf
36. Schiesser W (1976) DSS/2 introductory programming manual. Lehigh University, Bethlehem

37. Schiesser WE (1991) The Numerical Method of Lines. Academic Press, San Diego
38. Sincovec R, Madsen N (1975) Software for nonlinear partial differential equations. ACM Transact. Math. Software 1: 232-260
39. Sinha NK (1991) Linear Systems. Wiley, New York
40. Starzyk J (1982) FORSIM VI. A program for solving ordinary and partial differential equation systems (in Polish). Cyfronet, Kraków
41. Strong BR, Slagis GC (1981) TRANS2A, An unconditionally stable code for transient stress analysis in piping. Transactions of the ASME, J. Pressure Vessel Technology 103 (1): 50-58
42. Taler J, Przybyliński P (1982) Heat transfer through circular fins with variable conductivity and non-uniform heat transfer coefficient. Chemical and Process Engineering 3 (3-4): 659-676
43. Taler J (1998) Solving non-linear inverse heat conduction problems (in Polish). In: X Mass and Heat Transfer Symposium. Part 2. Scientific Papers of the Institute of Heat and Fluid Mechanics of Wroclaw University of Technology 53. Series: Conferences 9, pp. 831-841
44. Taler J, Taler D (1999) Experimental verification of space-marching methods used for solving inverse heat conduction problems (in Polish). In: XVII Congress of Thermodynamics Engineers, Conference materials Vol. 4, pp. 1443-1456
45. Taler J, Duda P (2001) Solution of non-linear inverse heat conduction problems. The method of lines. Heat and Mass Transfer 37: 147-155
46. Taler J, Duda P (2000) General method for solving inverse heat conduction problems that occur during thermal stresses monitoring in power plants (in Polish). Opole University of Technology, Series: Electrical Engineering 49 (255), pp. 421-437
47. Taler J, Duda P (2000) Numerical solution of inverse problems for monitoring non-stationary thermo-strength state of a boiler. Scientific Papers. Institute of Heat and Fluid Mechanics of Wroclaw University of Technology, Conferences 10, pp. 351-360
48. Taler J (1995) Theory and practice of heat transfer (in Polish). Ossolineum, Wrocław
49. Tautz H (1971) Wärmeleitung + Temperaturausgleich. Verlag Chemie, Weinheim
50. Timoshenko S (1951) Theory of Elasticity. McGraw-Hill, New York
51. Zill DG (1986) Differential Equations with Boundary-Value Problems. Prindle, Weber & Schmidt, Boston
52. Pekhovitch AI, Zidkih WM (1976) Calculation of transient heat conduction In solids (in Russian). 2nd edition, Energija, Leningrad

24 Solving Inverse Heat Conduction Problems by Means of Numerical Methods

This chapter presents simple algorithms for solving inverse heat conduction problems, the so called *space-marching methods.*

If one assumes that temperature is measured at the inner point $x = E$, in which also the heat flux $\dot{q}_E(t)$ is known, one is able to determine temperature $T_s(t)$ and heat flux $\dot{q}_s(t)$ on the surface $x = 0$.

It is more difficult to solve inverse problems than simple problems (direct). Usually measurement data $f(t_i)$, $i = 1, 2, ...$ is burdened with random measurement errors, which are strengthened during step-marching from a measurement point to body surface. The method for solving inverse transient heat conduction problems cannot be too sensitive to random measurement errors, since the determination of temperature or heat flux on the body surface can prove to be unstable.

In contrast to direct problems, the implicit finite difference method is less stable than the explicit method, since explicit methods are better at allowing for the phenomenon of delay in temperature changes at the measurement point $x = E$ than they are in the case of changes on the surface $x = 0$.

Exercise 24.1 Numerical-Analytical Method for Solving Inverse Problems

Describe a semi-numerical method for solving one-dimensional transient heat conduction problems, in which only derivatives are discreticised by means of the control volume method. Measure temperature of the insulated back plate surface f_i, $i = 1, 2, ...$ The unknowns to be calculated are temperature and heat flux on the plate front face and temperature distribution across the plate thickness.

Solution

Assume that thermo-physical properties of the plate material are constant. Two conditions at the point $x = E$ are known: temperature $f(t)$ and heat flux $\dot{q}_E(t)$. On the plate front face (Fig. 24.1), the boundary condition is unknown. A more comprehensive engineering problem will be solved in this exercise. Temperature measurement point $0 \le E \le L$ can lie at an arbitrary point across the plate thickness; in special case, on the back surface $E = L$. When the back surface is thermally insulated, then $\dot{q}_E = 0$.

In general, the boundary condition on the back plate surface is known. The plate, therefore, can be divided into two regions: inverse and direct.

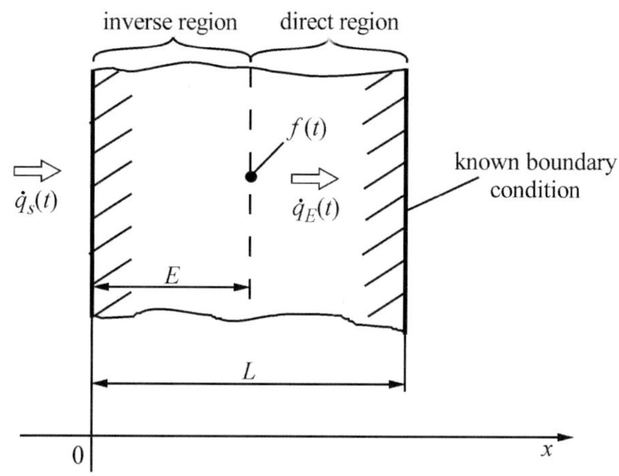

Fig. 24.1. Inverse heat conduction problem for a plate

Heat conduction equation has the form

$$\frac{1}{a}\frac{\partial T}{\partial t} = \frac{\partial^2 T}{\partial x^2} . \qquad (1)$$

Furthermore, two conditions are known at the temperature measurement point

$$T(E,t) = f(t), \qquad (2)$$

$$-\lambda \left.\frac{\partial T}{\partial x}\right|_{x=E} = \dot{q}_E(t) . \qquad (3)$$

Both, temperature $f(t)$ and heat flu $\dot{q}_E(t)$ are known at the discrete time points t_j, $j = 1, 2, ..., N$, where N is the number of temperature measurement points. In order to solve the inverse problem, one should determine the solution of the direct problem first in the region $0 \le x \le L$ with a boundary condition of the first kind (2) and with the assigned boundary condition on the back surface $x = L$. Once the direct (simple) problem is solved, the heat flux $\dot{q}_E(t)$ is calculated

$$\dot{q}_E(t) = -\lambda \left. \frac{\partial T}{\partial x} \right|_{x=E} . \tag{4}$$

If the thermocouple lies on an insulated back surface ($E = L$), then it is not necessary to solve the simple problem. Heat flux $\dot{q}_E(t)$ in such a case equals zero.

Inverse problems (1)–(3) will be solved by the step-marching method. Spatial derivatives will be approximated with difference quotients by means of the control volume method. Partial parabolic (1) will be approximated in this way by the ordinary differential equation system. Such method of solving partial equations is called the *method of lines* [4].

In the space-marching method, calculations begin from point $x = E$ ($i=M$), and then temperature at point $i = M - 1$ is determined. Next, temperature in node $i = M - 2$ is calculated from the heat balance equation for node $i = M - 1$.

Temperature calculations are carried out in the subsequent points, space-marching in the direction of surface $x = 0$. From the energy balance equation for node $i = 1$, the expression for plate surface heat flux $\dot{q}_s(t)$ is obtained.

In order to derive heat balance equations for individual nodes, the inverse region should be divided into control volumes. Nodes are also located on the boundaries $x = 0$ and $x = E$, since heat flux can be accurately determined at such points. Control volume boundaries divide the distance between adjacent nodes in half. In order to obtain the appropriate calculation accuracy, the node number should be sufficiently large.

In direct problems (simple), the node number M within the limits $10 \le M \le 20$ is adequate. In inverse problems, the node number can be much smaller, since the input data $f(t)$ and $\dot{q}_E(t)$ is attenuated and increasing node number M does not improve calculation accuracy. Usually a rather small node number $3 \le M \le 5$ ensures high calculation accuracy. In order to evaluate the effect the node number M has on the accuracy of the obtained results, various divisions of the inverse region, presented in Fig. 24.2, will be evaluated below.

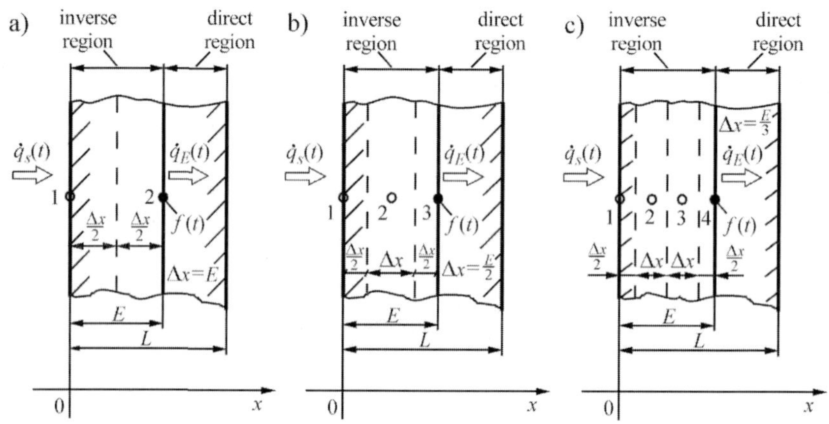

Fig. 24.2. Division of the inverse region into two ($M = 2$), three ($M = 3$) and four ($M = 4$) control volumes; $x = E$ is a temperature measurement point

a. *Division of an inverse region into two control volumes (Fig. 24.2a)*

Heat balance equation for nodes *2* and *1* has the form:

- node *2*

$$c\rho \frac{E}{2}\frac{df}{dt} = \frac{\lambda}{E}\left(T_1 - f\right) - \dot{q}_E, \tag{5}$$

- node *1*

$$c\rho \frac{E}{2}\frac{dT_1}{dt} = \dot{q}_s + \frac{\lambda}{E}\left(f - T_1\right). \tag{6}$$

According to the earlier described method, temperature T_1 is determined from (5) first

$$T_1 = f + \frac{\dot{q}_E E}{\lambda} + \frac{1}{2}\frac{E^2}{a}\frac{df}{dt}. \tag{7}$$

Next, heat flux $\dot{q}_s$ is calculated from (6), once (7) is substituted into (6)

$$\dot{q}_s = \dot{q}_E + \frac{\lambda E}{a}\left(\frac{df}{dt} + \frac{E^2}{4a}\frac{d^2 f}{dt^2}\right) + \frac{1}{2}\frac{E^2}{a}\frac{d\dot{q}_E}{dt}. \tag{8}$$

Equations (7) and (8) allow to calculate temperature and heat flux on the plate front face.

b. *Division of an inverse region into three control volumes (Fig. 24.2b)*

Increasing the number of control volumes results in the appearance of a higher order time derivatives in temperature and heat flux solutions of functions $f(t)$ and $\dot{q}_E(t)$.

Heat balance equations for individual control volumes have the following form (Fig. 24.2b):

- node *3*

$$c\rho\frac{\Delta x}{2}\frac{df}{dt}=\frac{\lambda}{\Delta x}\left(T_2-f\right)-\dot{q}_E\,, \tag{9}$$

- node *2*

$$c\rho\Delta x\frac{dT_2}{dt}=\frac{\lambda}{\Delta x}\left(f-T_2\right)+\frac{\lambda}{\Delta x}\left(T_1-T_2\right), \tag{10}$$

- node *1*

$$c\rho\Delta x\frac{dT_1}{dt}=\frac{\lambda}{\Delta x}\left(T_2-T_1\right)+\dot{q}_s\,, \tag{11}$$

where $\Delta x = E/2$.

First (9) is solved with respect to T_2

$$T_2=f+\frac{\left(\Delta x\right)^2}{2a}\frac{df}{dt}+\frac{\dot{q}_E\Delta x}{\lambda}. \tag{12}$$

Once (12) is substituted into (10) and solved with respect to T_1, one has

$$T_1=f+2\frac{\left(\Delta x\right)^2}{a}\frac{df}{dt}+\frac{1}{2}\frac{\left(\Delta x\right)^4}{a^2}\frac{d^2f}{dt^2}+2\frac{\dot{q}_E\Delta x}{\lambda}+\frac{\left(\Delta x\right)^3}{\lambda a}\frac{d\dot{q}_E}{dt}. \tag{13}$$

Once we account that $\Delta x = E/2$, (13) can be written in the form

$$T_1=f+\frac{1}{2}\frac{E^2}{a}\frac{df}{dt}+\frac{1}{32}\frac{E^4}{a^2}\frac{d^2f}{dt^2}+\frac{\dot{q}_E E}{\lambda}+\frac{1}{8}\frac{E^3}{\lambda a}\frac{d\dot{q}_E}{dt}. \tag{14}$$

Heat flux $\dot{q}_s$ is determined from (11) once (12) and (14) are substituted into this equation

$$\dot{q}_s = \dot{q}_E + \lambda\left(\frac{E}{a}\frac{df}{dt} + \frac{3}{16}\frac{E^3}{a^2}\frac{d^2 f}{dt^2} + \frac{1}{128}\frac{E^5}{a^3}\frac{d^3 f}{dt^3}\right) + \frac{1}{2}\frac{E^2}{a}\frac{d\dot{q}_E}{dt} + \frac{1}{32}\frac{E^4}{a^2}\frac{d^2\dot{q}_E}{dt^2}. \tag{15}$$

Solutions (7), (14) and (8) and (15) depict temperature and heat flux dependence at the plate surface on the derivatives with respect to time from functions that represent a measured temperature transient $f(t)$ and heat flux $\dot{q}_E(t)$ at point $x = E$. The order of the highest derivative depends on the number of control volumes. The order of the highest derivative from function $f(t)$ equals the number of control volumes in the expression for heat flux $\dot{q}_s(t)$.

c. *Division of an inverse region into four control volumes (Fig. 24.2c)*

Heat balance equations for control volumes no. *4*, *3*, *2* and *1* have, respectively, the following form

$$c\rho\frac{\Delta x}{2}\frac{df}{dt} = \frac{\lambda}{\Delta x}(T_3 - f) - \dot{q}_E, \tag{16}$$

$$c\rho\Delta x\frac{dT_3}{dt} = \frac{\lambda}{\Delta x}(T_2 - T_3) + \frac{\lambda}{\Delta x}(f - T_3), \tag{17}$$

$$c\rho\Delta x\frac{dT_2}{dt} = \frac{\lambda}{\Delta x}(T_1 - T_2) + \frac{\lambda}{\Delta x}(T_3 - T_2), \tag{18}$$

$$c\rho\frac{\Delta x}{2}\frac{dT_1}{dt} = \frac{\lambda}{\Delta x}(T_2 - T_1) + \dot{q}_s, \tag{19}$$

where $\Delta x = E/3$.

Once (16)–(19) are solved successively, the following expressions for the plate surface temperature and heat flux are obtained:

$$T_1 = f + \frac{1}{2}\frac{E^2}{a}\frac{df}{dt} + \frac{1}{27}\frac{E^4}{a^2}\frac{d^2 f}{dt^2} + \frac{1}{1458}\frac{E^6}{a^3}\frac{d^3 f}{dt^3} + \frac{\dot{q}_E E}{\lambda} + \frac{4}{27}\frac{E^3}{\lambda a}\frac{d\dot{q}_E}{dt} + \frac{1}{243}\frac{E^5}{\lambda a^2}\frac{d^2\dot{q}_E}{dt^2}, \tag{20}$$

$$\dot{q}_s = \dot{q}_E + \lambda\left(\frac{E}{a}\frac{df}{dt} + \frac{19}{108}\frac{E^3}{a^2}\frac{d^2 f}{dt^2} + \frac{2}{243}\frac{E^5}{a^3}\frac{d^3 f}{dt^3} + \frac{1}{8748}\frac{E^7}{a^4}\frac{d^4 f}{dt^4}\right) + \\ + \frac{1}{2}\frac{E^2}{a}\frac{d\dot{q}_E}{dt} + \frac{1}{27}\frac{E^4}{a^2}\frac{d^2\dot{q}_E}{dt^2} + \frac{1}{1458}\frac{E^6}{a^3}\frac{d^3\dot{q}_E}{dt^3}. \tag{21}$$

Formulas presented in this exercise for T_1 and $\dot{q}_s$ resemble in form the exact solution presented in papers [2, 3] and in Ex. 23.1. Once only the first three terms are accounted for in an infinite series, the exact solution has the following form:

$$T_1 = f + \frac{1}{2}\frac{E^2}{a}\frac{df}{dt} + \frac{1}{24}\frac{E^4}{a^2}\frac{d^2 f}{dt^2} + \frac{1}{720}\frac{E^6}{a^3}\frac{d^3 f}{dt^3} + \frac{\dot{q}_E E}{\lambda} + \\ + \frac{1}{6}\frac{E^3}{\lambda a}\frac{d\dot{q}_E}{dt} + \frac{1}{120}\frac{E^5}{\lambda a^2}\frac{d^2\dot{q}_E}{dt^2} + \frac{1}{5040}\frac{E^7}{\lambda a^3}\frac{d^3\dot{q}_E}{dt^3}, \tag{22}$$

$$\dot{q}_s = \dot{q}_E + \lambda\frac{E}{a}\left(\frac{df}{dt} + \frac{1}{6}\frac{E^2}{a}\frac{d^2 f}{dt^2} + \frac{1}{120}\frac{E^4}{a^2}\frac{d^3 f}{dt^3}\right) + \\ + \frac{1}{2}\frac{E^2}{a}\frac{d\dot{q}_E}{dt} + \frac{1}{24}\frac{E^4}{a^2}\frac{d^2\dot{q}_E}{dt^2} + \frac{1}{720}\frac{E^6}{a^3}\frac{d^3\dot{q}_E}{dt^3}. \tag{23}$$

One can see that exact and approximate solutions have a similar form. From the comparison of (20) with (21), one can see that the approximate solution (20) does not contain the term $d^3\dot{q}_E/dt^3$. Approximate solution (21), which allows to calculate heat flux $\dot{q}_s$ on the plate surface, contains term d^4f/dt^4, which does not appear in the exact solution. Other space-marching methods, used for solving one-dimensional inverse problems, are presented in references [5, 9, 10], while for two-dimensional problems in [8].

Exercise 24.2 Step-Marching Method in Time Used for Solving Non-Linear Transient Inverse Heat Conduction Problems

Describe the method for solving inverse heat conduction problems in a flat, cylindrical and spherical wall, which accounts for the dependence of a material's thermophysical properties on temperature [9, 10]. On the basis of the wall temperature measurement at an inner point r_E determine temperature T_1 and heat flux $\dot{q}_s$ on an inner wall surface while assuming that the boundary condition on an outer surface is known.

Solution

From the two methods presented in papers [9, 10], the Method 2 will be discussed here, in which the derivative after time is discretisized. Temperature is measured at point $r = r_E = r_N$ (Fig. 24.3).

Temperature distribution in the inverse region is defined by the heat conduction equation

$$c(T)\rho(T)\frac{\partial T}{\partial t} = -\frac{1}{r^m}\frac{\partial}{\partial r}\left(r^m \dot{q}\right), \tag{1}$$

where heat flux $\dot{q}$ is defined by Fourier Law

$$\dot{q} = -\lambda(T)\frac{\partial T}{\partial r}. \tag{2}$$

Exponent m depends on the shape of a wall and is

- $m = 0$ for a flat wall,
- $m = 1$ for a cylindrical wall,
- $m = 2$ for a spherical wall.

Thermophysical properties are functions of temperature and can be defined by different dependencies, for example:

$$\begin{aligned} c(T) &= c_0\left(1 + a_1 T + a_1 T^2 + a_3 T^3\right), \\ \rho(T) &= \rho_0\left(1 + b_1 T + b_2 T^2 + b_3 T^3\right), \\ \lambda(T) &= \lambda_0\left(1 + c_1 T + c_2 T^2 + c_3 T^3\right), \end{aligned} \tag{3}$$

where a_1, a_2, a_3, b_1, b_2, b_3, c_1, c_2 and c_3 are constants.

Quantities c_0, ρ_0 and λ_0 are, respectively, specific heat, density and thermal conductivity at temperature $T = 0°C$.

In order to simplify the notation, dimensionless variables are introduced

$$\begin{aligned} &C = \frac{c}{c_0}, \quad \Omega = \frac{\rho}{\rho_0}, \quad \Lambda = \frac{\lambda}{\lambda_0}, \quad X = \frac{r_E - r}{r_E - r_W} = \frac{r_E - r}{E}, \\ &Fo = \frac{a_0 t}{E^2}, \quad \theta = \frac{T\lambda_0}{\dot{q}_0 E}, \quad H = \frac{\dot{q}}{\dot{q}_0}, \end{aligned} \tag{4}$$

where $a_0 = \lambda_0/(c_0\rho_0)$ is thermal diffusivity at temperature $T = 0°C$.

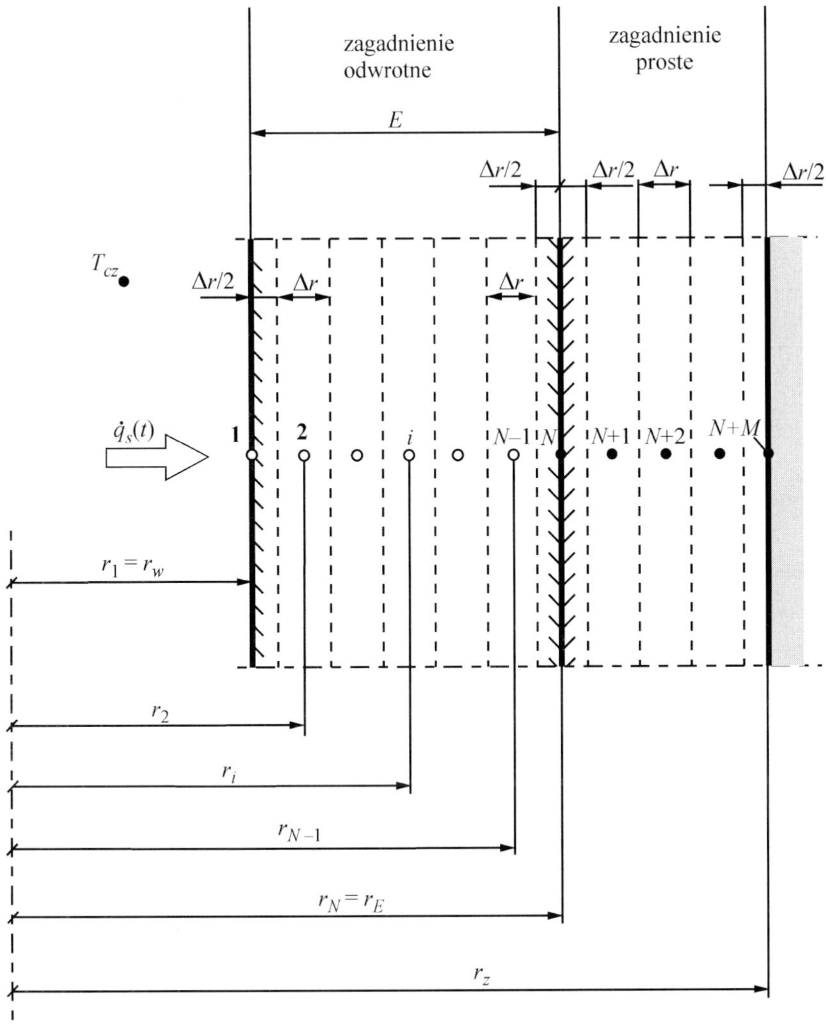

Fig. 24.3. Division of an inverse region into N control volumes; r_E – temperature measurement point, $r_N \le r \le r_{N+M}$ – simple solution region

Equation systems (1) and (2) can be written in the form

$$\frac{\partial H}{\partial X} = C_1 H + C(\Theta)\Omega(\Theta)\frac{\partial \Theta}{\partial Fo}, \tag{5}$$

$$\frac{\partial \Theta}{\partial X} = \frac{H}{\Lambda(\Theta)}, \tag{6}$$

where

$$C_1 = \frac{mE^*}{m - XE^*},$$

$$E^* = \frac{E}{r_E}.$$

Initial conditions have the form

$$\left[\Lambda(\Theta) \frac{\partial \Theta}{\partial X} \right]\bigg|_{X=0} = H_E(Fo), \tag{7}$$

$$\Theta\big|_{X=0} = F(Fo), \tag{8}$$

where

$$H_E(Fo) = \frac{\dot{q}_E(Fo)}{\dot{q}_0}, \qquad F = \frac{f(Fo)\lambda_0}{\dot{q}_0 E}. \tag{9}$$

In the initial condition (7), heat flux $\dot{q}_E$ appears at temperature measurement point $r = r_E$; one can determine temperature once the direct problem is solved in the region $r_E \le r \le r_z$ (Fig. 24.3). Once we determine temperature distribution in the direct solution region $r_E \le r \le r_z$, we can calculate heat flux by means of a formula with the accuracy of a second order. The first dependence, which can be applied for the calculation of $\dot{q}_E$ has the form

$$\dot{q}_E(Fo) = -\left[\lambda(T) \frac{\partial T}{\partial r} \right]\bigg|_{r=r_E} \approx -\lambda(f) \frac{4T_{N+1} - T_{N+2} - 3T_N}{2\Delta r}. \tag{10}$$

The second formula for $\dot{q}_E$ can be obtained from the energy balance equation for a control volume that lies in a direct region, on the boundary on which node N lies (Fig. 24.3)

$$c(T_N)\rho(T_N) \frac{\left(r_E + \frac{\Delta r}{2} \right)^{m+1} - r_E^{m+1}}{m+1} \frac{dT_N}{dt} = \\ = r_E^m \dot{q}_E + \left(r_E + \frac{\Delta r}{2} \right)^m \frac{\lambda(f) + \lambda(T_{N+1})}{2} \frac{T_{N+1} - f}{\Delta r}. \tag{11}$$

From (11), one obtains an expression for $\dot{q}_E$

$$\dot{q}_E = c(T_N)\rho(T_N)\frac{\left(r_E + \frac{\Delta r}{2}\right)^{m+1} - r_E^{m+1}}{(m+1)r_E^m}\frac{dT_N}{dt} - \tag{12}$$
$$-\frac{\left(r_E + \frac{\Delta r}{2}\right)^m}{r_E^m}\frac{\lambda(f)+\lambda(T_{N+1})}{2}\frac{T_{N+1}-f}{\Delta r}.$$

In order to solve the inverse problem, the derivative with respect to time $\partial\Theta/\partial Fo$, which occurs in (5) is approximated by means of the difference quotient P_j

$$\left.\frac{\partial\theta}{\partial Fo}\right| Fo_j \approx P_j\,. \tag{13}$$

Once we denote $H_j = H(X,Fo_j)$ and $\Theta_j = \Theta(X,Fo_j)$, the (5) and (6) together with conditions (7) and (8) are reduced to an initial problem for the ordinary differential equation system

$$\frac{dH_j}{dX} = C_1 H_j + C(\theta_j)\Omega(\theta_j)P_j\,, \tag{14}$$

$$\frac{d\theta_j}{dX} = \frac{H_j}{\Lambda(\Theta_j)}, \quad j = 0,\ \ldots,\ NT\,, \tag{15}$$

where $(NT + 1)$ is the total number of time points, in which the temperature distribution was determined. Initial conditions have the form

$$H_j\big|_{X=0} = H_E(Fo_j), \tag{16}$$

$$\theta_j\big|_{X=0} = F(Fo_j), \quad j = 0,\ \ldots,\ NT\,, \tag{17}$$

where

$$Fo_j = j(\Delta Fo) = j\frac{a_0(\Delta t)}{E^2}, \quad (NT+1) \text{ is the number of time steps.} \tag{18}$$

One can see, therefore, that the discretizations of the time derivative in the heat conduction equation leads, in the case of the inverse problem, to the initial problem for the ordinary differential equation system with respect to a coordinate X. Equations (14) and (15) together with the initial conditions (16) and (17) can be solved by means of the Euler method, Runge-Kutta method or by means of many other widely available methods. Additional attribute is a good knowledge of calculation results at different integration steps ΔX. Results obtained for the integration interval $0 \le X \le 1$ when $0.01 \le \Delta X \le 0.25$ do not differ much. Furthermore, programs developed for the purpose of integrating equation systems by means of the Runge-Kutta method are available in number of commercial software packages.

The accuracy of the obtained results depends, to a large degree, on the method used to approximate time derivative P_j –(13).

Stable solution algorithm is obtained once the time derivative is approximated by the central difference quotient

$$P_j = \frac{T\left(X, Fo_{j+1}\right) - T\left(X, Fo_{j-1}\right)}{2\left(\Delta Fo\right)}. \tag{19}$$

The results from paper [9] show that the optimum time step ΔFo lies in the interval $0.03 \le \Delta Fo \le 0.05$, although for larger ΔFo the calculation error is insignificant. When $\Delta Fo \le 0.07$, then mean square temperature and heat flux calculation error is not much greater than it is with the optimum time step. For the first and last time step respectively, the approximation with a difference quotient is carried out backwards and forwards due to the lack of measurement data

$$P_0 = \frac{T\left(X, \Delta Fo\right) - T\left(X, 0\right)}{\Delta Fo}, \tag{20}$$

$$P_{NT} = \frac{T\left(X, Fo_{NT}\right) - T\left(X, Fo_{NT-1}\right)}{\Delta Fo}. \tag{21}$$

Difficulties in making the approximation of the time derivative in the first and last time step by means of quotient (19) can be avoided by adding one fictional measurement point at the beginning and end of the analyzed interval.

Time derivative P_j can also be approximated by means of digital filters presented in paper [7]. When calculating P_j using local polynomial of the second degree, which approximates seven consecutive time points $T(X,t_i)$, one has

$$P_1 = \frac{-39T_1 - 6T_2 + 15T_3 + 24T_4 + 21T_5 + 6T_6 - 21T_7}{84(\Delta Fo)},$$

$$P_2 = \frac{-29T_1 - 6T_2 + 9T_3 + 16T_4 + 15T_5 + 6T_6 - 11T_7}{84(\Delta Fo)},$$

$$P_3 = \frac{-19T_1 - 6T_2 + 3T_3 + 8T_4 + 9T_5 + 6T_6 - T_7}{84(\Delta Fo)},$$

$$P_j = \frac{-3T_{j-3} - 2T_{j-2} - T_{j-1} + T_{j+1} + 2T_{j+2} + 3T_{j+3}}{28(\Delta Fo)}, \quad j = 4 ,..., NT - 3,$$

$$P_{NT-2} = \frac{1}{84(\Delta Fo)}\big(T_{NT-6} - 6T_{NT-5} - 9T_{NT-4} - 8T_{NT-3} - \tag{22}$$

$$- 3T_{NT-2} + 6T_{NT-1} + 19T_{NT}\big),$$

$$P_{NT-1} = \frac{1}{84(\Delta Fo)}\big(11T_{NT-6} - 6T_{NT-5} - 15T_{NT-4} - 16T_{NT-3} -$$

$$- 9T_{NT-2} + 6T_{NT-1} + 29T_{NT}\big),$$

$$P_{NT} = \frac{1}{84(\Delta Fo)}\big(21T_{NT-6} - 6T_{NT-5} - 21T_{NT-4} - 24T_{NT-3} -$$

$$- 15T_{NT-2} + 6T_{NT-1} + 39T_{NT}\big),$$

where

$$T_j = T(X, Fo_j).$$

Due to the application of filter (22), the accuracy of the method for small time steps increases, since the method becomes less sensitive to random temperature measurement errors.

To approximate the first derivative $P_j = \left.\dfrac{\partial \Theta}{\partial Fo}\right|_{Fo_j}$ one can also use filters based on 9 or 11 points $T(X,Fo_j)$.

Exercise 24.3 Weber Method Step-Marching Methods in Space

Describe step-marching methods, which are based on the finite difference method. Derive computational formulas that are applied to Weber method [11]. Also derive formulas by approximating time and space derivative in the heat conduction equation with the following finite differencies:

$$\left(\frac{\partial T}{\partial t}\right)_i^n \approx \frac{-3T_i^n + 4T_i^{n+1} - T_i^{n+2}}{2(\Delta t)}, \tag{1}$$

$$\left(\frac{\partial^2 T}{\partial x^2}\right)_i^n \approx \frac{T_{i-1}^n - 2T_i^n + T_{i+1}^n}{(\Delta x)^2}. \tag{2}$$

Assume that temperature $f(t)$ is measured on the back plate surface with the assigned heat flux $\dot{q}_E(t)$.

Solution

First *Weber method* will be discussed [11]. The starting point of the analysis is the hyperbolic heat conduction equation

$$\tau\frac{\partial^2 T}{\partial t^2} + \frac{\partial T}{\partial t} = a\frac{\partial^2 T}{\partial x^2}, \tag{3}$$

where τ is the relaxation time.

This equation was selected due to a greater numerical calculations stability, when $\tau > 0$. Once derivative after time and space is approximated by central difference quotients, (3) has the form

$$\frac{\tau}{(\Delta t)^2}\left(T_i^{n-1} - T_i^n + T_i^{n+1}\right) + \frac{1}{2(\Delta t)}\left(T_i^{n+1} - T_i^{n-1}\right) = \\ = \frac{a}{(\Delta x)^2}\left(T_{i+1}^n - 2T_i^n + T_{i-1}^n\right). \tag{4}$$

Next, temperature in node $(i-1)$ is determined from (4)

a)

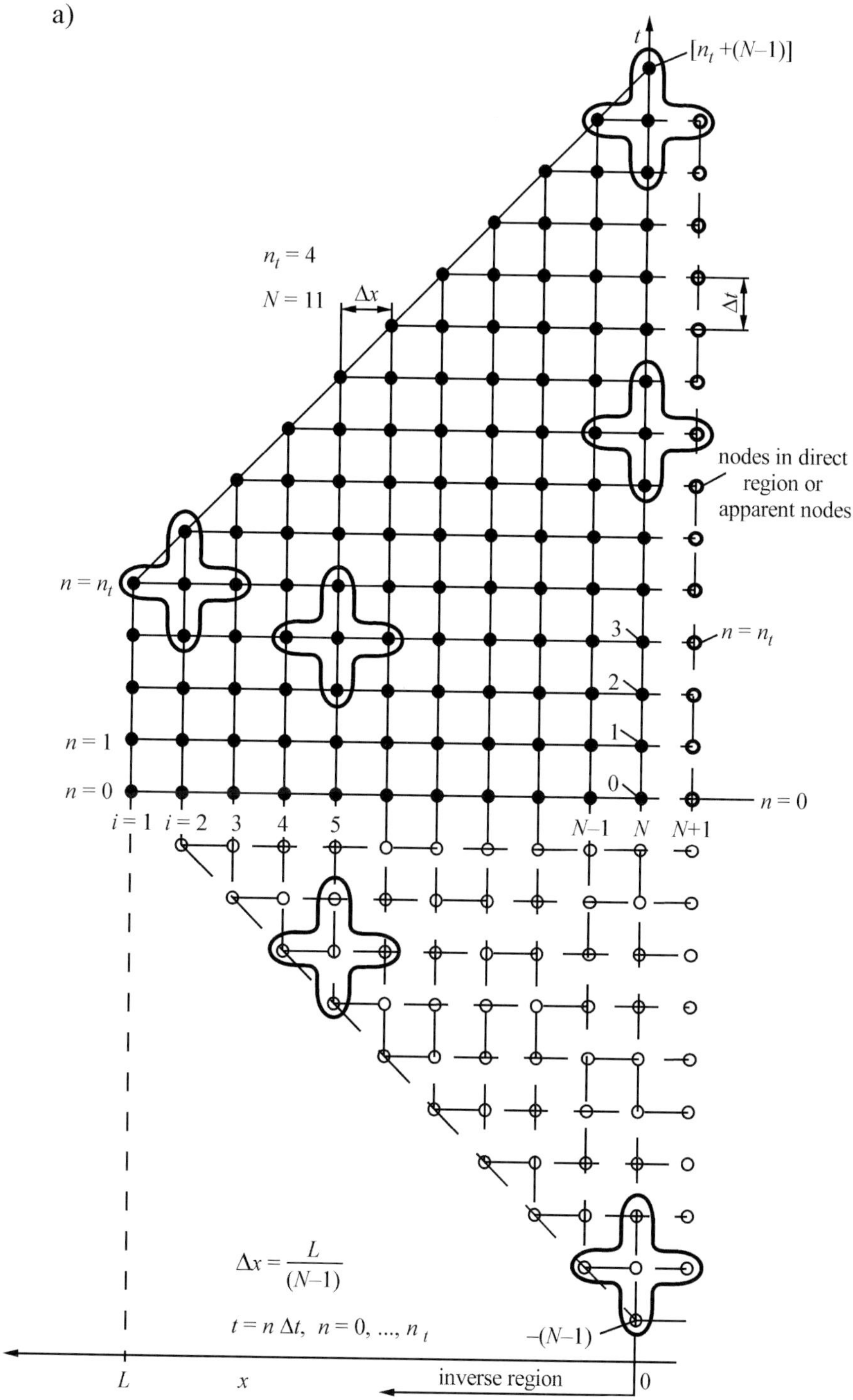

t
$[n_t+(N-1)]$
$n_t = 4$
$N = 11$
Δx
Δt
nodes in direct region or apparent nodes
$n = n_t$
$n = n_t$
3
2
1
0
$n = 1$
$n = 0$
$n = 0$
$i = 1$ $i = 2$ 3 4 5 $N-1$ N $N+1$
$\Delta x = \frac{L}{(N-1)}$
$t = n\,\Delta t,\ n = 0, ..., n_t$
$-(N-1)$
L
x
inverse region
0

long, then the 3rd or 2nd degree polynomial, which approximates back surface temperature changes, is not able to accurately approximate real measured temperature changes.

Averaging of temperature or heat flux changes on the front face is too strong and so the method becomes less accurate, i.e. δT increases.

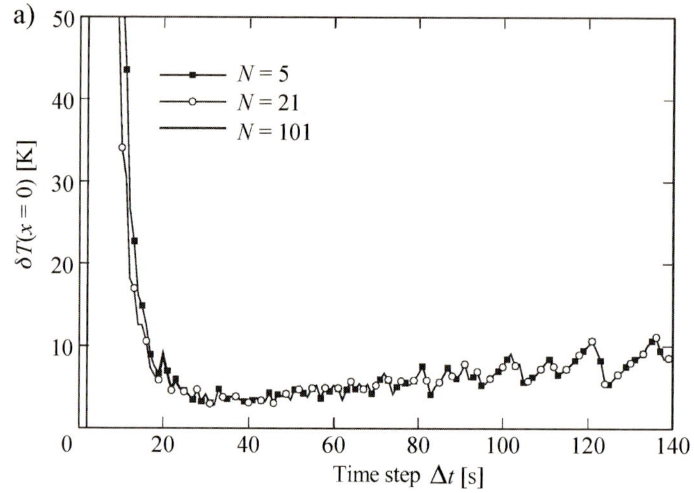

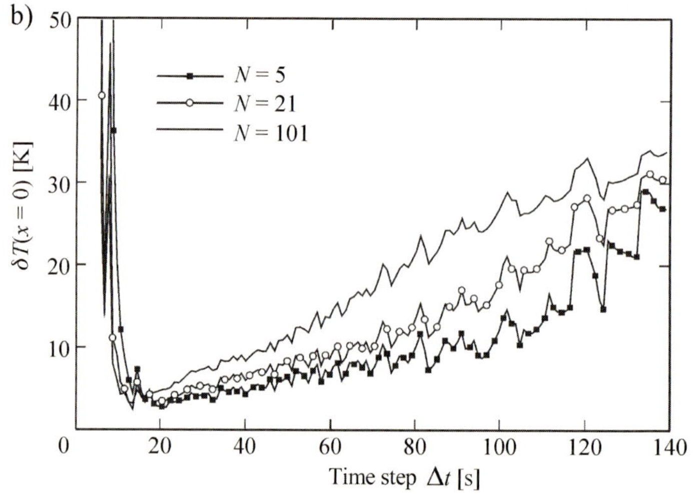

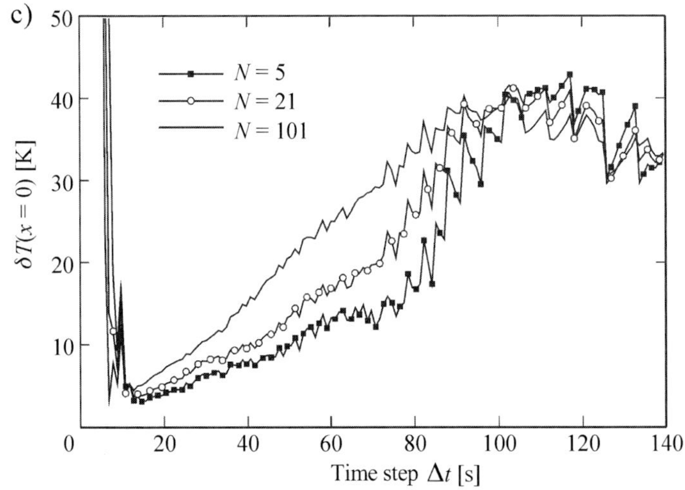

Fig. 24.7. Mean square temperature determination error of the front faces δT in time step function Δt for 4, 20 and 100 spatial steps Δx, for measurement data disturbed by random errors from interval ±0.3 K: a) time derivative approximation by means of the central difference quotient, b) time derivative approximation by means of the digital seven-point filter based on the polynomial of 2nd degree, c) time derivative approximation by means of the nine-point filter based on the polynomial of 3rd degree

With a larger value of Δt_{total}, the ability of the method to reconstruct temperature changes on the front face diminishes. From Fig.24.7 it follows that the optimum value of time step Δt is as follows:

- $\Delta t_{opt} \approx$ 27–40 s (ΔFo = 0.027–0.04), for the approximation of the time derivative by means of the central difference quotient,
- $\Delta t_{opt} \approx$ 18–20 s (ΔFo = 0.018–0.02), for the approximation of the time derivative by means of the seven-point filter, which is based on the polynomial of 2nd degree,
- $\Delta t_{opt} \approx$ 8–12 s (ΔFo = 0.008–0.012), for the approximation of the time derivative by means of the nine-point filter, which is based on the polynomial of 3rd degree.

Fig. 24.8 presents the history of temperature of the plate front face determined on the basis of a measured temperature transient on the back surface, disturbed by random errors from interval ±0.3 K.

a)

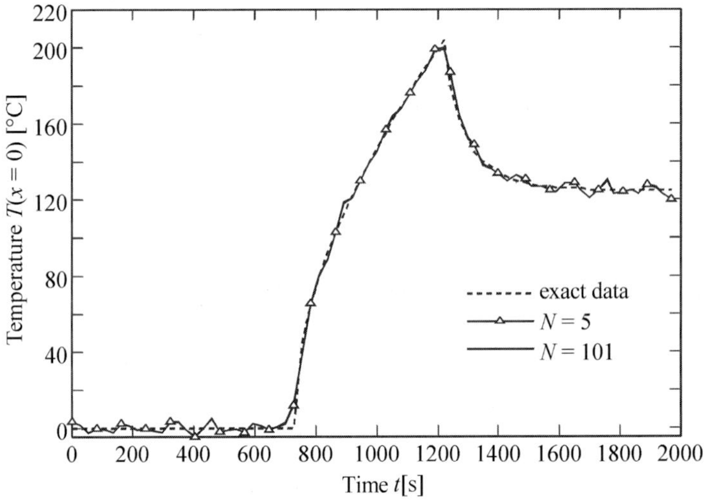
220
200
160
120
80
40
0
Temperature $T(x = 0)$ [°C]
0 200 400 600 800 1000 1200 1400 1600 1800 2000
Time t[s]
exact data
$N = 5$
$N = 101$

b)

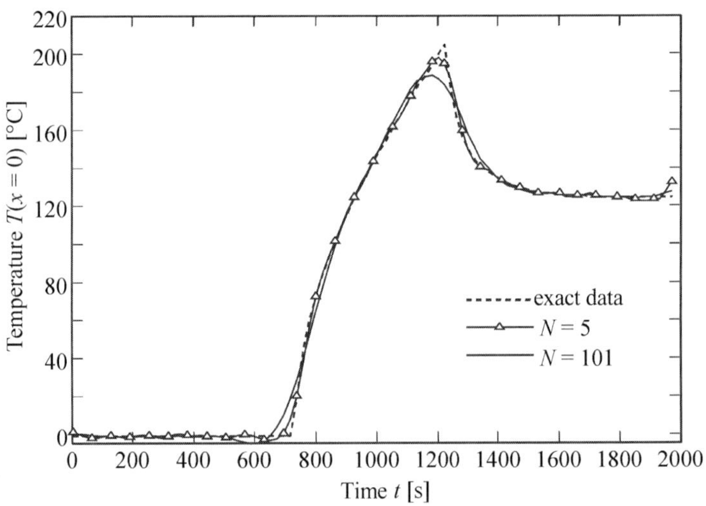
220
200
160
120
80
40
0
Temperature $T(x = 0)$ [°C]
0 200 400 600 800 1000 1200 1400 1600 1800 2000
Time t [s]
exact data
$N = 5$
$N = 101$

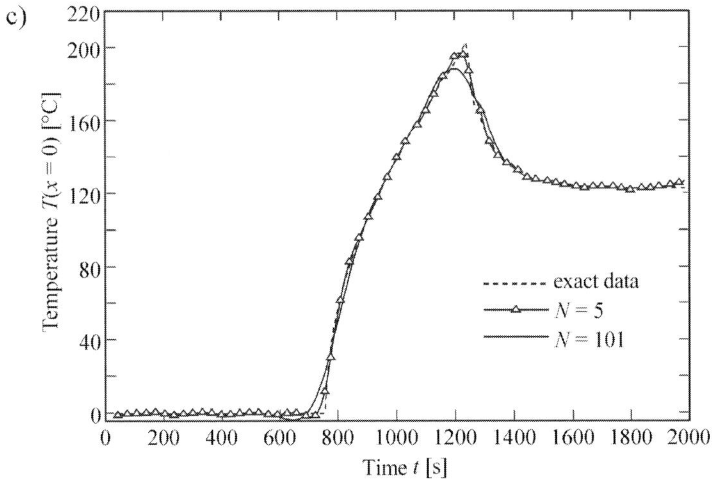

Fig. 24.8. Comparison of the plate front face temperature, determined on the basis of a temperature measurement taken on the back surface, with the exact solution of the direct problem; the measurement data was disturbed by random errors from interval ±0.3 K: a) approximation of the time derivative by the central difference quotient when $\Delta t = 27$ s; $\delta T = 3.03$ K for four spatial steps ($N = 5$) and $\delta T = 3.02$ for hundred spatial steps ($N = 101$), b) approximation of the time derivative by means of the seven-point filter $\Delta t = 21$ s; $\delta T = 2.34$ K for four spatial steps ($N = 5$) and $\delta T = 4.96$ K for hundred spatial steps ($N = 101$), c) approximation of the time derivative by means of the nine-point filter $\Delta t = 11$ s, $\delta T = 2.05$ K for four spatial steps ($N = 5$) and $\delta T = 4.19$ K for hundred spatial steps ($N = 101$)

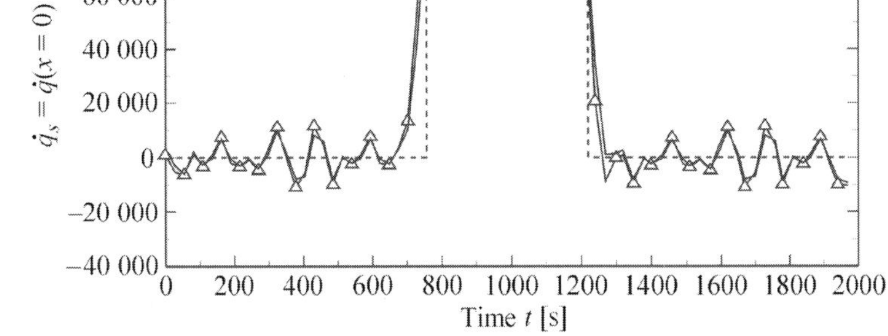

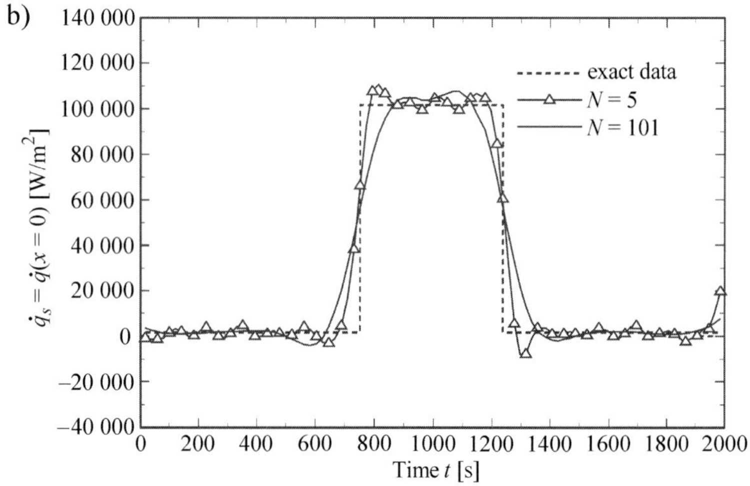

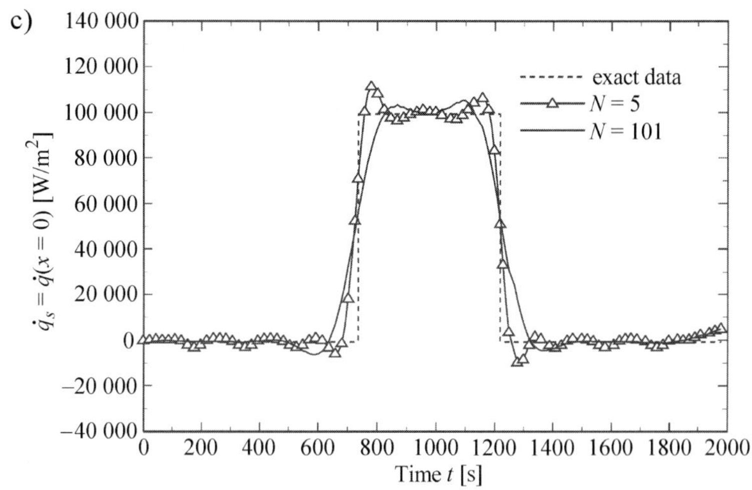

Fig. 24.9. Comparison of the plate front face heat flux, determined on the basis of a temperature measurement taken on the back surface, with the exact solution of the direct problem; the measurement data was disturbed by random errors from interval ±0.3 K: a) approximation of the time derivative by the central difference quotient when $\Delta t = 27$ s; $\delta\dot{q} = 16880.6$ W/m^2 for $N = 5$ and $\delta\dot{q} = 15784.2$ W/m^2 for $N = 101$, b) approximation of the time derivative by means of the seven-point filter $\Delta t = 21$ s, $\delta\dot{q} = 12262.5$ W/m^2 for $N = 5$ and $\delta\dot{q} = 14538.6$ W/m^2 for $N = 101$, c) approximation of the time derivative by means of the nine-point filter when $\Delta t = 11$ s; $\delta\dot{q} = 10381$ W/m^2 for $N = 5$ and $\delta\dot{q} = 12688.2$ W/m^2 for $N = 101$

From the analysis of Fig. 24.8 it follows that for optimum time steps, the accuracy of the method is very good, regardless of how the time derivative is approximated. In contrast to simple (direct) problems, increasing the number of spatial steps does not improve the accuracy of the inverse solution. The accuracy of the temperature and heat flux determination on the plate front face is already very good when the plate is divided into four control volumes Δx ($N = 5$). Similar results were obtained in the case of the plate surface heat flux (Fig. 24.9).

From the analysis of Fig. 24.9 it follows that in the case of the time derivative approximation with the central difference quotient, the method is less sensitive to the value of spatial step Δx. Results obtained for $N = 5$ and $N = 101$ do not differ much. The method, however, is more sensitive to random temperature measurement errors. This is particularly discernible in those time intervals in which the flow $\dot{q}_s$ on the plate surface equals zero.

Exercise 24.5 Determining Temperature and Heat Flux Distribution in a Plate on the Basis of a Temperature Measurement on an Insulated Back Plate Surface; Heat Flux is in the Shape of a Triangular Pulse

Determine heat flux and temperature of the plate front face on the basis of temperature measured on an insulated back surface. Use for the calculation identical data to the data used in Ex. 18.5: L = 0.01 m, $\dot{q}_N$ = 100000 W/m2, λ = 50 W/(m·K), a = $1 \cdot 10^{-5}$ m^2/s, T_0 = 20°C, $Fo_m = at_m/L^2$. Apply measurement data presented in Table 18.1, Ex. 18.5. Apply to the calculation both step-marching methods discussed in Ex. 24.3. Present calculation results, temperature $T_s(t)$ and heat flux $\dot{q}_s(t)$ on the plate front face in the tabular and graphical form. Using diagrams, compare the obtained results with transient $\dot{q}_s(t)$, which was assumed for the solution of the direct problem, and with surface temperature $T_s(t)$ calculated from the exact solution of the direct problem.

Solution

Back plate surface is thermally insulated; therefore, $\dot{q}_E(t) = 0$. Calculation results of surface temperature $T_s(t)$ and heat flux $\dot{q}_s(t)$ determined by means of the Weber method are presented in Table 24.1, while by means of the second method are presented in Ex. 24.3 and described by (15)–(18) in Table 24.2.

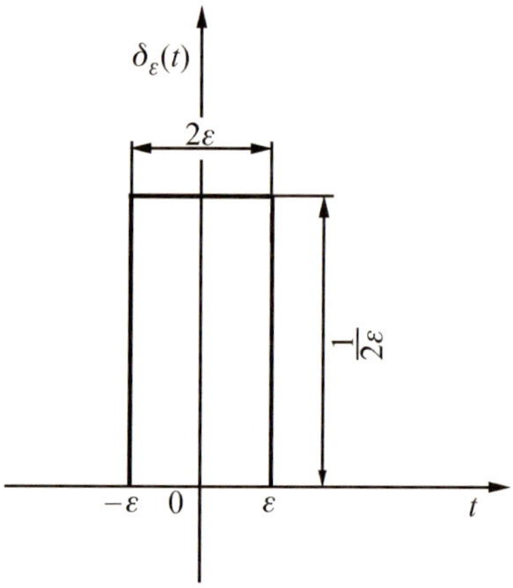

Fig. 25.1. Approximation of the Dirac function $\delta(t)$ by means of a rectangular pulse with a width of 2ε

$$\int_{-\infty}^{\infty} Q_0 \delta(t)\, dt = Q_0 \,. \tag{5}$$

Laplace transform is frequently used to determine transient temperature distribution. Laplace transform from the Dirac function $\delta(t - t_0)$ has the form [11, 31]

$$\mathcal{L}\{\delta(t - t_0)\} = e^{-st_0} \,. \tag{6}$$

When the impulse appears at the initial moment $t_0 = 0$, then

$$\mathcal{L}\{\delta(t)\} = 1 \,. \tag{7}$$

If temperature history $h(t)$ is known at a given spatial point $\boldsymbol{r}_0$ in the impulse function, then one can determine temperature transient $T(t)$ produced by the time-dependent impulse function $u(t)$

$$T(t) = \int_0^t h(t)\, u(t - \tau)\, d\tau \,. \tag{8}$$

Once the analyzed time interval is divided into steps under Δt and $t = k\Delta t$, (8) can be written in the form

$$T(k\Delta t) = \Delta t \sum_{i=0}^{k} h(i\Delta t)\, u(k\Delta t - i\Delta t), \tag{9}$$

where $k = 0, 1, 2, \ldots$

In the case of heat sources, which are continuously active within the whole body volume, the heat conduction equation has the form

$$c(T)\rho(T)\frac{\partial T}{\partial t} = \nabla \cdot \left[\lambda(T)\nabla T\right] + \dot{q}_v(\mathbf{r},t,T), \tag{10}$$

where $\dot{q}_v(\mathbf{r},t,T)$ is the density of the generated thermal power (heat flow), i.e. thermal power generated per unit of body volume.

Good examples of thermal energyemission within a whole body volume are exothermic chemical and nuclear reactions, inductive and resistant electric heating and heat dissipation processes during a liquid flow. The determination process of temperature fields in the presence of heat sources active within the whole body volume is not, in essence, significantly different from the process of solving source-free problems and will not be discussed separately.

Exercise 25.1 Determining Formula for Transient Temperature Distribution Around an Instantaneous (Impulse) Point Heat Source Active in an Infinite Space

Determine transient temperature distribution around an instantaneous point heat source, which is active in an infinite space. Thermo-physical properties of the body are temperature independent. Initial temperature T_0 is constant and independent of position.

Solution

Laplace transform will be used to determine temperature distribution around the point heat source. First, we will determine temperature field around the spherical void inside the infinite space (Fig. 25.2).

Thermal energy $Q_0 = 4\pi r_w^2 q_w$ is emitted in an impulse form on an inner surface of the void at a time $t = 0$, where q_w is the thermal energy per unit of void surface.

Temperature distribution around the void is described by the equation

$$\frac{\partial \theta}{\partial t} = a\left(\frac{\partial^2 \theta}{\partial r^2} + \frac{2}{r}\frac{\partial \theta}{\partial r}\right), \tag{1}$$

by boundary conditions

$$-\lambda \left.\frac{\partial \theta}{\partial r}\right|_{r=r_w} = q_w \delta(t), \tag{2}$$

$$\theta\big|_{r\to\infty} = 0 \tag{3}$$

and by initial condition

$$\theta\big|_{t=0} = 0, \tag{4}$$

where

$\theta = T - T_0$, $q_w = Q_0/(4\pi r_w^2)$,

T_0 is the initial temperature, $\delta(t)$ is the Dirac function.

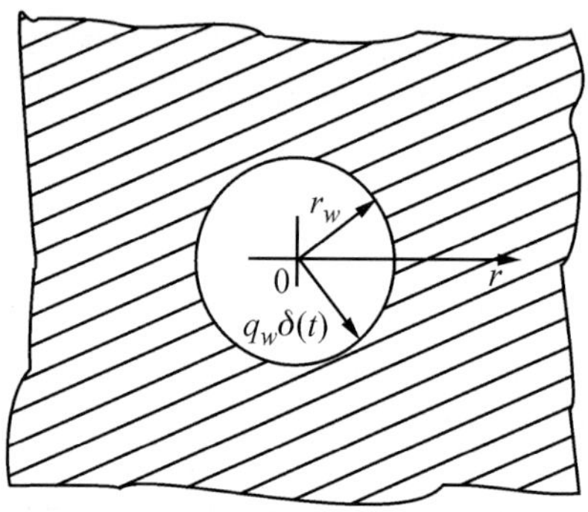

Fig. 25.2. Spherical void in an infinite space

Once the Laplace transform is applied to (1), conditions (2) and (3), one has

$$\frac{d^2\overline{\theta}}{dr^2} + \frac{2}{r}\frac{d\overline{\theta}}{dr} - q^2\overline{\theta} = 0, \tag{5}$$

$$-\lambda \frac{\partial\overline{\theta}}{\partial r}\bigg|_{r=r_w} = q_w, \tag{6}$$

$$\overline{\theta}\big|_{r\to\infty} = 0, \tag{7}$$

where $q^2 = s/a$.

The solution of (5) has the form

$$\overline{\theta} = A\frac{e^{qr}}{r} + B\frac{e^{-qr}}{r}. \tag{8}$$

Once condition (7) is accounted for, one obtains $A = 0$ and (8) is simplified to

$$\overline{\theta} = B\frac{e^{-qr}}{r}. \tag{9}$$

Once constant B is determined from boundary condition (6) and substituted into (9), one has

$$\bar{\theta} = \frac{Q_0}{4\pi\lambda r}\frac{e^{-q(r-r_w)}}{1+qr_w}. \tag{10}$$

If $r_w \to 0$, then (10) has the form

$$\bar{\theta} = \frac{Q_0}{4\pi\lambda r}e^{-qr}. \tag{11}$$

Accounting that (Table 14.2, Ex. 14.1)

$$\mathcal{L}^{-1}\left(e^{-qr}\right) = \frac{r}{2t\sqrt{\pi at}}e^{-r^2/(4at)} \tag{12}$$

one obtains

$$\theta(r,t) = \mathcal{L}^{-1}\left(\bar{\theta}\right) = \frac{Q_0}{c\rho(4\pi at)^{3/2}}e^{-r^2/(4at)}, \tag{13}$$

where Q_0 is the energy emitted by the point heat source and expressed in J.

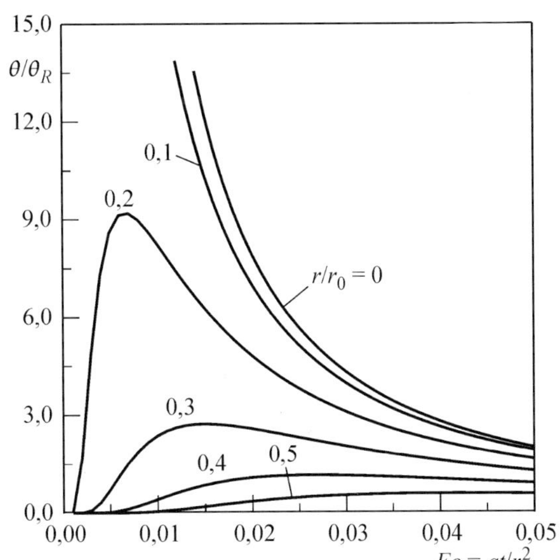

Fig. 25.3. Dimensionless temperature transient by thermal impulse point source calculated from (14) and; radius r_0 is an arbitrary distance from $r = 0$, where the heat source is located

If reference temperature $\theta_R = Q_0/(c\rho r_0^3)$, dimensionless radius r/r_0 and the Fourier number $Fo = at/r_0^2$ are introduced, then (13) can be written in the form

$$\frac{\theta(r,t)}{\theta_R} = \frac{1}{(4\pi Fo)^{3/2}} e^{-\left(\frac{r}{r_0}\right)^2 \frac{1}{4Fo}}. \tag{14}$$

Quantity r_0 is the arbitrarily chosen radius.

Figure 25.3 shows dimensionless temperature transient in the dimensionless radius function (r/r_0) and the Fourier number $Fo = at/r_0^2$. From the analysis of Fig. 25.3, it is evident that temperature decreases more rapidly the greater the distance from point $r = 0$, in which the impulse point source lies. Maximum temperature occurs at time $t_{max} = r^2/(6a)$. The maximum temperature value quickly decreases the greater the distance from $r = 0$.

Exercise 25.2 Determining Formula for Transient Temperature Distribution in an Infinite Body Produced by an Impulse Surface Heat Source

Determine transient temperature distribution in an infinite body produced by an impulse surface heat source. Heat source emitted energy per unit of surface area is $q_s = Q_0/A$ and is expressed in J/m^2. Initial temperature T_0 is constant.

Solution

The position of the surface heat source and the assumed coordinate system is given in Fig. 25.4.

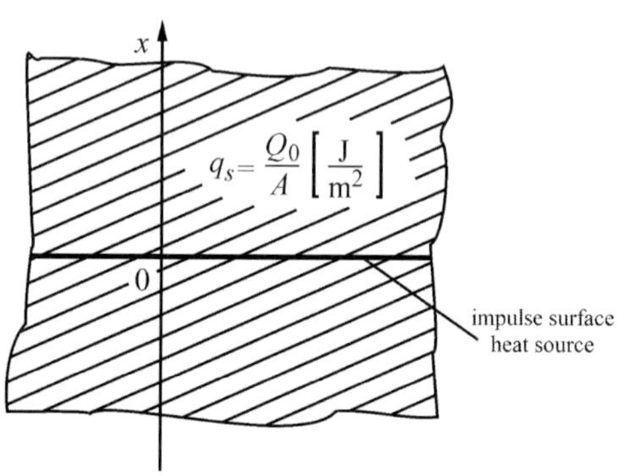

Fig. 25.4. Impulse surface heat source in an infinite space

Temperature distribution in the infinite space is only a function of the coordinate x and is defined by the heat conduction equation

$$\frac{\partial \theta}{\partial t} = a \frac{\partial \theta}{\partial x^2}, \tag{1}$$

by boundary conditions

$$-\lambda \frac{\partial \theta}{\partial r}\bigg|_{x=0} = \frac{1}{2} q_s \delta(t), \tag{2}$$

$$\theta\big|_{x \to \infty} = 0 \tag{3}$$

and by initial condition

$$\theta\big|_{t=0} = 0, \tag{4}$$

where $\theta = T - T_0$.

Boundary condition (2) contains $q_s/2$, since a half of the generated heat flows to an upper section of the half space, while the second half to a lower section due to the symmetry of the temperature field with respect to $x = 0$.

Once Laplace transform is applied to (1) and conditions (2) and (3), one has

$$\frac{d^2 \bar{\theta}}{dx^2} - q^2 \bar{\theta} = 0, \tag{5}$$

$$-\lambda \frac{d\bar{\theta}}{dx}\bigg|_{x=0} = \frac{1}{2} q_s, \tag{6}$$

$$\bar{\theta}\big|_{x \to \infty} = 0, \tag{7}$$

where $\bar{\theta}$ is the Laplace transform from θ, $q^2 = s/a$.

The solution of (5)–(7) has the form

$$\bar{\theta} = A e^{qx} + B e^{-qx}, \tag{8}$$

where from, after accounting for condition (7), one has

$$\bar{\theta} = B e^{-qx}, \tag{9}$$

since $A = 0$.

Once constant B is determined from (6) and substituted into (9), one obtains

$$\bar{\theta} = \frac{1}{2}\frac{q_s}{\lambda}\frac{e^{-qx}}{q}. \tag{10}$$

Once the inverse Laplace transform is determined (Table 14.2, Ex. 14.1), one obtains

$$\mathscr{L}^{-1}\left(\frac{e^{-qx}}{q}\right) = \sqrt{\frac{a}{\pi t}}e^{-x^2/(4at)}. \tag{11}$$

After substituting (11) into (10), a formula for temperature distribution in infinite space is obtained

$$\theta(x,t) = \frac{q_s}{2c\rho\sqrt{\pi a t}}e^{-x^2/(4at)}. \tag{12}$$

The (12) resembles in form (13), Ex. 25.1, which describes temperature distribution around the thermal impulse point source in infinite space.

Exercise 25.3 Determining Formula for Transient Temperature Distribution Around Instantaneous Linear Impulse Heat Source Active in an Infinite Space

Determine transient temperature distribution around the instantaneous linear impulse heat source, which is active in an infinite space. Heat source emitted energy per unit of length is $q_l = Q_0/L$ [J/m]. Initial temperature of the body T_0 is constant.

Solution

Fig. 25.5 shows the position of the linear heat source and the assumed coordinate system.

Heat conduction equation, which describes transient temperature field has the form

$$\frac{\partial \theta}{\partial t} = a\left(\frac{\partial^2 \theta}{\partial r^2} + \frac{1}{r}\frac{\partial \theta}{\partial r}\right), \tag{1}$$

where $\theta = T - T_0$.

The to-be-determined temperature distribution should satisfy the following boundary conditions:

$$-\lambda \frac{\partial \theta}{\partial r}\bigg|_{r=0} = q_l \delta(t), \tag{2}$$

$$\theta\big|_{r \to \infty} = 0 \tag{3}$$

and initial condition

$$\theta\big|_{t=0} = 0. \tag{4}$$

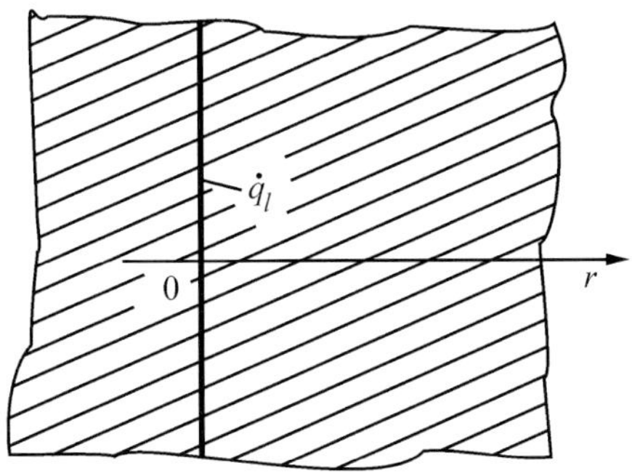

Fig. 25.5. Linear heat source in infinite space

Once Laplace transform is applied to (1) and then solved, one obtains

$$\overline{\theta} = AI_0(qr) + BK_0(qr), \tag{5}$$

where $\overline{\theta}$ is the Laplace transform from θ, $q = \sqrt{s/a}$.

It is difficult to find the inverse Laplace transform, since the modified zero order Bessel functions of the first and second kind, i.e. $I_0(qr)$ and $K_0(qr)$ do not approach zero, when $r \to 0$.

One can, however, rather easily find a solution to problems (1)–(4), if it is observed that Eqs. (13) in Ex. 25.1 and (12) in Ex. 25.2 are similar in form, which in the case of the linear source can be expressed as follows:

$$\theta(r,t) = H(t)\exp\left(-\frac{r^2}{4at}\right). \tag{6}$$

Once we assume that the heat source has a length L, we can calculate the energy emitted from the source to surroundings. When the analyzed spatial interval is infinitely long, the energy Q_0 emitted by the source equals the energy absorbed by the surroundings

$$Q_0 = \int_0^\infty c\rho 2\pi r L\,\theta(r,t)\,dr = 2\pi L c\rho \int_0^\infty r\,\theta(r,t)\,dr. \tag{7}$$

By substituting (6) into (7), one obtain

$$Q_0 = 2\pi L c\rho H(t) \int_0^\infty r \exp\left(-\frac{r^2}{4at}\right) dr . \tag{8}$$

If a new variable is introduced

$$\zeta = \frac{r}{\sqrt{4at}} , \tag{9}$$

Equation (8) can be written in the form

$$Q_0 = 8\pi L c\rho H(t) at \int_0^\infty \zeta e^{-\zeta^2} d\zeta , \tag{10}$$

hence, allowing that the integral equals1/2, one can determine $H(t)$

$$H(t) = \frac{Q_0/L}{4\pi c\rho at} = \frac{Q_0}{4\pi L\lambda t} = \frac{q_l}{4\pi\lambda t} . \tag{11}$$

By substituting (11) into (6), one obtains an expression that defines the temperature field

$$\theta(r,t) = \frac{q_l}{4\pi\lambda t} \exp\left(-\frac{r^2}{4at}\right) , \tag{12}$$

where $q_l = Q_0/L$ [J/m].

The determined temperature distribution (12) satisfies the differential (1) and conditions (2)–(4).

Exercise 25.4 Determining Formula for Transient Temperature Distribution Around a Point Heat Source, which Lies in an Infinite Space and is Continuously Active

Determine transient temperature distribution around a point heat source, which lies in an infinite space and is continuously active. Thermo-physical properties of the medium are constant and temperature independent. Initial temperature of the body T_0 is constant.

Solution

The superposition method (Chap. 16) will be applied to determine temperature distribution for a continuously active point source and the solution determined in Ex.25.1 for a thermal impulse point source.

Heat source activity, with a time-variable power, can be presented as the sum of all activities of an infinite number of impulse sources, which are successively being added at an infinitely small time intervals. Assume that new impulse heat sources emit Q_i energy at the successive time points t_i, $i = 0, 1, \ldots$ The first heat source added at the moment $t_0 = 0$ emits Q_0 energy. Temperature distribution can be calculated from (13), Ex. 25.1; the calculation is limited to interval $0 \le t \le t_1$

$$\theta_0(r,t) = \frac{Q_0}{c\rho(4\pi a t)^{3/2}} \exp\left(-\frac{r^2}{4at}\right), \qquad 0 \le t \le t_1 . \tag{1}$$

The second impulse heat source is added at the moment $t = t_1$; it emits Q_1 energy. To determine temperature distribution in interval $t_1 \le t \le t_2$, one can use (13) from Ex. 25.1 by counting time from the moment the source was switched on, from $t = t_1$. This means that t should be replaced in (13) by $(t - t_1)$

$$\theta_1(r,t-t_1) = \frac{Q_0}{c\rho\left[4\pi a(t-t_1)\right]^{3/2}} \exp\left[-\frac{r^2}{4a(t-t_1)}\right], \qquad t_1 \le t \le t_2 . \tag{2}$$

In accordance with the superposition method (Chap. 16), temperature distribution around the source is defined by the following expressions:

$$\theta(r,t) = \begin{cases} \theta_0(r,t), & 0 \le t \le t_1 \\ \theta_0(r,t) + \theta_1(r,t-t_1), & t_1 \le t \le t_2 . \end{cases} \tag{3}$$

In the case of a successively switched n heat sources, temperature distribution at time $t = t_n$ is formulated as

$$\theta(r,t) = \theta_0(r,t) + \theta_1(r,t-t_1) + \ldots + \theta_{n-1}(r,t-t_{n-1}) + \theta_n(r,t-t_n) . \tag{4}$$

If a heat source with power $\dot{Q}(t)$ is continuously active, then the thermal energy emitted at interval Δt is

$$\Delta Q(t_i) = \dot{Q}(t_i)\Delta t = \dot{Q}_i \Delta t . \tag{5}$$

When $\Delta t \to 0$, while the number of time intervals is $n \to \infty$, then the sum in (4) can be replaced by an integral and that gives

$$\theta(r,t) = \int_0^t \frac{\dot{Q}(\tau)}{c\rho\left[4\pi a(t-\tau)\right]^{3/2}} \exp\left[-\frac{r^2}{4a(t-\tau)}\right] d\tau . \tag{6}$$

By substituting

$$\zeta = \frac{r}{\sqrt{4a(t-\tau)}} \tag{7}$$

and accounting that

$$d\zeta = \frac{2ard\tau}{\left[4a(t-\tau)\right]^{3/2}}, \tag{8}$$

$$d\tau = \frac{\left[4a(t-\tau)\right]^{3/2}}{2ar} d\zeta \tag{9}$$

Equation (6) can be transformed into a form

$$\theta(r,t) = \frac{1}{4\pi\lambda r}\frac{2}{\sqrt{\pi}}\int\limits_{\zeta=r/\sqrt{4at}}^{\infty} \dot{Q}(\zeta)e^{-\zeta^2}d\zeta . \tag{10}$$

If the heat source power is constant

$$\dot{Q} = \dot{Q}_0 , \tag{11}$$

then, accounting that [16, 28]

$$\operatorname{erfc}\zeta = 1 - \operatorname{erf}\zeta = \frac{2}{\sqrt{\pi}}\int\limits_{\zeta}^{\infty} e^{-x^2}dx, \qquad \operatorname{erfc}(\infty) = 0 \tag{12}$$

Equation (10) assumes the form

$$\theta(r,t) = \frac{\dot{Q}_0}{4\pi\lambda r}\operatorname{erfc}\left(\frac{r}{\sqrt{4at}}\right). \tag{13}$$

In a steady-state, when $t \to \infty$ and erfc(0) = 1, one has from (13)

$$\theta(r) = \frac{\dot{Q}_0}{4\pi\lambda r}. \tag{14}$$

Due to the fact that (14) has a simple form and power $\dot{Q}_0$ can be easily measured, (14) is used to measure thermal conductivity λ of a medium in which a continuously active point heat source is located. Aside from $\dot{Q}_0$, the value of temperature $\theta(r)$ is also measured at one or few points.

Exercise 25.5 Determining Formula for a Transient Temperature Distribution Triggered by a Surface Heat Source Continuously Active in an Infinite Space

Determine transient temperature distribution due to a surface heat source, which is continuously active in an infinite space. Heat flow generated by this source is $\dot{q}_s(t)$ per unit of source surface. Thermo-physical properties of the medium are temperature independent. Initial temperature of the medium is T_0.

Solution

Continuously active surface heat source can be described as the sum of an infinitely large number of active impulse sources that individually generate thermal energy (per unit of source surface) equal to

$$\Delta q_s = \dot{q}_s \Delta t , \tag{1}$$

where Δt is the infinitely small time interval.

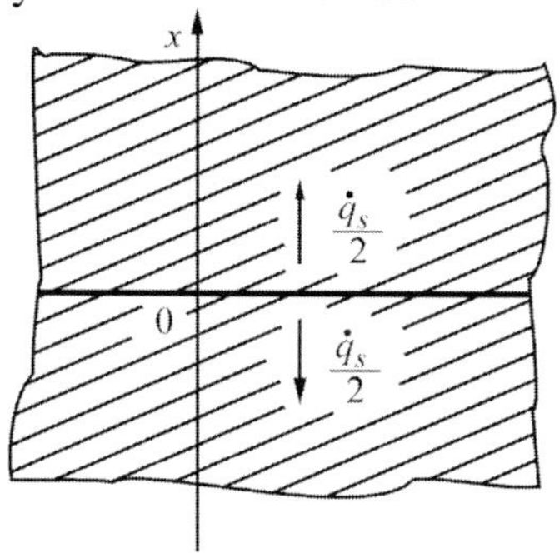

Fig. 25.6 Continuously active surface heat source in an infinite space

On the basis of solution (12), from Ex. 25.2, the superposition method can be applied as it was in Ex. 25.4 and the following expression for temperature distribution can be obtained

$$\theta(x,t) = \int_0^t \frac{\dot{q}_s(\tau)}{2c\rho\left[\pi a(t-\tau)\right]^{1/2}} \exp\left[-\frac{x^2}{4a(t-\tau)}\right] d\tau . \tag{2}$$

In order to calculate integral (2), a new variable is introduced

$$\zeta = |x| \Big/ \sqrt{4a(t-\tau)} . \tag{3}$$

Once (3) is differentiated on both sides, one has

$$d\tau = \frac{\left[4a(t-\tau)\right]^{3/2}}{2a|x|} d\zeta \,. \tag{4}$$

Moreover, if

$$\tau = 0, \quad \text{to} \quad \zeta = \frac{|x|}{\sqrt{4at}}, \tag{5}$$

$$\tau = t, \quad \text{to} \quad \zeta \to \infty, \tag{6}$$

and if one accounts for (3) and (4) in (2), one has

$$\theta(x,t) = \frac{|x|}{2\lambda\sqrt{\pi}} \int_{\zeta=|x|/\sqrt{4at}}^{\infty} \frac{\dot{q}_s(\zeta) e^{-\zeta^2}}{\zeta^2} d\zeta \,. \tag{7}$$

When $\dot{q}_s =$ const, then (7) is simplified to a form

$$\theta(x,t) = \frac{\dot{q}_s |x|}{2\lambda\sqrt{\pi}} \int_{\zeta=|x|/\sqrt{4at}}^{\infty} \frac{e^{-\zeta^2}}{\zeta^2} d\zeta \,. \tag{8}$$

Because

$$\int \frac{e^{-\zeta^2}}{\zeta^2} d\zeta = -\frac{e^{-\zeta^2}}{\zeta^2} - 2\int e^{-\zeta^2} d\zeta \tag{9}$$

and

$$\operatorname{erfc}\zeta = \frac{2}{\sqrt{\pi}} \int_{\zeta}^{\infty} e^{-x^2} dx, \qquad \operatorname{erfc}(\infty) = 0 \,, \tag{10}$$

then (8) assumes the form

$$\theta(x,t) = \frac{\dot{q}_s}{c\rho}\sqrt{\frac{t}{\pi a}} \exp\left(-\frac{x^2}{4at}\right) - \frac{\dot{q}_s |x|}{2\lambda} \operatorname{erfc}\left(\frac{|x|}{\sqrt{4at}}\right). \tag{11}$$

Equation (11) contains a module from coordinate x, since it can assume positive and negative values (Fig. 25.6). One can notice that identical formula is obtained when determining temperature distribution in a half space, either heated or step-cooled by a heat flow with a density $\dot{q}_s / 2$ (Ex. 14.3, (11)).

Exercise 25.6 Determining Formula for a Transient Temperature Distribution Around a Continuously Active Linear Heat Source with Assigned Power $\dot{q}_l$ Per Unit of Length

Determine transient temperature distribution around a continuously active linear heat source and a power $\dot{q}_l$ assigned per unit of length. Thermo-physical properties of the medium are temperature independent. Initial temperature of the medium is T_0.

Solution

Continuously active linear heat source can be presented as the sum of infinitely large number of active impulse sources, which individually generate thermal energy (per unit of source length) equal to

$$\Delta q_l = \dot{q}_l(t)\Delta t\,, \tag{1}$$

where Δt is the infinitely small time interval.
Once solution (12), from Ex. 25.3, and the superposition method are applied, the way they are in Ex. 25.4, the following expression, which describes temperature distribution, is obtained

$$\theta(r,t) = \int_0^t \frac{\dot{q}_l(\tau)}{4\pi\lambda(t-\tau)} \exp\left[-\frac{r^2}{4a(t-\tau)}\right] d\tau\,. \tag{2}$$

Once the new variable is introduced

$$u = \frac{r^2}{4a(t-\tau)}, \tag{3}$$

and both sides are differentiated, one has

$$d\tau = \frac{4a(t-\tau)^2}{r^2} du\,. \tag{4}$$

Accounting for (3) and (4), (2) has the form

$$\theta(r,t) = \frac{1}{4\pi\lambda} \int_{r^2/(4at)}^{\infty} \dot{q}_l(u) \frac{e^{-u}}{u} du\,. \tag{5}$$

If $\dot{q}_l$ = const, then (5) is simplified to

$$\theta(r,t)=\frac{\dot{q}_l}{4\pi\lambda}\int_{r^2/(4at)}^{\infty}\frac{e^{-u}}{u}du. \tag{6}$$

Once the definition of the exponential integral function is accounted for [16, 28]

$$Ei(-\xi)=-\int_{\xi}^{\infty}\frac{e^{-u}}{u}du \tag{7}$$

Equation (6), which defines temperature distribution, can be written in the form

$$\theta(r,t)=-\frac{\dot{q}_l}{4\pi\lambda}Ei\left(-\frac{r^2}{4at}\right). \tag{8}$$

Exponential integral function $Ei(-\xi)$ can be calculated from [16, 28]

$$Ei(-\xi)=0{,}577216+\ln\xi+\sum_{n=1}^{\infty}(-1)^n\frac{\xi^n}{n\cdot n!}. \tag{9}$$

Selected values of function $-Ei(-\xi)$ are listed in Table 25.1.

Table 25.1. The values of exponential integral function $-Ei(-\xi)$

ξ	$-Ei(-\xi)$	ξ	$-Ei(-\xi)$	ξ	$-Ei(-\xi)$	ξ	$-Ei(-\xi)$
0.00	∞	0.15	1.4645	0.70	0.3738	1.50	0.10002
0.01	4.0379	0.20	1.2227	0.75	0.3403	1.60	0.08631
0.02	3.3547	0.25	1.0443	0.80	0.3106	2.00	0.04899
0.03	2.9591	0.30	0.9057	0.85	0.2840	2.50	0.02499
0.04	2.6813	0.35	0.7942	0.90	0.2602	3.00	0.01305
0.05	2.4679	0.40	0.7024	0.95	0.2387	3.50	0.00697
0.06	2.2953	0.45	0.2849	1.00	0.2194	4.00	0.00378
0.07	2.1508	0.50	0.5598	1.10	0.1859	4.50	0.00207
0.08	2.0269	0.55	0.5034	1.20	0.1584	5.00	0.00115
0.09	1.9187	0.60	0.4544	1.30	0.13545	6.00	0.00035
0.10	1.8229	0.65	0.4115	1.40	0.11622	7.00	0.00012

If time t is sufficiently long or radius r is small, when $r^2/(4at) << 1$, one can neglected the infinite series in (9) and the (8) for temperature distribution has the form

$$\theta(r,t)=\frac{\dot{q}_l}{4\pi\lambda}\left[\ln\left(\frac{4at}{r^2}\right)-0.577216\right],\qquad \frac{r^2}{4at}<<1. \tag{10}$$

Equation (10) can be applied when it is necessary to experimentally determine the thermal conductivity λ, on the basis of measured temperature value $\theta\,(r,t)$ at one or more points, and the measured value $\dot{q}_l=\dot{Q}_0/L$ [W/m^2], where L is the length of the heating duct.

Exercise 25.7 Determining Formula for Quasi-Steady-State Temperature Distribution Caused by a Point Heat Source with a Power $\dot{Q}_0$ that Moves at Constant Velocity v in Infinite Space or on the Half Space Surface

Determine quasi-steady-state temperature distribution produced by a point heat source with a power $\dot{Q}_0$ that moves at constant velocity v in infinite space or on the half space surface. Also evaluate quasi-steady-state solution, which describes temperature distribution in a thin plate due to a linear heat source that moves at constant velocity v. Thermo-physical properties of both mediums are temperature invariant.

Solution

In order to find the solution, assume that the coordinate system moves at a constant velocity v with a point of origin 0_1 inside the moving source (Fig. 25.7).

Transient conduction equation

$$\frac{\partial^2\theta}{\partial x^2}+\frac{\partial^2\theta}{\partial y^2}+\frac{\partial^2\theta}{\partial z^2}=\frac{1}{a}\frac{\partial\theta}{\partial t} \tag{1}$$

will be transformed once coordinate ξ is entered into the mobile coordinate system, which originates at point 0_1 and moves at constant velocity in the direction of x axis

$$\xi = x - vt\,. \tag{2}$$

Accounting that

$$\begin{gathered}\frac{\partial\theta}{\partial x}=\frac{\partial\theta}{\partial\xi},\\ \frac{\partial^2\theta}{\partial x^2}=\frac{\partial^2\theta}{\partial\xi^2},\\ \frac{\partial\theta}{\partial t}=\frac{\partial\theta}{\partial\xi}\cdot\frac{\partial\xi}{\partial t}+\frac{\partial\theta}{\partial t}=-v\frac{\partial\theta}{\partial\xi}+\frac{\partial\theta}{\partial t}\end{gathered} \tag{3}$$

Equation (1) can be written in the form

$$\frac{\partial^2\theta}{\partial\xi^2}+\frac{\partial^2\theta}{\partial y^2}+\frac{\partial^2\theta}{\partial z^2}=-\frac{v}{a}\frac{\partial\theta}{\partial\xi}+\frac{1}{a}\frac{\partial\theta}{\partial t}\,. \tag{4}$$

a)

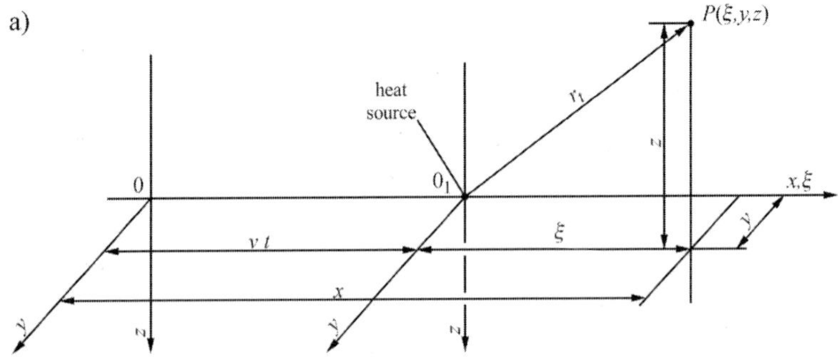

b)

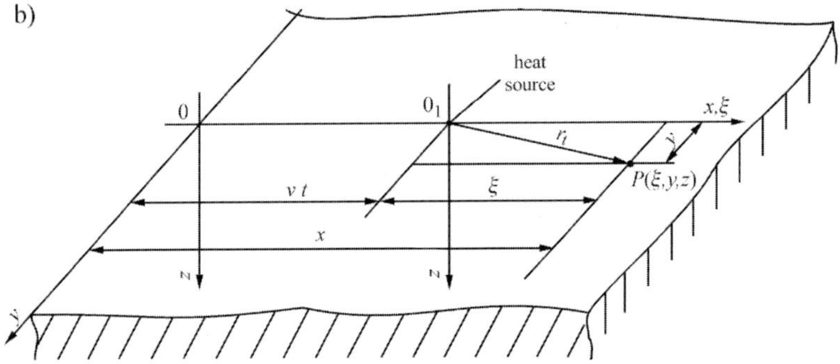

Fig. 25.7. Point heat source, which moves at constant velocity v along x axis: a) in infinite space, b) on the half space surface; 0 is the point of origin of the immobile system of coordinates x,y,z; 0_1 is the point of origin of the mobile system of coordinates ξ,η,z that moves along with the heat source

Symbol $\theta = T - T_{cz}$ is the excess of body temperature T above the medium's temperature T_{cz}.

In the case of thin plates with thickness δ, (4) can be simplified while neglecting the temperature drop across its thickness. If we assume that the coordinate of the lower plate surface is $z = -\delta/2$, while of the upper surface $z = \delta/2$, then (4) can be integrated over coordinate z

$$\int_{-\delta/2}^{\delta/2}\left(\frac{\partial^2\theta}{\partial\xi^2}+\frac{\partial^2\theta}{\partial y^2}+\frac{\partial^2\theta}{\partial z^2}\right)dz=\int_{-\delta/2}^{\delta/2}\left(-\frac{v}{a}\frac{\partial\theta}{\partial\xi}+\frac{1}{a}\frac{\partial\theta}{\partial t}\right)dz\,. \tag{5}$$

From the definition of average temperature

$$\bar{\theta}=\frac{1}{\delta}\int_{-\delta/2}^{\delta/2}\theta dz \tag{6}$$

one has

$$\overline{\theta}\delta = \int_{-\delta/2}^{\delta/2} \theta dz \,. \tag{7}$$

By accounting for (7), (5) has the form

$$\delta \frac{\partial^2 \overline{\theta}}{\partial \xi^2} + \delta \frac{\partial^2 \overline{\theta}}{\partial y^2} + \left.\frac{\partial \theta}{\partial z}\right|_{z=-\delta/2} - \left.\frac{\partial \theta}{\partial z}\right|_{z=\delta/2} = -\frac{v\delta}{a}\frac{\partial \overline{\theta}}{\partial \xi} + \frac{\delta}{a}\frac{\partial \overline{\theta}}{\partial t} \,. \tag{8}$$

If we account for the boundary conditions on the lower and upper plate surface

$$\lambda \left.\frac{\partial \theta}{\partial z}\right|_{z=-\delta/2} = \alpha \left.\theta\right|_{z=-\delta/2} , \tag{9}$$

$$\lambda \left.\frac{\partial \theta}{\partial z}\right|_{z=\delta/2} = \alpha \left.\theta\right|_{z=\delta/2} \tag{10}$$

and assume that $\overline{\theta} \approx \theta$ due to the insignificantly small temperature drop across the plate thickness, then (8) can be written in the form

$$\frac{\partial^2 \theta}{\partial \xi^2} + \frac{\partial^2 \theta}{\partial y^2} + \frac{2\alpha}{\lambda\delta}\theta = -\frac{v}{a}\frac{\partial \theta}{\partial \xi} + \frac{1}{a}\frac{\partial \theta}{\partial t} \,. \tag{11}$$

In a quasi-steady-state, one should assume that $\partial\theta/\partial t = 0$ in (4) and (11). Temperature distribution in the infinite space and the quasi-steady-state is described by the simple function [7, 15]

$$\theta = \frac{\dot{Q}_0}{4\pi\lambda r_t} \exp\left[-\frac{v}{2a}\left(\xi + r_t\right)\right], \tag{12}$$

where

$$\xi = x - vt, \qquad r_t = \sqrt{\xi^2 + y^2 + z^2} \,. \tag{13}$$

Once (12) is substituted into (4) and all the mathematical operations carried out, it becomes clear that solution (12) is correct; however, one should allow that $\partial\theta/\partial t = 0$. Equation (12) can also be obtained once the transient problem (4) is solved and the quasi-steady state is analyzed, when $t \to \infty$. Such method of determining (12) is thoroughly discussed in Ex. 25.8.

If the heat source moves around the half space surface, then the entire heat flow moves over to the half space, i.e. temperature excess $\theta(\xi, r_t)$ is twice as large as the excess obtained from (12) and has the form

$$\theta(\xi, r_t) = \frac{\dot{Q}_0}{2\pi\lambda r_t} \exp\left[-\frac{v}{2a}(\xi + r_t)\right], \tag{14}$$

where $r_t = \sqrt{\xi^2 + y^2 + z^2}$.

Symbol $\dot{Q}_0$ is the source power expressed in W. The remaining notations are shown in Fig. 25.7. Equation (14) can be applied in the case of thick plate welding, as the plates can be treated as semi-infinite bodies.

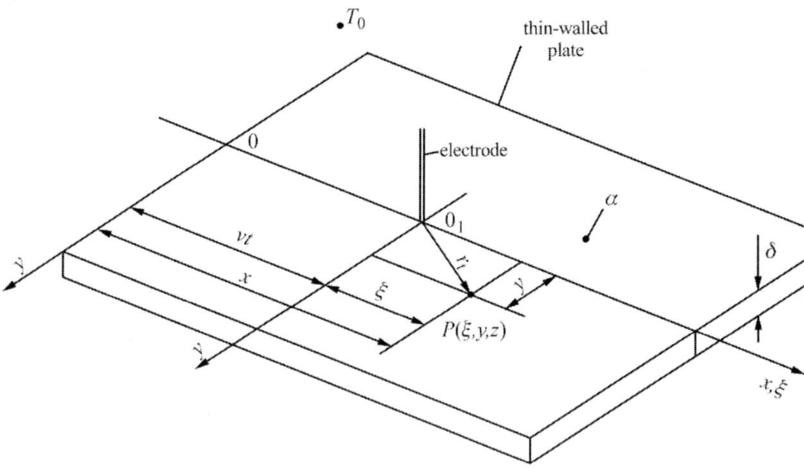

Fig. 25.8. Diagram of a linear heat source that moves at constant velocity v along x axis

Quasi-steady state solutions are frequently used to determine temperature fields during the welding processes. A good example of such a solution is the expressions that defines temperature distribution in a thin-walled plate with thickness δ (Fig. 25.8). Electric arc is treated as a linear heat source, whose power per unit of length is $\dot{q}_l = \dot{Q}_0 / \delta$. The solution has the form [12, 22, 25, 27, 30, 32] (Fig. 25.8)

$$\theta(\xi, y) = \frac{\dot{Q}_0}{2\pi\lambda\delta} K_0(u) \exp\left[-\frac{v\xi}{2a}\right], \tag{15}$$

where

$$u = r_t \sqrt{\left(\frac{v}{2a}\right)^2 + \frac{m\alpha}{\lambda\delta}}, \tag{16}$$

$$r_t = \sqrt{\xi^2 + y^2}\,. \tag{17}$$

One can easily check that (15) is the solution of (11) by simply substituting (15) into (11) and carrying out all the transformations. Equation (15) can be also obtained by determining the solution of the transient problem first, using the method described in Ex. 25.8, and by assuming next that $t \to \infty$.

If the plate gives off heat on one of its surfaces to surrounding air at temperature T_{cz}, while the opposite surface is thermally insulated, then one should assume in (16) that $m = 1$. When the plate is two-sided cooled, then $m = 2$. If the plate is insulated on one side, then one assumes that $m = 0$ in (16). Symbol α in (16) is the heat transfer coefficient on the plate surface, while K_0 a modified Bessel function of the second kind of zero order. The values of function K_0 are tabulated [16], but one can also calculate them by means of the library programs, such as for example [10].

Exercise 25.8 Determining Formula for Transient Temperature Distribution Produced by a Point Heat Source with Power $\dot{Q}_0$ that Moves at Constant Velocity v in Infinite Space or on the Half Space Surface

Determine transient temperature distribution due to a point heat source with power $\dot{Q}_0$, which moves at constant velocity v in an infinite space or on the half space surface. Thermo-physical properties of the medium are temperature independent, while the medium's temperature T_0 is constant.

Solution

A diagram of the point heat source that moves at constant velocity is presented in Fig. 25.9.

The superposition method will be used to determine temperature distribution; the method was already discussed in Chap. 16 and applied in Ex. 25.4 in order to determine temperature distribution around a stationary continuously active point source.

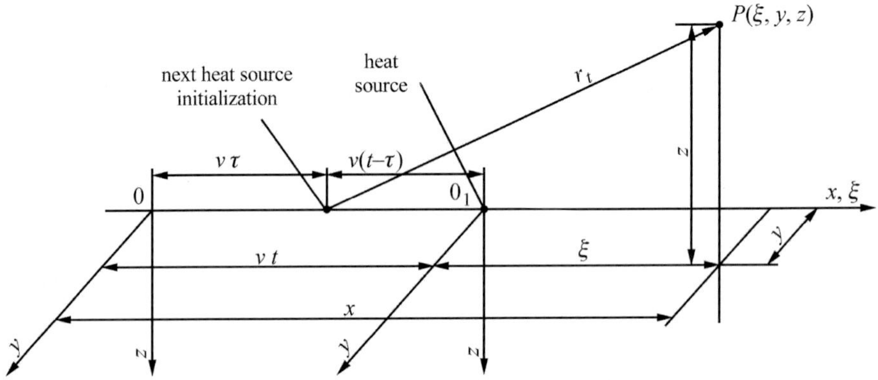

Fig. 25.9. A point heat source, which moves at constant velocity v along x axis; 0 is the starting point of the stationary system of coordinates x,y,z; 0_1 is the starting point of the moving system of coordinates ξ,y,z that moves at constant velocity v

A continuously active heat source can be treated as the sum of an infinite number of impulse sources successively activated at an infinitely small time intervals. In terms of the moving source, one should also account for the time-changing distance r_t (Fig. 25.9) between the instantenous heat source switched at time τ and the point $P = P(x,y,z) = P(\xi,y,z)$. Temperature at time t, when the source traversed vt, is the effect of an infinite number of active impulse sources that are being successively switched on at time interval $[0,t]$. In terms of the impulse source activated at time t, the temperature at point P can be calculated from (13) derived in Ex. 25.1

$$\theta_P = \frac{\dot{Q}_0}{c\rho\left(4\pi a\tau\right)^{3/2}}\exp\left[-\frac{\left(x-v\tau\right)^2+y^2+z^2}{4a\tau}\right]. \tag{1}$$

Distance r_t between the impulse heat source at point $(v\tau,0,0)$ and the point $P(x,y,z)$ (Fig. 25.9) is

$$r_t = \sqrt{\left(x-v\tau\right)^2+y^2+z^2}\,. \tag{2}$$

In accordance with the superposition method, the temperature distribution produced by the moving heat source is formulated as [13, 22, 27] (also Ex. 25.4)

$$\theta\left(x,y,z,t\right) = \int_0^t \frac{\dot{Q}_0}{c\rho\left[4\pi a\left(t-\tau\right)\right]^{3/2}}\exp\left[-\frac{\left(x-v\tau\right)^2+y^2+z^2}{4a\left(t-\tau\right)}\right]d\tau\,. \tag{3}$$

By introducing new variables

$$\psi = t-\tau, \qquad \xi = x-vt\,, \tag{4}$$

from which the following relationships result

$$d\psi = -d\tau\,, \tag{5}$$

$$x - v\tau = v\left(t-\tau\right) + \xi = v\psi + \xi\,, \tag{6}$$

$$\tau = 0 \rightarrow \psi = t\,, \tag{7}$$

$$\tau = t \rightarrow \psi = 0\,, \tag{8}$$

one is able to write (3) in the form

$$\theta\left(\xi, y, z, t\right) = \frac{\dot{Q}_0}{c\rho\left[4\pi a\right]^{3/2}} \int_0^t \left\{ \frac{1}{\psi^{3/2}} \exp\left[-\frac{\left(\xi + v\psi\right)^2 + y^2 + z^2}{4a\psi} \right] \right\} d\psi\,. \tag{9}$$

Once simple transformations are carried out, (9) assumes the form

$$\begin{gathered} \theta\left(\xi, y, z, t\right) = \frac{\dot{Q}_0}{c\rho\left[4\pi a\right]^{3/2}} \exp\left(-\frac{\xi v}{2a} \right) \times \\ \times \int_0^t \left[\frac{1}{\psi^{3/2}} \exp\left(-\frac{\xi^2 + y^2 + z^2}{4a\psi} \right) \exp\left(-\frac{v^2\psi}{4a} \right) \right] d\psi. \end{gathered} \tag{10}$$

In order to simplify (10), one should introduce a new variable [27]

$$\omega = \frac{v^2\psi}{4a}\,. \tag{11}$$

Once we account for (11) and carry out appropriate transformations, (10) has the form

$$\theta\left(\xi, y, z, t\right) = \frac{\dot{Q}_0 v}{16\lambda a \pi^{3/2}} \exp\left(-\frac{\xi v}{2a} \right) \int_0^{\frac{v^2 t}{4a}} \frac{1}{\omega^{3/2}} \exp\left(-\omega - \frac{u^2}{4\omega} \right) d\omega\,, \tag{12}$$

where

$$\begin{gathered} u = \frac{r_t v}{2a}, \\ r_t = \sqrt{\xi^2 + y^2 + z^2}\,. \end{gathered} \tag{13}$$

Radius r_t is the distance of the heat source from point P at the moment t. The integral in (12) is the dimensionless coefficient and resembles the modified Bessel function of the second type of zero order

$$K_0(u) = \frac{1}{2}\int_0^{\infty}\frac{1}{\omega}\exp\left(-\omega - \frac{u^2}{4\omega}\right)d\omega . \tag{14}$$

The longer the heating time t with the moving heat source is, the higher the value of the integral in (12) is and the higher the temperature at point P. The integral in (12) does not have the analytical solution and is numerically calculated. Quasi-steady state solution is obtained when time $t \to \infty$. It is clear from the calculation, however, that when the upper integration limit from (12) satisfies condition [13]

$$\frac{v^2 t}{4a} \geq 5 , \tag{15}$$

the temperature field is already quasi-steady.

One can assume, therefore, on the basis of condition (15) that the quasi-steady state of temperature distribution is already formed for time

$$t \geq \frac{20a}{v^2} . \tag{16}$$

It is clear, therefore, that the quasi-steady state is formed very quickly. Using the superposition method and solution (1), one can determine temperature distribution formulas for sources of different shapes and heat fluxes on the surface [13] that also move at constant velocity v.

Equation (12) was derived for the infinite space. In the case of the source with the same power $\dot{Q}_0$ that moves on the half space surface in (12), number 8 appears instead of number 16, since the total power $\dot{Q}_0$ emitted by the source moves to the half space. In terms of the infinite space, the heat source power disintegrates into two equal parts. The condition (16) is especially important in the case where integral (12) is numerically calculated by means of, for example, trapeze or rectangle method. One can determine, however, the analytical solution for $t \to \infty$ using (10). A new variable should be introduced

$$u^2 = r_t^2 / (4a\psi) \tag{17}$$

the following relationships accounted for

$$\psi = r_t^2 / (4au^2) , \tag{18}$$

$$d\psi = -\frac{r_t^2}{2au} du , \tag{19}$$

and, next, (17)–(19) accounted for in (10), to have

$$\theta(\xi, y, z, t) = \frac{\dot{Q}_0}{2\pi^{3/2}\lambda r_t} e^{-\frac{v\xi}{2a}} \int_0^{\infty} e^{-u^2 - \frac{p^2}{u^2}} du, \tag{20}$$

where

$$p = \frac{v r_t}{4a}. \tag{21}$$

Once we account that

$$\int_0^{\infty} e^{-u^2 - \frac{p^2}{u^2}} du = \frac{\sqrt{\pi}}{2} e^{-2p} \tag{22}$$

Equation (20) has the following form:

$$\theta(\xi, y, z, t) = \frac{\dot{Q}_0}{4\pi\lambda r_t} e^{-\frac{v}{2a}(\xi + r_t)}. \tag{23}$$

By comparing (23) to (12) from Ex. 25.7, one can see that both formulas are identical. In the case of the heat source, which moves on the half space surface, temperature distribution is formulated in (14) from Ex. 25.7, i.e. temperature excess θ is twice as large as the excess calculated from (23).

Exercise 25.9 Calculating Temperature Distribution along a Straight Line Traversed by a Laser Beam

During a welding process, a laser moves over the surface of a thick steel plate at a temperature of 20°C. Plate absorbed heat flow (power) is 2800 W. Laser moves at a speed of 3.3 mm/s. Calculate temperature distribution along the straight line, traversed by the laser beam. Assume that the thermo-physical properties of the plate are as follow: λ = 50 W/(m·K), ρ = 7800 kg/m^3, c = 460 J/(kg·K). Treat the plate as a semi-infinite body, while the temperature calculate under the assumption that the temperature distribution is in a quasi-steady state.

Solution

For the calculation, apply (14) from Ex. 25.7

$$\theta(\xi, r_t) = \frac{\dot{Q}_0}{2\pi\lambda r_t} \exp\left[-\frac{v}{2a}(\xi + r_t)\right], \tag{1}$$

where

$$r_t = \sqrt{\xi^2 + y^2} \,. \tag{2}$$

Equation (1) is valid for ((16), Ex. 25.8)

$$t \geq \frac{20a}{v^2} \,. \tag{3}$$

The point of origin of coordinates ξ,y (Fig. 25.8, Ex. 25.7) moves along with the heat source (the laser beam).

Accounting that in the given case $y = 0$ and $r_t = |\xi|$ (Fig. 25.10), (1) assumes the form

$$\theta(\xi) = \frac{\dot{Q}_0}{2\pi\lambda|\xi|} \exp\left[-\frac{v}{2a}\left(\xi + |\xi|\right)\right]. \tag{4}$$

Temperature diffusivity is

$$a = \frac{\lambda}{c\rho} = \frac{50}{460 \cdot 7800} = 1.3935 \cdot 10^{-5} \text{ m}^2/\text{s} \,.$$

Quasi-steady state is formed after time

$$t \geq \frac{20 \cdot 1.3935 \cdot 10^{-5}}{\left(3.3 \cdot 10^{-3}\right)^2} = 25.6 \text{ s} \,.$$

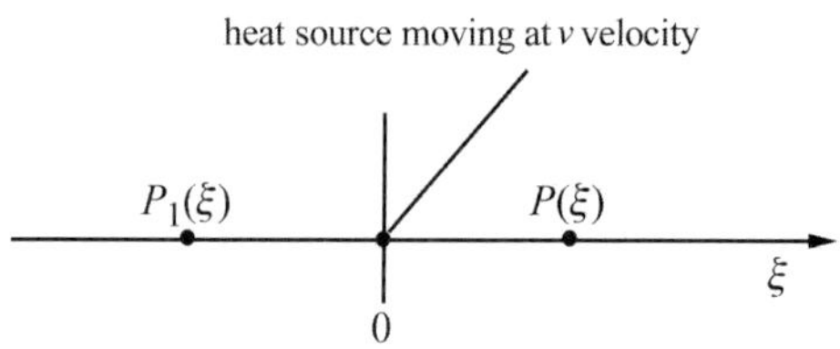

Fig. 25.10. A diagram that illustrates the movement of a laser along axis ξ

After substitution, (4) assumes the form

$$\theta(\xi) = \frac{2800}{2\pi \cdot 50|\xi|} \exp\left[-\frac{3.3 \cdot 10^{-3}}{2 \cdot 1.3935 \cdot 10^{-5}}\left(\xi + |\xi|\right)\right],$$

$$\theta(\xi) = \frac{8.9127}{|\xi|} \exp\left[-118.407\left(\xi + |\xi|\right)\right].$$

Temperature distribution is presented in Table 25.2 and Fig. 25.11.

Temperature $T(\xi)$ is formulated as

$$T(\xi) = T_0 + \theta(\xi) = 20 + \theta(\xi).$$

In the vicinity of the heat source, temperature is much higher than it is at steel melting; therefore, in this region the solution is strictly theoretical in character, since the formula employed for the calculation does not allow for steel melting and the subsequent heating in the liquid state.

Table 25.2. Temperature distribution along axis ξ

ξ [m]	T [°C]	ξ [m]	T [°C]	ξ [m]	T [°C]	ξ [m]	T [°C]	ξ [m]	T [°C]
–0.1	109.1	–0.015	614.18	–0.004	2248.18	0.001	7053.35	0.006	378.74
–0.075	138.84	–0.01	911.27	–0.003	2990.9	0.002	2795.14	0.007	262.65
–0.05	198.25	–0.008	1134.09	–0.002	4476.35	0.003	1479.98	0.008	187.55
–0.025	376.51	–0.006	1505.45	–0.001	8932.7	0.004	884.09	0.009	137.53
–0.020	465.64	–0.005	1802.54	0.000	∞	0.005	565.51	0.010	103.47

From the analysis of results presented in Table 25.2, one can see that the temperature behind the moving heat source decreases at much slower rate than it does in the region located in front of the source that moves at constant velocity.

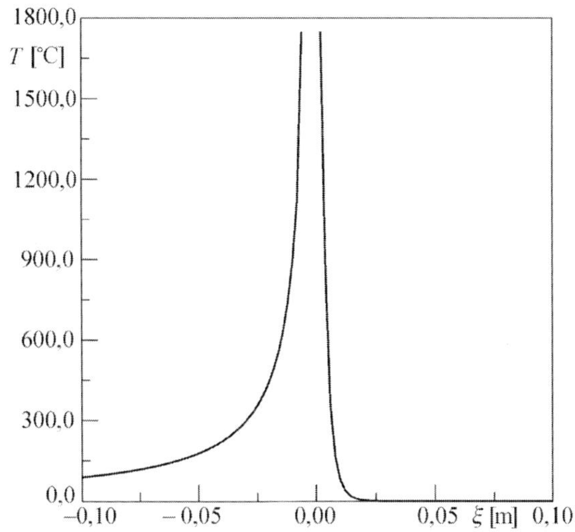

Fig. 25.11. Temperature distribution along the laser-traversed straight line

Figure 25.11 presents calculated temperature distribution in the distance function of a given point from the heat source. One can see that the plate temperature abruptly decreases as it moves away from the source, especially in the region located in front of the source.

Exercise 25.10 Quasi-Steady State Temperature Distribution in a Plate During the Welding Process; A Comparison between the Analytical Solution and FEM

Determine distribution of isotherms in quasi-steady state during butt welding of metal plates, which is in thickness $\delta = 6$ mm. The electrode moves at 1 mm/s. The welding parameters are as follow: $U = 20$ V, current intensity $I = 180$ A, heating efficiency $\eta = 0.6$. Thermo-physical constants of the steel are $\lambda = 41{,}9$ W/(m·K), $c = 570$ J/(kg·K), $\rho = 7350.8$ kg/m^3. Heat exchange with surroundings takes place through both plate surfaces when the heat transfer coefficient is $\alpha = 4.19$ W/(m^2·K). Initial temperature of the metal plate T_0 equals the temperature of the medium $T_{cz} = 20$°C. Also determine temperature distribution along the straight line traversed by the electrode and along the parallel line to a welding axis, which lies at a distance $y = 1$ cm. Use FEM and the approximate (15) from Ex. 25.7 to determine temperature distribution.

Solution

ANSYS programs will be employed to determine the isotherms. Thermal arc power is

$$\dot{Q} = UI\eta = 20 \cdot 180 \cdot 0.6 = 2160 \text{ W}. \tag{1}$$

Temperature distribution in a quasi-steady state is defined by (15) from Ex. 25.7, which in the given case assumes the form

$$\theta = T - T_{cz} = \frac{\dot{Q}}{2\pi\lambda\delta} e^{-\frac{v\xi}{2a}} K_0\left(r_t \sqrt{\frac{v^2}{4a^2} + \frac{2\alpha}{\lambda\delta}} \right), \tag{2}$$

where $r_t = \sqrt{\xi^2 + y^2}$. Once (2) is replaced, one has

$$\theta \quad \frac{2160}{2 \cdot 3.14 \cdot 41.9 \cdot 0.006} e^{\frac{-0.001\xi}{2 \cdot 1 \cdot 10^{-5}}} K_0\left(\sqrt{\xi^2 \quad y^2} \sqrt{\frac{0.001^2}{4 \cdot \left(1 \cdot 10^{-5}\right)^2} + \frac{2 \cdot 4.19}{41.9 \cdot 0.006}} \right),$$

$$\theta = 1368.135 \cdot e^{-50\xi} K_0\left(50\,332 \sqrt{\xi^2 + y^2}\right), \tag{3}$$

where x, y and ξ are expressed in m. Temperature excess θ [°C] is presented in Table 25.3. Temperature excess θ was calculated by means of the Mathcad program [19]. Next, temperature distribution was determined using the ANSYS program [1].

Table 25.3. Temperature excess in the plate θ [°C]

ξ [cm]	−20	−15	−10	−5	−3	−2	−1	0	1	2	3
y = 0	498.70	583.11	720.28	1012.7	1281.3	1528.2	2014.0	2693.3	781.76	210.62	63.21
y = 1	492.2	572.87	700.96	956.12	1158.8	1308.0	1473.5	1269.4	575.81	181.45	57.44

Calculations were done under the assumption that temperature field is three-dimensional. It was assumed that the length of the plate is $H = 0.6$ m, while the width $W = 0.6$ m. Electric arc was treated as a linear heat source with a power $\dot{q}_l = \dot{Q}_0/\delta =$ 2160/0.006 = 360000 W/m. It has been assumed, for the calculation with FEM, that the arc is a volumetric source, which measures 5×5×6 mm and whose center of gravity is located at the point of origin of coordinates (0,0,0). Once the new coordinate is introduced

$$\xi = x - vt \tag{4}$$

the transient heat conduction (4) from Ex. 25.7 has the form

$$c\rho\left(\frac{\partial T}{\partial t} - v\frac{\partial T}{\partial \xi}\right) = \lambda\left(\frac{\partial^2\theta}{\partial \xi^2} + \frac{\partial^2\theta}{\partial y^2} + \frac{\partial^2\theta}{\partial z^2}\right). \tag{5}$$

In the new coordinate system (ξ,y,z), the electrode is stationary, while the plate moves in the direction of ξ axis at a speed of $-v$ (Fig. 25.12).

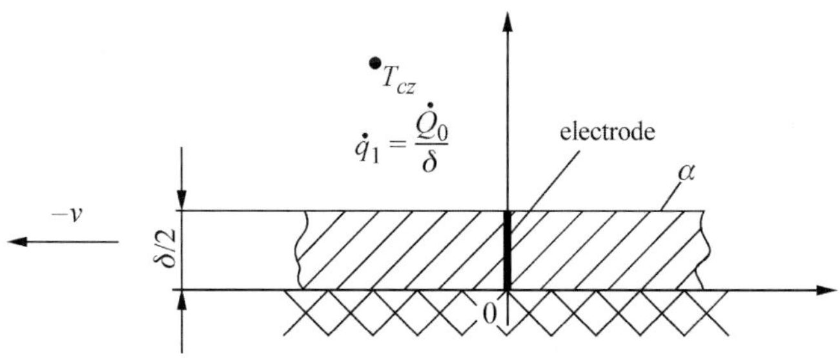

Fig. 25.12. Longitudinal cross-section of a plate with an assumed coordinate system

Calculation results for temperature distribution on the plate surface are presented in Table 25.4. Due to the symmetry, the temperature field was determined in 1/4 of the plate volume, with dimensions $H\times W/2\times\delta/2$ = 0.6×0.3×0.003 m.

a)

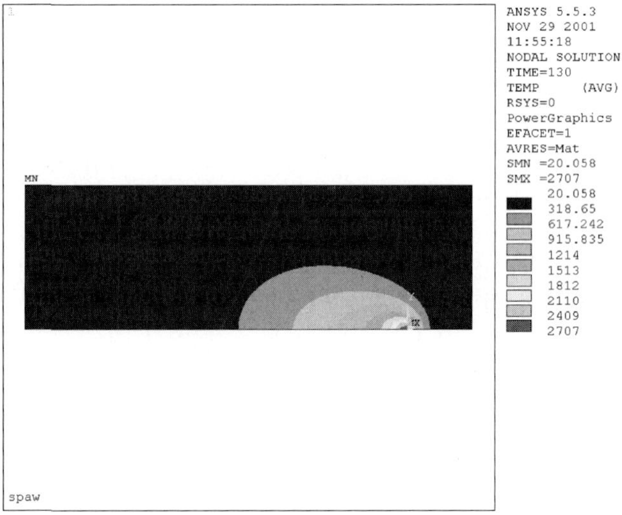

b)

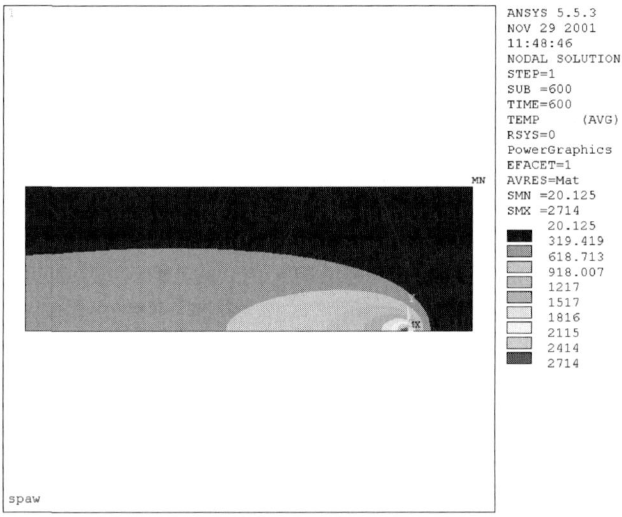

Fig. 25.13. Distribution of isotherms around an electrode: a) t = 130 s; b) t = 600 s – quasi-steady state; the presented plate fragment measures 0.35 × 0.11 m; the electrode is located on the bottom edge at a distance of 0.05 m from the right-hand side

Table 25.4. Plate temperature excess θ [°C] determined by means of FEM for a quasi-steady state; z = 0,003 m

ξ [cm]	−20	−15	−10	−5	−3	−2	−1	0	1	2	3
y = 0	498.70	583.11	720.28	1012.7	1281.3	1528.2	2014.0	2693.3	781.76	210.62	63.21
y = 1	492.2	572.87	700.96	956.12	1158.8	1308.0	1473.5	1269.4	575.81	181.45	57.44

Temperature distribution was calculated in a quasi-steady time by assuming that $\partial T/\partial t = 0$. Results given in Table 25.4 are for t = 600 s when $\partial T/\partial t \to 0$. The goodness of fit for the results obtained by means of (2) and FEM is very good (Tables 25.3 and 25.4). The distribution of isotherms on the plate surface, determined by means of FEM, is presented in Fig. 25.13.

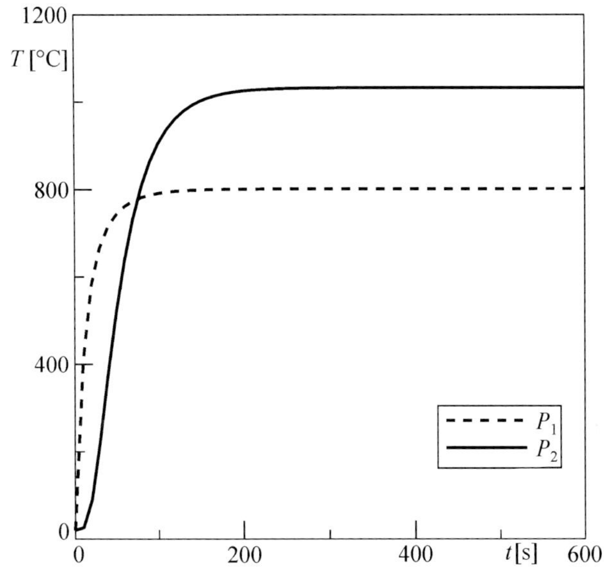

Fig. 25.14. Plate temperature history at points P_1 and P_2 located, respectively, at a distance ξ = 1 cm and ξ = −5 cm from the electrode axis, $y_1 = y_2 = 0$

The analyzed region $H \times W/2 \times \delta/2$ was divided into 7200 (120×60×1) elements. To evaluate how quickly the quasi-steady state is formed, calculations were also carried out during the transient state, assuming that T_{cz} = 20°C, $T(\xi, y, z, 0) = T_0 = T_{cz}$ = 20°C. The history of temperature excess $\theta = T - T_{cz}$ at points

$$P_1(\xi_1, y_1, z_1):\ \ \xi_1 = 0.01\ \text{m};\quad y_1 = 0.0\ \text{m}\ \ \text{and}\ \ z_1 = \delta/2 = 0.003\ \text{m},$$

$$P_2(\xi_2, y_2, z_2):\ \ \xi_2 = -0.05\ \text{m};\quad y_2 = 0.0\ \text{m}\ \ \text{and}\ \ z_2 = \delta/2 = 0.003\ \text{m},$$

in the function of time is presented in Fig. 25.14. From the analysis of temperature time transients at points P_1 and P_2 presented in Fig. 25.13, one can conclude that the quasi-steady state of temperature is already formed at time $t > 200$ s.

Literature

1. ANSYS 5.5.3(1998). ANSYS Inc., Urbana
2. Baehr HD, Stephan K (1994) Wärme-und Stoffübertragung. Springer Berlin
3. Barber JR (1967) Distribution of heat between sliding surfaces. J. Mech. Eng. Sci. 9: 351-354
4. Becker M (1986).: Heat Transfer. A Modern Approach. Plenum Press, New York - London
5. Becker M (2000) Nonlinear transient heat conduction using similarity groups. Transactions of the ASME, J. of Heat Transfer 122: 33-39
6. Bos J, Moes H (1995) Frictional heating of tribological contacts. Transactions of the ASME, J. Tribology 117: 171-217
7. Carslaw HS, Jaeger JC (1959) Condution of Heat in Solids. Ed. 2 Oxford University Pres, London
8. Dowden JM (2001) The Mathematics of Thermal Modeling. An Introduction to the Theory of Laser Material Processing. Chapman & Hall/CRC, Boca Raton-London
9. Francis HA (1970) Interfacial temperature distribution within a sliding hertzian contact. ASLE Trans. 14: 41-54
10. Goldak J, Chakravarti A, Bibby M (1984) A new finite element model for welding heat sources. Metall. Trans. 15B(6): 299-305
11. Goode SW (2000) Differential Equations and Linear Algebra. Prentice Hall, Upper Saddle River
12. Havalda A (1963) Thermal processes in electrical welding (in Polish). WNT, Warsaw
13. Hou ZB, Komanduri R (2000) General solutions for stationary/moving plane heat source problems in manufacturing and tribology. Int. J. of Heat and Mass Transfer 43: 1679-1698
14. Industrial Thermics. User's Guide (1996). Visual Analysis GmbH, München
15. Jaeger JC (1942) Moving sources of heat and the temperature at sliding contacts. Proc. Roy. Soc. NSW pp. 203-224
16. Janke E, Emde F, Lösch F (1960) Tafeln höherer Funktionen. Issue 6. Teubner Verlag, Stuttgart
17. Kasuya D, Yurioka N (1993) Prediction of welding thermal history by a comprehensive solution. Weld. Journal 72(3): 107-115
18. Ling FF (1973) Surface Mechanics. Wiley–Interscience, New York
19. MathCad 2000 (2000), MathSoft, Cambridge

20. Moore JE, Bibby MJ, Goldak JA, Santyr S (1986) A comparison of the point source and finite element schemes for computing weld cooling. In Nippes EF, Ball DJ (eds) Welding Research: The State of the Art. ASM, Miami, pp. 1-9
21. Myers PT, Uyehara OA, Borman GL (1967) Fundamentals of heat flow in welding. Welding Research Council Bulletin 123: 1-46
22. Myśliwiec M (1972) Thermo-mechanical fundamentals of welding (in Polish). WNT, Warsaw
23. Pilarczyk J (1983) Thermal phenomena in the process of welding (in Polish). In: Engineer's Guide. Welding Technology Vol. I. WNT, Warsaw, pp. 32-55
24. Radaj D (1992) Heat Effects of Welding. Springer, New York
25. Schneider PJ (1973) Conduction. Section 3. In: Rohsenow WM, Hartnett JP (eds) Handbook of Heat Transfer. McGraw-Hill, New York
26. Shaw MC (1984) Metal Cutting Principles. Oxford University Press, Oxford
27. Służalec A (1969) Metallurgic and thermal welding processes. Part II. Thermal welding processes (in Polish). Częstochowa University of Technology, Częstochowa
28. Thompson WJ (1997) Atlas for Computing Mathematical Functions. Wiley-Interscience Publication, New York
29. Tian X, Kennedy FE (1994) Maximum and average flash temperatures in sliding contact. Transactions ASME, J. Tribology 116: 167-174
30. Węgrzyn J (1971) Thermal and metallurgical welding processes (in Polish). Silesian University of Technology, Gliwice
31. Zill DG (1986) Differential Equations with Boundary-Value Problems. Prindle Weber & Schmidt, Boston
32. Baghranskij KW, Dobrotina ZA (1968) Theory od welding processes. Kharkov Univ. Press, Kharkov

26 Melting and Solidification (Freezing)

Melting and *solidification* are the two phenomena that frequently occur in nature and in many technological processes. Good examples of the phenomena that occur in nature are ice melting and water freezing, ground freezing (the uppermost surface layer), solidification of vulcanic lava and the melting processes that evolve deep under the earth surface. Examples of the phenomena that occur in many technological processes are the freezing and thawing of food products, casting, production of plastic products, welding, electrolytic machining and thermal energy accumulators, which make use of the metal or wax melting or freezing phenomena. In the cases when pure metals, ice or eutectic alloys undergo a phase change, one can observe a clear cut line between a liquid and a solid and a definite melting point t_m.

In glass materials, the transition from a liquid to a solid phase is gradual, since a sudden change in properties does not occur during the transition from one phase to the other. Figure 26.1 gives examples of phase changes that occur on a half space surface. For instance, solidification process during ice production or cast freezing is shown in Fig. 26.1a.

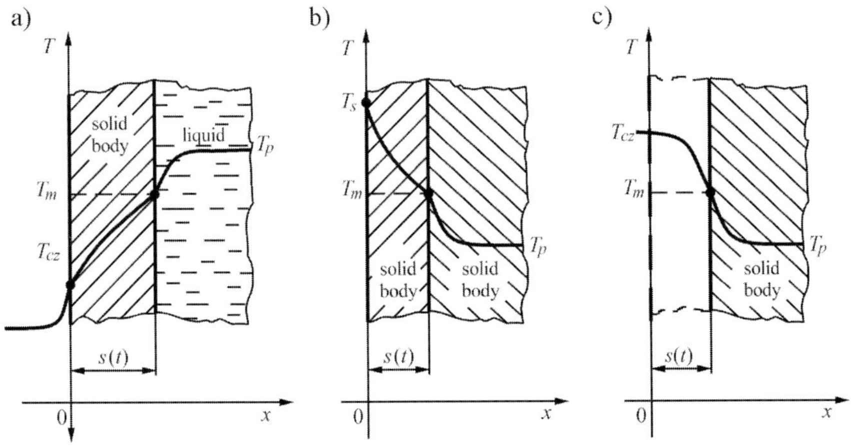

Fig. 26.1. Phase changes on the surface of a solid: (**a**) solidification (freezing), (**b**) decomposition, (**c**) ablation

In terms of the thermal or chemical decomposition, the layer of the degraded material is not removed from a body surface (Fig. 26.1b). A good example of decomposition is the carbonization of wood or fibre-strengthened plastics. Fig. 26.1c presents a case of ablation, when a layer of a melted material is being immediately removed from a surface of a solid (e.g. glass melting or Teflon sublimation).

It is assumed that in all three cases presented in Fig. 26.1, the phase change occurs at constant temperature T_m, as it does, for instance, in the cases of pure metals or water. During the phase change, a substancial change in the enthalpy of the substance occurs. If the enthalpy of a solid phase is designated by h_s [J/kg], while a liquid phase by h_l [J/kg], then a change in enthalpy during the transition from a solid to a liquid phase is

$$h_{sl} = \left(h_l - h_s\right) > 0 \,. \tag{1}$$

A change in enthalpy h_{sl} is also called the latent heat of melting and frequently denoted by a symbol L. During the melting of a material, the heat must be carried to a body, while during solidification (freezing), carried away from the body. Heat is absorbed or given off on a phase boundary, whose position changes in time. In the case of freezing, a liquid changes into a solid and the change in enthalpy is

$$h_{ls} = h_s - h_l = -h_{sl} \,. \tag{2}$$

Negative values of h_{ls} mean that at the phase boundary heat moves towards the body, which is being solidified or undergoes freezing. If L is the latent heat of melting or freezing, then

$$h_{sl} = -h_{ls} = L > 0 \,. \tag{3}$$

Heat balance equation, which describes the solidification process (freezing), has the following form:

$$\lambda_l \frac{\partial T_l}{\partial x} A \bigg|_s - \lambda_s \frac{\partial T_s}{\partial x} A \bigg|_s = \frac{d\left(m_{ls} h_{ls}\right)}{dt} , \tag{4}$$

where $dm_{ls} = \rho_s A ds$ is the increment of the frozen mass from the liquid phase, which occurred at time dt, while symbol A is the surface area on which the phase change occurs. If we allow that

$$\frac{d\left(m_{ls} h_{ls}\right)}{dt} = h_{ls} \frac{dm_{ls}}{dt} = h_{ls} \frac{d\left(\rho_s A s\right)}{dt} = \rho_s h_{ls} A \frac{ds}{dt} , \tag{5}$$

then (4) can be written in the form

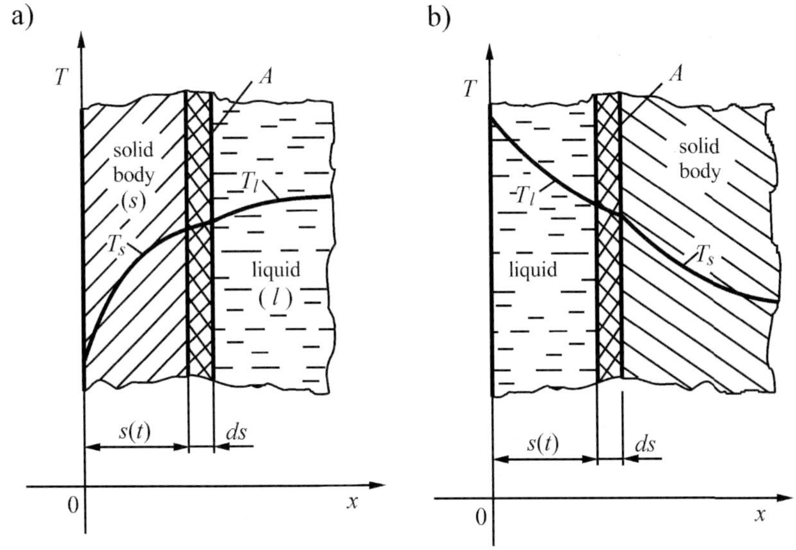

Fig. 26.2. Diagram of solidification (**a**) and melting (**b**) of semi-infinite body; $s(t)$ – location of a phase boundary, ds – increment of a solid phase (**a**) or a liquid phase (**b**) in time $(t, t + dt)$

$$\lambda_l \left.\frac{\partial T_l}{\partial x}\right|_{x=s} - \lambda_s \left.\frac{\partial T_s}{\partial x}\right|_{x=s} = \rho_s h_{ls} \frac{ds}{dt}. \tag{6}$$

By means of (6), we can determine the position of boundary $s(t)$ between the solid and the liquid phase. Later, latent heat of melting $h_{sl} = = L > 0$ will be applied, irrespectively of the process of solidification or melting. By allowing for (3) in (6), one obtains

$$-\lambda_l \left.\frac{\partial T_l}{\partial x}\right|_{x=s} + \lambda_s \left.\frac{\partial T_s}{\partial x}\right|_{x=s} = \rho_s h_{sl} \frac{ds}{dt}. \tag{7}$$

If we are going to move the first term of the equation above to its right-hand-side, this will signify that the heat flow, which moves towards the solid phase equals the sum of the heat flow, which moves towards the liquid and the heat flow that develops as a result of liquid freezing (Fig. 26.2a). In the case of melting (Fig. 26.2b), the heat flow, which moves towards the solid, consists of a heat flow that moves from the liquid side minus the heat flow used for the melting of the solid

$$-A\lambda_l \left.\frac{\partial T_l}{\partial x}\right|_{x=s} = -A\lambda_s \left.\frac{\partial T_s}{\partial x}\right|_{x=s} - A\rho_s h_{sl} \frac{ds}{dt}. \tag{8}$$

After transformations of (8), one has

$$-\lambda_l \left.\frac{\partial T_l}{\partial x}\right|_{x=s} + \lambda_s \left.\frac{\partial T_s}{\partial x}\right|_{x=s} = -\rho_s h_{sl} \frac{ds}{dt}. \tag{9}$$

From the comparison of (7) with (9), it follows that both, the process of solidification (freezing) and the process of melting can be described using a single formula

$$-\lambda_l \left.\frac{\partial T_l}{\partial x}\right|_{x=s} + \lambda_s \left.\frac{\partial T_s}{\partial x}\right|_{x=s} = \pm\rho_s h_{sl} \frac{ds}{dt}, \tag{10}$$

where symbol "+" refers to solidification, while symbol "–" to melting.

The phenomenon of liquid freezing and melting on the half space surface was the subject of work of J. Stefan [12]. Stefan has determined the thickness of the frozen liquid layer $s(t)$ by assuming that half space surface temperature $T_0 < T_m$ is constant. Additional assumption, which simplified the problem, was that constant temperature of a liquid equaled the melting point T_m. Heat conduction phenomenon in the liquid phase was taken into account by F. Neumann [5].

In many engineering problems, it is important to determine the time, after which the total solidification or freezing of a product occurs. Such problem was solved by R. Plank [10], who assumed that in the layer of a solidified or a melted material the temperature distribution is in a steady-state, i.e. he neglected the heat accumulation in the developing layer of a solid or liquid phase with thickness $s(t)$.

K. Nesselmann [9] and H. Martin [8] also worked on the same type of quasi-steady solutions. The accuracy of quasi-steady solutions is higher for larger enthalpies h_{sl}, when heat released during liquid freezing or absorbed during the melting of a solid plays a greater role in the course of the occurring phenomenon than the transient processes of heat accumulation in the developing layer of a solid or a liquid phase. Such situation occurs during water freezing, when water displays a high latent heat of freezing h_{ls}.

Ablation on the half space surface is analyzed in paper [13] with a convective boundary condition of 2nd kind, i.e. when there is a step-increase in heat flux. Ablation is used in thermal protection systems of bodies that are exposed to very high temperatures such as, for instance, space shuttles or jet nozzles. Large heat flows are transferred by a melting material, which is being immediately removed from the surface of an element due to large shear forces caused by objects or combustion gases, which move at high speed. The insulation of other construction elements is ensured at the expense of losing parts of thermal shields. Heat transfer problems that arise during melting or solidification (freezing) processes are the object of discussion in references [1, 3, 4, 6, 11, 14].

In this chapter, we will discuss in greater detail the process of solidification (freezing) on a half space surface, assuming that the temperature of the inner, deeper buried layers of a medium remains constant for $x \geq s(t)$ and equals the melting temperature $T_{m,.}$ We will also discuss a more commonplace problem, in which the process of heat conduction in a solid and liquid phase will be accounted for as well as the approximate quasi-steady state method for solving heat conduction problems, in which a phase change occurs. Using the second of the aforementioned methods, we will determine the freezing times for simple-shape food products and basic parameters, which characterize the ablation process with boundary conditions of 2nd and 3rd kind.

Exercise 26.1 Determination of a Formula which Describes the Solidification (Freezing) and Melting of a Semi-Infinite Body (the Stefan Problem)

Derive formulas, which describe solidification (freezing) and melting of a semi-infinite body (the Stefan problem [12]), assuming that surface temperature is T_0. In the case of solidification, assume that the temperature of a liquid is constant and equals the melting temperature T_m; do the same in the case of melting for the solid phase. Derive formulas for temperature distribution in a solid phase and liquid phase for the processes of solidification and melting, respectively. Also derive dependencies for the thickness of a solidified or melted layer in the function of time and the rate of phase boundary relocation.

Solution

Solidification and melting of semi-infinite body is illustrated, respectively in Fig. 26.3a and 26.3b.

a. *Solidification (freezing) of a semi-infinite body (Fig. 26.3a)*

Temperature distribution in a solid phase is described by the heat conduction equation

$$\frac{\partial T_s}{\partial t} = a_s \frac{\partial^2 T_s}{\partial x^2}, \quad 0 < x < s(t), \quad t > 0, \tag{1}$$

by boundary conditions

$$T_s\big|_{x=0} = T_0, \quad t > 0, \tag{2}$$

$$T_s\big|_{x=s} = T_m, \quad t > 0\,, \tag{3}$$

by initial condition

$$T_s\big|_{t=0} = T_m, \quad x > 0\,, \tag{4}$$

and the condition at the solid-liquid boundary

$$\lambda_s \frac{\partial T_s}{\partial x}\bigg|_{x=s} = \rho_s h_{sl} \frac{ds}{dt}, \quad t > 0\,, \tag{5}$$

while

$$s\big|_{t=0} = 0\,. \tag{6}$$

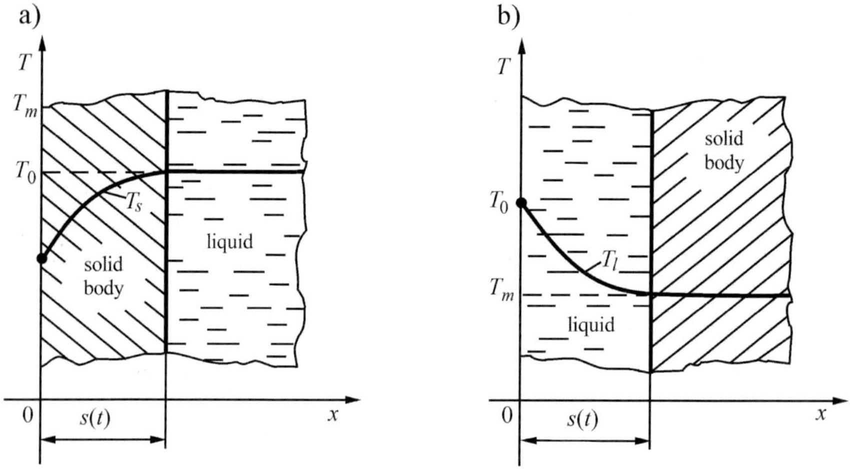

Fig. 26.3. Solidification (a) and melting (b) of a semi-infinite body, $s(t)$ – the thickness of a solidified (frozen) or melted layer

The solution of the heat conduction (1) for a half space is the function

$$T_s = T_0 + C\mathrm{erf}\left(\frac{x}{2\sqrt{a_s t}}\right), \tag{7}$$

which satisfies condition (2). By substituting (7) into condition (3), one obtains

$$T_m = T_0 + C\mathrm{erf}\left(\frac{s}{2\sqrt{a_s t}}\right). \tag{8}$$

If temperatures T_m and T_0 in (8) are constant and do not depend on time, then the argument of function erf must be independent of time. If the argument of function erf is equal to a certain constant β

$$\frac{s}{2\sqrt{a_s t}} = \beta , \tag{9}$$

hence,

$$s = 2\beta\sqrt{a_s t} , \tag{10}$$

then (8) can be written in the form

$$T_m = T_0 + C\mathrm{erf}\beta . \tag{11}$$

Once constant C is determined from (11) and substituted into (7), the temperature distribution in the solid phase is formulated as

$$T_s(x,t) = T_0 + (T_m - T_0)\frac{\mathrm{erf}\left(x/\sqrt{4a_s t}\right)}{\mathrm{erf}\beta} = T_0 + (T_m - T_0)\frac{\mathrm{erf}(\beta x/s)}{\mathrm{erf}\beta}, \tag{12}$$

while constant β is determined from (5).

From (10) it follows that

$$\frac{ds}{dt} = \beta\sqrt{\frac{a_s}{t}} . \tag{13}$$

It is easy to calculate heat flux on the phase boundary $\lambda_s(\partial T_s/\partial x)|_{x=s}$ once we account that

$$\frac{d(\mathrm{erf}x)}{dx} = \frac{2}{\sqrt{\pi}}e^{-x^2} . \tag{14}$$

Accounting for expressions (12) and (14), one has

$$\lambda_s\left(\frac{\partial T_s}{\partial x}\right)\bigg|_{x=s} = \lambda_s(T_m - T_0)\frac{2e^{-\beta^2}}{\sqrt{\pi}\,\mathrm{erf}\beta}\frac{\beta}{s} = \frac{\lambda_s(T_m - T_0)e^{-\beta^2}}{\sqrt{\pi}\sqrt{a_s t}\,\mathrm{erf}\beta} . \tag{15}$$

Substituting (13) and (15) into (5) yields

$$\beta e^{\beta^2}\mathrm{erf}\beta = \frac{Ste_s}{\sqrt{\pi}} , \tag{16}$$

where Ste_s is the *Stefan number*

$$Ste_s = \frac{c_s(T_m - T_0)}{h_{sl}} . \tag{17}$$

In order to determine β for the assigned value of Stefan number Ste_s, one should solve a non-linear algebraic (16) with one of the known methods, e.g. interval search method, interval bisection method, or secant method. Transient β in function Ste_s is presented in Fig. 26.4.

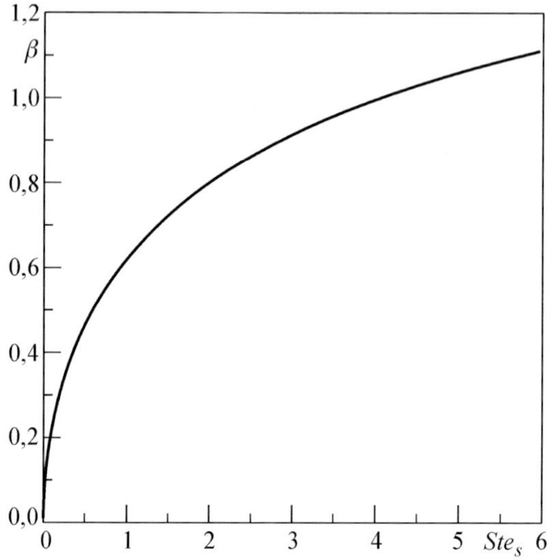

Fig. 26.4. The root of (16) in the Stefan number function Ste_s

Knowing the value of β, one can determine temperature distribution $T_s(x,t)$ and $s(t)$. One can also calculate solidification (freezing) time for a layer with the assigned thickness $s = d$. Therefore, from (10), one has

$$t_d = \frac{d^2}{4\beta^2 a_s}. \tag{18}$$

Time t_d is directly proportional to the square of the layer's thickness d and inversely proportional to the temperature diffusivity of the solid.

b. *Melting (Fig. 26.3b)*

Temperature distribution in the liquid phase is described by the heat conduction equation

$$\frac{\partial T_l}{\partial t} = a_l \frac{\partial^2 T_l}{\partial x^2}, \qquad 0 < x < s(t), \tag{19}$$

boundary conditions

$$T_l\big|_{x=0} = T_0, \qquad t > 0, \tag{20}$$

$$T_l\big|_{x=s} = T_m, \qquad t > 0, \tag{21}$$

initial condition

$$T_l\big|_{t=0} = T_m, \qquad x > 0 \tag{22}$$

and the condition at the liquid-solid boundary ((10), Chap. 26)

$$\lambda_l \frac{\partial T_l}{\partial x}\bigg|_{x=s} = \rho_l h_{sl} \frac{ds}{dt}, \qquad t > 0, \tag{23}$$

while

$$s\big|_{t=0} = 0 . \tag{24}$$

Once problems (19)÷(24) are solved, as in the case of solidification, one obtains

$$T_l(x,t) = T_0 - (T_0 - T_m)\frac{\operatorname{erf}\left(x/\sqrt{4a_l t}\right)}{\operatorname{erf}\beta}, \qquad t > 0, \tag{25}$$

where β is the root of equation

$$\beta e^{\beta^2} \operatorname{erf}\beta = \frac{Ste_l}{\sqrt{\pi}}, \tag{26}$$

while

$$Ste_l = \frac{c_l (T_0 - T_m)}{h_{sl}} . \tag{27}$$

The thickness of the liquid layer $s(t)$ is calculated from formula

$$s(t) = 2\beta\sqrt{a_l t} . \tag{28}$$

It is easy to determine from (28) the melting time of a layer with a thickness $s = d$

$$t_d = \frac{d^2}{4\beta^2 a_l} . \tag{29}$$

One can observe that (16) and (26) have a similar form. Once the Stefan number Ste_l is calculated from (27), the value β can be read out from Fig. 26.4 or Table 26.1.

Table 26.1. β of (16) or (26) in the Stefan number function Ste_s

β	Ste_s	β	Ste_s	β	Ste_s
0.01	0.0002	0.39	0.3370	0.77	1.7870
0.03	0.0018	0.41	0.3765	0.79	1.9240
0.05	0.0050	0.43	0.4189	0.81	2.0700
0.07	0.0098	0.45	0.4644	0.83	2.2250
0.09	0.0163	0.47	0.5130	0.85	2.3910
0.11	0.0244	0.49	0.5650	0.87	2.5690
0.13	0.0342	0.51	0.6205	0.89	2.7580
0.15	0.0457	0.53	0.6798	0.91	2.9610
0.17	0.0589	0.55	0.7431	0.93	3.1770
0.19	0.0740	0.57	0.8107	0.95	3.4080
0.21	0.0908	0.59	0.8827	0.97	3.6560
0.23	0.1096	0.61	0.9595	0.99	3.9210
0.25	0.1303	0.63	1.0410	1.01	4.2040
0.27	0.1531	0.65	1.1290	1.03	4.5080
0.29	0.1780	0.67	1.2220	1.05	4.8340
0.31	0.2050	0.69	1.3210	1.07	5.1830
0.33	0.2343	0.71	1.4260	1.09	5.5580
0.35	0.2660	0.73	1.5390	1.11	5.9600
0.37	0.3002	0.75	1.6590		

Exercise 26.2 Derivation of a Formula that Describes the Solidification (Freezing) of a Semi-Infinite Body Under the Assumption that the Temperature of a Liquid is Non-Uniform

Initial temperature of a liquid measures T_p and is higher than the melting temperature T_m. The surface temperature of a semi-infinite space was suddenly lowered to the temperature T_0 (Fig. 26.5), lower than the temperature T_m. Determine temperature distribution in the solid and liquid phase and the position of the boundary between the solid and liquid phase.

Solution

This problem was solved by F. Neuman [5] and also presented in the work by H. S. Carslaw and J. C. Jaeger [2]. Temperature distribution in the solid phase is described by the heat conduction equation

$$\frac{\partial T_s}{\partial t} = a_s \frac{\partial^2 T_s}{\partial x^2}, \qquad 0 < x < s(t), \qquad t > 0 \tag{1}$$

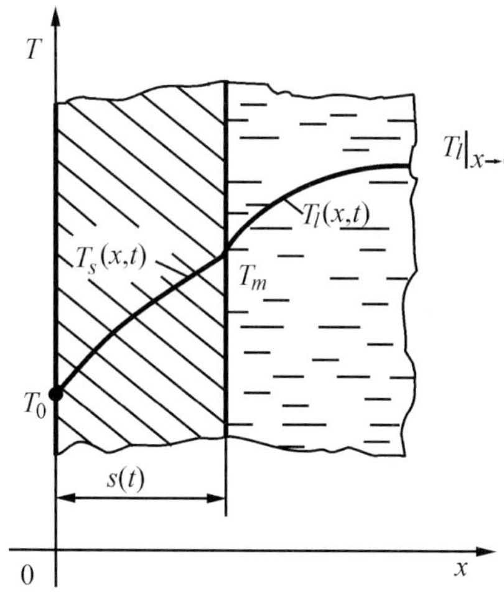

Fig. 26.5. Solidification (freezing) on the half space surface

and boundary conditions

$$T_s\big|_{x=0} = T_0, \qquad t > 0, \tag{2}$$

$$T_s\big|_{x=s} = T_m, \qquad t > 0. \tag{3}$$

Temperature distribution in the liquid phase is also described by the heat conduction equation

$$\frac{\partial T_l}{\partial t} = a_l \frac{\partial^2 T_l}{\partial x^2}, \qquad s(t) < x < \infty, \qquad t > 0, \tag{4}$$

by boundary conditions

$$T_l\big|_{x=s} = T_m, \qquad t > 0, \tag{5}$$

$$T_l\big|_{x=\infty} = T_p, \qquad t > 0 \tag{6}$$

and by initial condition

$$T_l\big|_{t=0} = T_p, \qquad x > 0. \tag{7}$$

Heat balance equation is also satisfied at the phase boundary $x = s$

$$\lambda_s \frac{\partial T_s}{\partial x}\bigg|_{x=s} - \lambda_l \frac{\partial T_l}{\partial x}\bigg|_{x=s} = \rho_s h_{sl} \frac{ds}{dt}. \tag{8}$$

Temperature distribution in the solid phase, which satisfies boundary condition (2), has the form

$$T_s(x,t) = T_0 + A\mathrm{erf}\left(x/\sqrt{4a_s t}\right), \tag{9}$$

while for liquids the solution, which satisfies condition (6), has the form

$$T_l(x,t) = T_p + B\mathrm{erfc}\left(x/\sqrt{4a_l t}\right). \tag{10}$$

From condition (3) it follows that solution (9) can satisfy such condition, if the argument of function erf is a constant

$$\beta = \frac{s(t)}{2\sqrt{a_s t}}, \tag{11}$$

hence, the expression that defines the position of the phase boundary is obtained

$$s(t) = 2\beta\sqrt{a_s t}. \tag{12}$$

Once we account for (9) and (11), condition (5) assumes the form

$$T_0 + A\mathrm{erf}\beta = T_m, \tag{13}$$

while after allowing for (10) and (11)

$$T_p + B\mathrm{erfc}\left(\beta\sqrt{\frac{a_s}{a_l}}\right) = T_m. \tag{14}$$

Constants A and B, determined respectively from (13) and (14), have the form

$$A = \frac{T_m - T_0}{\mathrm{erf}\beta}, \tag{15}$$

$$B = \frac{T_m - T_p}{\mathrm{erfc}\left(\beta\sqrt{\frac{a_s}{a_l}}\right)}. \tag{16}$$

By substituting constant A into (9) and constant B into (10), the expression for temperature distribution in the solid and liquid phase is obtained

$$T_s(x,t) = T_0 + (T_m - T_0)\frac{\operatorname{erf}\left(x/\sqrt{4a_s t}\right)}{\operatorname{erf}\beta}, \tag{17}$$

$$T_l(x,t) = T_p + (T_m - T_p)\frac{\operatorname{erfc}\left(x/\sqrt{4a_l t}\right)}{\operatorname{erfc}\left(\beta\sqrt{\dfrac{a_s}{a_l}}\right)}. \tag{18}$$

Once (12), (17) and (18) are substituted into (8) and all transformations carried out, one has

$$\frac{e^{-\beta^2}}{\operatorname{erf}\beta} + \frac{\lambda_l}{\lambda_s}\sqrt{\frac{a_s}{a_l}}\frac{T_m - T_p}{T_m - T_0}\frac{e^{-\beta^2(a_s/a_l)}}{\operatorname{erfc}\left(\beta\sqrt{\dfrac{a_s}{a_l}}\right)} = \frac{\beta\sqrt{\pi}}{Ste_s}, \tag{19}$$

where the Stefan number Ste_s is formulated as

$$Ste_s = \frac{c_s(T_m - T_0)}{h_{sl}}. \tag{20}$$

The appropriate value of constant β is obtained for the assigned value Ste_s once the non-linear algebraic (19) is solved. Knowing the value of β, one can calculate $s(t)$, $T_s(x,t)$ and $T_l(x,t)$ successively.

Exercise 26.3 Derivation of a Formula that Describes Quasi-Steady-State Solidification (Freezing) of a Flat Liquid Layer

Determine basic dependencies that describe the solidification (freezing) process in a quasi-steady-state of a liquid layer with thickness d. Assume that the plate is being symmetrically cooled on both sides. Heat transfer coefficient on the plate surface α is constant.

Assume that temperature distribution in the solid phase is in a steady state. Temperature of the liquid equals the temperature of melting T_m. Compare the obtained approximate solution with the analytical solution for $\alpha \to \infty$.

Solution

Such method of analysis for the processes of solidification and melting, based on the assumption that the solidified or melted body layer has a steady-state distribution, is described as a quasi-steady state. The accuracy of the quasi-steady-state analysis is usually very good, since the latent heat of melting h_{sl} or freezing h_{ls} is much higher, with respect to the absolute value, than the specific heat of a liquid or a solid. Therefore, transient heat accumulation in a layer with thickness $s(t)$ does not play a greater role. Moreover, solidification or freezing are usually long-lasting processes; therefore, temperature distribution in a solidified layer of a material resembles a steady-state distribution.

A diagram of the analyzed problem is presented in Fig. 26.6. Due to the symmetry, only a half of the plate will be analyzed below.

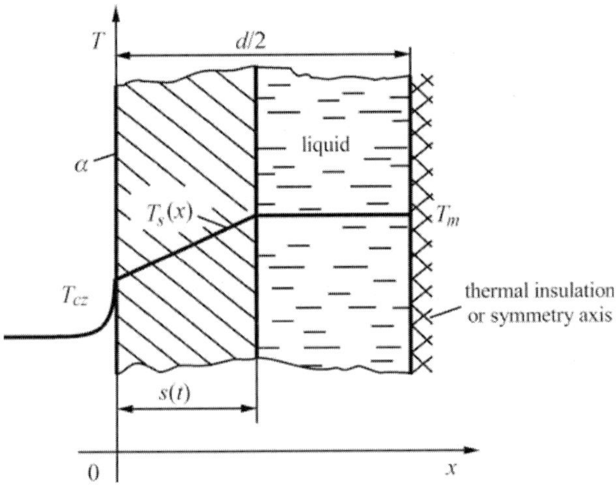

Fig. 26.6. Solidification (freezing) of a flat liquid layer; temperature distribution in the solid phase is in a quasi-steady-state

Temperature distribution is defined by the heat conduction equation

$$\frac{d^2T_s}{dx^2} = 0, \tag{1}$$

when boundary conditions are

$$\lambda_s \left.\frac{\partial T_s}{\partial x}\right|_{x=0} = \alpha\left(T_s\big|_{x=0} - T_{cz}\right), \tag{2}$$

$$T_s\big|_{x=s} = T_m, \tag{3}$$

and heat balance equation on the phase boundary $x = s$ is

$$\lambda_s \left.\frac{\partial T_s}{\partial x}\right|_{x=s} = \rho_s h_{sl} \frac{ds}{dt}, \tag{4}$$

$$s|_{t=0} = 0 .$$

Once (1) is twice integrated, one has

$$T_s(x) = A + Bx , \tag{5}$$

while, after (5) is substituted into conditions (2) and (3), one obtains

$$A = T_m - Bs,$$

$$B = \frac{\alpha(T_m - T_{cz})}{\lambda_s + \alpha s}. \tag{6}$$

If one accounts for constants (6) in (5), then temperature distribution in the solidified layer is formulated as

$$T_s(x) = T_m + B(x - s) = T_m + \frac{\alpha}{\lambda_s + \alpha s}(T_m - T_{cz})(x - s). \tag{7}$$

By substituting (7) into (4), the following differential equation is obtained

$$\rho_s h_{sl} \frac{ds}{dt} = \frac{\alpha(T_m - T_{cz})}{1 + \dfrac{\alpha s}{\lambda_s}}, \tag{8}$$

which, once a new variable is introduced,

$$z = 1 + \frac{\alpha}{\lambda_s} s \tag{9}$$

has the form

$$z\frac{dz}{dt} = \frac{\alpha^2(T_m - T_{cz})}{\lambda_s \rho_s h_{sl}},$$

$$\frac{d(z^2/2)}{dt} = \frac{\alpha^2(T_m - T_{cz})}{\lambda_s \rho_s h_{sl}}. \tag{10}$$

As a result of the separation of variables and the integration of the equation from $s = 0$ for $t = 0$ to $s = s(t)$ for t, one obtains

$$\left.\frac{z^2}{2}\right|_1^{1+\frac{\alpha}{\lambda_s}s} = \frac{\alpha^2\left(T_m - T_{cz}\right)}{\lambda_s \rho_s h_{sl}} t,$$

$$\frac{1}{2}\left(1+\frac{\alpha s}{\lambda_s}\right)^2 - \frac{1}{2} = \frac{\alpha^2\left(T_m - T_{cz}\right)}{\lambda_s \rho_s h_{sl}} t. \tag{11}$$

The thickness of the solidified (frozen) layer determined from (11) is expressed as

$$s(t) = \frac{\lambda_s}{\alpha}\left(\sqrt{1+\frac{2\alpha^2\left(T_m - T_{cz}\right)}{\lambda_s \rho_s h_{sl}} t} - 1\right). \tag{12}$$

When $\alpha \to \infty$, then boundary condition of 3rd kind changes into the condition of 1st kind; that is, the temperature of the half space surface T_{cz} is assigned. From (12), one obtains then

$$s(t) = \sqrt{\frac{2\lambda_s t}{\rho_s} \frac{\left(T_m - T_{cz}\right)}{h_{sl}}} = 2\sqrt{at}\sqrt{\frac{Ste_s}{2}}, \tag{13}$$

where $Ste_s = c_s(T_m - T_{cz})/h_{sl}$ is the Stefan number.

Accounting that ((9), Ex. 26.1)

$$\beta = \frac{s}{2\sqrt{a_s t}} \tag{14}$$

Equation (13) can be transformed into a form

$$\beta = \frac{s(t)}{2\sqrt{a_s t}} = \sqrt{\frac{Ste_s}{2}}. \tag{15}$$

The above formula allows to make an approximate determination of the root of (16) in Ex. 26.1

$$\beta e^{\beta^2} \operatorname{erf}\beta = \frac{Ste_s}{\sqrt{\pi}} \tag{16}$$

for small Stefan numbers: $Ste_s < 0.5$. Fig. 26.7 shows how the root β calculated from (16) compares with the root determined from the approximate (15). From the analysis of this figure, it is clear that condition $Ste_s < 0.5$ is well satisfied in the cases of water freezing or the freezing of substances, which contain a lot of water, such as food products.

If we assume the following data for water (l) and ice (s), $T_{cz} = -20°C$, $\rho_s = 917\ kg/m^3$, $\rho_l = 1000\ kg/m^3$, $h_{sl} = 333$ kJ/kg, $c_s = 2.05$ kJ/(kg·K), $\lambda_s = 2.2$ W/(m·K) and $T_m = 0°C$, then we will obtain the following values of Stefan number Ste_s:

$$Ste_s = \frac{c_s\left(T_m - T_{cz}\right)}{h_{sl}} = \frac{2.05\left[0-(-20)\right]}{333} = 0.1231.$$

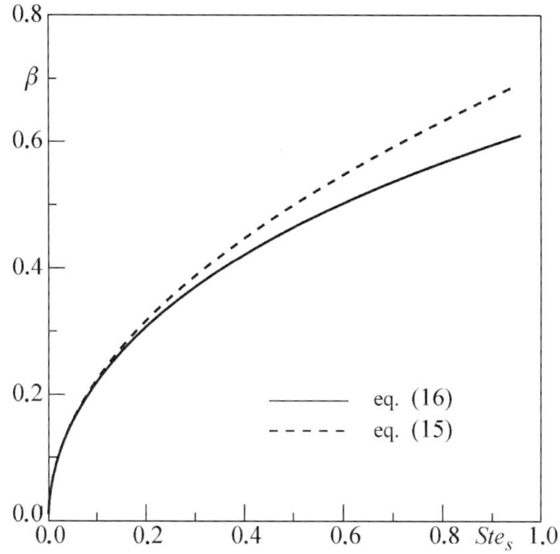

Fig. 26.7. Comparison of root β determined from the analytical (16) and approximate (15) equations

It is evident, therefore, that the quasi-steady-state analysis of the food freezing and defrosting phenomena is fully justified. The freezing time (solidification) t_z of a two-sided cooled product with thickness d (or with thickness $d/2$, cooled on one surface and insulated on the other) is determined from (11) once it is assumed that $s = d/2$

$$t_z = \frac{\rho_s h_{sl}}{T_m - T_{cz}}\left(\frac{1}{2}\frac{d}{\alpha} + \frac{1}{8}\frac{d^2}{\lambda_s}\right). \tag{17}$$

The product $\rho_s h_{sl} = q_z$ is a unit latent heat of freezing expressed in J/m^3, whose values for different food products can be found in literature on refrigerating engineering.

Exercise 26.4 Derivation of Formulas that Describe Solidification (Freezing) of Simple-Shape Bodies: Plate, Cylinder and Sphere

Derive formulas, which describe solidification (freezing) of simple-shape bodies (plate, cylinder, sphere) by assuming that temperature field is in a steady-state in a solidified (frozen) layer. Also assume that the product is enclosed in a thin-wall casing (package) with a thickness g, through which the heat is carried away to surroundings at temperature T_{cz} (Fig. 26.8).

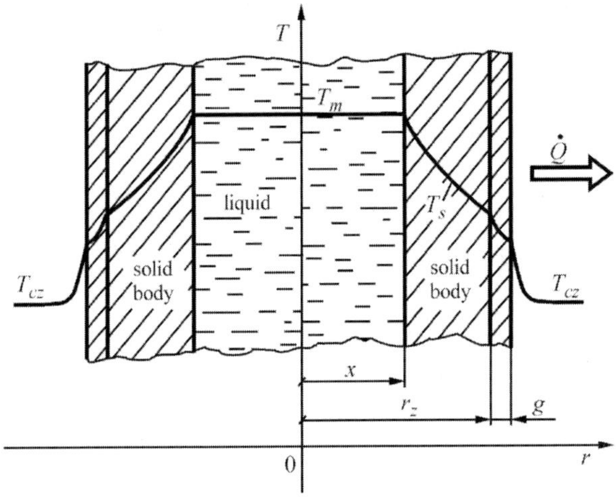

Fig. 26.8. Solidification of a body in a simple-shape casing with thickness g

Solution

Solidification process is treated as a steady-state process; therefore, the heat flow $\dot{Q}$, which develops during the solidification of a liquid at temperature T_m, passes to surroundings. The heat flow, which forms during the liquid solidification on the liquid-solid boundary $r = x$ can be written in the form

$$\dot{Q} = -\rho_s h_{sl} A(x) \frac{dx}{dt}. \tag{1}$$

Symbol "–" results from the negative value of the derivative dx/dt.

The heat flow $\dot{Q}$, which is carried away to surroundings, can also be calculated from formula

$$\dot{Q} = k_z A(r_z)(T_m - T_{cz}). \tag{2}$$

In (1), $A(x)$ is the surface area on the phase boundary. This area is formulated as

- $A(x) = HW$ for the plate (H – plate height, W – plate width),
- $A(x) = 2\pi xL$ for the cylinder (L – cylinder height),
- $A(x) = 4\pi x^2$ for the sphere.

Heat transfer coefficient k_z refers to the surface area $A(r_z)$, which is defined by the following formulas:

- $A(r_z) = HW$ for the plate,
- $A(r_z) = 2\pi r_z L$ for the cylinder,
- $A(r_z) = 4\pi r_z^2$ for the sphere.

Heat transfer coefficient k_z has the form

$$\frac{1}{k_z} = \frac{1}{k_x} + \frac{1}{k_{zew}}, \tag{3}$$

where

$$\frac{1}{k_{zew}} = \frac{g}{\lambda_{sc}} \frac{r_z^n}{(r_z + g/2)^n} + \frac{1}{\alpha} \frac{r_z^n}{(r_z + g)^n} \tag{4}$$

is the component of the heat transfer resistance ($r_{zew} = 1/k_{zew}$) that results from thermal wall resistance and thermal resistance from convective heat exchange on the outer surface.

The values of exponent n are as follow: $n = 0$ for the plate, $n = 1$ for the cylinder and $n = 2$ for the sphere.

Thermal resistance of the solidified (frozen) layer with thickness $(r - x)$ is

$$\frac{1}{k_x} = \frac{r_z}{\lambda_s}\left(1 - \frac{x}{r_z}\right) \quad \text{for the plate,} \tag{5}$$

$$\frac{1}{k_x} = \frac{r_z}{\lambda_s} \ln\left(\frac{r_z}{x}\right) \quad \text{for the cylindrical layer,} \tag{6}$$

$$\frac{1}{k_x} = \frac{r_z}{\lambda_s}\left(\frac{r_z}{x} - 1\right) \quad \text{for the spherical layer.} \tag{7}$$

Once we equate (1) and (2), we will obtain the following differential equation:

$$-\rho_s h_{sl} A(x)\frac{dx}{dt} = k_z A(r_z)(T_m - T_{cz}). \tag{8}$$

If we allow for (3)–(7) and corresponding surface areas $A(x)$ and $A(r_z)$ in (8), we will obtain the following differential equation:

$$-\rho_s h_{sl}\left(\frac{x}{r_z}\right)^{n+1}\frac{dx}{dt} = k_z(x)(T_m - T_{cz}), \tag{9}$$

where, $n = 0$ for the plate, $n = 1$ for the cylinder, $n = 2$ for the sphere.

Due to different forms of $k_z(x)$, the equation above will be separately integrated for each element.

a. *Plate*

Once (3) and (5) are accounted for in (9), the variables separated and then integrated within appropriate limits, one has

$$\int_{r_z}^{x}\left[\frac{r_z}{\lambda_s}\left(1-\frac{x}{r_z}\right)+\frac{1}{k_{zew}}\right]dx = -\int_0^t \frac{T_m - T_{cz}}{\rho_s h_{sl}}dt. \tag{10}$$

As a result of the integration, one has

$$\frac{r_z}{\lambda_s}\left(x-\frac{1}{2}\frac{x^2}{r_z}\right)+\frac{x}{k_{zew}}-\frac{1}{2}\frac{r_z^2}{\lambda_s}-\frac{r_z}{k_{zew}} = -\frac{T_m - T_{cz}}{\rho_s h_{sl}}t. \tag{11}$$

Solidification time (freezing) for the whole plate is obtained once $x = 0$ is substituted into (11)

$$t_z = \frac{\rho_s h_{sl}}{T_m - T_{cz}}\left(\frac{1}{2}\frac{r_z^2}{\lambda_s}+\frac{r_z}{k_{zew}}\right). \tag{12}$$

b. *Cylinder*

Once (3) and (6) are accounted for in (9), the variables separated and then integrated within appropriate limits, one obtains

$$\int_{r_z}^{x}\left[\frac{r_z}{\lambda_s}(\ln r_z - \ln x)+\frac{1}{k_{zew}}\right]\frac{x}{r_z}dx = -\int_0^t \frac{T_m - T_{cz}}{\rho_s h_{sl}}dt. \tag{13}$$

Once we determine integral

$$\int x \ln x dx = \frac{1}{2}x^2 \ln x - \int \frac{1}{2}x^2 \cdot \frac{1}{x} dx = \frac{1}{2}x^2 \ln x - \frac{1}{4}x^2 \quad (14)$$

and integrate (13) while accounting for (14), we have

$$\frac{1}{2}\left(\frac{r_z}{\lambda_s}\ln r_z - \frac{r_z}{\lambda_s}\ln x + \frac{1}{2}\frac{r_z}{\lambda_s} + \frac{1}{k_{zew}}\right)\frac{x^2}{r_z} - \frac{1}{2}\left(\frac{1}{2}\frac{r_z}{\lambda_s} + \frac{1}{k_{zew}}\right)r_z = -\frac{T_m - T_{cz}}{\rho_s h_{sl}}t. \quad (15)$$

We can determine solidification time (freezing) t_z for the whole cylinder by substituting $x = 0$ into (15)

$$t_z = \frac{1}{2}\frac{\rho_s h_{sl}}{T_m - T_{cz}}\left(\frac{1}{2}\frac{r_z^2}{\lambda_s} + \frac{r_z}{k_{zew}}\right). \quad (16)$$

c. *Sphere*

Once (3) and (7) are accounted for in (9), the variables separated and then integrated within appropriate limits, one has

$$\int_{r_z}^{x}\left[\frac{r_z}{\lambda_s}\left(\frac{r_z}{x} - 1\right) + \frac{1}{k_{zew}}\right]\frac{x^2}{r_z^2}dx = -\int_0^t \frac{T_m - T_{cz}}{\rho_s h_{sl}} dt. \quad (17)$$

After integration, one obtains

$$\left(\frac{1}{2}\frac{x^2}{\lambda_s} - \frac{1}{3}\frac{x^3}{\lambda_s r_z} + \frac{1}{3}\frac{x^3}{r_z^2 k_{zew}}\right) - \frac{1}{6}\frac{r_z^2}{\lambda_s} - \frac{1}{3}\frac{r_z}{k_{zew}} = -\frac{T_m - T_{cz}}{\rho_s h_{sl}}t. \quad (18)$$

If one assumes that $x = 0$ in (18), one obtains the freezing time for the sphere t_z

$$t_z = \frac{1}{3}\frac{\rho_s h_{sl}}{T_m - T_{cz}}\left(\frac{1}{2}\frac{r_z^2}{\lambda_s} + \frac{r_z}{k_{zew}}\right). \quad (19)$$

From the comparison of (12), (16) and (19), one can see that the freezing time for the plate ($n = 0$), cylinder ($n = 1$) and sphere ($n = 2$) can be expressed using a single formula only

$$t_z = \frac{1}{n+1}\frac{\rho_s h_{sl}}{T_m - T_{cz}}\left(\frac{1}{2}\frac{r_z^2}{\lambda_s} + \frac{r_z}{k_{zew}}\right). \quad (20)$$

Equation (20) can also be written in the form

$$t_z = \frac{1}{n+1}\frac{1}{Ste_s}\frac{r_z^2}{a_s}\left(\frac{1}{2} + \frac{1}{Bi_z}\right), \quad (21)$$

where

$$Ste_s = \frac{c_s\left(T_m - T_{cz}\right)}{h_{sl}}, \qquad Bi_z = \frac{k_{zew} r_z}{\lambda_s} \quad \text{i} \quad a_s = \frac{\lambda_s}{\rho_s c_s}.$$

In technical literature, for example in [4], (20) is usually written in the form given below, assuming that $g = 0$

$$t_z = \frac{\rho_s h_{sl}}{T_m - T_{cz}}\left(R\frac{d^2}{\lambda_s} + P\frac{d}{\alpha}\right), \tag{22}$$

where,

$$P = \frac{1}{2(n+1)}, \qquad R = \frac{1}{8(n+1)} \quad \text{and} \quad d = 2r_z. \tag{23}$$

Symbol d is the thickness of the two-sided cooled plate and the diameter of the cylinder or a sphere.

Equation (22) is of significance to package-free frozen products, when $g = 0$. The values of coefficients P and R are given in Table 26.2.

Table 26.2. Values of coefficients P and R calculated by means of (23)

Element	P	R
Plate	1/2	1/8
Cylinder	1/4	1/16
Sphere	1/6	1/24

For complex-shape products, appropriate values of coefficients P and R can be found in [4]. Solidification time (freezing) t_z determined from (22) is the minimum time, since the initial temperature of the body can be higher than the temperature of solidification T_m. Equation (22) is valid for smaller values of Stefan number Ste_s due to the fact that steady-state temperature distribution was assumed for the layer of the frozen product. For water and food products with a large water content, e.g. fruits, vegetables, meat, fish, fruit juices, ice cream, condition $Ste_s < 0{,}5$ is met and (22) yields satisfying results.

Exercise 26.5 Ablation of a Semi-Infinite Body

Present basic dependencies that describe the ablation process on the surface of a semi-infinite body, assuming that the boundary condition of 3rd and 2nd kind are assigned on the surface of a solid.

Solution

Initial temperature of the semi-infinite body T_p is lower than the temperature of melting T_m. Temperature distribution in a semi-infinite body with the boundary condition of 3rd kind is defined by the heat conduction equation (Fig. 26.9a)

$$\frac{\partial T}{\partial t} = a_s \frac{\partial^2 T}{\partial x^2}, \tag{1}$$

by boundary conditions

$$-\lambda_s \left.\frac{\partial T}{\partial x}\right|_{x=s(t)} = \alpha\left[T_{cz} - T\big|_{x=s(t)}\right], \tag{2}$$

$$T\big|_{x\to\infty} = T_p \tag{3}$$

and by initial condition

$$T\big|_{t=0} = T_p. \tag{4}$$

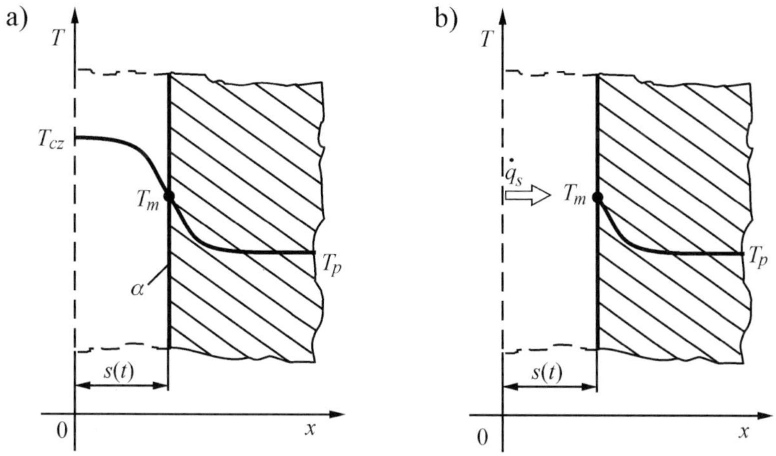

Fig. 26.9. Diagram of ablation on the surface of a semi-infinite body: a) boundary condition of 3rd kind, b) boundary condition of 2nd kind

The solution to problem (1)–(4) is given in paper [13]. First the body surface is heated from the initial temperature T_p to temperature T_m. The heating time t_m of the half space surface from temperature T_p to temperature T_m can be determined using (9) for the half space temperature derived in Ex. 14.4.

Time t_m is the solution of the non-linear algebraic equation

$$\frac{T_m - T_p}{T_{cz} - T_p} = 1 - \exp\left[\left(\frac{\alpha}{\lambda}\right)^2 a_s t_m\right] \operatorname{erfc}\left(\frac{\alpha}{\lambda}\sqrt{a t_m}\right). \tag{5}$$

For time $t >> t_m$, the ablation rate v is constant. Quasi-steady ablation rate, which follows from the balance equation of the heat flow on the surface $x = s(t)$, is calculated from formula

$$v_s = \frac{\alpha\left(T_{cz} - T_m\right)}{\rho\left[c_s\left(T_m - T_p\right) + h_{sl}\right]}. \tag{6}$$

Temperature distribution in the quasi-steady state can be determined from [13]

$$\frac{T - T_p}{T_m - T_p} = \exp\left[-\frac{v_s}{a_s}(x - s)\right], \qquad x \geq s. \tag{7}$$

When heat flux $\dot{q}_s$ is assigned on the half space surface [7], then the boundary condition on the surface $x = s(t)$ has the form

$$-\lambda_s \left.\frac{\partial T}{\partial x}\right|_{x=s(t)} = \dot{q}_s. \tag{8}$$

Temperature field is, therefore, expressed by (1), conditions (3) and (4) and by boundary condition (8). Time t_m, after which the initial surface temperature T_p of the semi-infinite body increases to the temperature of melting T_m is determined from (12) derived in Ex.14.3. Time t_m is calculated from formula

$$t_m = \frac{\pi}{4}\rho_s c_s \lambda_s \left(\frac{T_m - T_p}{\dot{q}_s}\right)^2. \tag{9}$$

Quasi-steady ablation rate $v_s = ds/dt$ follows from the balance equation of a heat flow on the surface $x = s(t)$

$$v_s = \frac{\dot{q}_s}{\rho_s\left[c_s\left(T_m - T_p\right) + h_{sl}\right]}. \tag{10}$$

Temperature distribution in the semi-infinite body for $t >> t_m$ is calculated from (7).

Ablation phenomenon is used in thermal shields of space shuttles and rockets [3]. Melted material transfers a lot of thermal energy and is immediately removed from the body surface as a result of an active aerodynamic shear forces. Materials used for the thermal shields are tantalum, wolfram, teflon, asbestos-phenol materials, cork, silicon-elastomers, nylon-phenol and quartz-phenol materials.

Exercise 26.6 Solidification of a Falling Droplet of Lead

A droplet of melted lead, which is in diameter $d = 3$ mm and whose initial temperature is $T_p = 400°C$, falls from a height of $h = 20$ m and solidifies at the moment it touches the ground. Air temperature T_{cz} is of 20°C. Use the following lead data for the calculation: temperature of solidification $T_m = 327°C$, specific heat $c = 130$ J/(kg·K), density $\rho = 11340$ kg/m^3 and latent heat of solidification $h_{sl} = 24.7$ kJ/kg. Determine heat transfer coefficient on the droplet surface.

Solution

The falling time of the lead droplet t will be calculated under the assumption that the droplet falls at a uniformly accelerated motion

$$h = \frac{1}{2} g t^2 , \tag{1}$$

$$t = \sqrt{\frac{2h}{g}} = \sqrt{\frac{2 \cdot 20}{9.81}} = 2.02 \text{ s} .$$

Heat flow emitted during the lead drop cooling is transferred to surroundings

$$-\frac{d(m_k c T)}{dt} = \alpha A_k (T - T_{cz}) . \tag{2}$$

Once variables are separated, one has

$$\frac{dT}{T - T_{cz}} = -\frac{\alpha A_k}{m_k c} dt . \tag{3}$$

Equation (3) can be also written in the form

$$\frac{d(T - T_{cz})}{T - T_{cz}} = -\frac{\alpha A_k}{m_k c} dt . \tag{4}$$

Once (4) is integrated, one obtains

Appendix A Basic Mathematical Functions

A.1. Gauss Error Function

Table A.1. Gauss error function

x	erf (x)	x	erf (x)	x	erf (x)	x	erf (x)	x	erf (x)
0	0	0.30	0.328627	0.60	0.603856	0.90	0.796908	1.20	0.910314
0.01	0.011283	0.31	0.338908	0.61	0.611681	0.91	0.801883	1.21	0.912955
0.02	0.022565	0.32	0.349126	0.62	0.619411	0.92	0.806767	1.22	0.915534
0.03	0.033841	0.33	0.359279	0.63	0.627046	0.93	0.811563	1.23	0.918050
0.04	0.045111	0.34	0.369364	0.64	0.634586	0.94	0.816271	1.24	0.920505
0.05	0.056372	0.35	0.379382	0.65	0.642029	0.95	0.820891	1.25	0.9229
0.06	0.067622	0.36	0.38933	0.66	0.649377	0.96	0.825423	1.26	0.925236
0.07	0.078858	0.37	0.399206	0.67	0.656627	0.97	0.82987	1.27	0.927513
0.08	0.090078	0.38	0.409009	0.68	0.663782	0.98	0.834231	1.28	0.929734
0.09	0.101281	0.39	0.418739	0.69	0.67084	0.99	0.838508	1.29	0.931898
0.10	0.112463	0.40	0.428392	0.70	0.677801	1.00	0.842701	1.30	0.934008
0.11	0.123623	0.41	0.437969	0.71	0.684665	1.01	0.84681	1.31	0.936063
0.12	0.134758	0.42	0.447467	0.72	0.691433	1.02	0.850838	1.32	0.938065
0.13	0.145867	0.43	0.456887	0.73	0.698104	1.03	0.854784	1.33	0.940015
0.14	0.156947	0.44	0.466225	0.74	0.704678	1.04	0.85865	1.34	0.941914
0.15	0.167996	0.45	0.475481	0.75	0.711155	1.05	0.862436	1.35	0.943762
0.16	0.179012	0.46	0.484655	0.76	0.717537	1.06	0.866143	1.36	0.945561
0.17	0.189992	0.47	0.493745	0.77	0.723821	1.07	0.869773	1.37	0.947312
0.18	0.200936	0.48	0.50275	0.78	0.73001	1.08	0.873326	1.38	0.949016
0.19	0.21184	0.49	0.511668	0.79	0.736103	1.09	0.876803	1.39	0.950673
0.20	0.222703	0.50	0.5205	0.80	0.742101	1.10	0.880205	1.40	0.952285
0.21	0.233522	0.51	0.529244	0.81	0.748003	1.11	0.883533	1.41	0.953852
0.22	0.244296	0.52	0.537899	0.82	0.753811	1.12	0.886788	1.42	0.955376
0.23	0.255023	0.53	0.546464	0.83	0.759524	1.13	0.88997	1.43	0.956857
0.24	0.2657	0.54	0.554939	0.84	0.765142	1.14	0.893082	1.44	0.958296
0.25	0.276326	0.55	0.563323	0.85	0.770668	1.15	0.896124	1.45	0.959695
0.26	0.2869	0.56	0.571616	0.86	0.7761	1.16	0.899096	1.46	0.961053
0.27	0.297418	0.57	0.579816	0.87	0.78144	1.17	0.902	1.47	0.962373
0.28	0.30788	0.58	0.587923	0.88	0.786687	1.18	0.904837	1.48	0.963654
0.29	0.318283	0.59	0.595936	0.89	0.791843	1.19	0.907608	1.49	0.964898

Table A.1. (cont.)

x	erf (x)	x	erf (x)	x	erf (x)	x	erf (x)	x	erf (x)
1.50	0.966105	1.80	0.98909	2.10	0.99702	2.40	0.999312	2.70	0.999866
1.51	0.967277	1.81	0.989524	2.11	0.997155	2.41	0.999346	2.71	0.999873
1.52	0.968413	1.82	0.989943	2.12	0.997284	2.42	0.999379	2.72	0.99988
1.53	0.969516	1.83	0.990347	2.13	0.997407	2.43	0.999411	2.73	0.999887
1.54	0.970586	1.84	0.990736	2.14	0.997525	2.44	0.999441	2.74	0.999893
1.55	0.971623	1.85	0.991111	2.15	0.997639	2.45	0.999469	2.75	0.999899
1.56	0.972628	1.86	0.991472	2.16	0.997747	2.46	0.999497	2.76	0.999905
1.57	0.973603	1.87	0.991821	2.17	0.997851	2.47	0.999523	2.77	0.99991
1.58	0.974547	1.88	0.992156	2.18	0.997951	2.48	0.999547	2.78	0.999916
1.59	0.975462	1.89	0.992479	2.19	0.998046	2.49	0.999571	2.79	0.99992
1.60	0.976348	1.90	0.99279	2.20	0.998137	2.50	0.999593	2.80	0.999925
1.61	0.977207	1.91	0.99309	2.21	0.998224	2.51	0.999614	2.81	0.999929
1.62	0.978038	1.92	0.993378	2.22	0.998308	2.52	0.999635	2.82	0.999933
1.63	0.978843	1.93	0.993656	2.23	0.998388	2.53	0.999654	2.83	0.999937
1.64	0.979622	1.94	0.993923	2.24	0.998464	2.54	0.999672	2.84	0.999941
1.65	0.980375	1.95	0.994179	2.25	0.998537	2.55	0.999689	2.85	0.999944
1.66	0.981105	1.96	0.994426	2.26	0.998607	2.56	0.999706	2.86	0.999948
1.67	0.98181	1.97	0.994664	2.27	0.998674	2.57	0.999722	2.87	0.999951
1.68	0.982493	1.98	0.994892	2.28	0.998738	2.58	0.999736	2.88	0.999954
1.69	0.983153	1.99	0.995111	2.29	0.998799	2.59	0.999751	2.89	0.999956
1.70	0.983790	2.00	0.995322	2.30	0.998857	2.60	0.999764	2.90	0.999959
1.71	0.984407	2.01	0.995525	2.31	0.998912	2.61	0.999777	2.91	0.999961
1.72	0.985003	2.02	0.995719	2.32	0.998966	2.62	0.999789	2.92	0.999964
1.73	0.985578	2.03	0.995906	2.33	0.999016	2.63	0.9998	2.93	0.999966
1.74	0.986135	2.04	0.996086	2.34	0.999065	2.64	0.999811	2.94	0.999968
1.75	0.986672	2.05	0.996258	2.35	0.999111	2.65	0.999821	2.95	0.99997
1.76	0.98719	2.06	0.996423	2.36	0.999155	2.66	0.999831	2.96	0.999972
1.77	0.987691	2.07	0.996582	2.37	0.999197	2.67	0.999841	2.97	0.999973
1.78	0.988174	2.08	0.996734	2.38	0.999237	2.68	0.999849	2.98	0.999975
1.79	0.98864	2.09	0.99688	2.39	0.999275	2.69	0.999858	2.99	0.999976

$$\operatorname{erf}(x) = \frac{2}{\sqrt{\pi}} \int_0^x e^{-t^2}\, dt,$$

$$\operatorname{erfc}(x) = 1 - \operatorname{erf}(x), \tag{A.1}$$

$$\frac{d}{dx}\left(\operatorname{erf}(x)\right) = \frac{2}{\sqrt{\pi}} e^{-x^2}.$$

Gauss error function can be calculated by means of series [1]

$$\operatorname{erf}(x)=\frac{2}{\sqrt{\pi}}\sum_{n=0}^{\infty}\frac{(-1)^n x^{2n+1}}{n!\,(2n+1)} \tag{A.2}$$

or approximate formula

$$\operatorname{erf}(x)=1-\left(c_1 y+c_2 y^2+c_3 y^3\right)e^{-x^2}, \tag{A.3}$$

where

$$y=\frac{1}{1+0,47047x}, \quad c_1=0,3480242, \quad c_2=-0,0958798, \quad c_3=0,7478556.$$

Maximum error in calculating function erf(x) by means of formula (A.3) is $2{,}5\cdot 10^{-5}$.

A.2. Hyperbolic Functions

Hyperbolic functions are defined as follows [1, 2]:

$$\sinh(x)=\frac{1}{2}\left(e^x-e^{-x}\right);$$

$$\cosh(x)=\frac{1}{2}\left(e^x+e^{-x}\right); \tag{A.4}$$

$$\tanh(x)=\frac{e^x-e^{-x}}{e^x+e^{-x}};$$

Derivatives of hyperbolic functions:

$$\frac{d}{dx}(\sinh u)=(\cosh u)\frac{du}{dx};$$

$$\frac{d}{dx}(\cosh u)=(\sinh u)\frac{du}{dx}; \tag{A.5}$$

$$\frac{d}{dx}(\tanh u)=\left(\frac{1}{\cosh^2 u}\right)\frac{du}{dx}.$$

Table A.2. Hyperbolic functions

x	sinh x	cosh x	tanh x	x	sinh x	cosh x	tanh x
0.0	0.0000	1.0000	0.0000	2.3	4.9370	5.0372	0.9801
0.1	0.1002	1.0050	0.0997	2.4	5.4662	5.5569	0.9837
0.2	0.2013	1.0201	0.1974	2.5	6.0502	6.1323	0.9866
0.3	0.3045	1.0453	0.2913	2.6	6.6947	6.7690	0.9890
0.4	0.4108	1.0811	0.3800	2.7	7.4063	7.4735	0.9910
0.5	0.5211	1.1276	0.4621	2.8	8.1919	8.2527	0.9926
0.6	0.6367	1.1855	0.5371	2.9	9.0596	9.1146	0.9940
0.7	0.7586	1.2552	0.6044	3.0	10.018	10.068	0.9951
0.8	0.8881	1.3374	0.6640	3.5	16.543	16.573	0.9982
0.9	1.0265	1.4331	0.7163	4.0	27.290	27.308	0.9993
1.0	1.1752	1.5431	0.7616	4.5	45.003	45.014	0.9998
1.1	1.3356	1.6685	0.8005	5.0	74.203	74.210	0.9999
1.2	1.5095	1.8107	0.8337	5.5	122.34	122.35	1.0000
1.3	1.6984	1.9709	0.8617	6.0	201.71	201.72	1.0000
1.4	1.9043	2.1509	0.8854	6.5	332.57	332.57	1.0000
1.5	2.1293	2.3524	0.9052	7.0	548.32	548.32	1.0000
1.6	2.3756	2.5775	0.9217	7.5	904.02	904.02	1.0000
1.7	2.6456	2.8283	0.9354	8.0	1490.5	1490.5	1.0000
1.8	2.9422	3.1075	0.9468	8.5	2457.4	2457.4	1.0000
1.9	3.2682	3.4177	0.9562	9.0	4051.5	4051.5	1.0000
2.0	3.6269	3.7622	0.9640	9.5	6679.9	6679.9	1.0000
2.1	4.0219	4.1443	0.9705	10.0	11013	11013	1.0000
2.2	4.4571	4.5679	0.9757				

A.3. Bessel Functions

Bessel functions $J_n(x)$ and $Y_n(x)$ are the solution of a differential equation

$$x^2 \frac{d^2}{dx^2} y + x \frac{d}{dx} y + \left(x^2 - n^2\right) y = 0, \quad \text{gdzie} \quad n = 0, 1, \ldots \tag{A.6}$$

where $J_n(x)$ and $Y_n(x)$ are the Bessel functions of the first and second kind of integer order n.

Modified Bessel functions $I_n(x)$ and $K_n(x)$ are the solution of the following equation:

$$x^2 \frac{d^2}{dx^2} y + x \frac{d}{dx} y - \left(x^2 + n^2\right) y = 0, \quad \text{gdzie} \quad n = 0, 1, \ldots \tag{A.7}$$

where $I_n(x)$ and $K_n(x)$ are modified Bessel functions of first and second kind of integer order n.

Literature

1. Abramowitz M, Stegun I (1965) Handbook of Mathematical Functions. Dover, New York
2. Thompson WJ (1997) Atlas for Computing Mathematical Functions. Wiley, New York

Appendix B Thermo-Physical Properties of Solids

B.1. Tables of Thermo-Physical Properties of Solids

Table B.1. Thermo-physical properties of metals

Metal	Atmospheric pressure								Liquid metal			
				at temp. 100 K		at temp. 298,15 K (25°C)			Pressure of saturated vapour, Pa			
	Temperature of melting	Temperature of boiling	Latent heat of melting	Thermal conductivity	Specific heat	Specific heat	Linear coeff. of thermal expans. (×10^6)	Thermal conductivity	Specific heat at temp. 2000 K	1.01×10^2	1.01×10^{-2}	1.01×10^{-4}
	°C		kJ/kg	W/(m·K)	kJ/(kg·K)	kJ/(kg·K)	K^{-1}	W/(m·K)	kJ/(kg·K)	Boiling point [K]		
Aluminum	660	2441	397.67	300	0.4814	0.9000	25	237	1.0884	1782	1333	1063
Stibium	630	1410	161.161	–	0.1674	0.2093	9	18.5	0.2595	1007	741	612
Beryllium	1285	2475	1356.264	–	0.2051	1.8251	12	218	3.2651	1793	1347	1085
Bismuth	271.4	1660	51.9064	–	0.1088	0.1256	13	8.4	0.1507	1155	851	677
Cadmium	321	767	55.2552	103	0.1967	0.2302	30	93	0.2637	655	486	388
Chromium	1860	2670	330.694	158	0.1926	0.4605	6	91	0.9377	1992	1530	1247
Cobalt	1495	2925	276.276	–	0.2386	0.4186	12	69	0.6865	2167	1652	1354
Copper	1086	2575	205.114	483	0.2553	0.3851	16.6	398	0.4939	1862	1391	1120
Gold	1063	2800	62.79	345	0.1088	0.1298	14.2	315	0.1486	2023	1510	1211
Iridium	2450	4390	138.138	–	0.0921	0.1298	6	147	0.1817	3253	2515	2062
Iron	1536	2870	272.09	132	0.2177	0.4521	12	80.3	0.8246	2093	1594	1297
Lead	327.5	1750	23.023	39.6	0.1172	0.1298	29	34.6	0.1381	1230	889	698
Magnesium	650	1090	368.368	169	0.0670	1.0172	25	159	1.3395	857	638	509
Manganese	1244	2060	267.904	–	0.2679	0.4772	22	–	0.8372	1495	1131	913
Mercury	–38.86	356.55	11.3022	–	0.1214	0.1381	–	8.39	–	393	287	227
Molybdenum	2620	4651	288.834	179	0.1381	0.2512	5	140	0.3726	3344	2558	2079
Nickel	1453	2800	297.206	158	0.2302	0.4437	13	89.9	0.7326	2156	1646	1343
Niobium	2470	4740	284.648	55.2	0.1884	0.2679	7	52	0.3474	3552	2721	2232
Osmium	3025	4225	142.324	–	–	0.1298	5	61	0.1633	–	–	–
Platinum	1770	3825	100.464	79	0.1005	0.1340	9	73	0.1800	2817	2155	1757
Plutonium	640	3230	12.558	–	0.0795	0.1340	54	8	0.1716	2200	1596	1252

Table B.1. (cont.)

Metal	Atmospheric pressure								Liquid metal			
				at temp. 100 K		at temp. 298,15 K (25°C)			Pressure of saturated vapour, Pa			
	Temperature of melting	Temperature of boiling	Latent heat of melting	Thermal conductivity	Specific heat	Specific heat	Linear coeff. of thermal expans. ($\times 10^6$)	Thermal conductivity	Specific heat at temp. 2000 K	1.01×10^2	1.01×10^{-2}	1.01×10^{-4}
	°C		kJ/kg	W/(m·K)	kJ/(kg·K)	kJ/(kg·K)	K^{-1}	W/(m·K)	kJ/(kg·K)	Boiling point [K]		
Potassium	63.3	760	60.697	–	0.6279	0.7535	83	99	–	606	430	335
Rhodium	3700	209.3	–	–	0.2428	8	8	150	0.3851	–	–	–
Selenium	217	700	66.976	–	–	0.3223	37	0.5	–	–	–	–
Silicon	1411	3280	1799.98	–	0.2595	0.7116	3	83.5	0.9084	2340	1749	1427
Silver	961	2212	110.929	450	0.1884	0.2386	19	427	0.2846	1582	1179	952
Sodium	97.83	884	113.022	–	0.9795	1.2265	70	134	–	701	504	394
Tantalum	2980	5365	171.626	59.2	0.1088	0.1423	6.5	54	0.1674	3959	3052	2495
Thorium	1750	4800	71.162	–	0.1005	0.1256	12	41	0.1967	2407	2407	1919
Tin	232	2600	59.0226	85	0.1633	0.2260	20	64	0.2428	1857	1366	1080
Titanium	1670	3290	418.6	31.2	0.3014	0.5233	8.5	20	0.7870	2405	1827	1484
Wolfram	3400	5550	192.556	235	0.0879	0.1340	4.5	178	0.1674	4139	3228	2656
Uranium	1132	4140	50.232	–	0.0921	0.1172	13.4	25	0.2009	2861	2128	1699
Vanadium	1900	3400	410.228	–	0.2553	0.4856	8	60	0.8665	2525	1948	1591
Zinc	419.5	910	113.022	132	0.2637	0.3893	35	115	–	752	559	449

Table B.2. Metals in a solid state – melting temperature and thermal properties at temp. 300 K

Metal	T [K]	ρ [kg/m³]	c [J/(kg·K)]	λ [W/(m·K)]	a [m²/s]×10⁶
Aluminium					
pure	933	2702	903	237	97.10
Duralumin (4.4% Cu, 1.0% Mg, 0.75% Mn, 0.4% Si)	775	2770	875	174	71.80
alloy 195, cast(4.5% Cu)	–	2790	883	168	68.10
Beryllium	1550	1850	1825	200	59.20
Bismuth	545	9780	122	7.90	6.59
Cadmium	594	8650	231	97	48.40
Copper					
Pure	1358	8933	385	401	117
electrolytic (Cu + Ag 99.90 minimum)	–	8950	385	386	112
bronze – alloy (10A1)	1293	8800	420	52	14.10
Brass (30% Zn)	1188	8530	380	111	34.20
New silver (15% Ni, 22% Zn)	–	8618	410	116	32.80
Constantan (40% Ni)	–	8920	420	22.7	6.06
constantan (45% Ni)	–	8860	–	23	–
Gold	1336	19300	129	317	127
Iron					
Pure	1810	7870	447	80.20	22.8
armco (99.75% Fe pure)	–	7870	447	72.70	20.7
Cast (4% C)	–	7272	420	51	16.7
Carbon steel					
AISI 1010 (0.1% C, 0.4% Mn)	–	7830	434	64	18.80
AISI 1042, annealed (0.42% C. 0.64% Mn, 0.063% Ni, 0.13% Cu)	–	7840	460	50	13.90
AISI 4130, hardened (0.3% C, 0.5% Mn, 0.3% Si, 0.95% Cr, 0.5% Mo)	–	7840	460	43	11.90
Stainless steel					
AISI 302 (18-8) (0.15% C, 2% Mn, 1% Si, 16–18% Cr, 6–8% Ni)	–	8055	480	15	3.88
AISI 304 (0,08% C, 2% Mn, 1% Si, 18–20% Cr, 8–10% Ni)	1670	7900	477	15	3.98
AISI 316 (0.08% C, 2% Mn, 1% Si, 16–18% Cr, 10–14% Ni, 2–3% Mo)	–	8238	468	13	3.37
AISI 410 (0.15% C, 1% Mn, 1% Si, 11.5–13% Cr)	–	7770	460	25	7.00
Lead	601	11340	129	35.30	24.10
Magnesium					
Pure	929	1740	1024	156	87.60
Alloy A8 (8% Al, 0.5% Zn)	–	–	–	–	–
Molybdenum	2894	10240	251	138	53.6

Table B.2. (cont.)

Metal	T [K]	ρ [kg/m³]	c [J/(kg·K)]	λ [W/(m·K)]	a [m²/s]×10⁶
Nickel					
Pure	1728	8900	444	91	23.00
Heat-resisting alloy –X750 (15.5% Cr, 1% Nb, 2.5% Ti, 0.7% Al, 7% Fe)	1665	8510	439	11.7	3.13
Nichrome (20% Cr)	1672	8314	460	13	3.40
Nimonic75 (20% Cr, 0.4% Ti)	–	8370	461	11.7	3.03
Hasteloy B (38% Mo, 5% Fe)	–	9240	381	12.20	3.47
Nickeline (50% Cu)	–	8800	421	19.5	5.26
chromel-P (10% Cr)	–	8730	–	17	–
alumel (2% Mn, 2% Al)	–	8600	–	48	–
Palladium	1827	12020	244	71.80	24.5
Platinum					
Pure	2045	21450	133	71.6	25.1
60 Pt-40 Rh	1800	16630	162	47	174
40 Rh	3	–	–	–	–
Silicon	1685	2330	712	148	89.2
Silver	1235	10500	235	429	174
Tantalum	3269	16600	140	57.50	24.70
Tin	505	7310	227	66.6	40.1
Titanium					
Pure	1993	4500	522	21.90	9.32
Ti-6Al-4V		4420	610	5.80	2.15
Ti-2AI-2Mn		4510	466	8.40	4.00
Tungsten	3660	19300	132	174	68.30
Tin	693	7140	389	116	41.80
Zirconium					
Pure	2125	6570	278	22.70	12.4
zircaloy-4 (1.2–1.75% Sn, 0.18–0.24% Fe, 0.07–0.13% Cr)	–	6560	285	14.2	7.6

Table B.3. Metals in solid state – temperature-dependent thermal conductivity λ [W/(m·K)]

Metal	Temperature [K]								
	200	300	400	500	600	800	1000	1200	1500
Aluminium									
Pure	237	237	240	236	231	218	–	–	–
Duralumin	138	174	187	188	–	–	–	–	–
alloy 195, cast	–	168	174	180	185	–	–	–	–
Copper									
Pure	413	401	393	386	379	366	352	339	–
bronze – alloy	42	52	52	55	–	–	–	–	–
Brass	74	111	134	143	146	150	–	–	–
New silver	–	116	135	145	147	–	–	–	–
Gold	323	317	311	304	298	284	270	255	–

Table B.3. (cont.)

Metal	Temperature [K] 200	300	400	500	600	800	1000	1200	1500
Iron									
armco	81	73	66	59	53	42	32	29	31
Cast	–	51	44	39	36	27	23	–	–
Carbon steel									
AISI 1010	–	64	59	54	49	39	31	–	–
AISI 1042	–	52	50	48	45	37	29	26	30
AISI 4130	–	43	42	41	40	37	31	27	31
Stainless steel									
AISI 302	–	15	17	19	20	23	25	–	–
AISI 304	13	15	17	18	20	23	25	–	–
AISI 316	–	13	15	17	18	21	24	–	–
AISI 410	25	25	26	27	27	29	–	–	–
Lead	37	35	34	33	31	–	–	–	–
Magnesium									
Pure	199	156	153	151	149	146	–	–	–
Alloy A8	–	–	84	–	–	–	–	–	–
Nickel									
Pure	105	91	80	72	66	68	72	76	83
Heat-resistant alloy-X-750	10.30	11.70	13.5	15.10	17.00	20.5	24.00	27.60	30.00
Nichrome	–	13	14	16	17	21	–	–	–
Platinum	73	72	72	72	73	76	79	83	90
Silver	420	429	425	419	412	396	379	361	–
Tantalum	58	58	58	59	59	59	60	61	62
Tin	73	67	62	60	–	–	–	–	–
Titanium									
Pure	25	22	20	20	19	19	21	22	25
Ti-6Al-4V	–	5.80	–	–	–	–	–	–	–
Tungsten	185	174	159	146	137	125	118	112	106
Zirconium									
Pure	25	23	22	21	21	21	23	26	29
zircaloy-4	13.30	14.20	15.2	16.20	17.20	19.20	21.20	23.20	–

Table B.4. Metals in a solid state – temperature-dependent specific heat *c* [J/(kg·K)]

Metals	Temperature [K] 200	300	400	500	600	800	1000	1200	1500
Aluminium									
Pure	798	903	949	996	1033	1146	–	–	–
Duralumin	–	875	–	–	–	–	–	–	–
alloy 195, cast	–	883	–	–	–	–	–	–	–
Copper									
Pure	356	385	397	412	417	433	451	480	–
bronze – alloy	785	420	460	500	–	–	–	–	–
Brass	360	380	395	410	425	–	–	–	–
New silver	–	410	–	–	–	–	–	–	–

Table B.4. (cont.)

Metals	Temperature [K]								
	200	300	400	500	600	800	1000	1200	1500
Gold	124	129	131	133	135	140	145	155	–
Iron									
armco	384	447	490	530	574	680	975	609	634
Cast	–	420	–	–	–	–	–	–	–
Carbon steel									
AISI 1010	–	434	487	520	559	685	1168	–	–
AISI 1042	–	–	500	530	570	700	1430	–	–
AISI 4130	–	–	500	530	570	690	840	–	–
Stainless steel									
AISI 302	–	480	512	531	559	585	606	–	–
AISI 304	402	477	515	539	557	582	611	640	682
AISI 316	–	468	504	528	550	576	602	–	–
AISI 410	–	460	–	–	–	–	–	–	–
Lead	125	129	132	136	142	–	–	–	–
Magnesium									
Pure	934	1024	1074	1170	1170	1267	–	–	–
Alloy A8	–	–	1000	–	–	–	–	–	–
Nickel									
Pure	383	444	485	500	512	530	562	594	616
Heat-resistant alloy–X-750	372	439	473	490	510	546	626	–	–
Nichrome	–	–	480	500	525	545	–	–	–
Platinum	125	133	136	139	141	146	152	157	165
Silver	225	232	239	244	250	262	277	292	–
Tantalum	133	140	144	145	146	149	152	155	160
Tin	215	227	243	–	–	–	–	–	–
Titanium									
Pure	405	522	551	572	591	633	675	680	686
Ti-6Al-4V	–	610	–	–	–	–	–	–	–
Tungsten	122	132	137	140	142	145	148	152	157
Zirconium									
Pure	264	278	300	312	322	342	362	344	344
zircaloy-4	–	–	300	314	327	348	369	–	–

Table B.5. Dielectric in a solid state – thermal properties

Dielectric	T [K]	ρ [kg/m^3]	c [J/(kg·K)]	λ [W/(m·K)]	a [m^2/s]$\times 10^6$
Aluminium oxide, Al_2O_3					
Sapphire	300	3970	765	46	15.20
Aluminium oxide	300	3970	765	36	11.90
	400	–	940	27	7.20
	600	–	1110	16	3.60
	1000	–	1225	7.60	1.60
	1500	–	–	5.40	–
Carbon					
Diamond (type IIb)	300	3300	510	1300	772
ATJ-S graphite	300	1810	1300	98	42

Table B.5. (cont.)

Dielectric	T [K]	ρ [kg/m^3]	c [J/(kg K)]	λ [W/(m·K)]	a [m^2/s]×10^6
	1000	–	1926	55	16
	2000	–	2139	38	9.80
	3000	–	2180	33	8.40
Pyrolytic graphite	300	2210	709	1950	1240
λ longitudinal to the layers	600	–	1406	892	287
	1000	–	1793	534	135
	2000	–	2043	262	58
λ perpendicular to the layers	300	2210	709	5.70	4.10
	600	–	1406	2.68	2.10
	1000	–	1793	1.60	1.67
	2000	–	2043	0.81	1.51
Epoxide graphite fibres	200	1400	640	8.70	9.70
(25% of volume) composite	300	–	935	11.10	8.50
λ longitudinal to the layers	400	–	1220	13.00	7.6
λ perpendicular to the layers	200	1400	640	0.68	0.76
	300	–	935	0.87	0.66
	400	–	1220	1.1	0.64
Carbon fibres	300	1860	810	110	73
	1000	–	1800	56	17
	2000	–	2140	36.50	9.20
	3000	–	2220	34.50	8.40
	4000	–	2260	34	8.10
	5000	–	2270	34	8.10
Ice	273	910	1930	2.22	1.26
Artificial materials	8	–	–	–	–
Acetylcellulose	300	1300	1510	0.24	0.12
neoprene rubber	300	1250	1930	0.19	0.079
phenol, bands	300	1760	1260	0.50	0.23
polyamide (nylon)	300	1140	1670	0.24	0.13
polyethylene (high density)	300	960	2090	0.33	0.16
Polypropylene	300	1170	1930	0.17	0.075
Polyvinyl chloride	300	1714.00	1050	0.092	0.051
Teflon	300	2200	1050	0.35	0.15
	400	–	–	0.45	0.19
Silicon dioxide, SiO_2	200	2650	–	16.40	–
Crystalline (quartz)	300	–	745	10.40	5.30
λ longitudinal to axis c	400	–	885	7.60	3.20
	600	–	1075	5.00	1.70
λ perpendicular to axis c	200	2650	–	–9.5	–
	300	–	745	6.20	3.1
	400	–	885	4.7	2.00
	600	–	1075	3.40	1.20

Table B.5. (cont.)

Dielectric	T [K]	ρ [kg/m^3]	c [J/(kg K)]	λ [W/(m·K)]	a [m^2/s]×10^6
Polycrystal (burned quartz glass)	300	2220	745	1.38	0.83
	400	–	905	1.51	0.75
	600	–	1040	1.75	0.75
	800	–	1105	2.17	0.88
	1000	–	1155	2.87	1.11
	1200	–	1195	4.00	1.51
Titanium dioxide, TiO_2 (ore)	300	4157	710	8.40	2.8
	600	–	880	5.00	1.40
	1200	–	945	3.30	0.84
Uranium dioxide, UO_2	300	10 890	240	7.90	3.00
	500	–	265	6.00	2.10
	1000	–	305	3.90	1.20
	1500	–	325	2.6	0.79
	2000	–	355	2.3	0.59
	2500	–	405	2.50	0.57

Table B.6. Insulating and construction materials – thermal properties

Material	T [K]	ρ [kg/m^3]	c [J/(kg K)]	λ [W/(m·K)]	a [m^2/s]×10^6
Asbestos- laminar and corrugated					
4 layers	300	190	–	0.078	–
	320	–	–	0.085	–
	340	–	–	0.091	–
	360	–	–	0.097	–
	380	–	–	0.101	–
8 layers	300	300	–	0.068	–
	320	–	–	0.073	–
	340	–	–	0.077	–
	360	–	–	0.080	–
	380	–	–	0.083	–
Brick					
B&W K-28 insulating	600			0.03	–
	1300	–	–	0.04	–
Chromited	400	3010	835	2.3	0.92
	800	–	–	2.5	–
	1200	–	–	2	–
Refractory	400	2645	960	0.9	0.35
	800	–	–	1.4	–
	1200	–	–	1.7	–
	1600	–	–	1.8	–
Simple	300	1920	835	0.72	0.45
Face brick	300	2083	–	1.3	–
Concrete					
Building stone 1-2-4 mixture	300	2100	880	1.4	0.75
Cork	300	160	1680	0.043	0.16

Table B.6. (cont.)

Material	T [K]	ρ [kg/m^3]	c [J/(kg K)]	λ [W/(m·K)]	a [m^2/s]×10^6
Cotton	300	80	1300	0.06	0.58
Glass:					
Quartz glass	300	2220	745	1.38	0.83
Pyrex glass	300	2640	800	1.09	0.51
Soda-calcium glass (25% Na_2O, 10% CaO)	300	2400	840	0.88	0.44
Foam glass	240	–	–	0.048	–
	260	–	–	0.051	–
	280	–	–	0.054	–
	300	145	–	0.058	–
	320	–	–	0.063	–
	340	–	–	0.067	–
Glass fibre	300	16	835	0.046	3.4
	300	40	–	0.035	–
	260	–	–	0.029	–
	280	–	–	0.033	–
	300	28	–	0.038	–
	320	–	–	0.043	–
	340	–	–	0.048	–
	360	–	–	0.054	–
	380	–	–	0.06	–
	400	–	–	0.066	–
Loose filling					
Cellulose, wood pulp,	290	–	–	0.038	–
Chemical paper pulp	300	45	–	0.039	–
	310	–	–	0.042	–
Vermiculite	240	–	–	0.058	–
	260	–	–	0.061	–
	280	–	–	0.064	–
	300	122	–	0.069	–
	320	–	–	0.074	–
	240	–	–	0.052	–
	260	–	–	0.056	–
	280	–	–	0.059	–
	300	80	–	0.063	–
	320	–	–	0.068	–
Mangan oxide 85%	300	270	–	0.062	–
	350	–	–	0.068	–
	400	–	–	0.073	–
	450	–	–	0.078	–
	500	–	–	0.082	–
Paper	300	930	2500	0.13	0.056
Polystyrene, rigid	240	–	–	0.023	–
	260	–	–	0.024	–
	280	–	–	0.026	–
	300	30–60	1210	0.028	0.4–0.8
	320	–	–	0.03	–

Table B.6. (cont.)

Material	T [K]	ρ [kg/m³]	c [J/(kg K)]	λ [W/(m·K)]	a [m²/s]×10⁶
Polyurethane resins	300	70	–	0.026	–
Rubber					
ebonite, hard rubber	270	1200	2010	0.15	0.062
Neoprene rubber	300	1250	1930	0.19	0.079
Rigid foam	260	–	–	0.028	–
	280	–	–	0.03	–
	300	70	–	0.032	–
	320	–	–	0.034	–
Snow					
Fresh	273	110	–	0.049	–
Dense	–	500	–	0.19	–
Soil					
Dry	300	1500	1900	1	0.35
Wet	300	1900	2200	2	0.5
Wood					
oak, longitudinal to fibres	300	820	2400	0.35	0.18
Perpendicular to fibres	300	820	2400	0.21	0.11
White pine, longitudinal to fibres	300	500	2800	0.24	0.17
Perpendicular to fibres	300	500	2800	0.1	0.071
Sheep wool	300	145	–	0.05	–

Table B.7. Tthermal conductivity λ [W/(m·K)] for selected materials at cryogenic temperatures

Material	Temperature [K] 5	10	30	100	200
Metals					
aluminium (2024-T4)	3.50	7.70	21	50	72
Brass	–	–	–	71	94
Copper (OFHC)	–	–	950	430	400
Carbon steel (1020)	4.30	12	34	64	70
Staainless steel (303)	0.29	0.71	3.50	9.00	12
Stainless steel (304)	0.16	0.82	3.30	9.50	13
Titanium	0.15	0.80	2.00	4.50	6.5
Non-metals					
Diamond	45	310	3300	10 000	–
Pyrex glass	–	–	–	0.56	0.89
Glass, Phoenix	–	0.11	0.17	0.55	–
nylon-66	0.018	0.033	0.21	–	–
Polyethylene	0.04	0.15	0.75	–	–
Silicon	350	2000	4700	900	–
Silicon rubber	–	–	–	0.18	0.21
Teflon	–	–	0.18	0.62	1
Vacuumgrease (Dow-Corning silicone)	0.021	–	–	–	–
Lacquer (G.E. #7031, stoving)	0.065	0.075	0.15	0.25	0.36

Table B.8. Composition of ferritic boiler steels; percentages (%) stand for mass fraction

Steel		Steel composition, %												
Symbol (PN)	Symbol (DIN)	C	Si	Mn	P	S	Al	Cr	Cu	Mo	Nb	Ni	V	W
K10	St 35.8	≤0.17	≤0.35	0.35–0.65	≤0.05	≤0.05	0.008	–	–	–	–	–	–	–
K18	St 45.8	≤0.22	0.10–0.35	0.4–1.0	≤0.05	≤0.05	0.009	–	–	–	–	–	–	–
16M	15Mo3	0.12–0.2	0.15–0.35	0.5–0.8	≤0.04	≤0.04	0.032	–	–	0.25–0.35	–	–	–	–
15HM	13CrMo44	0.1–0.18	0.15–0.35	0.4–0.7	≤0.04	≤0.04	0.04	0.7–1.0	–	0.4–0.5	–	–	–	–
10H2M	10CrMo910	≤0.15	0.15–0.50	0.4–0.6	≤0.04	≤0.04	0.007	2.0–2.5	–	0.9–1.1	–	–	–	–
13HMF	14MoV63	0.1–0.18	0.15–0.35	0.3–0.6	≤0.04	≤0.04	0.02	0.3–0.6	–	0.5–0.65	–	–	0.25–0.35	–
20H12M1F	X20CrMoV121	0.17–0.23	0.10–0.50	0.3–0.8	–	–	–	11.0–12.5	–	0.8–1.2	–	0.3–0.8	0.25–0.35	–
St36K	H I	≤0.16	≤0.35	≥0.4	≤0.05	≤0.05	0.036	–	–	–	–	–	–	–
St41K	H II	≤0.2	≤0.35	≥0.5	≤0.05	≤0.05	0.01	–	–	–	–	–	–	–
19G2	19Mn5	0.17–0.23	0.40–0.6	1.0–1.3	≤0.05	≤0.05	0.065	–	–	–	–	–	–	–
15NCuMNb	15NiCuMoNb5 (WB 36)	≤0.17	0.25–0.5	0.8–1.2	≤0.035	≤0.035	–	–	0.5–0.8	0.25–0.5	~0.02	1.0–1.3	–	–
	22NiMoCr37	0.17–0.23	0.1–0.35	0.5–1.0	≤0.025	≤0.025	–	0.3–0.5	–	0.5–0.8	–	0.6–1.0	–	–
	15MnNiMoV53	0.12–0.17	0.1–0.35	1.1–1.5	≤0.025	≤0.025	–	0.2–0.4	–	0.3–0.45	–	0.9–1.2	0.08–0.16	–
HCM2S (SUMITOMO company symbol)	7CrMoVTiB10-10 (Mannesmann company symbol)	0.04–0.10	≤0.5	0.1–0.6	≤0.03	≤0.01	≤0.03	1.9–2.6	–	0.05–0.3	0.02–0.08	–	0.2–0.3	1.45–1.75

Table B.9. Thermal conductivity λ [W/(m·K)] for selected ferretic boiler steels

Steel Symbol (PN)	Symbol (DIN)	Temperature [°C] 20	100	200	300	400	500	600	700
K10	St 35.8	57.5	56.4	53.6	49.8	45.4	41.1	37.6	–
K18	St 45.8	54.1	53	50.5	47	43.2	39.4	36.4	–
16M	15Mo3	48.5	48	46.5	44.2	41.3	38.2	35	–
15HM	13CrMo44	44.1	44.6	44	42.2	39.8	37	34.1	–
10H2M	10CrMo910	36	37.8	38.5	37.8	36.3	34.6	33.1	–
13HMF	14MoV63	48.5	48	46.5	44.2	41.3	38.2	35	–
20H12M1F	X20CrMoV121	24	24.3	24.8	25.2	25.6	26	26.4	–
St36K	H I	57.5	56.4	53.6	49.8	45.4	41.1	–	–
St41K	H II	57.5	56.4	53.6	49.8	45.4	41.1	–	–
19G2	19Mn5	43	44.1	43.7	42	39.5	36.6	–	–
15NCuMNb	15NiCuMoNb5 (WB 36)	39.5	41.1	41.5	40.5	38.5	36	–	–
	22NiMoCr37	43.1	45.7	47	46.4	44.7	42.3	39.9	–
	15MnNiMoV53	38.1	40.7	42	41.4	39.7	37.3	34.9	–
HCM2S (SUMITOMO company symbol)	7CrMoVTiB10-10 (Mannesmann company symbol)	–	37.1	37	36.3	35.2	34.3	31.9	30.3

Table B.10. Density ρ [kg/m^3]×10^{-3} for selected ferritic boiler steels

Steel Symbol (PN)	Symbol (DIN)	Temperature [°C] 20	100	200	300	400	500	600	700
K10	St 35.8	7.839	7.814	7.781	7.746	7.71	7.671	7.631	–
K18	St 45.8	7.834	7.809	7.777	7.743	7.706	7.668	7.628	–
16M	15Mo3	7.849	7.825	7.792	7.759	7.723	7.685	7.646	–
15HM	13CrMo44	7.849	7.825	7.793	7.759	7.724	7.687	7.648	–
10H2M	10CrMo910	7.834	7.811	7.781	7.75	7.716	7.682	7.645	–
13HMF	14MoV63	7.844	7.821	7.791	7.759	7.726	7.69	7.654	–
20H12M1F	X20CrMoV121	7.76	7.74	7.713	7.686	7.657	7.627	7.595	–
St36K	H I	7.839	7.816	7.784	7.752	7.717	7.681	–	–
St41K	H II	7.839	7.816	7.784	7.752	7.717	7.681	–	–
19G2	19Mn5	7.829	7.806	7.775	7.742	7.707	7.671	–	–
15NCuMNb	15NiCuMoNb5 (WB 36)	7.854	7.83	7.799	7.765	7.731	7.694	–	–
	22NiMoCr37	7.859	7.836	7.805	7.772	7.738	7.702	7.664	–
	15MnMoNiV53	7.854	7.831	7.8	7.767	7.733	7.698	7.661	–
HCM2S (SUMITOMO company symbol)	7CrMoVTiB10-10 (Mannesmann company symbol)	7.89	–	–	–	–	–	–	–

Table B.11. Specific heat capacity c [kJ/(kg·K)] for selected ferritic boiler steels

Steel		Temperature [°C]							
Symbol (PN)	Symbol (DIN)	20	100	200	300	400	500	600	700
K10	St 35.8	0.464	0.48	0.511	0.555	0.612	0.681	0.762	–
K18	St 45.8	0.464	0.48	0.511	0.555	0.612	0.681	0.762	–
16M	15Mo3	0.464	0.48	0.511	0.555	0.612	0.681	0.762	–
15HM	13CrMo44	0.464	0.48	0.511	0.555	0.612	0.681	0.762	–
10H2M	10CrMo910	0.464	0.48	0.511	0.555	0.612	0.681	0.762	–
13HMF	14MoV63	0.464	0.48	0.511	0.555	0.612	0.681	0.762	–
20H12M1F	X20CrMoV121	0.456	0.47	0.503	0.552	0.619	0.703	0.804	–
St36K	H I	0.464	0.48	0.511	0.555	0.612	0.681	–	–
St41K	H II	0.464	0.48	0.511	0.555	0.612	0.681	–	–
19G2	19Mn5	0.464	0.48	0.511	0.555	0.612	0.681	–	–
15NCuMNb	15NiCuMoNb5 (WB 36)	0.464	0.48	0.511	0.555	0.612	0.681	–	–
	22NiMoCr37	0.464	0.48	0.511	0.555	0.612	0.681	0.762	–
	15MnMoNiV53	0.464	0.48	0.511	0.555	0.612	0.681	0.762	–
HCM2S (SUMITOMO company symbol)	7CrMoVTiB10-10 (Mannesmann company symbol)	–	0.486	0.516	0.552	599	0.663	0.719	0.915

Table B.12. Average linear expansion coefficient $\overline{\beta}$ [1/K]×10^5 in an interval from 0°C to temperature T for selected ferritic boiler steels

Steel		Temperature [°C]							
Symbol (PN)	Symbol (DIN)	20	100	200	300	400	500	600	700
K10	St 35.8	1.23	1.28	1.35	1.41	1.46	1.51	1.55	–
K18	St 45.8	1.21	1.26	1.33	1.39	1.44	1.49	1.53	–
16M	15Mo3	1.19	1.25	1.32	1.38	1.43	1.47	1.5	–
15HM	13CrMo44	1.18	1.24	1.31	1.37	1.42	1.46	1.49	–
10H2M	10CrMo910	1.13	1.18	1.24	1.29	1.33	1.37	1.4	–
13HMF	14MoV63	1.13	1.19	1.25	1.3	1.34	1.38	1.41	–
20H12M1F	X20CrMoV121	1.04	1.07	1.11	1.14	1.18	1.21	1.23	–
St36K	H I	1.15	1.21	1.28	1.33	1.38	1.42	–	–
St41K	H II	1.15	1.21	1.28	1.33	1.38	1.42	–	–
19G2	19Mn5	1.15	1.21	1.28	1.33	1.38	1.42	–	–
15NCuMNb	15NiCuMoNb5 (WB 36)	1.17	1.23	1.29	1.35	1.4	1.44	–	–
	22NiMoCr37	1.17	1.22	1.27	1.32	1.36	1.4	1.43	–
	15MnMoNiV53	1.17	1.22	1.27	1.32	1.36	1.4	1.43	–
HCM2S (SUMITOMO company symbol)	7CrMoVTiB10-10 (Mannesmann company symbol)	–	1.23	1.27	1.3	1.33	1.34	1.36	1.38

Table B.13. Linear expansion coefficient β [1/K]×10^5 at temperature T for selected ferretic boiler steels

Steel Symbol (PN)	Symbol (DIN)	Temperature [°C] 20	100	200	300	400	500	600	700
K10	St 35.8	1.24	1.35	1.48	1.58	1.66	1.72	1.76	–
K18	St 45.8	1.22	1.33	1.46	1.56	1.64	1.7	1.74	–
16M	15Mo3	1.2	1.32	1.45	1.54	1.62	1.66	1.68	–
15HM	13CrMo44	1.19	1.31	1.43	1.53	1.6	1.64	1.66	–
10H2M	10CrMo910	1.14	1.24	1.35	1.43	1.5	1.55	1.57	–
13HMF	14MoV63	1.15	1.25	1.36	1.44	1.51	1.56	1.58	–
20H12M1F	X20CrMoV121	1.05	1.11	1.18	1.24	1.3	1.35	1.39	–
St36K	H I	1.17	1.28	1.4	1.49	1.56	1.6	–	–
St41K	H II	1.17	1.28	1.4	1.49	1.56	1.6	–	–
19G2	19Mn5	1.17	1.28	1.4	1.49	1.56	1.6	–	–
15NCuMNb	15NiCuMoNb5 (WB 36)	1.19	1.3	1.41	1.51	1.57	1.62	–	–
	22NiMoCr37	1.18	1.27	1.38	1.46	1.52	1.57	1.59	–
	15MnMoNiV53	1.18	1.27	1.38	1.46	1.52	1.57	1.59	–

Table B.14. Longitudinal elasticity modulus (Young's modulus) E [GPa] for selected ferretic boiler steels

Steel Symbol (PN)	Symbol (DIN)	Temperature [°C] 20	100	200	300	400	500	600	700
K10	St 35.8	212.7	207	199.5	191.7	183.4	174.8	165.9	–
K18	St 45.8	212.7	207	199.5	191.7	183.4	174.8	165.9	–
16M	15Mo3	213.7	208	200.5	192.7	184.4	175.8	166.9	–
15HM	13CrMo44	214	208.7	201.7	194	185.9	177.2	168	–
10H2M	10CrMo910	214	208.7	201.7	194	185.9	177.2	168	–
13HMF	14MoV63	214	208.7	201.7	194	185.9	177.2	168	–
20H12M1F	X20CrMoV121	217.2	212.9	206.3	198.3	189.1	178.4	166.4	–
St36K	H I	212.7	207	199.5	191.7	183.4	174.8	–	–
St41K	H II	212.7	207	199.5	191.7	183.4	174.8	–	–
19G2	19Mn5	213.7	208	200.5	192.7	184.4	175.8	–	–
15NCuMNb	15NiCuMoNb5 (WB 36)	213.7	208	200.5	192.7	184.4	175.8	–	–
	22NiMoCr37	212.7	207	199.5	191.7	183.4	174.8	165.9	–
	15MnMoNiV53	212.7	207	199.5	191.7	183.4	174.8	165.9	–
HCM2S (SUMITOMO)	7CrMoVTiB10-10 (Mannesmann)	–	209	205	200	194	187	176	161

Table B.15. Poisson ratio ν for selected ferretic boiler steels

Steel		Temperature [°C]						
Symbol (PN)	Symbol (DIN)	20	100	200	300	400	500	600
K10	St 35.8	0.283	0.285	0.288	0.292	0.298	0.305	0.314
K18	St 45.8	0.283	0.285	0.288	0.292	0.298	0.305	0.314
16M	15Mo3	0.283	0.285	0.288	0.292	0.298	0.305	0.314
15HM	13CrMo44	0.283	0.285	0.288	0.292	0.298	0.305	0.314
10H2M	10CrMo910	0.283	0.285	0.288	0.292	0.298	0.305	0.314
13HMF	14MoV63	0.283	0.285	0.288	0.292	0.298	0.305	0.314
20H12M1F	X20CrMoV121	0.283	0.285	0.288	0.292	0.298	0.305	0.314
St36K	H I	0.283	0.285	0.288	0.292	0.298	0.305	–
St41K	H II	0.283	0.285	0.288	0.292	0.298	0.305	–
19G2	19Mn5	0.283	0.285	0.288	0.292	0.298	0.305	–
15NCuMNb	15NiCuMoNb5 (WB 36)	0.283	0.285	0.288	0.292	0.298	0.305	–
	22NiMoCr37	0.283	0.285	0.288	0.292	0.298	0.305	0.314
	15MnMoNiV53	0.283	0.285	0.288	0.292	0.298	0.305	0.314

Table B.16. Chemical composition of selected austenitic steels; percentages (%) stand for mass fractions

Steel symbol	Steel composition, %										
(DIN or ASME Code N47)	C	Si	Mn	P	S	Al	Cr	Cu	Mo	Ni	Ti
X8CrNiNb1613	≤ 0.1	0.3–0.6	1–1.5	0.019	0.01	–	15–17	–	1.6–2.0	12–14	–
X8CrNiMoNb1616	≤ 0.1	0.3–0.6	1–1.5	0.032	0.016	–	15.5–17.5	–	–	15.5–17.5	–
X10NiCrAlTi3220 (Incoloy)	0.08	0.05	0.7	0.011	0.002	0.321		0.55	–	32	0.36
TP316 (AISI 316)	0.08	1	2	–	–	–	16–18	–	–	10–14	–

Table B.17. Thermal conductivity λ [W/(m·K)] for selected austenitic steels

Steel symbol	Temperature [°C]							
(DIN or ASME Code N47)	20	100	200	300	400	500	600	700
X8CrNiNb1613	16.4	17.3	18.4	19.5	20.6	21.7	22.9	24.1
X8CrNiMoNb1616	15.7	16.6	17.8	19	20.2	21.4	22.6	23.8
X10NiCrAlTi3220 (Incoloy)	12.8	14	15.5	17	18.4	19.8	21.2	22.6
TP316 (AISI 316)	12.8	14.5	16.4	17.8	18.9	20.5	22.4	23.8

Table B.18. Density ρ [kg/m^3]×10^3 for selected austenitic steels

Steel symbol (DIN or ASME Code N47)	Temperature [°C] 20	100	200	300	400	500	600	700
X8CrNiNb1613	7.847	7.814	7.772	7.729	7.685	7.64	7.593	7.545
X8CrNiMoNb1616	7.917	7.884	7.842	7.799	7.755	7.709	7.662	7.614
X10NiCrAlTi3220 (Incoloy)	7.923	7.893	7.855	7.816	7.775	7.733	7.689	7.644
TP316 (AISI 316)	8.328	–	–	–	–	–	–	–

Table B.19. Specific heat capacity c [kJ/(kg·K)] for selected austenitic steels

Steel symbol (DIN or ASME Code N47)	Temperature [°C] 20	100	200	300	400	500	600	700
X8CrNiNb1613	0.504	0.52	0.539	0.558	0.576	0.594	0.611	0.628
X8CrNiMoNb1616	0.504	0.52	0.539	0.558	0.576	0.594	0.611	0.628
X10NiCrAlTi3220 (Incoloy)	0.459	0.474	0.494	0.515	0.536	0.558	0.58	0.603
TP316 (AISI 316)	0.463	0.497	0.523	0.543	0.559	0.574	0.587	0.599

Table B.20. Average linear expansion coefficient $\overline{\beta}$ [1/K]×10^5 in an interval from temperature 0°C to temperature T for selected austenitic steels

Steel symbol (DIN or ASME Code N47)	Temperature [°C] 20	100	200	300	400	500	600	700
X8CrNiNb1613	1.66	1.7	1.75	1.8	1.84	1.87	1.91	1.94
X8CrNiMoNb1616	1.65	1.69	1.74	1.79	1.83	1.86	1.9	1.93
X10NiCrAlTi3220 (Incoloy)	1.47	1.52	1.57	1.62	1.66	1.7	1.74	1.77
TP316 (AISI 316)	1.53	1.61	1.68	1.74	1.79	1.82	1.86	1.89

Table B.21. Linear expansion coefficient β [1/K]×10^5 at temperature T for selected austenitic steels

Steel symbol (DIN)	Temperature [°C] 20	100	200	300	400	500	600	700
X8CrNiNb1613	1.67	1.75	1.85	1.93	1.99	2.05	2.1	2.13
X8CrNiMoNb1616	1.66	1.74	1.84	1.92	1.98	2.04	2.09	2.12
X10NiCrAlTi3220 (Incoloy)	1.48	1.57	1.67	1.76	1.83	1.89	1.93	1.97

Table B.22. Longitudinal elasticity modulus (Young's modulus) E [GPa] for selected austenitic steels

Steel symbol (DIN or ASME Code N47)	Temperature [°C]							
	20	100	200	300	400	500	600	700
X8CrNiNb1613	197.5	190.5	181.9	173.5	165.4	157.5	149.9	142.4
X8CrNiMoNb1616	197.5	190.5	181.9	173.5	165.4	157.5	149.9	142.4
X10NiCrAlTi3220 (Incoloy)	197.5	192	185.1	178	170.8	163.5	156	148.5
TP316 (AISI 316)	196	192	187.5	181	173	164.5	155	142

Table B.23. Poisson ratio ν for selected austenitic steels

Steel symbol (DIN or ASME Code N47)	Temperature [°C]							
	20	100	200	300	400	500	600	700
X8CrNiNb1613	0.278	0.282	0.287	0.292	0.298	0.305	0.312	0.319
X8CrNiMoNb1616	0.278	0.282	0.287	0.292	0.298	0.305	0.312	0.319
X10NiCrAlTi3220 (Incoloy)	0.278	0.282	0.287	0.292	0.298	0.305	0.312	0.319
TP316 (AISI 316)	0.278	0.282	0.287	0.292	0.298	0.305	0.312	0.319

Table B.24. Chemical composition of selected new martensitic steels that contain 9–13% Cr

Steel symbol	Steel composition, %											
	C	Si	Mn	P	S	Al	Cr	Cu	Mo	Nb	Ni	V	W
P 91 (H9AMFNb)	0.08 –0.12	0.2– 0.5	0.3 –0.6	≤0.02	≤0.015	≤0.04	8–9	≤0.25	0.85 –1.05	0.06 –0.1	≤0.4	–	–
P 92 (NF616)	0.07 –0.13	≤ 0.5	0.3 –0.6	≤0.02	≤ 0.01	≤0.04	8.5 –9.5	–	0.3 –0.6	0.04 –0.09	≤0.4	0.15 –0.25	1.5 –2.0
T122 (HCM12A)	0.07 –0.14	≤ 0.5	≤ 0.7	≤0.02	≤ 0.01	≤0.04	10 –12.5	0.3 –1.7	0.25 –0.6	0.04 –0.1	≤0.5	0.15 –0.30	1.5 –2.5

Table B.25. Thermal conductivity λ [W/(m·K)] of high-temperature-creep-resistance ferritic steel (martensitic)

Steel symbol	Temperature [°C]							
	20	100	200	300	400	500	600	700
P 91 (H9AMFNb)	28.7	29	29.6	29.7	29.7	29.6	29.3	28.4
P 92 (NF616)	27	27.8	28.2	28.5	28.8	29	27.8	27.4
T122 (HCM12A)	24.4	25.2	27.2	30	31.1	33	34.3	35

Table B.26. Specific heat capacity c [J/(kg·K)] of high-temperature-creep-resistance ferritic steels (martensitic)

Steel symbol	Temperature [°C]							
	20	100	200	300	400	500	600	700
P 91* (H9AMFNb)	448	486	523	565	620	703	804	–
P 92* (NF616)	448	486	523	565	620	703	804	–
T122 (HCM12A)	439.6	498.3	552.7	636.4	703.4	816.5	954.6	1235.2

* Values assumed for steel X20CrMoV121.

Steel density at temperature 20°C: for P91 – $\rho \approx 7750$ kg/m^3, for P92 – $\rho \approx 7750$ kg/m^3, for T122 – $\rho \approx 7870$ kg/m^3.

Table B.27. Average linear temperature expansion coefficient $\overline{\beta}$ [1/K]×10^5 of ferritic steel (martensitic) at an interval from temperature 20°C to temperature T

Steel symbol	Temperature [°C]						
	100	200	300	400	500	600	700
P 91 (H9AMFNb)	0.98	1.06	1.09	1.12	1.13	1.15	1.16
P 92 (NF616)	1	1.07	1.11	1.14	1.15	1.17	1.17
T122 (HCM12A)	0.94	1.06	1.07	1.1	1.12	1.15	1.16

Table B.28. Longitudinal elasticity modulus (Young's modulus) E [GPa] of high-temperature-creep-resistant ferritic steels (martensitic)

Steel symbol	Temperature [°C]							
	20	100	200	300	400	500	600	700
P 91 (H9AMFNb)	239	228	214	200	186	172	158	144
P 92 (NF616)	239	228	214	200	186	172	158	144
T122 (HCM12A)	216	211	204	196	185	174	158	132

B.2. Diagrams

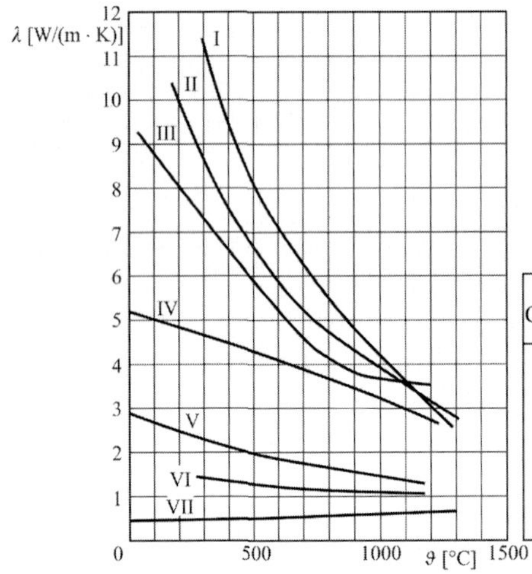

Curves	ρ	ψ	Al_2O_3	SiO_2	Fe_2O_3	CaO	MgO
	kg/m^3	–	%				
I	2850	0.171	1.6	2.9	5.3	2.0	88.3
II	2980	0.158	1.8	2.4	4.8	2.3	88.9
III	2400	–	0.4	5.0	1.6	1.7	92.1
IV	2340	0.34	6.5	2.7	4.4	–	86.4
V	2010	–	14.8	22.5	2.5	4.9	53.5
VI	2560	–	–	–	2.6	4.7	87.9
VII	2350	–	0.2	0.2	9.3	0.3	88.8

Fig. B.1. Thermal conductivity of magnesite [7]; ρ –density, ψ –volume fraction of piping

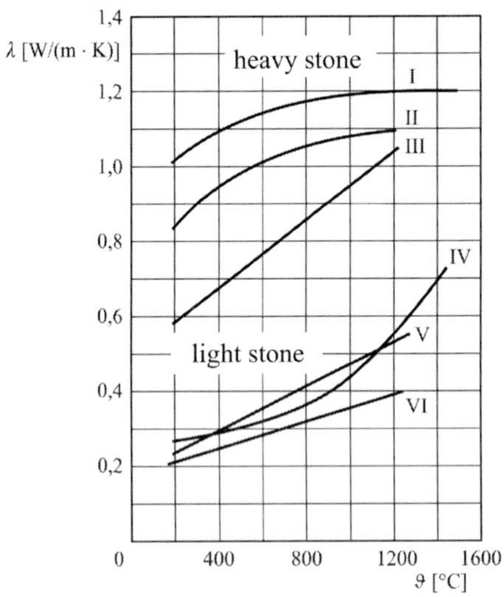

Curves	ρ	ψ
	kg/m^3	–
I	1900	0.27
II	1850	0.30
III	1750	0.33
IV	700 to 800	0.70
V	1000	–
VI	700	–

Fig. B.2. Thermal conductivity of chamotte [7]; ρ –density, ψ –volume fraction of piping

a)

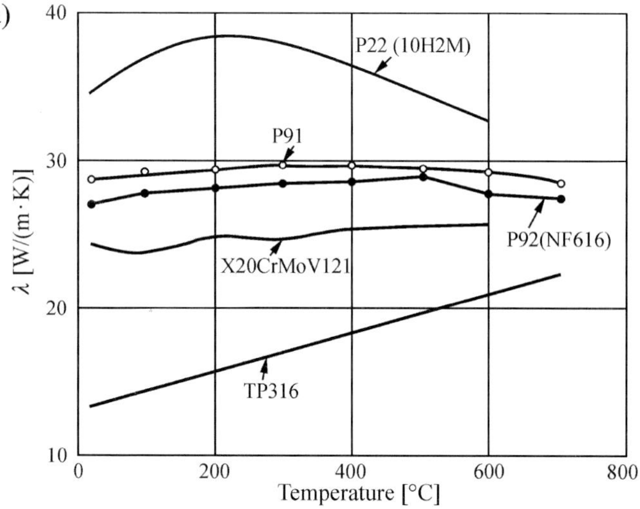
40
30
20
10
0
200
400
600
800
λ [W/(m·K)]
Temperature [°C]
P22 (10H2M)
P91
P92(NF616)
X20CrMoV121
TP316

b)

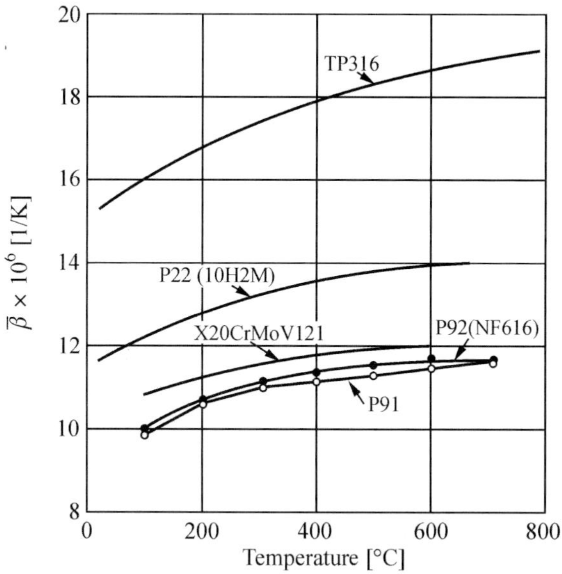
20
18
16
14
12
10
8
0
200
400
600
800
β̄ × 10⁶ [1/K]
Temperature [°C]
TP316
P22 (10H2M)
X20CrMoV121
P92(NF616)
P91

c)

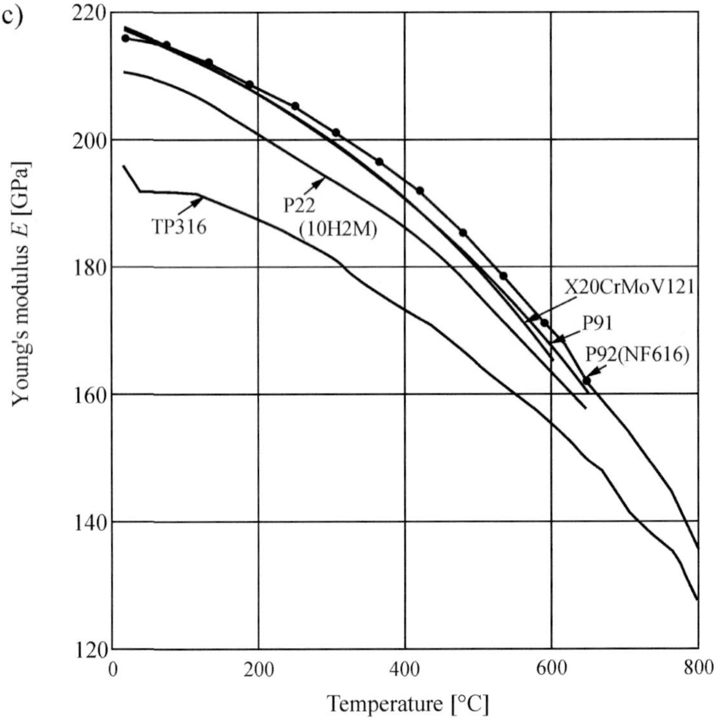

Fig. B.3. Comparison of the selected properties of the new high-temperature-creep-resistance martensitic. P91 and P92. steels [10] with standard steels: P22 – low-alloyed ferritic steel, whose equivalent is the Polish steel 10H2M or German 10CrMo910; X20CrMoV121 – German martensitic steel, TP316 – American austenitic steel: a) thermal conductivity λ, b) average linear expansion coefficient $\overline{\beta}$ from temperature of 20°C to given temperature, c) longitudinal elasticity modulus (Young's modulus) E

B.3. Approximated Dependencies for Calculating Thermo-Physical Properties of a Steel [8]

Density ρ at temperature 20°C

- carbon and low-alloyed steels. which contain Cr < 1%

$$\rho = (7.84 \pm 0.04) \cdot 10^3 \ \mathrm{kg/m^3}, \qquad \text{(B.1)}$$

• alloy steels. which contain 1–26% Cr, no Al

$$\rho = \left(7.82 - 0.008 \cdot \mathrm{Cr}[\%] \pm 0.06\right) \cdot 10^3 \ \mathrm{kg/m^3}, \tag{B.2}$$

• chromium steels, which contain 1–26% Cr and Al ≈ 1%

$$\rho = \left(7.74 - 1.2 \cdot 10^{-2} \cdot \mathrm{Cr}[\%] \pm 0.03\right) \cdot 10^3 \ \mathrm{kg/m^3}, \tag{B.3}$$

• austenitic steels

$$\rho = \left(8.001 - 0.1092 \cdot 10^{-2} \cdot \mathrm{Si}[\%] \pm 0.05\right) \cdot 10^3 \ \mathrm{kg/m^3}. \tag{B.4}$$

Thermal conductivity λ in a temperature function

• carbon and low-alloyed steels based on the example of steel St41K (H II) for $0 \le T \le 650^\circ\mathrm{C}$

$$\lambda = 52.1 - 0.0159 \cdot T - 1.86 \cdot 10^{-5} \cdot T^2 \pm 0.5 \quad \mathrm{W/(m \cdot K)}, \tag{B.5}$$

where T is expressed in [°C];

• steels that contain 5–26% Cr based on the example of steel X20CrMoV121 for $0 \le T \le 600^\circ\mathrm{C}$ (Cr = 11.8%)

$$\lambda = 23.9 + 0.41 \cdot 10^{-2} \cdot T + 0.01 \cdot 10^{-5} \cdot T^2 \pm 0.1 \quad \mathrm{W/(m \cdot K)}, \tag{B.6}$$

where temperature T is expressed in [°C];

• austenitic steels of type Cr-Ni and Ni-Cr for $0 \le T \le 1000^\circ\mathrm{C}$

$$\lambda = 14.6 + 1.27 \cdot 10^{-2} \cdot T \pm 0.3 \quad \mathrm{W/(m \cdot K)}. \tag{B.7}$$

where temperature T is expressed in [°C].

Specific heat capacity c in a temperature function

• carbon and low-alloyed steels for $0 \le T \le 650^\circ\mathrm{C}$

$$c = 422 + 0.931 \cdot T - 2.14 \cdot 10^{-3} \cdot T^2 + 2.64 \cdot 10^{-6} \cdot T^3 \pm 14 \ \ \mathrm{J/(kg \cdot K)}, \tag{B.8}$$

where temperature T is expressed in [°C];

• austenitic steels of type Cr-Ni and Ni-Cr for $0 \le T \le 1000^\circ\mathrm{C}$

$$c = 454 + 0.388 \cdot T - 3.22 \cdot 10^{-4} \cdot T^2 + 1.1 \cdot 10^{-7} \cdot T^3 \pm 12 \ \ \mathrm{J/(kg \cdot K)}, \tag{B.9}$$

where temperature T is expressed in [°C].

Longitudinal elasticity module (Young's modulus) E in function of temperature

- carbon and low-alloyed steels for $0 \le T \le 600°\text{C}$

$$E = \left(214 - 5.2 \cdot 10^{-2} \cdot T - 4.7 \cdot 10^{-5} \cdot T^2 \pm 5.6\right) \cdot 10^9 \text{ Pa}, \tag{B.10}$$

where temperature T is expressed in [°C];
- austenitic steels of type Cr-Ni and Ni-Cr for $0 \le T \le 800°\text{C}$

$$E = \left(200 - 8.3 \cdot 10^{-2} \cdot T \pm 4.2\right) \cdot 10^9 \text{ Pa}. \tag{B.11}$$

where temperature T is expressed in [°C].

Average temperature expansion coefficient $\overline{\beta}$ within temperature interval from 20°C to a given temperature T expressed in [°C]

- carbon and low-alloyed steels for $0 \le T \le 600°\text{C}$

$$\overline{\beta} = \left(11.4 + 8.23 \cdot 10^{-3} \cdot T - 4.8 \cdot 10^{-6} \cdot T^2 \pm 0.8\right) \cdot 10^{-6} \text{ 1/K}, \tag{B.12}$$

- chromium steels. which contain 5÷26% Cr for $0 \le T \le 600°\text{C}$

$$\overline{\beta} = \left(10.0 + 5.7 \cdot 10^{-3} \cdot T - 2.6 \cdot 10^{-6} \cdot T^2 \pm 1.0\right) \cdot 10^{-6} \text{ 1/K}, \tag{B.13}$$

- austenitic steels of type Cr-Ni for $0 \le T \le 700°\text{C}$

$$\overline{\beta} = \left(15.8 + 6.1 \cdot 10^{-3} \cdot T - 2.6 \cdot 10^{-6} \cdot T^2 \pm 0.8\right) \cdot 10^{-6} \text{ 1/K}. \tag{B.14}$$

Poisson ratio ν in function of temperature

- carbon and low-alloyed steels for $0 \le T \le 600°\text{C}$

$$\nu = 0.283 + 4.0 \cdot 10^{-5} \cdot T \pm 0.001, \tag{B.15}$$

where temperature T is expressed in [°C];
- austenitic steels of type Cr-Ni and Ni-Cr for $0 \le T \le 800°\text{C}$

$$\nu = 0.292 + 5.4 \cdot 10^{-5} \cdot T \pm 0.02, \tag{B.16}$$

where temperature T is expressed in [°C].

Numbers with a sign ± at the end of a (B.1)–(B.16) are a maximum error is determining a given quantity.

Literature

1. Cases of ASME Boiler and Pressure Vessel Code. Cases 2179. 2180 and 2199
2. French DN (1993) Metallurgical Failures in Fossil Fired Boiler. Ed. 2. Wiley, New York
3. Incropera FP, DeWitt DP (1996) Fundamentals of Heat and Mass Transfer. Ed. 4. Wiley, New York
4. Mills AF (1999) Basic Heat and Mass Transfer. Ed. 2. Prentice Hall, Upper Saddle River
5. Naoi H, Mimura H, Oghami M, Morimoto H, Tanaka T, Yazaki Y (1995) NF616 pipe production and properties of welding consumable development. In: Metcalfe E (ed.) The EPRI/ National Power Conference "New Steels for advanced plant up to 620°C"
6. Hernas A (1999) Creep-resistance steels and alloys (in Polish). Silesian Univ. Press, Gliwice
7. Pich R (1980) Werkstoffkennwerte. Dezember, EVT-Norm
8. Richter F (1983) Physikalische Eigenschaften von Stählen und ihre Temperaturabhängigkeit. Mannesmann Forschungsberichte. 930/1983. Sonderdruck Stahleisen-Sonderberichte 10. Verlag Stahleisen-m.b.H., Düsseldorf
9. Berechnungsblätter für den Wärmeübergang (1984) In: VDI-Wärmeatlas. VDI-Verlag, Düsseldorf
10. Wachter O (1995) Beschreibung der wolframhaltigen 9-bis 12% - Chromstähle unter besonderer Berücksichtigung des Einflusses der Hauptlegierungselemente, des Gefüges und der Wärmebehandlung auf die Gebrauchseigenschaften im Hinblick auf die Verwendung des Rohrleitungs- und Kesselbaustähle unter erhöhten Temperaturen, VGB-TW 508. Essen

Appendix C Fin Efficiency Diagrams (for Chap. 6, part II)

Efficiency of fins:

- straight (Fig. C.1),
- circular (Fig. C.2),
- square (Fig. C.3),
- hexagonal (Fig. C.4).

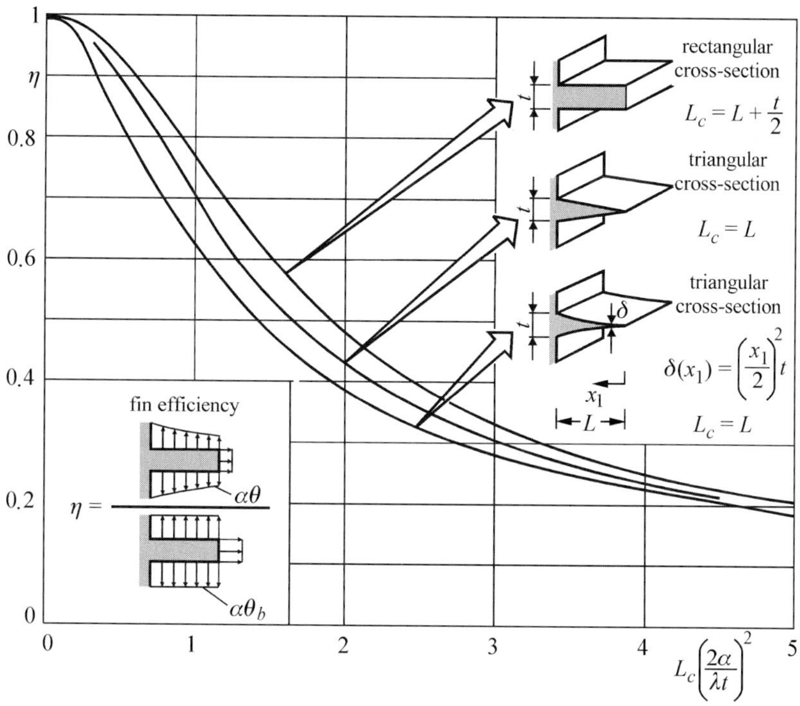

Fig. C.1. Efficiency of simple fins with different cross-sections [2]

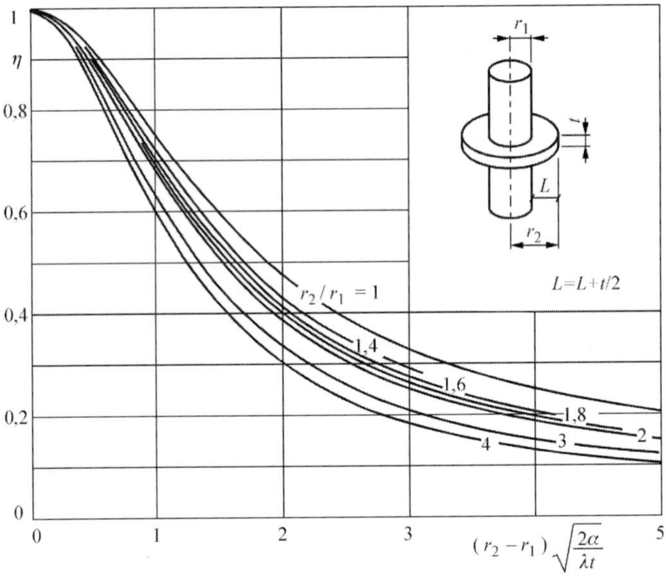

Fig. C.2. Efficiency of a circular fin of constant thickness [2]

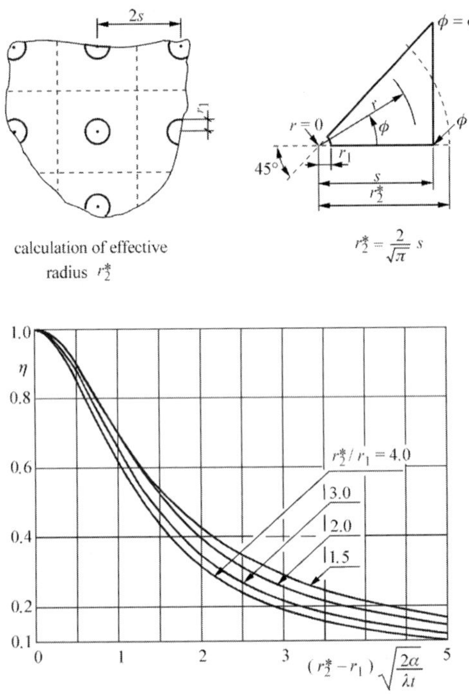

Fig. C.3. Efficiency of square fins [1]

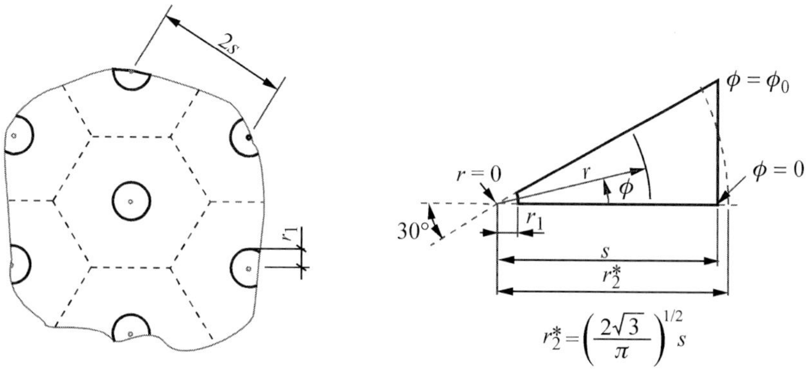

calculation of effective radius r_2^*

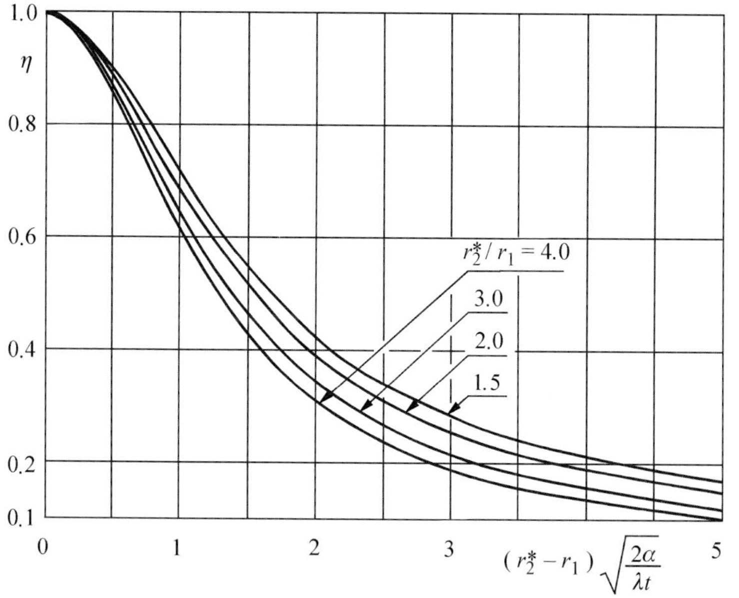

Fig. C.4. Efficiency of hexagonal fins [1]

Literature

1. Bejan A (1993) Heat Transfer. Wiley, New York
2. Sparrow EM, Lin SH (1964) Heat Transfer characteristics of polygonal and plate fins. Int. J. of Heat Transfer 7: 951

Appendix D Shape Coefficients for Isothermal Surfaces with Different Geometry (for Chap. 10, Part II)

Table D.1. Shape coefficients S for isothermal surfaces of different shapes; $\dot{Q} = S\lambda(T_1 - T_2)$

Shape	Shape coefficient S	Comments
1. Flat wall, lateral surfaces thermally insulated 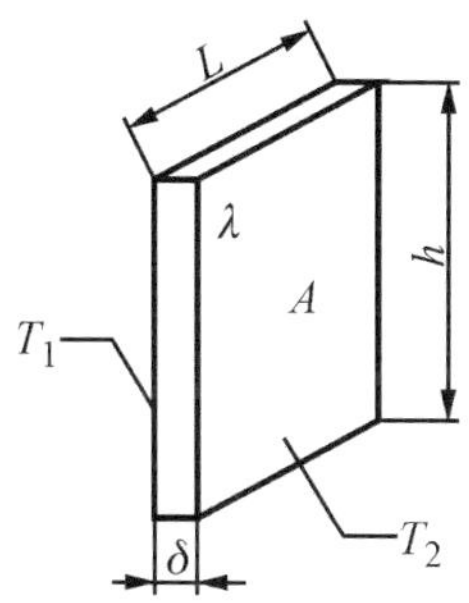	$S = \frac{A}{\delta} = \frac{hL}{\delta}$	If lateral surfaces are not thermally insulated, then $L > 5\delta$, $h > 5\delta$
2. Pipe with square cross-section and length L with insulated endings 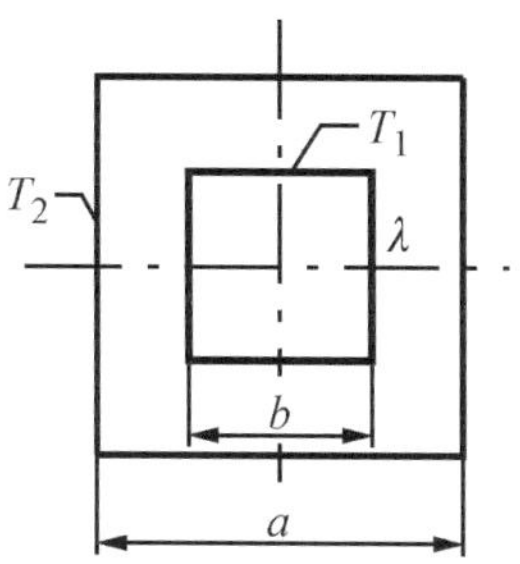	$S = \frac{2\pi L}{0.93 \ln \frac{a}{b} - 0.0497}$ or $S = \frac{2\pi L}{0.785 \ln \frac{a}{b}}$	$\frac{a}{b} > 1.4$, $\frac{a}{b} < 1.4$

Table D.1. (cont.)

Shape	Shape coefficient S	Comments
3. Long rectangular duct in a semi-infinite body 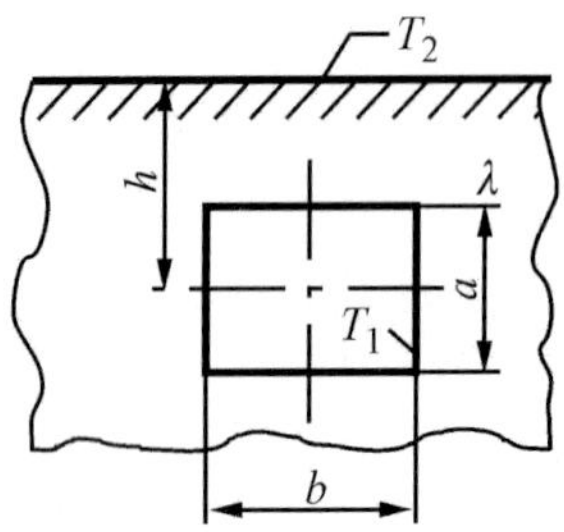	$S = \dfrac{L\left(5.7 + \dfrac{b}{2a}\right)}{\ln \dfrac{3.5h}{a^{0.75} b^{0.25}}}$	
4. Long beam that connects two flat walls with isothermic surfaces 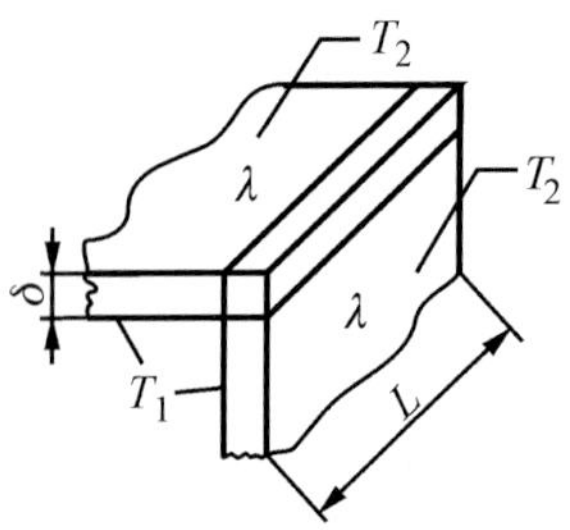	$S = 0.54L$	$L > 5\delta$, T_1 – inner surface wall temperature, T_2 – outer surface wall temperature and mount (ends) temperature of a beam
5. Corner hexahedron at the joint of three identically thick walls 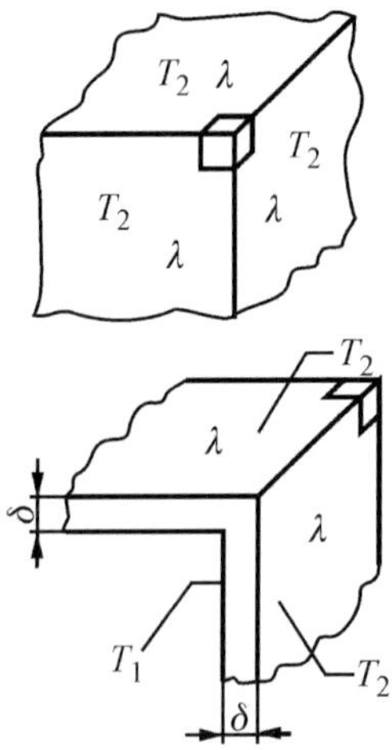	$S = 0.15\delta$	

Table D.1. (cont.)

Shape	Shape coefficient S	Comments
6. Single long pipe in a semi-infinite body 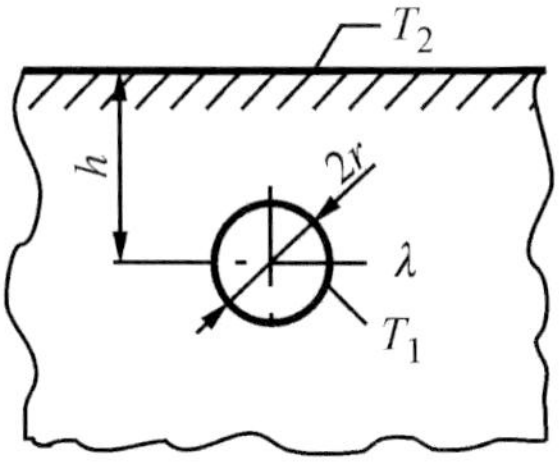	$S = \frac{2\pi L}{\ln\left[\frac{h}{r}+\sqrt{\left(\frac{h}{r}\right)^2-1}\right]}$, $h > r$; $S = \frac{2\pi L}{\ln\frac{2h}{r}}$, $h > 4r$	L – pipe length
7. Row of pipes spaced at equal intervals in a semi-infinite body 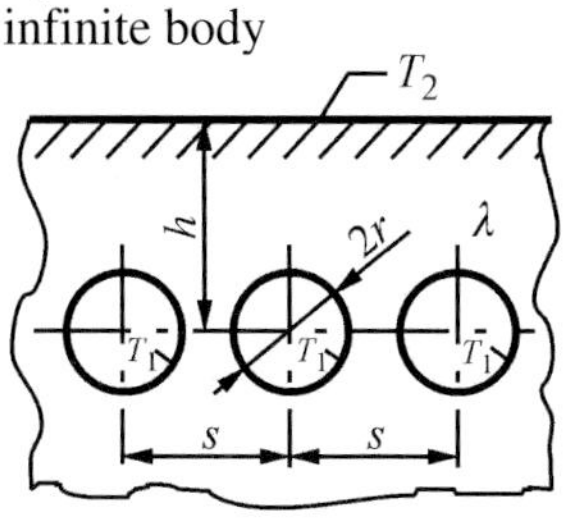	$S = \frac{2\pi L}{\ln\left[\frac{s}{\pi r}\sinh\left(\frac{2\pi h}{s}\right)\right]}$	L – pipe length $L \gg r$, $L \gg h$, $s \geq 3r$
8. Two pipes in semi-infinite body 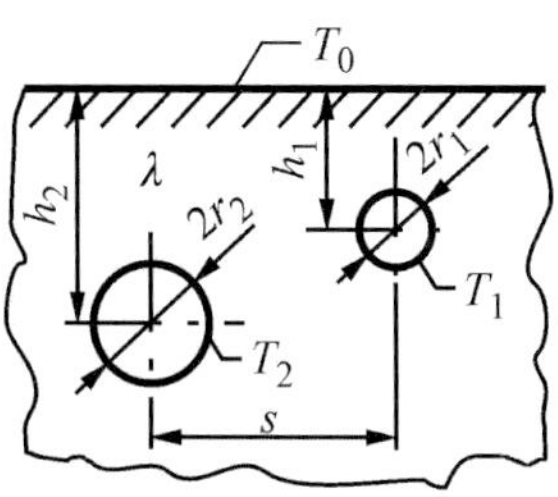	$S = \frac{2\pi L\left[\ln\frac{2h_2}{r_2}-\frac{\Delta T_2}{\Delta T_1}\ln\sqrt{\frac{s^2+(h_1+h_2)^2}{s^2+(h_1-h_2)^2}}\right]}{\ln\frac{2h_1}{r_1}\ln\frac{2h_2}{r_2}-\left(\ln\sqrt{\frac{s^2+(h_1+h_2)^2}{s^2+(h_1-h_2)^2}}\right)^2}$ $\Delta T_1 = T_1 - T_0$, $\Delta T_2 = T_2 - T_0$	L – pipe length $L \gg r_1, r_2$, $4r_1 \leq h_1 \leq h_2$
9. Two long pipes of different diameters in an infinite medium 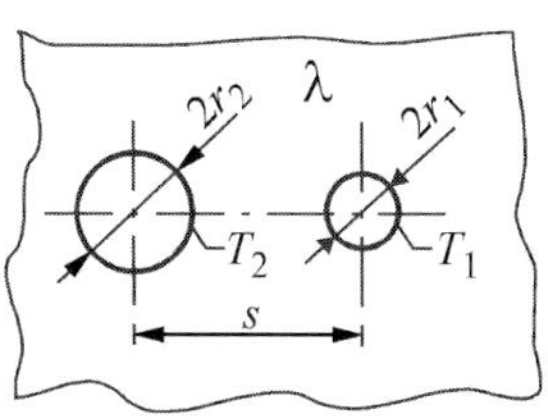	$S = \frac{2\pi L}{\operatorname{arcosh}\left(\frac{h^2-r_1^2-r_2^2}{2r_1r_2}\right)}$	L – pipe length $L \gg r_1, r_2$, $\operatorname{arcosh} x =$ $= \ln\left(x+\sqrt{x^2-1}\right)$, $x \geq 1$

Table D.1. (cont.)

Shape	Shape coefficient S	Comments
10. Long pipe inside a rectangular block 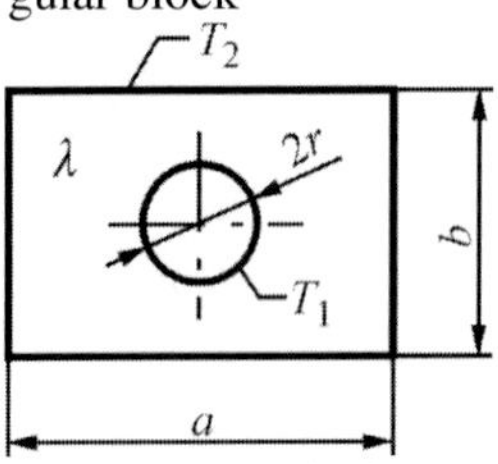	$S = \dfrac{2\pi L}{\ln\left[\dfrac{b}{r}\left(0.637 - 1.781 e^{-2.9\frac{a}{b}}\right)\right]}$	$L \gg a, b$, $a \ge b$
11. Long pipe inside a square-shaped block 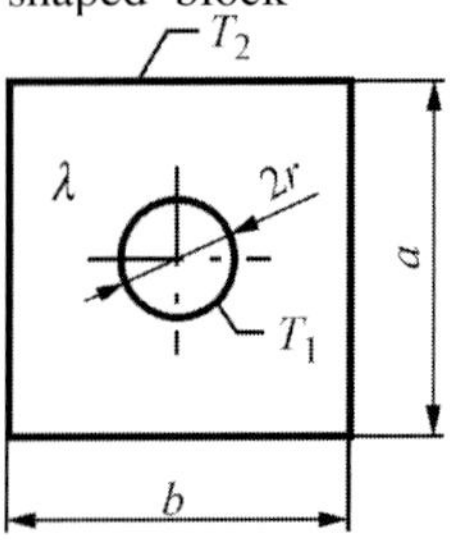	$S = \dfrac{2\pi L}{\ln\left(0.54\dfrac{a}{r}\right)}$	$L \gg a$
12. Long pipe perpendicular to semi-infinite body surface with a thermally insulated pipe end 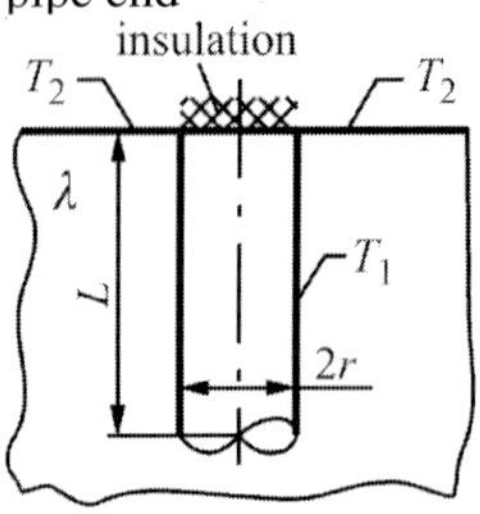	$S = \dfrac{2\pi L}{\ln\left(\dfrac{2L}{r}\right)}$	$L \gg r$
13. Two long off-centre cylinders 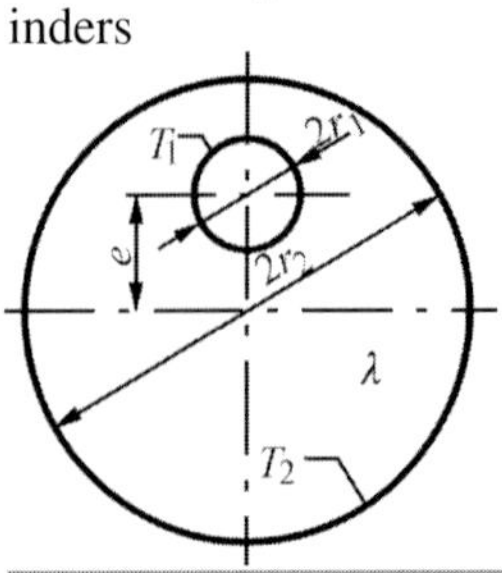	$S = \dfrac{2\pi L}{\operatorname{arcosh}\left(\dfrac{r_1^2 + r_2^2 - e^2}{2 r_1 r_2}\right)}$	$L \gg r_1$ – cylinder length, arcosh $x =$ $= \ln\left(x + \sqrt{x^2 - 1}\right)$, $x \ge 1$

Table D.1. (cont.)

Shape	Shape coefficient S	Comments
14. Long hollow cylinder 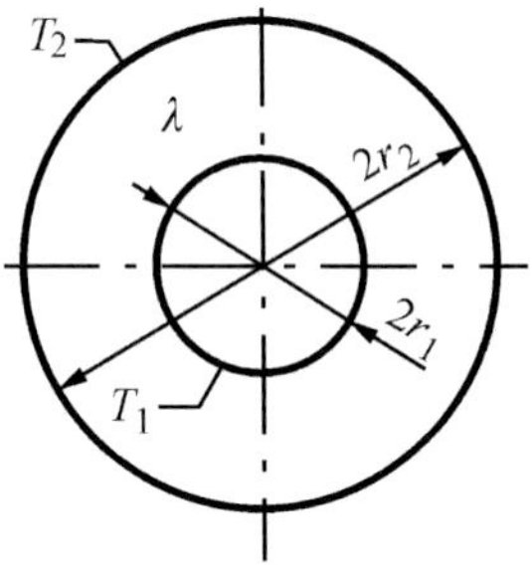	$S=\dfrac{2\pi L}{\ln\left(\dfrac{r_2}{r_1}\right)}$	L – pipe length, $L \gg r_1$
15. N number of long pipes spaced out at equal intervals inside a cylindrical area 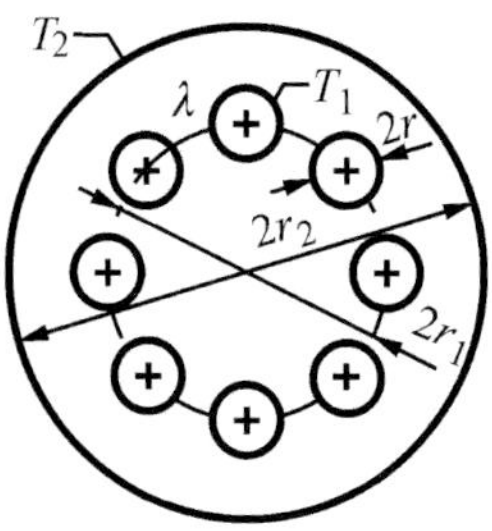	$S=\dfrac{2\pi L}{\ln\left(\dfrac{r_2}{r_1}\right)-\dfrac{1}{N}\ln\left(\dfrac{Nr}{r_1}\right)}$	L – pipe length, $r \ll r_1$, $N \gg 1$, N – number of pipe, Shape coefficient S refers to one pipe $\dot{Q}=NS\lambda(T_1-T_2)$
16. Long elliptical duct 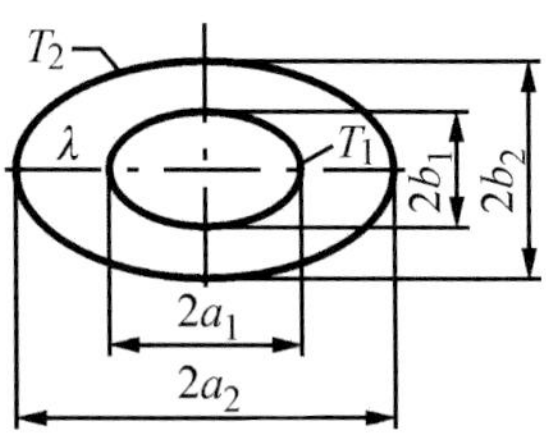	$S=\dfrac{2\pi L}{\ln\left(\dfrac{a_2+b_2}{a_1+b_1}\right)}$	$a_1^2-b_1^2=a_2^2-b_2^2$, $L \gg a_1$
17. Hollow sphere 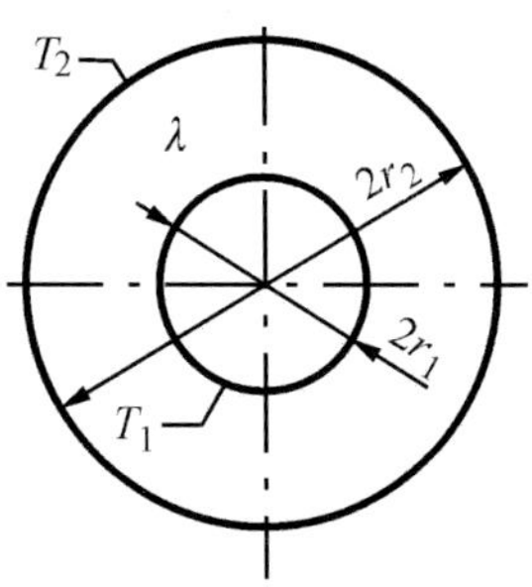	$S=\dfrac{4\pi}{\dfrac{1}{r_1}-\dfrac{1}{r_2}}$	

Table D.1. (cont.)

Shape	Shape coefficient S	Comments
18. Sphere in a semi-infinite body 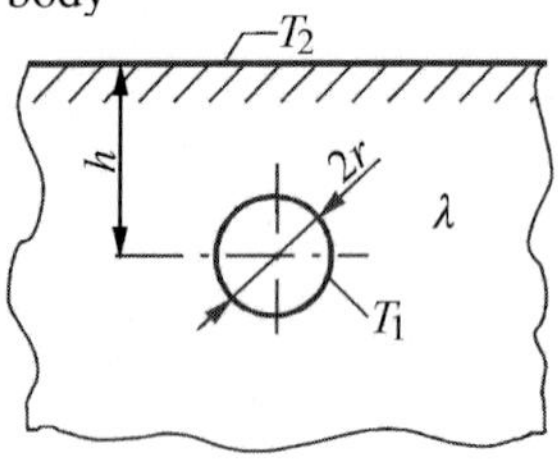	$S = \dfrac{4\pi r}{\left(1 - \dfrac{r}{2h}\right)}$	$h > r$
19. Hemisphere caved in a semi-infinite body 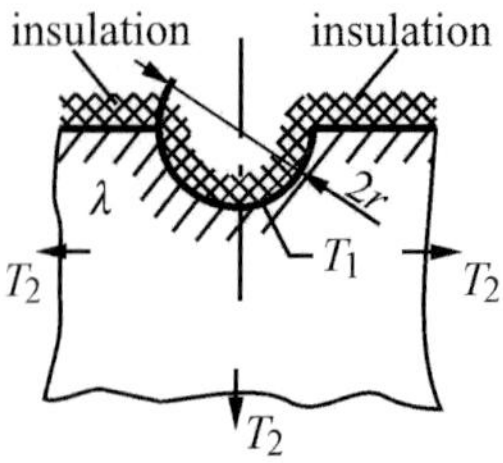	$S = 2\pi r$	The inner hemisphere surface at temperature T_1 is also thermally insulated
20. Sphere in a semi-infinite body with thermally insulated surface 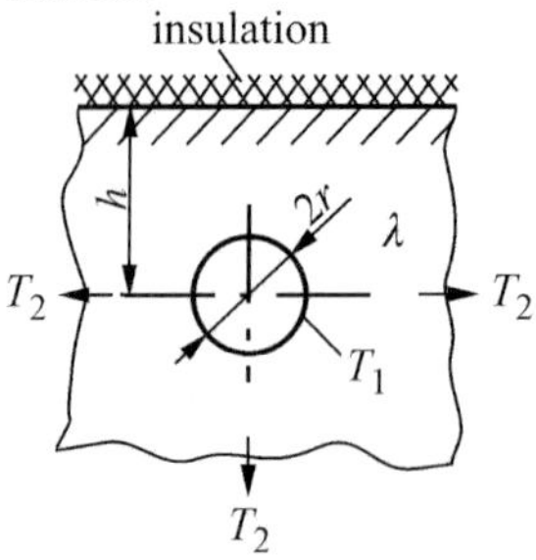	$S = \dfrac{4\pi r}{\left(1 + \dfrac{r}{2h}\right)}$	$h > r$
21. Two spherical cavities in an infinite medium 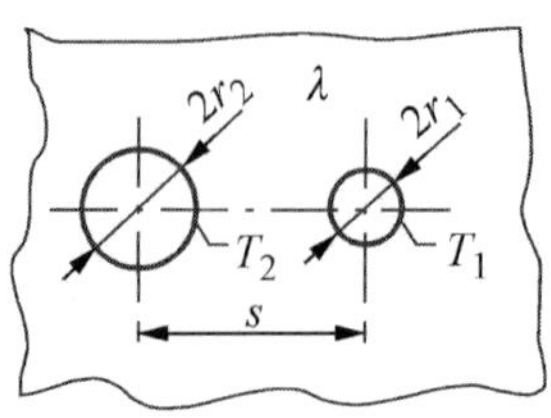	$S = \dfrac{2\pi r_2}{\dfrac{r_2}{r_1} + \left[1 - \dfrac{\left(\dfrac{r_2}{s}\right)^4}{1 - \left(\dfrac{r_2}{s}\right)^2}\right] - \dfrac{2r_2}{s}}$	$r_1 < r_2$, $\dfrac{s}{r_2} \geq 5$

Table D.1. (cont.)

Shape	Shape coefficient S	Comments
22. Two spherical cavities with identical diameters 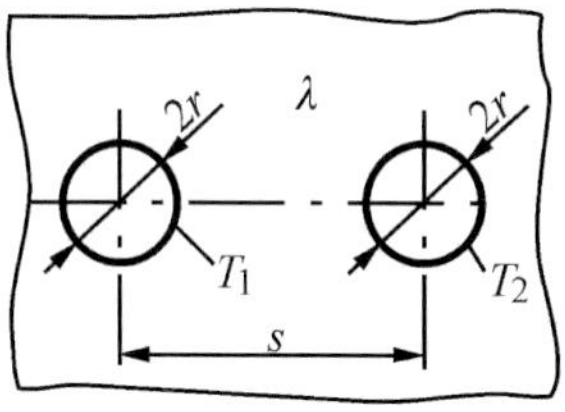	$S = \dfrac{4\pi r}{2\left(1-\dfrac{r}{s}\right)}$	$\dfrac{s}{r} \geq 5$
23. Two spherical cavities with identical diameters lo- 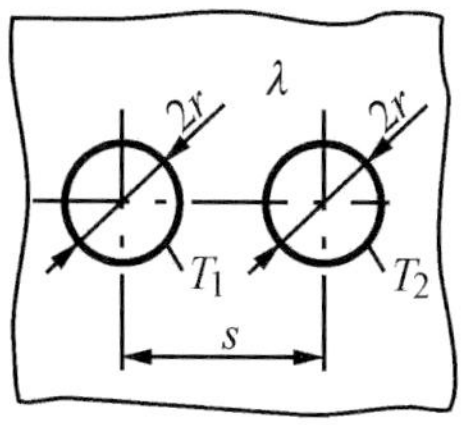	$S = \dfrac{4\pi r}{2}\left[1+\dfrac{r}{s}+\left(\dfrac{r}{s}\right)^2+\left(\dfrac{r}{s}\right)^3+ +2\left(\dfrac{r}{s}\right)^4+3\left(\dfrac{r}{s}\right)^5+\ldots\right]$	$2 < \dfrac{s}{r} < 5$
24. Long wires a) 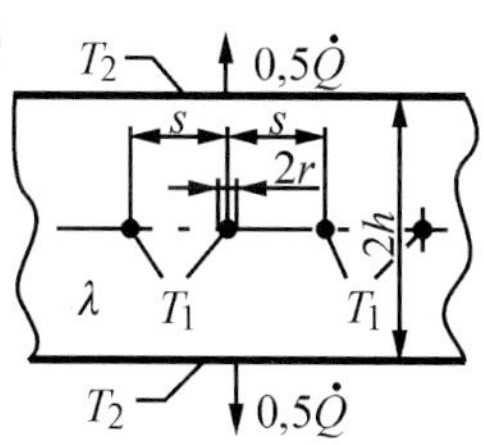 b) 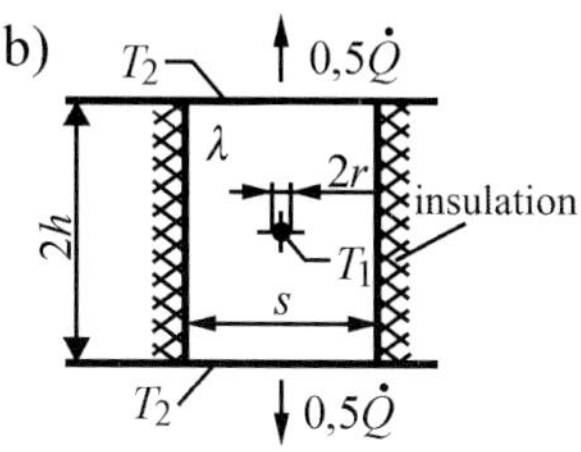 c) 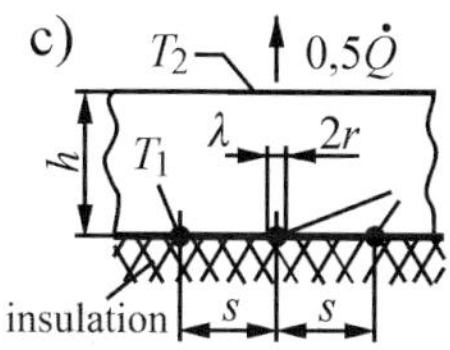	$S = \dfrac{2\pi L}{\dfrac{\pi h}{2s}+\ln\left(\dfrac{s}{\pi r}\right)}$	L – length, $L \gg r$, $r \ll h, s$, Shape coefficient S refers to a single wire, $Q = NS\lambda(T_1 - T_2)$, where N is the number of wires

Table D.1. (cont.)

Shape	Shape coefficient S	Comments
25. Long wires in a semi-infinite medium 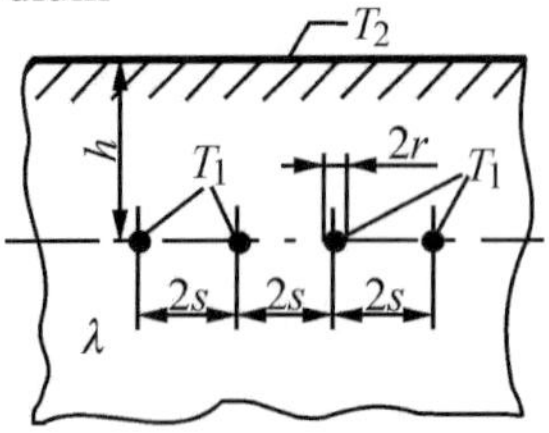 	$S = \dfrac{2\pi L}{\dfrac{\pi h}{s} + \ln\left(\dfrac{s}{\pi r}\right)}$	L – length, $L \gg r$, $r \ll h, s$, Shape coefficient S refers to a single wire, $\dot{Q} = NS\lambda(T_1 - T_2)$, where N is the number of wires
26. Long bar wires in a cross-section of a regular polygon with N sides 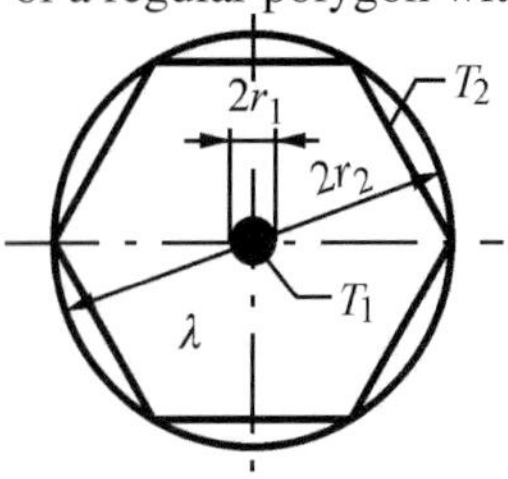 	$S = \dfrac{2\pi L}{\ln\left(\dfrac{r_2}{r_1}\right) - C}$	$\dfrac{r_2}{r_1} > 10$, $L \gg r_1$, N – number of sides N C 3 0.5696 4 0.2708 5 0.1606 6 0.1067 8 0.0570 10 0.0354 ∞ 0
27. Long rectangular bar with a wire inside 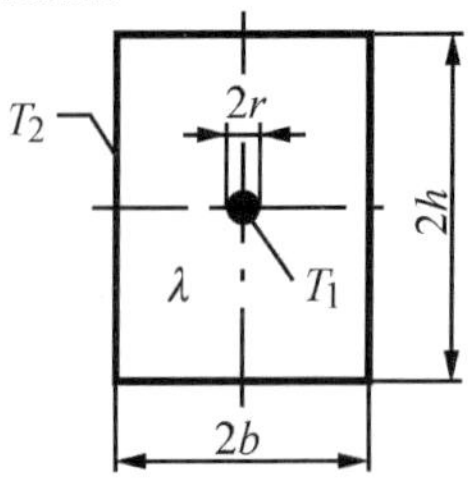 	$S = \dfrac{2\pi L}{\ln\left(\dfrac{4b}{\pi r}\right) - C}$	$L \gg r$, $\dfrac{b}{r} \geq 10$ h/b C 1.00 0.1658 1.25 0.0793 1.50 0.0356 2.00 0.0075 2.50 0.0016 3.00 0.0003 ∞ 0
28. A strip made of a thin long metal plate a) b) 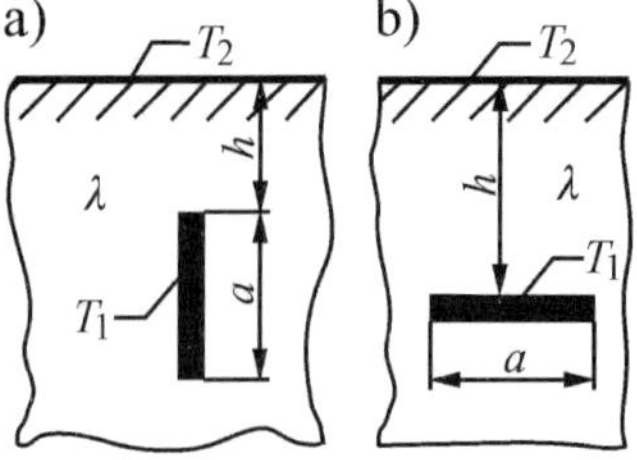 	a) $S = 2.38L\left(\dfrac{a}{h}\right)^{0.24}$ b) $S = 2.94\left(\dfrac{a}{h}\right)^{0.32}$	L – length of the strip, $L \gg a, h$, $0.5 < \dfrac{h}{a} < 12$

Table D.1. (cont.)

Shape	Shape coefficient S	Comments
29. Disk made of thin metal plate in a semi-infinite body 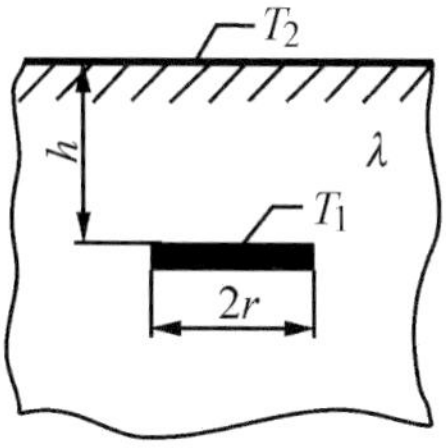	$S = \dfrac{4\pi r}{\dfrac{\pi}{2} - \arctan\left(\dfrac{r}{2h}\right)}$	$\dfrac{h}{r} > 2$
30. Two parallel disks in an infinite body 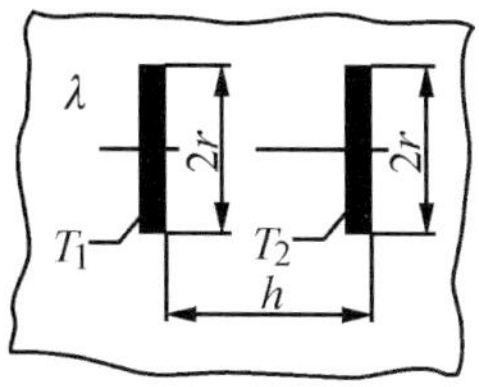	$S = \dfrac{2\pi r}{\dfrac{\pi}{2} - \arctan\left(\dfrac{r}{h}\right)}$	$\dfrac{h}{r} \geq 5$
31. Disk in a semi-infinite body 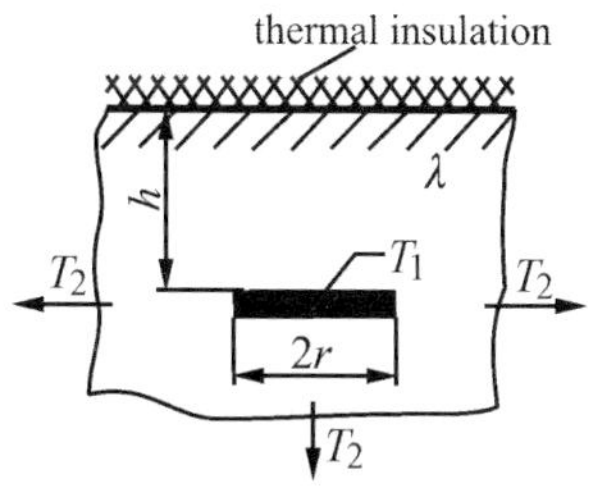	$S = \dfrac{4\pi r}{\dfrac{\pi}{2} + \arctan\left(\dfrac{r}{2h}\right)}$	$\dfrac{h}{r} > 2$
32. 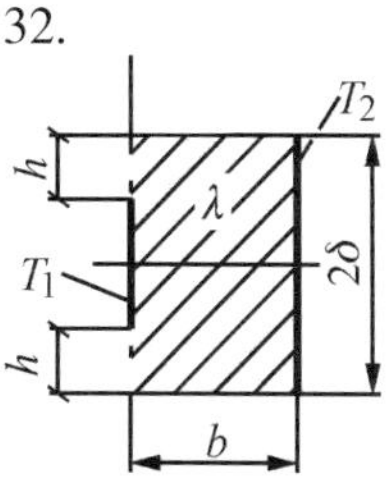	$S = \dfrac{2L}{\dfrac{b}{\delta} + S_{\dot{z}}}$	S – Fig. D.1, L – length measured perpendicularly to the diagram plane

Table D.1. (cont.)

Shape	Shape coefficient S	Comments
33. 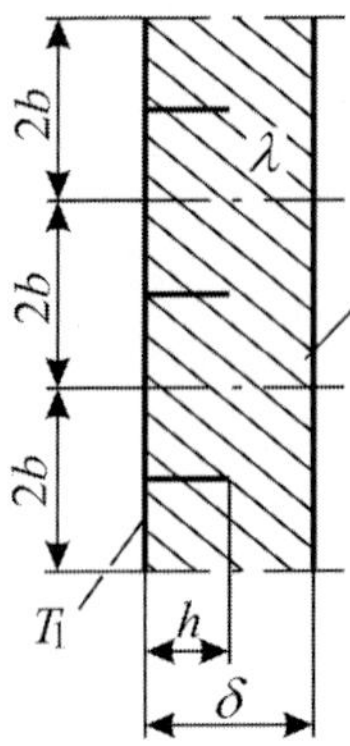	$S = N\left(\frac{2b}{\delta} + 2S_{\dot{z}}\right)L$	S – Fig. D.1, N – number of fins, L – length measured perpendicularly to the diagram plane
34. 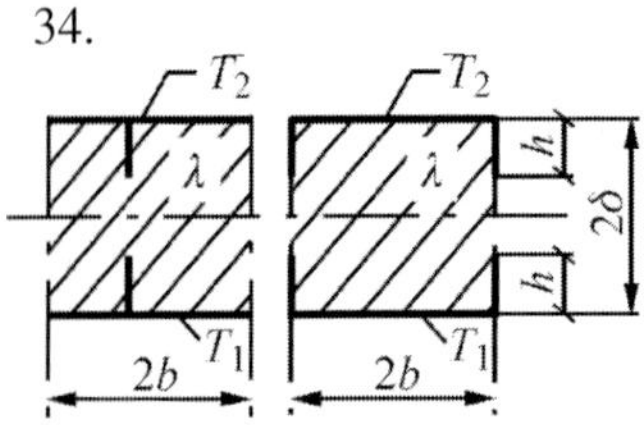	$S = N\left(\frac{b}{\delta} + 2S_{\dot{z}}\right)L$	as above
35. 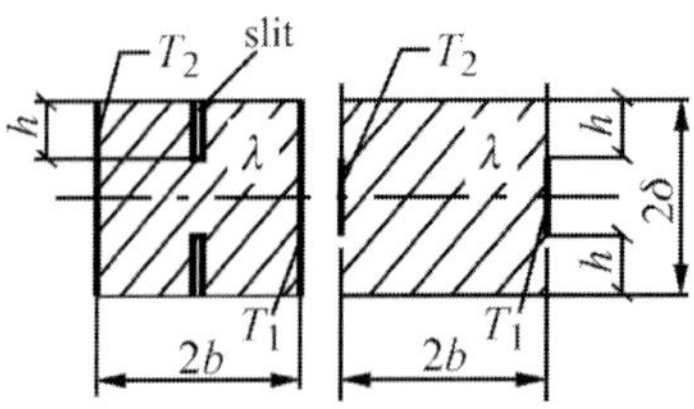	$S = \frac{L}{\frac{b}{\delta} + S_{\dot{z}}}$	as above
36. 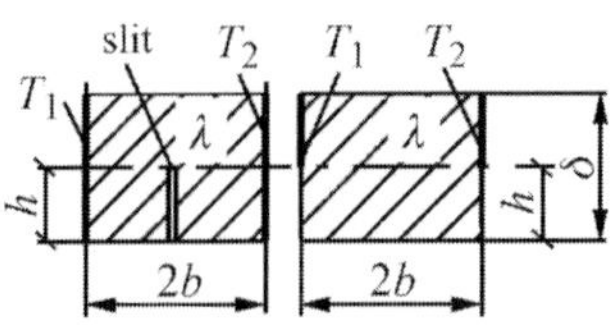	$S = \frac{L}{\frac{2b}{\delta} + 2S_{\dot{z}}}$	as above

Table D.1. (cont.)

Shape	Shape coefficient S	Comments
37. 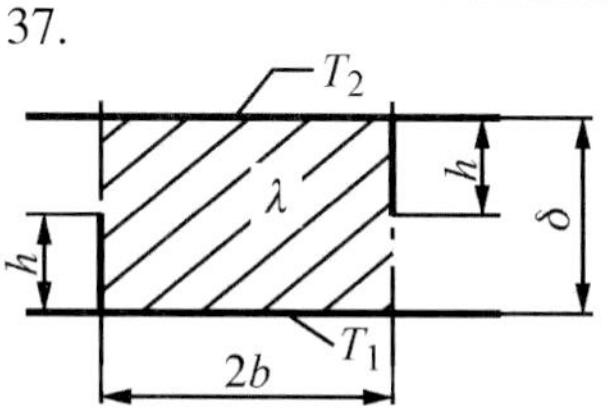	$S = \left(\frac{2b}{\delta} + 2S_{\dot z}^{p} \right) L$	$S_{\dot z}^{p}$ – Fig. D.2, L – length measured perpendicularly to the diagram plane
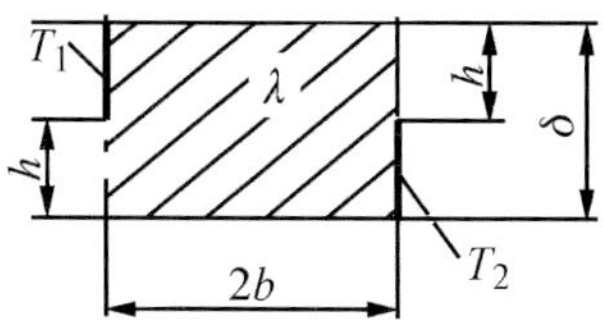	$S = \frac{L}{\frac{2b}{\delta} + 2S_{\dot z}^{p}}$	as above

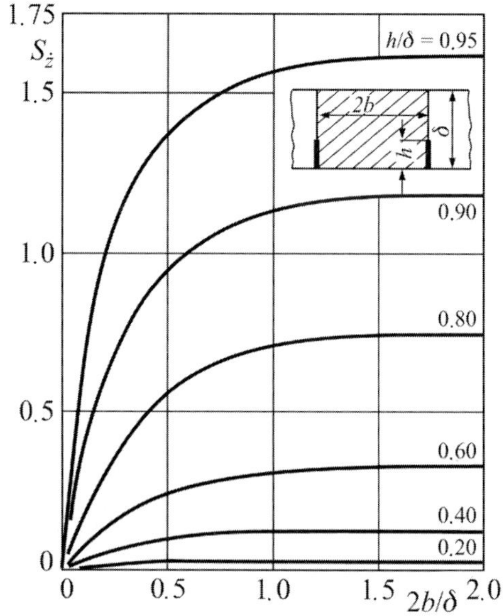

Fig. D.1. Shape coefficient S for unilaterally finned plates with respect to the unit of length measured perpendicularly to the diagram plane

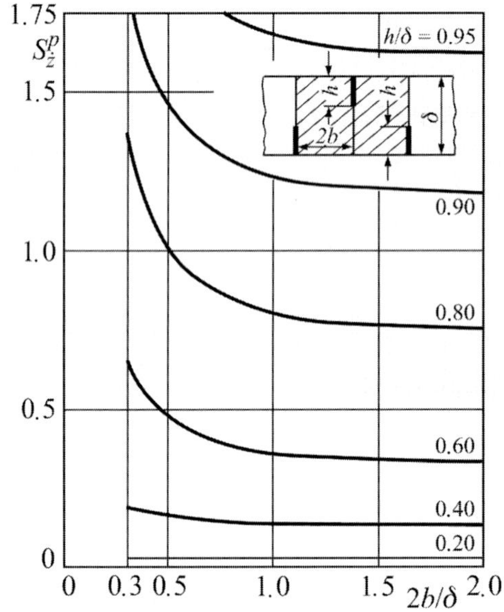

Fig. D.2. Shape coefficient S_z^p for unilaterally finned plates with respect to the unit of length measured perpendicularly to the diagram plane

Appendix E Subprogram for Solving Linear Algebraic Equations System using Gauss Elimination Method (for Chap. 6, Part II)

Subprogram for solving linear algebraic equations system using Gauss method

```
c   Gauss method
c   aa - matrix with coefficients, n - matrix dimension,
c   ainv - inverse matrix
        subroutine matinv(aa,n,ainv)
        dimension aa(50,50), ainv(50,50), a(50,100), id(50)
        nn=n+1
        n2=2*n
        do 100 i=1,n
        id(i)=i
        do 100 j=1,n
100     a(i,j)=aa(i,j)
        do 200 i=1,n
        do 200 j=nn,n2
200     a(i,j)=0.
        do 300 i=1,n
300     a(i,n+i)=1
        k=1
1       call exch(a,n,n,n2,k,id)
2       if (a(k,k)) 3,999,3
3       kk=k+1
        do 4 j=kk,n2
        a(k,j)=a(k,j)/a(k,k)
        do 4 i=1,n
        if(k-i) 41, 4, 41
41      w=a(i,k)*a(k,j)
        a(i,j)=a(i,j)-w
        if(abs(a(i,j))-0.0001*abs(w)) 42,4,4
42      a(i,j)=0.0
4       continue
         k=kk
        if(k-n)1,2,5
5       do 10 i=1,n
        do 10 j=1,n
```

```
      if (id(j)-i) 10,8,10
8     do 101 k=1,n
101   ainv(i,k)=a(j,n+k)
10    continue
      return
999   write(*,*)'Matrix is singular'
      return
      end

      subroutine exch(a,n,na,nb,k,id)
      dimension a(50,100),id(50)
      nrow=k
      ncol=k
      b=abs(a(k,k))
      do 2 i=k,n
      do 2 j=k,na
      if(abs(a(i,j))-b) 2,2,21
21    nrow=i
      ncol=j
      b=abs(a(i,j))
2     continue
      if(nrow-k)3,3,31
31    do 32 j=k,nb
      c=a(nrow,j)
      a(nrow,j)=a(k,j)
32    a(k,j)=c
3     continue
      if(ncol-k)4,4,41
41    do 42 i=1,n
      c=a(i,ncol)
      a(i,ncol)=a(i,k)
42    a(i,k)=c
      i=id(ncol)
      id(ncol)=id(k)
      id(k)=i
4     continue
      return
      end
```

Appendix F Subprogram for Solving a Linear Algebraic Equations System by Means of Over-Relaxation Method

Subprogram SOR section appendix f subprogram,for solving a linear algebraic equations system by means of over-relaxation method

```
      subroutine sor(a,nmax,mmax,n,xi,w,niter,toler,k)
      dimension a(nmax,mmax),xi(nmax)
      k=1
      err=1.
      do while ((k.le.niter).and.(err.gt.toler))
c     variable err is solution tolerance
            err=0.0
            do i=1,n
               s=0.0
                   do j=1,n
                      s=s-a(i,j)*xi(j)
                   enddo
                  s=w*(s+a(i,n+1))/a(i,i)
               err=err+s*s
               xi(i)=xi(i)+s
            enddo
            err=sqrt(err)
      k=k+1
      enddo
      k=k-1
      return
      end
```

Appendix G Subprogram for Solving an Ordinary Differential Equations System of 1st Order using Runge-Kutta Method of 4th Order (for Chap. 11, Part II)

Subprogram for solving an ordinary differential equations system of 1st order using Runge-Kutta method of 4th order

```
      SUBROUTINE RUNGE(N,T,F,TAU,DT,M,K)
      DIMENSION T(204),F(204),CAPY(204),P(204)
      M = M+1
      GO TO(1,2,3,4,5),M
    2 DO 12 I=1,N
      P(I) = F(I)
      CAPY(I)=T(I)
   12 T(I) = CAPY(I)+0.5*DT*F(I)
   11 TAU = TAU+0.5*DT
    1 K = 1
      RETURN
    3 DO 13 I=1,N
      P(I) = P(I)+2.0*F(I)
   13 T(I) = CAPY(I)+0.5*DT*F(I)
      K = 1
      RETURN
    4 DO 14 I=1,N
      P(I) = P(I)+2.0*F(I)
   14 T(I) = CAPY(I)+DT*F(I)
      GO TO 11
    5 DO 15 I=1,N
   15 T(I) = CAPY(I)+(P(I)+F(I))*DT/6.0
      K = 2
      M = 0
      RETURN
      END
```

Appendix H Determining inverse Laplace Transform (for Chap. 15, part II)

When applying Laplace transformation to the solving of transient heat conduction problems, the main difficulty lies in the fact that one needs to find inverse Laplace transform

$$\mathscr{L}^{-1}\left[F\right]=\frac{1}{2\pi i}\int_{\delta-i\infty}^{\delta+i\infty}F\left(s\right)e^{st}ds\,, \tag{H.1}$$

where $F(s)$ is the solution of the transient heat conduction problem in the image domain. Value δ is chosen in a way that all poles (roots of a denominator) $F(s)$ would lie on the left-hand-side of the path of integration, which lies on the right-hand-side of an imaginary axis [1–4].

The process of determining an integral in (H.1) is very complex and rarely applied when solving transient heat conduction problems. Below, we will describe the method for determining inverse Laplace transform usually applied in practice. Temperature distribution transform is obtained in the form [1, 2]

$$F\left(s\right)=\overline{T}\left(s\right)=\frac{g_1\left(s\right)}{g_2\left(s\right)}. \tag{H.2}$$

Function $g_2(s)$ in denominator has many zero points s_n , which satisfy equation

$$g_2\left(s\right)=0. \tag{H.3}$$

If $s \to s_n$, then function $F(s_n) \to \infty$; if, however, $s \neq s_n$, then the function has a finite value. Points s_n are called *poles* or *singularities*. Three cases will be analyzed here:

a) denominator $g_2(s)$ has individual poles, also called simple poles, which are other than zero
b) denominator $g_2(s)$ has a single pole $s = 0$ and many simple poles s_n, n = 1, 2, ...,
c) denominator $g_2(s)$ has k- number of zero poles $s = 0$ and many simple poles s_n, n = 1, 2, ...

Ad a) With different poles s_n, when each one is other than zero, inverse transform $\mathscr{L}^{-1}$ [$F(s)$] is formulated as

$$\mathscr{L}^{-1}\left[F(s)\right]=\sum_{n=0}^{\infty}\frac{g_1(s_n)}{g_2'(s_n)}e^{s_n t}, \tag{H.4}$$

where

$$g_2'(s_n)=\left[\frac{d}{ds}g_2(s)\right]_{s=s_n}. \tag{H.5}$$

Ad b) Denominator $g_2(s)$ of transform (H.2) can be in the given case written in the form

$$g_2(s)=sh(s). \tag{H.6}$$

Pole $s = 0$ is a single pole; other single poles s_n, $n = 1, 2, ...$ are determined from equation

$$h(s)=0. \tag{H.7}$$

Derivative $g_2'(s)$ is given by

$$g_2'(s_n)=\frac{dg_2(s)}{ds}=h(s)+sh'(s), \tag{H.8}$$

hence, for $s = 0$ one has

$$g_2'(s_n)=h(0), \tag{H.9}$$

while for $s = s_n$

$$g_2'(s_n)=s_n h'(s_n), \qquad n=1,\ 2,\ ... \tag{H.10}$$

Because for $s = 0$, $e^{st} = 1$; therefore, formula (H.4) assumes the form

$$\mathscr{L}^{-1}\left[F(s)\right]=\mathscr{L}^{-1}\left[\frac{g_1(s)}{sh(s)}\right]=\frac{g_1(0)}{h(0)}+\sum_{n=1}^{\infty}\frac{g_1(s_n)}{s_n h'(s)}e^{s_n t}. \tag{H.11}$$

It is a *Heaviside* formula.

Ad c) Inverse transform can be also calculated from formula

$$\mathscr{L}^{-1}\left[F(s)\right]=\frac{1}{2\pi i}\int_{\delta-i\infty}^{\delta+i\infty}F(s)e^{st}ds=\sum_{n=0}^{\infty}Res\left[F(s_n)e^{s_n t}\right]. \tag{H.12}$$

Residuum, which occurs under the summation sign, has the form

$$Res\left[F(s_n)e^{s_n t}\right]=\frac{1}{(k-1)!}\frac{d^{k-1}}{ds^{k-1}}\left[(s-s_n)^k F(s)e^{st}\right]\Bigg|_{s=s_n}, \tag{H.13}$$

where k is the multiplication factor of a given pole.

If transform $F(s)$ is expressed by (H.2), then $g_2(s)$ can be expanded into a Taylor series around pole s_n

$$g_2(s)=g_2(s_n)+(s-s_n)g_2'(s_n)+\ldots \tag{H.14}$$

Once we account for characteristic equations (H.3) and (H.14) from (H.13), we have

$$Res\left[F(s)e^{st}\right]=\frac{g_1(s_n)}{g_2'(s_n)}e^{s_n t}, \tag{H.15}$$

therefore, formula (H.4).

If the multiplicity factor of pole s_n is k, then the determination of a residuum from (H.13) is rather complex. In such a case, function $F(s)e^{s_n t}$ is expanded into a Laurent series around pole s_n

$$F(s)e^{st}=\frac{g_1(s)}{g_2(s)}e^{st}=\frac{c_{-k}}{(s-s_n)^k}+\frac{c_{-k+1}}{(s-s_n)^{k-1}}+\ldots+\frac{c_{-1}}{(s-s_n)}+ \tag{H.16}$$
$$+c_0+c_1(s-s_n)+\ldots,$$

where k is a multiplication factor of pole s_n.

Residuum of function $F(s_n)e^{s_n t}$ is equal, in such a case, to coefficient c_{-1}

$$Res\left[F(s_n)e^{s_n t}\right]=c_{-1}. \tag{H.17}$$

Inverse Laplace transform is, thus, formulated as

$$\mathscr{L}^{-1}\left[F(s)\right]=c_{-1}+\sum_{n=1}^{\infty}\frac{g_1(s_n)}{g_2'(s_n)}e^{s_n t}. \tag{H.18}$$

It is an expanded Heviside formula. In order to determine coefficient c_{-1}, functions $g_1(s)$, $g_2(s)$ and e^{st} are expanded into Taylor series around a multiple pole s_n. Allowing that $s_0 = 0$ is the multiple pole in heat conduction problems, the corresponding Taylor series have the form

$$g_1(s) = \sum_{m=0}^{\infty} \frac{1}{m!}(s-s_0)^m \frac{d^m g_1(s_0)}{ds_0{}^m} = A + Bs + Cs^2 + ..., \tag{H.19}$$

$$g_2(s) = \sum_{m=0}^{\infty} \frac{1}{m!}(s-s_0)^m \frac{d^m g_2(s_0)}{ds_0{}^m} = s^k\left(D + Es + Fs^2 + ...\right), \tag{H.20}$$

$$e^{st} = \sum_{m=0}^{\infty} \frac{1}{m!}(s-s_0)^m \frac{d^m\left(e^{s_0 t}\right)}{ds_0{}^m} = 1 + st + \frac{s^2t^2}{2} + ... , \tag{H.21}$$

where $s_0 = 0$.

For $s_0 = 0$ transform $F(s)e^{st}$ can be written as follows:

$$F(s)e^{st} = \frac{g_1(s)}{g_2(s)}e^{st} = \frac{A + Bs + Cs^2 + ...}{s^k\left(D + Es + Fs^2 + ...\right)}\left(1 + st + \frac{s^2t^2}{2} + ...\right), \tag{H.22}$$

where from, once we carry out the operations, we get

$$\begin{aligned} F(s)e^{st} &= \frac{1}{s^k}\left[\frac{A}{D} + \frac{1}{D}\left(At + B - \frac{AE}{D}\right)s + \right. \\ &\left. + \frac{1}{D}\left(C + Bt - \frac{AE}{D}t + \frac{A}{2}t^2 - \frac{AF}{D} - \frac{BE}{D} + \frac{AE^2}{D^2}\right)s^2 + ...\right]. \end{aligned} \tag{H.23}$$

It is easy to determine coefficient c_{-1} present in formula (H.18) by means of (H.23). What follows from formula (H.23) is that for a single pole ($k = 1$) present in $s = 0$, coefficient c_{-1} is

$$c_{-1} = \frac{A}{D}, \tag{H.24}$$

when $k = 1$.

For a double pole ($k = 2$) present in $s = 0$ from (H.23) one has

$$c_{-1} = \frac{1}{D}\left(At + B - \frac{AE}{D}\right). \tag{H.25}$$

Formulas for c_{-1} when $k = 3$ and higher multiplication factors of poles in $s = 0$ can be derived in a similar way.

Literature

1. Tautz H (1971) *Wärmeleitung + Temperaturausgleich.* Verlag Chemie, Weinheim
2. Privalov CC (1977) Introduction into theory of complex function (in Russian). Moscow, Nauka